# GEOTECHNICAL ENGINEERING

## SOIL AND FOUNDATION PRINCIPLES AND PRACTICE, FIFTH EDITION

## ABOUT THE AUTHORS

**Richard L. Handy, Ph.D.,** is a Distinguished Professor Emeritus in the Department of Civil, Construction, and Environmental Engineering at Iowa State University. He is also the founder of Handy Geotechnical Instruments, a company that manufactures innovative soil testing devices. Dr. Handy is the author of *The Day the House Fell* and co-author of the Third and Fourth Editions of *Soil Engineering*. Recognized as a scientist as well as an engineer, he is a Fellow in the Geological Society of America and also in the American Association for the Advancement of Science.

**M. G. Spangler** *(deceased)* was a Research Professor at Iowa State University and is well-known internationally as the author of the "Marston-Spangler theory for loads on underground conduits." He also conducted seminal research on pressures on retaining walls and many other topics. He was a recipient of the Marston Medal at Iowa State University and was an Honorary Member of ASCE.

# GEOTECHNICAL ENGINEERING

## SOIL AND FOUNDATION PRINCIPLES AND PRACTICE, FIFTH EDITION

R. L. Handy
M. G. Spangler

New York   Chicago   San Francisco   Lisbon   London   Madrid   Mexico City
Milan   New Delhi   San Juan   Seoul   Singapore   Sydney   Toronto

**The McGraw·Hill Companies**

McGraw-Hill books are available at special quantity discounts to use as premiums and sales promotions, or for use in corporate training programs. For more information, please write to the Director of Special Sales, Professional Publishing, McGraw-Hill, Two Penn Plaza, New York, NY 10121-2298. Or contact your local bookstore.

**Geotechnical Engineering: Soil and Foundation Principles and Practice, Fifth Edition**

1 2 3 4 5 6 7 8 9 0 DOC/DOC 0 1 9 8 7 6

ISBN-13: 978-0-07-148120-5
ISBN-10: 0-07-148120-6

**Sponsoring Editor**
  Larry Hager
**Editorial Supervisor**
  Janet Walden
**Project Manager**
  Andy Baxter
**Copy Editor**
  Alan Hunt
**Proofreader**
  John Bremer
**Indexer**
  Yvonne Dixon

**Production Supervisor**
  George Anderson
**Composition**
  Keyword Group Ltd
**Illustration**
  Keyword Group Ltd
**Art Director, Cover**
  Anthony Landi

# Contents at a Glance

# Contents

# Preface

Over 50 years ago the late Professor M. G. Spangler put this train on the track with the first edition of *Soil Engineering*—at a time when the engineer's fast track was a slide rule. The goal, stated in the Preface of the first edition, was to present material "in the simplest possible manner" and "to lead the reader through appropriate phases of the basic sciences of geology, pedology, soil physics, and physical chemistry and to illustrate the applications of these sciences in soil mechanics and soil engineering."

The present senior author has attempted to continue to fulfill those goals in this, the fifth edition, while introducing many new concepts and methods developed over the ensuing decades. There has been a shift in emphasis away from highway engineering applications, earth dams, and underground conduits, and more emphasis on foundations, retaining walls, landslides, and in-situ testing, topics that form the nucleus of the geotechnical engineer's repertoire. Materials that are unique to this textbook should make it equally useful for practicing engineers and for classroom instruction.

# Acknowledgments

The author wishes to acknowledge the many individuals who have given freely of their time and advice while bearing none of the responsibility. Dr. Alan Lutenegger, University of Massachusetts at Amherst, contributed to the outline, and the manuscript was reviewed by Dr. David White, Iowa State University. Others who have read selected sections or chapters include Dr. John Schmertmann, Dr. Nathanial Fox, Dr. Kord Wissman, Mr. Michael Lustig, Mr. Max Calhoon, Mr. Dwayne McAninch, Mr. Roger Failmezger, and Attorney Richard Garberson. Many other colleagues and former students have made valuable suggestions; they know who they are. Most important was my wife, Kathryn, for her unique combination of patience and encouragement. A special thanks goes to the editor Larry Hager, the project manager Andy Baxter, and many others who helped to pull it all together.

# SI and Conversion Factors

In this edition dimensions are given in both English and SI (Système International) units. Problems may be worked in either system. While SI is being adopted in academic circles, U.S. federal agencies, and professional societies, a working knowledge of the English system will be required in the U.S. for some years to come, so students and engineers caught in the transition must of necessity learn both.

## Mass vs. Force

In the SI the unit of mass is the kilogram; in the English system it is the slug. Corresponding designations for forces are the newton (SI) and the pound (English). Engineering emphasize forces, not masses; hence our basic units in the two systems are the newton and the pound. Grams and kilograms, being units of mass, require conversion to force by being multiplied by the acceleration of gravity. Weight is a force developed by a certain mass in a certain gravitational field. Another distinction is that force is a vector quantity and has direction, whereas mass does not. Thus a mass acted on by gravity gives a gravitational force that is directed downward.

## Density vs. Unit Weight

An important fundamental property of soil is its mass per unit volume, or density. However, since engineering deals with forces, the force equivalent of density is *unit weight*, in newtons per cubic meter or pounds per cubic foot. Unit weight is the preferred measure and should be used and designated as such. Units of density are $g/cm^3$ (grams per cubic centimeter) or $kg/m^3$ (kilograms per cubic meter), and it is incorrect to refer to a "density" in pounds per cubic foot because pounds are force. Unit weight, being a force term, has direction, and acts downward. It should be emphasized that the terms "density" and "unit weight" strictly speaking are not interchangeable. Density is of course convertible to unit weight by Newton's law: force $=$ mass $\times g$ (acceleration of gravity). On a unit volume basis, unit weight $=$ density $\times g$.

## Stress

Stress is defined as a force per unit area. Thus stress is in $N/m^2$ (newtons per square meter), which also is called the pascal in SI, or it may be in $lb/ft^2$ (psf or pounds per square foot) or $lb/in.^2$ (psi or pounds per square inch). Stress is not in units of $kg/cm^2$, which is mass per unit area and has no physical meaning. Therefore, even though $kg/cm^2$ (kilograms per square centimeter) is the common metric unit in the older European literature and still is widely used, it is incorrect and should be converted to kilopascals: $1 \, kg/cm^2 = 100 \, kPa$.

## Summary

Some mass- and force-derived units commonly used in geotechnical engineering are shown below. Inconsistencies occur in both the English and cgs systems due to use of force terms for mass and vice versa.

|  | English | SI | cgs |
|---|---|---|---|
| Mass, $M$ | slug | kilogram | gram |
| Force, $F$ | pound, ton | newton | gram force, tonne[a] |
| Stress, $FL^{-2}$ | pound/inch$^2$ | newton/meter$^2$ | gram force/centimeter$^2$ |
| Unit weight, $FL^{-3}$ | pound/foot$^3$ | newton/meter$^3$ | gram force/centimeter$^3$ |
| Density, $ML^{-3}$ | pound mass/foot$^3$ | kilogram/meter$^3$ | gram/centimeter$^3$ |

[a]Use of a force term for mass or a mass term for force is applicable only with a specific gravitational field.

The conversions in this book are based on the mean acceleration of Earth's gravitational field, $g = 9.80665 \, m/s^2$.

## Preferred SI Prefixes

The SI uses preferred multipliers in steps of $10^3$:

$G = giga = 10^9$

$M = mega = 10^6$

$k = kilo = 10^3$

$(No \; prefix) = 10^0 = 1$

$m = milli = 10^{-3}$

$\mu = micro = 10^{-6}$

$n = nano = 10^{-9}$

$p = pico = 10^{-12}$

The prefixes mega, kilo, milli, and micro are the most commonly used in geotechnical engineering. Prefixes are selected so that numbers are between 0.1 and 1000, except in tabular presentations where the rule may give way to consistency. Example: 0.7 m or 700 mm are acceptable; 0.0007 km is not.

## Allowable Units

The above arrangement becomes a bit fanciful when one considers appropriate units for area or volume, since $10^3$ steps in linear dimensions translate into $10^6$ steps in area dimensions and $10^9$ steps in volume dimensions. That is, $1 mm^3 = 10^{-9} m^3$, and $1{,}000{,}000 mm^3$ equals only $0.001 m^3$. Therefore, for area or volume, $cm^2$ and $cm^3$ are allowed, 1 cm equaling $10^{-2} m$. Then $1{,}000{,}000 mm^3 = 1000 cm^3 = 0.001 m^3$.

The preferred unit for angle is radians, but degrees and fractions (not minutes and seconds) are allowed and will be used in this text unless specifically noted.

## Computers and Significant Figures

In spite of their apparent infallibility, modern digital calculators and computers automatically contribute an error that all too frequently is copied into answers—the error being to show too many digits in the output. The output therefore is not the answer, but must be interpreted. For example, unless directed otherwise, a digital calculator will solve the following problem thusly:

$9.71 \times 1.3 = 12.623$    *Answer (incorrect)*

*In an engineering report this answer would be incorrect and should be marked wrong,* because it implies five-figure precision but one of the multipliers, 1.3, shows only two-figure precision. The answer therefore should show two significant figures:

$9.71 \times 1.3 = 13.$    *Answer*

Multiplication, division, and trigonometric functions do not change the number of valid significant figures. Summing or averaging increases the number of significant figures and is a basis for statistical measurements, whereas subtraction decreases the number of significant figures by removing from the front:

$129.71 - 129.11 = 0.60$

(5 significant figures reduced to 2 significant figures)

This is important in engineering measurement because when an accurate difference is wanted it should be directly measured, not obtained by subtracting two measured values.

The number of significant figures in an answer is a rough indication of measurement precision:

| Example | Implied measurement precision |
|---------|-------------------------------|
| 2 | ±50% = 1 significant figure = estimation or approximate measurement |
| 2.2 | ±5% = 2 significant figures = rough measurement |
| 2.22 | ±0.5% = 3 significant figures = ordinary measurement |
| 2.222 | ±0.05% = 4 significant figures = precise measurement |
| 2.2222 | ±0.005% = 5 significant figures = very precise measurement, usually attainable only by repetitive determinations and statistical averaging |
| 2.22222 | ±0.0005% = 6 significant figures = the highest order of precision |

The number of valid figures depends on variability of the property measured, high variability giving few legitimate significant figures. Soils are quite variable.

**Example**
Three soil samples from the same location are found to contain 9.6%, 6.3%, and 4.0% gravel. What is the average percentage of gravel?

Calculator answer: $(9.6 + 6.3 + 4.0) = 6.63333333\ldots\%$

If one were to screen 100 tons of this soil, it is very unlikely that 6.63333333...tons would be gravel. Inspection of the raw data shows there is no more than one significant figure, and the average therefore should be reported as 7%. This implies that about 7 tons, or about 6 to 8 tons, may be expected to be gravel. For a better estimate, a statistical procedure may be used.

## Precision and Accuracy

The above discussions of significant figures and rounding relate only to precision, or exactness of a measurement or test. Precision is evaluated by re-measuring several times under identical conditions and noting the reproducibility of the data. However, because a measurement is precise does not mean that it is accurate. In fact, many precise measures are inaccurate because "accuracy" implies the measurement accurately evaluates that which is intended to be evaluated. Income tax, for example, may be precisely calculated to the nearest penny and still can be inaccurate enough to constitute a felony.

## Rounding

An extra figure should be carried in calculations to reduce rounding error, but the final answer must be rounded to reflect the minimum number of significant figures of the contributing data. The standard recommended by

ASTM is to round a final digit 5 (five) upward if the preceding digit is odd. For example, $15.50 \times 12.10 = 187.55$ rounds to 187.6. Many calculators *always* round 5 upward.

## Conversion Factors

### Length

| | |
|---|---|
| 1 in. $= 25.4$ mm | 1 mm $= 0.0397$ in. |
| 1 ft $= 0.3048$ m | 1 m $= 3.281$ ft |
| 1 mile $= 1.609$ km | 1 km $= 0.6214$ mile |

### Area

1 in.$^2 = 645.2$ mm$^2$       1 mm$^2 = 0.00155$ in.$^2$
     $= 6.452$ cm$^2$       1 cm$^2 = 0.155$ in.$^2$
1 ft$^2 = 0.0929$ m$^2$       1 m$^2 = 10.764$ ft$^2$
1 acre $= 4047$ m$^2$       1 m$^2 = 0.000247$ acre

### Volume

1 in.$^3 = 16,390$ mm$^3$       1 mm$^3 = 0.0000610$ in.$^3$
     $= 16.39$ cm$^3$       1 cm$^3 = 0.0610$ in.$^3$
1 ft$^3 = 0.02832$ m$^3$       1 m$^3 = 35.31$ ft$^3$
1 yd$^3 = 0.7646$ m$^3$       1 m$^3 = 1.308$ yd$^3$

### Force

| | |
|---|---|
| 1 lb $= 4.448$ N | 1 N $= 0.2248$ lb |
| 1 kip $= 4.448$ kN | 1 kN $= 0.2248$ kip |
| 1 ton $= 8.896$ kN | 1 kN $= 0.1124$ ton |
| 1 dyne $= 10 \, \mu$N | |

### Force per Unit Length

| | |
|---|---|
| 1 lb/ft $= 14.59$ N/m | 1 N/m $= 0.0685$ lb/ft |
| 1 dyne/cm $= 1$ mN/m | |

### Stress (Force per Unit Area) ($1$ Pa $= 1$ N/m$^2$)

1 lb/in.$^2$ (psi) $= 6.895$ kPa       1 kPa $= 0.1450$ lb/in.$^2$
1 lb/ft$^2$ (psf) $= 47.88$ Pa       1 kPa $= 20.89$ lb/ft$^2$
1 ton/ft$^2$ (Tsf) $= 95.76$ kPa       1 Pa $= 0.0104$ ton/ft$^2$
1 atm $= 101.3$ kPa       1 kPa $= 0.009869$ atm
1 bar $= 100$ kPa       1 kPa $= 0.0100$ bar

### Unit Weight (Force per Unit Volume)

1 lb/ft$^3 = 0.1571$ kN/m$^3$       1 kN/m$^3 = 6.366$ lb/ft$^3$

## Mass-Force and Density-Unit Weight Conversions on Earth

$1 \, \text{kg} = 2.205 \, \text{lb} = 9.807 \, \text{N}$

$1 \, \text{tonne} \ (1000 \, \text{kg}) = 2205 \, \text{lb} = 1.102 \, \text{tons} = 9.807 \, \text{kN}$

$1 \, \text{kg/m}^2 = 0.2048 \, \text{lb/ft}^2$

$1 \, \text{kg/cm}^2 = 14.22 \, \text{lb/in.}^2 = 1.024 \, \text{tons/ft}^2 = 98.07 \, \text{kPa}$

$1 \, \text{Mg/m}^2 = 204.8 \, \text{lb/ft}^2 = 9.807 \, \text{kPa}$

*Water*

$1 \, \text{g/cm}^3 = 1 \, \text{Mg/m}^3 = 62.4 \, \text{lb/ft}^3 = 9.807 \, \text{kN/m}^3$

# 1 Introduction

## 1.1 BRANCHES OF MECHANICS

"Mechanics" may be defined as that part of physical science that treats of the action of forces on masses. There are many branches of mechanics, each applicable to a particular kind of mass or classification of matter. Mechanics pertaining to astronomical bodies is called celestial mechanics. The mechanics of gases is called pneumatics; of water, hydraulics; of solid bodies in motion, dynamics; of solid bodies at rest, statics; of heat-energy transfer, thermodynamics; and so on. Similarly, the branch of mechanics that deals with the action of forces on soil masses is called soil mechanics, and on rocks, rock mechanics.

## 1.2 BRANCHES OF GEOTECHNICAL ENGINEERING

The practice of engineering that involves application of the principles of soil mechanics is called *soil engineering*. Similarly, the practice of engineering that involves application of the principles of rock mechanics may be called rock engineering. Because rock mechanics for the most part grew out of soil mechanics, a close relationship exists between these engineering disciplines, which now are collectively referred to as *geotechnical engineering*.

*Foundation engineering* is the application of geotechnical engineering for the design of foundations for structures including buildings, walls, and embankments; the total load supported by a foundation obviously must not exceed the supporting capacity of the underlying soil. Less obvious, but also of critical importance, is that settlement of the completed structure must not be excessive or uneven.

Geotechnical engineering also is involved in highway engineering and in engineering for dams.

A more recent application of geotechnical engineering is *geoenvironmental engineering*, which involves assessment, prevention, and mitigation of ground and surface water pollution from landfills, lagoons, and hazardous waste sites.

## 1.3 WHY SOIL IS DIFFERENT

It is important to recognize that soils are different from all other construction materials. A list of top 11 differences is as follows, ranging from the obvious to the esoteric. An understanding and utilization of these characteristics occupy a major portion of this book, and of geotechnical engineering.

1. Soil is cheap—usually—and therefore is our most abundant construction material. Railway and highway embankments comprise the longest man-made structures in the world, and earth dams are the largest, many exceeding the bulk of the Great Pyramid. The interior of the Great Wall is soil.

2. The weight of the soil itself is a major factor in design, in some instances being so large that it cannot even support itself, so we have a landslide. Virtually every structure ultimately derives its support from soil or from rock; those that don't either fly, float, or fall over.

3. Soils are extremely variable, ranging from a harsh jumble of angular rock fragments on a steep mountain slope to massive billows of sand in dunes, to gentle ripples of sand on a beach, to free-standing hills of loessial silt, to soft, viscous gumbo that can mire a tank.

4. Whereas most construction materials are specified and manufactured to a given purpose, soils are simply there, to be either used or avoided depending on the good or bad qualities they may possess. A geotechnical engineer should have a sufficient understanding of geology and soil science to reliably identify, test, and evaluate the relevant soil properties and property variations at a site. For example, many soils become stiffer and stronger with increasing depth because of compression under the weight of the overburden. This property is utilized by building foundations on piles that transfer load downward to stiffer soils or strata.

5. Soils are not homogeneous solids, but are composed of mixtures of discrete particles that are in contact and surround jagged, spider-like voids that are filled with varying amounts of water and air. The properties of soils depend in part on packing of the particles, which is why soils often are compacted prior to being used to support pavements or foundations.

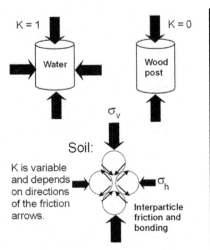

**Figure 1.1**
Soils are
particulate and
respond to
external load by
developing
friction at the grain
contacts. As a
result the $K$ ratio of
horizontal to
vertical stress is
quite variable, and
depends on the
stress history and
loading conditions.

6. When soil particles are disturbed they can exhibit a kind of mob behavior, jamming into each other like elbows on a New York subway. This can either markedly increase or decrease the soil strength. A decrease in strength explains why landslides continue to move until they find a more level geometry. Soils can be tricky.

7. Horizontal pressure from soil is an important consideration for design of retaining walls, but also is highly variable. Whereas the $K$ ratio of horizontal to vertical stress in a liquid is 1.0, and for a rigid solid is 0, in soils the ratio depends on the resistance of particles to sliding. This is illustrated in Fig. 1.1. $K$ for soils typically varies from 0.2 to 0.5, but can be much higher if high horizontal stress is inherited from overburden that has been removed by erosion.

8. Horizontal pressure depends on whether a soil is pushing or is being pushed. For example, soil piled against a retaining wall usually will exert much less pressure than if it is being pushed by a bulldozer. The reason for this is illustrated by changing directions of the friction arrows in the lower part of Fig. 1.1—whichever stress is higher, vertical or horizontal, determines the directions of the friction arrows and the $K$ ratio. A bulldozer that is designed on the basis of the wrong $K$ will have a built-in anchor. Soils can be very tricky.

9. An unsaturated soil derives part of its strength from the pull or suction of capillary water—the same pull that draws water up into a fine straw as shown at the right in Fig. 1.2. Then when the soil becomes saturated it loses this part of its strength. Some wind-blown loess soils that have never been saturated therefore collapse under their own weight if they become saturated.

10. Compressing a saturated soil causes stress to be carried in part by grain-to-grain contact and in part by pressure from the pore water, as

**Figure 1.2**

Pressure in pore water can be positive or negative, respectively decreasing or increasing frictional strength.

Water pressure pushes soil grains apart,

Capillary suction pulls grains together

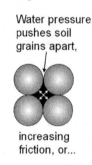

increasing friction, or...

decreasing friction.

Water

Air

shown at the left in Fig. 1.2. As pore water pressure increases, friction between the grains decreases and the soil becomes weaker. Measurement and control of pore water pressure therefore can be a major factor influencing construction.

11. Certain soil clay minerals blow up like an accordion when wet with water, and can exert enough pressure to lift pavements, floors, and foundations, and push in walls. This may come as a surprise, but the cost of repairs of damages caused by expansive clays exceeds that from landslides and earthquakes, and runs to billions of dollars per year.

This list is not complete, partly because geotechnical engineering is relatively new and the era of discovery is not over. Soils are complex materials, and tests are required to identify, evaluate, and effectively use them. Curtailing an investigation to save money up front will increase overall cost as the design then must be overly conservative to allow for any unrecognized problems. A competent geotechnical investigation is essential if problems are to be avoided, and it is axiomatic that the investigation that is not conducted prior to construction probably will be conducted later, when repairs are needed or the matter goes to court. The problems depicted in Fig. 1.3 went to court.

## 1.4   SITE INVESTIGATION

### 1.4.1   Site Geology

The first step in a site investigation is to identify the geological conditions, which can profoundly influence the plan and methods of the investigation. For example, borings performed on a hill whose flanks reveal a persistent rock outcrop will be far different from those performed in a river floodplain where rock may be tens or hundreds of meters deep. A preliminary assessment is made through recognition of surface features, reference to geological and agricultural soil maps, and records from prior borings in the vicinity.

**Figure 1.3**
This could have been avoided. Fill soil was identified by drilling, so part of the house was supported on piles. What was not recognized was that the fill was on top of an old landslide that was reactivated by weight of the fill.

## 1.4.2 Site Walkabout

A site visit by a geotechnical engineer may be somewhat bewildering to spectators as the engineer pokes around in adjacent hillsides or ravines, or even goes across the street or around the corner to observe existing structures. Such a seemingly casual inspection of nearby areas can provide important clues to site conditions and geology. For example, soil or rock layers exposed in a nearby ravine may be the same as those underlying the site—but don't count on it without other confirmation.

Often the most important clues to landslides or expansive clays are signs of distress in existing structures. Such clues can be as subtle as a pavement heaved up in the middle, or a scarp or step that may be devoid of vegetation or may support a line of younger trees extending laterally across a hillside. A series of shallow depressions can warn of the existence of subterranean caverns, or a mine or tunnel. Such observations can be extremely important because the causes can be missed by routine pattern boring; borings can reveal only what is encountered by drilling, not what is in between. Regardless of how many holes are bored, soil conditions between the borings are interpolated based on circumstantial evidence, a knowledge of the local geology, and general observations.

## 1.4.3 Drilling Program

The purpose of exploration drilling is to identify, sample, and test the soils. Drilling programs sometimes are laid out without a preliminary site visit but only if the engineer is familiar with and experienced in the area. Sometimes an initial investigation is made to select or evaluate different possible building sites and give information relative to design and construction. A design not only must be safe, it must be "buildable" within the economic and time constraints of a project.

## 1.5  SOIL TESTS

Testing is required to evaluate soils at a particular site, and normally are conducted as part of the boring program. Relatively undisturbed samples are taken for laboratory testing, and in recent years there has been a rapid increase in the use of rapid tests that are performed in situ in order to save time and money.

The footprint and complexity of a structure help to define appropriate depths and patterns of borings, and the tests that should be performed. Statistical methods are used to evaluate the results, and show that the more variable the soil, the more data are required for an accurate characterization of its properties. It also is important to define and identify different soils at a site in order to evaluate the properties of each instead of mixing apples and random dog turds.

One challenge for geotechnical engineers is to not only convince clients and their representatives of the need for a proper investigation, but also to heed the results. For example, when a pile foundation was recommended for an elaborate marble-faced building, the architect objected that he was not going to approve putting money into the ground "where it would not show." An argument was made that without an adequate foundation the consequences indeed would show, and the building was put on piles.

## 1.6  SOIL WATER

Architects or structural engineers who are not accustomed to dealing with the vagaries and inconsistencies of nature may regard soil as perverse and lacking in manners. What other material is so variable and so dependent on its environment? Even a change in the elevation of the groundwater table can have adverse effects. An important clue that may be overlooked is the soil color, gray indicating seasonal wet conditions.

## 1.7  SOIL VARIABILITY AND THE "FACTOR OF SAFETY"

Factor of safety is defined as a ratio between a design value and a failure value. For example, if a rope breaks under load $X$, the allowable load must be reduced by a factor of safety or the rope will break. Factors of safety in geotechnical engineering tend to be generous, often 3 to 5, to ensure that everything remains on the safe side. The more variable the soil, the less able one is to accurately define its average or most critical properties, and the higher the factor of safety. Thus "factor of safety" therefore in part is a "factor of ignorance," although it seldom is stated in this way, at least to a client. The less variable the soil and the more comprehensive the investigation, the lower can be the factor of safety.

Increasing a factor of safety by as little as 0.5 can add substantially to cost, so it often is advantageous to try and trim the factor of safety by having more soil tests and improving statistical reliability of the results. Thus, an inexpensive test that is performed many times may be more accurate and create more confidence than an expensive and more sophisticated test that can be performed only a few times for the same amount of money.

The factor of safety also may be lower where occasional failures are acceptable. An example is highways, where periodic repairs of weak spots is more cost-effective than overdesign of an entire project. On the other hand, an earth dam whose failure would endanger thousands of lives obviously requires a more reliable and conservative investigation and factor of safety.

## 1.8  ANCIENT APPLICATIONS OF GEOTECHNICAL ENGINEERING

The first human uses of soil and rock as engineering materials is lost in antiquity. Neanderthal or his predecessors may have been the first to recognize the advantage of structural engineering as they used a log to bridge a stream, but most effort probably was focused on simply staying alive. As glaciers receded, climatic changes raised lake levels, so people of the early Iron Age supported their lakeside dwellings on piles. Paved highways existed in Egypt several thousand years B.C.E., and were used by the pyramid builders for transportation of the construction materials. Remnants of underground cisterns, drains, tunnels, and aqueducts, and many other structures involving soil, have been unearthed at the sites of early Middle Eastern civilizations. Ancient engineers encountered and solved many practical problems in soil engineering, based on experience and trial and error.

Some primitive structures reveal an unexpected level of sophistication. Early stone arches as well as Inuit (Eskimo) igloos follow the ideal shape of a catenary similar to the St. Louis Arch, so sides do not sustain any bending moment, compared with the circular arches and domes of classical European architecture that required lateral support from columns and flying buttresses. There is a Darwinian factor in engineering, survival of the fittest.

## 1.9  EARLY LITERATURE ON SOIL ENGINEERING

In 1687, a French military engineer named Vauban set forth empirical rules and formulas for the design and construction of revetments to withstand lateral soil pressures, and nearly 200 years later, Wheeler, in his *Manual of Civil Engineering*, recommended the rules of Vauban for U.S. Military Cadets.

In 1691 Bullet, of the French Royal Academy of Architecture, presented the first theory of lateral earth pressures based on the principles of mechanics. It was he who introduced the concept of a "sliding wedge" of soil against a retaining wall. He assumed that the slip angle was 45°, which later was shown to be an oversimplification.

Coulomb in 1773 applied the principle of maxima and minima to the sliding-wedge theory to determine the most critical wedge angle, and showed that it depends in part on internal frictional properties of the soil. His formulas, while now recognized as simplifications, still are in use today and are the basis for many computer programs. It therefore is important to know the assumptions and limitations of a computer program prior to committing it to use.

Another important contribution by Coulomb was to recognize the concept that shearing resistance of soil involves two separate components, cohesion and friction. An empirical formula embodying these components now is almost universally accepted and used in geotechnical engineering practice, and is referred to as the Coulomb equation.

Coulomb's interest and insight into soil problems were inspired by his profession as a military engineer. After retiring as a result of ill health he invented the torsion balance while seeking a prize for inventing a frictionless navigational compass. He did not win the prize, but found a better use for his instrument by measuring the faint attractive and repulsive forces caused by electrical charges. He at first assumed that the attractions were inversely proportional to the separation distance, but his experiments then established that they relate to the square of the distance. Couomb's inverse square law governs not only electrostatic attractions, but also gravitational forces and, ironically, navigation of orbiting satellites.

## 1.10 NINETEENTH-CENTURY DEVELOPMENTS

In 1856 Rankine, in his treatise "On the Stability of Loose Earth," employed the concept of soil internal friction to retaining wall problems. His analysis gave a quasi-hydrostatic distribution of pressure that agreed with Coulomb's analysis based on a sliding wedge, but by application of the theory of conjugate stresses Rankine concluded that the resultant pressure on a wall acts parallel to the surface of the backfill instead of horizontally. The contributions of Rankine and Coulomb are regarded as classic and have served engineers well over the years, but are now known to be simplifications that are not precisely realized in engineering practice. For example, both theories predict that the resultant of pressure on a retaining wall acts at one-third of the height of the wall, whereas measurements indicate that it is higher because of partial support of the soil by wall friction and arching action. The consequent increase in overturning moment is covered by the factor of safety.

Also in 1856, two other concepts were introduced that play an important role in soil engineering. These are Darcy's Law defining gravitational flow of water through porous media such as soils, and Stokes' Law describing the equilibrium velocity of solid particles settling in liquids. Stokes' Law is used for measurement of fine soil particle sizes from their settlement rates.

Another important contribution in the field of soil pressures is that of Sir Benjamin Baker in 1881. Baker observed that the slip surface for a bank failure is not planar as indicated by the Rankine and Coulomb analyses, but incorporates vertically oriented cracks at the upper end. Decades earlier, in 1846, the internal structure of landslides was investigated in the field by Alexandre Collin, whose detailed cross-sections showed a curved instead of a flat slip surface. Collin's work unfortunately escaped the attention or was ignored by later workers, who perhaps were more taken with the classical theory, and was rediscovered only in recent years.

In 1871, Otto Mohr devised a simple graphical means for presentation and analysis of stress data, and "Mohr's circle" has become an invaluable tool for the modern geotechnical engineer. Mohr later confirmed and generalized the Coulomb failure criterion by relating to experimental findings, and Coulomb's earlier description of components of shearing strength sometimes is referred to as the "Mohr-Coulomb failure envelope."

Another nineteenth-century contribution that was destined to become extremely useful in modern soil engineering was the solution by a mathematician, Boussinesq (pronounced Boo-sin-esk). By use of elastic analysis, in 1885 he showed that stresses from a point load on the surface of soil should dissipate in three-dimensional space much like ripples from a stone thrown in water. Although soil is far from being an ideal elastic material, pressure measurements indicate that the Boussinesq solution is appropriate for determining pressures from foundations and for computing lateral pressures on retaining walls from loads applied at the surface of the soil backfill.

## 1.11  SOIL MOISTURE STUDIES

Early in the twentieth century soil scientists employed by the U.S. Department of Agriculture were active in the study of the mechanics of soil moisture. Among these was Briggs, who suggested a classification of soil moisture. Concurrently Buckingham proposed the concept of capillary potential and conductivity, which has led to a better understanding of the forces responsible for the retention and movement of capillary water in soils. Haines and Fisher developed an important concept of soil cohesion resulting from capillary forces or suction of water in soils that tends to pull soil grains together and increase friction.

In 1911 a Swedish scientist, Atterberg, by observing the properties of soils being molded in the hands, suggested two simple tests to determine upper and lower limiting moisture contents through which a soil exhibits the properties of a plastic solid. These tests for the "liquid limit" and "plastic limit," respectively, now form the basis for most engineering classification systems of soils.

## 1.12 MARSTON'S EXPERIMENTS

In the early 1900s at Iowa State College (now University), Anson Marston began full-scale studies of soil loads on pipes in ditches. He later extended the studies to include soil loads on culverts under soil embankments, which can result in a many-fold increase in vertical pressure on the pipe. Marston's work was extended by Spangler to include the response of flexible culvert pipes to these kinds of loads.

Marston's research perhaps is most notable because he introduced full-scale experimentation, and then attempted to explain the results from theoretical considerations. Then, perhaps most importantly, he presented the results in a simplified form that could be readily understood and used by practicing engineers.

## 1.13 TERZAGHI'S CONTRIBUTIONS

As a professor at Robert College in Turkey, Karl Terzaghi investigated a variety of soil problems and proposed the term "Erdbaumechanik" (soil mechanics) in 1925. For this and subsequent contributions Terzaghi is considered the "father of modern soil mechanics."

Terzaghi's innovative concepts derive in part from his educational background in both mechanical engineering and geology. His classical and widely used theory to explain the time-rate of consolidation of saturated soils is an adaptation of heat flow theory of thermodynamics. Terzaghi also proposed a theory for friction in soils that is used in mechanical engineering for friction in bearings.

An important part of Terzaghi's work was to verify his theories experimentally. He devised and constructed the first consolidometer, a device that now is commonplace in soil mechanics laboratories.

Again utilizing his background in mechanical engineering, Terzaghi proposed a widely used theory for foundation bearing capacity that is an adaptation of Prandtl's theory for a metal punch. He then reduced complicated mathematical relationships to a form that is readily understood and used by engineers. The relation between a diversified background and creativity is the topic of a book by Arthur Koestler (1964), who defined it as a "juxtaposition of conflicting matrices."

Terzaghi also conducted full-scale studies of pressures on retaining walls in order to test the Coulomb and Rankine theories, and dared to suggest that they might be oversimplified by ignoring an influence from soil arching.

**What Is the Classical Approach?**

The research methods of Marston, Terzaghi, and many others may be regarded as the "classical approach" in geotechnical engineering research. This involves: (1) field observation by a trained eye, (2) development of a theory to try and explain the observations, (3) experimentation to test and if necessary modify the theory, and (4) simplifications to put the new-found knowledge into practice. A fifth element is to maintain a positive outlook and never give up. Research that omits one or more of these steps may be fatally flawed. For example, it recently has become commonplace to publish computer-based analyses without experimental verifications in unintended support of the adage, "garbage in, garbage out."

In 1943, Terzaghi summarized classical soil mechanics theories in a book appropriately titled *Theoretical Soil Mechanics*, which still is in print. In 1948 he collaborated with a former student, Ralph Peck, to stress practical applications in another important book, *Soil Mechanics in Engineering Practice*, which remains in print in revised editions.

## 1.14 SOIL AS A CONSTRUCTION MATERIAL

In 1906 C. M. Strahan, a county engineer in Georgia, began a systematic study of the distribution of particle sizes in gravel-road surfaces in relation to road quality and performance. Strahan's conclusions from these correlative studies provided the basis for later research in granular soil stabilization that now plays an important role in the design and construction of highways and airport runways.

A contribution of outstanding importance based purely on experimentation is that of Proctor, who in 1933 defined modern principles of soil compaction by showing a relationship between compaction energy, moisture content, and density of a compacted soil. Proctor's test and its derivatives now are standards used in construction of virtually all soil structures including earth embankments, levees, earth dams, and subgrades for foundations or pavements.

## 1.15 SOIL CLASSIFICATION

In the 1920s, Terzaghi and Hogentogler introduced a scheme for soil classification that became the basis for the "AASHTO classification" used in highway work. In the 1940s, Arthur Casagrande of Harvard University introduced a soil classification for use by the U.S. Army in World War II, later named the "Unified Classification," and now used by most foundation engineers. Casagrande also made improvements in laboratory tests, including a mechanical device to measure the liquid limit that is based on a cog-wheel invention by Leonardo da Vinci. This device is now standard equipment in all soil mechanics laboratories.

## 1.16    LANDSLIDES

Field investigations, soil sampling, and testing gained new impetus and respect after a series of landslides was investigated by a committee appointed by the Swedish Royal Board of State Railways and chaired by Wolmar Fellenius. The results, published in 1922, included a simple method of analysis that remains the basis for a variety of modern computerized methods for evaluation of slope stability. In 1948, a textbook by MIT professor Donald W. Taylor contained considerable original material for the analysis of slope stability, with charts that were developed with the aid of his graduate students.

Starting in the 1950s, contributions towards a better understanding of clays, soil compressibility, and landslides came from Norway and the Norwegian Geotechnical Institute, led by Laurits Bjerrum and N. Janbu. In England, A. W. Bishop and A. W. Skempton respectively presented a new slope stability model and a theory explaining how slope failures in clays may be delayed for many decades. Purdue University pioneered the use of computers to solve slope stability problems, publishing a computer program that is widely used and is the basis for later copyrighted programs.

Engineering geologists use a different but more general approach to landslides by using their occurrences as a basis for landslide susceptibility maps that are particularly useful in planning. The geotechnical engineer should be aware of the availability of these maps in order to perform an intelligent investigation of a specific site.

## 1.17    FOUNDATIONS

As previously mentioned, while many bearing capacity theories have been proposed and used, the theory that forms the basis for most modern investigations is that of Terzaghi. In Canada, G. G. Meyerhoff extended and modified Terzaghi's bearing capacity theory, and at Duke University, A. Vesic suggested modifications based on model studies with sands. In the 1960s, T. W. Lambe of MIT introduced a new approach to the prediction of settlement called the "stress path method." One difficulty was the inability to accurately measure lateral soil stresses in the field, and the 1970s saw the introduction of new methods including self-boring "pressuremeters" in France and in England, and more recently the "$K_o$ Stepped Blade" in the U.S. As demonstrated by Terzaghi with his consolidometer, new instrumentation can lead to new discoveries.

The problem of sample disturbance was addressed by John Schmertmann in connection with settlement preditions, and by C. C. Ladd in connection with shear strength testing. An approach that is gaining favor is to test the soil in situ, with a variety of electronically instrumented cone-tipped penetration devices such as the "piezocone," which also monitors pore water pressure. The spade-shaped "Dilatometer" developed in Italy by Marchetti is used to measure modulus and predict settlement. The "Borehole Shear Test" developed in the U.S. measures drained or effective stress shear strength in situ, thereby avoiding disturbances from sampling.

Two important textbook references that emphasize a scientific approach to geo-technical engineering include *Soil Mechanics* by T. W. Lambe and R. V. Whitman, and *Fundamentals of Soil Behavior* by J. K. Mitchell.

## 1.18 SOIL DYNAMICS AND COMPUTER MODELING

Certain behaviors of soils in earthquakes, such as the development of quicksand or sand "liquefaction," contribute much of the damage to buildings. Studies of soil dynamics in relation to earthquake damage were pioneered by H. Bolton Seed and his associates at the University of California, Berkeley, and later by T. L. Youd. Influences of machine vibrations were studied by D. D. Barkan in Russia, and more recently by F. E. Richart at the University of Michigan.

Pile driving also involves soil dynamics, and procedures have been developed based on computer modeling of soil reactions during pile driving, a concept introduced in 1957 by a practicing foundation engineer, E. A. L. Smith.

The computer revolution also led to computer modeling of complex soil mechanics problems by finite element analysis. This requires mathematical modeling of soil strength and volume change behavior, which is difficult as these tend to be discontinuous functions. Emphasis has been directed toward refined laboratory testing to define idealized "constitutive equations" to describe soil behavior under widely varying stress environments, a problem that may be open-ended without some guiding theory.

As pile driving has gone out of favor in populated areas, alternatives have been introduced including drilled-and-filled concrete shafts. Major advances in design were made by Lyman Reece and Michael O'Neill in Texas, who developed design procedures based on full-scale load tests. A more recent and rapidly growing method involves replacing the concrete with aggregate that is rammed in place

(Rammed Aggregate Piers$^{TM}$) in order to increase lateral stress and strength of the surrounding soil.

## 1.19  GEOTEXTILES, GEOMEMBRANES, GEONETS

Another recent innovation in geotechnical engineering is the use of geosynthetic materials for drains, filters, lagoon linings, or tensile reinforcement within soil. Each use requires its own material properties, such as permeability (or impermeability), tensile strength, and toughness. Current uses of geosynthetics include acting as a separator or filter under landfills or between different soil layers, and as tensile members to improve foundation bearing capacity or stability of slopes and retaining walls.

## 1.20  ON BEING A GEOTECHNICAL ENGINEER

As implied by the above discussions, geotechnical engineering involves a broad knowledge base that includes soil mechanics and geology, and to a lesser degree groundwater hydrology, soil science, mineralogy, and statistics. This complexity precludes a "handbook approach" except at a technician level, and the M.S. or M.E. normally is considered the entry-level degree for a geotechnical engineer. One of the attractions of geotechnical engineering is that every new assignment is essentially a research project that will require a written report, so the preferred graduate degree is with thesis. Supporting course work includes courses in geology, agronomic soil survey, engineering mechanics, statistics, soil physics, clay mineralogy, and groundwater hydrology.

The demand for geotechnical engineers continues to increase as building expands into difficult or marginal sites. In many areas an environmental as well as an engineering assessment of a building site is required before building permits are granted. Insurance policies normally do not cover damages from wars or ground movements and leave it to owners and builders to try and avoid all wars and ground movements.

Most geotechnical engineers are consulting engineers, and others are employed by government agencies, highway departments, and universities. Doctorate degrees are mandatory in academia, and are increasingly common at higher levels in consulting.

Regardless of the level of academic training, the beginning consulting geotechnical engineer often starts by performing mundane chores in the field, identifying and describing soil samples as they come from borings. The next steps are to prescribe and/or perform appropriate soil tests, interpret and summarize the test results, and write reports under the direction of a senior engineer. All reports preferably will be reviewed by another geotechnical engineer. Communication skills obviously are essential. Computers and word processing are standard and make it easy to incorporate standardized "boiler plate" that can include discussions of area geology, and disclaimers prepared with the assistance of a lawyer.

## 1.21 GEOTECHNICAL ENGINEERS AND ENGINEERING GEOLOGISTS

Whereas geotechnical engineering is engineering applied to geological materials, engineering geology is geology applied to engineering problems. The two professions often share professional activities. The difference is mainly one of emphasis. Engineering geologists usually are trained first as scientists so they may approach a problem systematiclly, identifying all aspects and determining their relevance through classifications and correlations. Geotechnical engineers usually are trained first as engineers, and may prefer to look at a problem in simplified terms in order to solve it mathematically and arrive at an answer.

Engineers sometimes view geologists as being ready to devote a lifetime of study to a problem when the report must go out tomorrow, and geologists may see engineers as oversimplifying the problem to the point where important aspects are ignored or overlooked, and not fully appreciating the consequences from meddling with Mother Nature. Some consequences are readily predictable—straightening a river increases its gradient and erosion potential and takes out bridges. Sometimes where a geologist sees a river, an engineer sees a dam and reservoir. The geologist sees a reservoir that eventually and inevitably will be clogged with sediment. By combining the strengths from various specialties, engineers and scientists can form an alliance that can solve problems in a timely manner and hopefully minimize future screw-ups. Then all they have to do is convince the politicians.

## Problems

1.1. Define: (a) mechanics; (b) soil mechanics; (c) soil engineering; (d) geotechnical engineering.

1.2. Name some applications of geotechnical engineering.

1.3. Look up C. A. Coulomb in an encyclopedia and indicate whether the writer of that article was well informed concerning his engineering contributions.

1.4. List the names of six scientists and engineers whose contributions currently are being used in geotechnical engineering practice, and state the nature of each.

1.5. Who is considered the father of modern soil mechanics? Why?

1.6. Discuss features of geotechnical engineering that distinguish this field from structural engineering.

1.7. Which approach, that of the geotechnical engineer, the engineering geologist, or both, is most appropriate to solve the following problems:
   (a) Preparing a map of landslide hazards for zoning purposes, based on landslide occurrence in relation to rock type.
   (b) Figuring out how best to stop a landslide and restore an appropriate factor of safety.
   (c) Quantifying factors involved in a foundation failure, to be used as a basis for prorating damages in a lawsuit.
   (d) Identifying potential leaky bedrock around the edge of a water storage reservoir.
   (e) Designing erosion protection for a beach.

1.8. Look up information on Three Gorges Dam on the Yellow River in China and suggest some methods that might be used to reduce or prevent sedimentation of the reservoir.

1.9. What difficulty might be experienced as a result of straightening a river immediately upstream from a bridge? Why?

## Selected References

Atterberg, A. (1911). "Über die physikalische Bodenuntersuchung, und Über die Plastizität der Tone." *Internationale Mitteilungen für Bodenkunde* 1 (Part 1): 10.

Baker, Benjamin (1981). "The Actual Lateral Pressure of Earthwork." *Proc. Inst. Civ. Engrs.* (*London*) 65 (Part 3).

Boussinesq, J. (1885). *Application des Potentiels à l'Étude de l'Équilibre et du Mouvement des Solids Élastiques.* Gauthier-Villars, Paris.

Briggs, Lyman J., and McLane, John W. (1907). "The Moisture Equivalent of Soils." Bull. 45, U.S. Bureau of Soils, Washington, D.C.

Buckingham, E. (1907). "Studies on the Movement of Soil Moisture." Bull. 38, U.S. Bureau of Soils, Washington, D.C.

Casagrande, Arthur (1948). "Classification and Identification of Soils." *Trans. Am. Soc. Civ. Eng.* 113, 901.

Collin, Alexandre (1846). *Landslides in Clay.* Transl. from French by W. R. Schriever. University of Toronto Press, Toronto, 1956.

Coulomb, C. A. (1976). "Essai sur une Application des Règles de Maximis et Minimis à Quelques Problèmes de Statique, Relatifs à l'Architecture." *Mémoires de Mathématique & de Physique, présentés à l'Académie Royale des Sciences par divers Savans, & lus dans ses Assemblées,* 7, 1773, pp. 343–82, Paris. English transl. by J. Heyman, *Coulomb's Memoir on Statics*, Cambridge University Press, Cambridge, 1972.

Darcy, H. (1856). "Les Fontaines Publiques de la Ville de Dijon." Dijon, Paris.

Koestler, Arthur (1964). *The Act of Creation.* The Macmillan Co., New York.

Lambe, T. W., and Whitman, R. V. (1959). *Soil Mechanics.* John Wiley & Sons, New York.

Marston, A., and Anderson, A. O. (1913). "The Theory of Loads on Pipes in Ditches and Tests of Cement and Clay Drain Tile and Sewer Pipe." Bull. 31 of the Iowa Engineering Experiment Station, Ames, Iowa.

Mitchell, J. K. and Soga, K. (2005) *Fundamentals of Soil Behavior,* 3rd ed. John Wiley & Sons, New York.

Mohr, Otto (1871,2). "Berträge zur Theorie des Erddruckes." *Z. Arch. u. Ing. Ver. Hanover.* 17:344, and 18:67 and 245.

Proctor, R. R. (1933). "Fundamental Principles of Soil Compaction." *Engineering News-Record* (Aug. 31 and Sept. 7, 21, and 28).

Rankine, W. J. M. (1857). "On the Stability of Loose Earth." *Phil. Trans. Royal Soc., London.*

Stokes, C. G. (1851). "On the Effect of the Internal Friction of Fluids on the Motion of Pendulums." *Trans. Cambridge Phil. Soc.* 9 (Part 2): 8–106.

Strahan, C. M. (1932). "Research Work on Semi-Gravel, Topsoil and Sand-Clay, and Other Road Materials in Georgia." Bull. Univ. of Georgia 22, No. 5-a.

Taylor, D. W. (1948). *Fundamentals of Soil Mechanics.* John Wiley & Sons, New York.

Terzaghi, Karl (1925). *Erdbaumechanik auf bodenphysicalischer Grundlage.* Franz Deuticke, Leipzig and Vienna.

Terzaghi, Karl, and Peck, Ralph (1967). *Soil Mechanics in Engineering Practice*, 2nd ed. John Wiley & Sons, New York.

# 2 Igneous Rocks, Ultimate Sources for Soils

## 2.1 GEOLOGY AND THE GEOLOGICAL CYCLE

### 2.1.1 Knowing the Materials

Regardless of the field of specialization, an engineer should know and understand the materials with which he or she will be working. Naturally occurring soils and rocks are complex, and an appreciation of their particular properties can be critical to the success of a project. Formal course work in geology therefore is a prerequisite to the study of geotechnical engineering. The purpose of these introductory chapters is to serve as a reminder and to add emphasis to points that are most relevant in engineering. Students also are urged to re-read their basic text in physical geology. Large or complicated projects usually utilize the services of one or more professional engineering geologists.

### 2.1.2 Igneous Rocks as Primordial Sources for Soils

As shown in Fig. 2.1, the originating sources for soils are igneous rocks, that is, rocks that have solidified from molten material, or *magma*. Although most igneous activity occurred in past geological eras, active volcanoes are evidence that such activity continues, being concentrated along weaker zones of the Earth's crust at plate margins.

### 2.1.3 Weathering, Saprolite, and Residual Soils

The conversion of rock to soil involves physical and chemical processes of weathering. Generally, the farther a rock is removed from its nascent environment, the more vulnerable it is to changes in the environment. Thus, a rock that has formed under extremes of heat and pressure will be relatively immune

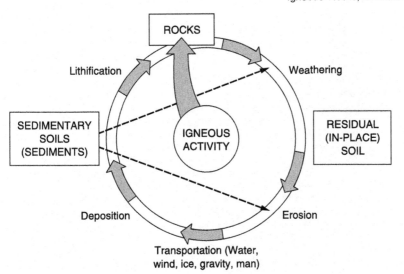

**Figure 2.1**
The geological cycle with shortcuts and subcycles.

to changes caused by heat and pressure, but if the same rock is exposed to air, acid rain, and frost action, it weathers.

*Saprolite* is from the Greek for decayed stone, and designates weathered rock that still retains the original rock structure. Saprolite looks like rock but has a distinguishing feature in that it falls apart when struck with a shovel.

Weathering proceeds from the ground surface downward, creating a mantle of *residual soil* that with increasing depth is transitional to saprolite and to unweathered rock. Borings in residual soil generally show a gradual transition from soil to soil containing rock fragments, to saprolite and solid rock.

## 2.1.4   Boulders Cut and Shaped by Weathering

It sometimes is assumed that large boulders are round because they have a history of rolling along in a stream, which might sound reasonable except that it is not easy to move a boulder that is as big as a house. Stone Mountain in Georgia is round and at last report was not going anywhere, so another explanation should be sought for rounding. Observations of granite rock outcrops reveal boulders that have been isolated and shaped by chemical weathering penetrating along intersecting fractures. This is illustrated in Fig. 2.2. The round shape is encouraged because weathering penetrating the surface causes the rock to expand and flake off like layers of an onion. This origin of boulders is important because borings intended to establish a depth to competent rock may instead hit a boulder that is floating in a soil matrix. More than one boring is required, or the boring should be extended deep by coring to assure that the rock is connected. Soils derived from igneous rocks are discussed later in this chapter.

**Figure 2.2**

Weathering of granite proceeds faster along fractures, leaving an array of rounded outcrops and boulders.

### 2.1.5   Sedimentary Soils

Residual soils are readily removed and redeposited through actions of wind, moving water, or glacial ice, to become *sedimentary soils* or *sediments*. Engineers frequently refer to these materials simply as "soils," but the term "sedimentary soils" is more descriptive and helps to flag major differences in properties and in behavior from residual soils. The properties of sedimentary soils relate to the source rock and to the transporting agent, whether it is gravity, water, wind, glacial ice, or activities of scrapers, bulldozers, and pocket gophers. For example, if the source rock is granite, granite disintegrates into sand having the same mineralogical composition as the granite; then as transportation proceeds by wind or water, softer minerals are worn away and concentrate harder minerals, in particular quartz.

### 2.1.6   Recycling of Sediments

Sedimentary soils readily can become eroded, transported, and redeposited as new sediments that reflect the source material and the most recent mode of origin. As an example, a glacial deposit that is a mixture of sand, silt, and clay can be a source for alluvium where the sand, silt, and clay are segregated.

### 2.1.7   Sediments into Rocks

Sediments that are buried for long periods of time can succumb to pressures from overlying sediments, and/or cementation by secondary minerals such as carbonates. Sediments that are cemented become sedimentary rocks—*shale, sandstone*, and *limestone*. Sedimentary rocks in turn can weather and support weathering to residual soils that can re-enter the geological cycle.

## 2.1.8  Random Fill vs. Engineered Fill

Soils that have recently been moved from their place of origin are redeposited as *fill*. The soil loses most of its identifying features and substitutes other characteristics that are forensic clues that it is recent fill, such as containing random bits of bricks or glass. Identification of random fill is critical because it has not undergone geological processes of compression and cementation that give a soil strength. Random fill therefore includes some of the weakest, most compressible, trashiest, noxious, combustible, and least predictable materials that can be encountered by engineers and builders. There are better places to build than on top of old newspapers and bedsprings. More details on recognition of random fill are presented in the next chapter and are among the more important tools used by the geotechnical engineer.

Engineered fill is selected and compacted under controlled conditions to increase its strength and decrease its compressibility. Fill that has been placed under these controlled conditions in layers is a dense, competent construction material used for foundations, highway embankments, and earth dams.

## 2.1.9  Relative Abundance of Sedimentary Rocks and Sediments

Although sedimentary materials comprise only a thin skin around an igneous earth, that skin covers most of the exposed areas of continents. As a general picture, a deep igneous rock complex is overlain by sedimentary rocks that occur in discrete layers. The top layer can be residual soil developed on either of these types of rocks, or may consist of a blanket of geologically younger clayey, silty, and sandy sediments that have been deposited by wind, water, or glacial ice.

In the last century soil erosion has been rapidly accelerated by cultivation, such that the upper layer of sediment on river floodplains often can be identified as "post-cultural." The soil that presently is being carried down streams and rivers is rapidly filling reservoirs behind dams, or building deltas extending outward into the sea. Post-cultural eroded sediment often contains agricultural chemicals that are a threat to fisheries and can create "dead spots" in the sea close to the delta.

## 2.2  WEATHERING AND SOIL MINERALS

### 2.2.1  Inherited Minerals

The mineralogical composition of sediments relates to the source rock and weathering. For example, the composition of a pile of rock fragments that have fallen off a cliff reflects the composition of the rocks in the cliff. Rock fragments making up rockfalls are released as a result of *physical weathering* by cyclical heating and cooling, wetting and drying, and freezing and thawing. The loose rock is then transported by gravity and a little bouncing to the bottom of the cliff.

### 2.2.2 Making New Minerals

*Chemical weathering* involves modification or destruction of the original minerals. Thus, if the rockfall of the preceding paragraph has been modified by chemical weathering, then eroded, transported, and redeposited as a sediment, the composition of the sediment will reflect its entire history, including composition of the parent rock, the extent to which the minerals have been modified by weathering, the transporting agent, and the mode of deposition. Because of the importance of sedimentary soils for engineering, the various possibilities are discussed in the next chapter.

Chemical weathering creates a new suite of soft, hydrated minerals called *clay minerals*. As the name implies, the fine-grained portion of soil that is clay mostly consists of these special clay minerals. Clay mineral particles are ionically active and carry a pattern of electrical charges that contribute to a distinctive and often bad behavior for engineering. When dried, a soil containing clay minerals will become a hard solid that can support a skyscraper, but the same soil when wet may be as soft as a bowl of gelatin. Clay soils are commonly at the base of landslides and are problem soils for foundations and other uses.

### 2.2.3 Conflicting Definitions of Soil

The geological term for soil is *regolith*, which is from the Greek for blanket-stone. Regolith signifies the mantle of loose, incoherent rock material at or near the Earth's surface, regardless of whether that material has been eroded and redeposited, or has been left at its place of origin from weathering. Thus, both residual soils and sediments are considered part of the regolith.

Engineers regard soil as unconsolidated surficial material that does not require blasting for removal. The engineer's soil therefore is the approximate equivalent of the geologist's regolith, except that with the development of ever-heavier earth-moving equipment the engineer's definition of soil changes, and what in the days of horse-drawn scrapers was rock now is soil. An exact definition based on physical properties therefore is important for cost estimates and bid documents, and has been central to more than a few lawsuits, as excavating costs are multiplied roughly by a factor or 10 if blasting is required. Generally the engineering definition of soil now relates to the ease of removal by a ripping tooth pulled behind the largest readily available bulldozer.

Soil scientists are agronomists whose main focus is on agriculture. The agronomic definition of soil therefore is that it is the unconsolidated surface layer that can support plant life. Thus the layer of rust on steel exposed to weathering is a soil if it supports a growth of moss or lichens.

## 2.3 GEOMORPHOLOGY AND AIRPHOTO INTERPRETATION

### 2.3.1 Geomorphology

Geomorphology is the study of landforms, which provides important clues to the composition of rocks and soils making up a landform. For example, a conical shaped hill is the mark of a volcano, and if one is aware of the mechanics of volcanoes, one can expect to find sloping layers of lava mixed with volcanic rubble and ash.

### 2.3.2 Airphoto Interpretation

Geomorphic patterns readily can distinguish different soil parent materials such as the type of bedrock, alluvial sands and clays, dune sands, etc., because where the composition of these soils is closely related to the pattern of the landform, that in turn relates to its method of formation. For example, collapsed caverns have considerable engineering significance and identify the rock as most likely being a limestone. The existence of a collapsed cavern or *sink* also is revealed by stream drainage that disappears into the sink.

The study of geomorphology is aided by *airphoto interpretation*, which literally gives a big picture. In recent years high-level satellite photography has become readily available on the internet, and gives an even bigger picture but with less detail.

Airphotos conventionally are photographed in strips along a flight path, with the camera synchronized with ground speed to create about a 60 percent overlap between adjacent photos. This offers an advantage for airphoto interpretation because overlapping pairs can be viewed with a stereoscope, the left eye seeing the left picture and the right eye seeing the right. This in effect puts the eyes where the camera was, and gives an exaggerated three-dimensional view of the ground surface, like peering down at a scale model.

Aerial photos usually are black and white, but also may be in color or made with special films such as infrared, which shows green vegetation as white and water as black. "False color" combines infrared and color photography to make infrared reflection more obvious by printing it in red. This is useful for identifying open water and assessing crops, and for military purposes such as to differentiate between living vegetation and camouflage.

The LANDSAT group of satellites gives rapid coverage and spectral diversity useful for computerized inventorying of resources. This group of satellites is in north-south polar orbits, covering areas from 120 to 200 m wide on single passes. Repetitions every few days help penetrate a problem of traditional aerial photography, cloud cover. Satellite imagery now is capable of fine resolution, and can be pointed fore and aft to generate stereo pairs for determining ground elevations relative to known benchmarks. Satellite imagery also offers a highly sophisticated selection of light wavelengths that are useful for land utilization studies and military uses.

Satellite photos of the entire world can be viewed with Google after clicking on Google Earth. This installs a computer program that enables one to move about and zoom in on a particular city or locality. In selected areas the detail is such that the view can be tilted to give a three-dimensional effect. This valuable function is offered at no charge and requires a high-speed internet connection.

The interpretation of aerial imagery should be gist for a computer, but computerized interpretation has not yet caught up with the human *photointerpreter*. However, a computer can readily compare images of the same area to detect changes, such as development of a nuclear facility.

The geotechnical engineer and geomorphologist are most closely attuned to the relationship between landscape appearance and process. A photographer sees hills where the geomorphologist sees valleys between the hills and looks for evidence of fractures, faults, tilted sedimentary rock layers, etc., as they affect the drainage pattern. Most hills consist of remnants left after erosion, although some such as volcanoes are constructive.

Airphoto interpretation always should be supplemented with geological maps and on-site studies to supply "ground truth." Examples of airphoto interpretation are shown later in this and in the next chapter.

## 2.4  THE HUMAN STORY OF PLATE TECTONICS

### 2.4.1  Continental Drift

Scientists long speculated on reasons why volcanoes tend to line up along arcs that define continental margins. Examples are the Aleutian Islands, Japan, and a "ring of fire" enclosing the Pacific Ocean basin. An explanation that was championed by a German meteorologist and explorer, Alfred Wegener, is called *continental drift*.

Wegener was not by training a geologist, and his book, *The Origins of Continents and Oceans*, was first published in 1915 and viewed with what might conservatively be called raw skepticism. Nevertheless, unlike geological concepts that derive from religious beliefs or intuition, the hypothesis of continental drift was based on hard evidence, and attracted sufficient attention that in 1928 Wegener's proposals were the subject of a symposium attended by many of the most prominent geologists of the day. Wegener's proposals were slammed verbally and in print, and his credentials were questioned. However, with the persistence of a bear smelling a hot apple pie, Wegener carried his book through three more editions.

Wegener died at the age of 50 while attempting to save a fellow Arctic explorer on the Greenland icecap. Had he lived into his eighties, he would have witnessed the most remarkable turnaround in scientific opinion since the Earth stopped being flat. Wegener's concepts were confirmed by drilling, sampling, and radio-active dating of rocks from the sea floor. Contrary to what might be assumed, the youngest rocks in the Atlantic Ocean basin are not close to the continents but are out in the middle in what is called a *rift zone*, where the ocean floor is pulling apart and pushing the continents away from each other. The rift zone also is the site of extensive volcanic activity such as in Iceland. The major rift zone in the Atlantic Ocean is called the Mid-Atlantic Ridge.

Wegener's ideas thus sparked a revolution in geological thinking that rivals an earlier revolution in biological thinking by Charles Darwin, but the matter has been much less controversial because people do not take it personally.

Wegener got his ideas from observing a geometrical "fit" of the eastern edge of the Americas to the western edges of Europe and Africa, and noting that fossils are identical up until the age of the reptiles and then diversified.

The concept of sea-floor spreading is analogous to two opposing conveyor belts that are carrying the continents apart. Many moving fragments have been discovered that are bounded by faults, and because the drifting plates are not limited to continents the continental drift hypothesis has progressed to a more general theory called *plate tectonics*.

## 2.4.2 Paleomagnetism a Digital Switch

Not only does magnetic north move about, the magnetic poles periodically switch, north for south and south for north. This rather astonishing conclusion was verified by magnetometer measurements and radioactive dating of the sea floor at different distances from the Mid-Atlantic Ridge, which show a regular progression of flip-flops in magnetism with time, with the youngest rocks closest to the Ridge.

As minerals in an igneous rock cool below a temperature called the Curie point, they record and retain a record of Earth magnetism at the time of cooling. Measurements of magnetism of the sea floor indicate a consistent symmetrical pairing of magnetic patterns on both sides of rifts such as the Mid-Atlantic Ridge. These patterns have helped to locate many more plate separations, and the oscillations confirm that paired rock strata on two sides of a rift are the same age, an additional evidence for plate drift.

As shown at the right in Fig. 2.3, the Mid-Atlantic Ridge cuts through Iceland, then runs southward like baseball stitching, parallel to the curves of Africa and following the approximate middle of the Atlantic Ocean. The location in Iceland is shown in Fig. 2.4 and explains volcanic activity and hot springs in that island. The Pacific-Indian Ocean Ridge extends southward from the San Andreas Fault through the Gulf of California, past Easter Island, westward between Australia and Antarctica, and bends back north through the Red Sea. Thus, if the marches continue, both the Gulf of California and the Red Sea will become sites of volcanic activity and eventually enlarge to become oceans.

### 2.4.3  Three Kinds of Plate Bashing

The highest mountains in the world, the Himalayas, are being created by the collision of two continental plates. At the other extreme, where two oceanic plates collide they both plunge downward, forming a deep ocean trench flanked by a line of volcanic islands or *island arc*. Examples are the Aleutians, Japan, Indonesia, the Philippines, and the Solomon Islands.

Where a continental plate collides with a thinner oceanic plate, as along the western edge of the Americas, the continental plate tends to override the oceanic

**Figure 2.3**

Plate margins, volcanoes, and earthquakes. At the left the boundary of the Pacific Plate defines the "Ring of Fire." Zig-zagging of faults was first described by H. Tuzo Wilson of Canada. (U.S. Geol. Survey.)

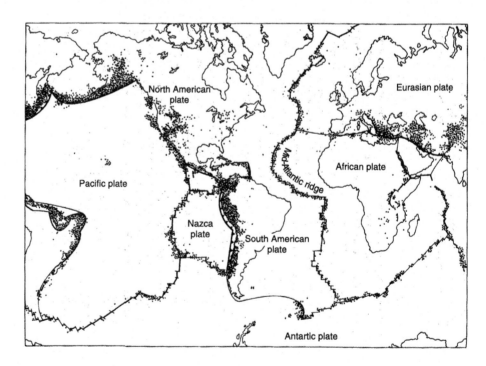

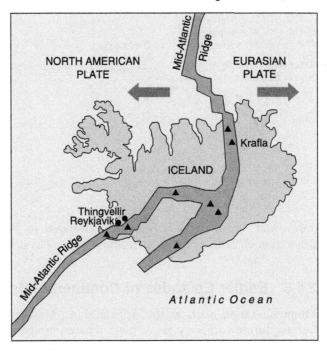

**Figure 2.4**

Plate separation creates hot spots in Iceland. (From Kious and Trilling, 1996.)

plate, which is composed of denser rocks. This pushes up mountain ranges along the leading edge of the continent, coupled with volcanism and flanked by an offshore ocean trench.

## 2.4.4   Plate Tectonics and Earthquakes

The movement of one plate over or past another seldom is smooth, but involves a frictional phenomenon called *stick-slip*. A common example is the unseemly screech of chalk on a blackboard. Sticking at plate margins does not prevent plates from moving, so elastic energy is stored in rocks on both sides of the fault. When slip finally does occur that energy is released like a spring and causes an earthquake. Since the stress relief is local, at what is termed the *epicenter*, stress is transferred to flanking areas that did not slip, making them more susceptible to slipping and causing an earthquake. Hence a ground rule is that the longer the period of inactivity, the more stress builds up, and the more severe the earthquake will be when it comes. The most dangerous areas are not where an earthquake has recently occurred, but along active faults where they have not recently occurred, in what is termed a "window" of inactivity.

## 2.4.5   What Drives Drift?

The deep interior of the earth is largely a mystery, with clues coming from the velocity of sound waves following a direct path from an earthquake epicenter through layers of the globe. The interior of the earth is hot because of radioactive

**Figure 2.5**

Convection current hypothesis to explain continental drift. The driving force may be heat generated from deep-seated radioactivity.

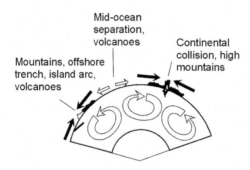

decay, and one hypothesis, which is illustrated in Fig. 2.5, is that very slow convection currents occur in soft, near-liquid rocks deep within the earth.

### 2.4.6   Earlier Episodes of Continental Drift

Mountain ranges such as the Appalachians, Urals, and Alps appear to have formed during earlier cycles of plate separation and collision. The most stable areas are located away from active plate margins in South America, Africa, Australia, and Eurasia.

## 2.5   VOLCANIC ROCKS AND VOLCANOES

### 2.5.1   Occurrence of Volcanoes

As previously indicated, volcanoes cluster along lines of plate collisions, and along lines of plate separations at mid-ocean ridges. Volcanoes outline oceanic plates of the Pacific to define the famous "Ring of Fire" that includes the Aleutian Islands and Japan. Such strings of volcanic islands are referred to as *island arcs*. One of the largest eruptions in recent history was when the island of Katmai in the Aleutians erupted violently in 1912 and blew out several cubic kilometers of material. Probably the most famous volcanic explosion is that of Krakatoa in the Strait of Java, which in 1883 shattered eardrums, blew away most of the 2300 ft (700 m) high island, created a huge tsunami ("tidal" wave), and created lasting red sunsets around the world with the dust (Winchester, 2001). (Tsunamis also are created by fault-related earthquakes, as witness the devastating 2004 tsunami in the Indian Ocean.)

Mt. Vesuvius in Italy is perennially active and is over a subduction zone as the African plate is moving northward at about one inch (2–3 cm) per year, slowly closing the Mediterranean basin while being pushed underneath the Eurasian plate.

**Figure 2.6**
One way to enforce a no-passing zone, Hawaii. Basalt is a dense, black, fine-grained rock. Light areas in the photo still are red-hot.

Many volcanic features occur in the United States. The Cascade Range includes Mts. Shasta, Rainier, and St. Helens. The Hawaiin Islands are volcanoes, and Alaska forms a margin of the "Ring of Fire." Other examples are San Francisco Mountain, Mt. Taylor, and the San Juan Mountains in Utah, Arizona, Colorado, and New Mexico.

## 2.5.2 Shield Volcanoes and the Strange State of Hawaii

Lava that has a relatively low viscosity can flow out to form domes instead of the cones. Hawaii has the largest assemblage of volcanoes having this shape, referred to as shield volcanoes.

The island of Hawaii, or the "Big Island," is the largest of the chain and the newest that extends above sea level. Farther west the islands are progressively smaller and older, finally culminating in a series of tiny islands called atolls. This is explained by a deep, stationary lava source that occasionally cuts a hole in an overlying Pacific oceanic plate to create a volcanic island. The next island already is forming to the east of Hawaii and consists of a huge volcanic mountain extruding deep under the sea. The rate of plate movement deduced from the successive ages of the volcanoes is as much as 5 cm (2 in.) per year.

Volcanic eruptions associated with Yellowstone Park also indicate plate movement over a stationary lava source. This started in what now is southeastern Oregon about 16.5 million years ago. As the continent continued to move WSW, eruptions occurred in southern Idaho and northwestern Wyoming. The last explosive eruption, 2.1 million years ago, resulted in ash being carried as far as western Iowa.

### 2.5.3   Atolls

Atolls are doughnut-like remnants of volcanic islands. Their origin was hypothesized by Charles Darwin: after a volcano dies and the island slowly sinks into the sea floor, a ring of coral growing on the flanks of the volcano continues to grow upward until all that is left is the atoll of living coral.

### 2.5.4   Craters and Calderas

Extrusion of lava relieved internal pressure in a volcano, and as cooling begins the vent area at the apex of a volcanic cone tends to subside and form a *crater*. *Calderas* are similar areas of subsidence but on a much larger scale, sometimes extending kilometers in diameter. Calderas may be a location of continuing volcanic activity, as in Hawaii. Crater Lake in Oregon actually is a caldera.

### 2.5.5   Basalt

Basalt, which comprises the bulk of most volcanoes, is an *extrusive rock*, meaning that it has solidified after having been extruded at the ground surface. Extrusion and rapid cooling freezes the rock into a fine crystal structure, and very rapid cooling creates a volcanic glass called *obsidian*. Lava that flows into the sea is quenched and may shatter and contribute to a black beach sand composed of obsidian.

One consequence of extrusion is a sharp reduction in fluid pressure that results in exsolution of gases, creating round bubble holes in the rock. The resulting *vesicular basalt* sometimes is used for landscaping. An extreme example of bubbling is *pumice* or rock froth, which is so vesicular it may float on water. Pumice can be pulverized to form a fine sanding agent for furniture and an ingredient in mechanic's soap.

### 2.5.6   Some Geomorphic Features of Lava Flows

Slow cooling of a flow after it has stopped moving can result in a hexagonal array of shrinkage cracks that create columnar joining, Fig. 2.7. If a surface crust cools sufficiently to solidify while the lava underneath is moving, the result is a harshly jumbled surface of jagged rocks. Near the borders of flows, molten lava underneath may run out, leaving a *lava tube*. After the lava has solidified, lava flows typically are very permeable so there is little or no surface runoff of rainwater, which drains through the tubes. Lava tubes exiting at sea level can collect waves and create dramatic waterspouts in the sea cliff.

### 2.5.7   Volcanic Necks

Erosion of volcanic cones gradually takes away the cone and leaves only hard rock comprising the neck as a monument. Examples are prominent in Western movies.

**Figure 2.7**
Contraction during cooling of a broad lava flow can create a distinctive pattern of vertical cracks and columnar jointing. (Author's photo from fishing trip north of Lake Superior in Canada.)

## 2.5.8 Pyroclastic Rocks

The explosive nature of some volcanic eruptions creates rocks that are igneous in origin and sedimentary in deposition, as materials are blown into the air and then settle out near and far from the volcano. "Pyro" means fire; "clastic" means fragmented. The greatest danger to life from volcanoes is not from lava flows, which usually are slow enough that people can get out of the way, but violent explosions that throw huge masses of incandescent debris, or pyroclastics, high into the air. As the particles are quite heavy they settle and form a density current that sweeps down the mountainside, leaving a burned-out trail of fire and devastation.

### Cones

Volcanic cones consist of layers of relatively soft pyroclastics interpersed with layers of hard lava. Pyroclastics are classified by size into bombs, cinders, ash, and dust. Rarely, fine stringers of glass are formed called Pelle's hair.

### Bombs

Are incandescent masses of lava that are tossed into the air and come down with a mushy thump close to the exploding crater, sometimes catching in the crotches of trees.

### Volcanic Cinders

Are much more abundant than bombs. They also fall close to the vent and annihilate all life and vegetation. Cinders often comprise the bulk of volcanic cone mountains, and small cones called *cinder cones* are almost entirely made up of cinders.

### Volcanic Ash

Being finer than cinders, can be carried tens of kilometers, and constitutes the most lethal part of most eruptions. Deposits of ash may be meters thick, particularly close to the erupting volcano. The ancient Roman towns of Pompeii and Herculaneum were buried by volcanic ash.

**Pozzolans**

Volcanic ash is important as a *pozzolan*, named after the town of Pozzuoli located on the Bay of Naples, Italy. Volcanic ash can occur wherever there has been volcanic activity; for example, the eruption that created what is now Yellowstone Park threw out ash that was carried eastward by prevailing winds for many hundreds of kilometers. As the ash cooled rapidly it is mainly composed of glass that the Romans discovered reacts with hydrated lime, $Ca(OH)_2$, to make a hydraulic cement—that is, a cement that sets under water. The famous Roman aqueducts were cemented with pozzolanic mortar, and the still-standing roof of the Parthenon is composed of pozzolanic concrete. Any fine siliceous material such as industrial fly ash that reacts with lime now is referred to as a pozzolan.

Pozzuoli, Naples, and the ancient Roman city of Pompeii are close to the still active Mt. Vesuvius, and in recent years the town of Pozzuoli has risen approximately 4 m (13 ft) due to underground magmatic activity, causing the town to be relocated.

## 2.5.9   Bentonite and Montmorillonite

Volcanic ash is highly susceptible to chemical weathering because of the disordered atomic arrangement of the constituent glass. Chemical weathering can convert entire layers of volcanic ash into a white, sticky, highly expansive clay called *bentonite*. Because of its large water-absorbing properties bentonite is utilized in drilling mud and as a sealant for lagoons and basements. The dominant clay mineral in most bentonite is *montmorillonite*, named after the town of Montmorillon in France.

Montmorillonite is one of a group of minerals called *smectites* that form in many soils and can be very troublesome in engineering. Smectites lose part of their crystalline water on drying and reabsorb it upon wetting, causing cycles of shrinking and swelling that can lift and destroy vulnerable structures. *More geotechnical engineering problems are caused by smectite-rich expansive clays than by any other type of soil.* On the other hand, small percentages of smectite in a soil can benefit plant growth by acting as temporary storage for water and plant nutrients.

## 2.5.10   Plateau Basalts

On an area basis, the most abundant lava flows are not volcanic cones, but occur where lava emerges from a fissure and can cover thousands of square kilometers. The Columbia Plateau of Washington, Oregon, and Idaho consists of repeated lava flows averaging about 30 m (100 ft) thick, and covering an area of about $520,000 \, km^2$ $(200,000 \, m^2)$. Another prominent example is the Deccan Plateau of west central India.

## 2.5.11 Weathering of Basalt

Even though the geological origins of the Columbia Plateau in the U.S. and the Deccan Plateau in India are similar, the soils developed by weathering in these two areas are very different because of the marked differences in climate and weathering. Lava flows in dry climates are not highly weathered. The contrast due to weathering is exemplified by the island of Hawaii, where lava flows on the climatically dry side of the island remain as jagged rock, while those that are younger but occur on the wet side have weathered and support lush vegetation.

Weathering of basalt is aided by its composition of high-temperature minerals and its small crystalline grain size. Generally, minerals that crystallize at high temperatures, such as those in basalt, are more readily weathered than those that crystallize at lower temperatures, such as those in granite. Quartz, as a low-temperature mineral in granite, survives and is concentrated in sands.

## 2.5.12 Dikes

Dikes are wall-like masses of igneous rock that have intruded into fractures in older rock. Dikes associated with volcanoes often form a radiating array extending outward from the volcanic neck. Dikes also are associated with granite intrusions (see below) and may extend through cracks in sedimentary rock layers, in which case they usually consist of the last minerals to be crystallized, quartz, feldspar, and mica, sometimes enriched with rare minerals including gold. Because rock in a dike often is harder than the rock into which it is intruded, erosion causes dikes to stand above the surrounding ground level.

## 2.6 OCCURRENCES OF GRANITE

### 2.6.1 Intrusive Igneous Rocks

Whereas volcanoes are composed of *extrusive* rocks, meaning that they formed from magma that was extruded at the ground surface, another category of igneous rocks is called *intrusive*, meaning that the magma has cooled more slowly under a cover of older rocks. With more time for crystallization the individual grains become larger, so individual crystals are visible with the unaided eye.

### 2.6.2 Occurrences of Granite

Granite and closely related rocks are the most abundant intrusive igneous rocks, making up the cores of most mountain ranges, where the granitic mass is called a *batholith*, which literally translates to "deep rock." As geological erosion works to pare down a mountain range, the removal of load allows the earth's crust to "float" back up, with consequent shifting and rubbing of earth crustal layers that are sources for occasional earthquakes. Granitic mountains in the U.S. include the Colorado Front Range, Sangre de Cristo Mountains, Bighorns, Beartooth, Belt, Idaho Batholith, Sierra Nevada, and the Coastal Range of southeast Alaska and British Columbia.

Granite mountains often are flanked by once-horizontal layers of sedimentary rocks that are tilted to lap up on the sides of the mountain range. The more resistant layers of sedimentary rock become high ridges called *hogbacks*. The Black Hills are a relatively small granite body that is surrounded by hogbacks.

### 2.6.3  Composition of Granite

Granite and closely related rocks such as granodiorite generally are light gray or pink in color, with a scattering of dark minerals such as biotite, or black mica, creating a "salt and pepper" effect. The main components in granite are *feldspars*, which are light-colored gray or pink minerals that show cleavage, meaning that the grains break along flat fracture surfaces. The existence of cleavage in minerals was an early clue that there might be an orderly internal arrangement of atoms comprising a crystalline structure.

Feldspars are readily weathered, forming clay minerals that incorporate $OH^-$ ions from water into their crystalline structure. The resulting increase in volume of feldspar crystals causes granite to fall apart into sand. Granitic rocks therefore are the ultimate sources for most natural sands that occur in dunes and river bars.

The clear, colorless mineral that may be seen in a broken face of granite is *quartz*. Quartz is readily recognized from its clear, glass-like appearance and the lack of any cleavage. Quartz is the last mineral to crystallize from a granitic magma, and therefore fills voids between other crystals. Quartz grains therefore initially are somewhat rounded, and become more rounded as they are repeatedly transported and deposited, then eroded, transported, and redeposited as sand, an action that can be observed with every wave washing up on a beach. The large quartz crystals that are found for sale in rock shops and are cut into slices for controlling frequencies in electronic devices are not igneous in origin, but have formed in caverns through slow accretion of ions from cold solutions.

### 2.6.4  Magmatic Differentiation

Whereas basalt is a heavy, black rock composed of high-temperature minerals, granite is a light-colored rock mainly composed of low-temperature minerals. According to a theory called *magmatic differentiation*, granite can form from a basalt magma if the heavy, high-temperature minerals crystallize first and sink. Field evidence confirming this has been discovered in thick horizontal ledges of intruded rocks called *sills*.

On the other hand, a theory that works qualitatively can suffer irreparable damage when numbers are put to it, a precaution that must be remembered. There is too much granite in the world for it all to have been derived from magmatic differentiation. Another hypothesis called "granitization" suggests that the high temperatures and solutions associated with intrusion of magma can cause it to react with and incorporate existing siliceous rocks and limestone. Evidences for this include the limestone layered structure that is incorporated into granite. Granite that originates by this means is not strictly an igneous rock but is a *metamorphic rock*, discussed later.

### 2.6.5 Older Mountains and Continents

Granite and related rocks also outcrop in geologically older and less topographically rugged mountains such as those in the Appalachian system, Piedmont Plateau, and the Blue Ridge Mountains. Granite and associated metamorphic rocks are prominent in Stone Mountain, Georgia, the White Mountains in New England, and a small area designated the St. Francis Mountains in southeast Missouri.

### 2.6.6 The Really Old Shield Areas

As erosion proceeds through hundreds of millions of years, the exposed base of a granitic mountain range broadens, so more and more granite and associated metamorphic rocks are exposed. This results in broad *continental shield areas*.

**Figure 2.8**
Aerial photo of a granitic shield area scoured clean by glacial erosion. A subtle criss-cross pattern of streams and more heavily vegetated areas defines vertical fractures. (USDA photo, Northern Minnesota.)

The complex nature of rocks exposed in a shield relates to its origin as the core of a former mountain range. In North America the *Canadian Shield* constitutes most of eastern Canada and extends southward into New York as the Adirondack Mountains, and into northern Minnesota and Wisconsin as the Superior Upland. This shield also has been eroded and stripped of much of its residual soil by continental glaciation.

### 2.6.7   The Basement Complex

The elevated positions of continents relative to oceans suggests that they are underlain by rocks that are less dense than basalt. Deep drilling has confirmed the existence of granitic rocks underlying the sedimentary layers. Another contributing factor may be a higher temperature from innate radioactivity, which would result in a lower density. Because of its unknown features it sometimes is referred to as the *basement complex*.

### 2.6.8   Geomorphology of Granite Areas

Exposures of granite, whether in a mountain range or shield area, are marked by a distinctive criss-cross network of fractures that are readily seen from the air and on airphotos, Fig. 2.8. Because weathering and erosion proceed faster along the fractures, they also influence the locations of streams, and the resulting *drainage pattern*. The tendency for streams to bend at nearly right angles when they follow fractures gives a *rectangular drainage pattern*. Rectangular drainage also can be apparent on topographic maps.

Other important clues can be seen from a car window. The tendency for weathering to proceed along intersecting fractures creates round boulders embedded in a matrix of sand, so as the sand washes away the boulders become exposed as rounded knobs. Boulders that eventually are freed to roll down the hill are sometimes mistakenly assumed to have been rounded by rolling along in streams.

As the surface of a granite mass weathers and expands, it can create enough shearing stress that surface layers will become loose and scale off in a process called "exfoliation," leaving a rounded exposure. An example is Stone Mountain, Georgia.

### Problems

2.1.  What is meant by the "Ring of Fire"?

2.2.  Describe three classes of volcanoes. Would you expect their mineralogical compositions to be the same or different?

2.3.  Look up Alfred Wegener in an encyclopedia or other article or book. Does the article mention the shabby treatment that he received from the scientific hierarchy?

2.4. Contact the USGS web site for a map showing recent earthquake and volcanic activity around the world, and see if it relates to plate tectonics.

2.5. The rate of plate movement varies between about 1 and 10 cm/yr. What are the corresponding upper and lower limits for how far two plates may have separated in 4.5 billion years? How does this compare with the circumference of the Earth at the equator, where the diameter is 12,757 km?

2.6. Name two major occurrences of granite-like rocks. How are the occurrences related?

2.7. Rounding of granite boulders sometimes is attributed to rolling by running water. Is there any other explanation? Which explanation is more likely to dominate?

2.8. Most sands are believed to derive from disintegration of granite that is about one-fourth quartz and three-fourths feldspar. Explain why these percentages typically become reversed in river and beach sands. Which do you expect is the harder mineral, quartz or feldspar?

2.9. A slow-moving lava flow threatens a village. What steps might be taken to try and stop it? (Human sacrifice is not an acceptable answer.)

2.10. Name three widely divergent ways that a volcanic eruption can take human life. Which takes the most?

## References and Further Reading

Attewell, P. B., and Farmer, I. W. (1976). *Principles of Engineering Geology.* Chapman & Hall, London.

Fristrup, Borge (1966). *The Greenland Ice Cap.* Rhodos, Copenhagen, and University of Washington Press.

Kehew, Alan B. (1995). *Geology for Engineers and Environmental Scientists,* 2nd ed. Prentice-Hall, Englewood Cliffs, N.J.

Kious, W. J., and Trilling, R. I. (1996). *This Dynamic Earth.* U.S. Government Printing Office, Washington, D.C. Online version at: pubs.usgs.gov/gip/dynamic/dynamic.html

Winchester, Simon (2003). *Krakatoa.* HarperCollins, New York.

# 3
## Special Problems with Sedimentary Rocks

## 3.1 SEDIMENTARY ROCKS

Rocks that have been weathered, eroded, and transported are first deposited as loose sediments that given time and favorable conditions eventually will solidify into sedimentary rocks. Sediments and sedimentary rocks form a relatively thin blanket covering about three-fourths of the area of the continents—a blanket has some holes where igneous mountains and shield areas stick through.

Regardless of their present positions on continents, most sedimentary rocks are marine in origin, having been deposited in former shallow seas. Only limited areas of continents were submerged at any one time, leaving adjacent higher ground as the sources for sediments.

The advances and retreats of shallow seas were relatively slow, and enabled a sandy beach deposit to be transformed into a continuous layer of sand. As a sea retreated, the layer of sand would cover a layer of clay that was being deposited in deeper water offshore. In this way a series of alternating layers of sand and clay would be deposited.

Other layers are composed of chemically precipitated calcium carbonate that is deposited farther offshore and is destined to become limestone. Other types of limestone originate in shallow water as accumulations of shells. Sedimentary rock layers therefore tend to be sequential, involving varying amounts of sandstone, limestone, and shale.

Sedimentary rocks also vary widely in resistance to weathering and erosion, so harder rocks tend to cap flat hills as softer rocks are exposed on the sides. This is the classic form called a mesa, or tableland. If layers of sedimentary rocks are tilted on edge, the harder rock layers will form high ridges and softer rocks will occupy valleys in between. The relative hardness and softness therefore is indicated by the landform and the geomorphology.

## 3.2 THE GEOLOGICAL TIME SCALE

The major subdivisions of geological time is into eras, periods, and epochs. These separations are based on fossil evidence and on extinctions. For example, the Mesozoic era ended with an abrupt termination of the dinosaurs, now usually attributed to a severe climate change following the impact of a meteor. Breaks in deposition of sediments create erosional surface "unconformities."

The "Paleozoic era" literally means "ancient-life-time" and was dominated by invertebrates including many varieties of now-extinct shellfish. The "Mesozoic era," or "middle-life-time," was a time of dinosaurs and the earliest mammals. The "Cenozoic era," or "recent-life-time," is dominated by mammals.

Unconformities are very common and subdivide a rock column into "formations" of consistent geological age. "Formation" is not a time designation but a rock designation, and a single formation may incorporate a variety of rock layers. Formations for the most part are identified from fossils, which is very important in exploration drilling for oil.

Radioactive dating depends on constant rates of radioactive decay of certain isotopes, and indicates that the majority of geological time was without life except for single-celled plants, or algae. These rocks generally have been buried under younger rocks and constitute part of the "basement complex." They may be lumped together in age as "pre-Cambrian," which means that they originated prior to the earliest, Cambrian, period of the Paleozoic era. Granitic shield areas often are referred to as "pre-Cambrian shield."

## 3.3 CLASSIFICATION OF SEDIMENTARY ROCKS

The three major groups of sedimentary rocks are *shale, sandstone*, and *limestone*. A fourth less common group is called evaporites and includes rock salt and gypsum.

### 3.3.1 Shale

Shale is by far the most abundant sedimentary rock. Although strictly speaking shale is a rock, it is relatively weak and soft, and engineers consider it to be a soil. However, shale differs from most soils in a number of ways that affect its engineering uses. Shale typically is dense and has a very low content of voids, causing it to be relatively impermeable to water. Second, shale typically has a finely layered bedding structure and readily shears along these planes. Third, shale used as a fill soil gradually degrades and reverts to clay, causing delayed difficulties and failures. While engineers may be content to call shale a soil, there are abundant reasons why it also should carry a designation as shale.

The cause of the finely layered "shaley structure" of shale is compression under large amounts of overburden coupled with geological time. This process of densification is called *consolidation*, similar to the consolidation that occurs in clay under a foundation load but to a much larger degree. Thus, whereas the void content of soils is measured in tens of percent, the void content of a shale may be only a few percent. As shale consolidates, the clay mineral platelets tend to become oriented flat and parallel to one another and create the shale structure. *Claystones* and *siltstones* are rocks having a similar grain-size composition that are not so thinly bedded.

"Consolidation" originally was proposed by an English geologist, Lyell, to explain the compression of sediments in the sea. Nevertheless, petroleum geologists refer to this process as "compaction," even though in engineering "compaction" refers to mechanically induced densification, and differs from consolidation because it squeezes out air instead of water.

### Shales and Landslides
The planes of weakness in shales are highly conducive to landslides, particularly if the rock layers have been tilted from deformations of the Earth's crust. This occurs as a result of plate collisions, faulting, folding, and mountain building. Sometimes the Earth's crust has been compressed so that rock layers are pushed at the edges to form a series of symmetrical folds like ripples on water. Some very large and tragic landslides are attributed to shale.

### Shale Rebound from Unloading
Another property of shale that contributes to a catalogue of lamentable surprises is a tendency to rebound and expand when pressure is removed by excavations. Expansion is triggered if water is available to weaken the electrostatic bonding in the clay. Excavations in shale therefore should not be allowed to remain open any longer than absolutely necessary, and should be kept dry. This can be accomplished by use of a thin layer of asphalt or Portland cement concrete.

Geologically old shale does not contain expansive clay minerals, and while pressures from elastic rebound are relatively moderate, they can lift lightly loaded floors while leaving nearby bearing walls intact. Damages are reduced if floors and walls are kept separated. However, interior partition walls founded on a heaving floor slab can be troublesome. Shale deposits that are Paleozoic or older generally have been chemically modified or "deweathered" to more stable clay minerals.

## Expansive Shale

Geologically young shale, younger than Paleozoic, often contains expansive clay minerals, or minerals that expand on wetting and shrink on drying. Tests described later in this book can determine the presence and expansive potential of expansive clay.

*Expansive clays* constitute one of the most difficult engineering problems, inflicting extensive damages on highways, floors, walls, and foundations. Several control measures may be used. A common approach is to try and keep the water out, but this may be difficult because of the high affinity of an expansive clay for water. The attraction can be reduced with *chemical stabilization*, which is most commonly accomplished using hydrated chemical lime, $Ca(OH)_2$.

## Weathering of Shale

Shale weathers by simply disintegrating back into clay. As weathering is most intense near the ground surface, several distinctive layers may form that are called a "soil profile." A typical soil profile consists of an organic-rich topsoil that is weak and should not be used in engineering construction, over a subsoil in which clay has been concentrated. The details of soil profiles depend on many factors including climate, moisture and temperature environment, erosion rate, and time, and are treated in Chapter 5.

## Geomorphology of Shale

Geomorphology is the study of landforms as they relate to geological materials and processes. Streams cutting headward into shale are like fingers randomly reaching out for water, and create a *dendritic drainage pattern*. Shale soils have a low permeability and are readily erodible, which results in a finely sculptured drainage pattern. An example is shown in Fig. 3.1. The total stream length per unit of area is defined as the "drainage density," which is high for shale.

## Why Are Stream Valleys Wider at the Top?

Unless they are cut into hard rock, valleys of small tributary streams normally are V-shaped, that is, wider at the top. It is unrealistic to suppose that streams initially were wide and became narrower as they cut downward. Instead, other processes must kick in to widen the valleys. The processes depend on slopes of the valley walls and can be sequential:

1. A vertical cut in weak soil is unstable, so the cut develops a landslide that tends to level things out.
2. A flattened slope gathers more rain so it is subjected to increased slope wash erosion. Eroded slopes tend to be concave as the elevation at the lower end reaches a local base level such as a stream or river that restricts downcutting.
3. Weathered or loose soil on a slope tends to move downhill by gravity, a slow process that is called *soil creep*. Soil creep gradually rounds the edges of unprotected hilltops.

**Figure 3.1**

Airphoto of shale area showing a fine, dendritic drainage pattern. A river cuts across the corner at the left. (USDA photo.)

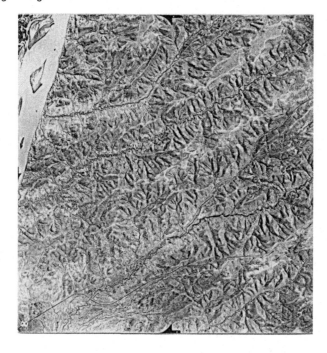

### Why Does Erosion Stop Near the Crest of a Hill?

Whereas downcutting at the toe of a slope is controlled by a local base level, the width at the crest of a hill also is subject to a different kind of control. As slopes on opposing flanks of a hill work headward and approach one another, the area of hilltop collecting rain diminishes so slope retreat slows down and may come to a stop at some "critical distance" that defines the widths of ridges. The explanation was first proposed by a hydraulic engineer, Robert Horton, who used his background in engineering to become a pioneer in the field of quantitative geomorphology. As the critical distance is relatively constant for a given soil under given conditions, this leads to a fairly constant width of hilltops, or interfluves. The edges of the interfluves are rounded by weathering and by soil creep.

### Why Are Some Hills Flat on Top?

As previously mentioned, a flat-topped hill or mesa indicates a protective layer of harder rock lying horizontal across the top. The situation changes if the rock layer is inclined, in which case it outcrops along the high side, creating landforms that are discussed later in this chapter.

Observations and explanations from geology can be quite helpful in geotechnical engineering, as a steep slope should be investigated for landslides, and a shallower slope for soil creep. Creep, although slow, eventually can multiply the soil pressure against a retaining wall, and explains why older low retaining walls that were not engineered gradually tilt and eventually may fall over.

*Question:* In Fig. 3.1 identify (a) a dendritic drainage pattern; (b) a base level for the dendritic drainage; (c) slope retreat and a critical distance; (d) soil creep; (e) a narrow, flat-topped hill that is capped with a harder rock.

### Shale as a Construction Material

The hardness of dense shale can be deceiving, because when it is broken up and used as fill it may slowly absorb water and soften into clay. The moisture content of unweathered shale can be as low as 2 or 3 percent compared with 25 to 50 percent or more in more ordinary clayey soils. Gradual softening of shale can result in slope failures mysteriously occurring years after an embankment is completed. Considerable caution should be exercised before using an untried shale in construction.

## 3.3.2 Sandstone

Sandstone is less abundant and less troubling than shale, being composed of sand grains that are cemented by clay, carbonates, iron oxides, or silica. Most sandstone is relatively soft, but some is hard enough that it can be used as coarse aggregate.

### Feldspathic Sandstone

Sandstone that is close to a granitic source can have a granite-like composition that is dominated by pink feldspars. These are referred to as "arkose." An example is rocks of the Garden of the Gods in Colorado.

### Quartz Sandstone

Far more common is sand that has been transported sufficiently to wear down feldspars and leave quartz, which is a much harder mineral. This occurs regardless of whether the sand is carried by wind or by water, as it moves by *saltation*, or bouncing along. The dominant mineral in most sandstone is quartz, and extensive reworking as a beach sand in geological time eventually can result in a nearly pure quartz sandstone that is used for manufacture of glass. "Ottawa sand" is from a sandstone of Ordovician age that has highly rounded grains and is used as a glass sand and as a reference sand in engineering tests.

### Graywacke

This is gray, fine-grained sandstone that originally was deposited in deltas. Being in relatively thick, deep deposits it is much more likely to be encountered in deep borings for petroleum than in geotechnical engineering.

### Geomorphology of Sandstone

Sandstone, being more permeable than shale, has a lower runoff and a lower drainage density, or total stream length per unit area (Fig. 3.2). Sandstone layers usually have variable cementation, such that harder layers are more resistant to erosion. Horizontal layers of harder rock therefore cap and create flat-topped hills or mesas.

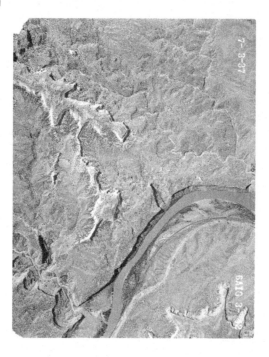

**Figure 3.2**
Mesa County, Colorado. Hard sandstone caps the flat hills, and a more resistant layer also is seen in the high bank along the river. These features are obvious in this airphoto because there is no vegetative cover, but the same relationship also exists in forested or cultivated areas. Width, 2.9 km (1.8 miles). (USDA photo.)

*Mesas* are obvious where vegetation has been held in check by a dry climate, but also occur in humid climates, where the edges are muted. The more readily erodible rock exposed on the sides of a mesa typically is shale or a softer sandstone. Therefore a search for hard aggregate will lead to the top layer of a mesa and not the sides.

### Cementation and Weathering of Sandstone

Sandstones weather back to loose sand as cementing materials are weakened or leach out. The cements that bond sand into sandstone are, in increasing order of strength, clays, iron oxides (in which case the sandstone is red), calcium carbonate, and quartz. Quartz-cemented sand in particular can be quite hard as quartz grains grow by crystallization so the enlarged crystals interlock.

### 3.3.3  Limestone

Limestone is the light-colored rock that occurs in ledges that outcrop in roadcuts or along stream valleys. Limestone is quarried and is a valuable aggregate used in engineering. However, most limestone is not satisfactory for such purposes, and limestone bedrock has some peculiarities that can present serious problems for foundation engineers.

## Why Call It Limestone?

The name limestone comes from its use to make lime for mortar, which dates from ancient times. The dominant mineral in limestone is *calcite*, or calcium carbonate, $CaCO_3$. When calcite is heated to several hundred degrees Celsius it decomposes and releases carbon dioxide, $CO_2$, leaving quicklime, $CaO$. Quicklime retains a pebble shape until it is hydrated by adding water to make hydrated lime, $Ca(OH)_2$. The name "quick" can be attributed to the heat released by the rapid hydration.

## Other Minerals in Limestone

While the most common calcium carbonate mineral in limestone is calcite, another mineral with the same chemical composition accumulates in shells of invertebrate marine animals. This mineral is "aragonite," which can be much harder than calcite due to crystallite orientations and possible interactions with organic matter. "Coquina" is soft limestone that is composed almost entirely of shells and may be used for surfacing roads. Coral reefs are another example of limestones that are made up of skeletons of marine animals.

## Dolomite

This is a closely related carbonate rock in which some of the calcium has been replaced during geological time by magnesium to make $Ca.Mg(CO_3)_2$. Dolomite generally is older than limestone and less desirable as concrete aggregate. Dolomite is both a rock name and a mineral name.

## Concretions and Caliche

Calcium carbonate is slightly soluble in water that is made acidic by the presence of $CO_2$. Carbon dioxide released by combustion of carbon compounds dissolves in atmospheric water to make the acid rain that is so devastating to monuments made of limestone, and contributes to global warming. Carbon dioxide also is released into topsoil by microorganisms. Rain infiltrating downward through topsoil therefore can become acidic and dissolve carbonate minerals that then are redeposited as rounded nodules or "concretions." In arid and semiarid climates the occasional movement of water upward and subsequent evaporation leaves a crusty deposit of calcium carbonate near the ground surface called *caliche*. Caliche sometimes is used as gravel on roads.

## Residual Soils Developed in Limestone

Competent limestone rock is suitable for supporting building foundations, but a major goal of exploration is to establish its competence. Weathering of limestone, in contrast to weathering of shale or sandstone, leaves relatively little residue after the calcium carbonate is simply dissolved away. Surface soils on limestone typically are clayey, and incorporate increasing percentages of rock fragments with depth, being transitional to solid rock. Foundation borings in limestone normally are continued deep enough into the rock to ensure that it is continuous downward, and not a disconnected piece of rock or "float" resting in clay.

### Clay Pockets in Limestone

Surface water penetrating limestone follows along vertical fractures, creating solution cavities. As shown in Fig. 3.3, these often are filled from above with residual soil. The vertical orientation of clay pockets and their random nature is such that they may be missed by exploration drilling borings and come as an unpleasant surprise when excavations are opened for construction of foundations. The surprise often can be averted by inspecting nearby outcrops and roadcuts, as in Fig. 3.3. Failing that, geophysical measurements such as earth resistivity or the use of ground-penetrating radar can be quite helpful. Such tests may be performed by geotechnical engineers or by geologists.

---

### *Case History*

A tunneling machine essentially is a horizontal drill that is pushed forward by hydraulic pistons that push against opposing plates expanded against the sides of the completed tunnel boring. Therefore in order for the machine to advance it must have competent rock to push against. Exploration borings were conducted every 50 ft (15 m) along a tunnel route and showed no clay pockets, but the machine found one that was between the borings. The machine was stopped and has no capability for backing up, so a tunnel was started from the other end so that the machine could be pulled forward, and the tunnel completion date was moved back one year. A geophysical survey probably would have prevented this.

---

### Figure 3.3

A thin layer of residual soil covers this limestone, but weathering has proceeded along vertical fractures to make solution cavities that are infiltrated with soil, creating pockets of soft clay that will not support a foundation.

## Caverns and Sinks

Whereas clay pockets penetrate from the ground surface downward, caverns occur at the level of a groundwater table and remain as open voids until the top caves in. Caverns are difficult to locate as they can readily be missed by borings, and geophysical seismic and resistivity measurements focus on what is there and not what is gone. Ground-penetrating radar would be most useful but the penetration depth is limited, particularly in clay.

Caves are of obvious concern in foundation engineering, as a roof collapse can drop part or all of a building or other structure into the ground without advance warming. The consequences can be devastating. A cavern collapse under a railroad track in South Africa left a train suspended in midair by welded rails.

A cavern that already has collapsed is a *sink*, and cavernous ground often can be recognized from the occurrence of sinks. Sinks may be obvious, as in Fig. 3.4, or also can be detected from streams that disappear and have no visible outlet.

Sinks also are a major concern for groundwater supplies, as they are direct conduits leading into aquifers. Whatever falls or is thrown into a sink therefore may be only a step away from the well, and the once-common practice of dumping everything into a sink from bald tires to dead dogs and old automobile batteries must stop.

Sinks also are a challenge for investigators because a floor may only conceal part of a deeper cavern. A hint as to consequences of a brash action may be found in animal bones lying in the bottom of a sink.

Cavernous ground impacts the feasibility of a dam because of the potential for leakage. Sealing of leaky dam foundations and abutments can be attempted by

**Figure 3.4**

Collapsed caverns, or sinks, commonly dot the landscape in limestone areas. It is the caverns that are not known and have not yet collapsed that are more likely to create serious foundation problems.

grouting, and it is not unusual for the volume of grout pumped to exceed the volume of the dam itself. Also, groundwater flow through underground channels can wash away the grout as it is injected, in which case asphalt or fibrous materials may be added to try and seal things off. Experience indicates that the prognosis for success is marginal.

### Repair of Clay Pockets and Sinks
Clay pockets or sinks encountered at a building site can be dug out and filled with rubble and concrete, or if they are too deep, bridged over. Rubble or broken concrete is grouted to make a solid plug, or the sink can be bridged by steel mesh or girders that support rubble fill that then is grouted. The grouted mass should be bowl-shaped to prevent sinking into the sink.

### What Is Karst Topography?
Terrain that is dotted with sinks to the extent that they dominate the landscape is called "karst topography" and is easily recognized on airphotos (Fig. 3.5). The name comes from a locality in the Alps, which is called a "type locality." Examples in the U.S. are in Kentucky and southern Indiana.

The ultimate karst is where sinks overlap and leave only high, straight-sided knobby hills that are subjects of classic Chinese lithographs (Fig. 3.6).

### Discovering Caverns
Underground caverns may be missed by drilling, and do not show up well on resistivity or seismic surveys. Gravity contouring is not precise, and

**Figure 3.5**

Aerial photograph of karst topography in Kentucky. (USDA photo.)

**Figure 3.6**
Remnants of karst topography along the Yangtze River, China.

ground-penetrating radar, while definitive, is handicapped by low penetration in wet soils. Cave exploration and mapping by adventurous "spelunkers" may be used but there are obvious risks, one of the most serious being a rapid rise in the groundwater table in response to rain running into the sinkholes.

### Depths of Caverns

As $CO_2$-charged rain water infiltrates vertically down through cracks and channels, it reaches and seeps laterally toward a nearby stream valley, in the process eating away at the limestone to hollow out a cavern. A groundwater table is a muted expression of ground surface elevations, so caverns normally tend to dip downward toward a neighboring valley where the groundwater may emerge as springs.

As valleys cut deeper, groundwater levels are lowered and develop cavern systems at successively lower levels. Vertically oriented channels connecting the different levels are appropriately referred to by cave explorers as "glory holes." Such stacking aids drainage and exploration of the upper levels, but lowering of the groundwater table also reduces buoyancy and in effect increases the weight of the rocks and induces collapse, particularly of shallow caverns where the roof is thinnest.

### Tunneling in Limestone

Tunneling in limestone below a groundwater table can invite precipitous flooding by encountering an open fissure or cavern. This danger is reduced if the tunnel opening is sealed off and pressurized, but costs are much higher because without careful control workers may get the "bends," or nitrogen bubbles in the blood stream during repressurizing. A temporary seal can be achieved by ground freezing. Because of the high costs associated with such contingencies, small exploration tunnels often are driven to inspect for potential hazards.

### Cherty Limestone

The famous white cliffs of Dover contain generous amounts of flint, which in this case is a black, very hard silica rock. Flint is a synonym for chert, which can be any color, but usually is tan or gray and has a waxy appearance. Flint also can designate tools or arrowheads made by "flint knapping." Skillfully dressed "Folsom points" are evidence of early man in the Americas.

Chert occurs as hard nodules or nodular layers in limestone, and dissolution of the limestone therefore leaves a chert concentrate at the ground surface. Chert is microcrystalline quartz but also often contains a noncrystalline, amorphous silica component called *opal*. Opaline chert reacts with alkalis in Portland cement concrete to make an expansive, destructive, oozing gel, and concrete aggregate containing opaline chert is called *reactive aggregate*. The reaction can be controlled by use of low-alkali cement or by the addition of a fine-grained pozzolan such as fly ash to react preferentially with the alkali.

## 3.3.4  Evaporites

A class of sedimentary rocks that is even more soluble than limestone is called *evaporites*. As the name implies, these are rocks that have been deposited through evaporation of water. Because of their high solubility, evaporites rarely outcrop at the ground surface, and then only in arid or semiarid regions. Two common minerals in evaporites are gypsum and halite, which is rock salt. Gypsum is used to make plasterboard.

Gypsum is both a rock name and a mineral name, where it identifies a hydrated calcium sulfate. Gypsum crystals occur as secondary deposits in other sedimentary rocks and soils. Sulfates in soils or in sea water cause unfavorable expansive reactions with Portland cement concrete that can be minimized or controlled by the use of sulfate-resistant cement.

Caverns form much more rapidly in evaporites compared with limestone. While rare in soils, about one-third of the U.S. is underlain by evaporite deposits, mostly in semiarid areas, and solution caverns occur in western Kansas and Texas, New Mexico, Michigan, and in Nova Scotia.

### Salt Domes

Salt layers under pressure from the overburden become plastic, and being less dense than other rock tend to squeeze into large blobs that slowly rise and punch through overlying rock layers to make *salt domes*. Salt domes are commercial sources for salt and for elemental sulphur, and rock layers penetrated by the salt can form traps for petroleum.

> **Case History**
>
> Drilling for oil into a salt dome turned out rather badly when drilling encountered a working salt mine and caused it to suddenly become flooded with water. The miners worked a hasty retreat and a fishing boat on a nearby lake nearly went down the drain.

### 3.3.5  Geomorphology of Tilted Sedimentary Rock Layers

The influence of geomorphology becomes particularly significant where rock layers are pushed up, folded, and tilted. This is common in or near mountainous areas due to plate movements. High-angle tilting results in parallel ridges of more resistant rock layers called *hogbacks* (Fig. 3.7). Tilting at a lower angle results in lower hills called *cuestas* that are steep on one side and flat on the other. Cuestas in turn are transitional to relatively flat-lying mesas.

Because hogbacks and cuestas create resistant ridges, rocks between the ridges obviously must be softer, more readily eroded, and have gentler slopes. Streams follow elongated valleys between the ridges with periodic crossings that developed while the land was being tilted. The result is a characteristic *trellis drainage pattern*. Such a pattern on maps or airphotos is a sure sign that there are tilted rock layers of varying hardness.

## 3.4  METAMORPHIC ROCKS

**Definition of Metamorphism**

Metamorphism literally means "change in form." Metamorphic rocks are igneous or sedimentary rocks that have been transformed by heat, pressure, and/or chemical solutions. A guiding principle of metamorphism, as in the case of weathering, is that the farther a material is removed from its nascent environment, the more vulnerable it is to change. Thus, igneous rocks formed at high temperatures may be only slightly modified by metamorphism at a high temperature, whereas shale subjected to the same temperature will be completely transformed. Moderate metamorphism transforms shale into slate, used for roofs and

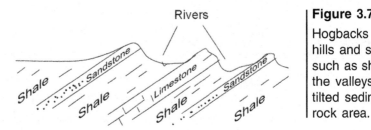

**Figure 3.7**

Hogbacks form the hills and soft rocks such as shale form the valleys in a tilted sedimentary rock area.

blackboards before plastics. More severe metamorphism can change shale into *schist*, a rock that glitters from its content of mica, which is commonly mistaken for gold.

Metamorphic rocks are less abundant than igneous or sedimentary rocks, occurring close to igneous activity. Large spotty areas are exposed in continental shield areas. Metamorphism can be expected to occur close to any igneous intrusions, including those from volcanoes.

### Mica Schist

As shale is the most abundant sedimentary rock and is easily metamorphosed, the most common metamorphic rock is mica schist. Heat and pressure transform the clay minerals into mica, which has a similar crystal structure but a much larger grain size, so individual flackes of mica are visible to the unaided eye. Mica, being close to clay minerals in composition, is fairly resistant to weathering, so soil developed from weathering of mica schist often contains abundant mica that is springy and difficult to compact. Such soils therefore are given a separate engineering classification based on their fractious physical properties.

### Marble

True marble is metamorphosed limestone. Most is imported from Italy. Some commercial "marble" is not metamorphic but is dense limestone that takes a polish.

### Quartzite

One of the hardest natural stones is sandstone that has been metamorphosed into a dense pattern of interlocking quartz grains, appropriately called quartzite. Quartzite is highly resistant to weathering. It forms a superior concrete aggregate, but crushing is expensive and wears on equipment.

### Metamorphosed Igneous Rocks

The most common metamorphic rock from an igneous source is "gneiss" (pronounced "nice"), which looks like granite but has a flow structure. Some granite also is considered to be metamorphic because it appears to retain some sedimentary rock structure. Basalt that undergoes metamorphism can recrystallize into a dark-colored layered rock that also is called schist, but does not contain abundant mica.

## 3.5 FAULTS AND EARTHQUAKES

### Definition

A *fault* is a fracture along which there has been slipping. Faults can occur in any kind of rock. Grinding action along a fault creates soil-like *fault gouge*. The more grinding that has occurred, the thicker the zone of pulverized and broken rock material that is designated a *fault zone*. Active faults are acknowledged sources of earthquakes.

A *fracture* is similar to a fault except that there is no relative movement and no gouge. Fractures are common in granite areas where as previously mentioned they influence drainage patterns, and criss-crossing of fractures allows weathering into boulders.

Because fault zones are weak and easily eroded, they strongly influence stream patterns as sections of streams follow the fault lines. Strike-slip faults where one side moves laterally with respect to the other cause obvious offsets of linear features including streams, hogbacks, roads, fences, and the pattern of trees in orchards. The San Andreas fault is a strike-slip fault that eventually will put Los Angeles close enough to San Francisco that they may be able to share the same fire department.

## Normal Faults

Faults where one side is elevated relative to the other are revealed by *fault scarps* that represent an exposed surface of the fault itself (Fig. 3.8). The high side of a fault scarp is quickly invaded by gullies, giving the scarp a triangular or trapezoidal facet that can aid in its identification. A similar topographic feature occurs where toes of hills are truncated by lateral erosion by a river.

Dating of a fault is important to establish whether a fault is active or not, and to establish earthquake recurrence intervals. Indirect evidence for recent activity includes (a) freshness of the fault scarp, (b) ages of trees growing on the exposed scarp, (c) ages of sediments cut by the scarp, and (d) age of earthquake-related features such as sand boils, determined from age of the overlying sediments.

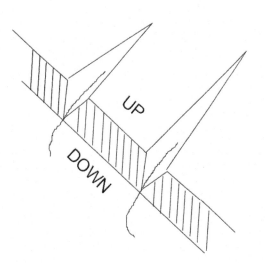

**Figure 3.8**
Diagram showing stream erosion of a fault scarp.

### Sand Boils

Sand boils are, as the name suggests, sand that has boiled up and spilled out at the ground surface, usually creating a small cone shaped like a miniature volcano. They have a considerable significance in engineering because the sand has become fluidized by vibration, and buildings that are supported on sand that liquefies as the result of an earthquake either sink into the ground or topple over. The potential of a sand to liquefy is an important consideration in earthquake-prone areas, and will be discussed in some detail later in this book. Radioactive dating of sediments covering prehistoric sand boils can provide important clues to the recurrence interval of major earthquakes—whether in tens, hundreds, or thousands of years.

## 3.6  SUBSIDENCE OVER UNDERGROUND MINES

### Mine Shafts

The most dangerous aspect of an abandoned mine is not ground settlement or subsidence, but the vertical shafts that were used for ventilation and to go in and out of the mine. Often the shafts are covered by a few loose boards and concealed in weeds, ready to victimize anything or anybody who should happen to take a false step. When an old mine shaft is discovered it can be sealed in the same manner as a sink, with steel rails or beams laid across the opening to support concrete rubble, which then is grouted to form a solid plug.

### Mining Methods

Mines in igneous rock generally pose less of a problem for cave-ins than those in sedimentary rocks that are more likely to have weak layers. Mines usually tunnel into discrete rock layers or seams. In the case of coal or limestone mines entire areas can be mined out in a waffle-like pattern of tunnels called *room-and-pillar*. This method commonly takes out about 40 to 50 percent of a rock layer and leaves the rest to support the mine roof. Additional support may be provided with timbers, or the roof can be reinforced with *rock bolts* that are installed into holes bored into the ceiling. The bolts themselves are steel rods that are grouted or epoxied in, covered with a steel plate that acts as a washer, and tightened to create an arch.

### Longwall Mining

A relatively recent innovation in coal mining first developed in England is to remove a long section of an entire coal seam with a cutter that runs back and forth on a track. The roof behind the machine is temporarily supported by hydraulic jacks that are individually released and moved forward to keep up with the mining machine. The cutter is pushed into the seam with horizontal jacks, and as mining proceeds the roof is let down, leaving little or no void. This procedure is called *longwall mining*. While longwall mining is a near-ideal arrangement from the standpoint of percent extraction, it causes differential settlement of the overlying rocks that can make an area at least temporarily unusable for support of structures.

## Abandoned Coal Mines

Most underground mines are below the natural groundwater table, so when pumping stops they become flooded with water. This is advantageous for several reasons: (1) the weight of a submerged rock mass is reduced by about one-half, which can almost make up for the stress concentration from room-and-pillar mining and helps to explain why more mines have not caved in; (2) submergence protects support timbers from decay; and (3) submergence reduces the likelihood of spontaneous underground burning. These advantages are compromised by lowering the groundwater table or if the mine is drained, which can lead to spontaneous combustion in the mine that is very difficult to control.

Flooding of a coal mine unfortunately does not completely prevent underground burning because of air pockets and roof collapses, and a pervasive coal-smoke odor often can be detected in mined-out areas. In extreme cases underground fires warm the earth up to the ground surface, causing snow melt and contributing noxious-fumes to the atmosphere. Because of the trapped heat a fire quickly rekindles as soon as any air becomes available. Dry ice may be used to cool a mine fire and put it out with $CO_2$. Another approach that may be more permanent is to pump in water and fly ash, in effect recycling the unburnable part of the coal back into the mine.

## Difficulties Locating Collapsed Mine Tunnels

Collapse of a mine does not occur all at once, but is localized where roof support is weakest. Because room-and-pillar mining follows a rectangular pattern, a hint to a collapsing mine can be a series of shallow ground depressions following a straight or intersecting lines.

Even where collapse features are evident at the ground surface, finding or isolating the cause can be difficult because rubble from a caving mine roof occupies more space than the rock did prior to caving. Caving therefore proceeds upward by a process called "stoping" (rhymes with groping). As stoping proceeds upward the cross-sectional area of a tunnel is reduced and can disappear, so collapse will never reach the ground surface. It therefore is the shallow mines that are of the greatest concern.

## Example 3.1

Assume that after roof rock falls into a mine tunnel the rock contains 10 percent voids compared to zero percent before caving. The mine tunnel is 2 m (6 ft) high and 50 m (150 ft) deep. Will stoping extend to the ground surface?

*Answer:* At the end of collapse the depth X that equals the initial height of the tunnel will contain 10 percent voids. Therefore

$$0.10 \, X = 2 \, m$$
$$X = 20 \, m \, (60 \, ft)$$

As $20 \, m < 50 \, m$ depth, the answer is no.

Boring into a collapsed mine tunnel therefore may only mean boring into some loose rock, which may be difficult to detect. The best procedure is to use a recirculating fluid such as a *drilling mud* to carry cuttings up out of the hole, because if loose rock is encountered there will be a drop in pressure of the drilling fluid.

### Mine Maps

The most useful tool for discovery of an abandoned mine should be a map made at the time of mining, but records often are inaccurate and may no longer exist. As many small mines were marginal operations using mules or horses to take out the coal or other rock, they may not have included the luxury of accurate mapping. The main sources of maps are state and federal geological surveys and bureaus of mines, and mining companies if they still exist.

### Drilling in Rock

The most common drill bit used for production drilling in rock is the Hughes tricone bit that has three toothed rollers that impact the rock. In softer rocks such as shale a hardened steel "fishtail" bit may be used. Where rock cores are required for identification or testing, a hollow-core bit laced at the end with industrial diamonds is used.

Regardless of the type of bit, some type of circulating fluid is required to cool the bit and bring up cuttings. Most common is a mud made by mixing *bentonite*, which is expansive clay mineral, with water. This is down the hollow "drill stem" to come out at the drill bit. The mud is pumped down under pressure and carries drill cuttings upward in the annulus between the drill pipe and the wall of the boring.

A bentonite mud is used because in the event of an interruption in drilling, it is "thixotropic," or touch-sensitive, so it sets into a weak gel that keeps the rock particles suspended instead of sinking down on top of the drill bit. The same muds are used in construction of retaining walls by excavating vertical slots in the soil, called *slurry trenching*.

The next chapter deals with sedimentary soils that are not cemented into rocks and comprise the bulk of engineering soils, and Chapter 5 deals with the weathering changes that result in a soil profile that can include expansive clays.

# Problems

3.1. Which is more important, recognizing a geological hazard such as an old landslide or using a sophisticated computer program for the analysis?

3.2. What engineering problems are associated with the use of shale (a) as a fill material, (b) as a foundation soil, (c) in a cut slope?

3.3. What weathering products would you expect from the following: (a) granite, (b) basalt, (c) sandstone, (d) shale, (e) limestone containing chert nodules?

3.4. What engineering problems can be associated with foundations on limestone?

3.5. Give two alternatives when a foundation excavation encounters a clay pocket in limestone.

3.6. (a) What is fault gouge, and how might it be identified from airphotos or from borings? (b) What is its significance if any for building foundations? (c) What is its significance if any for earth dams?

3.7. A house is settling and cracking, and is in line with several ground depressions in an area known to have been mined a century or more ago. (a) What may be happening and why did it take so long to happen? (b) Give two alternatives for remedying the situation.

3.8. Local residents have been using a limestone sink as a dump. You are asked to make a presentation at a meeting of the county supervisors. Outline your presentation and prepare sketches that should help to make it more convincing.

3.9. Sulphurous odors are detected in a neighborhood that is known to have old coal mines. Should there be any concern for carbon monoxide poisoning in a house built in this area? What might be done to prevent it?

3.10. Discuss problems associated with extinguishing underground coal mine fires, and keeping them extinguished.

3.11. How does the elevation of a groundwater table influence mine cave-ins? Is the same condition likely to occur with natural caverns? Why (not)?

3.12. A topographic map reveals a trellis drainage pattern. Where is the best area to explore for hard rock for aggregate?

3.13. Why are shallow mines more likely to cause problems than deep mines?

3.14. How may lowering of the groundwater table influence the stability of structures located over an abandoned room-and-pillar mine?

3.15. What is longwall mining and where was it invented?

3.16. (a) In Fig. 3.9 what is schist and what is implied by it being next to the granite? (b) The sedimentary rock layers in Fig. 3.9 are parallel but the limestone layer cuts across the schist layers. Explain. What is the geological name for the boundary between the limestone and the schist?

**Figure 3.9**

The Black Hills area of South Dakota is a classic dome that illustrates many geological relationships including hogbacks, metamorphic rock close to a granite intrusion, and escalation of erosion from uplift. The width of the dome is about 70 km (45 miles). (USGS Guidebook 25, Int. Geol. Congress, 1933.)

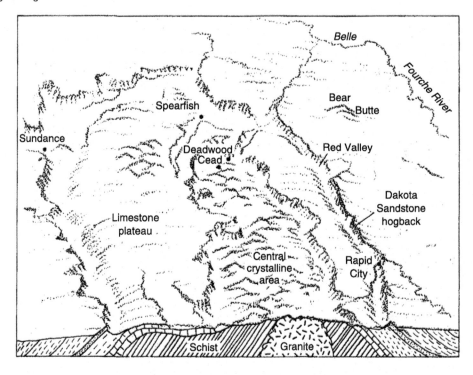

## Further Reading

Jackson, Donald D. (1982). *Underground Worlds*. Time-Life Books, Alexandria, Va.

Jackson, Julia A., ed. (1997). *Glossary of Geology*, 4th ed. American Geological Institute, Alexandria, Va.

Kehew, Alan E. (1995). *Geology for Engineers and Environmental Scientists*, 2nd ed. Prentice-Hall, Englewood Cliffs, N.J.

Martinez, J. D., Johnson, K. S., and Neal, J. T. ( 1997). "Sinkholes in Evaporite Rocks." *American Scientist*, 86(1), 38–51.

Miller, Russell (1983). *Continents in Collision*. Time-Life Books, Alexandria, Va.

Way, Douglas S., and Everett, John R. (1997). Ch. 3, "Landforms and Geology," in *Manual of Photographic Interpretation*, 2nd ed. American Society for Photogrammetry and Remote Sensing, Bethesda, Md.

# 4
## Soils That Are Sediments

## 4.1 SEDIMENTS IN ENGINEERING

### 4.1.1 Overview

Most soils used in engineering actually are sediments instead of residual soils that have been developed in place by weathering. Their properties therefore relate to sedimentation processes instead of weathering.

Sediments are geologically young compared with sedimentary rocks, and have not undergone significant lithification or hardening. Sand is a sediment, sandstone a rock. Sedimentary soils may subsequently be altered by weathering, which is the topic of the next chapter.

Sand and gravel deposited in river bars are readily recognized as sediments, as is sand that has been shaped by wind into dunes. However, other sediments that are less obvious are much more plentiful and play a more significant role in engineering. These include broad swaths of soils left plastered across the northern parts of the continents by continental glaciers and blankets of dust that were carried and deposited by winds.

An understanding of sedimentary processes helps to define and explain differences that impact engineering uses. A sediment that has been crushed under the weight of a continental glacier most likely will be an excellent foundation soil, whereas the sand that is loosely deposited in a dune is not. Some soils make excellent fill materials; others do not. Some soils expand and other soils collapse when wet with water. Sediments are the most variable of all engineering materials, and geotechnical engineers must be able to recognize, know, and appreciate their properties—the good, the bad, the indifferent, and the impossible. In some cases the best way to deal with a difficult soil is to replace it.

### 4.1.2  Transportation and Sorting

There are five common transporting agents for sediments: wind, water, ice, gravity, and animals including man.

*Sorting* refers to a tendency for sediments to be separated into uniform grain sizes, which varies according to the transporting agent. Landslide and glacial till soils exhibit little or no sorting action, particularly when compared with water- and wind-deposited soils.

A stream or a river is capable of excellent sorting action, with gravel being deposited by rapidly flowing current, sand by a moderate current, silt in a slow current, and clay with practically no current at all. Common sediments and their degrees of sorting are listed in Table 4.1.

## 4.2  SEDIMENTS AND GEOGRAPHY

### 4.2.1  Location, Location

The fact that many towns and cities were founded near rivers or seacoasts emphasizes the relevance of sedimentary soils in engineering. The reason of course is accessibility for transportation and trade. Sediments in alluvial and coastal areas often are recently deposited and relatively soft, having settled to an equilibrium density under their own weight. This means that if an additional weight such as a foundation is added, it will compress the soil and settle. The amount of settlement that can be allowed is the most common criterion used in foundation design. The soil density and stiffness tend to increase with depth, a factor that is utilized when founding structures on piles.

### 4.2.2  Settlement

Sediments that are in equilibrium with their own weight are said to be *normally consolidated*. Venice is an example of a city that is situated on a soil that still is normally consolidating, which explains why Venice is sinking. Both the amount and rate of settlement are important considerations for any design.

The position of a groundwater table also affects settlement because of the buoyant reduction in soil weight—the lower the weight, the less a soil will tend to consolidate under its own weight. For example, consider the weight of a bucket of sand both in and out of the water. According to a principle first put forth by Archimedes, the weight of the submerged object is reduced by the weight of the water that it displaces. This difference is readily calculated by knowing the density of the soil and of the water.

| Mechanism | Sediment name (sorting) | Characteristics |
|---|---|---|
| Gravity | Talus or scree (poor) | Rockfalls: steep cone-shaped deposits of loose rock fragments at the base of cliffs or mountains |
| | Landslides (poor) | Downslope sliding of a mass of soil. Usually caused by erosion and removal of soil at the toe and triggered by wet conditions |
| Water and gravity | Slope wash or colluvium (moderate) | Loose mix of soils collecting near the base of gentle hillslopes |
| Water | Alluvium (excellent) | Wide variety of soil materials from boulders to clay, deposited by streams and rivers |
| | Beach (excellent) | Loose gravel or sand deposits reworked by wave action |
| | Offshore deposits (moderate) | Soft silts and clay deposited in relatively quiet water |
| Ice | Glacial drift (variable) | Wide variety of deposits from glacial melting, including glacial till and glacio-fluvial deposits |
| | Till (poor) | Heterogeneous mix of all sizes of soil materials, sometimes firmly compacted under the weight of the ice |
| Water and ice | Glacio-fluvial deposit (good) | Sand and gravel deposited from glacial meltwater. Deposits extend down river valleys far from the glacial source |
| Wind | Dune sand (excellent) | Loose fine-grained sand |
| | Loess (good) | Thick deposits of silt that may collapse if wet, transitional to thinner deposits of silty clay with increasing distance from a source area |
| Water and plants | Peat (poor) | Fibrous plant material. Generally considered the worst soil for engineering purposes but good for potting plants |
| Man | Fill (poor) | The most random and potentially dangerous of all sediments. We know who is to blame |
| | Engineered fill (selected) | Selected soils that have been compacted under strict guidelines for use as foundation soils, earth embankments, and lagoon linings |

Table 4.1 Common sedimentary soils and their transporting agents

Buoyancy therefore is an important consideration in geotechnical engineering, and simply lowering the groundwater table, for example by pumping from wells, will increase the weight of the emerging soil by approximately a factor of two. The additional weight leads to additional compression of the underlying soil. This is dramatically illustrated in Mexico City, where pumping from wells has resulted in settlement that in some areas is measured in tens of meters.

Another option in areas of soft soils is to support structures on piles that extend down to underlying hard strata or bedrock. In this case settlement of adjacent unsupported areas can create the illusion that buildings are rising up out of the ground. In places like Mexico City a common countermeasure is to support a structure on hydraulic jacks that can be lowered to keep pace with the settlement.

### 4.2.3  Medieval Construction

Many medieval structures have survived not because the builders were better engineers, but because manpower was many times slower than machines, and centuries often were required to complete a castle or cathedral. Typically, the foundation was simply large stones laid flat, and structures settled as they were being built. Corrections then were made during building to compensate for uneven settlement and tilting.

For example, courses of masonry in the famous Leaning Tower indicate that as tilting occurred during construction and was compensated by increasing the height on the low side. Tilting then pursued in a different direction because, as soil under the low side was compressed, it became stronger and less compressible. Had the corrections not been made, the increasingly eccentric loading would have caused the tower to topple, and after each correction was made a new cycle of tilting proceeded in a different direction. It was only after the tower was completed that there was no further correction and an increasing danger of falling over. The latest correction devised by English and Italian geotechnical engineers involved adding a temporary surcharge load to the high side, which arrested further tilting, and then augering soil out from under the high side to bring it back to a safer position. What makes this tower so unique is because it survived; most medieval towers collapsed.

### 4.2.4  Surcharging

Fourteenth century construction scheduling does not fit a modern mold, and a simple procedure that can be used to reduce or control settlement is to preload the soil with a weight and pressure that are equal to that of a proposed structure, allow time for the soil to compress, and then remove the weight and build the structure. Extra weight may be used to hurry things along, and normally does no harm so long as it does not punch down into the soil sufficiently to cause a shear failure. This method of construction is called *surcharging*.

Surcharge loads usually are piles of soil that can be moved to successive building sites in a complex. Another option is to store all building materials on a site footprint slab; then as the structure is built the weight stays the same. Surcharging is common where a new highway must cross soft, compressible terrain. Surcharging can save money but requires time, and owners see time as money. It therefore is important to be able to predict the time that will be required and monitor the progress of settlement. A method described later in this book shows how to predict future settlement from measurements made over a period of time.

## 4.3 GRAVITY DEPOSITS

### 4.3.1 Talus

Steeply sloping heaps of rock fragments at the bases of rock outcrops are called *talus* or *scree*. The rubble deposits have fallen off and slid down to a marginally stable, relatively steep slope angle, Fig. 4.1. The deposits tend to be cone-shaped with the apex pointing to the source.

The stability of a talus slope is readily measured by simply walking on it, as the weight of a foot can start things sliding. For obvious reasons such a slope is not a satisfactory foundation but it sometimes cannot be avoided, particularly when building roads in mountainous areas. In that case vibration or blasting may be used to cause the deposit to settle into a more comfortable slope angle, while recognizing that more scree will be on the way. As mountainous areas usually are earthquake areas, a good shake may accomplish the same thing until the next shake that may be bigger.

**Figure 4.1**
Road constructed across a talus slope where there was not much alternative, Karakorum Highway, Pakistan.

The marginally stable slope angle of gravity deposits is called the *angle of repose*. It typically is around 35°, but may be as high as 45° in coarse, angular rubble. The angle of repose also may be observed where sand, gravel, rock, or grain are dumped onto a pile from a conveyor belt, and occurs in nature on the back sides of active sand dunes.

### 4.3.2 Rockfalls and Rock Avalanches

Rockfalls are as the name implies, and pose obvious dangers for anybody or anything that lies in their path. Methods of protection include cutting benches to catch the falls, covering steep slopes in loose rock with steel mesh, or building a roof to support sliding rocks so that the end appearance is that of an open-faced tunnel.

A *rock avalanche* involves a mass movement of loose rock. A rock avalanche is analogous to a snow avalanche but is potentially bigger and more devastating. Velocities can attain 100 km/hr (60 mph). An advance warning can be obtained by monitoring the creep rate of a potentially unstable rock mass.

### 4.3.3 Creep

Creep is an imperceptibly slow downhill movement. Creep occurring in advance of a rock avalanche can be monitored by careful measurements of the ground movement, or may be indicated by fences that gradually move out of line. Special microphones may be used to detect and monitor subaudible "rock noises" that come off as minute clicking noises associated with stick-slip. The more scientific name is *acoustic emissions*, and an increase in the rate of occurrence is a precursor to mass movement.

Curiously, rock noises apparently are audible to animals, which become visibly agitated and may attempt move off of a slope before it fails.

Creep of soil instead of rock is common on hillslopes and is not necessarily a precursor of a landslide. Creep is aided by wetting and drying cycles, and by freezing and thawing of near-surface soil. Visible indications of soil creep are slow tilting of retaining walls, fence posts, and grave markers. Another common indication is curving tree trunks because a tree corrects for early tilting by growing upright. Then as a tree becomes large enough to become firmly anchored, soil flows around it, piling up on the uphill side and leaving a shallow cavity in the tree shadow on the downhill side.

Creep is important to recognize because it increases lateral soil pressures against retaining walls and foundations.

*Colluvium* is soil that has been moved to the toe of a slope by creep or by a combination of creep and periodic alluvial activity.

### 4.3.4 Landslides

Landslides progress more slowly than rock avalanches and faster than creep. It must be emphasized that landslides are *not* the same as erosion, but involve mass downslope movements of soil or rock masses along discrete shear zones. As in the case of faults, landslides tend to grind and stir soils along the slip surfaces, creating thickened slip zones.

Landslides occur when acting forces from the weight of a soil mass equal the maximum resistance to shearing. They usually occur after prolonged periods of rain, as frictional restraint is reduced by buoyancy from a rising groundwater table. The causative factors, analysis, prevention, and repair are major concerns of the geotechnical engineer. Structures involved in landslides often are a total loss, Fig. 4.2. Insurance companies put losses from ground movements in the same category as those from nuclear war, in that they are not covered by ordinary homeowner insurance policies.

### 4.3.5 Recognizing Landslides

Engineering geologists specialize in recognizing old landslides, which is important because an old landslide readily can reactivate whenever conditions are right. One obvious clue to a landslide that can be overlooked by the unwary is a *scarp*, which is a bare, exposed part of the shear surface that defines the upper boundary of the slide. The scarp usually is steeply inclined and may be nearly vertical, and forms the riser of an arcuate or curved step. Landslide scarps become less obvious when they are overgrown with vegetation. In this case an important clue is a row of smaller trees, bush, or shrubs that follows a contour around a hillside. These are the newcomers that have grown since the scarp was first exposed.

**Figure 4.2**
Split level in Tarzana, Cal.: the house was level and then it split. A landslide not only can pull down structures, the sites are destroyed.

Counting tree rings therefore can help date the first start of a landslide. Large trees that move with a landslide usually die if their tap roots are broken.

Soil underneath a landslide is weakened by sliding so one scarp frequently leads to another farther up on a hillside. Because the slip surface typically is concave, each new sliding block tends to tilt back in the direction of the scarp and trap rainwater in shallow ponds. Infiltrating water then further aggravates sliding. Pockets of standing water therefore should be drained as a first step toward stabilizing a landslide.

Soil in a shear zone is weakened because cohesive bonds are disrupted, and as soil grains roll over one another they dilate and increase in volume, and suck in more water. As this process continues, the basal soil in a landslide can turn into a viscous mud that squeezes out in the toe area and can make access very difficult. Test boring often requires the use of track-mounted drilling machines, and even then the toe areas may be inaccessible.

Creeks or rivers running at the toe of a landslide often are pinched and may be temporarily dammed by the moving mass of soil. Since the dam is loose, unconsolidated soil, as soon as it is overtopped by flowing water, it will be breached by erosion, causing a flood downstream. Cutting a channel to prevent damming also is risky because removal of restraint at the toe will allow more sliding.

### 4.3.6   Landslide Abuse and Retaliation

When a landslide occurs the reaction is automatic—all one has to do is hire a bulldozer and put the soil back. This is about the worst thing that can happen unless one makes a hobby of nurturing and aggravating landslides. If the soil was not stable the first time it slid, it certainly will not be stable the next time around. As previously mentioned, soil in the slip zone is remolded so cohesion is lost, and it sucks in water to further reduce its strength. Pushing soil back up the slope makes reactivation of the landslide a virtual certainty.

Another way to abuse a landslide is to remove soil from the toe, because it is the weight and resistance from the toe area that stops sliding. This caper often is accomplished when the landslide is old and covered with vegetation so that its existence is not recognized or appreciated. Landslides also are triggered by natural removal of soil from the toe by erosion, in particular by lateral cutting by a river.

### 4.3.7   Concealing a Landslide

A landslide decimates property values because stopping it and repairing a structure often costs more than the structure is worth. Residential lots on landslides are virtually worthless, and laws require that such deficiencies be revealed to a potential buyer. A few developers or real estate salespeople who are either unscrupulous or blissfully ignorant will cover or smooth over a landslide scarp with fill and patch or conceal cracks in houses or other structures in order to make a sale. Any attempt to conceal a landslide is fraudulent and gives cause to nullify a sale and require a full refund, and bring the agent's license into critical review.

### 4.3.8  Precursors of Landslides

Landslides often begin with soil creep that relieves stress at the top of a slope and results in a vertical tension crack. Surface runoff water then enters the crack and helps to saturate the soil and promote more aggressive sliding. The deeper the crack, the higher the hydraulic pressure acting to move the soil. The downhill force is a product of both the water pressure and the crack area, and therefore is the square of the crack depth. A first step toward preventing sliding therefore is to seal open ground cracks to prevent entry of surface water.

---

*Case History*

Creep was observed but was a futile warning for a huge landslide that occurred in 1964 in northern Italy. The slide completely filled the reservoir above Vaiont Dam and created such a monstrous splash that the wind broke windows over a mile (1.6 km) away. The wave overtopped the dam by about 100 m (300 ft) and washed down through the valley, taking the lives of over 2000 people. Creep was observed and monitored prior to the slide, and an attempt was made to drain the reservoir, but drainage did not keep up with the rate of soil creep so the lake level kept rising. As buoyancy reduced friction at the toe, creep turned into a landslide encompassing an area of about 1.6 × 2.4 km (1 × 1.5 miles). Some of the engineers at the dam were convicted of negligence because of failure to initiate a timely evacuation.

---

### 4.3.9  Landslides and Earthquakes

Earthquakes trigger landslides by adding the effect of oscillating horizontal movements, rather like shaking peas in a pan. As the ground moves back and forth the corresponding accelerations cause horizontal forces in accordance with Newton's Law, $f = ma$: force equals mass times acceleration. The design of soil and rock slopes in earthquake-prone areas must take these forces into account.

## 4.4  DEPOSITS FROM ICE

### 4.4.1  An Outrageous Proposal

The suggestion made in 1840 by a Swiss naturalist, Louis Agassiz, that large portions of the northern continents at one time were covered by

**Figure 4.3**
Boulders that have been faceted and striated are evidence for dragging by moving ice. The grinding action of continental glaciers was the ultimate source for most agricultural soils, whether deposited by ice or by wind or water.

glaciers, may have made some people question his sobriety. Aggasiz's specialty at the time was the study of fossil fish. However, he was a competent and critical observer, and he saw similarities between deposits in North America and glacial deposits in the Alps. In particular he saw linear drag marks with a roughly north-south orientation scored into bedrock, which reminded him of home.

Boulders within glacial deposits also show scrape marks and often are flattened on one or more sides, Fig. 4.3. Agassiz became a professor at Harvard and revolutionized the teaching of natural sciences by emphasizing field study, an emphasis that also has a home in geotechnical engineering.

### 4.4.2  Extent

Prior to continental glaciation the Missouri River flowed north into what now is Hudson Bay, which occupies a basin that was pushed down below sea level by the weight of the glacial ice.

Approximately 30 percent of the continental land mass has been covered at one time or another by continental glaciers. Nearly 10 percent still is covered, including Greenland, Antarctica, and the northern islands of Canada. The top of the Greenland icecap is at an elevation of over 3000 m (10,000 ft) and the bottom is below present sea level, so the maximum thickness of continental glaciers would be measured in kilometers or miles. Further glacial melting therefore is a major concern because of the rise in sea level.

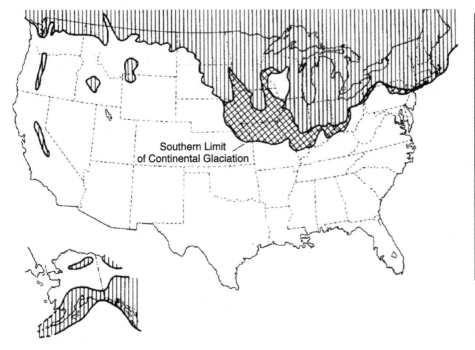

**Figure 4.4**
Glaciated areas in the U.S. Vertical shading indicates younger (Wisconsin age) glacial advances that are dominated by glacial features. Cross-hatching shows earlier glacial deposits that are incised by streams and covered with wind-blown silt or loess.

Continental glaciers once covered broad northern areas of the continents. In Europe, most of the British Isles, Scandinavia, and northern parts of Germany, Poland, and Russia were glaciated. Glaciation in North America pushed the Missouri and Ohio Rivers to their present locations and left broad areas of glacially derived sediments shown in Fig. 4.4.

Glaciation was a temporary inconvenience for mankind, but after the development of agriculture it has been a huge plus because it created a mantle of fertile soil. Several soil deposits are associated with continental glaciation—soil deposited from the ice itself, soil deposited by water from melting ice, and soil picked up by winds crossing exposed river bars that deposited dust across broad areas of uplands.

## 4.4.3  Glacial Erosion

Alpine glaciers are confined along the edges by mountains, and scoop out characteristic U-shaped valleys. In contrast, continental glaciers are free to spread, stripping soil from bedrock, leveling hills, plucking out lake basins, and carrying the glacial sediment southward to be deposited wherever the ice runs out and melts. It is no coincidence that the Great Lakes of the north-central U.S. are elongated in the directions of former glacial movement. Farther north, large seas such as Hudson Bay and the Baltic Sea owe their existence to depression of the Earth's crust under the weight of glacial ice, and even now, thousands of years after the ice has melted, the crust still is rebounding upward.

### 4.4.4  Beaches and Strandlines

The extent of depression of the earth's crust is indicated by the inclination of uplifted beaches that originally were level. Such beaches are called *strandlines*. Strandlines bordering Hudson Bay, the Great Lakes, and the Baltic Sea now are tilted in directions consistent with a hypothesis of crustal rebound. The directions of bedrock striations indicate that the Hudson Bay area was at the center of the Laurentide ice sheet, and measurements and dating of strandlines show that the Gulf of Bothnia in the northern part of the Baltic Sea is rising at a rate of about 1 cm (0.5 in.) per year.

Tilted strandlines also indicate rebound around Great Salt Lake as the lake level lowered as a result of desiccation. The ancestral lake, known as Lake Bonneville, was about 330 m (1000 ft) higher during the Pleistocene, and the land has risen about 70 m (230 ft).

### 4.4.5  Mechanics of Glacial Sliding

Rock debris concentrated in basal ice is a kind of "glacial sandpaper." The drag marks, deep longitudinal gouges left in bedrock, are called *striations*. Nevertheless the overall frictional resistance to sliding had to be low to carry the ice for such long distances with a low slope angle, and must have been aided by pressure-melting of ice at the bottom of the glacier. This would transfer the weight to liquid water trapped between the ice and the soil, thereby creating a *positive pore-water pressure* sufficient to support the ice. A similar decrease in friction from positive pore-water pressures also plays an important role in landslides.

### 4.4.6  Life in the Pleistocene

The time of glaciation is referred to as the *Pleistocene epoch* of the *Cenozoic (recent life) era*, and extended from about 2.5 million to 10,000 years ago. Subarctic cold and a generous food supply led to many experimental models of cats, cave bears, sloths, mammoths, and mastodons. A few mammoths survived in miniature on islands off the northern coast of Siberia almost into historical times, 5000 years ago. Mammoths found in permanently frozen ground, or *permafrost*, yield frozen tissues that are being studied for their DNA.

The early Pleistocene saw the emergence of man, who now probably would be referred to as intellectually challenged. The stocky and large-brained Neanderthal man appeared about 100,000 years ago, and the taller and equally large-brained Cro-Magnon man first appeared about 35,000 ybp (years before present) and threw his weight around. Modern man is Cro-Magnon with shoes and a haircut.

## 4.4.7    River Valleys in the Pleistocene

An indirect consequence of continental glaciation was a lowering of sea level, as much water was locked up in cold storage in the glaciers. The most telling evidence that this occurred is thick deposits of sediments in modern river valleys, particularly close to their outlets into the sea. These sediments only could have been deposited if sea level were lower because of a basic requirement for water to run downhill.

This influence of continental glaciation was world-wide because as sea level lowered, rivers were free to cut downward and create deep, entrenched river valleys; then as the glaciers melted and sea level rose, river valleys close to the sea became drowned as *estuaries* or, if scoured out by glacial ice, *fiords*.

## 4.4.8    Sedimentation of River Valleys

Rivers draining glaciated areas carried large amounts of sediment, a process that still can be observed in Alaska. During Pleistocene periods of glacial retreat the availability of this *glacio-fluvial* sediment and the simultaneous rise in sea level caused valleys to be filled with sediment, creating the base for broad floodplains that now extend hundreds of miles upstream from the sea. The broad valley of the Lower Mississippi River south of Cairo, Illinois, is an example. The thick sediment fill in river valleys directly influences foundation designs for bridges, etc., because of the large depth to bedrock.

---

*Question:* What volume of glacial ice would be required to lower sea level 100 m? The total surface area of the Earth is approximately 509,600,000 km², of which oceans cover about 71 percent.

*Answer:* 36,200,000 km³ = 8,700,000 cubic miles. This does not include displacement as the weight of the ice pushed down the Earth's crust.

---

## 4.4.9    Glacial Drift, the Deposit

Deposits from glaciers are collectively called *glacial drift*. The rock and mineral composition of drift reflects the source area from which it came, which in the case of continental glaciation may be only a few hundred kilometers to the north. Where glacial erosion attacked mainly granite, the resulting glacial drift will contain large quantities of sand, gravel, and boulders. Where the glacial ice gouged into shale, the resulting drift has a high percentage of clay. Incorporation of ground-up limestone by glaciers causes much glacial drift to be *calcareous*, meaning that it contains calcium carbonate. Copper and traces of silver and gold have been found in glacial deposits, and back-tracking has led to the discovery of valuable diamond-bearing kimberlite rocks in the Northwest Territories.

### 4.4.10 Glacial Outwash

Glacially derived water deposits that are carried beyond the limits of glacial advances are called *outwash*. These are mainly sand and gravel and are important sources of aggregate for use on roads and in concrete.

### 4.4.11 Glacial Till

The most abundant glacial deposit is *glacial till*, which is deposited by slow melting of ice such that there is little sorting by running water. Till deposited by modern glaciers is a mud that readily flows under its own weight, and gradually settles into a solid mass as the the soil loses water and consolidates.

*Subglacial till* has been been run over and compressed into a hard mass by the weight of the glacier. In engineering terms such a soil is said to be *overconsolidated*, meaning that it has been consolidated under a pressure that is in excess of that which exists today. The pressure involved in overconsolidation is called the *overconsolidation pressure*, also called the *preconsolidation pressure*. This is an important measure in foundation engineering because it represents a pressure that can be replaced without causing appreciable settlement. It is surcharge imposed by the weight of a glacier.

### 4.4.12 Overconsolidation Pressure and Ice Thickness

The overconsolidation pressure determined from laboratory tests may not represent the maximum pressure imposed by the weight of glacial ice because of restricted drainage as water is trapped between soil on the bottom and ice on the top. This is another evidence for the existence of positive pore-water pressure that aided glacial movement. The overconsolidation pressure determined from laboratory testing may better reflect the maximum ice thickness where the glacier has overridden a porous rock such as limestone.

### 4.4.13 Retreat of an Ice Front

Retreat of a glacier does not mean that the glacier backed up, but signifies retreat of a glacial *margin* when the rate of melting exceeds the rate of ice advance. During the transition from advance to retreat, the rate of advance temporarily equals the rate of retreat, so the ice front is stationary even though the ice still is moving. This causes a large pile-up of sediment called a *terminal moraine*. Periodic surges during the final retreat result in hills called *recessional moraines* that have a similar origin.

During the final retreat large blocks of ice may stagnate and be incorporated into the moraines. Later when the blocks of ice melt they leave steep-sided depressions called "kettle lakes." Kettle lakes are common in the northern U.S. and in the British Isles.

## 4.4.14 Two Kinds of Till

Terminal and recessional moraines are not overridden by a glacier so they are not overconsolidated. Glacial till that is not overconsolidated sometimes is called *superglacial till*, not because it is super but because it was deposited by melting from the top. A close association with water draining from the melting ice causes superglacial till to be more sandy than subglacial till, and to contain irregular pockets and layers of sand.

Superglacial till, being less dense than subglacial till, more readily weathers, so a distinction also may be made on the basis of soil color, brown on top of gray.

## 4.4.15 Ground Moraine

The uniform layer of glacial sediment left during a steady retreat of an ice front is called a *ground moraine*. Ground moraines have a gently to moderately rolling topography with internal drainage, meaning that streams collect into pools and have no exit. The resulting wet conditions in the swale areas give a ground moraine a mottled appearance on aerial photographs. Ground moraines also can show fingerprint-like patterns caused by seasonal retreats of the ice front.

Erosion by surface runoff water gradually removes soil from the shallow hills of a ground moraine and deposits it in adjacent swales, where the soil tends to be wet, clayey, and highly compressible. Swale soils also can contain expansive clay minerals, so drainage may dry them out and make them vulnerable to later rewetting.

## 4.4.16 Peat in the Swales

Peat is vegetation that has grown and died in bogs, and is protected from decay by being under water. Peat is common in ground moraine areas. As peat is mostly water, it is a very difficult soil for the engineer. Piles or piers are used to support structures such as bridges, and road embankments for roads may be built high enough that they can sink and displace the peat until it reaches solid soil at the bottom. The process is speeded up by drilling through the embankment and placing dynamite charges in the peat layer.

Another approach is to float an embankment on foamed plastic. The least expensive alternative may be to simply go around the bog. Most important is to recognize areas of peat in time to influence location, design, and construction.

## 4.5 GLACIO-FLUVIAL DEPOSITS

### 4.5.1 A Mixed Breed

Glacio-fluvial deposits are sediments deposited by water from melting glaciers. They extend beyond the margins of glaciers, but also are deposited within moraine areas as the glacier retreats.

Although glacial melting normally proceeds from the glacier surface downward, the water released by melting readily infiltrates downward through cracks and flows as a river underneath the ice. After the ice has melted the alluvial deposit remains as a ridge of sand and gravel, called an *esker*, as shown in Fig. 4.5.

*Kames* are sand-gravel mounds that accumulated in pockets in the ice. They often occur in association with kettle lakes. *Kame terraces* are deposited along edges of glaciers confined in valleys, and show evidence of collapse after the ice in the valley melted. Eskers and kames may be used as local sources for sand and gravel, and appear as light areas on airphotos because of good drainage.

### 4.5.2 Outwash

Glacial sediment carried down river valleys is referred to as *outwash* and typically consists of a wide range of coarse particle sizes, from sand up to gravel and even boulders. Outwash-carrying streams do not meander lazily down their floodplains, but race downhill in a wild series of interconnecting, rapidly shifting channels called a *braided stream*, as shown in Fig. 4.6. The rapid current in a braided stream leaves deposits of sand and gravel that may be covered with silt during waning stages of the river.

**Figure 4.5**

Subglacial stream actively forming an esker and emerging at the terminus of the Matanuska glacier, Alaska. Note the heavy concentration of sediment in the basal ice.

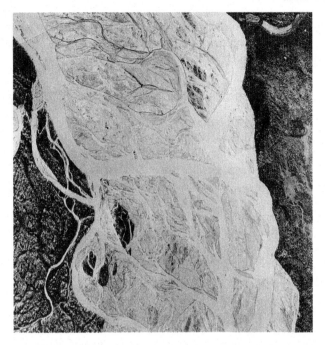

**Figure 4.6**
Aerial photograph of a braided river carrying glacial outwash in Alaska.

*Outwash plains* are created where outwash streams spill out over their banks to deposit a broad deposit of sand and gravel. Outwash plains tend to be fan-shaped as they spread out from a glacial front. Long Island and Cape Cod contain a series of such fans.

*Outwash terraces* are alluvial terraces that are elevated remnants of former floodplains that were abandoned when the river entrenched or cut downward. Outwash terraces are common along rivers that carried glacial outwash because, as the nature of the sediment being carried by the river changed, its downslope gradient changed. Generally the downhill gradient of a river carrying a full load of coarse granular outwash is steeper than that of the modern floodplain, so outwash terraces are high close to a glacial front and decrease in relative elevation with distance downstream, eventually merging into and plunging below the modern floodplain.

As a glacial front does its final retreat and the river starting point moves northward, the river normally will entrench and leave a series of step-like terraces, the oldest being the highest and the most extensively dissected by later stream erosion.

Outwash terraces are prime sources for sand and gravel. Intelligent prospecting for gravel in these areas requires an appreciation of the landforms and recognition of stream terraces. These generally appear lighter on airphotos and are confirmed with test borings.

As will be pointed out, braided streams are not limited to glacial rivers, but also occur in arid and semiarid climates where sediment is abundant and water limited.

### 4.5.3  Special Alluvial Features of Valley or Alpine Glaciers

Although continental glaciers did not invade high mountains, valley glaciers in these areas advanced during the Pleistocene and left soil deposits in lower parts of their valleys. Glacially sculpted valleys have a characteristic U-shape, compared with a V from stream downcutting, and uneven gouging often leaves a series of small lakes arranged like beads down a valley.

Another feature of glacial valleys that differentiates them from stream valleys is that tributary glaciers have elevations that are even at the top of the ice, whereas tributary streams cut down to elevations dictated by the bottom. Therefore after the valley glaciers have melted, tributary valleys are "hanging valleys" marked by waterfalls.

### 4.5.4  Glacial Damming

An interesting and significant byproduct of glaciation is temporary damming of large river valleys by ice. The scale of damming created by continental glaciation can be huge. The largest example in North America is Lake Agassiz, which was created by damming of a Missouri River that originally flowed northward through what is now the Red River valley into Hudson Bay. The Lake Agassiz plain extends from Canada southward into eastern North Dakota and western Minnesota. It is known for its flatness, its fertility for agriculture, and its deposits of expansive clay. Glacial Lake Agassiz was larger than the combined areas of all of the present Great Lakes.

### 4.5.5  Varved Clays

The rate of filling of glacial marginal lakes was seasonal, with more rapid melting during summers contributing layers of relatively light-colored silt, followed by a more gradual sedimentation of darker-colored and finer clay during winter months. The seasonal cycling results in *varved clays*, and the number of years during which sedimentation was active can be determined by simply counting the layers or varves.

### 4.5.6  Causes of Continental Glaciation

Continental glaciation was cyclical, resulting in a series of glacial advances separated by mild interglacial periods A mathematical theory developed by a Serbian engineer, M. Milankovich, and published in 1920, related cold periods to periodic changes in eccentricity of the Earth's orbit around the Sun. In the 1950s radiocarbon dates of glacial sediments challenged the theory, but detailed studies of cores of deep-sea sediments in the 1970s reveal better agreement with the predicted dates. This is the "Milankovich" or "astronomic" hypothesis.

Many other explanations also have been suggested, including climatic oscillations caused by surges of glaciers in Antarctica, or changes in reflectance of the Sun's energy as snow cover accumulated during a series of harsh winters. The Ewing-Donn hypothesis, first proposed in the 1950s and modified in the 1960s, suggested that melting of the Arctic ice pack created a source for snow that then accumulated sufficiently to depress global temperatures, which led to freezing over of the source for snow, thereby setting up a cycle.

The discovery of lithified glacial till called "tillite" indicates that continental glaciation also occurred during earlier geological eras. The causal factors affecting such dramatic climatic changes are relevant to the interpretation of global warming.

## 4.6 ALLUVIAL DEPOSITS

### 4.6.1 Down the River

Rain falling on a slope initially runs off as sheet wash, then concentrates into parallel channels or "rills" that can be observed on bare roadcuts. Rills randomly intersect and combine downslope into streams that in turn connect with one another into rivers. The random branching resembles limbs in a tree, and is called a *dendritic drainage pattern.*

Streams that are aggressively downcutting are considered "youthful." Youthful streams have relatively steep downhill gradients that enable them to flow rapidly, and therefore erode and move large particles. The flow may only be intermittent after periods of rain. During waning stages of each cycle, sand and gravel accumulate in the stream bottom. Prospectors for gold or other heavy minerals therefore pan their way upstream until traces of gold run out, and then up the adjacent hillsides.

Alluvial deposits from youthful streams are confined to narrow valleys and generally are thin and temporary, being washed away by the next major runoff event. They often co-mingle with colluvial soil brought down by gravity and sheet wash from adjacent slopes. The combined deposit can arbitrarily be called "local alluvium."

Headward erosion by youthful streams slows down when the collection area for rainfall diminishes, and essentially stops at the previously mentioned "critical distance." This distance defines the width of intervening hilltops, or "interfluves."

### 4.6.2 Playfair's Law and Horton's Demonstration

This relationship between stream size and valley size was noted by John Playfair, a Scottish mathematician and philosopher, in 1802, and he reasoned that streams therefore must cut their valleys. A relationship to geomorphic landform was formulated in 1945 by an engineer, R. E. Horton. Horton defined the smallest headwater streams as *first-order* streams. Two or more first-order streams then combine to make a *second-order stream*; two or more second-order streams combine to make a *third-order stream*; and so on. The highest-order stream in North America is the Mississippi River, which is tenth order. Through this simple numbering system Horton showed that the higher the stream order, the larger the drainage basin.

### 4.6.3 Base Level

The term "base level" was proposed in 1875 by an American geologist, John Wesley Powell, a civil war veteran whose expedition was the first to traverse the Grand Canyon in boats. Powell saw firsthand that the depth of cutting by a stream is governed by thresholds of hard rock that create a *base level*. The ultimate base level is sea level; the bottom of a river bed can erode below sea level, but this depth is limited because the water must run downhill. Localized base levels can occur anywhere along the length of a river or stream where it encounters harder rock, or where an excess of sediment is carried in and deposited by a tributary stream.

Because of the existence of an ultimate base level, the gradient or slope of rivers generally increases upstream, and in general the lower the stream order, the higher the stream gradient.

### 4.6.4 Meandering Streams

As downward erosion is halted by a base level, excess energy becomes directed toward *meandering*, or lateral erosion into a series of sinusoidal loops that lengthen the river channel and therefore decrease its gradient, which in turn slows the flow and decreases erosion. Meanders normally are spaced at an interval of about five to seven river widths, so small rivers have small meanders and large rivers have large meanders.

Meandering streams still erode, as the momentum of water flowing around a meander loop carries it to the outside of the bend, where it erodes the channel wider and deeper. The main thread of flow, called a "thalweg," moves back and forth across a meandering river in order to impinge on the outside of each meander. The straight section between adjacent meanders is a relatively shallow, sandy *reach*. The changing depths of a meandering river channel become obvious to the boater as a boat drags bottom in the reaches. Prior to the use of bridges and ferries, reaches were sought out for fording, but can be areas of quicksand, which is discussed in Section 14.9.

Meandering presents a dynamic equilibrium between erosion and deposition. A stream that has reached this balance is said to be *mature*. There is an important implication in this, because straightening a meandering river makes it shorter, and increases its gradient and potential for erosion. Straightening a river upstream from a bridge therefore can be a bad idea unless the river banks are protected from erosion. This normally is accomplished by driving steel sheet pile or placing large rocks called "rip-rap."

## 4.6.5 River Meandering and Property Lines

Shifting river channels undercut soil along the outside of a meander so that it slides off into the river, and simultaneously deposits sand bars on the inside of the meander. Courts make a legal distinction between a rapid change caused by channel abandonment, referred to in legal terms as an "avulsion," and a slow, gradual shift due to bank erosion, referred to as an "alluvion." An avulsion occurs when one meander loop catches up with another, and leaves legal boundaries intact, whereas a slow channel migration takes property lines with it. An avulsion in 1898 left the town of Carter Lake, Iowa, on the Nebraska side of the Missouri River, so Omaha residents drive through Iowa to get to their airport, which would seem an ideal place to require a toll.

## 4.6.6 Floodplains of Meandering Rivers

Lateral meandering creates relatively level floodplains that, as the name implies, are subjected to flooding. Floodplains nevertheless are favored industrial sites because they are close to river transportation.

River meandering and associated sorting activities create a host of different sedimentary floodplain soils that vary from relatively clean gravel and sand to heavy, soft clay. Most deposits are readily identifiable from their landforms and their appearance on airphotos, as shown in Fig. 4.7. The most important deposits are as follows:

### Point Bars
Probably the most conspicuous deposit of a meandering river is sand that occupies the inner area of each meander loop. The term "point bar," like many other terms used to describe rivers, comes from river navigation. As meander loops migrate downstream the point bars are like footprints forming a line down the river. Because the point bar inside one meander loop is directly across the river from the next one, they form a continuous band of sand that is criss-crossed by the river channel. Bridges across meandering rivers therefore, at least in part, are supported on point bars. After they are deposited, point bars soon are covered by vegetation, but their identity still can be determined from their position relative to existing or former channels of a meandering river.

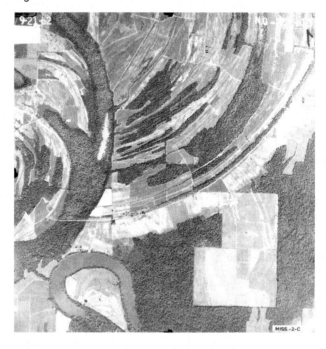

**Figure 4.7**

Filled former river channels (clay plugs) occur at curved outer margins of the extensive point bar sand deposits. A newer channel has remnants of an oxbow lake. The downstream progress of this meander was halted by the older clay plug, which caused the cutoff. The smaller oxbow probably involves a tributary river. At the lower right, patterns on older parts of the floodplain are obscured by trees and surficial backswamp clay deposits. The lighter area rimming the larger clay plug has a road and is a natural levee silt deposit. (USDA photo.)

Meander migration is not smooth owing to periodic bank caving and flood surges, so point bars typically contain arcuate ridges that represent former river margins. During high river stages the shallow channels across a point bar may be reoccupied and can even develop their own scaled-down meander patterns. Riverboat pilots called these channels "chutes," and used them as shortcuts during high water. The risk was in getting stuck when the river level went down.

### Oxbow Lakes and Clay Plugs

If for any reason the downstream progress of a meander is impeded, the next one upstream may catch up, causing a *neck cutoff*. This leads to a sequence of events that is very important in geotechnical engineering. First, the abandoned river channel rapidly becomes plugged at the ends to create an *oxbow lake*. Because the oxbow is isolated from the river, fine sediment carried into the lake during high river stages is trapped and slowly settles out to form a clay deposit called a *clay plug*. Clay plugs are poor foundation materials and are readily identified from airphotos. Where clay plugs cannot be avoided, special measures such as temporary surcharging will be required to reduce or control settlement.

Particularly devastating is if a clay plug is not recognized so that only part of a structure settles.

The thickness of a clay plug depends on the depth of the river channel *at the time of the cutoff*, which, as it occurred during a period of high water, will tend to be deeper than the existing channel. Clay plugs are thicker in the central area of the meander where the river channel was deeper.

### 4.6.7 Clay Plugs and Cutoffs

Even though clay plugs are relatively soft clay, they are slow to erode and therefore can hold back the downstream progress of a meander loop, which allows the next loop to catch up and create a cutoff, which in turn initiates another cycle of oxbow lake, clay plug, and cutoff. A clay plug that is not otherwise visible from the ground or from the air may be revealed by its effect on a river channel.

### 4.6.8 Meander Belt

The part of a mature river floodplain that is subjected to active meandering often is confined by lines of clay plugs that have the appearance of uneven parentheses running down the floodplain. This is a *meander belt*, Fig. 4.8. Outside of the meander belt, periodic flooding and deposition of clay gradually obscures the older meander patterns and oxbows so that they can only be detected with borings.

Occasionally a river will escape from its meander belt and start a new series of cutoffs and clay plugs that will define a new meander belt. This has occurred

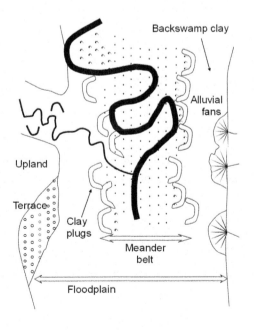

**Figure 4.8**

Diagram showing deposits on a floodplain and associated terrace of a meandering river. Most of the area within the meander belt is point bar.

repeatedly in the Mississippi River above New Orleans, each time forming a new delta. As the delta extends outward, the river level must rise in order to preserve a downhill gradient, so the system becomes unstable and the river will seek to escape the confines of its meander belt and levees. This is the situation at New Orleans.

### 4.6.9 Overbank Deposits

Water spilling out over the banks of a river slows down and deposits sediment. Fine sand and silt are deposited close to the river channel in low ridges called *natural levees*, which are preferred building sites. Exploration borings must extend through the natural levee because it normally will be on top of softer floodplain clay and may conceal a buried clay plug.

Some floodplain areas will have a lighter color on airphotos and appear to have a braided pattern indicating a sand deposit, but this is misleading because the sand may only be a meter or so thick. This is a *sand splay*, which occurs where a levee has been breached at some time in the past.

Clay carried overbank or through a breach is slow to settle out, and forms a continuous blanket on top of older deposits. This clay is called a "backswamp" or "slackwater" deposit, and can be meters thick. It frequently is expansive clay and therefore can be of critical importance in engineering. Desiccation at the surface creates a harder crust that can be utilized for supporting lightly loaded foundations.

### 4.6.10 Natural and Artificial Levees

The first response to a flood is to build or increase the height of a levee. This confines the river from a breakout and increases its water level and gradient. Breaching a levee on one side of the river can save the other side, which is why levees are patrolled during flood stages to prevent the use of dynamite.

As a delta builds out into a sea, the river is extended and natural levees are built up higher during flood. A river level that is higher than the adjacent floodplain is a recipe for disaster. The most famous example of delta extension is the heavily loess-laden Yellow River (Huang Ho) in China, considered to be the muddiest major river in the world. In some locations the delta has extended as much as 8 km (5 miles) *in one year*, and natural levees have been built up to the extent that the river as much as 21 m (70 ft) above the adjacent floodplain. Sudden breakouts are almost impossible to contain, and have killed hundreds of thousands of people, giving the river its name "River of Sorrow." Dams are being constructed to control flooding, but a river carrying such a heavy load of sediment will quickly silt up the reservoirs and reduce their effectiveness. One plan is to allow the river to flow on through during periods when it is most heavily laden with silt.

## 4.6.11  Alluvial Terraces

Terraces are formed when a mature river is *rejuvenated*, that is, when it again becomes youthful as a result of lowering of its base level. This can occur when the base level of harder rock is cut through or a dam breaks. Then instead of cutting laterally, the river cuts downward, or *entrenches*. Meanders upstream then cut deep scrolls in the former floodplain.

River entrenchment was world-wide during the Pleistocene glaciation as sea level was lowered over 100 m as water was held as ice. Then as glaciers melted and sea level rose, the deeply entrenched river valleys were drowned to become estuaries. Rivers carrying glacial outwash filled their valleys with alluvial gravel and sand. Glacial sand and gravel outwash deposits extend over 100 m (300 ft) below the modern floodplain, indicating the extent to which sea level was lowered.

Deposits in alluvial terraces reflect their origin: for example, terraces of rivers that did not carry glacial outwash may have approximately the same composition as the modern floodplain, whereas those associated with glacial outwash mainly contain sand and gravel. Older terraces that were formed prior to loess deposition may be covered by wind-blown loess, discussed in a later section.

The youngest terraces may be so low that they are best seen from the ground by observing slight differences in ground elevation. Low terraces also may be subject to flooding, then being referred to as "second bottoms."

## 4.6.12  Meandering and Tributary Entrenchment

The position of river meanders on a floodplain also affects the base level of tributary streams crossing the floodplain to enter the river. As meanders swing from one side to the other, tributary base levels are alternately raised and lowered, causing erosion and problems with bridges.

## 4.6.13  Braided Streams

Some streams are so loaded with sediment that there is no energy left over for pattern meandering. They nevertheless can aggressively and randomly erode their banks as channels are plugged and diverted by sediment. The channels of a braided stream divide and recombine to enclose almond-shaped sand bars. The channels are ever-changing because of local encounters with tree stumps or boulders. There usually is one dominant channel that can shift during periods of high water.

Two common occurrences of braided streams are: (1) in arid/semiarid areas where there is a shortage of water, and (2) as previously indicated, as glacial outwash in which case there is an excess of sediment, Fig. 4.9. In arid areas streams are intermittent and may flow only briefly after a rain, but the rare heavy rain can

**Figure 4.9**

A braided stream carrying glacial outwash and a source for loess. A dust cloud is silhouetted against the mountain. Matanuska River, Alaska.

create a "flash flood" and channel instability that washes out roads and bridges. Heavy rains therefore can create extremely hazardous driving conditions in the desert, particularly at night when vision is obscured.

### 4.6.14   Deposits from Braided Streams

Braided stream gradients and flow rates generally are considerably higher than those for meandering streams of a comparable size, which is consistent with their higher load-carrying capability. The high gradient and flow rate keeps small particles in suspension, so deposits are mainly coarse-grained, sand and gravel. The erratic shifting of channels creates uneven lenses and beds that are *cross-bedded* as one channel cuts across another that has been filled with sediment, as illustrated in Fig. 4.10.

Each cycle of deposition following a decline of high water involves a gradually decreasing flow rate, so large particles are deposited first, followed by progressively finer particles. A single bed therefore may contain gravel at the bottom, grading upward into finer gravel, sand, and finally silt, clay being washed out. The bottom-to-top, coarse-to-fine transition is called *graded bedding*. Gravel layers can indicate how deep the river has scoured during past periods of high water, which is important for the design of bridge pier foundations, as scour can extend even deeper because of diversion of current and a concentrated flow.

### 4.6.15   Silt and Wind Erosion

During waning river stages the water velocity may slow down sufficiently to deposit silt on top of sand bars of both braided and meandering streams. In braided streams the lack of protective vegetation and large exposed areas are

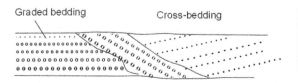

**Figure 4.10**
Illustration of graded bedding and cross-bedding.

Graded bedding      Cross-bedding

**Figure 4.11**
The large alluvial fan at the center has been cut off at the toe by the river. A large, steep talus slope is at the right of the fan, and a road may be seen running across the fan. Indus River, Karakorum Range, Pakistan.

susceptible to wind erosion, and silt may be observed billowing upward from exposed sand bars in Alaska. The silt then is deposed as loess.

Terraces left by braided streams are important sources for aggregate, and may be recognized on airphotos from darker, branching channels enclosing lighter almond-shaped sand bars.

## 4.6.16 Alluvial Fans

A sediment-laden stream can become braided if there is a sudden decrease in gradient, the most common example being when a tributary stream flows out onto a floodplain. Sediment is deposited and clogs the stream channel so it finds another route, which then becomes clogged. This process is repeated and develops a radial pattern that builds up a fan-shaped deposit appropriately called an *alluvial fan* (Fig. 4.11). As alluvial fans are relatively coarse, open-graded material, in arid areas they are important reservoirs for groundwater.

Coalescing fans from adjacent streams constitute an *alluvial plain* or "bahada." This can be a preferred site for a road or other construction, as it is not subject to flooding by the river. However, it still is subject to flash flooding and erosion by the tributary.

In desert areas where there is no outlet, fans can build up until they submerge bases of mountains. Some formidable examples in the U.S. are the Basin and Range Province of Nevada, Utah, and southern California, and Death Valley in California. Alluvial fan deposits reflect their localized sources but also exhibit some sorting action because as a fan builds outward, coarser particles are deposited first and fines are carried farther out. In closed desert basins the fine particles that are not deposited in fans are carried into an intermittent lake or *playa*, to build up a clay deposit that may be alkaline and highly expansive.

Braided streams and alluvial fans may be seen in miniature in roadside ditches after rain.

## 4.7 SEA AND LAKE DEPOSITS

### 4.7.1 Deltas

Deltas were named by Herodotus in the fifth century B.C.E., from the shape of the Nile delta, like a Greek Δ with the apex pointing upstream. However, most deltas extend outward so a delta shape is not necessarily an identifying feature. A delta is deposited as a river or stream flowing into a lake or ocean loses velocity and deposits its sediment.

As in the case of alluvial fans, coarser particles carried into a delta are deposited first and finer materials are carried farther out to constitute "bottomset beds." As the delta builds outward, the bottomset beds are covered with "foreset beds" that are deposited on a steeper slope, as illustrated in Fig. 4.12. The last materials to be deposited are the "topset beds."

### 4.7.2 Freshwater vs. Saltwater Deltas

Freshwater deltas differ markedly from those deposited in salt water because of the flocculating effect of salt on suspended clay. As clay enters salt water the

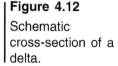

**Figure 4.12**
Schematic cross-section of a delta.

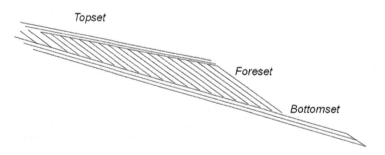

Topset

Foreset

Bottomset

electrical charge that normally keeps particles repellent to one another changes so they attract one another; the clay becomes *flocculated* into silt-size particles that settle out and make up most of the foreset beds. In fresh water, clays remain suspended and slowly settle out into the lakebed and as bottomset beds in a delta. Foreset beds in a freshwater delta are mainly sand and silt.

Topset beds in both types of deltas are similar to river floodplain deposits, being extensions of the channel sands, natural levees, and backswamp clays. A "bird's foot" delta, such as that of the Mississippi, is defined by natural levees that continue and channelize the river for a distance out into the sea.

### 4.7.3 Estuarine Deltas

An *estuary* is a river valley that has been drowned by the postglacial rise of sea level and where tides are present. Sedimentation that is confined within an estuary constitutes "estuarine delta," which is mainly tidal mud flats.

### 4.7.4 Deltas in Artificial Lakes

The several benefits from dams, namely flood control, irrigation, power generation, recreation, and navigation, are not without environmental consequences. Some are predictable from the river dynamics: a dam is the base level for the portion of the river that is upstream, causing the stream to pond as a reservoir. Delta building then initiates where the river enters the lake, and people who are in the market for a lakeside lot are advised to avoid the upper reaches of the reservoir.

As the delta grows and the reservoir becomes silted in, uses of the lake gradually will be compromised. A clue is growing mud flats that eventually will become floodplain. In the U.S., a minimum reservoir design life of 100 years has been considered acceptable, but sedimentation rates often show that estimates that were used as a basis for dam building were overly optimistic. Reducing soil erosion in the drainage area has obvious benefits, and where feasible smaller check dams can be built upstream from a main reservoir to catch sediment in areas that can more easily be cleaned out. Adding height to a dam is far more difficult, and dredging is expensive and may cost more than the dam while posing a problem of where to put the spoil.

The seriousness of the coming problem is shown by reservoirs that already have become silted in 15 years after construction. By the time a dam was completed on the Yellow River in China in the 1970s, the capacity of the reservoir for power generation had been reduced by three-fourths. If sedimentation continues at the present rate, the reservoir will be completely filled with loess-derived silt by 2050, 80 years after completion of the dam

The most promising solution to the sedimentation problem may be to bypass sediment-laden water, particularly during floods that carry the bulk of the sediment. After completion of the Three Gorges Dam on the Yangtze River in China in 2009, the plan is to draw the reservoir down during the flood season, which will allow 90 percent of the annual sediment load to pass through with 60 percent of the annual inflow of water. The procedure undoubtedly will be rife with controversy, because managing a dam to prevent sedimentation also means managing it to allow flooding. Even though it should extend the reservoir life by a factor of 10, the anticipated life still is only 100 years. Thus, while hydroelectric power is widely regarded as a renewable resource, the reservoirs created by damming are not readily renewable.

### 4.7.5  Downstream from a Dam

Removal of sediment makes a river more erosive, so the channel immediately downstream from a dam generally deepens and works with increased energy against its banks, increasing the need for bank protection to prevent landslides.

Sediment depletion by a reservoir also impacts the delta downstream that holds a tenuous balance against wave erosion. The delta of the Colorado River guards the Imperial Valley of California, one of the richest agricultural areas in the world, against invasion by the sea, but the sea has been advancing since completion of the Hoover Dam in 1934.

Wave erosion of deltas is the primary source for beach sand that is distributed by longshore currents. Thus as a delta becomes depleted, beaches may require protection from wave erosion. Current policy in many areas is to prohibit building closer to the sea than the anticipated beach position after 50 or 100 years.

In summary, changes that are worked on a river, whether by shortening, confining it with levees, damming, or other measures, all affect the natural balance of the river system, its delta, and the associated beaches. This does not mean that there is no net gain, but only that engineers and planners should be fully aware of the eventual consequences and incorporate them into their feasibility studies.

### 4.7.6  Beach Deposits

After they are derived from deltas or other sources such as erosion of headlands, beaches are subjected to continuous, unending wave action that grinds away soft minerals and concentrates hard minerals, in particular quartz. Beach sediments also are well sorted (all one size), well rounded, and can vary in size from gravel to sand.

Longshore currents that carry sands along the lengths of beaches are products of winds blowing toward the beach at an angle. With certain wind directions two opposing longshore currents can come together in a bay and create a dangerous *rip tide* that flows offshore. Swimmers or boaters caught in a rip tide should move parallel with the shore instead of trying to fight against it, which is a losing battle.

### 4.7.7 Breakers and Barrier Beaches

The drag of wind on the water causes a near-circular motion where the diameter of the circle represents the height of the waves. As a wave progresses into shallow water, frictional drag at the bottom distorts the circular orbit and reduces the wavelength so that water piles up. The wave breaks over at the top when the depth is about 1.5 times the wave height.

Breaking dissipates energy, so sand deposition occurs to the landward side of the break zone and builds a submerged offshore bar. The bar can be built above sea level during storms, in which case it becomes a *barrier beach* that is separated from the shore by a *lagoon*. The lagoon then tends to trap fine-grained sediments brought in by streams and rivers.

Barrier beaches are common along many coastlines, including the eastern and southern coasts of the United States, and are shown on maps. They tend to erode on the seaward side and build up on the landward side, so a beach slowly migrates landward and covers the associated clayey lagoon sediments. The clay helps to prevent intermingling of fresh and salt water in the groundwater supply, but also forms a soft zone for foundations.

### 4.7.8 Fossil Beaches

In the long term most land masses are not stable relative to a sea level that also is not stable. An emerging coastline may preserve beaches as part of the upland. Such beaches often are marked by sand dunes. A submergent coast is more likely to be marked by narrow beaches, rugged cliffs, and estuaries.

## 4.8 EOLIAN SANDS

### 4.8.1 Dunes

Sand dunes are among the most easily recognized sedimentary deposits because of their sweeping curves and frequently blowing sand. Less obvious are sand dunes that no longer are active and are referred to as *stable dunes*. Stable dunes support protective vegetation but nevertheless often are pock-marked with wind-eroded *blowouts*.

Active dunes require a continuous source of sand, and therefore occur adjacent to beaches or alluvial sand, particularly along braided rivers. Desert dunes derive from the extensive alluvial fans in deserts. Contrary to popular conceptions, sand dunes cover only about one-fourth to one-third of desert areas, the rest being mainly exposed rock and alluvium.

### 4.8.2   Movement of Sand

Whereas silt is fine enough to be carried suspended in air as dust, sand grains quickly settle out and bounce along on the ground, sand-blasting exposed rocks and ankles, and clipping off old fence posts close to the ground. The transport mechanism is called *saltation*, and requires that sand dunes be connected to a source. The connection can either be more dunes or a barren "desert pavement" that is covered with wind-polished stones. Desert pavements may show thin, straight sand strips or streaks after a single storm.

### 4.8.3   Cliff-Head Dunes

Eddies created by winds crossing a river floodplain can carry sand to the top of the adjacent river bank where it is deposited as "cliff-head dunes." These dunes are analogous to snow drifts, and normally remain anchored to their source areas, growing ever larger and extending inland. In arid areas, isolated dunes called "barchans" may break away and migrate downwind as the "children" of cliff-head dunes.

### 4.8.4   Dune Shapes

Dunes may exhibit many shapes indicative of the directions of prevailing winds. Most common where there is a generous supply of sand are *transverse dunes* that resemble gigantic ripples. Where there is less abundant sand, transverse dunes may become partitioned into *barchans*, which are the classic crescent-shaped dunes that are a favorite with photographers. The tails or horns of a barchan sweep off downwind, a tail from one dune leading to the head of the next.

A British Army engineer, R. A. Bagnold, made an extensive study of dunes in the Sahara during World War II. He found that with two instead of one dominant wind direction, one tail of a barchan tends to grow longer and give a classic "seif" dune, named for its shape like an Arabian sword. The long tails of seifs may link into a continuous chain or "longitudinal" dune with a succession of peaks and saddles. Whereas the maximum size of barchans is limited to about 30 m (100 ft) because of blowing off at the crest, longitudinal dunes, being alternately blown at from both sides, may continue to build to a height of 200 m (650 ft), and become a very prominent landform extending downwind for many kilometers.

### 4.8.5   Dune Slip Face and the Angle of Repose

A characteristic of all active dunes regardless of shape is the occurrence of a *slip face* on the leeward side, as illustrated in Fig. 4.13. The slip face is the main area of deposition, as sand grains bouncing up the windward side drop into the wind shadow behind the dune. Sand grains sprinkling down a slip face adjust to a constant angle that depends on the frictional characteristics of the sand.

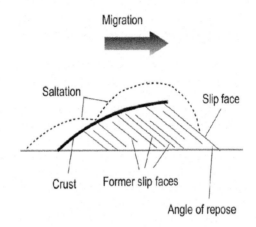

**Figure 4.13**
Anatomy of a sand dune.

This is the *angle of repose*, and normally is less than 35°, although it can give the impression of being much steeper.

### 4.8.6    Internal Structure of a Sand Dune

Continual bombardment of the windward side of a dune by saltating sand grains develops a crust that slowly creeps up over the dune, sometimes extending out over the edge of the slip face. The dune itself also migrates as sand grains are dislodged from the windward side and carried across to the slip face. The migrating dune therefore encompasses former slip surfaces, as shown in Fig. 4.13. The interior of a dune therefore remains relatively soft and is an invitation for vehicles to become stuck if tires break through the surface crust.

### 4.8.7    Migration Rate

As dunes migrate they cover and encapsulate whatever is in the way—trees, roads, houses, whatever. As a general rule one should not place anything close to the lee side of an active sand dune unless it is on wheels.

The rate of dune migration is predictable based on empirical measurements. Large dunes migrate more slowly than small ones because more sand is required to extend the slip face. While data are limited, measurements of several barchans in Egypt gave an average rate of advance of about $R = 180/H \leq 20 \, \text{m/y}$, where $R$ is meters per year and $H$ is the dune height in meters.

**Example 4.1**
Your brother-in-law asks to borrow money to invest in a beach motel that has a swimming pool and tennis court, and is about 200 m downwind from a beautiful, active, 10-m-high sand dune. How long will he have before the motel is part of the scenery and he has to apply for government aid?

*Answer:* $R = 180/10 = 18$ m/y. To go 200 m will require $200/18 = 11$ years. The "For Sale" sign probably will appear after about 10 years, as soon as they get the drift.

### 4.8.8 Stabilizing Sand Dunes

Dune migration is a challenge. One response that is used to keep roads open is to keep a road grader handy, push the sand off, and wait for more. A more reasoned approach is to stop encroachment by cutting off the source of sand. This usually involves anchoring the sand with vegetation that can grow to keep pace with the rate of sand accumulation. Migration also can be slowed by covering the windward side to prevent erosion or creep of the surface layer. If all else fails, a cover can be constructed over a road or railroad so that the dune can walk over the top.

## 4.9 EOLIAN SILT, OR LOESS

### 4.9.1 Definition

Loess is eolian dust that, as shown in the background of Fig. 4.9, still may be seen blowing off glacial outwash to be deposited on nearby terraces and upland. Loess is mainly silt, having grain sizes that are finer than sand and for the most part are coarser than clay. The name is Anglicized from the German löss, which literally means loose. The German pronunciation is approximated by "lerse" but more common pronunciations are "luss," "less," and "lo-ess."

While loess is mostly silt, it also can contain some clay and minor amounts of fine sand. The silt and sand are mainly quartz and feldspars, and the clay fraction often consists of expansive clay minerals that usually are not in a sufficient amount to make the soil expansive. However, weathering processes discussed in the next chapter can turn it into expansive clay.

### 4.9.2 Geography of Loess

Silt and clay are carried aloft as clouds of dust that is carried in suspension and spread across many tens of kilometers. In the U.S. and in Europe, most loess was derived by winds blowing across exposed bars of braided, outwash-carrying rivers during Pleistocene continental glaciation. The loess was deposited during the Pleistocene between 25,000 and 10,000 years ago and therefore was witnessed by early man. A native American name for the Missouri River valley is "valley of smoke."

Loess deposits are thickest close to source areas and thin exponentially with distance. A common assumption is that this reflects a prevailing wind direction, but that does not explain deposition on both sides of a source. A more logical

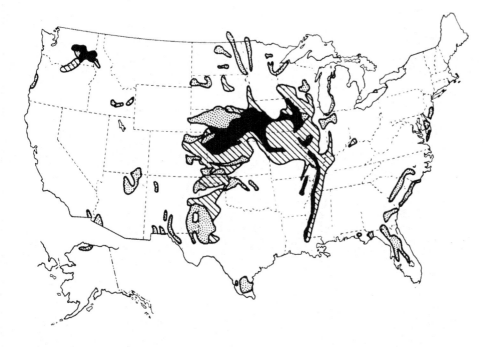

**Figure 4.14**

Major eolian surface deposits in the U.S.

explanation is based on river valleys being approximately linear source areas, so the thickest deposits should be from winds blowing nearly parallel with a linear source. Then, as the wind direction approaches a right angle, the silt is carried farther. The same pattern of distribution has been observed in dust deposits derived from gravel roads.

The maximum loess thickness in the U.S. is about 50 m (150 ft), and the loess cover extends from Nebraska and Kansas eastward to southern Ohio, and southward along the Mississippi floodplain into Tennessee and Mississippi. Other areas include the Palouse loess in southeastern Washington. Loess deposits in the U.S. are shown in Fig. 4.14. The approximate extent of loess deposits in Europe is shown in Fig. 4.15.

The thickest loess deposits in the world are in China, where the soil is believed to have been blown from high Asian desert areas over a period of 2.4 million years. The total thickness exceeds 120 m (370 ft), and the ease of erosion of loess is a major factor contributing to the silt content, color, and difficulties experienced with the Yellow River.

### 4.9.3 Stable Slope Angles

Gully walls and man-made cuts in loess, such as shown in Fig. 4.16, can have a height and steepness that seems to be out of character for such a soft deposit. Artificial cuts may be inclined at 4 vertical to 1 horizontal in order to minimize erosion from rain. It once was believed that calcium carbonate cemented the

**Figure 4.15**

Approximate extents of glaciation and of loess deposition in Eurasia. Loess map is from Russian sources, reprinted in *Loess Letter 48*, Nottingham Trent University, Ian Smalley, ed.

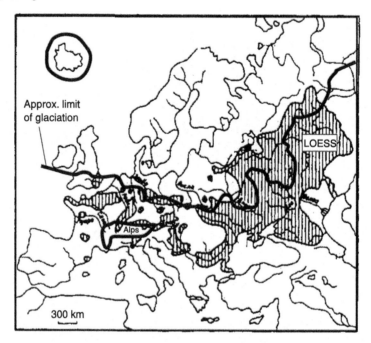

**Figure 4.16**

Intrepid student engineer suspended on a bosun's chair in order to collect samples of loess in western Iowa.Here the soil is so porous that it will collapse if it becomes saturated with water. A vertical tension crack and landslide are at the left.

soil grains together, but this is not borne out by microscopic examination or by the fact that loess close to a source can collapse under its own weight if wet with water. The cohesive mechanism therefore appears to involve capillary forces or "suction" of water enhanced by clay surface activity and that is reduced by wetting.

### 4.9.4 Collapsible Loess

A common problem associated with many loess deposits is *collapsibility*, meaning that it collapses when wet with water. This is consistent with the hypothesized eolian origin from wind such that it never has become saturated. Ground-water tables in collapsible loess are perched near the bottom of the deposit, and soil below a groundwater table is dense and not collapsible.

The problem for the engineer is to keep a collapsible loess from ever becoming saturated. Saturation can occur at leaky joints in pavement gutters, near leaky water or sewer pipes, or near downspouts. Close attention must be paid to surface drainage to prevent ponding, and roof water should be carried several meters away from foundations.

Damages from collapse also can be prevented by presaturating in order to induce collapse prior to construction, or by founding structures on deep foundations such as pile. Presaturation is the obvious choice for dams and canals, but after collapse the wet soil remains soft and compressible and may require surcharging.

### 4.9.5 Changes with Distance from a Source

Collapsible loess occurs close to a source area. Farther away, loess deposits are thinner, more clayey, more likely to have been saturated, and less likely to be collapsible. As the loess was carried by wind, the larger silt grains settled more rapidly, allowing clay to become more concentrated farther away, where its sedimentation was aided by rain. A slower rate of deposition also provides more opportunity for weathering during deposition.

Because the properties of a loess deposit gradually change with increasing distance from a source, loessial soils exhibit a spectrum of behavior from collapsible close to a source to relatively stable to expansive farther away. This is referred to by geologists as a "facies change" (*facies* from Latin for face). Expansiveness or collapse can be measured in a laboratory compression test or can be indicated from the soil density and classification that are discussed in later chapters.

## 4.10  MAN-MADE FILL

### 4.10.1  Overview

As the best building sites become used and re-used, random fill becomes more commonplace and can contain everything from concrete rubble to tree limbs and old refrigerators. The recognition of landfill can be one of the easiest but nevertheless most critical duties of the geotechnical engineer. Random fill usually is not compacted, and even if it is compacted it can contain open voids.

Surcharging can be beneficial but remains an uncertain remedy because of gradual disintegration or decay of components in the fill. Another method of treatment is to repeatedly drop a large weight using a crane, a process called *deep dynamic compaction*. Driven piles can be used as foundation elements, but even this can be difficult if the fill contains structural elements such as concrete rubble that also makes the fill difficult to investigate by drilling.

### 4.10.2  Clues

Fill soil is out of place—that is, it does not fit a natural order. This requires an appreciation for the natural order, which is a reason for emphasis in this and the next chapter. Fill soil does not follow the dictates or conditions imposed by geology or by weathering. Fill soil is a mix. A common evidence for fill is a buried layer of topsoil that under normal circumstances would be on top.

Other critical evidence is bits of glass, concrete, brick fragments, bones, coal ashes, lumber, bedsprings, soup cans, etc. Such clues to landfill *must be recorded in boring logs and soil reports*, which are legal records and can be subject to subpoena.

Fill containing organic material or garbage can generate methane and if allowed to accumulate inside a building can cause an explosion. Buildings should be separated from such fill with a plastic liner and/or ventilation system.

Landfills that include toxic or radioactive wastes add another dimension to the problem, and are addressed by specialists in geoenvironmental engineering.

The failure to recognize random fill and take it into account in design almost inevitably leads to future difficulties.

## 4.11  SUMMARY

Most geotechnical problems are not the result of mathematical errors or use of the wrong formulas, although such errors obviously must be avoided. Some problems result from poor or inappropriate construction practices, but most derive from an inadequate appreciation for the soil, its properties, and its variability. A geotechnical engineer learns to have a critical eye in order to see problems before they happen.

### Problems

4.1. Explain how the weight of a continental glacier can affect properties of the underlying soil deposits. How may this relate to allowable foundation pressure to minimize settlement?

4.2. How can a continental glacier move so far on such a low slope angle?

4.3. Explain how one might identify the following from airphotos, and discuss engineering features:
   (a) Alluvial terrace composed of sand and gravel deposited by a braided river.
   (b) Peat bog.
   (c) Terminal moraine.
   (d) Esker.
   (e) Stable sand dunes.
   (f) Point bar and clay plug.
   (g) Landslides.
   (h) Alluvial fans.
   (i) Talus.

4.4. Differentiate between a youthful stream, a braided stream, and a meandering stream.

4.5. What confines a meander belt within a broad floodplain? Which are most effective for containment, clay plugs or natural levees? Why? Draw a cross-section showing a clay plug and associated point bar, backswamp clay, and a natural levee.

4.6. Sketch a cross-section for a stream valley that has a succession of progressively lower floodplain levels. What are these levels called? Which is the oldest?

4.7. The transportation agent for sand dunes and for loess silt is the same, wind. Which deposit do you expect to be closer to a source? Why?

4.8. What changes occur in loess with increasing distance from a source? Explain the changes.

4.9. How do deposits from a braided stream differ from those from a meandering stream? Make a list of deposits from a meandering stream.

4.10. Radiocarbon dating shows that Wisconsin-age loess in the United States was deposited during the period between 29,000 to 14,000 ybp (years before present). How does the average thickness accumulation per year of the thickest loess deposits compare with the accumulation of dust alongside a dusty road, measured to be as much as 2.5 mm/y (0.1 in./y)?

4.11. Suggest ways to keep sand dunes from encroaching on a housing development.

4.12. The edge of a field of barchans 6 m high is located 2.4 km (1.1 miles) upwind from an express highway. Estimate the time of arrival.

4.13. Would you expect to find any gravel terraces along the Mississippi River floodplain? Why?

4.14. If you did find gravel terraces along the Mississippi River floodplain, would they be loess covered? Why?

4.15. Which should give the better prediction of maximum scour depth that might be anticipated during a flood, the deepest water depth in the river, or the maximum clay thickness in a nearby oxbow? Why?

4.16. Where would you prospect for sources of coarse aggregate in Fig. 4.9?

4.17. An association of property owners around a small artificial lake has come to the conclusion that the lake is filling with sediment. What are some options?

4.18. How can extension of a river delta lead to more flooding upstream?

4.19. How does a dam contribute to erosion by a river downstream from the dam? To the reduction of the river delta?

4.20. Why and at what depth do waves break?

4.21. In the open sea a tsunami has a very long wavelength and is barely noticeable. Why is it so devastating when it reaches land?

4.22. Install the free version of Google Earth on your computer.

    (a) Go the Grand Canyon, tilt the image, and see if there is evidence for sedimentary rock layers overlying an igneous rock complex.

    (b) Zoom in on your area and browse for some of the features listed in Problem 4.3.

    (c) Check out Greenland for evidence of global warming.

    (d) Find a mountain range that has the appearance of having been eroded by streams. What mountains are constructive?

    (e) It may be a small world but it's complicated. Play.

## Further Reading

Bagnold, R. A. (1941). *The Physics of Blown Sand and Desert Dunes*. Methuen, London. Reproduced by Dover, New York, 1965.

Chorlton, Windsor (1983). *Planet Earth Ice Ages*. Time-Life Books, Alexandria, Va.

Flint, Richard Foster (1971). *Glacial and Pleistocene Geology*, 3rd ed. Wiley, New York.

Google Earth (high-speed internet connection is required).

Handy, Richard L. (1972). "Alluvial Cutoff Dating from Subsequent Growth of a Meander." *Geol. Soc. Amer. Bull.* 83, 475–480.

Handy, Richard L., et al. (1975). "Unpaved Roads as Sources for Fugitive Dust." *Transportation Res. Bd. News*, 60, 6–9.

Handy, Richard L. (1976). "Loess Distribution by Variable Winds." *Geol. Soc. Amer. Bull.* 87, 915–927.

Jackson, Julia A., ed. (1997). *Glossary of Geology*, 4th ed. American Geological Institute, Alexandria, Va.

Kehew, Alan E. (1995). *Geology for Engineers and Environmental Scientists*, 2nd ed. Prentice-Hall, Englewood Cliffs, N.J.

Kiersch, George A. (1964). "Vaiont Reservoir Disaster." *ASCE Civil Engineering* 34(3), 32–39.

Leopold, Luna B., Wolman, M. Gordon and Miller, John P. (1964). *Fluvial Processes in Geomorphology*. Freeman, San Francisco.

Miller, Russell (1983). *Planet Earth Continents in Collision*. Time-Life Books, Alexandria, Va.

Philipson, Warren, ed. (1997). *Manual of Photographic Interpretation*, 2nd ed. American Society for Photogrammetry and Remote Sensing, Bethesda, Md.

Strahler, Arthur N. (1971). *The Earth Sciences*, 2nd ed. Harper and Row, New York, 1971.

# 5

## The Soil Profile

## 5.1  SOIL PROFILES IN ENGINEERING

### 5.1.1  Occurrence

A soil profile designates near-surface soil layers that have been modified by processes associated with weathering. Weathering not only changes some soil minerals and removes others, it concentrates and relocates soil particles in characteristic layers. In lay terms the layers have names like "topsoil," "subsoil," and "hardpan." Many soil profiles contain concentrations of expansive clay minerals, and soil profiles develop in rocks as well as in sediments. As sedimentary soils normally have been recycled from earlier stages of weathering, they tend to be less susceptible to further change.

The most important function of a soil profile is for agriculture. Soils that are good for agriculture generally are poor for engineering and vice versa, making the uses complementary instead of competitive. For example, black topsoil that is excellent for plant growth has poor properties for engineering, and therefore is normally stripped off from construction sites and saved for later spreading and landscaping. In fact, the topsoil that is left in place and buried under fill soil eventually can cause serious problems including foundation failures and landslides.

### 5.1.2  Relation to Erosion

A soil profile represents a balance between weathering and erosion, weathering developing the profile while erosion simultaneously strips away the upper part. A steep eroded hillside may have little or no development of a soil profile while nearby flat areas on material of the same age can have a deep and strongly developed weathering profile.

The time required for development of a soil profile is measured in hundreds of years. Erosion not only strips away valuable topsoil, it fills nearby road ditches and increases the rate of accumulation of sediments on floodplains and in lakes and deltas. Agricultural fertilizers that are carried along with eroded topsoil can impact aquatic life, and have created an infamous "dead zone" where the Mississippi River enters the Gulf of Mexico.

### 5.1.3 Relation to Climate

Weathering and soil profiles are closely related to climatic conditions, for example being weakly developed in the Arctic, and in areas where there is little rain. On the other hand, soil profiles developed in tropical climates can extend tens of meters deep into the in-place rock or sediment.

Soil profiles in temperate zones frequently contain an expansive clay layer subject to volume changes during normal cycles of wetting and drying. Such soils usually are easily recognized upon close inspection from crack patterns in the soils. Expansive clays exert pressures sufficient to lift pavements, floors, and foundations, so their recognition is of paramount importance in geotechnical engineering. Expansive clays in soil profiles have for the most part been ignored in the engineering literature, and on maps that are indicated to show the distribution of expansive clays. Such maps may show only geological occurrences such as in shale, alluvium, and coastal plain deposits. Ironically, the latter occurrences owe their origin to expansive clays developed in weathered soil profiles that have been eroded, transported, and deposited many hundreds of kilometers from their places of origin. Therefore tropical areas where expansive clay is not part of the weathered soil profile often have extensive deposits in alluvial deltas and rice paddies.

## 5.2 DEVELOPMENT OF A SOIL PROFILE

### 5.2.1 Weathering

Weathering almost inevitably involves chemical changes that contribute to expansion and disintegration. For example, exfoliation or flaking-off of the surface of a granite boulder, and its eventual disintegration into sand, are the result of hydration reactions that cause a volume expansion of the grains of feldspar.

Chemical weathering converts rock minerals into new minerals, in particular a distinct class of minerals called clay minerals. Water plays a key role in the conversions, as $OH^-$ ions are essential building blocks in clay mineral crystal structures. As indicated in the preceding chapter, clays can be concentrated by erosion and deposition as sediments, but with rare exceptions their ultimate genesis is by weathering.

0
1
2
3
4 ft

0
0.5
1.0 m

Houston Black series    Holdrege series    Tunbridge series

**Figure 5.1**

Three examples of soil profiles. In all cases, A horizons should be removed prior to use in engineering. The Houston Black is an expansive clay with a B horizon that extends to a depth of about 2.6 m (9 ft) and causes extensive engineering problems. A somewhat shiny appearance is caused by slickensides or smearing of clay along shear surfaces. Tunbridge is officially designated the Vermont state soil. R denotes rock. (From a calendar publication of the Soil Science Society of America, 2002.)

Because weathering is most intense at the ground surface, there is a transition from highly weathered soil to unconsolidated "parent material" that may be either a weathered rock or a sediment. Different layers in the weathered zone constitute the soil profile. The layers are called "horizons" and are designated by capital letters, the most common from the ground surface down being A, B, and C. Various subscripts are used to designate specific properties. For example a subscript $A_1$ designates a dark-colored topsoil that develops under grassy vegetation whereas $A_2$ signifies a light gray or white layer that develops under forest. Some examples of soil profiles are shown in Fig. 5.1, where the Tunbridge series shows a strong $A_2$ horizon. The zone of maximum development of the B horizon is called the $B_2$.

## 5.2.2 The A Horizon

The *A horizon* is the common "topsoil." (The scientific name is "epipedon," which means over-soil.) A horizons often are rich in humus and organic plant residues and have a loose, loamy texture preferred by gardeners. An A horizon usually is darker in color than the underlying soil except under forest conditions, where intense leaching by humic acids derived from forest litter creates an ash-gray or white layer. This is particularly true under conifer forest, where Russian soil scientists who pioneered the classification of soils named these "podzols," which means ash. A white or ashy layer therefore is diagnostic of development under forested conditions and may be designated an E (for eluviated) horizon. An example is shown at the right in Fig. 5.1.

The thickness of the A horizon typically ranges from a few centimeters (inches) to about 0.6 m (2 ft). Because of its high content of organic matter the A horizon usually exhibits undesirable engineering characteristics such as high compressibility and rebound, low shear strength, resistance to compaction, and variable plasticity. In construction operations the A horizon generally is removed and saved for topping-out and seeding, sodding, or planting for erosion control or aesthetic purposes.

The A horizon also is the site for intensive weathering to form clay minerals, but these tend to be carried downward by percolating water to be deposited in the next lower layer, the B horizon. This process is called *eluviation.*

In engineering the A horizon is the layer to avoid. A horizon soil that is left on a slope and buried under fill can cause a landslide, no matter how well the fill soil has been compacted. Acceptable construction practice involves removing all vegetation and A horizon material, and cutting horizontal steps prior to placing soil fill on a hillside. *Exploration borings must penetrate all of the way through fill material to ensure that there is no buried topsoil, trash, or organic material that may cause future problems.* Any construction on such a site also must bear an additional cost for removal of the poor material. Structures that are built on top of weak or compressible layers may suffer severe damage and require expensive remedial measures.

### 5.2.3   B Horizon

In humid or subhumid climates, clay minerals are concentrated in the B horizon by eluviation. Most clay-rich B horizons form a valuable reservoir for water for plants during periods of drought, and clay minerals act as temporary holding sites for fertilizers.

Two expansive clay B horizons are shown in Fig. 5.1, the highly expansive Houston and the moderately expansive Holdrege soil series. The Holdrege also has a transitional BC horizon.

As shown in Fig. 5.2, because soil horizons follow contours of hills, they are transected by horizontal floors and foundations, resulting in differential uplift or settlement. Expansive clay B horizons often are removed and replaced with nonexpansive soil. This can be the same clay stabilized by mixing with a few percent hydrated lime. Expansive B horizon clay can readily push a basement wall off its foundation, as shown at the bottom of Fig. 5.3. A failure to recognize and deal with expansive clay is an invitation for future problems and lawsuits.

#### Blocky Structure
Soil containing expansive clay typically develops vertical shrinkage cracks during periods of dry weather. Such cracks can extend a meter or more deep, extending to a nonexpansive soil layer or to a depth below which the moisture content remains stable.

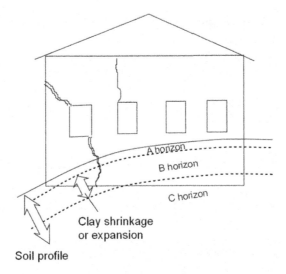

**Figure 5.2**
Typical damages from an expansive clay B horizon. This crack pattern indicates shrinkage.

**Figure 5.3**
Basement wall and sewer pipe being pushed in by wet-dry cycling of an expansive clay B horizon. This may happen years after construction, as cycling is required to compress loose soil backfill placed against the outside of the wall. Also a severe problem with expansive shale C horizons as in the Denver area.

Moisture cycling of a B horizon containing expansive clay creates a "subangular blocky" structure with the individual blocks or *peds* measuring a centimeter or so across. Clay coatings on the outside of the individual peds are planes of weakness that maintain the blocky structure. *A blocky structure therefore is diagnostic of expansive clay.*

### Slickensides

In climates with extended wet and dry periods, severely expansive B horizon clays develop inclined shear planes. These result from lateral stresses that are high enough to shear the soil after filling of open ground cracks during dry cycles. Slipping along these planes causes clay particles to be smeared to form *slickensides*. Soils that are so severely stressed as to form slickensides are notoriously bad actors. *Slickensides are diagnostic of highly expansive clays.*

## 5.2.4   C Horizon and Saprolite

Unconsolidated soil material underneath the B horizon is called the *C horizon*. This may either be sediment such as loess, glacial till or alluvium, or weathered rock. A C horizon in rock usually is transitional to an underlying *saprolite*, which is defined as weathered rock that still retains the rock appearance and structure. A rock that appears to be sound and falls apart when hit with a shovel most likely is saprolite, which is from the Greek for "rotten rock."

## 5.2.5   Caliche and Laterite

In arid and semiarid areas where evaporation exceeds infiltration of water, the net upward movement can carry with it calcium carbonate that cements the soil into *caliche*. Caliche can occur in A, B, or C horizons. It sometimes is used for surfacing unpaved roads.

Another hard subsoil that occurs in tropical climates is called *laterite*, which is soil that is cemented with iron oxides released by weathering. Lateritic layers often form a mesa-like protective cap on hills, but are differentiated from bedrock because they cut across bedrock layers.

## 5.2.6   Subhorizons

Subscripts on A, B, and C horizons further define important characteristics. Some of the more important subscripts for engineers are as follows. Older soil surveys may use different subscripts, which will be noted in the survey report.

| | |
|---|---|
| b | buried soil or paleosol (formerly p) |
| f | permanently frozen soil (permafrost) |
| g | gleying, gray color indicating unoxidized soil due to poor drainage conditions |
| k | accumulation of carbonates (formerly ca) |
| o | accumulation of iron and aluminum oxides |
| p | plowed layer |
| r | weathered rock or saprolite, shale, or dense glacial till |
| ss | slickensides, indicating expansive clay |
| t | accumulation of clay |
| x | frangipan, indicating high density and poor seasonal drainage conditions |

Y       accumulation of gypsum
Z       accumulation of soluble salts, for example in dried lakebeds

Many of these substrips have important implications in engineering and in geology. For example, an $A_b$ horizon is a buried topsoil, and buried topsoils can be conducive to landslides.

---

*Question:* Which other subscripts can have engineering significance?

---

## 5.3 FACTORS INFLUENCING A SOIL PROFILE

### 5.3.1 Climate

Climate, in particular the patterns of temperature and precipitation, exerts a major influence on weathering. For example, soils in the humid tropics are far more deeply and severely weathered than those in arctic or desert conditions.

**Weathering Stages of a Soil**
Climate strongly affects clay mineralogy because of selective leaching and reconstitution of mineral ingredients. A guide to intensity of weathering is the *weathering stages* for the fine-clay size fraction by M. L. Jackson and his associates at the University of Wisconsin (Table 5.1).

As might be expected, gypsum and its occasional companion, halite or rock salt, are relatively soluble and leach out first, so a soil containing gypsum is not highly weathered unless the gypsum came in later as a secondary mineral.

| Stage | Mineral | Formula or common occurrence |
|-------|---------|------------------------------|
| 1 | Gypsum | $CaSO_4 \cdot 2\,H_2O$, an evaporite |
| 2 | Calcite | $CaCO_3$, limestone |
| 3 | Hornblende etc. | Dark minerals in granite, basalt |
| 4 | Biotite | Black mica in granite |
| 5 | Feldspars | Dominant in granite, basalt |
| 6 | Quartz | $SiO_2$ abundant in granite |
| 7 | Muscovite | K-rich white mica, granite |
| 8 | Illite, vermiculite, chorite | Mica clay mineral, shale |
| 9 | Smectite | Expansive clay mineral, soil |
| 10 | Kaolinite | Nonexpansive China clay |
| 11 | Gibbsite | $Al(OH)_3$, aluminum ore |
| 12 | Hematite | $Fe_2O_3$, red iron oxide |
| 13 | Anatase | $TiO_2$, soil |

**Table 5.1**

Jackson-Sherman (1953) sequence indicated by minerals remaining in the fine-clay fraction. Arrows indicate progressive removal of potassium (K) in the series muscovite to smectite, and of silica ($SiO_2$) in the series smectite to gibbsite

The next weathering stage is represented by calcite, or calcium carbonate. Thus, a soil that contains calcite still is in an early stage of weathering.

The presence of quartz and loss of feldspars represents an intermediate stage of weathering. Quartz, although highly resistant to weathering in sand and silt sizes, is more aggressively attacked in the fine fraction because of surface energy considerations.

A more advanced intermediate stage includes a weathering sequence *mica-illite-smectite* that reflects a progressive leaching of K, or potassium. These clay mineral groups are discussed in more detail in Chapter 6. They dominate soil profiles in most of the U.S. Smectite is expansive clay. Smectite clays occur in soil profiles in the central and south-central U.S., and in soils formed in basic (volcanic) rocks of the far west.

More intense weathering is characterized by a sequence *smectite-kaolinite-gibbsite* that shows progressive removal of silica, or $SiO_2$, from the mineral structures. *Kaolinite represents a more advanced stage of weathering than smectite.* Kaolinite dominates soils in warm climates with an annual rainfall exceeding about 100 cm (40 in.), such as in the southeastern U.S. and in subtropical areas. This means that these areas have little expansive clay except for that which may be brought in by wind or water.

Kaolinite is named for a mountain in China and is used in China and elsewhere to make porcelain. Weathering to gibbsite requires tropical or subtropical climates with high rainfall. Gibbsite deposits are sources for aluminum ore.

### Significance of a Red Soil Color

Smectite clays usually contain iron as part of their somewhat casual crystalline structure, and weathering to well-crystallized kaolinite ejects the iron, which oxidizes to give soil a characteristic red color. Tropical soils therefore generally are rust-red, which also identifies them as being nonexpansive.

## 5.3.2  Mathematical Expression for Soil Profile

The natural factors influencing soils were formulated by a Russian soil scientist, Dokuchaev, in 1898. His formula was modified by Hans Jenny in 1941 to an algebraic notation:

$$s = f(cl, o, r, p, t) \tag{5.1}$$

This formula is shorthand for soil (s) as a function of climate (cl), organisms (o), topography (r), parent material (p), and time (t). The formula means that if any one of the factors changes, the soil will be different.

The significance of climate "cl" already has been discussed.

"o" for organisms includes vegetation. A most prominent example is forest versus grassland: forested soil profiles are thinner and have the gray, ash-like layer that is acidic, whereas grassland soils typically have a thick black or brown A horizon and a near-neutral pH. A low pH can contribute to corrosion of metal pipes buried in soil.

"r" designates topography, in particular as it affects the rate of erosion: steep slopes show little or no weathering to form a soil profile.

"p" relates to parent material, whether granite or basalt, sandstone, limestone, or shale. Basalt as in lava flows contains relatively high-temperature minerals that are readily weathered compared with granite, which undergoes only moderate weathering before disintegrating into sand. Shale minerals already are far down on the weathering scale and show only moderate changes, mainly by disintegrating into clay. Sandstone disintegrates into sand, and limestone dissolves so that there is little left to make residual soil. Figure 5.4 shows a generalized map of soil parent materials in the U.S.

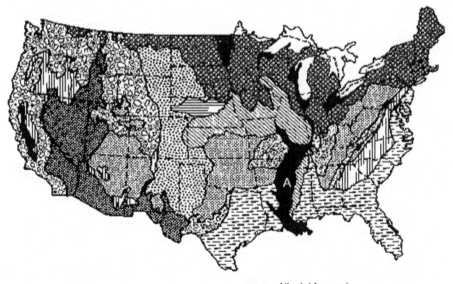

**Figure 5.4**

Generalized map of soil parent materials in the mainland U.S. Except at A alluvium is not shown. (Modified from Thompson and Troeh, 1973.)

| | |
|---|---|
| Marine | Alluvial fan and desert accumulations |
| Alluvial | Complex mountainous areas and outwash |
| Limestone | Soft sedimentary rock |
| Sandstone and shale | Lacustrine |
| Crystalline rocks | Loess |
| Glacial drift | Sand |

"t" means age of the parent material as modified by the rate of erosion. For example, an active sand dune has excellent drainage and no time for development of a soil profile, so the profile is nil. A bog may be thousands of years old but have no soil profile because continuously wet conditions have prevented oxidation and weathering.

---

*Question:* What are the soil parent materials in your area? Information that is not shown in Fig. 5.4 is available from geological maps and from the internet.

---

## 5.4  IMPLICATIONS FROM BURIED SOIL PROFILES (PALEOSOLS)

### 5.4.1  Origin and Occurrence

Paleosols (ancient soils) are soil profiles that have been buried underneath more recent deposits. The clay layer that occurs underneath a layer of coal, referred to by miners as "underclay," is an ancient paleosol. More common and closer to the ground surface are paleosols that separate different episodes of sediment deposition from wind, water, or ice. Paleosols are important geological markers because they separate stratigraphic units of different ages. They also are important in engineering because they create perched groundwater tables in otherwise well-drained soils and can cause landslides. They also can have their good points, as a perched groundwater table can reduce or prevent seasonal cycling of expansive clays.

The time interval during which a paleosol has developed may be much longer than for the modern soil profile, so as a result paleosols tend to be more highly weathered and have a more pronounced soil profile. The same horizon designations are used, but with a subscript b: $A_b$, $B_b$, $C_b$, indicating that this is a buried soil.

As in the case of modern soils, the properties of the paleosols depend on factors involved in their origin, which can be quite different from those that exist today. For example, a paleosol developed in basalt and subsequently covered by a later lava flow will be baked like a clay pot. A lateritic paleosol formed under tropical conditions can occur in today's temperate climate or even in Antarctica. Paleosols in alluvial deposits often show a marked increase in the rate of deposition in recent decades following the introduction of agriculture.

### 5.4.2  Perched Groundwater

Perched groundwater conditions result when downward percolating rain water encounters an obstacle such as a clay-rich layer in a paleosol. However, the perched water will drain out where a paleosol is exposed on a slope by erosion,

**Figure 5.5**

Quicklime stabilization of a highly expansive $B_b$ horizon with a series of soil borings. The proximity to the valley slope allowed drainage of the perched groundwater table and aggravated the condition.

so this is the location where expansive clay problems are most likely to occur. The house in Fig. 5.5 is on an eroded hillside where the exposed paleosol clay undergoes wet-dry cycling. The condition was further aggravated by a leaky floor drain.

## 5.5 GROUNDWATER AND SOIL COLOR

### 5.5.1 Definition of a Groundwater Table

It sometimes is assumed that a groundwater table represents the saturation elevation in a soil, but that is not correct, because soil above the water table also can be saturated by capillary action, as water can be drawn up into a small glass tube. The amount of capillary rise is small in sand, but can be substantial when pores are finer, as in clay. The groundwater table is defined as the level to which groundwater will rise in an open boring where capillary action has no effect.

### 5.5.2 Importance of a Groundwater Table

Which weighs more, a bucket of sand suspended in air or the same bucket of sand suspended in water? The answer is obvious, or at least it was to Archimedes, who noted that the weight of an object submerged in water is reduced by the weight of the water that is displaced. Similarly, buoyant forces on soil that is submerged below a groundwater table cause it to weigh considerably less than the same soil that is not submerged.

If the level of the groundwater table changes, so does the weight of the soil. That is why lowering a groundwater table by pumping from wells can cause areas to subside, as the increased weight of the soil compresses soil that is underneath. A notable example is Mexico City, where the amount of subsidence is measured in meters or even tens of meters.

Conversely, raising a groundwater table reduces the weight of soil, which is a reason why abandoned mines should be flooded, to reduce the likelihood that they will collapse. Pumping out an abandoned mine can be extremely unwise as it can double the weight supported by the mine.

### 5.5.3 Locating the Groundwater Table

When infiltrating rain water reaches a groundwater table it moves laterally to the nearest exit, normally a stream valley. The groundwater table therefore represents a subdued version of the ground surface, being higher under hills and exiting at or near the ground surface in valleys.

The groundwater elevation is not constant, but varies seasonally depending on the rate at which water infiltrates into the ground from rain or snow melt. A low position of the groundwater table is no guarantee of future immunity from problems such as basement flooding. The most important clue to transitory elevations of a natural groundwater table is soil color, discussed below.

### 5.5.4 Dowsing

Few things can inspire the confidence of a water dowser more than a $50 bill—unless it is a $100 bill, in which case he will try twice as hard. A dowser holds a forked stick that mysteriously points down when over water. Skeptics have tested the system by blindfolding the dowser and walking him or her over a bridge, but that is not a fair test, as water has to be surrounded by soil in order to give off the proper aura. A variation for buried water pipes or electric lines is two steel rods bent in L-shapes and held so that the long ends are horizontal and point forward. They then swing together over the buried object.

There is another reason why water witching is successful: groundwater can be found anywhere, even in a desert, if the hole is drilled deep enough. A boring eventually is likely to encounter some kind of permeable water-bearing stratum. If it doesn't, you get your $50 back.

Some dowsing concepts are simply incorrect, such as groundwater occurring in "veins" or "rivers" instead of as pore fluids. The only underground lakes and rivers are those in abandoned mines and limestone caverns. For a history and status of dowsing see Legget (1962). On the other hand, if a good client expresses a belief in water dowsing it can be unwise to argue with a religion.

### 5.5.5 Drilling to Locate a Groundwater Table

Groundwater elevations are routinely measured and recorded during geotechnical exploration drilling, by measuring water levels after they stabilize in open borings. Even this is not foolproof because if a boring penetrates into a zone of artesian pressure, the water level not only will rise higher in the hole but will shoot up as a geyser. If a boring encounters and punches through a perched groundwater table, the boring may drain before the water level can be measured. Special techniques, namely the use of expandable packers, can be used to seal off artesian or perched water. Moisture contents and densities of soil samples provide information on the level(s) of saturation.

### 5.5.6 Relation of Groundwater Level to Soil Color

Engineers must design for worst anticipated groundwater conditions, so past positions are as important as current positions. If groundwater elevations are assigned too low, a rise in the groundwater table can readily float an empty swimming pool or underground tank.

Below a permanent groundwater table iron oxides usually are in a reduced ($Fe^{2+}$) state because of deoxidation by organic matter. The soil color is a shade of gray and may have a greenish or bluish tint. A horizon subscript g, for *gleying*, signifies gray colors in a soil profile. Above the groundwater table, tan, brown, or rusty red colors indicate oxidation of the iron compounds ($Fe^{3+}$) by infiltrating oxygen-charged rain water.

Gray colors may dominate in a zone of capillary saturation, but there will be vertical fingers of tan or brown where oxygen-charged water has penetrated downward through cracks and empty root channels. Soluble iron compounds migrating to these sites can precipitate and make concretionary "pipestems."

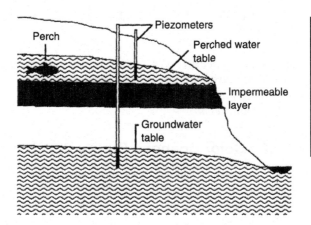

**Figure 5.6**
Perched groundwater table. (From *The Day the House Fell* by R. L. Handy, with permission of the American Society of Civil Engineers.)

Mottled or mixed soil colors often indicate a fluctuating groundwater table, in which case the highest elevation of the gray represents the highest former groundwater level. This is particularly useful for determining a need for tile drains, for example to prevent a landslide or flooded basements.

---

*Question:* What color sequence might be expected with a perched groundwater table?

*Hint:* Is the soil below a perched table saturated?

---

## 5.5.7 Minerals that Give Soil Its Color

Soil colors mainly depend on the colors of the coatings of the mineral grains rather than the colors of the grains themselves. Humic acids give the A horizon a dark color. Iron oxide coatings give soil a red, red-brown, or yellowish-brown color. Only a small percentage of iron compounds is sufficient to give a red or yellow-brown color. Red-brown color is attributed to an oxide mineral, *hematite*, and yellowish brown to a hydroxide mineral, *goethite*. Reddish-brown colors indicate another iron hydroxide mineral, *lepidocrocite*. Black specks are mainly manganese oxides.

Reduction of iron from the ferric to ferrous state requires dismantling the mineral crystal structure and removal of oxygen, which can occur only in a strong reducing environment. That is why red colors can persist in rocks for thousands or millions of years. Reduction can occur rapidly in the presence of organic matter, and occurs under asphalt pavements and in the presence of natural gas leaks. This can have forensic significance in the case of a fire or explosion, by indicating whether a gas pipe has recently been broken or if it has been leaking for some time.

Gray or white colors indicate that a soil has been leached of its iron oxide and hydroxide coatings so that one sees mineral colors that normally are dominated by quartz.

In arid and semiarid areas a white, crusty appearance indicates secondary deposits of calcium carbonate or, in arid climates, gypsum or rock salt. These deposits also can develop artificially in irrigated areas such that they eventually become sterile for agriculture.

## 5.5.8 Color Changes During Soil Sampling

Gray, greenish, or bluish hues indicating reducing conditions rapidly change to light gray or brown when exposed to air. Color identifications therefore should be made immediately after sampling in the field. Samples that are color-coded in the laboratory should be broken open to examine the interiors.

### 5.5.9 Color Standards

Most geotechnical reports describe soil color in everyday terms based on visual impressions, such as light or dark gray, black, tan, brown, buff, red-brown, etc. A more systematic notation was introduced by Albert Munsell in 1905 and is widely used in many different fields. Geotechnical engineers should at least be familiar with the shorthand of the Munsell system, and may wish to use it.

Colors are characterized by letters and numbers indicating (1) a number and capital letter for hue, or spectral color, (2) a number for value, for the degree of lightness or darkness, and (3) separated by a slash, a number for chroma, or intensity of the color. In the Munsell system the three components are sequential: hue, value, and chroma.

Hues most applicable for soils are designated by R for red, YR for yellow-red, and Y for yellow, all indicative of oxidizing conditions. GY signifies green-yellow, G is for green, BG is for blue-green, B is for blue, and N is for neutral or colorless. G, B, and N hues indicate unoxidized or gleyed soils that have been influenced by a groundwater table. An example is 10YR 5/4, for a color on the 10YR (yellow-red) chart with a medium value of 5 and a medium chroma of 4.

The color of a soil will change as much as two or three steps in value and chroma depending on the moisture condition. For this reason, and because of the possibility for change after more than a few minutes exposure to air, color identifications should be made on field-moist samples prior to air-drying. Dry soils are moistened prior to a color determination but may not regain their color.

## 5.6 AGRONOMIC SOIL MAPS

### 5.6.1 Availability of Soil Maps

Governmental agencies in many countries are actively engaged in mapping soils as an aid to land evaluation and to improve soil uses in agriculture and in engineering. In the U.S., soil surveys and maps are prepared on a county-by-county basis and are available from state universities or county or state offices of the USDA-NRCS (U.S. Department of Agriculture, Natural Resources Conservation Service). Most soil surveys stop at city limits, although cities may contract the studies as an aid to planning. County soil surveys nevertheless show what can be expected in a given area.

### 5.6.2 Soil Series

The basic agronomic soil mapping unit is the *soil series*. Each soil series is given a name, usually the name of a lake, city, stream, or other geographical entity near the place where the series has been described and mapped. Examples are the Mankato series, named after Mankato, Minnesota; and the Hilo series named after Hilo, Hawaii. After a soil has been described and named, that name is applied to all soils having essentially the same weathering profile and genetic history.

A soil series represents a single solution to eq. (5.1); that is, all soils within a series have the same number of horizons of similar depth, developed under the same climate and same kinds of vegetation, with essentially the same slope and landscape factors, from the same kind of parent material, and of approximately the same age. An allowable range is included in each of the factors, but the consistency is sufficient to allow assignment of a particular soil to a particular series. As one soil series will intergrade into another, analyses are made to define cutoff points.

### 5.6.3  Soil Association Areas

Soil series that dominate or characterize a particular area define a *soil association area*. Block diagrams showing soil association areas are included in county soil survey reports and are useful for visualizing and interpreting soils in a given area. A representative block diagram is shown in Fig. 5.7.

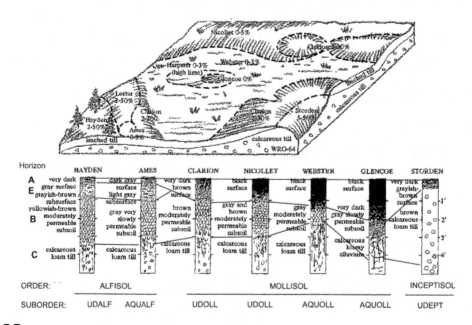

**Figure 5.7**

Block diagram showing topographic relationships of soil series in a soil association area where the parent materials are glacial till and slope wash deposits. Alfisols are developed under forest and Mollisols under grass. "UD" in the suborder designation stands for a humid climate, and "AQU" for wet conditions. The latter series also have expansive clay minerals concentrated in their B horizons. (USDA.)

### 5.6.4 Soil Classification

Soil classification follows a biological model, from top down consisting of order, suborder, great soil group, family, and series.

Classical names for great soil groups like Podzol, Brunizem, and Chernozem are Russian in origin, and respectively mean ashy, brown earth, and black earth, in reference to the color of the topsoil. These soil groups are called "zonal soils" because they bear close relationships to climatic zones.

More relevant in engineering is a system developed multinationally by soil scientists, called soil taxonomy. This system is based on measured soil properties, and new words were coined to try and convey the most significant properties. The names therefore are useful if one can penetrate their complicated system and polyglot origins.

The ten soil orders are described in the following paragraphs, arranged more or less according to increasing degree of weathering. A hypothetical relationship of soil orders to climate is shown in Fig. 5.8.

### 5.6.5 Ordering the Orders

Names for soil orders normally are not a major concern in engineering, but can be helpful for characterizing a particular soil series. For example, if a soil series is identified as a "Histosol" or "Vertrisol" it should raise some alarm bells for the engineer. Other soil orders are much less threatening, but the names nevertheless can reveal important properties.

*Histosol* means "tissue soil" and signifies bog soils, peat, and organic muck. In a soil classified as "Cryofibrist," "cryo" means cold, "fibr" implies fibrous, and "ist" is the designation for Histosol. The order is designated by the *last* three or four letters in a soil name, so when "ist" is appended to the end of a soil name one

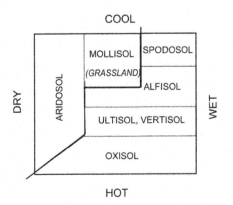

**Figure 5.8**
Generalized relationship between soil orders of the Seventh Approximation and climate.

can expect severe engineering problems, namely wet conditions and highly compressible soil such as peat or muck.

*Entisol* means "recent soil." Entisols have little or no genetic horizonation—that is, they have an AC profile, and sometimes a very weak B horizon. Suborders are defined for wet soils (Aquents), sandy soils (Psamments), arid soils (Ustents), and humid soils (Udents). Under Aquents, the great group Cryaquent (cold-wet-recent) includes tundra soils, indicative of permafrost, with attendant engineering problems.

*Inceptisol* means "beginning soil." Horizons in Inceptisols mainly reflect changes in color and pH, but there is little textural change due to clay movement. Suborders are defined for wet soils (Aquepts), predominantly volcanic ash or allophane (amorphous clay) soils (Andepts), dark-colored A horizon soils (Umbrepts), and light-colored A horizon soils (Ochrepts). Most important in engineering are Aquepts because they are wet, and Andepts because they may be sensitive to shock or vibrations. Cryaquepts (cold-wet-beginning soil) are on permafrost.

*Aridisol* means desert soil. Aridisols have light-colored A horizons and a variety of B horizons containing secondary deposits of salt, gypsum, or calcium carbonate. Suborders are defined for nonclayey (Orthid) and clayey (Argid) soils; the latter contain highly expansive clays that afford severe engineering problems.

*Mollisol* means "soft soil," in reference to the dark-colored, crumbly A horizon that is excellent for agriculture. Mollisols develop on grasslands and often have clay-rich B horizons containing expansive clay minerals. Suborders indicate if they have developed on limestone (Rendolls), or have a white A2 horizon (Albolls), show wetness plus gleying or mottling (Aquolls), develop in cool or high altitudes (Altolls), develop in humid climates (Udolls), or in hot, dry climates (Ustolls and Xerolls). For engineering purposes, the A horizon is stripped away and saved, and the B horizon should be recognized as probably being expansive. Soils in this order formerly were called Brunizem, Chernozem, Chestnut, and related soils.

*Alfisol* means "aluminum-iron soil." These soils have a light-colored A2 horizon over a clayey B. They develop under boreal or deciduous forest, usually in calcareous deposits of Pleistocene age (i.e., those containing calcium carbonate). Suborders are defined for wet conditions (Aqualfs), high, cool climate (Altalfs), humid climate (Udalfs), and hot, dry climate (Ustalfs and Xeralfs). The B horizons are well developed, blocky, and montmorillonitic or expansive. Formerly these were called Gray-Brown Podzolic, Gray Wooded, and related soils.

*Vertisol* means "inverted soil," that is, turned upside down due to churning action of expansive clay. Vertisols have a black A1 horizon that sometimes extends to 1 to 1.5 m (3 to 5 ft) depth. Vertical mixing occurs when shrinkage cracks fill in from above so that they cannot close when the clay re-expands. The resulting high horizontal stresses surpass the strength of the soil, so it fails by shearing along inclined shear planes. Smearing of clay (slickensides) along the shear planes imparts permanent weakness. Mixing prevents the development of a normal soil profile and arrests weathering at the smectite stage instead of allowing it to proceed to more stable, less expansive clay minerals that are more in keeping with the climate.

The disruption of the soil mass by shearing sometimes creates a hummocky microrelief called *gilgai*, tilting trees and fenceposts. High lateral soil stresses make basements uneconomical because walls push in, and can exhume graves so that burials are in soil mounds placed above the natural ground surface. The parent material for Vertisols typically is alluvial clay or highly weathered limestone or basalt. The climate may be temperate to tropical, but must include a dry season for cracking and vertical mixing to occur. Severe engineering problems include tilting and tearing apart of buildings, upward pulling of pile foundations, and distress to highways. In the older literature these soils are called Grummosol, Regur, or Black Cotton Soils.

*Spodosol* means "wood ash (Greek, *spodes*) soil" and contains an accumulation of free oxides or humus or both. Formerly these were called Podzols, Brown Podzolics, and Ground Water Podzols. They have an ash-colored E horizon (hence the name). Spodosols are soils of the humid, usually coniferous forest areas. They typify sandy parent materials in these locations, and do not develop on parent material containing more than about 30 percent clay. Suborders are defined for wet conditions (Aquods), humus accumulation (Humods), iron plus humus accumulation (Orthods), and iron accumulation (Ferrods). The humic acids in Spodosols can prevent setting of Portland cement.

*Ultisol* means "ultimate soil" and has a variable A horizon over a red or yellow clayey B, which often contains iron-emented nodules or an iron-cemented layer called *plinthite* (hard layers also are called laterite). Ultisols are highly weathered acidic soils developed under forest, grass, or marsh in pre-Pleistocene deposits. B horizons are very well developed, and contain nonexpansive kaolinite clay mineral plus some montmorillonite.

*Oxisol* means "oxide soil" characterized by a red B horizon of low density, weakly cemented silt, and clay that is mostly aluminum and iron oxide and hydroxide minerals plus kaolinite clay. More readily weathered minerals including quartz and montmorillonite have been degraded or removed. The occurrence of Oxisols is restricted to tropical and subtropical climates and older (pre-Pleistocene) land surfaces. They often have a soft or hard iron accumulation layer or plinthite. Suborders are defined for wet conditions (Aquox), extreme weathering (Acrox), humid climate (Udox), hot, dry climate (Ustox), and arid climate (Idox). Aluminum ore, or bauxite, mainly derives from Oxisols. The weathering stage is kaolinite or gibbsite.

### 5.6.6 Deciphering the Names

Soil taxonomy names start with a prefix designating some key soil property used in classification, followed by more general names. For example, "Argalboll" should be read as Arg-al-boll, for having a clay layer (arg), a white layer (alb), in a soft (Mollisoll soil order) soil. The taxonomic name therefore carries more information than the former name for this soil, which was "Planosol," or flat soil.

Soils are not isolated species, but form a continuum and are transitional one to another. Subgroup names are added in front to indicate whether a soil fits the central concept for that group, in which case it is called "Orthic." If it represents an intergrade into another great group, that group will be indicated. This kind of detail is required to satisfy the sensibilities of soil scientists but ordinarily is of little concern in engineering.

### 5.6.7 Important Soil Orders in Engineering

The most critical soil orders for engineers are as follows:

*Histosol*, because it designates peat that is weak, usually wet, highly compressible, and a nuisance in engineering.

*Vertisol*, because it designates a considerable depth of highly expansive clay that can cause severe foundation and other problems.

*Aridosol*, because it means desert conditions and may include highly expansive clay concentrated in shallow lakebeds or "playas," and may include a layer cemented by secondary carbonates, "caliche."

*Mollisols* and *Alfisols*, because they are likely to contain expansive clay B horizons (subsoils).

*Oxisols and Ultisols*, because they are leached to such a low density that they may collapse under a foundation load, the clay mineral is not expansive but nevertheless can contribute to landslides, and there may be a surficial hard, laterite layer.

### 5.6.8 Soil Database

Identifications, descriptions, and engineering data for soil series are included in maps published by the USDA-NRCS. A soil series database is maintained at http://www.statlab.iastate.edu/soils/index/html/

### 5.6.9 The Geotechnical Report

Information, data, analyses, recommendations, and conclusions resulting from an investigation ordinarily are presented in a written report that is discussed in more detail in the last chapter of this book. However, it is important to emphasize that a brief review of the site geology and soils is presented as introductory material. Geological and pedological information helps to define and optimize the approach used in the investigation. For example, if the site is on an upland having a uniform geology, the boring and testing program may be reduced if results are consistent with expectations. On the other hand, if the site is on a floodplain of a meandering stream, borings must be close enough that they will not miss the clay-filled oxbow that may be hidden away underneath the ground surface.

In some cases, for example in a limestone sink or undermined area, no reasonable amount of drilling will provide absolute assurance, and a geophysical investigation may be warranted. The geotechnical report must emphasize that it can describe only what has been encountered in the borings and not what lies between. The more detailed the investigation, the more reliable the results but the higher the cost. It is important to maintain communications between all parties so that there are no misunderstandings. A credible discussion of site geology and soils is basic to any site investigation.

## Problems

5.1. Weathering to form clay minerals is most intense in the A horizon, so where does the clay go? Give evidences that support your answer.

5.2. Strength tests show that a dry, clayey $B_{2t}$ behaves like a gravel. (a) With reference to par. 5.2.6 what is the meaning of the subscript "t"? (b) Explain the peculiar behavior of this clay. (c) How might clay coatings on the peds affect this behavior?

5.3. A soil is described as having A over $C_g$ horizons. (a) What can you infer regarding groundwater conditions? (b) What are the order and suborder?

5.4. A $C_k$ horizon occurring in the southwestern United States is crushed and used on gravel roads. (a) What is the cementing component in this material? (b) What is another name for this horizon? (c) What is the order?

5.5. The iron-cemented or laterite layer of tropical soils is frequently used as road metal. (a) What is the horizon designation? (b) What is the soil order?

5.6. The next chapter describes the clay mineral that dominates in Vertisols. What is it?

5.7. Identify hue, value, and chroma for (a) 10YR 5/4; (b) 10YR 2/2; (c) N 9/0.

5.8. Borings indicate the following sequences of horizons from the top down. Give one or more possible explanations for each sequence: (a) A, C; (b) A, B, C, B, A; (c) A, B, C, A, B, C; (d) B, C.

5.9. Compare the suitability for foundation bearing: $B_{2t}$, $B_x$, $B_{ss}$.

5.10. (a) Examine soil horizons exposed in a roadcut in your area and identify A, B, and C horizons. (b) Do the same from soils obtained from a boring made with a hand auger.

## References and Further Reading

Bigham, J. M., and Ciolkosz, E. J., ed. (1993). *Soil Color*. Special Publication 31, Soil Science Society of America, Madison, Wis.

Dinauer, R. C., ed. (1977). *Minerals in Soil Environments*. Soil Science Society of America, Madison, Wis.

Handy, R. L. (2002). "Geology, Soil Science, and the Other Expansive Clays." *Geotechnical News* 20(1), 40–45.

Jackson, M. L., and Sherman, G. R. (1953). "Chemical Weathering of Minerals in Soils." *Advances in Agronomy* 5, 221–338.

Jenny, Hans (1980). *The Soil Resource*. Springer-Verlag, New York.

Legget, R. F. (1962). *Geology and Engineering*, 2nd ed. McGraw-Hill, New York.

Soil Science Society of America (1997). *Glossary of Soil Science Terms 1996*. Soil Science Society of America, Madison, Wis.

Soil Science Society of America (2002). *Soil Planner 2002: State Soils and Protecting Important Farmland*. Soil Science Society of America, Madison, Wis.

Soil Survey Division Staff, USDA (1993). *Soil Survey Manual*. USDA Handbook No. 18. U.S. Government Printing Office, Washington, D.C.

Soil Survey Staff, Soil Conservation Service, USDA (1975). *Soil Taxonomy—A Basic System of Soil Classification for Making and Interpreting Soil Surveys*. U.S. Government Printing Office, Washington, D.C.

Thompson, L. M., and Troeh, F. R. (1973). *Soils and Soil Fertility*, 3rd ed. McGraw-Hill, New York.

# 6 Soil Minerals

## 6.1 INTRODUCTION

Minerals vary from the aristocratic brilliance of diamonds to the common jumble of quartz grains and clay minerals. All minerals have one property in common: they consist of regularly repeating arrays of atoms. The tendency for crystals to exhibit flat faces led to the first speculation that they are made up of a regular arrangement of atoms as early as the fifth century B.C.E. A simple analogy is to the repeated pattern of wallpaper, but in three dimensions. Glass, even a natural glass such as obsidian, is not a mineral because it does not have a regular array of atoms, and glass fractures along curved instead of planar surfaces. Glass is not crystalline; it is amorphous. Not all gemstones are minerals: opal is amorphous silica, and amber is amorphous and petrified tree sap. Ice is a mineral; liquid water is not.

The crystalline nature of minerals affects the way in which they refract light, and the difference often is directional. For example, a crystal of calcite, which is the most common calcium carbonate mineral, has different refractive indices that depend on the polarization of the light waves penetrating through the crystal. A black dot observed through a crystal of calcite appears as two dots because the light rays become polarized, the same as in polarizing sunglasses but in two directions. The two rays transmit differently. This can be confirmed by viewing the dots through a polarizing filter and rotating the filter, in which case the two dots will take turns appearing and disappearing. This property of minerals is used in a polarizing microscope to help identify minerals.

Clay particles are too fine for their shapes to be resolved by a light microscope, so clay mineralogy was mainly a matter of guesswork based on chemical analyses until the introduction of X-ray diffraction in the early 1900s. Direct observation of clay particles later became possible with the higher resolving power of the electron microscope. More recent was the development of the scanning electron microscope, which is a kind of subminiature video camera and is particularly useful for viewing clay structures.

Both the electron and scanning electron microscope require that samples be in a vacuum that desiccates the clay and can alter its composition and structure.

The most useful tool for identification of clay minerals is X-ray diffraction, which does not require a vacuum and can be performed without pretreating the soil. Even though this tool is not commonly used for routine soil analyses, the student should know its capabilities. For example, a wet clay can be X-rayed as it is drying out to determine if it is expansive. Furthermore, the extent of expansion and contraction can be demonstrated to depend on the nature of the free ions that are associated with the clay, such that changing the ions can change the expansive properies. X-ray diffraction demonstrates that there is more to geotechnical engineering than boring or punching holes in the ground.

# 6.2 X-RAY DIFFRACTION AND GEOTECHNICAL ENGINEERING

## 6.2.1 Not Your Usual X-Rays

If you ask a medical technician what is meant by X-ray diffraction, you most likely will be met with a querulous look and be asked to repeat the question. Instead of using penetration of X-rays to observe an inner structure, X-ray diffraction defines relative positions of atoms in minerals or other solid objects.

The strongest diffractions are obtained from crystalline materials that have a regular array of atoms arranged in planes so that the X-rays in effect bounce off as if reflected from a mirror. A critical distinction is made from mirror reflections because X-ray diffraction can occur only at specific angles that depend on distances between the reflecting planes of atoms. By measuring the angles at which reflections occur, one can decipher the distances and identify the minerals. It also is possible to measure exactly how much a clay structure expands when it is wet with water.

## 6.2.2 Production of X-Rays

X-rays are generated by an electron beam striking a metal plate that is aptly called the "target." The voltage pulling electrons to the target is high enough to knock electrons out of target atoms, and as orbiting or shell electrons are replaced, energy is given off as X-rays. The target in an X-ray tube therefore glows with a brilliant but invisible light. The voltage is of the order of 50,000 volts. Such high voltages are strictly prohibited in television or computer screens to prevent the production of X-rays.

Replacement of orbiting electrons from outer shells in atoms involves a stepwise energy change, so X-rays have particular wavelengths. Other "white radiation" wavelengths also are produced and are used for medical X-rays, but for diffraction measurements only a single wavelength can be used, so the rest are filtered out.

The strongest monochromatic or single-wavelength X-rays are produced when a position in the innermost or "K" atomic shell is replaced by one falling in from the next outer shell, which generates "K$\alpha$ radiation." The most useful X-rays for mineral identification are from a copper target and are therefore referred to as "copper K$\alpha$ radiation."

### 6.2.3 Geometry of Diffraction

Figure 6.1 shows a *diffractometer* that is designed to measure diffraction angles from a sample placed in an X-ray beam. A beam is obtained by simply looking at the target through some narrow slits, and is directed toward the sample that is at the center of a circular goniometer. A counter then is moved around the outside of the circle to detect X-rays, and is synchronized with a chart recorder so that when diffraction occurs and the count increases, the angle can be measured. One of the most famous uses of X-ray diffraction was to determine the structure of DNA.

Unlike reflections in a mirror the diffraction angle is the same on both sides of the crystal and is designated by $\theta$. As shown in Fig. 6.1, the angle measured around the circumference of a diffractometer is $2\theta$.

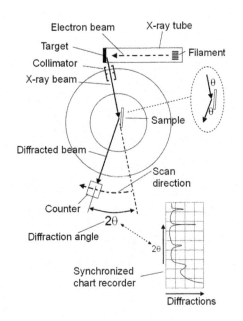

**Figure 6.1**

Schematic diagram of an X-ray diffractometer used for identifying minerals in soils and measuring clay mineral expansion.

### 6.2.4   Relating Diffraction Angles to Crystal Structure

The interpretation of X-ray diffraction would be relatively simple if each crystalline material had only one set of reflecting planes, and as a matter of fact analyses of clay minerals do emphasize only one set of planes. However, there also are other planes having different interatomic spacings, much as parallel lines drawn at various angles through figures in wallpaper have different spacings (Fig. 6.2). Each mineral species therefore presents a combination of diffraction angles and intensities that serves as a fingerprint to identify that mineral.

Thousands of diffraction patterns have been collected, categorized, and published for purposes of mineral or crystal identification by the American Society for Testing and Materials. Because of its speed and accuracy, X-ray diffraction has many uses including forensic investigations.

### 6.2.5   Bragg Made It Easy

The principles of X-ray diffraction were described by von Laue in the early 1900s, but a simplified model was needed before diffraction was easily understood and utilized. The model was devised by father-and-son English physicists William and Lawrence Bragg, who shared a Nobel Prize for their contribution. The relationship describing diffraction of X-rays from crystals is called Bragg's Law, but a more accurate description is "Bragg's reflection analogy."

**Figure 6.2**

Atoms in a crystal occur in a regular pattern like flowers in wallpaper, so distances between identical planes depend on the orientation of the planes and the size and arrangement of the flowers and caterpillars.

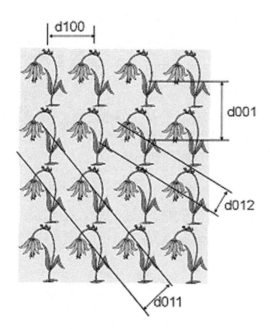

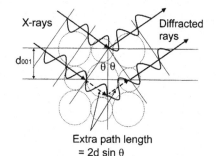

Bragg, father and son, suggested that even though X-rays actually come from point sources, namely electrons in crystalline atoms, the rays may be thought of as reflecting from atomic planes. As a wave front is reflected from adjacent planes, the interlayer distance imposes an extra path length that is shown in Fig. 6.3. In order for rays to reinforce, that extra path length must equal an exact number of wavelengths. This enables writing an equation:

$$n\lambda = 2d \sin \theta \tag{6.1}$$

where $n$ is a whole number, $\lambda$ is the X-ray wavelength, $d$ is the distance between identical crystal planes, and $\theta$ is the diffraction angle. This is the Bragg Law. When a reflection occurs, the interlayer distance $d$ is obtained by measuring the angle and knowing the X-ray wavelength.

If a reflection is obtained with $n=1$, according to eq. (6.1) another reflection should be obtained with $n=2$, or at exactly one-half of the $d$-spacing. This is called a *second-order reflection*. In general as the order increases the reflection becomes weaker, but that is not always the case. Reflection intensities are modeled with Fourier analyses for determination of crystal structures.

The unit of measure is the Ångstrom, or Å, which equals $10^{-8}$ cm or 10 nm (nanometers). As a reference, the diameter of an oxygen ion in crystal structures is 2.64 Å.

## Example 6.1

Mica, the shiny, flakey mineral in mica schist, is closely related to clay minerals and gives an X-ray diffraction angle $2\theta = 8.8°$ with an X-ray wavelength of 1.54 Å. At what angle may one expect a second-order reflection?

*Answer:* Equation (6.1) solved for the first-order reflection is

$$(1)(1.54) = 2(d) \sin (4.4°)$$

from which $d = 10.0$ Å. Substituting this value in eq. (6.1) and solving for $n = 2$ gives

$$(2)(1.54) = 2(10.0) \sin \theta$$
$$2\theta = 17.7°$$

A similar solution can be made for third-order and higher reflections. Note that diffraction angles always are expressed in terms of $2\theta$.

### 6.2.6 Defining Orientations of Crystal Planes

Planes in a crystal are defined on the basis of their relationships to *crystallographic axes*. These axes may or may not be orthogonal. In rock salt or diamond, which crystallize in the cubic system, the axes are at right angles. The smallest repeating pattern is called a unit cell, which in this case has the shape of a cube.

---

*Question:* Can you identify a unit cell in Fig. 6.1?

---

The orientation of a crystalline plane is described by a "Miller index," which is the number shown on the plane orientations in Fig. 6.1. The most useful Miller index in clay mineralogy is 001. The $d_{001}$ distance is the distance between repeating layers stacked like pages in a book. For mica, $d_{001} = 10$ Å, so each layer in a mica crystal is 10 Å thick, or about the diameter of three oxygen atoms.

## 6.3  CLAY MINERALOGY

### 6.3.1  Atoms and Ions

For most minerals the use of the name "atom" for individual components is not strictly correct because an atom has a neutral charge, which means that the number of protons, each with a $(+)$ charge, equals the number of electrons, each with a $(-)$ charge. Minerals for the most part are composed of *ions*, which are atoms that either have a surplus or a shortage of electrons. This occurs when one atom gives up an electron to another kind of atom in order to fill out a shell. For example, a sodium atom, Na, readily loses an electron from its outer shell to chlorine, Cl, which needs one to fill its outer shell. The result is $Na^+Cl^-$, sodium chloride or rock salt. The linkage between sodium and chlorine is not a true chemical bond, but an ionic attraction that is easily put off by an intruder such as water that has a high dielectric constant, so $Na^+Cl^-$ readily dissolves in water.

Because the crystal structure of crystalline minerals such as $Na^+Cl^-$ (the mineral name is halite) is continuous and ionic attractions have no particular directional orientation, there is no way to define a discrete molecule. In true molecules the bonds are covalent, which means that instead of electrons being transferred they are shared. Atoms of many gases, for example, pair off to make molecules such as $O_2$, $H_2$, and $N_2$.

Covalent bonds are directional, and generally are stronger than ionic bonds. Diamond and quartz are heavily imbued with covalent bonds, and these minerals are hard and obviously do not dissolve in water. Unlike ionic attractions, covalent bonds are highly directional as they involve orbiting electron shells.

Many minerals incorporate both ionic and covalent bonding. Calcium carbonate, for example, is composed of calcium ions, $Ca^{2+}$, which are held ionically to covalently bonded carbonate, $(CO_3)^{2-}$, to make $Ca^{2+}(CO_3)^{2-}$, or $CaCO_3$. The ionic bond is easily attacked by a weak acid so calcium carbonate dissolves, but the covalently bonded $CO_2$ stays together and escapes as gas.

## 6.3.2 Part-Time Ions

Atomic pairs that are most likely to lose or gain electrons are those at the opposite edges of the periodic table. Na and Cl, for example, are at the left and right edges respectively. Bonds between closer atom pairs are less ionic and more covalent, as in the case of Si and O, which make up $SiO_2$, quartz. Covalent bond percentages were quantified by Linus Pauling, who estimated the Si—O bond to be approximately 50 percent ionic and 50 percent covalent.

## 6.3.3 Some Unique Aspects of Water

Water is a unique molecule in that a hydrogen atom that loses an electron becomes a proton, a tiny, intense positive charge. A proton exposed on the surface of the oxygen atom in a water molecule is highly attractive to the oxygen atom of another water molecule. Even though the two molecules are electrically neutral, this can build a loose structure as a result of "hydrogen bonding."

Hydrogen bonding causes an open lattice structure when water freezes to ice. When ice melts this structure collapses, which is a unique property of water. Were it not for this unique property, polar ice would sink and accumulate at the bottom of the sea. Also, applying pressure to ice to increase the density can cause it to melt, which is important for moving glaciers, sleds, and ice skates.

Another factor affecting water is the angle between two hydrogen atoms, which is fixed by the orbital geometry at about 105°. Each water molecule therefore has a positive end and a negative end. This dipolar nature causes water to be attracted to other ions, particularly small, positively charged ions such as $Na^+$.

$Na^+$ and $Ca^{2+}$ ions are associated with clay minerals and carry with them adsorbed hulls of water that can influence the expansive properties. These ions also affect the net electrical charge on clay particles and affect whether they are attracted to one another, or *flocculated*, or repel and are *dispersed*. These properties can dramatically influence the engineering properties of a clay.

### 6.3.4   Atomic Size and Coordination

The largest ions have a net negative charge because orbital electrons repel one another. This also means that if an electron is lost, the size of an ion changes. The diameter of a ferrous iron ion, $Fe^{2+}$, is larger than that of its more positive alter ego, ferric iron or $Fe^{3+}$. Ion size also depends on the number of electron shells, which increases downward in the periodic table.

A basic building block in many minerals is one silicon ion, $Si^{4+}$, surrounded by four oxygen ions, each having two negative charges as $O^{2-}$. The oygen diameter is 2.6 Å, and that of a silicon ion is 0.82 Å, so one $Si^{4+}$ fits into holes between four oxygens. The shape of this building block is that of a tetrahedron, and silica is said to have a "tetrahedral coordination" with oxygen. Tetrahedral coordination is shown at the top in Fig. 6.4.

In quartz, which is $SiO_2$, each oxygen is shared by two tetrahedrons. This linkage creates a continuous three-dimensional structure, which is the reason why quartz is hard and has no natural cleavage direction: quartz crystals have flat faces because they grow that way. Ionic substitutions can change and weaken the three-dimensional structure, for example in feldspars, which are easier to weather than quartz and have cleavage.

Because of its abundance in rocks and minerals and its large diameter compared with positive ions, most of the Earth is composed of oxygen ions.

### 6.3.5   Other Coordinations

The outer ions in tetrahedral coordination can touch but not overlap, and if separated far enough will allow room for more oxygen ions. These principles are known as "Pauling's rules," and enabled him to work out the crystal structure of mica. If aluminum, $Al^{3+}$, substitutes for silicon, the diameter of an aluminum ion is 1.0 Å instead of 0.8 Å for silicon, which spreads the structure sufficiently to allow 6 oxygen ions instead of 4. The coordination number is 6, with oxygen ions arranged as corners of an octahedron to give *octahedral coordination*, shown at the middle in Fig. 6.4.

Pauling's rules govern stable coordinations and establish that it is not chemical composition but ionic sizes that control a crystal structure. This means that certain ions having about the same diameter can substitute others having approximately the same diameter without changing the crystal structure or the mineral name. For example, some $Mg^{2+}$ can substitute for $Al^{3+}$ without disrupting the crystal structure. Small changes in size are compensated by "puckering" of the tetrahedral layer, as shown near the middle of Fig. 6.4.

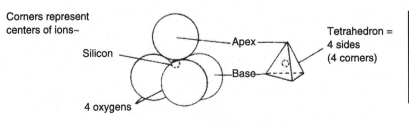

**Figure 6.4**
Tetrahedral and octahedral coordinated sheets that combine to make layers.

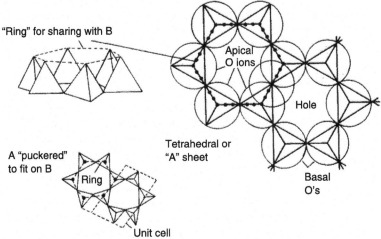

TETRAHEDRAL COORDINATION

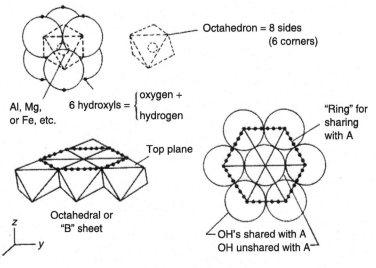

OCTAHEDRAL COORDINATION

### 6.3.6 Building a Layer Silicate

Mica shares silica tetrahedral in only two directions instead of three directions as in quartz, and therefore readily cleaves into flakes. Clay minerals also have shared bonding in only two directions, as silica tetrahedral link together at three corners instead of all four. This kind of sharing characterizes *layer silicates*. The sharing arrangement is shown in the top part of Fig. 6.4.

In addition, octahedrally coordinated aluminum ions can share corners to form a continuous layer, which is shown in the bottom part of Fig. 6.4.

### 6.3.7 Combining Layers

The final step is to make a sandwich from the two kinds of layers by sharing apical ions. One type is open-faced with one silica layer and one alumina layer. This is illustrated at the top in Fig. 6.5, where oxygen ions in the layers are shown touching. This is the mineral *kaolinite*, which is abundant in soils in tropical and semitropical climates.

Kaolinite is nonexpansive because a hydrogen ion, the dynamic proton, slips between two opposing oxygens and creates a hydrogen bond—the same kind of bond that occurs organized in ice and less well organized in liquid water.

A thicker sandwich is formed when two silica layers enclose one alumina layer, as shown at the bottom in Fig. 6.5. This is a three-layer silicate and will be the focus of attention as it includes expansive clays.

### 6.3.8 As Easy as AB

A silica tetrahedron viewed from the side somewhat resembles a letter "A" and an aluminum octahedron is kind of like a "B," so a graphic shorthand for the two classes of layer silicates are "AB" and "ABA." Kaolinite can be represented by AB • AB • AB . . . where the • represents hydrogen bonding.

Kaolinite is named for a mountain in China, and is the main constituent in China clay used for the manufacture of porcelain. Less pure deposits are used for stoneware or bricks. Kaolinite dominates the clay fraction of highly weathered soils in tropical and semitropical climates, but also is common in shales, particularly in tropical paleosols occurring underneath coal strata.

Kaolinite is identified by X-ray diffraction from its basal spacing $d_{001} = 7.14 \, \text{Å}$, which is the thickness of an AB layer. It is approximately the thickness of three oxygen atoms, $3 \times 2.6 = 7.8 \, \text{Å}$, with allowance for the stacking arrangement.

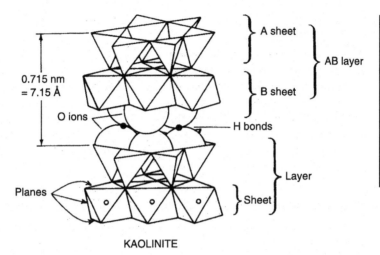

KAOLINITE

**Figure 6.5**

Two classes of clay minerals depending on the number of sheets per layer. One kind is used for making fine porcelain and the other can gradually create havoc.

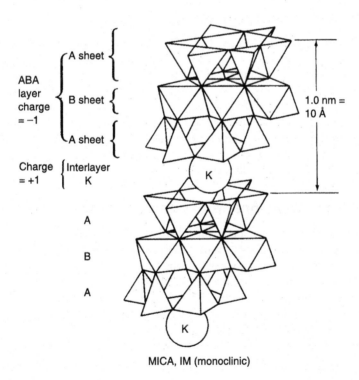

MICA, 1M (monoclinic)

### 6.3.9 ABA

The three-layer structure shown in the lower part of Fig. 6.5 is thicker than the AB by one oxygen ion, about 10 Å. In mica a potassium ion, $K^+$, has about the same size as an oxygen and fits into hexagonal openings in the faces of stacked A layers. These holes are shown at the upper right in Fig. 6.4. Potassium ions, being positive while the net charge on the layers is negative, hold the layers together, so mica can occur as large but relatively weak crystals. Mica is used commercially for heat-resistant insulating sheets in electrical appliances. A black variety of mica called biotite is common in granite.

*Question:* What is the coordination of the interlayer $K^+$ ion in mica?

*Answer:* It's not 6.

### 6.3.10 Take Away the K

Weathering strips away $K^+$ from mica so the crystal becomes weaker and less well organized. This is the *illlite* family of minerals, named for the state of Illinois. Illite is abundant in marine shales that tend to pick up $K^+$ ions from sea water.

When all of the $K^+$ is gone, the layers separate and the crystal structure takes in water and expands like an accordion. Even a change in relative humidity can cause expansion or contraction due to gain or loss of interlayer water. These are the ubiquitous and damaging expansive clays.

### 6.3.11 Influence from Interlayer Ions

In most naturally occurring expansive clays the place of $K^+$ is taken by calcium ions, $Ca^{2+}$, plus variable amounts of water. Calcium exerts some control on expansion, which when the clay is wet normally gives a maximum interlayer spacing of 19.6 Å. Of that about 10 Å is the mineral ABA layer and the rest is water molecules clustered around calcium ions and having an ice-like structure. Air-drying removes water and the spacing goes to about 15.4 Å. Heating to over 100°C is required to drive off all of the water.

*Question:* What volume expansion can be attributed to the change in $d_{001}$ of calcium expansive clay from air-dry to saturation?

*Answer:* $100(19.6 - 15.4)/15.4 = 27\%$ and the floor is slowly rising.

Calcium ions are attracted and held between layers of an expansive clay, with some additional ions held at the edges. Because the ions are loosely held, they are *exchangeable*, which means that other ions can be substituted and can markedly change the clay properties.

The changes in expansive clay *d*-spacing are stepwise and show *hysteresis*, which means that the humidity must be over 90 percent for a calcium clay to fully expand, and below about 50 percent humidity for it to dry to a spacing of 15.4 Å.

### 6.3.12 Can Foundation Pressure Prevent Expansion?

How much pressure can be exerted by hydration of an expansive clay? Is the pressure really enough to lift a building, and if so, how heavy?

To answer this a special compression device was mounted on a diffractometer, and it was found that re-expansion of an air-dry calcium montmorillonite from 15.4 to 19.6 Å was prevented with an applied pressure of 4 MPa (40 tsf). In contrast, a more ordinary foundation pressure of 0.4 MPa (4 tsf) allowed 90 percent expansion (Handy and Demirel, 1987). The relationship was linear, so the heavier the foundation pressure, the less the expansion. Therefore, under the same moisture conditions, floors are lifted more than adjacent load-bearing foundations.

One way to prevent future expansion is to pre-wet clay so that it reaches equilibrium expansion prior to construction. However, settlement then can occur when the clay dries out. One of the more common causes of damage is from drying shrinkage of soil around the edges of a floor or shallow foundation, particularly if the situation is aggravated by suction from tree roots. Damages may occur years or even decades after construction.

### 6.3.13 Sodium Clay and Lime Treatment

The expansive nature of clay is greatly increased if interlayer $Ca^{2+}$ ions are substituted by ions having a lower positive charge, in particular sodium ($Na^+$) or hydrogen ($H^+$). Sodium clay occurs naturally in alkaline conditions such as in playas, and hydrogen clay occurs in some forested A horizons with acid soil conditions. The severe expansive properties of a sodium clay can be reversed by adding a calcium salt, in particular hydrated lime, $Ca(OH)_2$, so that $Ca^{2+}$ ions replace $Na^+$. Lime treatment also benefits calcium clays for reasons discussed later in this chapter.

### 6.3.14 Expansive Clay Mineral Names

*Montmorillonite* is named for a village where it was first described in France. Later it was discovered that the chemical composition of expansive clays is quite variable, so expansive clays were referred to collectively as "montmor-illonite group minerals." *Smectite* now is used for these minerals. This is a very old name that Pliny used with reference to cleansing, probably because expansive clay is highly absorptive and is used as fuller's earth to remove lanoline from wool. Smectite clay has similar uses in geoenvironmental engineering.

### 6.3.15 Tests to Measure Clay Expandability

Because of the unruly nature of the $d_{001}$ spacing of smectite clays, they are treated with an organic chemical to stabilize the layer spacing before being examined by X-ray diffraction. The more common chemical is ethylene glycol, the main component in automotive antifreeze. This fixes the $d_{001}$ spacing at 17.6 Å. This treatment was used for Fig. 6.6.

Other tests to measure the physical character of clay at different moisture contents are more commonly used to characterize its expansive character. These tests, which are described later in this text, are a basis for classification of soils for engineering purposes.

Another approach is to measure the *cation exchange capacity* of a soil, which indicates the amount and kind of interchangeable cation. This is a wet chemical analysis, and the pH must be adjusted to neutral to obtain consistent results.

**Figure 6.6**

X-ray diffraction chart for a loess soil containing quartz and expansive clay mineral (smectite). The larger the individual grains the sharper the peaks.

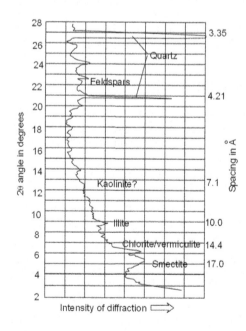

The most direct approach is to measure the expansion in a compression device that allows wetting of the soil. This method is discussed in Chapter 16 in this text.

## 6.3.16 Particle Size Analysis by X-ray Diffraction

Soil particle sizes ordinarily are measured by sieving or by monitoring the settlement rate in water, as described in the next chapter. However, the thickness of thin crystals can be measured from its effect on the sharpness of a diffraction peak, referred to as "line broadening." As a crystalline particle becomes thinner its orientation can be more variable with respect to the X-ray beam without causing annulment of waves. This is shown on the diffraction diagram in Fig. 6.6, where quartz shows very sharp diffraction peaks, illite and kaolinite are less sharp, and smectite peaks are quite broad. Even glass or liquids diffract a broad peak or halo, which indicates an average spacing of molecules even though the spacing is random.

While many other factors such as diffractometer alignment and mineral crystallinity are involved in peak broadening, a reasonable estimate of an equivalent crystallite thickness for a platy mineral can be made with the following formula, which applies for copper $K\alpha$ radiation:

$$D = \frac{150}{\beta \cos \theta} \tag{6.2}$$

**Example 6.2**
The diffraction peak width at half-intensity for the illite peak in Fig. 6.6 is approximately $0.3°$ and for the smectite peak $1.0°$ $2\theta$, respectively. What are the estimated average crystallite thicknesses, and how many unit cell layers do they represent?

*Answer:* For the illite, $D = 150/[0.3 \cos (8.8/2)] = 500$ Å, which represents $500/10 = 50$ layers. The solution for the smectite gives $150$ Å, or only about 8.5 layers.

## 6.3.17 Non-Clay Layer Silicates

Two classes of layer silicates have ABA-B crystal structures and a $d_{001}$ spacing of 14.4 Å, with the interlayer "B" representing a separate sheet that does not share ions with the other sheets. The layers are held together with hydrogen bonds. Chlorite minerals are common in shale, and the additional B has $OH^-$ corner ions. Vermiculite, which occurs in soils, has water in octahedral corner positions such that upon heating, the water is released explosively and each grain puffs up like popcorn. Expanded vermiculite is used as a fireproof building insulation, but some can contain asbestos, which if breathed can cause lung cancer. The mineral name comes from the shape of the expanded particles, *vermes* being Latin for worms.

## 6.4 PROBLEMS WITH AIR-DRYING SOIL SAMPLES

### 6.4.1 Hysterical about Hysteresis

As shown in Fig. 6.7, there is a considerable hysteresis or lag effect between expansion on wetting and shrinkage on drying of expansive clay. This can be attributed to drying constricting the interlayer spacing so that re-entry of water is slow. If the clay does not fully rehydrate, test results will be biased toward lower moisture contents. For example, the optimum water content for compaction can be 6 to 8 percent too low, which can have serious consequences.

Air-dried soils therefore should be thoroughly mixed with water and allowed to age overnight prior to testing. Atterberg limits used to classify soils also will be too low and can result in an erroneous classification that is on the unsafe side.

### 6.4.2 A Special Problem Clay

X-ray diffraction shows that special precautions are necessary to prevent drying of soils containing "halloysite" clay mineral. This is a kind of kaolinite that has formed in a continuously wet environment, and has a layer of water sandwiched between AB layers. This increases the basal spacing to 10 Å, the same as the ABA spacing of mica and illite, but when the clay is air-dried the spacing permanently decreases to 7.2 Å and the crystals roll up like tiny newspapers. Re-wetting therefore cannot replace the lost water.

Soils containing this mineral therefore should not be air-dried prior to testing. The problem can be detected by comparing Atterberg limit tests on a soil before and after air drying: if the limits are different, soils from the site should

**Figure 6.7**

Stepwise hydration and dehydration of Ca-montmorillonite and relation to layers of water.

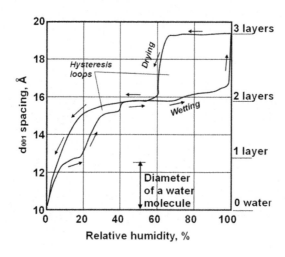

not be allowed to air-dry and should be stored in sealed plastic bags or other containers.

Whenever soils are air-dried prior to testing, this and the procedure used to rehydrate should be stated in the test report.

## 6.5 LIME STABILIZATION OF EXPANSIVE CLAY

### 6.5.1 Overview

Mixing hydrated lime, $Ca(OH)_2$, with soil to make it stronger dates from prehistory, and lime mortar was used for masonry walls well into the twentieth century. In modern times lime is mixed with clay soil and compacted for road bases. A more recent adaptation of this procedure is to simply drill holes into an expansive clay layer and either mix the soil with hydrated lime or fill the open borings with quicklime. It therefore is important to recognize the chemical reactions that take place.

### 6.5.2 Flocculation with Lime

An immediate reaction between expansive clay and lime will be observed as the clay appears to dry up and becomes crumbly, even though there is no change in the water content. The reason for this is that the clay particles become stuck together, and the clay is said to be *flocculated*. This stabilizes the expansive nature and greatly increases the clay strength. A simple method for "drying" expansive clay in a mudhole without really drying it is to dump in a bag or two of hydrated lime and mix it in.

This change occurs even though the clay already is calcium-saturated, so one hypothesis to explain the sudden flocculation is that $OH^-$ ions from dissolved lime pull $H^+$ ions from within the clay crystals to make water, and in doing so increase the negative charge on the clay. This is called a "pH-dependent charge." The negative charge is balanced by the $Ca^{2+}$ ions that remain outside of the clay layers and link clay particles together.

### 6.5.3 Why Use Quicklime?

For safety reasons hydrated lime is preferred for mixing with soil for road bases, but quicklime is preferred for filling borings in a procedure called "drilled lime." There are several reasons for this.

First, the quicklime quickly extracts water from the soil, hydrates, and approximately doubles in volume. The drying action can quickly stop a landslide, and as the lime hydrates and expands, the surrounding expansive clay shrinks and develops radial cracks as water is extracted from the soil. The expanding lime paste then injects into the cracks and disperses into the soil. Meanwhile the net overall volume change is nil.

Another benefit is that as quicklime hydrates, it releases considerable heat that helps to dissolve the lime and speed up the chemical reactions. Drilled quicklime is used to stabilize expansive clays in landslides and under foundations, as shown in Figs. 5.5 and 21.1.

### 6.5.4 Difficulty with Clay that Is Not a Mineral

The importance of the crystalline nature of clay minerals for engineering is emphasized by problems that can occur with a noncrystalline clay called *allophone*. Allophane clays derive from easily weathered volcanic basalt in areas of heavy rainfall where drying is not an option. Allophane is a gel, a colloid that has no regular arrangement of ions and can contain more water than solids. In fact, the weight of water per unit volume may be as high as 3 to 6 times the weight of the solids. Occurrences are on islands of the Pacific, including areas of Japan and Hawaii.

Allophane soils are difficult because they are sensitive, meaning that disturbing them destroys what little structure they have and turns them into soup. Lime stabilization requires modifying ionic charges on crystals and is not effective. These soils sometimes are dried in a kiln and crushed to make aggregate.

## 6.6  NON-CLAY MINERALS

### 6.6.1  Importance of Non-Clay Minerals

While most emphasis is placed on clay minerals in soils, non-clay minerals such as quartz play an important role and dominate granular soils, especially sand. Whereas most of the strength of a clayey, cohesive soil is strongly influenced by interactions of clays with water, most of the strength of granular soils relates to grain sizes, shapes, and packing, and the mineral composition is less significant. The discussion of non-clay minerals therefore is closely tied to grain size.

### 6.6.2  Minerals in Gravel, Cobbles, and Boulders

Large soil particles are essentially pieces of rocks. The particles are too large to be moved by wind, and in a desert subjected to wind erosion they are left behind as a lag concentrate or "desert pavement." Transportation by water results in rapid rounding-off of corners because of the mass and corresponding energy dissipated during bed impacts, so the dominant particle shape is highly rounded. This is in contrast to particle shapes produced in a rock crusher, which are highly angular.

Angular shapes are preferred for most engineering uses because of improved interlocking of individual particles.

### 6.6.3  Minerals in Sand

Sand by definition is finer than gravel and coarser than silt. A natural boundary exists between sand and gravel because sand grains tend to be composed of single mineral species instead of being fragments of rocks.

The ultimate source for most sand is disintegration of granite by weathering, so it initially is about 25 percent quartz and the remainder mainly feldspars with a sprinkling of mica and iron-rich dark minerals. Transportation by wind or water then wears away the feldspars and concentrates quartz, so typical river sand is about 75 percent quartz and the remainder feldspars and other minerals. Transportation also rounds the grains.

### 6.6.4  Silt

A natural boundary also exists between sand-size and silt-size particles because silt grains are small enough that they are carried in suspension instead of impacting one another, so there is little or no impact rounding. A visual distinction can be made between sand and silt in that individual silt grains are difficult to discern with the unaided eye, although they readily can be seen with a 10× or 20× lens.

As particle sizes further decrease and surface areas increase, non-clay minerals become less resistant to weathering so soils are dominated by the clay minerals.

### 6.6.5  Identifying Non-Clay Minerals

Table 6.1 lists some diagnostic X-rays for clay and non-clay mineral particles. Non-clay minerals are characterized by color, hardness, and cleavage. Cleavage refers to a tendency to fracture along discrete atomic planes. Mica, for example, has excellent cleavage along basal crystallographic planes. Feldspars have only moderate cleavage but quartz has no cleavage, so crystals of quartz that have flat faces grew that way. Cleavage can occur in more than one crystallographic direction, which affects the shapes of individual mineral grains.

Hardness is quantified by determining the scratch order of different minerals. This order is called the Mohs hardness scale, which has numbers from 1 for the softest to 10 for the hardest mineral, which is diamond. Gypsum (hardness 2) can be scratched with a fingernail, calcite (hardness 3) with a penny, and feldspar (hardness 6) with a hard knife. Quartz has a hardness 7.

The presence of calcite indicates a low degree of weathering and is determined by touching soil with a few drops of dilute hydrochloric acid (1:10 in water) and

**Table 6.1**

X-ray diffraction spacings used in mineral identification

| Mineral | Structure | Spacing, Å | Remarks |
|---|---|---|---|
| Kaolinite group | AB | 7.14 | Nonexpansive |
| Halloysite | AB+H | 10.0 | Must keep wet |
| Mica clay minerals | ABA | 10.0 | Nonexpansive |
| Smectites | ABA+$n$H | 10+, variable | Expansive |
| Treated with ethylene glycol | | 17.7 | For identification |
| Air-dry, Ca-saturated | | 15.4 | Two water layers |
| Wet, Ca-saturated | | 19.6 | Three water layers |
| Air-dry, Na-saturated | | 12.2 | One water layer |
| Wet, Na-saturated | | Unlimited expansion under zero loading | |
| Vermiculite | ABA+H | 14.4 | Nonexpansive |
| Chlorite | ABA+B | 14.4 | Nonexpansive |
| Gypsum | $CaSO_4(2H_2O)$ | 4.29 | Sulphate mineral |
| Quartz | 3D linked A | 4.21, 3.26 | Often dominant |
| Feldspars | 3D linked A | 3.2 | Many varieties |
| Calcite | $CaCO_3$ | 3.03 | Calcium carbonate |

Note: $1\ \text{Å} = 10^{-8}\ \text{cm} = 0.1\ \text{nm}$ (nanometer).

watching for a fizz. Portland cement concrete also reacts because hydrated cement compounds gradually become carbonated by $CO_2$ from air.

### 6.6.6 Geoenvironmental Problems with Silt-Size Particles

Prolonged inhaling of any dust particles that are silt size and smaller can lead to serious health problems.

Black lung disease that leads to early retirements of coal miners involves inhaling large quantities of relatively inert carbon particles that plug lung passages. Kaolinite dust from processing of china clay, metal dusts from grinding, and organic dusts such as fungal spores ("farmers' lung") cause similar problems. An air-borne fungus indigent to desert areas causes lung scarring and flu-like symptoms referred to as "valley fever."

Dusty environments are common near unpaved roads and construction sites unless there is active control, and dusts can become highly concentrated in confined areas such as in tunnels and mines.

Mineral dusts vary in toxicity depending on their composition and age. Silicosis develops from inhaling silica (quartz) dust, but not all silica dust is equally toxic. Quartz has no natural cleavage, so grinding exposes highly active unsatisfied chemical bonds that take $OH^-$ ions from water and leave a surplus of $H^+$ ions with instantly acidic conditions. Breathing freshly pulverized quartz dust therefore is like putting a quick shot of acid into the lungs. As mineral surfaces age they are deactivated by adsorbing water and gas molecules.

Asbestos defines a group of toxic minerals in which silica tetrahedra are shared in only one direction so that particles are shaped like needles. Asbestos once was widely used in fireproof insulation, fiberboard, and roofing materials, and is present in many buildings constructed prior to the 1970s. About 95 percent of the commercial asbestos is composed of the mineral chrysotile, in which each fiber is a rolled-up tube with a diameter of the order of 250 Å and a length of up to 0.25 mm. When inhaled, the tiny, rigid, spike-like particles remain as a permanent irritant that eventually can cause cancer. The cancer risk is greatly increased by smoking, and the disease can develop many years after the exposure to asbestos dust. Although dust-related respiratory diseases cause several thousand deaths per year in the U.S., many times more fatalities are linked to tobacco smoking.

## Problems

6.1. What is a hydrogen bond? How does it help to explain the properties of water?

6.2. Kaolinite sometimes is called a 1:1 clay mineral, and smectite a 2:1. Why?

6.3. Which is more expansive, Na-smectite or Ca-smectite? How might one change smectite from the sodium to the calcium form?

6.4. Which is more dangerous to breathe, freshly ground quartz in a mine or pulverized limestone in road dust? Why?

6.5. Coal miners sometimes make a choice between the use of face masks and early retirement. Explain.

6.6. Based on an X-ray basal spacing of $Ca^{2+}$ smectite of 15.4 Å and a mineral ABA layer thickness of 10 Å, what is the percent water held in the interlayer? Assume that the specific gravity of ABA is 2.65 and of the interlayer water 1.0.

*Answer:* The ABA mass is $10 \times 2.65\,g = 26.5$, and of water $5.4 \times 1.0 = 5.4\,g$. This gives $5.4/26.5 = 20\%$ water on an oven-dry weight basis.

6.7. Which is more effective to stabilize expansive clay: (a) drilling holes and filling them with quicklime; (b) drilling holes and filling them with hydrated lime; (c) drilling holes and mixing hydrated lime with the soil, referred to as "lime columns"?

6.8. Do you expect that agricultural lime, which is calcium carbonate, would be effective to stabilize expansive clay? Why (not)?

6.9. What is the weight percent interlayer water in a saturated $Ca^{2+}$ smectite?

*Answer:* 36%.

6.10. How might two clays having the same chemical composition have very different physical properties?

## References and Further Reading

Grim, R. E. (1968). *Clay Mineralogy*. McGraw-Hill, New York.

Handy, R. L., and Williams, W. W. (1967). "Chemical stabilization of an active landslide." *ASCE Civil Engineering*, 37, (8).

Handy, R. L., and Demirel, T. (1987). "Swelling Pressures and d-Spacings of Montmorillonite." *Proc. 6th Int. Conf. on Expansive Soils* 1, 161–166.

Mitchell, J. K. (1993). *Fundamentals of Soil Behavior*, 2nd ed. John Wiley & Sons, New York.

Moore, D. M., and Reynolds, R. C. (1997). *X-Ray Diffraction and the Identification and Analysis of Clay Minerals*. Oxford University Press, New York.

Velde (1992). *Introduction to Clay Minerals: Chemistry*. Springer, New York.

# 7

# Particle Size and Gradation

## 7.1 GRAIN SIZES AND ORDERS OF MAGNITUDE

### 7.1.1 Size Ranges or Grades

The individual particles that make up a soil vary in size by orders of magnitude. For example, the size difference between a 0.002 mm clay particle and a 2 m diameter boulder is 6 orders of magnitude, or about the same as between a Volkswagen and the Moon. It therefore is convenient to define *particle size grades* by defining discrete ranges in particle sizes that define clay, silt, sand, gravel, cobbles, and boulders. Each size grade covers a range in particle sizes—that is, all gravel particles obviously are not the same size. "Clay" thus defined relates to a range in particle sizes without regard to their mineralogy. However, because of a relationship between weatherability of different minerals and particle size, most clay-size particles are composed of the special group of minerals designated as clay minerals. A particular soil therefore will consist of varying percentages of clay, silt, and sand sizes with occasional coarser material.

## 7.2 GRADATION CURVES

### 7.2.1 Logarithmic Grain-Size Scale

Because of the broad range in particle sizes that can make up a particular soil, sizes are conveniently plotted on a logarithmic scale. The advantage becomes apparent by comparing Fig. 7.1, where sizes are plotted on a linear scale, with Fig. 7.2, where the size distribution for the same glacial till soil is plotted logarithmically. Figure 7.2 also shows the huge variations in particle sizes between some common soils deposited by wind, water, and ice.

**Figure 7.1**

Plotting particle sizes to a linear scale emphasizes the wrong end of the size scale—the gravel and not the clay.

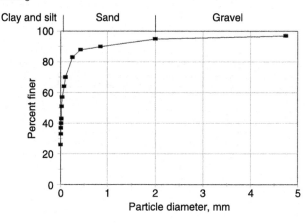

**Figure 7.2**

Semilogarithmic graph of the same particle size data for the glacial till soil and for several other soils.

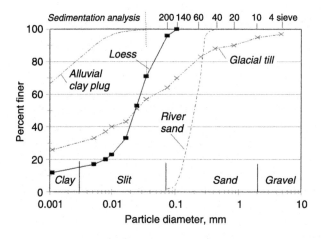

## 7.2.2  Particle Size Accumulation Curves

The graphs in Figs. 7.1 and 7.2 show particle size data as "percent finer" than each size on a dry-weight basis. This is a *particle size accumulation curve*.

Figure 7.3 shows the relationship between an accumulation curve and a bar graph or histogram representation of the same data. The data are obtained by passing soil through a succession of progressively finer sieves and weighing the amount retained on each sieve. The bar heights in the upper graph show each of these amounts. Mathematically the upper graph is the differential or slope of the lower graph, which is the particle size distribution curve. Conversely, the lower graph represents the integral of the upper graph.

The median or average grain size can be read directly from a particle size accumulation curve, as shown by the arrows in Fig. 7.3. The median grain size is defined on the basis that 50 percent of a soil by weight is finer, and 50 percent is

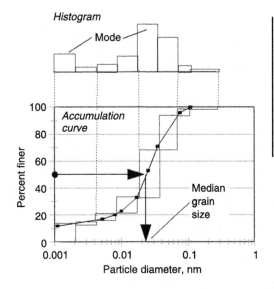

**Figure 7.3**

Relation between a particle size accumulation curve showing a median grain size and a histogram showing modal sizes.

coarser. In Fig. 7.3 this percentage occurs at 0.021 mm, which is in the size range for silt. The median grain size is designated by $D_{50}$. Another reference size that has been found to relate to the permeability or hydraulic conductivity of soils is $D_{10}$.

**Example 7.1**

What is $D_{10}$ for the soil in Fig. 7.3?

*Answer:* Slightly smaller than 0.001 mm.

## 7.2.3 Modes

The highest bar on a histogram data plot indicates a dominant particle size, which is designated the *mode*. Although a mode is not the same as a median size, in Fig. 7.3 the two are close because of the symmetrical shape of the major portion of the histogram. This symmetry reflects a statistical normal distribution, not of particle sizes, but of *logarithms* of the particle sizes because particles settle out of a suspension according to the square of their diameter instead of their diameter.

In Fig. 7.3 another mode occurs in the clay size range smaller than 0.002 mm, probably due in part to clay adhering to coarser grains when they settled out. Two or more modes also can indicate soil mixtures, as when two strata are combined in one sample or sand has infiltrated into interstices in a gravel deposit. B horizon soils are bimodal because of infiltration by clay from the A horizon. Engineered soils often are mixtures in order to improve their engineering properties.

While a histogram is instructive, an accumulation curve is easier to plot and is almost universally used in engineering. Modes occur on an accumulation curve where slopes are steepest, and component soil percentages are indicated where the curve flattens out.

### Example 7.2

Large samples of glacial till often contain a mix of different component soils. What are component percentages in the glacial till in Fig. 7.2?

*Answer:* The first steep section of the curve is at 41%, which therefore represents one component soil. The second break is at 60% so the difference is $60 - 41 = 19\%$, which represents a second component. Similarly, the third break at 90% defines $90 - 60 = 30\%$ for a third component, and a fourth component makes up the remaining 10%. The three components percentage are $41 + 19 + 30 + 10 = 100\%$. The respective soils are (a) mainly clay plus some silt, (b) all silt, (c) mainly fine sand, and (d) a mixture of coarse sand and gravel.

## 7.3 DEFINING SIZE GRADES

### 7.3.1 Making the Grades

Not all sand particles are exactly the same size, which means that "sand" must cover a range of particle sizes, the only requirement being that they are smaller than gravel and larger than silt grains. Natural size boundaries occur between gravel and sand, between sand and silt, and between silt and clay, but the boundaries are transitional and somewhat arbitrary, and different organizations have adopted different definitions.

Gravel particles require a higher water velocity to be moved than sand, and wind does not move them at all. Sand particles move by bouncing, or saltation, and silt grains are mainly carried in suspension, as the mud in muddy water or the dust in air. Clay particles are so fine that they are very slow to settle out of suspension and consist of separate mineral species, the clay minerals.

### 7.3.2 Sieve Sizes

Soils are separated into size grades by sieving, or sifting through a series or "nest" of standardized wire mesh sieves arranged from the coarsest down to the finest. Common sieve sizes used in engineering are listed in Table 7.1.

A sieve is a wire fabric, so the sieve number does not describe the size of the opening but designates the number of wires per inch or millimeter. As a matter of convenience some size grades are defined on the basis of standard sieve sizes: gravel, for example, commonly designates particles that are coarser than 2 mm, which is the size of the opening in a No. 10 (wires to the inch) sieve.

| No. (wires per inch) | Opening, mm | Comment | |
|---|---|---|---|
| (Lid) | — | | **Table 7.1** |
| 4 | 4.75 | *Gravel* | Standard sieve sizes |
| 10 . . . . . . . . . . . . . . . . . . . . | 2.0 . . . . . . . . | Size separating gravel and sand | used in geotechnical |
| 20 | 0.85 | ↑ | engineering |
| 40 | 0.425 | | |
| 60 | 0.25 | *Sand* | |
| 140 | 0.106 | ↓ | |
| 200 . . . . . . . . . . . . . . . . . . . . | 0.075 . . . . . . . . | Size separating<br>sand from silt | |
| (Pan) | — | *Silt and clay fall on*<br>through and collect in the pane | |

One complication is that sieve openings are not round; they are approximately square. Spherical particles can pass through regardless of their orientation, but few soil grains are spheres. Sieves therefore are vigorously shaken or vibrated for a prescribed time in a sieve shaker in order to achieve reproducibility of the data.

### 7.3.3 Details of the Gravel-Sand Size Boundary

Although the most common size boundary between sand-size and gravel-size particles is 2 mm, this size separation is not universal, even within geotechnical engineering.

The Unified Soil Classification System used in earth dam and foundation engineering makes the separation at the No. 4 (3/16 in.) sieve, and material from 4.76 to 2 mm in diameter is considered "very coarse sand." These and other size boundaries are indicated in Fig. 7.2 Because the boundaries differ, it is important that they be defined or included on graphs showing the particle size distribution, as indicated by the vertical lines and grade names across the bottom in Fig. 7.2.

### 7.3.4 The Sand-Silt Size Boundary

As silt particles are fine enough to be carried in suspension they show little or no rounding of corners, whereas sand particles typically are abraded and rounded at the corners and edges from having been transported and bounced along by wind or water. However, the boundary is transitional, and for convenience it often is defined on the basis of a sieve size. In geotechnical engineering practice the boundary between sand and silt usually is that of a 200-mesh sieve opening, 0.075 mm or 75 μm (micrometers). The earlier designation was "microns.") Sand therefore presents a range in particle sizes between 0.074 and 2 mm diameter, a size ratio of 27.

Soil scientists prefer to make the separation between sand and silt at 0.020 mm or 20 μm. However, as shown by the loess and sand soils in Fig. 7.2, the natural boundary may be closer to the No. 200 sieve (0.074 mm) or even slightly larger. Geologists sometimes use 1/16 mm = 0.067 mm, sometimes rounded off to 0.06 mm. However, the occurrence of a natural break in the general vicinity tends to diminish the influence on constituent percentages.

### 7.3.5    The Silt-Clay Size Boundary

The most widely accepted size definition of clay is particles that are finer 0.002 mm or 2 μm. An earlier definition was based on the resolving power and eyepiece calibration of a light microscope at the U.S. Bureau of Soils, and set the boundary at 0.005 mm (5 μm). Later mineralogical investigations showed that this boundary is too high, but meanwhile it became established and still is occasionally used in geotechnical engineering. The 0.005 mm size also requires less interpolation from measurements that routinely are made after 1 hour and 1 day testing time. This is discussed in more detail in section 7.4.6.

### 7.3.6    Silt-Clay Boundary Based on Physical Properties

Another approach is to define clay on the basis of its plasticity or moldability with water, as silt is crumbly while clay is sticky and can be molded into different shapes. These relationships are quantified by two simple tests called *Atterberg limits*. These tests and the relationship to engineering soil classifications are discussed in Chapter 12. The limits define a moisture content range over which a soil can be molded. This range is the *plasticity index*, which is a fundamental soil property in geotechnical engineering.

In order to avoid possible confusion between the two approaches, a clay content based on particle size may be referred to as *clay-size* material.

## 7.4    MEASURING PARTICLE SIZES

### 7.4.1    General Approach to Size Measurement

Some shortcuts are in order because so many particles must be measured in order to obtain statistical reliability. One shortcut is to use sieves and screen a representative soil sample. Another approach that is used for particle sizes too small to be separated on sieves is to disperse the soil in water and make a determination based on the sedimentation rate, with largest particles settling the fastest.

One of the most important steps in analysis is obtaining a representative soil sample. Large samples are spread out on a flat surface and "quartered," that is, cut into four pie-shaped sectors and then combining opposing sectors and returning the other half of the sample to the bag. This procedure is repeated until the soil sample is small enough to be managed. A more rapid method for quartering uses a "riffle-type" sample splitter that has parallel shuts, with half directing the sample one way and the other half the other. Soils are air-dried prior to quartering and sieving, but as discussed in Chapter 6, if a soil contains halloysite clay mineral, it should be saved and sealed against drying.

## 7.4.2  Sedimentation Analysis

Sieving is appropriate for measuring the amounts of sand and gravel in a soil, but silt and clay sizes are too small to be separated by sieving. Also, clay particles tend to be aggregated together into coarser particles and to occur as coatings on coarser particles. Gravel is removed by sieving, and the rest of the soil normally is soaked in water and then agitated and dispersed using a chemical dispersing agent. The suspension then is tested by measuring sedimentation rates, and finally the part of the soil that is retained on a fine sieve is dried and analyzed by sieving.

The general procedure is as follows. After sieving to remove gravel and coarser particles, the soil is soaked in water containing a small amount of a chemical dispersing agent, usually sodium hexametaphosphate, a water softener that is available in the detergent department of a supermarket. The dispersing agent forces substitution of sodium ions for exchangeable calcium ions on the clay by creating an insoluble phosphate precipitate.

The suspension then is agitated for a set amount of time with a standardized mechanical or air-jet stirring device. Ideally this will separate but not break individual soil grains. The soil suspension is diluted to 1 liter in a vertical flask and stirred in preparation for starting the test.

The starting time is noted and the suspension is allowed to settle for various time intervals. After each time interval, the density of the suspension is determined at a particular depth with a hydrometer. An alternative method is to sample the suspension with a pipette, then dry and weigh the sample.

The larger the weight of particles remaining in suspension, the denser the liquid, and the higher the hydrometer will float. An engineering hydrometer is calibrated to read directly in grams of soil per liter of suspension. Readings normally are taken after 1 minute and at various time intervals to 1 hour and then after 24 hours.

After the sedimentation analysis is completed, the soil is washed on a fine sieve to remove the silt and clay particles, then dried and the sand fraction analyzed by passing through a series of sieves.

**Figure 7.4**

Sampling theory
in sedimentation
analysis: at a
particular
sampling depth
the suspension
contains a
representative
sample of all sizes
smaller than the
size that will settle
to that depth.

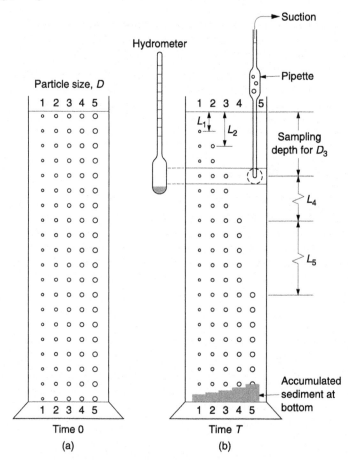

### 7.4.3 Sedimentation and the Percent Finer

A sedimentation analysis automatically measures the amounts finer than a specific grain size. This is illustrated in Fig. 7.4: after a certain time all particles larger than a certain depth have settled a calculated distance and therefore cannot occur at depths shallower than that distance. On the other hand, finer particles remain suspended and therefore are measured.

After each hydrometer reading the hydrometer is removed so that particles will not settle on the bulb. Removal stirs a small portion of the upper part of the suspension, but the effect is small so long as particles move horizontally and not vertically relative to the suspension—as the instrument is removed, the level of the suspension goes down, and when it is replaced the level goes back up. The depth to the center of volume of the submerged part of the hydrometer is the effective sampling depth that is used in the calculations, and depends on the depth of sinking. This depth is obtained from a calibration chart or table, Table 7.2.

| Hydrometer reading, g/l | Depth, mm | Table 7.2 |
|:---:|:---:|:---|
| 5 | 155 | Depth to hydrometer |
| 10 | 147 | center of volume |
| 15 | 138 | |
| 20 | 130 | |
| 25 | 122 | |
| 30 | 114 | |
| 35 | 106 | |
| 40 | 97 | |
| 45 | 89 | |
| 50 | 81 | |

Note: Adapted from ASTM Designation D-422.

Temperature also must be controlled and measured to enable correction for changes in the fluid viscosity.

## 7.4.4   Stokes' Law of Sedimentation

In 1851 a British mathematician, G. G. Stokes, solved for the settlement velocity of spherical particles in a suspension by equating their buoyant weight to viscous drag on the outer surfaces. Surface area increases in proportion to the radius while weight increases as the radius cubed, so the larger the particle, the faster it will settle. The classic derivation for Stokes' formula in the cgs system is

$$R = 6\pi r \eta v \tag{7.1}$$

where $R$ is the resisting force in g cm/s$^2$, $r$ is the particle radius in cm, $\eta$ is the fluid viscosity in poise or g-cm$^{-1}$s$^{-1}$, and $v$ is the settlement rate in cm/s. Equating to the buoyant weight of a spherical soil grain gives

$$6\pi r \eta v = \frac{4}{3}\pi r^3 (\gamma - \gamma_w) g \tag{7.2}$$

where $\gamma$ and $\gamma_w$ are respectively the density of the soil grain and that of water, and $g$ is the acceleration of gravity. Solving for velocity $v$ gives

$$v = \frac{2(\gamma - \gamma_w)g r^2}{9\eta} \tag{7.3}$$

Thus, the settlement rate $v$ depends on the square of the particle radius $r$.

Experiments have confirmed the validity of the formula for particles between 0.001 and 0.10 mm in size, that is, for silt and most clay particles. Sand sizes are influenced by mass displacement considerations that slow their rates of sinking, and sizes smaller than about 0.001 mm settle more slowly. In 1827 an English botanist, Robert Brown, noticed that pollen grains suspended in water jiggled about when observed in a microscope, a movement that now is called *Brownian motion*. This grabbed the attention of an employee of the Swiss patent office, who wrote a brief paper attributing it to random molecular bombardment. The employee's name was Albert Einstein, who later became famous for another matter. Particles smaller than about 0.001 mm tend to remain in suspension and are referred to as *colloidal size* particles.

According to eq. (7.3) the rate of settling depends on the specific gravity $\gamma$ of the particles, which varies depending on the mineral. Because sedimentation is a bulk test, an average specific gravity is used in the calculations for particle size. A method for measuring average specific gravity is described later in this chapter. However, the assumption that all grain densities are average means that particles of dense minerals will be reported as larger than their true dimensions because they settle faster.

Sedimentation rate is influenced by the fluid viscosity, $\eta$, which in turn depends on temperature. A standardized temperature of 20°C (68°F) is used for laboratory analyses. Other temperatures may be used with appropriate correction factors based on viscosity tables.

An obvious limitation of Stokes' Law is that it applies only to spherical particles, whereas silt grains are angular and clay particles flat. Particle sizes determined from sedimentation rates often are reported in terms of "equivalent particle diameters."

## 7.4.5 Simplifying Stokes' Law

In eq. (7.3), a particle radius in cm equals the diameter $0.05D$ in mm. The settling velocity in cm/s equals $600L/T$, where $L$ is the settling distance in mm and $T$ is time in minutes. Substituting values for the acceleration of gravity and the viscosity gives

$$D = K\sqrt{L/10T} \tag{7.4}$$

where $D$ and $L$ are in mm and $T$ is in minutes. $K$ depends on the specific gravity of the soil and temperature of the solution; with a representative soil specific gravity of 2.70, and a standardized temperature of 20°C, $K = 0.01344$. Other values for this coefficient for different specific gravities and temperatures are given in ASTM Designation D-422.

**Example 7.3**

A soil suspension is prepared containing 50 g/l. After 60 minutes the hydrometer reads 22 g/l. The temperature is controlled at 20°C. (a) What particle diameter is being measured, and (b) what is the percent of particles finer than that diameter?

*Answer:* (a) The effective depth of the hydrometer is obtained by interpolation of data in Table 7.2, which gives $L = 127$ mm. From eq. (7.4),

$$D = 0.01344\sqrt{127/10 \times 60} = 0.0062\,\text{mm} = 6.2\,\mu\text{m}.$$

(b) $P = 100 \times 22/50 = 44\%$.

### 7.4.6 Interpolating the Percent 2 μm Clay from Hydrometer Analyses

The sedimentation time for a hydrometer analysis to measure $2\,\mu\text{m}$ clay is approximately 8 hours, which is inconvenient with an 8-hour working day. However, as this part of the accumulation curve often is approximately linear on a semilogarithmic plot, the percent $2\,\mu\text{m}$ clay can be estimated from a proportionality of the respective logarithms. As an approximation,

$$P_{002} = 0.4P_{001} + 0.6P_{005} \tag{7.5}$$

## 7.5 USES OF PARTICLE SIZE DATA

### 7.5.1 Median Grain Size

As previously mentioned, the size that defines 50 percent of the soil as being finer and 50 percent coarser is the *median grain size*, designated as $D_{50}$, and is read from the intersection of the particle size distribution curve with the 50 percent line, as shown in Fig. 7.3. The median approximates but is not the same as a mean or average particle size, which would be very difficult to determine because it would involve measuring many individual particles and calculating an average.

### 7.5.2 Effective Size and Uniformity Coefficient

A measurement that often is made for sand is the effective size, $D_{10}$, or the size whereby 10 percent of the particles are finer, and was shown by an engineer, Allen Hazen, to correlate with the permeability of filter sands. Hazen defined the uniformity coefficient, $C_u$, as the ratio $D_{60}/D_{10}$. The uniformity coefficient can be as low as 1.5 to 2 for washed sands that are nearly all one size. For engineering uses a soil is said to be "well graded" if it contains a wide range of particle sizes. A well-graded sand-gravel mixture may have a uniformity coefficient of 200–300.

#### Example 7.4
The sand in Fig. 7.2 has approximate values of $D_{10} = 0.12$ mm and $D_{60} = 0.20$ mm, from which $C_u = 1.7$. For engineering purposes this soil would be described as "poorly graded."

Because $D_{10}$ is off the chart for fine-grained soils, another measure for degree of uniformity suggested by a geologist, Trask, is the "sorting coefficient," $S_o$, which

| Table 7.3 Mechanical analysis data and determinations of weight percents finer than sizes indicated | Sieve number (particle diameter in mm) | Weight percent retained on each sieve | Weight percent finer |
|---|---|---|---|
| | Sieve analysis: | | |
| | No. 4 (4.76) | 0 | 100 |
| | No. 10 (2.0) | 4 | $100-4=96$ |
| | No. 20 (0.84) | 4 | $96-4=92$ |
| | No. 40 (0.42) | 3 | $92-3=89$ |
| | No. 60 (0.25) | 7 | $89-7=82$ |
| | No. 100 (0.147) | 4 | $82-4=78$ |
| | No. 200 (0.075) | 13 | $78-13=65$ |
| | Sedimentation analysis: | | |
| | (0.025) | | Hydrometer reading $=52$ |
| | (0.010) | | "   31 |
| | (0.005) | | "   21 |
| | (0.001) | | "   8 |

is defined as $(D_{75}/D_{25})^{1/2}$. A more complicated calculation also may be made to obtain a statistical standard deviation.

### 7.5.3 Example of Mechanical Analysis

Measurement of soil particle sizes is called a "mechanical analysis." Data from a mechanical analysis are shown in Table 7.3.

The percent 0.002 mm clay is estimated from eq. (7.4), which gives $D_{002} = 0.4 \times 8 + 0.6 \times 21 = 16$ percent finer than 0.002 mm. The various size grades are as follows:

| Size grade | | Calculated percent by weight |
|---|---|---|
| Gravel (retained on No. 10 sieve) | | 4 |
| Sand (retained on No. 200 minus % gravel) | $(100-65)-4=$ | 31 |
| Silt (coarser than 0.002 mm minus % gravel and sand) | $(100-16)-4-31=$ | 49 |
| Clay (finer than 0.002 mm) | | 16 |
| Colloidal clay (finer than 0.001 mm) | (8) | |
| | Total | 100 |

### 7.5.4 Granular vs. Fine-Grained Soils

Concrete mixes are designed based on a concept that largest particles are touching, and progressively finer particles fill in the voids. The same concept applies to soils, and a broad range of particle sizes is considered to be "well graded." If coarse grains are in contact and voids between them are filled with smaller particles, the soil must increase in the volume, or *dilate*, in order to shear. This adds appreciably to the shearing resistance.

In many soils the silt and clay content are high enough to separate larger soil grains so that shearing can occur through the silt-clay matrix without dilatancy, which causes a marked reduction in the soil shearing strength. Artificial mixtures of sand plus clay show that this property change occurs at about 25 to 30 percent clay. The two distinct modes of behavior distinguish "granular soils" from "fine-grained soils."

### 7.5.5  Soil Mixtures

In Fig. 7.5 a poorly graded silt soil is combined with a poorly graded sand to obtain a more uniform grading. In this example the mix is 50–50, and the construction lines are shown dashed. A better grading could be obtained by reducing the percentage of A and increasing that of B. The effectiveness of an improved grading can be determined with strength tests. Geologists refer to a well-graded soil as being "poorly sorted," which means the same thing even though the connotations are different.

Flat portions of a particle size accumulation curve indicate a scarcity of those sizes, and a soil showing this attribute is said to be "gap-graded." Gap grading tends to give lower compacted densities and strength, and higher permeability.

### 7.5.6  Soil as a Filter

Filters are barriers that can transmit water while retaining soil particles that otherwise would be carried along in the water. Filter soils usually are sands. A common use of a filter is in the toe drainage area in an earth dam, where control

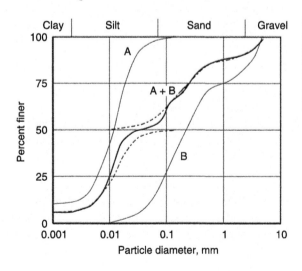

**Figure 7.5**

Combining two poorly graded soils A and B to obtain a more uniform grading A + B.

of seepage is important to prevent water from emerging on the earth slope where it might lead to piping and failure. Geotextile filters generally are more expensive but are easier to install than are layers of sand, and are less likely to be damaged or compromised during construction.

### Design

Protective filters act as a drain while resisting clogging by fine particles. They also cannot permit a breakthrough, and may be required to provide insulation against frost action. The finer sizes of particles in a soil filter tend to control its performance. Generally the filter $F_{15}$ size is compared with the $D_{85}$ size for the base soil. (To avoid confusion the filter size is designated with $F$ instead of $D$.) A conservative and acceptable guide for design is $F_{15}/D_{85} < 5$.

An additional requirement for the retention of clay, for example in the core of an earth dam, is that $F_{15} < 0.5$ mm.

### Example 7.5

Is the sand in Fig. 7.2 an appropriate filter for an earth dam constructed from the glacial till in the same figure?

*Answer:* The sand has $F_{15} = 0.12$ mm, and the till has $D_{85} = 0.4$ mm. Then $0.12/0.4 = 0.3 < 5$, so the filter should perform adequately. In addition $F_{15} < 0.5$ mm so there should be little or no clay penetration.

*Question:* What if the dam is constructed from the loess in the figure?

## 7.5.7   Geotextile Filters

The apparent opening size (AOS) of geotextile fabrics is defined as $O_{95}$, which is the size for 95 percent of glass beads of a particular size grade to pass through during sieving (ASTM Designation D-4751). One criterion in regard to filtration of soil is that $O_{95}/D_{85} < 2$ or 3, where $D_{85}$ is for the soil.

## 7.5.8   Grouting

Grouting is pumping of a fluid under pressure into a soil so that it either (a) permeates the soil, referred to as *injection grouting*, or (b) displaces the soil, called *compaction grouting*. The determination of whether a grout will inject into the soil pores or displace the soil is mainly dependent on the relations between the respective particle sizes.

Injection grouting is a common remedial treatment used to solidify loose foundation soil and rock underneath buildings, dams, and other structures. Injection grouting also is used to seal leaks under dams or lagoons, to seal off and contain buried hazardous wastes, and to seal off the groundwater aquifers in preparation for tunneling.

Compaction grouting is a relatively new procedure that normally is intended to laterally compact and densify loose soil to reduce settlement under a foundation load.

Regardless of the grouting procedure the maximum grouting pressure is limited by the overburden pressure of the soil, or lateral planar injections can lift the soil. When this occurs, pumping pressure should decrease while the pumping rate increases, referred to as the grout "take."

If the lateral stress existing in the soil is lower than the vertical pressure from overburden, the pumping pressure at which the "take" occurs is that which causes vertical radial cracking and is used as an approximate measure of lateral stress in the soil. This is called "hydraulic fracturing." It was first developed in the petroleum production industry to increase the flow of oil into oil wells.

### Grout Materials

The most common grout materials for rocks and soils are aqueous suspensions of Portland cement and/or fly ash. Sand-cement mortar may be used for grouting rubble that has large voids. Bentonite sometimes is used as a sealing grout, but has the disadvantage that it will shrink and crack when it dries out.

The first injection grout was developed by Joosten in Germany and uses chemical solutions of sodium silicates and calcium chloride, which react to make insoluble calcium silicate and sodium chloride. Some more recent chemical solution grouts have been removed from the market because of potentially toxic effects on groundwater. Emulsions of asphalt in water are sometimes used as grout for sealing cracks and joints in basements.

### Soil Groutability

For injection grouting the particle size ratio is reversed from that used designing filters, $D_{15}$ for the soil and $G_{85}$ for the grout. To ensure success, the ratio should be substantially higher than the corresponding ratio of 5 used for filters. Tests by the U.S. Army Corps of Engineers suggest that the ratio of soil $D_{15}$ to cement $G_{85}$ should be a minimum of 20. $G_{85}$ for Portland cement typically is about 0.040 to 0.050 mm. The smaller figure represents high-early strength cement, and also is fairly representative of fly ashes. Specially ground cements may have $G_{85}$ of only 0.005 mm. Bentonite is composed of montmorillonite particles that expand on wetting, with an effective hydrated $G_{85}$ of about 0.030 mm.

### Example 7.6

Can any of the soils of Fig. 7.2 be injection grouted with cement grout?

*Answer:* The soil with the largest $D_{15}$ is the sand, with $D_{15} = 0.12$ mm. For cement, assume $G_{85} = 0.050$ mm. Then $D_{15}/G_{85} = 2.4 \ll 20$, so this sand cannot be injected with cement grout. The sand still may be a candidate for compaction grouting or injection grouting with chemical solutions, depending on the properties that are required.

As a general guide:

- Gravel or very coarse sand can be injection grouted with cement and/or fly ashs.
- Medium to fine sand can be compaction grouted with cement/fly ash or injection grouted with sodium silicate or specially ground fine cement.
- Silt can be compaction grouted.
- Clay cannot be grouted, but expansive clay can be stabilized by a diffusion process of hydrated lime, which is much slower than the other processes.

Partly because of the difficulty in controlling injection grouting and knowing where the grout goes, compaction grouting has become increasingly popular in recent years.

# 7.6 DESCRIBING PARTICLE SHAPE

## 7.6.1 Particle Shape and Engineering Behavior

The shapes of soil grains can influence engineering behavior, as round grains obviously are more likely to slip and roll than angular fragments that mesh or interlock together. For this reason crushed rock normally creates a stronger surface of a "gravel" road than do the more rounded particles of gravel. On the other hand gravel, having been through many cycles of pounding against a beach or river bottom, is more likely to be harder and less likely to degrade into dust.

The main effect of angularity is harshness, or the tendency for the soil to dilate or increase in volume during shearing, a matter that can be quantified with strength tests.

Grain shapes closely relate to their mineralogy and origin; quartz sand grains derived from disintegration of granite tend to be round, whereas grains of feldpar derived from the same rock are more angular, and grains of mica are flat. Alluvial gravel generally is well rounded, sand less so, and silt not at all. Dune sand not only shows rounding, but the grain surfaces are etched from repeated impacts.

The measurement of shapes of individual grains can be time-consuming, but measurement of grain profiles can be digitized and automated. A chart that can be used to estimate shape, or "sphericity," is shown in Fig. 7.6. Sphericity theoretically is the ratio of a grain surface area to that of a sphere, but can be approximated by dividing the intermediate grain width by its length. As this does

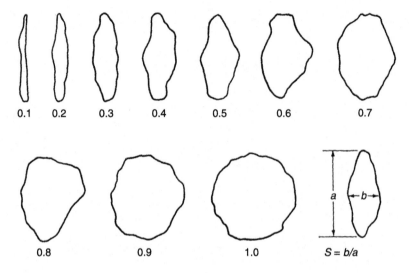

**Figure 7.6**
Chart for evaluating the shapes of individual soil grains from their profiles, 1.0 representing the approach to a sphere.

not take into account the shortest grain dimension, it tends to overestimate sphericity of flat particles such as mica.

### 7.6.2 Special Problems with the Shape of Mica Grains

Especially troublesome, is that mica particles are flat and also are springy, so compacting a soil with a high content of mica is like trying to compact a bucket of springs. Although micaceous soils are not common, their behavior is such that they are given a special category in some engineering classifications, and the glitter is not gold.

## 7.7 TEXTURAL CLASSIFICATION OF SOILS

### 7.7.1 Describing Different Proportions of Sand + Silt + Clay

The first step in characterizing grain sizes in a soil is to take the soil apart and assign the component parts to size grades, namely gravel, sand, silt, and clay. Next let us describe the products when we put them back together. A naturally occurring gravel deposit almost inevitably will contain some sand, and a naturally occurring silt deposit almost inevitably will contain some clay, so when does it stop having "silt" for a soil name and start being a "clay"?

"Clay" therefore can mean either (a) clay mineral, (b) clay size, or (c) a deposit or soil that is mainly clay but also contains other minerals and grain sizes. Engineers tend to use a term such as "clay" interchangeably for its several meanings, and should be certain that it is used in a context that ensures that everybody will know what it means.

**Figure 7.7**

A soil textural
chart based on
the 0.075 mm
definition of silt
size and
the 0.002 mm
definition of clay
size.

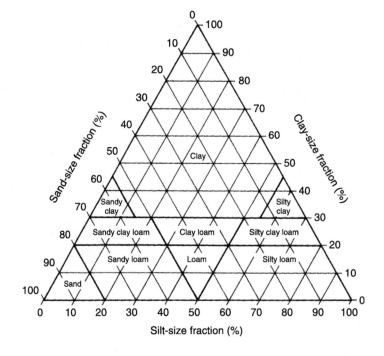

### 7.7.2    Soil Textures and Particle Sizes

Soil scientists who do soil mapping in the field originally proposed the term
"texture" to describe the "feel" of moist soil squeezed with the fingers. A soil
might have a gritty or sandy feel, or it might have a smooth feel, more like
modeling clay. "Loam" came to mean a somewhat loose and crumbly feel that is
great for agriculture.

Soil textures are quantified by relating them to the percentages of sand, silt, and
clay. The various ranges are shown on a triangular "textural chart" such as
Fig. 7.7. Boundaries on textural charts have been changed from time to time as
size definitions have changed, but the concept remains valid and useful.

The textural chart is read by entering any two of the three percentages and moving
onto the chart in the directions of the corresponding short lines around the edges.
For example, the boundary between clay and clay loam is at 30 percent clay-size
material. It will be seen that a clay texture can contain as much as 55 percent
sand. However, to qualify as a sand *texture* the soil must contain over 80 percent
sand.

Textural terms apply to the non-gravel portion of a soil, so the percentages are
adjusted for gravel content. If the gravel content exceeds 10 percent the soil is
"gravelly."

**Example 7.7**

What is the textural classification for the soil in Section 7.5.3?

*Answer:* The soil contains 31% sand and 49% silt. These figures are adjusted for the 4% gravel content: $31/0.96 = 32.6\%$ sand and $49/0.96 = 51.0\%$ silt. The texturally is "silty clay loam."

# 7.8 SPECIFIC GRAVITY OF SOIL PARTICLES

## 7.8.1 Definition and Use

Specific gravity is defined as the density of a material divided by the density of water at 4°C, which is water at its densest. According to eq. (7.3) the specific gravity is required in order to interpret settlement analyses. Some representative specific gravities for different minerals are shown in Table 7.4. Most sands have a specific gravity of 2.65–2.68; most clays, 2.68–2.72.

## 7.8.2 Measurement

A common method for measuring the specific gravity of a large object is to weigh it in air and then submerge it in water. The difference equals the weight of the water displaced, a discovery made by Archimedes in his search for a way to determine the purity of gold. The weight divided by the weight lost therefore

| | | | |
|---|---|---|---|
| Gold | 19.3 | Terribly expensive | **Table 7.4** |
| Silver | 10.5 | Pocket change | Specific gravities of |
| Galena (PbS) | 7.5 | Cubes that look like silver but aren't | some selected solids |
| Pyrite ($FeS_2$) | 5.0 | Cubes that look like gold but aren't | |
| Hematite ($Fe_2O_3$) | 4.9–5.3 | Red iron oxide in soils | |
| Limonite ($Fe_2O_3 \bullet nH_2O$) | 3.4–4.3 | Yellow or brown iron oxides in soils | |
| Iron silicate minerals | 2.85–3.6 | Dark minerals in basalt, granite | |
| Calcite ($CaCO_3$) | 2.72 | Most abundant mineral in limestone | |
| Micas | 2.7–3.1 | Flakey | |
| *Quartz($SiO_2$)* | *2.65* | *Most abundant mineral in soils* | |
| *Feldspar (Na and Ca silicates)* | *2.55–2.65* | *Most abundant mineral in rocks* | |
| *Kaolinite* | *2.61* | *Clay mineral* | |
| *Smectites* | *2.2–2.7* | *Expansive clay minerals* | |
| Glass | 2.2–2.5 | Lead glass = 3 | |
| Halite (NaCl) | 2.1–2.3 | Rock salt | |
| Liquid water ($H_2O$) | 1.00 | At its densest, 4°C | |
| Ice ($H_2O$) | 0.918 | Floats on liquid water | |

represents the weight divided by the weight of an equal volume of water, which by definition is the specific gravity:

$$G = \frac{W}{W - W_b} \qquad (7.6)$$

where $G$ is the specific gravity and $W$ and $W_b$ are the weight and buoyant weight respectively.

A slightly different procedure is used for soils and is a bit more tricky. A flask is filled with water and weighed; call this $A$. Then $W$, a weighed amount of soil, is put into the flask and displaces some of the water, giving a new total weight, $C$. As shown in Fig. 7.8, the weight of the water displaced is $(A + W - C)$. Hence,

$$G = \frac{W}{A + W - C} \qquad (7.7)$$

Experimental precision is unhappy with subtracting a weight from the denominator, so measurements are exacting. Recently boiled or evacuated distilled water ensures that there is no air that might come out of solution to make bubbles, and clay soils are not previously air-dried. Less critical is a temperature-dependent correction for the specific gravity of water, which at 20°C is 0.99823. (Specific gravities are reported to three significant figures.) Details are in ASTM Designation D-854. It will be noted that weights and not masses are measured, even though the data are usually recorded in grams.

**Example 7.8**

A flask filled to a reference mark with water weighs 690.0 g on a laboratory scale. When 90.0 g of soil are added, the filled flask weighs 751.0 g. The water temperature is 20°C. (a) What is $G$? (b) What effect will the temperature correction have? (c) What if as a result of measurement error the soil weight is 1 g too high, an error of 1.1%?

Fig. 7.8 Using a pycnometer to measure specific gravity.

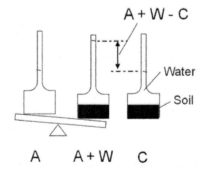

*Answer:*

(a) $G = 90/(690.0 + 90.0 - 746.0) = 2.65$.

(b) Dividing by 0.998 to correct for water temperature does not affect the answer.

(c) $G' = 91/(690 + 91 - 746) = 2.53$. A suggested assumed value would be more accurate.

## Problems

7.1. Plot a particle size accumulation curve for soil No. 4, Table 7.5, by entering the data on a computer spreadsheet and selecting the logarithmic option for the particle sizes. (Optionally this can be done manually using 5-cycle semilogarithmic paper.) (a) Evaluate the effective size and uniformity coefficient. (b) What is the median grain size? (c) Defining clay as <0.002 mm, silt as 0.002–0.074 mm, sand as 0.074–2.0 mm, and gravel as >2.0 mm, what are the percentages of clay, silt, sand, and gravel?

7.2. Classify soil No. 4 according to the chart in Fig. 7.7 after adjusting the percentages for gravel content.

7.3. Plot a particle size accumulation curve for soil No. 1, Table 7.5. (a) Identify the median and mode(s). (b) If there are two modes, what is the approximate percentage of each soil in the mixture? (c) Using the size grades defined in Problem 7.1, find the percentages of clay, silt, sand, and gravel. (d) Adjust the grade percentages for gravel and classify the soil by the chart in Fig. 7.7.

7.4. Calculate the effective size and uniformity coefficient for soil No. 1.

*Answer:* $D_{10} = 0.0039$ mm, $C_u = 192$.

7.5. By inspection indicate which of the soils in Table 7.5 should be designated as gravelly.

7.6. For the first five soils in Table 7.5 compare measured 0.002 mm clay contents with those interpolated from the 0.001 mm and 0.005 mm clay contents by eq. (7.5).

7.7. What is meant by a "well-graded" soil? What is the reason for considering such a soil to be well graded?

7.8. Which soils in Table 7.5 can be injection-grouted with a mixture of Portland cement, fly ash, and water?

7.9. Soil No. 12 in Table 7.5 is to be separated from No. 14 by means of a filter. From the two particle size accumulation curves, define (a) desirable characteristics of a geotextile filter, (b) the gradation(s) required for soil filter(s): if a single filter layer is not adequate, use two. (c) Select appropriate soil(s) from the table to use as filter(s).

**Table 7.5**

Mechanical analysis of soils: percentage passing various sieve sizes

| Soil No. | Sieve number and size of opening (mm) | | | | | | | | | Sedimentation size (mm) | | | | LL | PI | Soil No. |
|---|---|---|---|---|---|---|---|---|---|---|---|---|---|---|---|---|
| | 1 in. 26.7 | ¾ in. 18.8 | ⅜ in. 9.4 | No. 4 4.75 | No. 10 2.00 | No. 40 0.42 | No. 60 0.25 | No. 100 0.149 | No. 200 0.074 | 0.050 | 0.005 | 0.002 | 0.001 | | | |
| 1 | 100 | 90 | 80 | 72 | 67 | 56 | 44 | 34 | 24 | 21 | 11 | 7 | 4 | 29 | 7 | 1 |
| 2 | | 100 | 99 | 98 | 97 | 96 | 91 | 80 | 71 | 63 | 34 | 25 | 18 | 39 | 14 | 2 |
| 3 | | | | 100 | 100 | 99 | 96 | 92 | 80 | 73 | 41 | 31 | 23 | 76 | 21 | 3 |
| 4 | | | | 100 | 97 | 84 | 66 | 50 | 32 | 24 | 5 | 4 | 4 | 54 | 16 | 4 |
| 5 | | | | | 100 | 96 | 85 | 61 | 34 | 31 | 13 | 10 | 7 | 69 | 7 | 5 |
| 6 | | | | | | 100 | 95 | 88 | 80 | 54 | 25 | 14 | 5 | 35 | 10 | 6 |
| 7 | | | | | | | 100 | 99 | 98 | 95 | 9 | 7 | 6 | 80 | 9 | 7 |
| 8 | | | | | | 100 | 97 | 76 | 60 | 45 | 35 | 21 | 10 | 35 | 17 | 8 |
| 9 | | | | | 100 | 95 | 88 | 81 | 65 | 59 | 18 | 11 | 6 | 27 | 5 | 9 |
| 10 | | 100 | 99 | 97 | 93 | 70 | 58 | 56 | 44 | 42 | 24 | 17 | 11 | 41 | 12 | 10 |
| 11 | | 100 | 98 | 92 | 82 | 50 | 42 | 35 | 28 | 25 | 12 | 8 | 5 | 38 | 16 | 11 |
| 12 | 100 | 93 | 77 | 64 | 48 | 24 | 20 | 16 | 12 | 11 | 8 | 7 | 6 | 13 | 4 | 12 |
| 13 | | | | | 100 | 99 | 84 | 48 | 12 | 8 | — | — | — | — | N.P. | 13 |
| 14 | | | | | | 100 | 100 | 99 | 95 | 93 | 68 | 49 | 34 | 86 | 49 | 14 |
| 15 | | | | | 100 | 79 | 60 | 48 | 34 | 30 | 14 | 11 | 9 | 24 | 8 | 15 |
| 16 | | | | 100 | 94 | 89 | 87 | 85 | 81 | 73 | 39 | 24 | 12 | 59 | 43 | 16 |
| 17 | | | 100 | 98 | 94 | 48 | 42 | 33 | 26 | 22 | 13 | 11 | 10 | 47 | 24 | 17 |
| 18 | 100 | 98 | 97 | 96 | 94 | 91 | 90 | 89 | 86 | 80 | 42 | 27 | 16 | 45 | 17 | 18 |
| 19 | | | | | 100 | 45 | 38 | 30 | 22 | 20 | 10 | 7 | 4 | 16 | 6 | 19 |
| 20 | | | | | 100 | 98 | 96 | 95 | 93 | 89 | 64 | 44 | 29 | 84 | 53 | 20 |

7.10. Combine soils 1 and 3 in Table 7.5 in such proportions that the resulting mixture contains 20 percent 5 μm clay. Draw the particle size accumulation curve of the mixture.

## References and Further Reading

Grim, Ralph E. (1962). *Applied Clay Mineralogy*. McGraw-Hill, New York.

Koerner, Robert M. (1990). *Designing with Geosythetics*, 2nd ed. Prentice-Hall, Englewood Cliffs, N.J.

Mitchell, J. K. (1993). *Fundamentals of Soil Behavior*, 2nd ed. John Wiley & Sons, New York.

Sherard, J. L., Dunnigan, L. P., and Talbot, J. R. (1984). (a) "Basic Properties of Sand and Gravel Filters," and (b) "Filters for Silts and Clays." *ASCE J. Geotech. Engr. Div.* 110(6), 684–718.

Sherard, James L. (1987). "Lessons from the Teton Dam Failure." *Engng. Geol.* 24, 239–256. Reprinted in G. A. Leonards, ed., *Dam Failures*, Elsevier, Amsterdam, 1987.

# 8

## Soil Fabric and Structure

## 8.1 SOME IMPORTANT RELATIONSHIPS

### 8.1.1 Relevance in Engineering

Probably no characteristic of soils is more changeable and has a greater influence on engineering properties than fabric, which is the spatial arrangement between soil particles and soil voids. Larger aspects of fabric such as bedding, cross-bedding, shrinkage cracks, and shear surfaces are defined as soil structure, or more specifically, macrostructure. However, structure and fabric often are used interchangeably. Mitchell (1993) defines structure as fabric that includes its strength and stability.

While the engineering importance of fabric has been recognized for many decades, the difficulty of measurement and of defining meaningful measures has put fabric almost in a category of afterthought, and it is only recently that marked changes in soil behavior have been observed that have brought fabric to a microscopic center-stage.

For example, the most common rudimentary treatment of soil, compaction, requires that the existing soil structure must be broken down in order to expel air and push grains together. Compaction, if properly performed, makes soil stronger and less compressible under load. However, if the soil is compacted too wet so that voids are full of water, remolding can cause the soil to dramatically become weaker. If a soil is compacted too dry it can retain an open structure that will collapse upon wetting even though a specified minimum density has been achieved. Both of these problems involve soil fabric.

Fabric is closely related to particle size and mineralogy, and therefore is conveniently considered in relation to whether a soil is granular, cohesive, or has attributes of both categories of soil.

## 8.1.2  Overview of Granular Soil Fabric

The fabric of granular soils can be investigated by impregnating the soil with plastic and viewing cut faces. Another preparation procedure is to cut and grind thin sections that may be examined under a polarizing microscope, which enables identification of the minerals.

Such studies often reveal a preferred horizontal orientations of pores and flat or elongated grains. More recent studies indicate vertical stacking of grains on top of one another, which is in contrast to the widely hypothesized close-packing arrangements that are analogous to crystal structure.

## 8.1.3  Overview of Cohesive Soil Fabric

Because their small particle size is difficult to observe under a microscope, clay fabric mainly has been defined by inference from crystal structure and physical properties. For example, clay crystals are known to be flat, and stirring a suspension of clay in water makes it more liquid, like stirring a milk shake. If the liquid is allowed to stand it sets or stiffens. Then, like a milk shake, it will respond and turn back into a liquid with stirring. The inference is that there must be some interparticle bonding or optimization of particle contacts that is disrupted by stirring. In a milk shake, tiny ice crystals can temporarily become disorganized; in a clay suspension, it is the clay particles that are temporarily disorganized. This property is called *thixotropy*, which is from Greek for touch-sensitive.

Clay fabric is directly observed under a scanning electron microscope, but the requisite vacuum dehydrates the clay and can change its structure. Freeze-drying in which a sample is first frozen and then evacuated for a long period of time can help to preserve the structure.

Soil mixtures containing both granular and cohesive components combine both elements of fabric, but normally one will dominate. The transition occurs at about 25 to 30 percent clay, which is sufficient to separate and prevent contact between larger soil particles.

## 8.2  GRANULAR SOIL FABRIC

### 8.2.1  Single-Grained Structure

A soil that shows no discrete relationship between one grain and another is said to have a *single-grained structure*. A random single-grained structure is shown in Fig. 8.1(a). However, in many instances elongated particles tend to be aligned parallel and horizontally as illustrated in Fig. 8.1(b). In this case the pores also are oriented horizontally, reducing shearing strength in a horizontal direction. Preferred orientation of flat particles occurs during sedimentation because it is unlikely that particles will stand on end.

**Figure 8.1**

Single-grain soil structures that influence engineering properties.

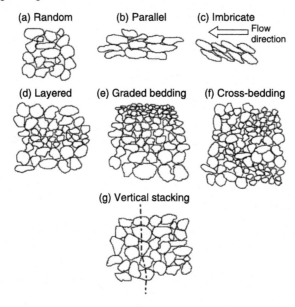

(a) Random  (b) Parallel  (c) Imbricate
Flow direction

(d) Layered  (e) Graded bedding  (f) Cross-bedding

(g) Vertical stacking

Flat or elongated particles deposited from flowing water often become arranged like shingles in what geologists refer to as "imbricate structure." This is illustrated in Fig. 8.1(c), where the arrow indicates the direction of fluid flow. Horizontal bedding as in Fig. 8.1(d) is common as a result of changing sedimentation conditions and can create horizontal zones that are stronger and weaker than the average.

A cycle of sedimentation often involves a waning stage that results in a gradual decrease in the content of coarse particles. As a result a particular layer may show an increase in coarseness with depth (Fig. 8.1(e)), called "graded bedding." "Cross-beddding" (Fig. 8.1(f)) is a product of dynamic and random cycling of erosion and deposition in braided stream deposits.

## 8.2.2 Grain Stacking

A preferred vertical orientation of grain-to-grain contacts was discovered by Oda (1978), who allowed sand grains to settle in water, then gently tapped the sides of the container. The sands then were impregnated with plastic, sectioned, and the grain orientations and contacts measured under a microscope. Stacking occurred regardless of grain shapes that included both flat particles and spheres.

Stacking is illustrated in Fig. 8.1(g) and has been confirmed by other studies, but remained a bit of a mystery until it was found to be related to sedimentation. Figure 8.2 shows a stacking structure resulting from dropping two different sizes of particles at random positions along a line as those landing on top of one another tend to stack instead of sliding off. The same tendency was confirmed with photoelastic studies of plastic disks (Mitchell, 1993; Santamarina et al., 2001).

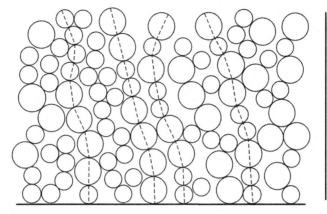

**Figure 8.2**
Simulated random sedimentation of spherical particles tends to create a stacking arrangement and vertical orientation of voids, resulting in low lateral stress and high horizontal compressibility.

Although not yet confirmed by measurements, an open arrangement from particle stacking may help explain low lateral stresses and collapsibility of loess that suddenly densifies under its own weight when it becomes saturated with water (Handy, 1994). It also may increase the responsiveness to horizontal compaction that occurs during compaction grouting, pile driving, and the construction of Rammed Aggregate Piers.

## 8.2.3  Granular Soils Containing Clay

Although the single-grained structure applies only to particles that are silt-size and larger, microscopic examinations show that coarse particles commonly are coated with clay oriented flat against the grain surfaces. The presence of clay coatings is revealed under a polarizing microscope that can block out light coming through the host grains so that the clay coatings appear as bright fringes. In tropical soils the coatings are iron oxides and hydroxides, which gives these soils their red color.

Although clay coatings probably decrease the coefficient of sliding friction between coarser grains, the clay also increases contact pressures because the affinity of clay for water can create suction. The amount of suction is sensitive to the water content, which explains why the strength of clayey granular soils is reduced by wetting.

## 8.2.4  Clay Bridges

Even a moderate clay content causes clay to become concentrated at the particle contacts, forming clay connectors or bridges. These are illustrated in Fig. 8.3(a). This phenomenon was originally described by Grim for molding sands, which are mixtures of sand and clay used in metal casting. Clay is added to increase cohesive strength because sand without clay readily crumbles.

**Figure 8.3**

(a) Clay bridges
add cohesive
strength to
granular soils,
but (b) grain
separation occurs
at about 25 to
30 percent clay.

(a) Clay bridges

(b) Clay matrix

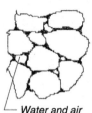

— *Water and air*

The molding sand studies showed that with an increasing clay content there becomes a point where the coarser grains no longer are in contact, and the soil behaves more like clay than a sand or silt. That percentage varies depending on the gradation of the coarse material, but usually is about 25–30 percent clay as the sand changes from granular to cohesive. The separation of sand particles is illustrated in Fig. 8.3(b).

**Loess Studies**

Loess is almost entirely silt and clay, with the clay content gradually increasing with distance from a source. Loess containing less than 16 percent 0.002 mm clay has over a 90 percent probability of being collapsible, which is a phenomenon that is limited to low-density granular soils, but if the clay content exceeds 32 percent the soil is very unlikely to be collapsible but behaves as a moderately expansive clay (Handy, 1973).

# 8.3  COHESIVE SOIL FABRIC

## 8.3.1  Cardhouse Structure

Clay particles in suspension in water tend to combine into flocs as a result of edge-to-face and edge-to-edge attractions between individual crystals, creating an open, "cardhouse" structure (Fig. 8.4(a)). Slow sedimentation may result in stacking similar to that which occurs with granular soils and a loose "honeycomb" structure (Fig. 8.5), and scuba divers describe a transition from water to soil such that they have difficulty feeling where the water leaves off and the soil begins.

## 8.3.2  Clay Particle Orientation under Overburden Pressure

As the cardhouse structure of Fig. 8.4(a) becomes compressed under load, the clay particles tend to become oriented flat, as in Fig. 8.4(b). The most extreme examples of pressure-induced orientation are shales.

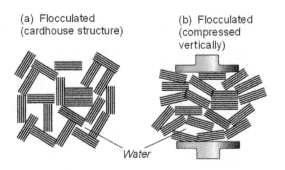

(a) Flocculated
(cardhouse structure)

(b) Flocculated
(compressed
vertically)

Water

**Figure 8.4**
Two-dimensional representations of flocculated and dispersed clay structures.

(c) Mechanically dispersed
(often thixotropic)

Water

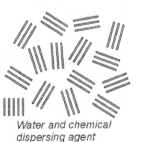

(d) Chemically dispersed

Water and chemical
dispersing agent

**Figure 8.5**
Simulated random settlement of clay flocs creating a "honeycomb" structure.

Even though clay particles are too small to be individually resolved in a light microscope, a parallel orientation can be inferred from viewing vertically oriented thin sections in a polarizing microscope. Clay mineral grains become dark when their optical axes are aligned with the polarizing axes of the microscope, so the degree of preferred orientation can be obtained by measuring light transmission as the stage is rotated.

### 8.3.3 Shear-Induced Orientation of Clay Particles

Shearing of clay, either in a landslide or as a result of horizontal pressures in expansive clays, creates shiny surfaces that indicate that the clay particles have become oriented parallel with those surfaces. These "slickensides" are permanently weak contact surfaces.

The goal of compaction is to push soil grains together, but if compaction is continued too long or if the soil is too wet, compaction pressure is transferred to the soil water that has no resistance to shearing. As a result the soil shears, or slips internally, reorienting and smearing clay particles along the shear planes. Such a soil is said to be *overcompacted*, and is permanently damaged by shear planes and slickensides. Before it can be reused and recompacted it must be dried and pulverized to destroy the sheared structure.

### 8.3.4 Pedologic Structures

As mentioned in Chapter 5, clay-rich B horizons often develop a blocky structure from a three-dimensional array of tension cracks developed from seasonal drying shrinkage. This structure is shown in Fig. 8.6 and with increasing depth is transitional to a columnar structure between vertically oriented shrinkage cracks. Under low pressures the blocky structure can cause clay to behave more like a granular soil than a clay. The blocky structure is preserved by shiny oriented

**Figure 8.6**

A blocky "B horizon" structure that is transitional to a columnar structure with depth. Soil above the dotted line is fill. The soil color is a mottled gray and brown color indicating wet conditions prior to exposure by erosion and sliding.

"clay skins," technically called "argillans," that cover surfaces of the individual blocks or peds.

## 8.4 SENSITIVITY

### 8.4.1 Sensitive Clays

Sensitivity is more than skin-deep, as all clays are somewhat sensitive, which means that they lose a substantial portion of their strength when their flocculated structure is disturbed. This is illustrated in Fig. 8.4(c). However, flocculation still may triumph in the long run if the soil then is left undisturbed. A loss of strength upon remolding or stirring and regain with aging is called *thixotropy*.

Thixotropic behavior explains the "setup factor" that occurs from pile driving in cohesive soils, such that if driving stops for a few hours penetration cannot easily be restarted. After a day or so the pile-bearing capacity commonly is increased by a factor of two, so load tests normally are conducted after sufficient time has elapsed to allow thixotropic setting.

As previously discussed, thixotropy also is a vital attribute of drilling muds that are circulated down through a drill pipe and come back up in the annular space between the pipe and the edges of a borehole. Similar muds are used to hole slurry trench excavations open for constructing cast-in-place retaining walls.

Muddy streams and rivers usually supply sufficient agitation to keep clays dispersed. They then can slowly settle out in quiet water such as in a lake. This is in contrast to settling in salt water, where the clay is flocculated and quickly settles. Deltas in freshwater lakes therefore are sandy while those in sea water contain more clay.

### 8.4.2 Measuring Sensitivity

Sensitivity is defined as the ratio of strength measured before remolding to that measured after remolding at the same soil moisture content and density.

Sensitivity normally is determined from laboratory unconfined compression tests. An undisturbed cylindrical sample is obtained by pushing a thin-walled steel tube; it is then extruded, and loaded vertically in a compression machine until it fails. The soil sample then is remolded without allowing any loss of water, pressed into a cylindrical mold so that it is the same size as the original specimen, and again tested in unconfined compression. and the ratio of the initial strength to the remolded strength is the *sensitivity*. The strength after remolding is called a *residual strength*.

| Table 8.1 | Sensitivity | Descriptive term |
|---|---|---|
| The Rosenqvist terms for sensitivity | <2 | Insensitive |
| | 2–4 | Moderately sensitive |
| | 4–8 | Sensitive |
| | 8–16 | Very sensitive |
| | 16–32 | Slightly quick |
| | 32–64 | Medium quick |
| | >64 | Quick |

Sensitivity also can be quickly measured in the field with the *vane shear test*. In this test, two vertical blades arranged to make an X are pushed into soil and twisted. The torque to cause failure is measured, and twisting continued to obtain a residual shear strength and strength ratio.

The sensitivity of clays typically ranges from about 2 to 4, and emphasizes the importance of avoiding unnecessary disturbance. Terms used to describe sensitivity are shown in Table 8.1.

### 8.4.3  Chemically Dispersed Clays

Clays are routinely dispersed in the laboratory as part of the procedure for particle size analysis, discussed in the preceding chapter. In this case the clay particles are separated as in Fig. 8.4(d) and the electrical charge is modified so that the particles repel instead of attracting one another. The details of this action are described in Chapter 10.

Dispersed clays can remain dispersed in a laboratory flask for months or years without settling out, but opposing forces of gravity and Brownian motion eventually cause the suspension to develop a density gradient. There normally is no discrete X-ray diffraction spacing because the clay particles are randomly separated and too thin.

Naturally dispersed clay can occur in alkaline playa lakes where the clay mineral is sodium-saturated smectite. This is the most expansive of naturally occurring expansive clays. Drying draws particles together and results in shrinkage cracks, and wetting re-expands the clay.

### 8.4.4  Quick Clays

A particularly troubling but fortunately unusual clay is called *quick clay*. Unlike quicksand, which is a condition brought on by rising seepage water, quick clay is a clay that is exceptionally sensitive and turns to liquid if disturbed.

Quick clay is clay that has been deposited and flocculated in salt water, usually in estuaries, and then leached of excess salt so that the flocculated structure becomes

**Figure 8.7**
Mudslide in quick clay annexing the town of Nicolet, Quebec. (National Research Council, Canada.)

unstable. When such a structure is stressed, as by a tremor or other vibration, the clay can suddenly turn into a liquid with devastating consequences. Quick clays occur in terraces along the St. Laurence River in Canada and in Norway, partly as a result of rebound of the earth's crust following melting of continental glaciers.

A quick clay can be stable for decades and then, when sharply vibrated, as by an explosion, ground tremor, or when a truck goes by, can suddenly become a rapidly flowing liquid mudslide, as shown in Fig. 8.7.

## 8.4.5   Dispersive Clays

Weakly flocculated soils can become dispersed from the action of flowing water. This is particularly important in earth dams and levees, where a small leak can quickly become larger and cause eventual failure by piping.

A "pinhole test" was developed by engineers of the USDA Soil Conservation Service to detect dispersive clays. The test involves drilling a 1 mm diameter hole through a compacted soil specimen and running water through the hole under standardized conditions. The hole enlargement after a particular time is a measure of the soil dispersivity (Sherard et al., 1976). Extensive testing has shown that clay minerals in dispersive soils usually are smectites, with over 60 percent of their exchangeable cations being sodium.

Another test for dispersive clay uses the hydrometer test described in the preceding chapter both with and without a chemical dispersing agent. The soil that can be dispersed without the addition of a dispersing agent is assumed to be dispersive (ASTM Designation D-4221).

### 8.4.6 Dispersive Silt

The pinhole test also determines erosion-prone silty soils that strictly speaking are not dispersive. Tests of the loess soil used for the core of the ill-fated Teton Dam did not pass the pinhole test, but did pass the double-hydrometer test (Sherard, 1987). The dam failed by piping, that is, by enlargement of seepage channels under and through the dam until the flow reached the category of a deluge shown in Fig. 8.8. Warnings were given in time to evacuate most of the people from the area downstream but nevertheless 14 lives were lost. This tragic misadventure cast a long shadow over the future of big dam building in the U.S., but earth dams continue to be built in developing countries of the world.

## 8.5 SOIL MACROSTRUCTURE

### 8.5.1 Overview

*Macro* means large, and *mega* implies larger by a factor of a million or so. For the present purpose *macrostructure* may be defined as soil or rock structure that is large but far enough apart that it may be missed by soil borings. Pockets of clay such as shown in Fig. 3.3 come under this definition of macrostructure, whereas the sink in Fig. 3.4 may be considered *megastructure*. Megastructure such as a cavern or landslide should not be missed by borings so long as they are correctly interpreted.

**Figure 8.8**

Teton Dam, Idaho, 5 June 1976, the highest dam in the world to fail, so far. (Photo © Mrs. Eunice Olson, used with permission.)

Macrostructure that is missed by soil borings becomes a matter for speculation, inference, and expectation. For example, vertical ground cracks that open during a dry season are macrostructure that will be missed by borings but nevertheless are readily observable by an experienced field person who knows what to look for.

A failure to recognize megastructure can lead to embarrassment and possible legal difficulties. The engineer who fails to take notice of shrinkage cracks, sagging ground, a landslide in the back yard, or distressed structures across the street, may be forced into a learning experience. Attempts to conceal troublesome megastructure such as a landslide at best is gross ignorance, and at worst, fraud.

Some common descriptive terms for soil macrostructure are listed in Table 8.2. The engineer will be familiar with these terms and know their significance.

## Problems

8.1. Whereas the soil fabrics shown in Figs. 8.1 and 8.3(a) show grains touching, in an actual cross-section obtained by sawing across a soil specimen only a few will touch. Explain.

8.2. What is the meaning and significance of thixotropy?

8.3. Give an example of a mechanically dispersed clay and a chemically dispersed clay.

8.4. What is the X-ray diffraction spacing of a sodium smectite dispersed in water? Of a flocculated calcium smectite in water?

8.5. What is the difference between a quick clay and a dispersive clay? Which might be most likely to influence a landslide? The stability of a levee? Of a retaining wall where pressure is lowered as soil is partly restrained as a result of its internal shearing resistance?

8.6. Explain the geography of quick clay, taking into account that sea level rose over 100 m from melting of continental glaciers.

8.7. Explain why the last of a convoy of military vehicles passing over wet soil is more likely to get stuck than the first vehicles in the convoy.

8.8. What is meant by overcompaction?

8.9. Muddy water is observed flowing from a leak on the downstream face of a large earth dam. Consider and discuss consequences from the following options:
    (a) Run for high ground.
    (b) Try to seal the leak by boring and grouting.
    (c) Draw down the reservoir.
    (d) Dump soil into the cavity eroded by the leak.
    (e) Dump coarse granular material into the cavity.
    (f) Inform the authorities and warn of the possible consequences.

**Table 8.2**

Descriptive terms for macrostructure

| Descriptive term | Criteria | Significance |
|---|---|---|
| Homogeneous | Same color and appearance throughout | The "ideal" soil that seldom occurs in nature |
| Lensed | Inclusions of small pockets of different soils, such as small lenses of sand in a mass of clay | Typical of glacial and alluvial sediments, affects groundwater seepage, and leakage |
| Stratified | Alternating horizontal layers >6 mm thick | Typical of alluvial or lacustrine sediments |
| Laminated | Same, but layers <6 mm thick | Same as above. Varved clays of glacial lakebeds, reflecting annual melting cycles |
| Fissured or fissile | Breaks into thin layers | Characteristic of shales, indicating parallel orientation of clay particles The Landslide Invitational. Compacted embankments can degrade into clay and fail without warning |
| Columnar | Vertical cracks | Can indicate presence of smectite |
| Blocky | Cohesive soil that breaks into small, hard angular lumps or peds | From wet-dry cycling of smectite, especially in a B horizon |
| Slickensided | Inclined, shiny planes that usually are striated | Slip planes caused by seasonal high lateral stress in expansive clay. Also from overcompaction |
| Cemented | Hardened soil layer from secondary carbonates (white), iron oxides (red-brown), or other materials | Influences ease of excavation; may be used for road metal |
| Punky | Ground tremors when stomped on | Saturated, overcompacted clay |
| Viscous yielding | Does not support a person's weight; send student on ahead | Soft, saturated silt or clay |

Note: Adapted and expanded from ASTM Designation D-2488.

(g) Have a beer to calm the nerves while you wait and see what happens.

(h) Write a letter to the newspaper pointing out that operations and maintenance of the dam are not your responsibility.

8.10. Assign grades A through F to the macrostructures in Table 8.2 in accordance with their potential impact on engineering uses.

# References

Handy, R. L. (1973). "Collapsible Loess in Iowa." *Proc. Soil Sci. Soc. Amer.* 37(2), 281–284.

Handy, R. L. (1994). "A Stress Path Model for Collapsible Loess." In Derbyshire, ed., *Genesis and Properties of Cohesive Soils.* NATO ASI Series, Kluwer, Dordrecht, The Netherlands, pp. 33–49.

Mitchell, J. K. (1993). *Fundamentals of Soil Behavior*, 2nd ed. John Wiley & Sons, New York.

Oda, M. (1978). "Significance of Fabric in Granular Mechanics." *Proc. U.S.-Japan Seminar on Continuum Mechanics and Statistical Approaches in the Mechanics of Granular Materials*, pp. 7–26.

Santamarina, J. C., Klein, K. A., and Fam, M. A. (2001). *Soils and Waves.* John Wiley & Sons, New York.

Sherard, J. L., Dunnigan, L. P., Decker, R. S., and Steele, E. F. (1976). "Pinhole Test for Identifying Dispersive Soils." *ASCE J. Geotech. Eng. Div.* 102(GT1), 69–85.

Sherard, J. L. (1987). "Lessons from the Teton Dam Failure." *Enging. Geol.* 24, 239–256. Also in G. A. Leonards, ed., *Dam Failures*, Elsevier, Amsterdam.

# 9 Soil Density and Unit Weight

## 9.1 DENSITY AND MOISTURE CONTENT

### 9.1.1 Heavy Is the Soil

Soil density and unit weight are basic to many soil and foundation problems, as the weight of soil involved can eclipse that of other materials used to build structures. For example, the weight of soil removed for a basement usually exceeds the weight of the house that will stand in its place. This does not ensure that the house will be safe from settlement because its weight is not evenly distributed, but is mainly carried by the foundations.

The most critical component of a foundation is not concrete, which obviously is much stronger than the soil on which it rests. Nor does the weight of the building simply derive support from the soil underneath the foundation, as foundations often are extended to a width that enables the soil alongside to keep the soil underneath from squeezing out. Fig. 9.1 illustrates what can happen if the soil alongside is removed. Support for foundations therefore is not simply a matter of soil strength but also involves a calculated balance that takes into account both the weight and strength of the soil. In Fig. 9.1 a basement was excavated next to the foundation for a municipal fire station with predictable consequences.

Soil weight also contributes the "body forces" that are the driving forces for landslides, and exert pressures on buried pipes and tunnels. Lateral bulging of soil under its own weight is opposed by support from retaining walls.

Soil weight also can have a positive influence because it contributes to friction that helps to *restrain* a landslide and *support* a building foundation. Soil weight therefore can appear on both sides of the equation.

**Figure 9.1**
Collapse of a wall of a municipal fire station was triggered by removing the restraining effect of weight of the soil from alongside the foundation. Photo is courtesy of Prof. J. M. Hoover.

## 9.1.2   Variations in Weight Depending on Moisture Content

In some ways a soil resembles a sponge—light in weight when dry and heavy when wet. As the dry weight is constant, it is the reference for density and unit weight calculations. For example, if a sample of soil weighs 100 g wet and 75 g dry, the percent moisture is based on 75 g, not on the total weight of 100 g. The moisture content, instead of being 25 percent, is $100 \times (100 - 75)/75 = 33$ percent.

This convention can yield some rather curious results: For example, if a soil sample weighs 100 g wet and 50 g dry, its moisture content is $100 \times (100 - 50)/50 = 100$ percent. Some natural soils contain more water than solids, in which case the moisture content is over 100 percent! For this one needs a college education.

The relation between moisture content and degree of saturation can be critical, because a load applied to and acting to compress a saturated soil will be partly carried by the pore water, which has zero strength. The soil that is firm enough to support a heavy truck when dry can turn into a mudhole when the soil becomes saturated with water.

## 9.1.3   Influence of Density on Strength

Generally the more dense the soil, the stronger it is, although there are important exceptions. Compaction specifications define the kind of soil, its moisture content, and a target density. Other criteria such as "walk-out" of a sheepsfoot roller are not reliable because a roller can "walk-out" if the soil is dry but contains sufficient

**Figure 9.2**

The largest man-made structures are not earth dams or the Pyramids, but highways built on engineered and compacted soil. U.S. Interstate Highway system. Adapted from the U.S. Federal Highway Administration.

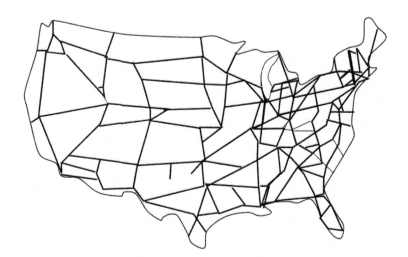

voids that it will turn to mush after rain, and a lightweight roller will walk out more readily than a heavy one.

Density and moisture content measurements are routinely conducted in all but the smallest compaction projects in order to ensure adherence to specifications. Modern instruments allow this to be done indirectly and quickly using non-destructive gamma-ray transmission and neutron tests that are discussed in Chapter 13. This chapter discusses the procedures used to calculate density relationships in a soil.

## 9.2 SOLID + LIQUID AND GAS

### 9.2.1 Three Phases

A block diagram such as shown in Fig. 9.3 is useful for dealing with many definitions related to soils. The total volume is 1.0, and portions of the volume made up by solids, water, and air are designated with subscripts s, w, and a.

### 9.2.2 Density and Unit Weight

Density is a mass property per unit volume that is the same regardless of the influence of gravity, whereas unit weight is the tug that is measured per unit volume in response to gravity. Nuclear instruments measure density that is converted to unit weight based on gravitational pull at the Earth's surface through

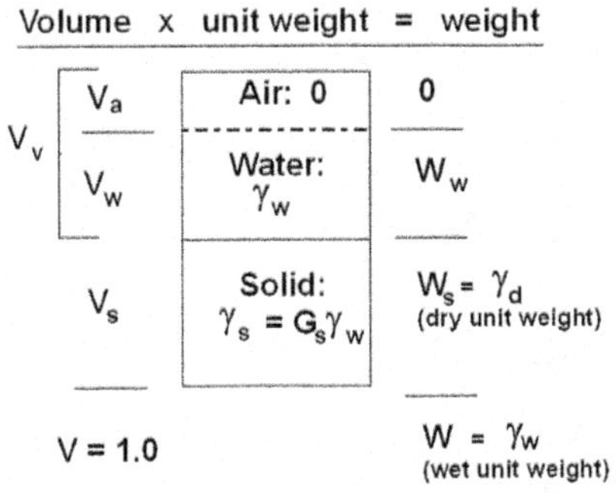

Volume x unit weight = weight

**Figure 9.3**

A block diagram simplifies calculations of moisture content and density relationships by filling in the blanks according to the formula shown at the top. The dashed line indicates variability of water and air contents as the total volume of voids, $V_v$, remains constant.

Newton's formula $f = ma$. On Earth the acceleration usually is denoted by $g$ and nominally is $9.81 \, \text{m/s}^2$ or $32.2 \, \text{ft/s}^2$. Gravitational acceleration varies with altitude and location relative to mountain ranges and ocean basins, but the amount of the variation on Earth does not significantly influence the determination of unit weights. In dealing with weights on Earth, *unit weight* is the term preferred in engineering.

## 9.2.3 A Benchmark for Unit Weight

The solid portion of soil shown in Fig. 9.3 is the base for calculating the *dry unit weight*. The moisture content may change but the dry unit weight remains the same (unless it is expansive clay). Including water gives a *wet unit weight*, which varies between limits that are fixed by the dry unit weight and by the unit weight when the soil is saturated with water.

The wet unit weight and moisture content are measured and then converted to dry unit weight, which reflects only the degree of packing of a soil.

**Example 9.1**

A cylindrical sample of undisturbed soil 3 in. (76.2 mm) in diameter and 6 in. (152.4 mm) long weighs 3.04 lb (1.38 kg mass) and 2.57 lb (1.166 kg mass) after oven-drying. What are (a) its wet and dry unit weights and (b) its densities?

*Answer:* (a) The volume of the cylindrical sample is 42.4 in.$^3$, or 0.0245 ft$^3$. The wet unit weight is $3.04 \div 0.0245 = 124 \, \text{lb/ft}^3$. The dry unit weight is $2.57 \div 0.0245 = 105 \, \text{lb/ft}^3$.

(b) The volume is 695 cm$^3$. The wet density is 1.99 g/cm$^3$ and the dry density is 1.68 g/cm$^3$. In SI these values are expressed as Mg/m$^3$, also called "bulk densities," the bulk density of water being 1.00.

**Example 9.2**
Calculate wet and dry unit weights using SI.

*Answer:* The wet and dry weights are 13.52 N and 11.43 N respectively. The sample volume is $695(10)^{-6}$ m$^3$. The respective unit weights are 19.5 and 16.4 kN/m$^3$.

### 9.2.4 The Void Ratio, *e*

A convenient measure of a soil dry density that is not affected by specific gravity of the soil particles or by the moisture content is the *void ratio*, *e*, which is defined as the volume of voids divided by the volume of solids. The volume of voids is the combined volume of water and air. Then by definition,

$$e = \frac{V_v}{V_s} \tag{9.1}$$

where *e* is the void ratio, $V_v$ the volume of voids, and $V_s$ the volume of voids. This is one of only a few formulas in geotechnical engineering that should be committed to memory.

### 9.2.5 Formula for Moisture Content, *w*

As previously noted, moisture content is defined as the weight (or mass) percent water referenced to the dry weight (or mass) of the soil. By definition,

$$w = 100 \times \frac{W_w}{W_s} \tag{9.2}$$

where *w* is the moisture content in percent, $W_w$ is the weight of water, and $W_s$ is the weight of the solids.

**Example 9.3**
Determine the moisture content in Example 9.1.

*Answer:* First subtract the dry weight from the wet weight to obtain the weight of water: $3.04 - 2.57 = 0.47$ lb. Then $w = 100 \times (0.47/2.57) = 18.3\%$. A similar calculation can be made using grams weight or Newtons.

## 9.3 DRY UNIT WEIGHT

One of the most frequently used formulas in geotechnical engineering converts a measured wet unit weight to a dry unit weight. With reference to Fig. 9.3, the definition for wet unit weight is

$$\gamma_w = \frac{W}{V} = \frac{W_s + W_w}{V} \tag{9.3}$$

where in the block diagram $V = 1$. Then

$$\gamma_w = W_s + W_w \tag{9.4}$$

Solving eq. (9.2) for $W_w$ gives

$$W_w = W_s \times \frac{w}{100} \tag{9.5}$$

where $w$ is the moisture content in percent. Substituting in eq. (9.4) gives

$$\gamma_w = W_s + W_s(w/100) = W_s(1 + w/100) = \gamma_d(1 + w/100) \tag{9.6}$$

$$\gamma_a = \frac{\gamma_w}{1 + w/100} \tag{9.7}$$

In other words, *the dry unit weight of a soil is equal to its wet unit weight divided by 1 plus the moisture content expressed as a decimal fraction.* This formula is used so often that it should be practiced and committed to memory.

### Example 9.4

Use eq. (9.7) to convert the wet unit weight in Example 9.1 to a dry unit weight.

*Answer:*

$$\gamma_d = \frac{\gamma_w}{1 + w/100} = \frac{124}{1 + 0.183} = 105 \text{ lb/ft}^3$$

The same formula, which is so astounding for its simplicity, also may be used to convert wet to dry density: $1.99/(1 + 0.183) = 1.68 \text{ Mg/m}^3$, or unit weights in SI: $19.5/1.183 = 16.5 \text{ kN/m}^3$.

## 9.4 SATURATED AND SUBMERGED UNIT WEIGHT

### 9.4.1 Saturated Unit Weight

The soil unit weight as measured gives only part of the story because it is the unit weight after the soil becomes saturated with water that may be most critical. This can be calculated by use of the block diagram since at saturation the voids are completely filled with water. The saturated unit weight is frequently encountered in landslides.

### Example 9.5

Calculate the saturated unit weight for Example 9.1 assuming that the specific gravity of the mineral portion equals 2.70.

*Answer:* The simplest procedure is to write known quantities into a block diagram and go from there. The known quantities are written as shown in Fig. 9.4. The wet unit weight or weight per unit volume, 124 lb, was measured, and the dry unit weight, or weight of the

**Figure 9.4**

A block diagram makes calculations easy and most density formulas superfluous: fill in the known values and calculate the unknowns.

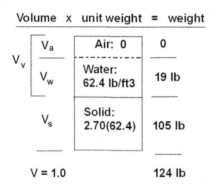

Volume x unit weight = weight

| $V_v$ | | Air: 0 | 0 |
|---|---|---|---|
| | $V_a$ | | |
| | $V_w$ | Water: 62.4 lb/ft3 | 19 lb |
| | $V_s$ | Solid: 2.70(62.4) | 105 lb |

V = 1.0                    124 lb

solids per unit volume, 105 lb, was calculated. The difference is the weight of the water, 19 lb. The respective unit weights for the solids and for the water are entered into the appropriate boxes. From the formula at the top, $V_s = 105/(3.70)(62.4) = 0.62$, which may be written on the diagram. As the total volume is 1.0, the volume of voids is $V_v = 1.0 - 0.62 = 0.38$. If that volume is filled with water, its weight will be $W_w = 0.38 \times 62.4 = 23.5$ lb, and the saturated unit weight will be $W_{sat} = 105 + 23.5 = 128$ lb/ft$^3$.

---

*Question:* What is the void ratio in the previous example?

*Answer:* The calculated values for $V_s = 0.62$ and $V_v = 0.38$ should be written at the left on the block diagram. Then $e = 0.38/0.62 = 0.61$.

---

## 9.4.2  Submerged Unit Weight

Unit weight is greatly influenced by the position of the groundwater table, which submerges and buoys up the soil. To illustrate this concept, imagine weighing a bucket of sand and then re-weighing it after it is suspended by a rope in water. It was Archimedes who argued that the reduction in weight equals the weight of the water displaced. Once a submerged unit weight has been calculated, determining a submerged unit weight is simplest of all. Simply subtract the unit weight of water:

$$\gamma_{sub} = \gamma_{sat} - \gamma_w \tag{9.8}$$

**Example 9.6**

What is the submerged unit weight from the preceding example?

*Answer:* $\gamma_{sub} = 128 - 62.4 = 65.5$ lb/ft$^3$.

Note that in order to calculate a submerged unit weight the saturated unit weight must be calculated first, unless for some indefensible reason a soil that is submerged is assumed to be unsaturated.

"Sudden drawdown" is when a submerged soil is left high and wet, as during receding flood levels. This converts a submerged unit weight to a saturated unit weight, which approximately doubles the effective weight of the soil. Collapse of river banks therefore occurs during waning stages of a flood and not during the flood. The same situation can occur in a lake if the level is rapidly lowered, which is an important consideration in reservoir management as a sudden drawdown can create sudden landslides.

### 9.4.3 Summary of Use of the Block Diagram

A general procedure is as follows:

- The wet unit weight and moisture content of a soil are measured.
- Dry unit weight is calculated using eq. (9.7).
- The two unit weights are written on a block diagram.
- A mineral specific gravity is assumed or measured and its unit weight calculated.
- The volume of solids is calculated from the dry unit weight and mineral unit weight.
- The volume of solids is subtracted from 1.0 to obtain a volume of voids.
- The void ratio is calculated from eq. (9.1).
- To predict the saturated unit weight, assume that all voids are filled with water and calculate its weight from the volume and unit weight of water, then add this weight to the dry unit weight to obtain a saturated unit weight.
- The submerged unit weight is calculated from eq. (9.8).

A block diagram can be used equally effectively with the cgs or SI systems of measurement with the unit weight of water equal to $1.0\,g/cm^3$ or $1\,Mg/m^3$, or $9.807\,kN/m^3$, respectively. Specific gravity values are dimensionless and the same in any system.

### 9.4.4 Other Measures Related to Density and Moisture Content

*Porosity* is defined as the volume of pores divided by the volume of voids, in percent:

$$n = 100\,\frac{V_v}{V} \tag{9.9}$$

Porosity is related to the void ratio by the following formula:

$$n = 100\,\frac{e}{1+e} \tag{9.10}$$

*Degree of saturation* is defined as the volume of water divided by the volume of voids, in percent:

$$S = 100\frac{V_w}{V_v} \tag{9.11}$$

A formula that may be used to determine saturated unit weight from the void ratio, specific gravity, and wet unit weight is as follows:

$$\gamma_{sat} = \gamma_w(G + e)/(1 + e) \tag{9.12}$$

## 9.5 VALUES FOR SPECIFIC GRAVITY

The specific gravity of the solid particles of a soil, represented by $G_s$ in Fig. 9.3, is the ratio of the average density of the solids to that of water. Specific gravities of different minerals vary widely. That of borax is only about 1.7, whereas that of the iron oxide mineral hematite is 5.3, and some less common minerals have much higher values. The usual preponderance of quartz, with a specific gravity of 2.65, and feldspars, with values of 2.55 to 2.75, narrows the range in most soils to between 2.6 and 2.7.

Some values of specific gravities are shown in Table 9.1. The specific gravity of volcanic ash, which is mostly glass, is low because glass has no regular crystal structure with requisite packing of atoms. That of the Oxisol is high because of its iron content.

## 9.6 FACTORS AFFECTING SOIL UNIT WEIGHT

The theoretical upper limit for unit weight can be calculated assuming a zero void ratio. With quartz as the soil mineral the upper limit is:

cgs system: Maximum $\gamma_d = 2.65\,\text{Mg/m}^3$

English units: Maximum $\gamma_d = 2.65 \times 62.4 = \underline{165.4\,\text{lb/ft}^3}$

SI: Maximum $\gamma_d = 2.65 \times 9.81 = \underline{26.0\,\text{kN/m}^3}$

| Table 9.1 | | |
|---|---|---|
| Specific gravities of solids in selected soils. | Volcanic ash, Kansas | 2.32 |
| | Kaolinite | 2.61 |
| | Alluvial smectite clay | 2.65 |
| | Platte River sand | 2.65 |
| | Loess from central U.S. | 2.70 |
| | Micaceous silt, Alaska | 2.76 |
| | Oxisol (latosol), Hawaii | 3.00 |

For comparison, the unit weight of Portland cement concrete, which contains water, is approximately $2.4\,Mg/m^3$, $150\,lb/ft^3$, or $23.6\,kN/m^3$. Heavier concretes used for radiological containment vessels or to contain gas pressures in deep oil wells require denser aggregates.

A theoretical lower bound with all particles touching can be obtained by assuming uniform spheres packed in a cubic arrangement (Fig. 9.5).

The void ratio of the cubic arrangement can be obtained from the volume of a sphere and the volume of the enclosed cube. If $V = 1$, $V_s = 0.5236$, and

$e = 0.928$

Multiplying $V_s$ times the mineral unit weight gives the dry unit weight, which with quartz spheres is

$$\gamma_d = 2.65 \times 0.5236 = 1.39\,Mg/m^3$$
$$\gamma_d = 1.39 \times 62.4 = 86.6\,lb/ft^3$$
$$\gamma_d = 13.6\,kN/m^3$$

Coincidentally silt soils that have a lower unit weight than about $90\,lb/ft^3$ ($1.44\,Mg/m^3$; $14.1\,kN/m^3$) generally are collapsible.

A rhombic arrangement is obtained by sliding one layer of spheres over interstices between the next lower layer of spheres, and is the densest packing arrangement for uniform spheres. This also is the pattern of a tetrahedral arrangement around the void, as each sphere is supported by and in contact with three spheres in the next lower layer. This arrangement gives a volume of solids $V_s = 0.75$, and a void ratio of

$e = 0.34$

The corresponding dry density is

$$\gamma_d = 2.65 \times 0.75 = 1.99\,Mg/m^3, (124\,lb/ft^3), 19.5\,kN/m^3$$

This is a high unit weight for a soil and will be decreased if there are voids where particles are not in contact. The density will be even higher as spaces are filled with progressively smaller particles, approaching that of Portland cement concrete.

Cubic          Rhombic

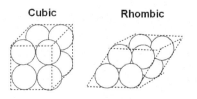

**Figure 9.5**

Packing arrangements for uniform spheres. Similar arrangements occur in crystals.

The average density from the cubic and rhombic packing arrangements is

$$\gamma_d = 2.65 \times 0.75 = 1.7\,\text{Mg/m}^3, (105\,\text{lb/ft}^3), 16.6\,\text{kN/m}^3$$

which is within the usual range of densities obtained after compaction.

### Example 9.7

Calculate the saturated and submerged unit weights for the cubic packing arrangement.

*Answer:* Subtracting the volume of solids from 1.0 gives $V_v = 0.4764$. The weight of water to fill all voids is $0.476\,\text{Mg/m}^3$, or $0.476 \times 62.4 = 29.7\,\text{lb/ft}^3$, or $0.476 \times 9.807 = 4.67\,\text{kN/m}^3$. This is added to the dry density to give $\gamma_{sat} = 29.7 + 86.6 = 116.3\,\text{lb/ft}^3$ ($8.3\,\text{kN/m}^3$). Then

$$\gamma_{sub} = 116.3 - 62.4 = 53.9\,\text{lb/ft}^3\,\text{or}\,18.2 - 9.81 = 8.5\,\text{kN/m}^3$$

## 9.7  MEASURING SOIL MOISTURE CONTENTS

Most but not all water in moist soil is "free water" that is liquid and occupies the open void spaces. Other water is adsorbed on mineral surfaces, particularly the clay mineral surfaces because of the fineness of clay and its surface activity, and in expansive clays part of the water is loosely held between the layers in coordination with exchangeable cations. In addition, OH ions that are part of the clay mineral crystal structure are released as water if the clay is heated to a high temperature.

These several categories of water, therefore, do not all boil off at 100°C, so a standardized drying temperature is used to determine moisture content. That temperature is $110 \pm 5°C$ (ASTM Designation D-2216). To determine moisture contents, samples of soils are weighed, dried overnight in an oven held at 105°C, removed, and allowed to cool in a desiccator, which is a sealed vessel containing calcium chloride, $CaCl_3$, as a desiccant. After cooling, the samples are removed and re-weighed. Controlled cooling is required or the soil clay can re-adsorb water from the atmosphere.

Nuclear instruments used to measure moisture contents in the field actually measure hydrogen atoms regardless of whether they are in water or are part of the clay mineral structure or the content of organic matter, and require calibrations to oven determinations for each soil type. The use of nuclear instruments and other means for measuring soil densities in the field is discussed in Chapter 13 on soil compaction.

### Example 9.8

A soil sample plus an aluminum dish weighs 195 g before oven-drying and 136 g afterwards. The dish weighs 15 g. (Most small laboratory balances weigh in grams.) What is the moisture content?

*Answer:* The difference in weight is the weight of the water: $196 - 136 = 59$ g. The weight of the dish, or tare weight, must be subtracted to obtain the dry weight of the soil: $136 - 15 = 121$ g. The moisture content is $w = 100 \times 59/121 = 49\%$.

## Example 9.9

A cylindrical soil sample 75 mm in diameter by 100 mm long weighs 690 g. After oven-drying it weighs 580 g. The specific gravity of the solid particles is $G = 2.70$. What are the (a) moisture content, (b) wet unit weight, (c) dry unit weight, (d) void ratio, (e) porosity, (f) saturated unit weight, (g) submerged unit weight?

*Answer:* The first step is to write known or given quantities in a block diagram as shown at the top in Fig. 9.6. The next step is to calculate and fill in the empty spaces as shown sequentially by the arrows starting at the lower right in the lower part of the figure. That is:

- The weight of water equals the difference between the wet and oven-dry weights, $690 - 580 = 110$ g.
- The volume of solids equals the weight of solids divided by $G_s = 580 \div 2.70 = 215$ cm$^3$.
- The volume of water equals the weight of water divided by $G_w = 110 \div 1.0 = 110$ cm$^3$.

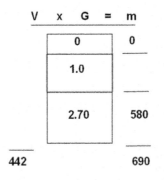

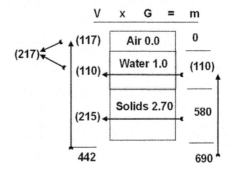

**Figure 9.6**

Block diagram for Example 9.9.

- The volume of air equals the total volume minus the volumes of water and solids: $442 - 215 - 110 = 117\,\text{cm}^3$.
- The volume of voids equals the volume of water plus the volume of air: $110 + 117 = 227\,\text{cm}^3$.

After the block diagram is completed it is a simple matter to calculate the required quantities:

(a) $w = 100 \times 110/580 = 19.0\%$.

(b) $\gamma_w = 690/442 = 1.56\,\text{g/cm}^3$; or $\times\ 62.4 = 97.4\,\text{lb/ft}^3$.

(c) $\gamma_d = 580/442 = 1.31\,\text{g/cm}^3$; or $\times\ 62.4 = 81.9\,\text{lb/ft}^3$.

(d) $e = 227/215 = 1.06$.

(e) $n = 100 \times (227/442) = 51.3\%$.

(f) For saturated unit weight the air voids are replaced by water that has a weight of $117 \times 1.0 = 117$. Then $\gamma_{sat} = (690 + 117)/442 = 1.83\,\text{g/cm}^3$ or $113.9\,\text{lb/ft}^3$.

(g) $\gamma_{sub} = 1.83 - 1.0 = 0.83\,\text{g/m}^3$, or $113.9 - 62.4 = 51.5\,\text{lb/ft}^3$.

## Problems

9.1. Convert the following wet densities and moisture contents to dry densities:

(a) $18.85\,\text{kN/m}^3$, 15%.

(b) $117\,\text{lb/ft}^3$, 17%.

(c) $21.21\,\text{kN/m}^3$, 21%.

(d) $15.87\,\text{kN/m}^3$, 5%.

(e) $120\,\text{lb/ft}^3$, 12%.

(f) $14.14\,\text{kN/m}^3$, 102%.

9.2. If 2.70, find the void ratio for each of the above.

9.3. Calculate the saturated and the submerged unit weights of each of the above.

9.4. An undisturbed sample of soil weighing 49.95 N (11.23 lb) is coated with paraffin and suspended by a string in water to give a submerged weight of 24.42 N (5.49 lb). If the sample contains 13.4% moisture and has a true specific gravity of 2.65, calculate the following properties: (a) wet density, (b) dry density, (c) void ratio, (d) porosity, (e) saturation, (f) percent water voids, (g) percent air voids, (h) percent solids.

9.5. If $171,900\,\text{m}^3$ ($224,800\,\text{yd}^3$) of soil are removed from a highway cut in which the void ratio is 1.22, how many cubic meters (yards) of fill having a void ratio of 0.78 could be constructed?

9.6. If the true specific gravity of the soil in Problem 9.5 is 2.70 and it contains 10% moisture, how many meganewtons (tons) of soil will have to be moved? How many meganewtons (tons) of dry soil will there be? How many meganewtons (tons) of water will there be in the soil?

9.7. (a) Explain how a moisture content can exceed 100%. (b) What soil minerals are indicated by moisture contents in excess of 100%? (c) Why is it advantageous to define moisture content on the basis of dry weight rather than total weight?

9.8. A proposed earth dam will contain $4{,}098{,}000\,m^3$ $(5{,}360{,}000\,yd^3)$ of soil. It will be compacted to a void ratio of 0.80. There are three available borrow pits, which are designated as A, B, and C. The void ratio of the soil in each pit and the estimated cost of moving the soil to the dam is as shown in the tabulation.

What will be the least earth-moving cost, and which pit will it be most economical to use?

| Pit | Void ratio | Cost of moving | |
|-----|-----------|----------------|-----------|
| | | per $m^3$ | per $yd^3$ |
| A | 0.9 | $1.67 | $1.28 |
| B | 2.0 | 1.19 | 0.91 |
| C | 1.6 | 1.56 | 1.19 |

9.9. Prepare a bar graph with vertical bar heights representing dry and submerged unit weights of (a) uniform quartz spheres in rhombic packing, (b) uniform quartz spheres in cubic packing, (c) solid quartz, (d) Portland cement concrete.

## Further Reading

American Society for Testing and Materials, *Annual Book of ASTM Standards*, Vol. 4.08: *Soil and Rock*.

Mitchell, J. K. (1993) *Fundamentals of Soil Behavior*, 2nd ed. John Wiley & Sons, New York.

# 10 Soil Water

## 10.1 THE HYDROLOGICAL CYCLE

The most obvious part of the hydrological cycle involves rainfall that runs downhill and accumulates into streams and rivers that combine and eventually reach the sea. Less evident is the evaporation that simultaneously takes place, rises to form clouds, and returns to the ground surface as rain or snow to rejoin the hydrological cycle. Even less evident is water that infiltrates into the soil where its influences are manifest in springs, wells, and green lawns. These factors plus the influence of transpiration from vegetation and recycling by animals complete the hydrological cycle.

The percentage of precipitation that infiltrates into the ground is close to 100 percent in cavernous limestone areas or desert sands and decreases to 10 to 20 percent on impervious soils such as in shale or granite. Much depends on the rate of rainfall and the moisture condition of the soil. The measurement and prediction of rainfall and runoff are the realm of the hydrologist, and measurement and flow of groundwater that of the groundwater hydrologist.

Runoff water is the main sculptor of landscape, and erosion patterns and the resistance to erosion are important clues to the compositions of rocks and soils, as discussed in earlier chapters. The management of infiltrating water is a major concern for geotechnical engineers as it influences such diverse factors as soil strength, reservoir levels, and soil conditions and erosion at construction sites. Water can reduce the bearing capacity of soils and is a key factor in landslides. Geotechnical engineers also are concerned with seepage of water through earth dams, levees, and soil lining irrigation ditches, as well as flow into underdrains and wells. It is important that an engineer has a working knowledge of the principles governing the flow and retention of water in soil, and the effect of water on strength and stability of this material.

## 10.2  GROUNDWATER

### 10.2.1  Overview

Soils and pervious rocks contain a vast subsurface reservoir of water that has percolated downward to a zone of saturation. Not only is groundwater a major source for water supplies, but it helps to maintain lakes and streams during periods of low rainfall or drought. Water stored in soil pores is essential for survival of terrestrial plants and animals. The importance of groundwater may not be fully appreciated until a supply becomes depleted, which currently is a matter of considerable concern in areas where more water is being extracted for irrigation than is being replenished by rainfall and snow melts. Climatic change as a result of deforestation also can have a devastating influence on water supplies, as in the constantly expanding Sahara. It is humbling to realize that North Africa once was the breadbasket for Rome.

### 10.2.2  Location of the Groundwater Table

Water infiltrating into soil is drawn downward by gravity until it reaches a zone of saturation, where it starts to seep laterally to the nearest outlet that normally is a stream valley. The zone of saturation extends above the groundwater table because of capillary forces similar to those that draw water up into a fine tube. The level of the groundwater table therefore is measured in an open boring that does not support capillary water. This level, also called the *phreatic surface*, represents the elevation at which the pressure in the soil water is equal to atmospheric pressure. Figure 10.1 is a schematic diagram showing the distribution of soil water.

The level of the groundwater table is measured and included in boring logs. Since saturated soil is removed during the boring operation, time is required for water

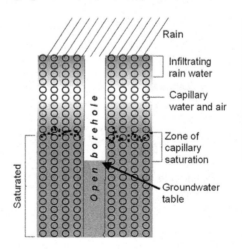

**Figure 10.1**

Rain infiltrating into soil recharges the groundwater.
In fine-grained soils a zone of capillary saturation occurs above the groundwater table.

to enter the boring and reach a stable level. Normally measurements are made after 1 hour, but they also may be repeated after 24 hours. The time required depends on the ease with which water flows through the soil, being less for highly permeable soils than for less permeable soils such as clays.

A groundwater table is *not an underground lake* unless it happens to be in a cavern, and it is *not flat like a table*. Instead, the water table is higher under hilltops and declines to the level of nearby lakes or stream valleys, where groundwater sometimes will be seen emerging as springs. Swimmers in lakes often will encounter zones of colder water caused by groundwater feeding into the lakes as springs.

The most apt description of a groundwater table is as a "subdued replica of the surface of the ground," as shown in Fig. 10.2. The reason for this is the cumulative nature of infiltration on a hillside, as infiltrating water that reaches the groundwater table must seep laterally to find an escape path.

### 10.2.3  Perched Groundwater Tables

If downward-infiltrating water encounters a relatively impermeable soil layer, the rate of infiltration may be slowed sufficiently to cause the water level to back up into the overlying soil, causing what is called a "perched" groundwater table. Perched water tables sometimes contribute to landslides, and can be a problem in excavations.

A perched water table is illustrated in Fig. 10.2. Such conditions often can be predicted from a knowledge of the soil strata and buried clayey paleosols, or ancient weathered soils that have a high content of clay.

A perched water table also may exist in loose fill soil that is placed on top of less permeable soil. Without the fill, the water would be free to run off. A sewage

**Figure 10.2**

Nomenclature descriptive of various groundwater conditions.

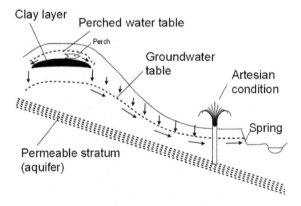

lagoon is an example of an intentionally perched water table, and a permeable landfill may create an unintentionally perched condition.

## 10.2.4  Artesian Conditions

Sometimes when a boring penetrates through an impermeable soil into a more permeable layer, water will rise up into the borehole and may even emerge or geyser out at the ground surface. This is an artesian condition, and its occurrence and the boring depth should be carefully noted in boring logs. The cause is illustrated in Fig. 10.2, and, as can be seen from the figure, it shows the level of the groundwater table at some distance uphill from the boring site.

Tapping into an artesian aquifer can go out of control if water emerges rapidly enough to erode and open the boring. The use of a drilling mud that is heavier than water can contain moderate artesian pressures, and if high artesian pressures are anticipated a special dense mud should be used. Artesian conditions can be expected on floodplains that are capped with clay and are flanked by flat terraces that gather water. An astute driller may keep a wood fencepost close at hand to drive down into a boring and prevent a nuisance from becoming a disaster. The plug can be made more permanent by ramming in dry concrete premix. Auger borings may even require the sacrifice of the auger if its withdrawal allows unimpeded erosive flow. Continuing escape of large amounts of artesian water will draw down the groundwater level and can dry up nearby wells as well as causing surface erosion and flooding.

A gaping hole, locally known as "Jumbo," was described as the "eighth wonder of the world" in 1886 when artesian flow went out of control at Belle Plaine, Iowa, after the wrong size of casing was inserted into a water well. Flow continued for over a year and finally was stopped by dumping in 40 railroad cars of rock, 25 tons of Portland cement, a diversity of scrap iron, plus unrecorded amounts of sand and clay.

## 10.2.5  Soil Color and the Groundwater Table

The level of a water table changes depending on rainfall, snow melting, and whether the ground is frozen or saturated when the snow melts. Curiously, rain can dry frozen soil by thawing and breaking through the frozen barrier that causes a perched water condition.

Typically the water table is highest during and following spring rains, and lowest during autumn and winter. Because soil water is continually in motion it seldom reaches a condition of equilibrium unless there is an impervious cover such as a pavement that prevents infiltration and evaporation. Prevention of surface evaporation can create saturated conditions immediately underneath a pavement, even in a dry climate, and an eventual change in soil color.

Permanent saturation tends to cause gray or even a bluish or green soil color indicative of reducing conditions. During periods when the groundwater table is low, infiltrating water carries oxygen into the gray soil so that the soil becomes mottled with brown colors. Water infiltrating through root channels draws soluble iron compounds toward the channels where the iron oxidizes and becomes concentrated as rust minerals. This also occurs where root channels penetrate a zone of capillary saturation. Soil color therefore is an important indicator of former levels of a groundwater table that may be experienced again in the future.

For example, landslides usually stop during a dry season and give a false sense of security when all that is needed for them to start again is more rain. The design of corrective measures should be based on the highest anticipated level of the groundwater table, which can be indicated by the soil color.

## 10.3  CLASSIFICATION OF SOIL WATER

*Gravitational water*, or free water, is water that flows in response to gravity. It is this water that flows into wells or drains.

*Capillary water* is held by surface tension forces in small pores or capillaries. One characteristic of capillary water is *negative pore pressure* relative to atmospheric pressure. Negative pore pressure equals suction, since it in effect sucks water up above the phreatic surface.

All capillary water, including that which occurs in the zone of capillary saturation, is said to be in the *capillary fringe*. The upper limit of the capillary fringe is indistinct and may be at the ground surface. It is emphasized that the water itself is not different; only the dominant forces retaining the water are different.

*Hygroscopic water* or adsorbed water is that which is held by adsorptive forces on mineral surfaces. Careful measurements indicate that sliding friction between mineral grains involves adsorbed water. This is reasonable because of the array of positive and negative forces exposed at the surface of an ionic crystal. Oxygen ions dominate mineral surface areas, which therefore tend to carry a negative charge. In clay minerals the negative charge is greatly increased by ionic substitutions within the crystal structure.

Hygroscopic water may be defined as water that is retained after air-drying and lost upon oven-drying at 105°C. The amount of hygroscopic water therefore varies depending on temperature and relative humidity.

Adsorbed water can be removed by heating or by a high vacuum. After evacuation incremental amounts of water are allowed back into the system, and highly sensitive weight measurements show a stepwise weight gain as surfaces become covered with discrete layers of molecules. Since the size of the water molecule is known, a simple calculation gives the surface area of the sample. This is the *Brunnaer-Emmet-Teller* or BET method and is used in the study of clays. Teller later was in charge of development of the hydrogen bomb.

*Interlayer water* is specific to expansive clay minerals, being water that is adsorbed between clay mineral layers. Considerable interlayer water remains in place after oven-drying at 110°C, which is sufficient to remove hygroscopic water. Temperatures in excess of 200°C may be required to remove all interlayer water.

*Structural OH water* is not water until clay minerals are heated sufficiently to break down their crystal structure, normally in the range 500–700°C, depending on the clay mineral.

*Crystalline water* is ice.

*Evaporated water* is part of the soil air, in which the relative humidity may be 100 percent. Drying at the ground surface creates a gradient that draws water up from below.

## 10.4  MOVEMENT OF WATER

Gravitational water moves only in response to gravity, which is vertically downward in an unsaturated medium because gravity has no horizontal component. As illustrated in Fig. 10.2, when gravitational water encounters a groundwater table that impedes downward seepage, it moves downward laterally toward an outlet that is lower in elevation. If a potential outlet is at the same elevation, gravitational water will not move.

Capillary water may move in any direction including horizontal in response to a gradient in capillary pressure, referred to as capillary potential or matric potential. Therefore when a gradient is induced by drying at the ground surface, capillary water is attracted toward the drier zone. Capillary water also moves in response to ground freezing, which by lowering the vapor pressure of water in effect dries a soil out. This movement of capillary water upward into a freezing zone results in the formation of pure ice crystals that cause frost heave, discussed in the next chapter.

Hygroscopic and interlayer water move in response to changes in temperature and relative humidity. Freezing therefore can simultaneously cause shrinkage of expansive clay by drawing out interlayer water, and expansion by attracting capillary water that contributes to frost heave.

## 10.5 THE DIPOLAR NATURE OF WATER

### 10.5.1 Geometry of a Water Molecule and Hydrogen Bonding

Each water molecule consists of two hydrogen ions that are positively charged and one oxygen that is negatively charged to form $H_2O$. The two hydrogen atoms are arranged to be complementary to orbiting electrons in the outer shell of the oxygen, and therefore are not located on opposite sides of the oxygen atom, but are at an apical angle of about 104°. A schematic representation is shown in Fig. 10.3(a).

The angle between the hydrogen ions dictates a relatively open hexagonal crystal structure of ice. When ice melts, the crystal structure collapses and hydrogen bonding becomes more random. This allows a closer fit of water molecules so liquid water is denser than ice, which is a unique property of water. Ice floats.

As previously discussed in relation to the kaolinite crystal structure, the loss of an electron from the hydrogen atom leaves a positively charged proton that is the source for *hydrogen bonding* (Fig. 10.3(b)). The resonant frequency of a hydrogen bond is absorptive to infrared radiation, so water appears black on infrared photos.

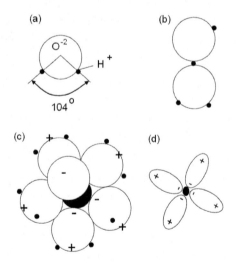

**Figure 10.3**

(a) A water molecule showing separation of the hydrogen ions; (b) hydrogen bond between two water molecules; (c) hydration water with negative ends of water molecules attracted to a $Ca^{2+}$ ion; (d) simplified representation of water dipoles attracted to a positive ion.

## 10.5.2 Polarity of the Water Molecule

Another important feature of the water molecule, aside from its tendency to participate in hydrogen bonding, is that the arrangement of hydrogen ions creates positive and negative ends to the molecule. Polarity makes water a good solvent for ionic crystals such as NaCl, by weakening the electrostatic charges that hold the crystal together. A measure of polarity is the *dielectric constant*, which is high for water. Water also decreases the surface hardness of minerals such as quartz that are only partly ionically bonded. For example, the scratch hardness of quartz is lower in water than in air. This also applies to glass, which is disorganized silica molecules, and is more easily scratched with a glass cutter if it is wet.

## 10.5.3 Interaction with Cations

Water dipoles are attracted to ions in solution, particularly to positive ions that are smaller than negative ions, where electrons in outer shells repel one another and increase the ion diameter. This adsorption is illustrated in Fig. 10.3(c), where negative ends of water molecules are attracted to a positive $Ca^{2+}$ ion in solution. This kind of attraction also exists between water and cations in the interlayer space of expansive clay minerals. Because $Ca^{2+}$ cations have a more positive hold on adsorbed water than single-valence $Na^+$ cations, calcium-smectite is much less expansive than sodium-smectite.

# 10.6 DIFFERENTIAL THERMAL ANALYSIS (DTA)

## 10.6.1 Overview

An early method for studying clay minerals involved measuring thermal delays occurring in small samples as they are heated. For example, calories are required to cook off free water at 100°C, which causes a lag in temperature. A clever arrangement was devised to amplify these differences, heating an inert reference sample and monitoring the difference in temperature between the test sample and the reference. This is easily accomplished with two opposing thermocouples attached in series so that a voltage difference occurs only when one is hotter than the other. This is the scheme for *differential thermal analysis* or *DTA*.

## 10.6.2 Uses of DTA

While X-ray diffraction has replaced DTA for clay mineral identifications, the reactions that occur during heating are an indication of the different energy levels for different classes of soil water. This is illustrated in Fig. 10.4, where a large peak above 100°C indicates the loss of hygroscopic water from a sample that previously was air-dried. Such a reaction is described as "endothermic" because it takes heat. This is followed by a smaller peak at higher temperature, from loss of water held

**Figure 10.4**

Differential
thermal analysis
of a smectite clay.

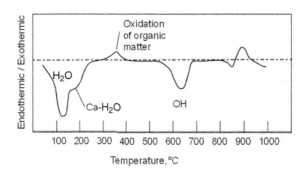

by interlayer $Ca^{2+}$ ions as shown in Fig. 10.3(c). The test therefore can differentiate between Ca- and Na-smectite. A small exothermic peak at 350°C indicates burning of organic material.

The higher temperature peaks in Fig. 10.4 show the loss of OH ions from the clay mineral structure, a final complete breakdown of the structure, and then recrystallization to form new minerals. Specialized instrumentation now allows simultaneous weighing and X-ray diffraction analyses of samples as they are being heated and analyzed.

## 10.7   COLLOIDAL FRACTION OF CLAY

### 10.7.1   Definition

The term "colloid" is derived from a Greek word meaning "glue-like." The colloidal fraction in soil is largely responsible for its plasticity when wet and hardness when dry. Colloids are defined as particles that are small enough that they remain in suspension in water, or smaller than about 0.1 μm and down to molecular size. Clay-size is defined as smaller than 2 μm, so fine clay that dominates smectites is colloidal-sized.

Colloids mixed with water are called *suspensions* and not solutions, which involve dispersion as ions or molecules. The two can be distinguished because a light beam passing through a solution is invisible, but can be seen passing through a suspension, referred to as the "Tyndall effect." Smoke is a colloidal suspension of particles in air. Milk is a colloid; carbonated water is a solution. Asphalt emulsions are colloidal suspensions in water.

Colloids have a large surface area per unit volume. This can be illustrated by dividing a cube into smaller cubes, as shown in Fig. 10.5. Dividing a unit volume into 5 slices each way increases the surface area from $6 \times 1 \times 1 = 6$ to $125 \times 6 \times 0.2 \times 0.2 = 30$, or by a factor of 5. Now let us similarly divide into 1000 slices on a side, and the surface area is increased by a factor of 1000.

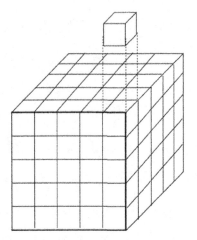

**Figure 10.5**
The finer the particles the larger the surface area.

Every surface has unsatisfied bonds that adsorb water and other molecules. Colloidal suspensions are highly surface-active, which means that they are sensitive to the presence of chemicals that can act as either dispersing agents or flocculating agents.

## 10.7.2 Double-Layer Theory

The stability of suspensions remained somewhat of a mystery prior to the development of double-layer theory, which was proposed in the nineteenth century by von Helmholtz and later modified independently by Guoy and Chapman. The theory is based on a supposition that a surface carries a negative charge. This can be attributed to the large size and larger surface exposure of oxygen ions in minerals, and to net charge deficiencies in expansive clay minerals that are balanced with interlayer cations. In a suspension, negative surfaces attract positive ions that are accompanied by hydration water as illustrated in Fig. 10.3(d), so they do not approach closely enough to neutralize the surface charge on the colloid. The combination of negative surface charge balanced by cations that can approach but do not attach constitutes a *diffuse double layer*. This concept is illustrated in Fig. 10.6.

There also are weak attractive forces between particles or molecules that are called *van der Waals attractions*. These forces are relatively constant while the diffuse double layer changes depending on the hydration, concentration, and electrical charges of the cations. Therefore by juggling cations in the diffuse double layer a colloid can be "stabilized," that is, remain dispersed, or it can become flocculated. Flocculation is the preferred attribute in soils used for engineering purposes, although dispersion is utilized in particle size analysis.

A high salt concentration or more highly charged positive ions such as $Ca^{2+}$ compress the double layer and cause flocculation, whereas lower charged and

**Figure 10.6**
Diffuse double-layer cartoon.

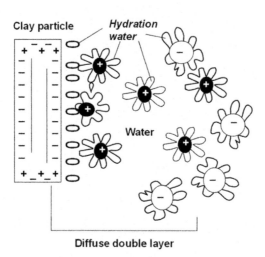

Diffuse double layer

highly hydrated ions such as $Na^+$ can cause a suspension to remain dispersed. Mathematical details of double-layer theory are treated by Mitchell (1993).

The chemistry of drilling muds used in oil well drilling is complicated by several factors including encountering water-soluble rock layers. "Salt domes" are created when a deep layer of rock salt becomes plastic under pressure and, being less dense that most rock, pushes upward through overlying strata, trapping oil in the tilted stata. Drilling muds are specially concocted to prevent swelling of shale and to contain high gas or fluid pressures encountered in deep drilling. Heavy minerals such as barite are added to increase the density of the mud to prevent blowouts. However, fluid pressures must not be so high as to cause hydraulic fracturing and loss of the drilling fluid. If gas dissolved in the drilling mud comes out of solution as the mud is pumped upward, this will decrease the mud density and increase the potential for a blowout. If more mud is observed coming out of the hole than is going in, drilling is immediately stopped and an expandable "blowout preventer" is used to seal the space between the well bore and the drill pipe. A denser mud then is cautiously pumped down the hole.

### 10.7.3   Electrophoresis and Electroosmosis

Electrophoresis is proof that colloidal particles carry an electrical charge, because they can be made to move by application of a voltage gradient. This is readily demonstrated with a U-tube as shown in Fig. 10.7. The direction of migration establishes whether the charge on the particles is positive or negative, and the rate of migration can be measured under a microscope to determine the amount of the charge on the particles, called the "zeta potential." Electrophoresis is used to separate DNA fragments for DNA fingerprinting, and is used medically to identify proteins in blood.

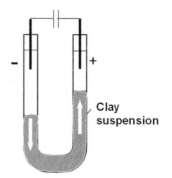

**Figure 10.7**
Electrophoresis moves negatively charged clay particles toward a positive electrode.

Clay suspension

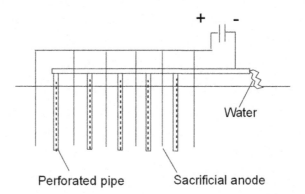

**Figure 10.8**
Setup for drainage by electroosmosis.

Water

Perforated pipe          Sacrificial anode

As negatively charged clay particles move, positively charged ions move in the opposite direction and take their water hulls with them. This is called *electroosmosis*, and was put to practical use for draining soils in 1939 by Leo Casagrande in Germany. Because electroosmosis moves water that is adsorbed in the double layer, it can extract water that cannot be drained by gravity.

Electroosmosis can be accomplished in the field using a d.c. welding generator to supply the voltage. Electrodes are spaced to give a maximum voltage gradient of about 1 volt per 25 mm (1 volt per inch) to avoid loss of energy by heating; thus a 40 volt generator translates into an electrode spacing of 40 in. or 1 meter. Perforated pipes are used as cathodes so cations and their associated water hulls reaching a cathode are plated out, releasing their water. A schematic diagram is shown in Fig. 10.8. Water flows up through the pipes to a manifold and is conducted to an external drain. The anode must supply cations to the system, so iron water pipe usually is used and must be replaced periodically.

Because electroosmosis mobilizes cations and water in the double layer it is more effective than gravity drainage in fine-grained soils. However, because of its high cost, electrical drainage is not used where well points or vacuum well points will do the job.

### 10.7.4   pH-Dependent Charge

The electrical charge or zeta potential of clays increases with increasing pH or alkalinity as hydrogen ions are removed from the clay to make water. A qualitative indication of pH of a solution of different salts can be obtained by considering their reactions with water: For example, $NaCl + H_2O = NaOH + HCl$, a strong acid and a strong base, so the solutions are near neutral. $CaCl_2 + H_2O = Ca(OH)_2 + 2HCl$, a strong acid and a weak base. Ice-melting salts that contain this chemical are highly corrosive to automobiles. (The same argument applies to $MgCl_2$.)

Hydrated lime, $CaCO_3 + H_2O = Ca(OH)_2 + H_2CO_3$, a weak acid and a strong base, which makes lime an effective stabilizing agent, particularly for expansive smectite clays. Furthermore $H_2CO_3 = H_2O + CO_2\uparrow$; as carbon dioxide gas escapes, the weak acid, which is the same as in carbonated soft drinks, turns to water.

## 10.8   SUMMARY OF ATTRACTIVE AND REPULSIVE FORCES INVOLVING WATER AND SOIL MINERALS

The several interactions involving water and soil minerals can be summarized as follows:

- Water molecules are linked by hydrogen bonds in ice. When ice melts the hydrogen bonds become more random, which allows closer packing and a higher density so that ice floats on liquid water. This property is unique to water.
- The geometry of the water molecule causes it to be dipolar, which in turn leads to clustering of water molecules around ions in solution.
- Mineral surfaces, in particular clay mineral surfaces, usually carry a negative charge that attracts water dipoles as adsorbed water, and cations with their associated water molecules that are part of the diffuse double layer.
- Ions and water in the double layer can be moved with d.c. electricity to remove water from difficult clays.
- Suppression of the double layer by drying or by the addition of multivalent ions allows particles to come close enough that they attract one another through van der Waals forces so that they flocculate.
- Lime stabilization combines the flocculating potential of $Ca^{2+}$ ions with an increase in the negative pH-dependent charge on smectite clay particles.

### Problems

10.1.   Draw a diagram of the hydrological cycle.

10.2.   What are three principal ways that precipitation is disposed of after falling to the earth?

10.3. Prepare a table outlining a classification of soil water and state the basis for each class.

10.4. How can one distinguish by means of a single boring between a main water table and a perched water table?

10.5. Name two ways in which a soil can be saturated above the water table.

10.6. Name two natural processes by which the elevation of a water table may be lowered.

10.7. How may a water table be lowered artificially?

10.8. What causes a water table to rise? Explain.

10.9. What clues would you look for to indicate former high elevations of a water table?

10.10. What happens when a dispersed clay dries out? Explain in terms of double-layer theory.

10.11. A soil is dispersed for particle size analysis by adding a small amount of a sodium salt. Explain how this works using double-layer theory.

10.12. A landslide in wet clay soil is taking the back yard of a multimillion-dollar home owned by your favorite movie star, and if it keeps on raining the house will be next to go. Wells will not drain the smectite clay and it is not practical to move the house. Two possibilities are suggested, electrical drainage, or boring holes in the clay and filling the holes with quicklime. Tell how each method will operate.

10.13. How will calcium carbonate, $CaCO_3$, affect the pH of water? Compare its effectiveness with hydrated lime for chemical soil stabilization.

10.14. What is the difference between a water dipole and hydrogen-bonded water? Which occurs where?

10.15. What is the zeta potential?

10.16. DNA fragments carry a negative charge and are separated using electrophoresis because smaller pieces move faster. Suggest a reason why they move faster.

## Further Reading

Baver, L. D., Gardner, W. H., and Gardner, W. R. (1972). *Soil Physics*, 4th ed. John Wiley & Sons, New York.

Bjerrum, L., Moum, J., and Eide, O. (1967). "Application of Electro-osmosis to a Foundation Problem in a Norwegian Quick Clay." *Geotechnique* 17, 214–235.

Grim, R. E. (1968). *Clay Mineralogy*, 2d ed. McGraw-Hill, New York.

Ho, Clara, and Handy, R. L. (1963). "Characteristics of Lime Retention by Montmorillonitic Clays." *Highway Research Record* 29, 55–69.

Mitchell, J. K. (1993). *Fundamentals of Soil Behavior*, 2nd ed. John Wiley & Sons, New York.

# 11 Pore Water Pressure, Capillary Water, and Frost Action

## 11.1  PUTTING IT TOGETHER

### 11.1.1  Of Soil and Water

The preceding chapters dealt with the nature of soil minerals and their relationships to water; this chapter emphasizes water pressures, both positive and negative, that either tend to push soil grains apart, reducing their contact friction, or pull them together, increasing friction. The first is called *positive pore water pressure*; the second, *negative pore water pressure* or *suction*. The reference pressure for both is atmospheric pressure, which tends to be equalized in soils above a groundwater table that are more or less permeable to air.

### 11.1.2  Positive Pore Water Pressure

Because positive pore water pressure acts to push soil grains apart and reduce their intergranular friction, it can be a key component of shearing strength. Positive pore water pressure plays a key role in landslides, which therefore usually occur after heavy rains when the groundwater table is high.

Positive pore water pressure also is subject to dynamic influences from compression, which creates a gradient that moves water out of the soil and allows it to compress. A gradient also exists across soil zones through which seepage occurs because of the viscous resistance to flow. These factors are discussed in detail in later chapters that deal with settlement of foundations and seepage rates through and under dams and levees, and into open excavations.

### 11.1.3  Negative Pore Water Pressure

Soil suction is what holds a child's sand castle together, until waves or tides produce a feeling of sad inevitability as negative pore pressure is relieved and the castle collapses. The sand must be moist in order to be able to build a sand castle in the first place. The transitory nature of the part of soil strength that depends on moisture content gives it the name "apparent cohesion."

Negative pore water pressure also sets up a pressure gradient that acts to suck in more water, which if available is responsible for frost heave. As this is closely related to capillarity it is discussed in this chapter, along with the depth of freezing and permafrost.

## 11.2  EFFECTIVE STRESS

### 11.2.1  Equilibrium Pore Water Pressure

A scuba diver knows from personal experience that water pressure increases with depth under water. The same increase exists in saturated soils, since voids are interconnected and transmit pressure. In a static situation, the pore water pressure is simply calculated from

$$u = \gamma_w h \tag{11.1}$$

where $u$ is pore water pressure, $\gamma_w$ is the unit weight of water, and $h$ is the depth below a groundwater table.

### 11.2.2  Equation for Effective Stress

One of the most important insights by Terzaghi was to recognize that, because water pressure on soil grains tends to push the grains apart, it can simply be subtracted from total stress to obtain what he defined as *effective stress*. This actually is a simplification because it assumes that water pressure acts across the entire cross-section of soil instead of only on the area represented by the voids. However, true grain contact areas are very small and make up only a small fraction of the entire cross-section. This assumption greatly simplifies calculations and, as it tends to overpredict the influence from positive pore water pressure, is on the safe side. By considering soil stress on a total area basis, one can write:

$$\sigma' = \sigma - u \tag{11.2}$$

where $\sigma'$ is the effective stress, $\sigma$ is total stress, and $u$ is the pore water pressure. This is the Terzaghi equation for effective stress, and even though it is deceptively simple, it is one of the most important relationships in soil mechanics.

**Example 11.1**
The unit weight of a saturated soil is $125 \, lb/ft^3$ ($19.6 \, kN/m^3$ or $2.0 \, Mg/m^3$). What are the total and effective stresses at a depth of $3.05 \, ft$ ($1 \, m$)?

*Answer:* Total stress is $\sigma = 3.05 \times 125 = 381\,\text{lb/ft}^2$ ($1 \times 19.6 = 19.6\,\text{kN/m}^2$ or kPa). The pore water pressure is $3.05 \times 62.4 = 190\,\text{lb/ft}^2$ ($1 \times 9.81$ kPa). Then

$$\sigma' = 381 - 190 = \underline{191\,\text{lb/ft}^2\ (19.6 - 9.81 = 9.79\,\text{kPa})}$$

In the cgs system $\sigma = 1 \times 2.0 = 2.0\,\text{Mg/m}^2$, $u = 1 \times 1.0$, and $\sigma' = 2 - 1.0 = 1.0\,\text{Mg/m}^2$.

It will be noted that the effective stress for a submerged soil having this unit weight is almost exactly one-half of the total stress, which is a useful approximation.

### Example 11.2

Repeat the above calculation for effective stress using the *submerged unit weight* of the soil.

*Answer:* As discussed in the preceding chapter, submerged unit weight at saturation equals total unit weight minus the unit weight of water: $\gamma_{sub} = 125 - 62.4 = 62.6\,\text{lb/ft}^3$ ($19.6 - 9.81 = 9.79$ kPa). Then

$$\sigma' = 3.05 \times 62.6 = \underline{191\,\text{lb/ft}^2\ (1 \times 9.79 = 9.79\,\text{kPa})}$$

The two answers are the same, and spreadsheet calculations sometimes are simplified by using submerged unit weight for the part of the soil that is below a groundwater table to calculate effective stress.

## 11.2.3   Does Pore Water Pressure Affect Shearing Stress?

Pore pressure acts normal or perpendicular to grain surfaces and therefore subtracts from compressive stress. How does pore water pressure influence shearing stress? Under static conditions water has no shearing resistance and is not influenced by pressure within the water. It is only by exerting pressure, which tends to separate the soil grains, that pore water pressure affects the shearing strength of soil.

## 11.2.4   Effective Stress at the Bottom of a Lake or Ocean

Let us assume that soil is submerged under 1000 m of water; does eq. (11.2) still hold true? It might intuitively be assumed that the weight of the water would push the soil grains together. On the other hand it can be argued that the high water pressure will push them apart. Actually, both occur and balance out. That is, pressure from the water column adds both to the total stress, $\sigma$, and adds equally to the pore water pressure, $u$, in eq. (11.2), so effective stress, $\sigma'$, remains the same. This explanation for low densities of sea bottom soils was first recognized by a geologist, Lyell, and later quantified by Terzaghi in the effective stress equation.

The same observation can be made by applying Archimedes' principle: as a buoyant force equals the weight of water displaced, it is not affected by the weight of water underneath or on top of a submerged object.

## 11.2.5  Excess Pore Water Pressure and Shear Strength

Pushing soil grains together, as occurs under a foundation load, reduces the pore space and causes a corresponding increase in the pore water pressure in a saturated soil, until the water can escape and the system returns to equilibrium. The temporarily increased pore pressure affects intergranular contact pressures and friction the same as does static pore water pressure, and eq. (11.2) still applies.

The response of pore water pressure to external loading is a critical relationship in geotechnical engineering. It explains why rapid loading of saturated clay soil, as under the wheel of a car, can result in the car being stuck in the mud, whereas if the same load is applied after the soil dries out there is no problem.

The rate of loading is particularly important for large structures such as earth dams, so the pore water pressure is monitored using observation wells or piezometers. If the pore water pressure is too high for the structure to be safe, construction is stopped until there has been sufficient time for part of the excess pore water pressure to drain out. The heavy stone castles of Europe survived because they were built over time spans of hundreds of years instead of being plunked down in a matter of months.

Settlement occurs as excess pore pressure declines and more load is transferred from the water to the soil grains, causing the soil to compress. This time-related compression is called *consolidation*. Consolidation of soil takes time, which depends in part on the permeability of the soil. The time function was analyzed by Karl Terzaghi and is presented in a later chapter.

Structures built on saturated sand are of less concern because sand has a sufficiently high conductivity for water that it can drain out as a load is applied.

Excess pore water pressure also develops during formation of a sedimentary deposit as more load is slowly applied at the surface. Very high pore water pressures have been measured in deep oil wells, still left over from sedimentation and enhanced by expansion of the pore water as a result of the increase in temperature with depth, called a "geothermal gradient."

## 11.2.6  Is an Artesian Condition the Same as Excess Pore Water Pressure?

This can be argued, because there is no question that an artesian condition represents an excess of water pressure that can be relieved by drainage. However, the cause is not compression of the soil voids, but water moving downward from an elevated position through a confined porous soil layer such as a sand or gravel, much as if the water were moving in a pipe. The source normally is a hill where the groundwater table naturally is higher. An artesian condition therefore does not qualify as excess pore pressure, because the pressure is in equilibrium with remotely connected groundwater. Artesian pressure is analogous to pressure in a water faucet that is remotely connected to a water tower.

# 11.3  NEGATIVE PORE WATER PRESSURE

### 11.3.1  Capillary Rise

A simple demonstration of negative pore water pressure is by analogy to water retained by capillary forces in a small clean glass tube, as shown in Fig. 11.1. Generally, the smaller the diameter of the tube, the higher the water will rise. A similar situation exists with soils; the finer the soil and the smaller the void diameter, the higher the capillary rise. As atmospheric pressure at the phreatic surface is the base for measurement, pressure in the capillary is negative and decreases with height above that surface, as shown in the figure.

### 11.3.2  Surface Tension

Capillary rise is attributed to surface tension of the water, which is the result of unbalanced attractive forces at the interface between water and air. This is illustrated in Fig. 11.2, where the low attractive force between water and air molecules results in higher dipolar attractions in the water surface.

Surface tension exists at any interface between solid, liquid, and gas, but its effects are most obvious at liquid-gas interfaces, where surface tension causes the surface of a liquid to behave as if it were covered with a tightly stretched membrane. The surface sometimes is considered a separate phase called a "contractile skin." A membrane analogy is an appropriate and useful concept in helping to explain

**Figure 11.1**

Capillary rise and negative pore pressure.

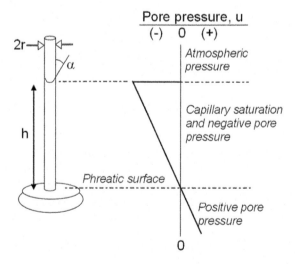

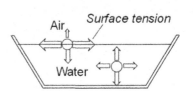

**Figure 11.2**
Unbalanced molecular attractions in a surface.

surface-tension phenomena, such as support for water insects that skate across and depress the water surface.

### 11.3.3 Surface Tension and Capillary Rise

In a capillary tube, surface tension forces attach to the glass and hold the water up, but the weight of the water causes the surface to sag downward in the middle, somewhat like a blanket that is being held at the edges. This creates a visible meniscus. The height of capillary rise usually is measured to the bottom of the meniscus, even though that is not strictly correct. The walls of the capillary must be clean and free of any hydrophobic material such as oil that would prevent the adsorption of water molecules to the glass.

If a fluid does not bond to glass, the $\alpha$ angle in Fig. 11.1 exceeds 90° and capillarity pushes downward so that the meniscus is reversed. A reversed meniscus may be observed at the top of a mercury column in a thermometer.

### 11.3.4 Oil on Troubled Waters

Oil is hydrophobic, or water-hating, and being less dense than water spreads over the surface in a thin film that diffracts light into rainbow-like colors. An oil film therefore creates two surface-tension interfaces, air-to-oil and oil-to-water. The net increase in total surface tension acts to smooth wave action and give meaning to the expression to "spread oil on troubled waters." Plutarch's explanation was that oil makes water more slippery so that wind slips on by, but he had no knowledge of surface tension. However, this misconception lives on in abortive experiments to develop an "airplane grease."

### 11.3.5 Surface Tension and Surface Energy

Surface tension is a force per unit length, $F/L$. Multiplying the numerator and denominator by $L$ gives $(FL)$ divided by $(F/L^2)$, which is work or energy per unit area. Surface tension therefore is synonymous with surface energy, the energy required to generate a surface. Reducing surface tension therefore reduces surface energy.

### 11.3.6 Soaping Surface Tension

The manipulation of surface tension is at the heart of the soap and detergent industries, which use organic compounds whose purpose is to lower the surface tension between the unclean and that which is to be cleaned. Although the amount of energy required to make new surfaces is reduced, energy still is required, whether by use of a stirring mechanism in a washer or pounding clothes on a rock.

Organic molecules in detergents are somewhat hydrophobic or water-hating so they concentrate in the water surface. There they interrupt and weaken water-to-water molecular attractions, making it easier to generate suds or foam. Only a tiny amount of organic chemical can produce this effect, so foam is an early indication of organic pollutants in a lake or river.

### 11.3.7 Measuring Surface Tension

A simple device that illustrates measurement of surface tension is shown in Fig. 11.3, where the pulling force is $2LT$, where $L$ is the length of the expanding surface and $T$ is surface tension in units of force per unit length. The force is $\times 2$ because two surfaces are involved.

Another way to measure surface tension is from the height of rise in a capillary tube, as in Fig. 11.1, if the inside diameter of the tube is known. The height of rise is obtained by equating the surface tension at the circumference with the weight of the water column in the tube:

$$2\pi r T \cos\alpha = \pi r^2 h \rho g \tag{11.3}$$

where $r$ is the radius of a circular capillary tube in units that are consistent with units of $T$,

> $T$ is the surface tension of the liquid, which can be in millinewtons per meter,
>
> $\alpha$ is the contact angle between the meniscus surface and the wall of the tube,
>
> $h$ is the height of rise of the liquid in meters,
>
> $\rho$ is the density of the liquid, in grams per cubic meter, and
>
> $g$ is the acceleration of gravity ($9.81\,\mathrm{m/s^2}$).

**Figure 11.3**

Measuring surface tension.

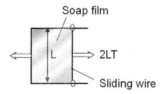

Soap film

2LT

L

Sliding wire

Solution of eq. (11.3) for $h$ yields

$$h = \frac{2T}{r\rho g}\cos\alpha \tag{11.4}$$

With perfect wetting, $\alpha = 0°$ so $\cos\alpha = 1.0$. Dimensions in the above equation will be seen to be consistent since by definition $1\,\text{N} = 1\,\text{kg-m/s}^2$.

### Example 11.3
Predict the height of rise of water with surface tension $72\,\text{mN/m}$, in a clean capillary tube $0.1\,\text{mm}$ in diameter.

*Answer:* From eq. (11.4),

$$h = \frac{2\times 72\,\text{mN/m}}{0.5(10)^{-4}\,\text{m} \times (10)^6\,\text{g/m}^3 \times 9.81\,\text{m/s}^2}\cos 0$$
$$= 0.294\,\text{m} = 294\,\text{mm}$$

In solving an equation such as this it is essential that units be included to avoid decimal error.

According to eq. (11.2) the height of capillary rise is inverse to capillary diameter, so as an approximation

$$d_e = \frac{0.03}{h} \tag{11.5}$$

where $d_e$ is defined as a *capillary equivalent pore diameter* in mm, and $h$ is in meters.

### Example 11.4
What is the capillary equivalent pore diameter in a soil where the height of capillary saturation above the groundwater table is $2\,\text{m}$?

*Answer:* Soil pores obviously are very irregular and do not have a uniform diameter. According to eq. (11.5) the equivalent diameter is $d_e = 0.03/2 = 0.015\,\text{mm}$ or $15\,\mu\text{m}$, which is silt-size.

## 11.3.8  The Vadose Zone

Because the height of the zone of capillary saturation is somewhat erratic, being dictated by the maximum pore size, this zone also is called the *closed capillary fringe*. Additional capillary water is held above that zone in an *open capillary fringe* that is held by smaller pores and can extend to the ground surface. Open and closed capillary fringes are illustrated in Fig. 11.4. The unsaturated zone of soil including the open capillary fringe is called the *vadose zone* (from the Latin *vadum* for shallow water, as in "wade").

Negative pore pressure in the closed capillary fringe is readily calculated from the distance above the groundwater table, but is less easily found in the open zone

**Figure 11.4**

Zones of capillary
water and the
vadose zone.

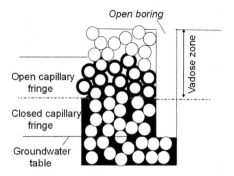

because part of the capillary cross-section is occupied by air. A correction
therefore must be made to the effective stress equation to account for this effect,
as discussed later in this chapter.

### Example 11.5

Soil at a depth of 3.05 ft (1 m) is in a zone of capillary saturation that extends 2.02 ft
(0.67 m) above the groundwater table. The unit weight of the soil is $125\,\text{lb/ft}^3$ ($19.6\,\text{kN/m}^3$).
(a) What are the total and effective stresses at this depth? (b) Is the soil stronger or weaker
than that at the level of the groundwater table?

*Answer:* Total stress is $3.05 \times 125 = 381\,\text{lb/ft}^2$ ($1 \times 19.6 = 19.6\,\text{kPa}$). Pore water pressure is
$-2.02 \times 62.4 = -126\,\text{lb/ft}^2$ ($-0.67 \times 9.81 = -6.6\,\text{kPa}$).

The effective stress is $\sigma' = \sigma - u = 381 - (-126) = \underline{507\,\text{lb/ft}^2}$ ($19.6 - \{-6.6\} = \underline{10.0\,\text{kPa}}$).
Because of the higher normal stress and intergranular friction the soil is stronger in the
capillary zone.

## 11.4  POSITIVE AND NEGATIVE PRESSURES FROM SURFACE TENSION

### 11.4.1  Pressure Inside a Soap Bubble

Calculating the pressure inside a soap bubble may seem a bit whimsical, but even
whimsy can have purpose. Furthermore the concept and calculation are so
simple, and the result is sure to enliven a flaccid conversation. Does the pressure
required to make a bubble increase, decrease, or stay the same as the bubble grows
larger?

From Fig. 11.5 the internal pressure over a diametric area equals 2 times the
surface tension around the circumference because both an inner and an outer
surface are involved:

$$\pi r^2 p = 2T\pi \times 2r \tag{11.6}$$

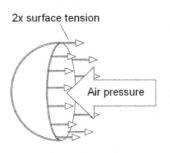

**Figure 11.5**
Force balance inside a soap bubble.

2x surface tension

Air pressure

where $r$ is the radius of the bubble, $p$ is the pressure inside the bubble, and $T$ is surface tension at a film-air interface. Hence

$$p = \frac{T}{r} \tag{11.7}$$

The internal pressure therefore decreases as the bubble grows larger. Soap lowers the surface tension and surface energy.

### 11.4.2 Capillary Rise and Surface Tension

The influence of surface tension on the height of rise in a cylindrical capillary can be solved by assuming that the half-bubble shown in Fig. 11.5 is rotated 90° to represent the meniscus in the tube. In this case only one surface is involved and $u$ can represent negative pressure in water immediately under the capillary surface:

$$u = \frac{-2T}{r} \tag{11.8}$$

This may be equated to the weight of the column of suspended water:

$$u = \frac{-2T}{h} = -\pi r^2 h \rho$$

from which

$$h = \frac{2T}{\pi r^2 \rho} \tag{11.9}$$

where $h$ is the height of capillary rise, $T$ is surface tension, $r$ is the radius of the capillary, and $\rho$ is the density of the fluid. The smaller the capillary, the higher the height of capillary rise.

### 11.4.3 Negative Pressure in Capillary Water Linking Two Soil Particles

Capillary water in unsaturated soil can take a doughnut or annular shape around particle contacts, as illustrated in Fig. 11.6. This problem is somewhat more complicated because the water-air interface has two radii, $r_1$ and $r_2$, one of which is convex and the other concave. The concave radius acts to reduce pressure in the

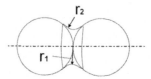

**Figure 11.6**

Annular ring of water linking two soil particles. The main attractive force is not from surface tension but from the negative pore pressure created by the surface tension.

water, whereas the convex radius tends to increase it. As the concave radius is smaller, the net effect is a decrease in pressure that pulls the grains together.

The equation for negative pressure in this case uses the Laplace function to define the surface:

$$u = -T\left[\frac{1}{r_1} + \frac{1}{r_2}\right] \tag{11.10}$$

in which $r_1$ and $r_2$ are radii of curvature of a warped surface where it intersects two orthogonal principal planes. The expression in the parentheses is called the total curvature of the surface. A negative radius creates a positive pore pressure, so in Fig.11.5, $r_1$, the internal radius of the ring of water, increases pressure in the water and therefore is negative.

The negative pore pressure in capillary water therefore equals the surface tension of the water times the *total curvature* of the meniscus or air-water interface. This relationship is important because, as the moisture content changes, both radii of curvature also change.

The influence of total curvature is illustrated in Fig. 11.7, where it will be seen that, if the moisture content increases such that the outer surface of the annular ring of connecting water becomes straight-sided (Fig. 11.7(a)), $r_1$ is infinite so its reciprocal makes no negative contribution to pore water pressure, which therefore is positive.

When a soil is dried (Fig. 11.7(b)), the smaller radius is positive so pore pressure becomes negative. Opposing this tendency is the cross-sectional area of the ring of water as the soil dries out (Fig. 11.7(c)), which eventually will result in a net decrease in attraction between the two particles. Finally, $r_1$ can become so small that the negative pressure equals the vapor pressure of water, causing the soil to spontaneously dry out. This does not occur in clay because its adsorptive properties permit a high negative pressure. When sand dries it crumbles whereas a clay keeps getting stronger.

(a)    (b)    (c)

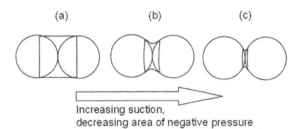

Increasing suction,
decreasing area of negative pressure

**Figure 11.7**
Influence of drying
on capillary forces
in soils: more
suction but over
a smaller area.

This relationship was studied in the early 1920s by Haines and Fisher, two soil scientists at the U.S. Bureau of Soils. A relationship for the negative pressure inside the water is:

$$u = \frac{T(3r_1 - 2a)}{r_1^2} \tag{11.11}$$

where $r_1$ is as shown in Fig. 11.6 and $a$ is the radius of the spheres. If $a$ is unity, $r_1 = 0.67$ for $u = 0$, which is between (a) and (b) in Fig. 11.7. The compressive force pulling the particles together is the sum of the surface tension and negative pressure times the cross-sectional area of the water ring, which can be shown to be (Kirkham and Powers, 1972):

$$f = \pi T(2a - r_1) \tag{11.12}$$

The total force per unit area is the sum of all such forces between particles. If $N$ is the number of particles per unit area, with cubic packing, $N = 1/4a^2$. Then

$$F = N \times f = f/4a^2$$

$$F = \pi T \left[ \frac{1}{2a} - \frac{r_1}{4a^2} \right] \tag{11.13}$$

This formula indicates that the smaller the particle size, the larger is the cohesive force per unit area resulting from capillary tension. The maximum force will be when $r_1$ is very small, that is, as the soil becomes drier so long as the moisture films remain intact.

## 11.5 SOIL MECHANICS OF UNSATURATED SOILS

### 11.5.1 A Shift in Emphasis

Much of the emphasis in geotechnical engineering has been on saturated soils, as this normally is the most critical condition when strength is lowest and unit weight is at a maximum, or is reduced to a minimum by submergence. However, many soils are not and may never become saturated. An important example is soil that has been mechanically compacted for support of a road or foundation, because compaction is halted when most *but not all* of the air has been squeezed out. A compacted soil therefore is not a saturated soil, as some air remains trapped even if the soil is submerged.

Unsaturated conditions dominate in soils such as dune sand and coarse loessial silt that are so permeable that they do not support a groundwater table. Surficial zones in most soils are not saturated, and an unsaturated zone can extend very deep in deserts and semiarid areas where there is little rainfall to replenish the groundwater supply.

As previously mentioned, a correction is required to the Terzaghi effective stress (eq. (11.2)) in unsaturated soils because pore water does not occupy the complete cross-sectional area of the soil pores. A simplified version of a correction originally proposed by Bishop is as follows (Fredlund and Rahardjo, 1993):

$$\sigma' = \sigma - \chi u \tag{11.14}$$

where $\chi$ is a number between zero and one, and the other variables are as defined in eq. (11.2).

When a soil is saturated, $\chi = 1$, and eq. (11.14) reduces to eq. (11.2). When a soil is completely dry, $\chi = 0$ and total stress equals effective stress, as obviously there is no influence from pore water pressure if there is no pore water. However, the relationship to percent saturation is not linear, and has been found to depend on soil type and even on the direction of the applied stress according to the void orientation and particulate packing arrangement.

The increase in cohesive strength on drying was shown by a classic experiment by a co-worker of Atterberg, whose contributions are presented in the next chapter. In 1914 in Germany, Johansson dried soil briquettes and measured the force necessary to split them with a wedge. His results are shown in Fig. 11.8, where strength steadily increases down to a moisture content of 14 percent at point A, where there is a break in the curve. At that point the soil suddenly became lighter in color, indicating the entrance of air into the pores and supporting the introduction of a $\chi$ factor as in eq. (11.14).

The complete Bishop equation includes pore air pressure, which under most field conditions is negligible. However, this inclusion is necessary for laboratory testing, where, in order to apply an external pressure to soil specimens, they are sealed in a rubber membrane such that included air cannot escape and also is compressed.

## 11.5.2  Pressure, Head, and Potential

In fluid mechanics pressure is one component of total head, which is the sum of pressure head, elevation head, and velocity head. Because fluid flow in soils is relatively slow, velocity head is negligible.

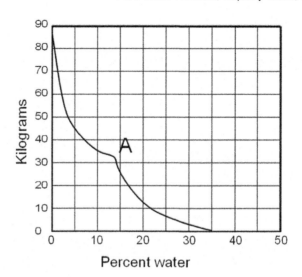

**Figure 11.8**
Change in penetration resistance of soil from the presence of air, from data of Johansson.

Pressure is a driving force analogous to voltage in an electrical system, and therefore is a *potential*. A capillary driving force was defined in the early 1900s as a positive quantity called "capillary potential," but as capillary pressures are negative relative to atmospheric pressure, the preferred term now is *matric potential*.

*Osmotic potential* refers to the attractive forces from salts and by clays, which can be many negative atmospheres.

*Total suction* is the sum of matric and osmotic potentials.

Matric and osmotic potential are used in the algebraic sense, that is, the potential increases when it becomes less negative. Moisture flow is toward drier soil, other factors being the same, because the potential is low, that is, more negative. In a wet soil the matric potential is high, and in a saturated soil or a soil below the water table the matric potential is zero, this being the highest possible value.

In thermodynamic terms, potential represents the free energy of soil water, or the amount of work required to remove a unit mass of water.

## 11.5.3 Units of Potential

Potential may be expressed in any units that are appropriate for indicating pressure, such as kilopascals, kilonewtons per square meter, bars, or pounds per square inch. Also, it may be convenient in certain cases to express the potential in terms of the height of a column of liquid such as meters or feet of water. Conversions are:

$$1\,\text{bar} = 100\,\text{kPa} = 100\,\text{kN/m}^2 = 14.5\,\text{lb/in.}^2 = 1.02\,\text{kg/cm}^2$$
$$= 750\,\text{torr} = 750\,\text{mm Hg} = 29.5\,\text{in.Hg} = 10.2\,\text{m}\,H_2O = 33.5\,\text{ft}\,H_2O$$
$$= 10\,\text{dyne/cm}^2 = 0.987\,\text{atmosphere}$$

Another term, $pF$, is no longer widely used, and is the logarithm of the height of capillary water in centimeters. The $p$ is analogous to pH and $F$ signifies free energy.

### Example 11.6

The zone of capillary saturation in a soil under equilibrium conditions extends 1 m above the water table. What is the indicated matric potential in meters of water?

*Answer:* 1 m.

## 11.5.4 Movement of Capillary Water

When capillary potential is in balance with gravitational potential, the soil moisture is in a state of static equilibrium so there is no upward or downward flow. Ordinarily, the matric potential is more negative in dry soils than in wet ones. However, this is not necessarily true for soils that are texturally different, as moisture may flow from a relatively dry sandy soil into a wetter clay if that is dictated by a potential gradient. In other words, the criterion for unsaturated moisture flow from one soil region to another is the relative potential of the two regions, and not their respective moisture contents.

In addition, moisture in a field soil seldom reaches a state of equilibrium because of the relatively slow rate at which capillary water moves in the soil compared with changing weather conditions. When a capillary tube is inserted into a vessel of water, the liquid rises in the tube almost immediately, but in the soil, considerable time is required for a similar movement to take place. Consequently, when rain falls on the surface, that portion of the precipitation that soaks into the soil causes a wave of wetness that migrates slowly downward. Before an equilibrium distribution of moisture is established, however, evaporation may dry soil near the surface so that the matric potential is reduced and moisture starts to move back up. Alternating rains and periods of evaporation, plus the effects of transpiration and variations in barometric pressure and soil temperature, keep capillary water in motion.

## 11.5.5 Moisture Contents Stabilized by an Impervious Cover

By preventing normal evaporation and transpiration, a building or pavement is conducive to establishing equilibrium with the water table below, and as a result, capillary moisture tends to accumulate in a pavement subgrade after the pavement is laid. Measurements indicate that such accumulation occurs slowly and that 3 to 5 years may elapse before the maximum accumulation is reached. The pavement or building also serves as an insulator that reduces the range and rate of change of temperature in the underlying soil. Even subgrades under pavements in arid regions have been known to become very wet, and the soil may lose a substantial part of its bearing capacity by accumulation of capillary moisture from a water table below. These factors point to the relevance of capillary characteristics of soils under buildings and pavements.

## 11.5.6 Capillary Siphoning

A siphon normally is thought of as a tube filled with fluid and connecting a higher with a lower fluid level so that the weight of the fluid in the lower part of the siphon causes it to flow out and draw more fluid in at the top. If a host of tiny capillaries is substituted for the tube, they also can act as a siphon, but the action is relatively slow.

Capillary siphoning can cause an unexpected lowering of the water level behind an earth dam if there is a continuous soil layer running over the top of the impervious core. The core material therefore is carried to the ground surface even though it is substantially higher than the highest water level behind the dam.

## 11.5.7 Effect of Temperature

The surface tension of water is inversely related to temperature, so cooling increases the attraction of soil for capillary water. Thus, cooling during fall and winter tends to draw more moisture into soil subgrades under pavements, whereas summer heating reduces the soil affinity for water. Emergency water supplies in deserts may be collected by excavating a shallow hole and covering it with a plastic film so that water moving upward toward the cooler night temperatures condenses on the underside of the film. The film is depressed by a weight in the middle to make a drip point into a cup placed in the hole.

## 11.5.8 Ground Freezing

A sudden and drastic influence from temperature occurs when water freezes, as the surface tension (surface energy) of ice is very high and the vapor pressure very low. Freezing temperatures therefore attract both capillary water and water vapor that migrate to a freezing zone to form ice crystals and lenses. These generally are oriented horizontally and lift the overlying soil due to *frost heave*. Water is replenished by upward movement of water from the groundwater table.

The amount of frost heave also depends on the capillary conductivity and matric potential of the soil, which in turn relate to its gradation. This relationship is discussed in the next chapter, which is on the engineering classification of soils. The most troubling soils for frost heave are those that combine permeability and capillary potential, which are silts containing small amounts of clay.

### 11.5.9  Effect of Dissolved Salts and Chemicals

An increase in the amount of dissolved salts in the soil water slightly increases its surface tension and thereby lowers the matric potential of the soil, but the effect is relatively minor. On the other hand a chemical additive that changes the wetting angle, such as a bituminous compound, can halt capillarity. In road embankments this also can be achieved by incorporating a layer of impermeable plastic, but that has the disadvantage of also preventing infiltration of rainwater.

### 11.5.10  Effect of Clay Mineralogy

Because of their high affinity for water and the associated diffuse double layer of ions and water molecules, active clay minerals such as smectite can profoundly influence matric potential. As the water content is decreased by drying, adsorptive forces increase, and the last water adsorbed is held by an osmotic potential of thousands of bars.

## 11.6  SORPTION CURVES

### 11.6.1  A Simple Model that Illustrates Capillarity in Soil

A simple and direct way to determine the equilibrium water content in soil in the capillary fringe or vadose zone is to fill a tube with soil and suspend the bottom in water. After a sufficient amount of time has elapsed the soil can be extruded, cut into sections, and the moisture content of each section measured and, if appropriate, its strength determined.

A typical result is shown in Fig. 11.9. The height above the free water surface represents the matric potential. The curves for till and loess are based on more precise measurements of matric potential, discussed below. As can be seen, soil layers tend to reach equilibrium moisture contents that are characteristic of each layer, and this system can model soil layers under a pavement. These experiments also demonstrate the difference in capillary conductivity of the two soils, since in the experiment shown in the figure it took approximately three times as long for the wetting front to rise in the glacial till as in the loess. Note also that the loess has a substantially higher capacity for capillary water, even though it contains less clay than the till.

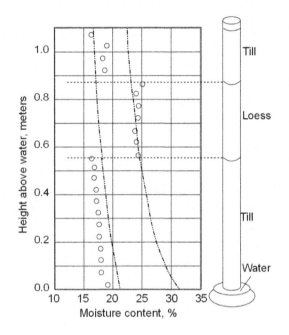

**Figure 11.9**
Moisture contents
sucked up by
different soil layers
in a tube with the
bottom in water to
simulate a
groundwater table.
Dashed lines are
from adsorption
measurements for
the two soils.

## 11.6.2 Hysteresis

If a sorption experiment such as that outlined above is performed from the wet side down instead of the dry side up, different curves are obtained because of the lag effect or *hysteresis*. A simple model for capillary hysteresis is called the "ink-bottle effect" and is illustrated in Fig. 11.10. In the left part of the figure, water rising into a capillary stops where the capillary is enlarged at the "bottle." However, if the bottle already is filled, as shown at the right, capillary attraction allows it to retain water. The ink-bottle therefore is reluctant to take in water, but one that is full also is reluctant to give it up.

The ink-bottle analogy also applies to expansive clay mineral particles as edges of particles pinch together during drying so that the interior tends to retain water, and then act to slow down re-entry of water during re-wetting.

## 11.6.3 Adsorption and Desorption Curves

Capillary hysteresis also becomes apparent by comparing the sorption curve obtained during draining, called a *desorption curve*, with that obtained during wetting, called an *adsorption curve*. Examples are shown in Fig. 11.11, where it also will be seen that upon repeating the experiment the second cycle does not exactly track the first. This is reasonable since the first adsorption-desorption

**Figure 11.10**

Ink-bottle
hypothesis to
explain capillary
hysteresis.

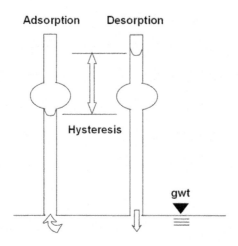

**Figure 11.11**

Adsorption and
desorption curves
for a loessial soil
in the Tama soil
series, after
Richards (1941).

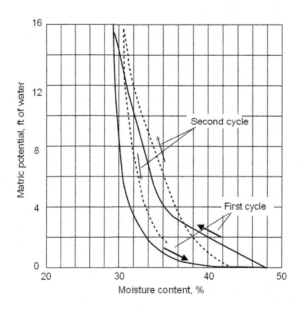

cycle can be expected to leave some of the "ink-bottles" full of water so that the
second cycle tends to retain higher moisture contents. Another factor may be
modification of the voids themselves, particularly as this soil contains expansive
clay mineral.

As a practical matter, water that is drawn up into soil underneath a pavement as
the result of a temperature gradient or freezing will tend to be retained according
to the desorption instead of the adsorption part of the cycle, resulting in higher
retained moisture contents.

## 11.7 MEASURING MATRIC POTENTIAL

### 11.7.1 Filter Paper Method

A simple procedure that is used in soil science and demonstrates the principles might be called the "limp potato chip" method, except that instead of potato chips it uses filter paper. A standardized filter paper is selected and must be calibrated by wetting and then re-weighed after equilibrating at different values of suction. These are provided by glass desiccators containing selected salt solutions.

A test is performed by placing three layers of oven-dried filter paper in contact with about 200 g of soil in a container that is sealed and allowed to equilibrate. The outer layers of paper protect the middle layer from becoming soiled, and it is the middle layer that is used in the determination. The filter paper is removed and weighed, then oven-dried and re-weighed. Then by reference to the filter paper calibration curve the matric potential is obtained at that particular soil moisture content. The soil also is weighed, oven-dried, and re-weighed to obtain its moisture content.

If the filter paper contacts the soil, the result represents combined matric suction and osmotic suction, the latter related to the dissolved salts and clay mineral surface activity.

### 11.7.2 Principle of a Tensiometer

A more rapid method for measuring matric potential of a soil either in the laboratory or in the field is with a *tensiometer*. The essential features are shown in Fig. 11.12. A porous ceramic cup is sealed to a glass or plastic tube that is connected to a vacuum gauge. At the top of the tube is a cap with an O-ring seal.

The tensiometer must first be prepared by soaking the ceramic cup in water that has been recently de-aired by boiling. Then a vacuum is pulled at the top to draw water up into the tube and de-air the gauge. The top is sealed off and the system tested by allowing water to evaporate from the surface of the ceramic, which should create a negative pressure that can be read on the gauge. The cup then is re-immersed in de-aired water and the gauge pressure should return to zero in one or two minutes. If it does not, the de-airing procedure is repeated and the gauge may have to be re-zeroed.

The saturated tensiometer cup then is imbedded into close contact with the soil that is to be measured. Unsaturated soil sucks water through the porous wall of the cup to create a negative pressure reading on the gauge. The end point, or the point of zero potential, will always be at the saturation moisture content of the soil. Several instruments based on this principle are described by Fredlund and Rahardjo (1993).

A limitation of a tensiometer is that tension cannot exceed 1 atmosphere because water vapor bubbles will form in the porous cup so that the instrument is no longer operative. Nevertheless, this device offers many possibilities for studying soil-moisture changes within its range of applicability.

### 11.7.3 Zero-Point Translation Devices

In order to avoid cavitation of water in a tensiometer, the entire system can be put under pressure. The most common method for doing this is to place saturated soil specimens on a porous ceramic plate that is saturated with water and has sufficient capillary retention that it does not allow entry of air (Fig. 11.13). Then air pressure is applied to push water out of the soil and into the plate. Equilibrium will be reached at a particular differential pressure that represents the matric

**Figure 11.12**

Schematic diagram of a soil tensiometer to measure capillary suction or matric potential.

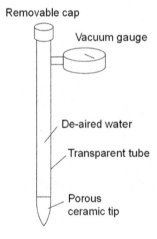

**Figure 11.13**

Pressure-plate apparatus for measuring negative pore water pressures >1 atmosphere.

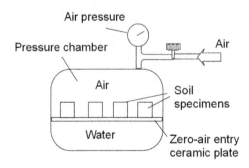

potential. The system then is depressurized and the moisture content of the soil determined. By performing the test with different values of differential pressure, one obtains a desorption curve that gives moisture-tension relationships in the approximate range 1 to 15 bars.

A modification of this system by Hilf of the U.S. Bureau of Reclamation uses a probe with a ceramic tip that is inserted into a soil specimen that is under external air pressure. In order to minimize water transfer, a null-type measuring system is used and connected to a vacuum gauge. As water is sucked out of the ceramic tip the gauge becomes negative, so to prevent cavitation the external air pressure is increased to a null point where water does not move into or out of the soil specimen. That pressure is the matric potential.

# 11.8  FROST HEAVE

## 11.8.1  Where to Look

There probably is no more dramatic illustration of the effects of movement of capillary water than frost heave, where the ground surface may spontaneously rise many inches or tens of millimeters in a single winter season. Frost heave normally is limited to northern climates where freezing occurs, but also can be a serious problem under cold-storage lockers, where installations often become unusable in a matter of one or two decades.

It once was assumed that the frost damage was caused by expansion of water upon freezing, but this 9 percent volume change is not sufficient to explain heaves measured in fractions of a meter. A close examination of the soil involved in frost heave will reveal a series of layers of pure ice separating soil layers, as shown in Fig. 11.14. The cumulative thickness of the ice layers represents the total amount of frost heave, which can lift and crack building floors and foundations and pavements, and prevent doors from opening.

## 11.8.2  Spring Breakup

Frost heave is only part of the problem because when the ice layers and lenses melt, the soil becomes overly saturated with water and can turn into mud. In unpaved roads such areas are called "frost boils." They also occur under paved roads where they become evident when a truck breaks through. It is for this reason that a road may be embargoed for heavy loads, until the ground has thawed and the excess water has drained out.

Thawing from the ground surface downward aggravates an already serious problem by preventing excess water from draining downward. Thawing that is hastened by rain can speed drying by allowing drainage downward.

**Figure 11.14**

Two cores showing ice lenses that lifted a seven-story building in Canada. (From Penner and Crawford, 1983, National Research Council of Canada.)

### 11.8.3 Three Requirements for Frost Heave

Obviously frost heave requires cold temperatures. A second requirement is the availability of unfrozen water, usually from a shallow groundwater table. A third requirement is for the soil to be sufficiently permeable to transmit a damaging amount of water during a single winter season.

Frost heave results from a potential gradient that draws water into the freezing zone. If the freezing front remains stationary, water will continue to be drawn up and freeze to make a layer of relatively pure ice. If the weather turns colder so that freezing penetrates downward, it encapsulates the ice layer and prevents further growth. Slow freezing is conducive to frost heave.

Soil and water conditions required for frost heave are as follows:

- The soil must have sufficient capillary tension to draw water into the freezing zone, which means that sands do not exhibit very much frost heave.
- The soil must be sufficiently permeable that a significant amount of water can be transferred during a winter season, which means that dense clays do not exhibit very much frost heave.
- There must be an available supply of water that normally will come from a groundwater table.

## 11.8.4 Control of Frost Heave

Frost heave can be reduced or prevented by interfering with any one of the required conditions. For example, freezing of foundation soils is prevented by placing the foundation below the maximum depth of annual freezing or by keeping a building heated during winter. Pavement can be laid on top of an insulating layer such as a few inches of Styrofoam®.

Because so many different factors contribute to frost heave it is not always predictable, and a road that suffers damage one year may not be damaged the next. The reason may be lowering of the groundwater table, or perhaps rapid freezing that does not give sufficient time for moisture to move to a rapidly progressing freezing front. Snow is an insulator that in theory can either contribute to frost heave by slowing freezing, or prevent frost heave by preventing ground freezing, depending on weather conditions.

## 11.8.5 Diary of a Frost Heave

As the frost line, or line separating frozen from unfrozen soil, penetrates downward, water in the soil pores freezes, which from the standpoint of matric potential dries the soil out. Liquid water therefore moves to the freezing front. Some water may also move as vapor, which creates a loose array of ice crystals inside voids called "frazzle ice," but this does not make a major contribution to heave.

As water reaching the frost line becomes supercooled, freezing is triggered by ice crystals already in the soil so that the crystals grow. The tenacity with which the molecules are held causes a rejection of foreign matter, so the ice is relatively clear. Heave-producing ice layers therefore grow along the frost line as long as they are fed from below by capillary water. The frost line and ice layers normally are parallel to the ground surface but can be inclined or vertical behind a wall.

As the weather becomes colder and freezing proceeds faster than capillary conductivity can supply water to the freezing front, freezing penetrates deeper and encapsulates the ice layer in frozen soil. If freezing slows or pauses, conditions are reinstated for production of a new ice layer. This cycle can continue and build layer upon layer of ice, as shown in Fig. 11.14. The total frost heave therefore is the sum of all ice layer thicknesses (Fig. 11.15).

With the onset of warmer weather, thawing for the most part takes place from the ground surface down, creating temporary boggy conditions until thawing progresses all of the way through the frozen soil layer so that excess water from thawing of the ice layers can drain downward. In areas of frost heave, roads often are protected during seasonal thawing by embargoing heavy truck traffic.

**Figure 11.15**

Frost heave causing muddy conditions when the ice melts.

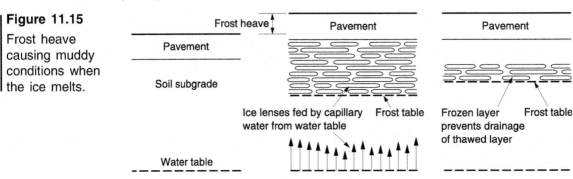

(a)  (b)  (c)

## 11.9  DEPTH OF FREEZING

### 11.9.1  Locating Foundations below the Depth of Freezing

The most obvious way to prevent damage from frost heave is to found structures below the maximum depth of freezing. A map showing maximum depths of annual freezing that may be expected in the U.S. is shown in Fig. 11.16.

Freezing depths are determined not only by air temperatures but also by thermal properties of soils and their cover. For example, a saturated soil will not freeze as deeply under a given set of conditions as an unsaturated soil because of the heat transfer required to freeze water. Also, soil under a thick layer of vegetation or under a snow blanket will not freeze as deeply as if the soil were exposed to the elements.

### 11.9.2  Weather Records

An evaluation of freezing conditions can be obtained from weather records. Maximum and minimum daily temperatures are averaged and subtracted from the nominal freezing temperature of water to obtain *degree-days below freezing*. The same general procedure is used for furnace and air conditioner design, only the reference temperature is the comfort zone for people and/or equipment. For example, if the temperature range for a particular day is from $-5°C$ to $-15°C$, the average is $-10°C$, or 10 degree-days below freezing.

### 11.9.3  The Stefan Equation

Freezing and thawing depths during a single winter season can be used to estimate the extent of freeze-thaw cycling, which is an important factor affecting

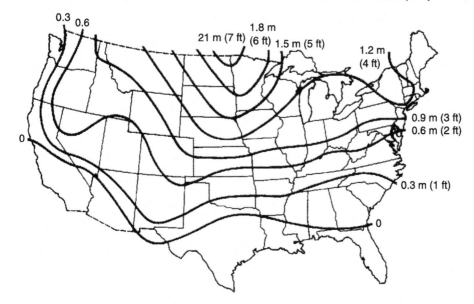

**Figure 11.16**
Maximum frost depths for design purposes.

deterioration of pavement materials. The Stefan equation is a simplification as it is based on the heat capacity of water in the soil and therefore overpredicts freezing depths in dry soils. Also, it does not take into account additional water drawn into the soil by frost heave, which can have a substantial effect on the rate of thawing. The equation is

$$Z = C_t \sqrt{\frac{L_t F}{L}} \qquad (11.15)$$

where $Z$ = depth of freezing, in meters (ft).

$C_t$ = a constant. In SI, $C_t$ = 13.1; in the English system, 6.93.

$k_t$ = thermal conductivity coefficient of frozen soil, W/m · K (watts per meter-Kelvin) or Btu/ft$^2$-hr-°F per ft.

$F$ = degree days below 0°C or 32°F.

$L$ = latent heat of fusion of water in soil, kJ/m$^3$ (kilojoules per cubic meter) or Btu/ft$^3$.

$L$ for water is 333.7 kJ/or 142.4 Btu/lb. $L$ for soil water therefore is

$$L = 340\, w\gamma_d \quad \text{or}$$
$$L = 1.43\, w\gamma_d \qquad (11.16)$$

where $w$ is the percent moisture and $\gamma_d$ is the dry unit weight of the soil, in kN/m$^3$ or lb/ft$^3$, respectively.

**Table 11.1**

Thermal conductivity
(from charts of
Kersten, 1952)

| Dry density | | Moisture content, % | $k_t$ (frozen soil) | | $k_u$ (unfrozen soil) | |
|---|---|---|---|---|---|---|
| kN/m³ | lb/ft³ | | W/m·K | Btu/ft²-hr/° F/ft | W/m·K | Btu/ft²-hr/°F/ft |
| Sandy soils: | | | | | | |
| 20.4 | 130 | 5 | 2.3 | 1.3 | 2.6 | 1.5 |
| | | 10 | 4.2 | 2.4 | 3.1 | 1.8 |
| 18.9 | 120 | 5 | 1.7 | 1.0 | 2.0 | 1.2 |
| | | 15 | 4.0 | 2.3 | 2.8 | 1.6 |
| Soils with >50% silt and clay: | | | | | | |
| 18.9 | 120 | 10 | 1.7 | 1.0 | 1.6 | 0.9 |
| 17.3 | 110 | 10 | 1.3 | 0.8 | 1.3 | 0.7 |
| | | 18 | 2.1 | 1.2 | 1.7 | 1.0 |
| 15.7 | 100 | 10 | 1.0 | 0.6 | 1.0 | 0.6 |
| | | 20 | 1.8 | 1.0 | 1.4 | 0.8 |
| 14.1 | 90 | 10 | 0.8 | 0.5 | 0.8 | 0.5 |
| | | 30 | 2.1 | 1.2 | 1.3 | 0.7 |

Selected thermial conversion factors:

| | |
|---|---|
| 1 Btu | = 1.055 kJ |
| 1 Btu/ft²-hr-°F/ft | = 1.730 W/m·K (watts per meter-kelvin) |
| 1 cal/cm·s·K | = 418.4 W/m·K |
| 1 Btu/ft² | = 37.26 kJ/m³ (kilojoules per cubic meter) |
| 1 cal/cm³ | = 4187 kJ/m³ |
| 1 Btu/lb | = 2.326 kJ/kg = 22.81 kJ/kN |
| 1 cal/g | = 4.1868 kJ/kg |
| 1 W (watt) | = 1 J/s |
| 1 J | = 1 N m |

Some representative values for $k_t$ are given in Table 11.1. Calculations for depth of thawing use $k_u$, which is lower than $k_t$ because the heat conductivity of liquid water is less than that of ice.

**Example 11.7**

The average temperature for a 10-day period is $-5°C$. Calculate the depth of freezing in a sandy soil with $\gamma_d = 18.9$ kN/m³ (120 lb/ft³) and a moisture content of 12%.

*Answer:* Interpolating in Table 11.1 gives $k_t = 3.3$ W/mK. $L = 340 \times 12 \times 18.9 = 77,000$ kJ/m³. From eq. (11.15),

$$Z = \sqrt[13.1]{\frac{3.3 \times 10 \times 5}{77,000}} = 0.6 \, m \, (24 \, in.)$$

(b) How many degree-days above 0°C will be required for complete thawing, assuming no change in moisture content and no direct heat from the sun?

*Answer:* From Table 11.1, $k_u = 2.6$ W/m·$K$, so

$$0.6 = {}^{13.1}\sqrt{\frac{2.6 \times F}{77,000}}, \quad \text{or} \quad F = 62 \text{ degree-days C.}$$

Because of the difference between $k_t$ and $k_u$, in this case it will require 62 degree-days of thawing to compensate for 50 degree-days of freezing. As previously mentioned, the calculations do not take into account additional water drawn into the soil by frost heave, which mainly will affect the rate of thawing.

An application of the Stefan equation is illustrated in Fig. 11.17 and shows several periods when thawed soil was on top of frozen soil that would prevent downward drainage. Also, by simply counting the thawing and refreezing periods at different depths it will be found that the number of freeze-thaw cycles is a maximum at the ground surface and decreases with depth, which is an important consideration for pavement design in freezing climates.

---

*Question:* In Fig. 11.17, during what months might it be expected that most of the frost heave would occur?

*Answer:* November and December. Why?

---

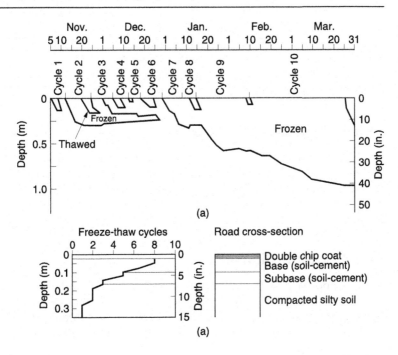

**Figure 11.17**

Freezing depths calculated using the Stefan equation indicate 8 cycles in the road base course during one winter season. (After K. P. George.)

### 11.9.4  Insulating Against Freezing

The most obvious control measure against frost heave is to prevent freezing. Foam plastic, usually high-density polystyrene called "Geofoam," has a thermal conductivity of about 0.026 W/K·m, which is about one-hundredth that of soil. However, after absorbing water the thermal effectiveness is estimated to be reduced to about 30 to 40 times that of soil. In the 1980s a 51 mm (2 in.) thick layer was successfully used to prevent frost heave on I-70 in Colorado where the freezing depth was 1.8 m (6 ft).

The Stefan equation is not applicable to predict foam thickness because $L$ for the foam depends on a variable and unknown content of water. However, as a rough guide the required thickness can be taken as 1/30 to 1/40 times the normal depth of freezing. In the example cited above the thickness would have been 45 to 60 mm (1.8 to 2.4 in.), which closely agrees with the case history.

Adding salt or calcium chloride to soil also can reduce or prevent frost heave by lowering the freezing point, but the effect may last only one or two seasons until the salts leach out and become pollutants.

### 11.9.5  Drainage

The most common methods used to control or reduce frost heave in roadbeds is to raise the grade by scooping out ditches, and provide a drainable subgrade soil. As a general guide, the distance from the bottom of a pavement to the highest probable elevation of the water table should not be less than 2 m (6 ft), and a larger distance is desirable if it can be obtained at a reasonable cost. Increasing the distance to a water table decreases the potential gradient and the rate of upward movement of capillary water. Another advantage of a high-crowned road grade is that snow blows off.

### 11.9.6  Granular Subbase

Another approach is to replace frost-susceptible soil with a coarse-grained soil that has a low capillary attraction for water. Ideally the subbase should extend to the maximum freezing depth, but as a practical matter this usually is not feasible. Therefore a granular subbase often is combined with drainage by ditches and/or tile lines.

### 11.9.7  Cutoff Blanket

A third approach is to replace upper subgrade soil with a nonexpansive clay having low permeability. This has the disadvantage that clay tends to be weak. An alternative is the use of an impermeable geomembrane. In either case it is important to also control and minimize infiltration of surface water by providing surface drainage.

## 11.10 PERMAFROST

### 11.10.1 Overview

Permanently frozen ground, termed "permafrost," underlies nearly one-fifth of the total land area of the world. Permafrost is most widespread in the northern hemisphere in the treeless high Arctic, referred to as *tundra*, but also can extend south of the tree line and as far south as the 50th parallel. The southern boundary is irregular, and patches of permafrost occur sporadically beyond the boundary. As may be expected, extensive permafrost occurs in Antarctica.

Permafrost occurs where the mean ground surface temperature is below 0°C (32°F), which in turn is influenced not only by the air temperature but also by ground cover, topography, and local climatic and soil conditions. The depth to which permafrost penetrates is limited by the geothermal gradient, as shown in Fig. 11.18. The thickness of permafrost therefore varies from a few meters to several hundred meters, in general being thicker along the coast of the Arctic Ocean, and diminishing in thickness farther south.

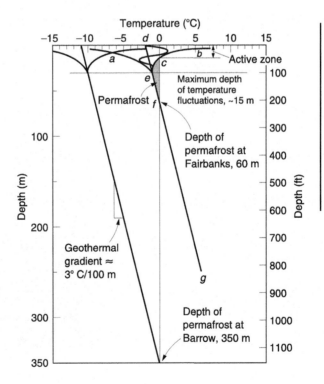

**Figure 11.18**

Permafrost and the geothermal gradient. The active zone undergoes annual changes in temperature, and the permafrost thickness is limited by the geothermal gradient.

### 11.10.2 Permafrost Active Zone

Seasonal changes in temperature from winter to summer cause thermal energy waves that slowly penetrate and are damped out with depth as the heat flux from each wave moves upward as well as downward. The result is a damped sinusoidal curve for ground temperatures at any given time, shown by the curved line between *d* and *e* in Fig. 11.18. These lines present an envelope of minimum and maximum ground temperatures, and extend downward into the permanently frozen zone.

The line for the maximum temperature defines the maximum depth of thawing, which is called the *active zone*. Ice lenses generally are abundant below the active zone and cause major ground subsidence if conditions change so that melting proceeds downward. According to the theory for frost heave the lenses may be "paleo-ice" formed during initial freezing.

### 11.10.3 Permafrost Landforms

Thermal cycling in frozen soil below the active zone causes contraction tensile cracks that fill with water trickling down from the thawed active zone. The water immediately freezes, creating permanently frozen vertical wedges of ice. The wedges then grow incrementally during subsequent thermal cyling, pushing the ground up between adjacent wedges. The cracks often follow a hexagonal pattern similar to a honeycomb, and create a characteristic feature of permafrost areas called *polygonal ground*, shown in Fig. 11.19. If soil immediately next to the wedges is pushed up, the result is "low-centered" polygons, or if the entire soil mass  between wedges is pushed up, the result is "high-centered" polygons.

**Figure 11.19**

Summer on the tundra. Water is trapped on the surface of Arctic tundra by permafrost. Vertical ice wedges have pushed up soil to form low-centered polygons that pond water and can cause partial melting of the permafrost to form large lakes.

Polygons may follow linear soil boundaries such as abandoned lakes or river channels, in which case they are rectangular.

### 11.10.4 Drainage in Permafrost Areas

Permanently frozen ground does not allow infiltration of surface water, so permafrost areas are poorly drained and marshy during melt seasons, even though the annual precipitation is that of a desert. During thaw seasons the tops of ice wedges melt down, and low-centered polygons fill with meltwater. Small streams tend to follow ice wedges, and where wedges meet they melt out pockets that may be meters deep and appear like beads—hence the name "beaded streams."

### 11.10.5 Artesian Pressure from Confined Drainage

Water seeping downslope through the active zone becomes confined when the top of the soil freezes during the onset of winter. This can lead to some extraordinary consequences. A pingo is a large mound, of the order of 30–45 m (100–150 ft) in diameter and 8–9 m (25–30 ft) high, where a frozen soil layer has been pushed up by hydrostatic pressure and contains a core of ice. A *palsa* is similar to a pingo, but consists mainly of peat soil and occurs most frequently near the southern fringe of the discontinuous permafrost.

An abandoned building can delay ground freezing so confined downslope seepage finds an outlet inside the building, which then fills with water so that ice cascades out of the windows.

### 11.10.6 Thawing of Permafrost

Permafrost is sensitive to changes in the thermal regime and, as shown in Fig. 11.20, will thaw underneath a heated building unless steps are taken to prevent it. Destruction of ground cover such as by clearing of trees promotes thawing, and tilting of the trees as ice wedges melt away creates an effect known as "drunken forest." Several years are required after clearing for ground temperatures to warm up sufficiently for agriculture.

Lakes dot the high Arctic and change the soil thermal regime sufficiently to thaw the upper part of the permafrost. This causes subsidence as ice lenses and wedges melt, so lakes are perpetuated. Large areas of the Arctic coastal plain of North America therefore present a patchwork of oblong thaw lakes that are oriented in the direction of summer prevailing winds. As the lakes migrate and find outlets they drain and the cycle starts over again.

Irregular pock-marked topography resulting from thawing of permafrost and associated ice masses is called "thermokarst," analogous with limestone sink areas.

**Figure 11.20**

Settlement from thawing of permafrost by a heated building, Big Delta, Alaska.

### 11.10.7  Premafrost and Engineering

The most direct approach to prevent problems from melting of permafrost is to keep it frozen, but this is becoming more difficult because of the sensitivity of northern climates to change, with the result that these areas are the first to feel major impacts from global warming. The moisture content at which a soil becomes liquid is called the "liquid limit," and when permafrost melts, the soil moisture content usually exceeds the liquid limit and is trapped until it can drain away or evaporate.

Seasonal thawing can be prevented by supporting structures on piles extending into the permafrost layer and allowing winter air to circulate underneath the structures in the winter. Another method uses ducts in soil under a structure in order to force-circulate cold winter air.

Oil pipelines must be protected from loss of support by thawing permafrost in order to prevent environmentally disastrous oil spills. The scheme shown in Fig. 11.21 was used for the Trans-Alaska pipeline, and employs a passive heat transfer mechanism with a vaporizing refrigerant. During winter when the soil is warmer than the air, refrigerant in the soil zone evaporates, taking heat with it, then rises and condenses in the colder ends of the tubes, releasing heat to the air. The refrigerant then flows down by gravity to replenish the supply at the bottom. During summer the cycle is broken because the refrigerant remains at the bottom of the tubes. A similar, less efficient, but less costly system uses convection, as air inside a pipe is warmed by the soil and moves up and out. The air is replenished by cold winter air from the ground surface and conducted downward through a center pipe.

Roads can pose a difficult problem in permafrost areas because, if they are built up for better drainage, permafrost can rise in the road embankment and create a

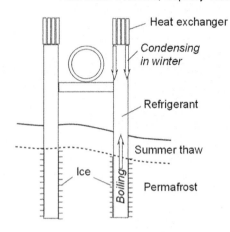

**Figure 11.21**

Schematic diagram showing method for preservation of permafrost under the Trans-Alaska pipeline.

barrier to water seeping through the active zone, leading to ponding on the uphill side of the road. Culverts should be large enough to allow flow of air and thawing during summer months.

Tundra of the high Arctic is marshy and supports a variety of mosses, lichens, flowers, and shrubs that are part of the eco-chain, are easily damaged, and are very slow to recover. Most land travel therefore is during winter when the tundra is frozen.

## 11.10.8 Cold-Storage Plants

Cold-storage facilities simulate Arctic conditions and gradually create a permafrost bulb and frost heave. An air space or auxiliary subfloor heating system separated from the cold room by insulation can avoid such eventualities.

Shutting down a frost-heaved cold-storage facility can create another set of problems as thawing of ice lenses leads to soft "frost boil" soil conditions under the floor and footings. It is not unusual to find older facilities for sale.

## 11.10.9 Insulation to Prevent Thawing or Freezing

If a mean temperature is above or below freezing, insulation alone will not prevent ground thawing or freezing, respectively, but will only delay it. Insulation can be effective during a single cycle, for example to prevent frost heave during a single winter season, but will not solve a permafrost problem if the overall direction of heat flow is not changed. A method that is successful for relatively small buildings is to place the structure on a raised gravel pad that has a relatively high thermal conductivity so that it cools from the outside. This and other arrangements are discussed by Phukan (1991).

# 11.11   SUMMARY

Positive pore water pressure weakens soil by reducing friction between soil grains. Pore pressure therefore is subtracted from total stress to obtain *effective stress*.

Capillary action draws water up into a fine-grained soil, creating a capillary fringe. Soil in the capillary fringe may be saturated, so saturation is not a criterion for locating a groundwater table. A groundwater table is measured in an open boring where there is no capillary effect. Pore water pressures in a capillary fringe are negative relative to atmospheric pressure.

Under static conditions, pore water pressure increases linearly with depth below a groundwater table, whereas soil weight-induced stresses increase with depth below the ground surface.

Compressing saturated soil increases its pore water pressure. The increase over static groundwater conditions is *excess pore water pressure*, which like static pore water pressure acts to decrease effective stress and friction.

Capillary action is a lead actor in frost heave.

Permafrost conditions require special considerations and will become more prominent as mineral resources are developed in Arctic areas.

## Problems

11.1. Why is the term "matric potential" now preferred over "capillary potential"?

11.2. A soil unit weight above the groundwater table averages $115 \, lb/ft^3$ and below the groundwater table averages $112 \, lb/ft^3$. The water table is 6 ft deep below the ground surface. Prepare a plot of total and effective stresses to a depth of 20 ft.

11.3. Distinguish between artesian water pressure and excess pore water pressure.

11.4. (a) Calculate the height of rise of pure water in a clean glass tube whose inside diameter is 0.15 mm, using $T = 7.75 \, mN/m$. (b) What is the height of rise if the tube is sufficiently dirty to produce a wetting angle of 15°?

11.5. If the outside diameter of the tube in Problem 11.4 is 0.35 mm, what is the compressive stress in the walls of the tube in part (a) of that problem?

What parallel can be drawn to compressive stress in the capillary zone of a soil?

11.6. Two soils have the sorption characteristics shown in the accompanying table. Draw the sorption curves with moisture content along the $x$-axis.

| | Moisture content (%) | |
| --- | --- | --- |
| Water, m | Soil A | Soil B |
| 0.10 | 47 | 32 |
| 0.316 | 33 | 17 |
| 1.0 | 22 | 12 |
| 3.16 | 17 | 9 |
| 10.0 | 13 | 7 |
| 31.6 | 10 | 5 |

11.7. Assume that two masses of soils A and B in Problem 11.6 are at the same elevation above a water table and are in intimate contact with each other. If soil A contains 17% moisture and B contains 12%, will capillary water flow from A to B or from B to A, or will it remain static?

11.8. A subgrade soil under an impervious pavement has the sorption characteristics of soil A in Problem 11.6. What will be the accumulated moisture content of the soil 0.3 m (1 ft) below the bottom of the pavement, if the water table is constant at an elevation 1.5 m (5 ft) below the bottom of the pavement?

11.9. Repeat Problem 11.8, using soil B in Problem 11.6.

11.10. What is meant by the zone of capillary saturation? Is the pressure in the soil water within this zone greater than, less than, or equal to atmospheric pressure?

11.11. In a certain locality the water table rises steadily from October to April and then falls steadily from May through September. Will the zone of capillary saturation extend a greater or lesser distance above the water table in July than in March?

11.12. The water in a capillary tube extends 0.9 m (3 ft) above a free water surface. Determine the pressure in the water in this capillary tube: (a) at the free water surface; (b) 0.3 m (1 ft) above the free water surface; (c) 0.6 m (2 ft) above it; (d) just below the meniscus; (e) just above the meniscus.

11.13. Explain to a 6-year-old why sand castles built at the water margin on a beach crumble when the tide comes in. Do they also crumble when they dry out?

11.14. What is the best time to drive a four-wheeler on a sandy beach, when the tide is coming in or going out? Why?

11.15. Soil immediately underneath a pavement is 6 ft above the groundwater table. If the sorption curve for this soil is as shown in Fig. 11.11, what will be the probable moisture content? Which is more appropriate, the desorption or adsorption curve? Why?

11.16. At what stage of frost action in the subgrade may it be necessary to impose a traffic embargo on a road or runway? How long should the embargo last?

11.17. Under what conditions will rain dry out a soil?

11.18. Name three basic conditions that must exist before ice lenses will grow in a soil subgrade.

11.19. Describe three principles that may be employed in pavement design to inhibit the growth of heave-producing ice lenses. Explain how each method accomplishes its purpose.

11.20. Should ice lenses in soil behind a retaining wall be oriented with their thin dimensions vertical or horizontal? Explain.

11.21. Explain how freezing an expansive clay can cause it to shrink.

11.22. What minimum foundation depths are recommended to prevent frost damage in the following cities: Chicago, New York, Denver, Atlanta, Minneapolis, Los Angeles, Seattle, Phoenix?

11.23. Maximum and minimum recorded daily air temperatures over a 10-day period are as follows. Predict the depth of freezing: (a) in a clay with a unit weight of $14.9 \text{ kN/m}^3$ ($95 \text{ lb/ft}^3$) and a moisture content of 20%; (b) in a still lake.

| Day | 1 | 2 | 3 | 4 | 5 | 6 | 7 | 8 | 9 | 10 |
|---|---|---|---|---|---|---|---|---|---|---|
| Min (°C) | 0 | −14 | 9 | −6 | 20 | −16 | −16 | −10 | −8 | −10 |
| Max (°C) | 12 | 2 | 1 | 3 | −12 | −8 | −6 | 2 | 4 | 10 |

11.24. Define and give reasons for permafrost, active zone, oriented lake, pingo.

11.25. What dictates the bottom elevation of permafrost?

11.26. Explain the development and significance of patterned ground.

11.27. A four-story apartment building located on permafrost is sinking into the ground. Is there any way to prevent further damage to this building?

11.28. Why, when permafrost thaws, does the ground usually settle, and why does it do so unevenly?

11.29. Suggest two methods to maintain permafrost under a heated building. How might these methods be modified to also apply to foundations for a cold-storage warehouse?

# References and Further Reading

Andersland, O. B., and Anderson, D. M. (1978). *Geotechnical Engineering for Cold Regions.* McGraw-Hill, New York.

Baver, L. D., Gardner, W. H., and Gardner, W. R. (1972). *Soil Physics*, 4th ed. John Wiley & Sons, New York.

Fredlund, D. G., and Rahardjo, H. (1993). *Soil Mechanics for Unsaturated Soils.* John Wiley & Sons, New York.

Jumikis, A. R. (1962). *Soil Mechanics.* Van Nostrand, Princeton, N.J.

Jumikis, A. R. (1966). *Thermal Soil Mechanics.* Rutgers University Press, New Brunswick, N.J.

Jury, W. A., and Horton, R. (2004). *Soil Physics*, 6th ed. John Wiley & Sons, New York.

Kersten, M. (1952) "Thermal Properties of Soils." Special Report No. 2, Highway Research Board, 161–166.

Kirkham, D., and Powers, W. L. (1972). *Advanced Soil Physics.* Wiley-Interscience, New York.

Penner, E., and Crawford, C. B. (1983). "Frost Action and Foundations." National Research Council, Canada, DBR Paper No. 1090.

Phukan, A. (1991). "Foundations in Cold Regions." In H.-Y. Fang, ed., *Foundation Engineering Handbook*, 2nd ed. Van Nostrand Reinhold, New York.

Richards, L. A. (1941). "Uptake and Retention of Water by Soil as Determined by Distance to a Water Table." *J. Am. Soc. Agron.* 33, 778–786.

Soil Science Society of America (1996). *Glossary of Soil Science Terms.* SSSA, Madison, Wis.

Spangler, M. G. (1952). "Distribution of Capillary Moisture at Equilibrium in Stratified Soil." Special Report No. 2, Highway Research Board.

Van Olphen, H. (1991). *An Introduction to Clay Colloid Chemistry.* Krieger Publ. Co., Melbourne, Fla. (reprint edition).

# 12

## Soil Consistency and Engineering Classification

## 12.1 CLASSIFICATION AND SOIL BEHAVIOR

### 12.1.1 Classification and Engineering Properties

Soil properties that are of most concern in engineering are strength and volume change under existing and future anticipated loading conditions. Various tests have been devised to determine these behaviors, but the tests can be costly and time-consuming, and often a soil can be accepted or rejected for a particular use on the basis of its classification alone.

For example, an earth dam constructed entirely of sand would not only leak, it would wash away. Classification can reveal if a soil may merit further investigation for founding a highway or building foundation, or if it should be rejected and either replaced, modified, or a different site selected. Important clues can come from the geological and pedological origin, discussed in preceding chapters. Another clue is the engineering classification, which can be useful even if the origin is obscure or mixed, as in the case of random fill soil.

### 12.1.2 Classification Tests

Engineering classifications differ from scientific classifications because they focus on physical properties and potential uses. Two tests devised in the early 1900s by a Swedish soil scientist, Albert Atterberg, are at the heart of engineering classifications. The tests are the *liquid limit* or *LL*, which is the moisture content at which a soil become liquid, and the *plastic limit* or *PL*, which is the moisture content at which the soil ceases to become plastic and crumbles in the hand.

Both limits are strongly influenced by the clay content and clay mineralogy, and generally as the liquid limit increases, the plastic limit tends to decrease.

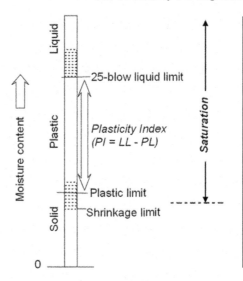

**Figure 12.1**
Schematic representation of transitions between solid, plastic, and viscous liquid behaviors defined by liquid and plastic limits. These tests are basic to engineering classifications and emphasize influences of clay mineralogy and capillarity.

The numerical difference between the two limits therefore represents a range in moisture contents over which the soil is plastic, and is referred to as the *plasticity index* or *PI*. By definition,

$$PI = LL - PL \tag{12.1}$$

where *PI* is the plasticity index and *LL* and *PL* the liquid limit and plastic limit, respectively. This relationship is shown in Fig. 12.1. Because the plasticity index is a difference in percentages and not in itself a percentage, it is expressed as a number and not a percent. Also shown in the figure is the shrinkage limit, which is discussed later in the chapter.

## 12.1.3 Preparation of Soil for Testing

As discussed in relation to clay mineralogy, drying a soil can change its adsorptive capacity for water and therefore can change the liquid and plastic limits. If the soil contains the clay mineral halloysite, dehydration from air-drying is permanent, so to obtain realistic data the soil must not be dried prior to testing. A similar change can occur in soils that have a high content of organic matter.

Air-drying nevertheless is still an approved method because it is more convenient for storing soil samples and for dry sieving, because only the portion of a soil passing the No. 40 (425 μm) sieve is tested. Also, many existing correlations were made on the basis of tests of air-dried samples. If a soil has been air-dried it should be mixed with water for 15 to 30 minutes, sealed and stored overnight, and re-mixed prior to testing. Details are in ASTM D-4318.

### 12.1.4 Liquidity Index

The liquidity index indicates how far the natural soil moisture content has progressed between the plastic and liquid limits. If the soil moisture content is at the plastic limit, the liquidity index is 0; if it is at the liquid limit, it is 1.0.

The formula for the liquidity index is

$$LI = \frac{w - PL}{LL - PL} \tag{12.2}$$

where $LI$ is the liquidity index, $w$ is the soil moisture content, $PL$ is the plastic limit, and $LL$ is the liquid limit. The liquidity index also is called the relative consistency.

## 12.2 MEASURING THE LIQUID LIMIT

### 12.2.1 Concept

The concept of the liquid limit is simple: keep adding water to a soil until it flows, and measure the moisture content at that point by oven-drying a representative sample. Two difficulties in application of this concept are (1) the change from plastic to liquid behavior is transitional, and (2) flow can be prevented by thixotropic setting.

In order to overcome these limitations, Atterberg suggested that wet soil be placed in a shallow dish, a groove cut through the soil with a finger, and the dish jarred 10 times to determine if the groove closes. While this met the challenge of thixotropy, it also introduced a personal factor. Professor A. Casagrande of Harvard University therefore adapted a cog arrangement invented by Leonardo da Vinci, such that turning a crank drops a shallow brass cup containing wet soil 10 mm onto a hard rubber block, shown in Fig. 12.2. The crank is turned at 2 revolutions per second, and the groove is standardized.

Casagrande defined the liquid limit as the moisture content at which the groove would close after 25 blows, which increased the precision of the blow count determination. Different amounts of water are added to a soil sample and stirred in, and the test repeated so that the blow counts bracket the required 25. As it is unlikely that the exact number will be achieved at any particular moisture content, a graph is made of the logarithm of the number of blows versus the moisture content, a straight line is drawn, and the liquid limit read from the graph where the line intersects 25 blows (Fig. 12.3).

### 12.2.2 Procedure for the Liquid Limit Test

A quantity of soil passing the No. 40 sieve is mixed with water to a paste consistency and stored overnight. It is then re-mixed and placed in a standardized

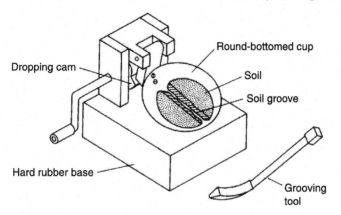

**Figure 12.2**
Casagrande-da Vinci liquid limit device.

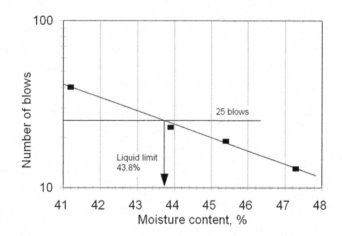

**Figure 12.3**
Semilogarithmic plot for determining a soil liquid limit.

round-bottomed brass cup, and the surface is struck off with a spatula so that the maximum thickness is 10 mm. The soil pat then is divided into two segments by means of a grooving tool of standard shape and dimensions. The brass cup is mounted in such a way that, by turning a crank, it can be raised and allowed to fall sharply onto a hard rubber block or base. The shock produced by this fall causes the adjacent sides of the divided soil pat to flow together. The wetter the mixture, the fewer shocks or blows will be required to cause the groove to close, and the drier the mixture, the greater will be the number of blows.

The number of blows required to close the groove in the soil pat is determined at three or more moisture contents, some above the liquid limit and some below it. The logarithm of the number of blows is plotted versus the moisture content and a straight line is drawn through the points, as shown in Fig. 12.3. The moisture content at which 25 blows cause the groove to close is defined as the liquid limit. Tests usually are performed in duplicate and average results reported.

A "one-point" test may be used for routine analyses, in which the number of blows is between 20 and 30 and a correction that depends on the departure from 25 is applied to the moisture content. See AASHTO Specification T-89 or ASTM Specification D-423 for details of the liquid limit test.

## 12.3 MEASURING THE PLASTIC LIMIT

### 12.3.1 Concept

Soil with a moisture content lower than the liquid limit is plastic, meaning that it can be remolded in the hand. An exception is clean sand, which falls apart on remolding and is referred to as "nonplastic." It is the plasticity of clays that allows molding of ceramics into statues or dishes. At a certain point during drying, the clay can no longer be remolded, and if manipulated, it breaks or crumbles; it is a solid. The moisture content at which a soil no longer can be remolded is the *plastic limit*, or *PL*.

The standard procedure used to determine the plastic limit of a soil is deceptively simple. The soil is rolled out into a thread, and if it does not crumble it is then balled up and rolled out again, and again, and again … until the thread falls apart during remolding. It would appear that a machine might be devised to perform this chore, but several factors make the results difficult to duplicate. First, the soil is continuously being remolded, and second, it gradually is being dried while being remolded. A third factor is even more difficult—the effort required to remold the soil varies greatly depending on the clay content and clay mineralogy. Despite these difficulties and the lack of sophistication, the precision is comparable to or better than that of the liquid limit test.

### 12.3.2 Details of the Plastic Limit Test

The plastic limit of a soil is determined in the laboratory by a standardized procedure, as follows. A small quantity of the soil-water mixture is rolled out with the palm of the hand on a frosted glass plate or on a mildly absorbent surface such as paper until a thread or worm of soil is formed. When the thread is rolled to a diameter of 3 mm ($\frac{1}{8}$ in.), it is balled up and rolled out again, the mixture gradually losing moisture in the process. Finally the sample dries out to the extent that it becomes brittle and will no longer hold together in a continuous thread. The moisture content at which the thread breaks up into short pieces in this rolling process is considered to be the plastic limit (Fig. 12.4). The pieces or crumbs therefore are placed in a small container for weighing, oven-drying, and re-weighing. Generally at least two determinations are made and the results

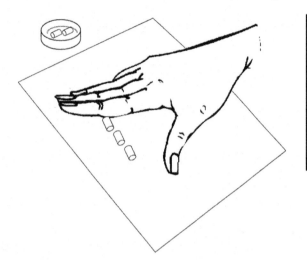

**Figure 12.4**
The plastic limit test. As the soil thread crumbs, pieces are collected in a metal container for oven-drying to determine the moisture content.

averaged. See AASHTO Specification T-90 or ASTM Specification D-424 for details of the plastic limit test.

## 12.4   DIRECT APPLICATIONS OF LL AND PL TO FIELD SITUATIONS

### 12.4.1   When a Soil Moisture Content Exceeds the Plastic Limit

The liquid limit and plastic limit tests are more diagnostic than descriptive of soil behavior in the field because the tests involve continual remolding. However, there are some important situations where remolding occurs more or less continuously in the field. One example is soil in the basal zone of a landslide. As a landslide moves, it shears and mixes the soil. This mixing action can occur if the soil moisture exceeds the plastic limit. If through chemical treatment such as with *drilled lime* (quicklime) the plastic limit is increased, the landslide stops.

### 12.4.2   When a Soil Moisture Content Exceeds the Liquid Limit

Exceeding the soil liquid limit in the field can generate harmful and potentially devastating results, as the soil may appear to be stable and then when disturbed can suddenly break away, losing its thixotropic strength and becoming transformed into a rapid churning, flowing mudslide that takes everything in its way. The rate of sliding depends on the slope angle and viscosity of the mud; the lower the viscosity and steeper the slope, the faster the slide. The most devastating mudslides in terms of loss of life therefore occur in mountainous terrain where the mud moves faster than people can get out of the way and escape almost certain death. Quick clays that appear stable can turn into a soup that can be poured like pancake batter.

### 12.4.3 Liquefaction

Another example where a liquid limit may be exceeded is when a saturated sand or silt suddenly densifies during an earthquake so that all of its weight goes to pore water pressure. This is *liquefaction*, which can cause a sudden and complete loss of shear strength so that landslides develop and buildings may topple. The consequences, diagnosis, and prevention of liquefaction are discussed in more detail in a later chapter.

## 12.5   THE PLASTICITY INDEX

### 12.5.1   Concept

The plasticity index, or PI, is the numerical difference between the liquid and plastic limit moisture contents. Whereas the two limits that are used to define a PI are directly applicable to certain field conditions, the plasticity index is mainly used to characterize a soil, where it is a measure of cohesive properties. The plasticity index indicates the degree of surface chemical activity and hence the bonding properties of clay minerals in a soil. The plasticity index is used along with the liquid limit and particle size gradation to classify soils according to their engineering behavior.

An example of a direct application of the plasticity index is as an indicator of the suitability of the clay binder in a soil mixture used for pavement subgrades, base courses, or unpaved road surfaces. If the PI of the clay fraction of a sand-clay or clay-gravel mixture is too high, the exposed soil tends to soften and become slippery in wet weather, and the road may rut under traffic. On the other hand, if the plasticity index is too low, the unpaved road will tend to "washboard" in response to resonate bouncing of wheels of vehicular traffic. Such a road will abrade under traffic and antagonize the public by producing air-borne dust in amounts that have been measured as high as *one ton per vehicle mile per day per year*. That is, a rural unpaved road carrying an average of 40 vehicles per day can generate up to 40 tons of dust per mile per year. Most collects in roadside ditches that periodically must be cleaned out.

### 12.5.2   A PI of Zero

Measurements of the LL and PL may indicate that a soil has a plasticity index equal to zero; that is, the numerical values of the plastic limit and the liquid limit may be the same within the limits of accuracy of measurement. Soil with a plasticity index of zero therefore still exhibits a slight plasticity, but the range of moisture content within which it exhibits the properties of a plastic solid is not measured by the standard laboratory tests.

### 12.5.3 Nonplastic Soils

Drying and manipulating a truly nonplastic soil such as a clean sand will cause it to abruptly change from a liquid state to an incoherent granular material that cannot be molded. If it is not possible to roll soil into a thread as small as 3 mm in diameter, a plastic limit cannot be determined and the soil is said to be *nonplastic*, designated as NP in test reports.

## 12.6 ACTIVITY INDEX AND CLAY MINERALOGY

### 12.6.1 Definition of Activity Index

The *activity index* was defined by Skempton to relate the PI to the amount of clay in a soil, as an indication of the activity of the clay and therefore the clay mineralogy: the higher the activity index, or AI, the more active the clay. The activity index was defined as the PI divided by the percent 0.002 mm (or 2 μm) clay:

$$AI = \frac{PI}{C_{002}} \tag{12.3}$$

where $AI$ is the activity index, $PI$ the plasticity index, and $C_{002}$ the percent 2 μm clay determined from a particle size analsysis. The basis for this relationship is shown in Fig. 12.5.

### 12.6.2 Relation to Clay Mineralogy

The relationship between activity index and clay mineralogy is shown in Table 12.1. Clay mineral mixtures and interlayers have intermediate activities. Data in the table also show how the activity of smectite is strongly influenced by the adsorbed cation on the plasticity index.

### 12.6.3 Modified Activity Index

The linear relationships in Fig. 12.5 do not necessarily pass through the origin. This is shown in Fig. 12.6, where about 10 percent clay is required to generate plastic behavior. This also has been found in other investigations (Chen, 1988).

It therefore is recommended that for silty soils eq. (12.3) be modified as follows:

$$A = \frac{PI}{C_{002} - k} \tag{12.4}$$

where $A$ is the activity, $C$ is the percent of the soil finer than 0.002 mm, and $k$ is a constant that depends on the soil type. For silty soils $k = 10$. If this equation

**Figure 12.5**

Linear relations between PI and percent clay. Ratios were defined by Skempton (1953) as activity indices, which are shown in parentheses. The Shellhaven soil clay probably is smectite.

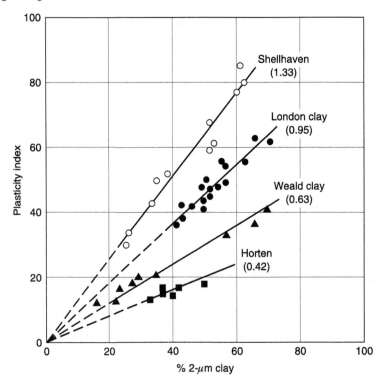

**Table 12.1**

Activity indices of selected clay minerals (after Grim, 1968)

| | |
|---|---|
| Na$^+$ smectite (montmorillonite) | 3–7 |
| Ca$^{2+}$ smectite (montmorillonite) | 1.2–1.3 |
| Illite | 0.3–0.6 |
| Kaolinite, poorly crystallized | 0.3–0.4 |
| Kaolinite, well crystallized | <0.1 |

is applied to the data in Fig. 12.6 with $k = 10$, then $A = 1.23 \pm 0.04$ where the $\pm$ value is the standard error for a number of determinations $n = 81$. This identifies the clay mineral as calcium smectite, which is confirmed by X-ray diffraction.

## 12.7 LIQUID LIMIT AND COLLAPSIBILITY

### 12.7.1 Concept

A simple but effective idea was proposed in 1953 by a Russian geotechnical engineer, A. Y. Denisov, and later introduced into the U.S. by Gibbs and Bara of

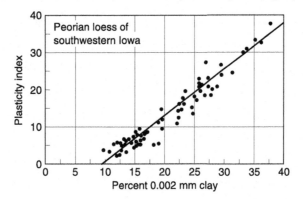

**Figure 12.6**

Data suggesting a modification to eq. (12.3) for silty soil.

the U.S. Bureau of Reclamation. Denisov argued that if the moisture content upon saturation exceeds the liquid limit, the soil should be collapsible—that is, it should collapse and densify under its own weight if it ever becomes saturated. The most common collapsible soil is loess, which is a widespread surficial deposit in the U.S., Europe, and Asia. Because loess increases in density with depth and with distance from a source, only the upper material close to a source may be collapsible, so this is a valuable test.

## 12.7.2  Moisture Content Upon Saturation

The soil unit weight and specific gravity of the soil mineral grains are required to calculate the moisture content upon saturation, which are entered into the following equations:

SI: $w_s = 100(9.807/\gamma_d - 1/G)$ $\qquad\qquad\qquad\qquad$ (12.5)

English: $w_s = 100(62.4/\gamma_d - 1/G)$ $\qquad\qquad\qquad\qquad$ (12.5a)

where $w_s$ is the percent moisture at saturation, $\gamma_d$ is the dry unit weight in kN/m$^3$ or lb/ft$^3$, and $G$ is the specific gravity of the soil minerals. Solutions of this equation with $G = 2.70$ are shown in Fig. 12.7.

## 12.8  CONSISTENCY LIMITS AND EXPANSIVE SOILS

### 12.8.1  Measuring Expandability

Expandability can be determined with a *consolidometer*, which is a device that was developed to measure compression of soil but also can be used to measure expansion under different applied loads. Samples are confined between porous ceramic plates, loaded vertically, wet with water, and the amount of expansion measured. An abbreviated test measures expansion under only applied pressures that can simulate a floor or a foundation load.

**Figure 12.7**

Denisov criterion for collapsibility with $G = 2.70$. Data are for loess at 3, 40, and 55 ft (1, 12, and 16.8 m) depths in Harrison County, Iowa. The deepest soil was mottled gray, suggesting a history of wet conditions, and is indicated to be noncollapsible.

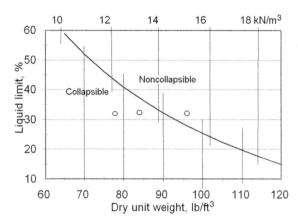

Many investigations have been made relating expansion to various parameters, including activity, percent finer than 0.002 mm, percent finer than 0.001 mm, plasticity index, and liquid limit. For the most part the studies have used artificially prepared soil mixtures with varying amounts of different clay minerals.

### 12.8.2  Influence of Surcharge Pressure

Generally the higher the vertical surcharge pressure, the lower the amount of expansion. This leads to a common observation in buildings founded on expansive clay: floors in contact with the soil are lifted more than foundations that are supporting bearing walls and columns and therefore are more heavily loaded. Partition walls that are not load-bearing are lifted with the floor.

### 12.8.3  Lambe's PVC Meter

A rapid method for measuring clay expandability was developed by T. W. Lambe and his coworkers at MIT. In this device, soil expands against a spring-loaded plate and the expansion is measured. Because the vertical stress increases as the soil expands, results are useful for classification but do not directly translate into expansion amounts that may be expected in the field.

### 12.8.4  Influence of Remolding

Chen (1988) emphasizes that expansion is much lower for undisturbed than for disturbed soil samples subjected to the same treatment, indicating an important restraining influence from soil fabric. Therefore the expansive clay that is inadvertently used for fill soil, as sometimes happens, may expand much more than if the clay were not disturbed. The reason for this has not been investigated, but an argument may be made for a time-related cementation effect of edge-to-face clay particle bonding, which would prevent water from entering and separating the clay layers.

### 12.8.5 Relation to PI

Figure 12.8 shows the conclusions of several researchers who related clay expandability to the plasticity index or PI. Curves A and B show results for remolded samples, and curves C and D are from undisturbed samples where surcharges were applied to more or less simulate floor and foundation loads, respectively. It will be seen that the lowest expandability is shown by curve D, which is for undisturbed soil under the foundation load.

### 12.8.6 Relation to Moisture Content

Generally expansion pressure decreases as the soil moisture content increases, and expansion stops when the smectite clay is fully expanded. This depends on the relative humidity of the soil air and occurs well below the point of saturation of the soil itself. This is illustrated in Fig. 12.9, where seasonal volume change occurs between 30 and 70 percent saturation. Expansion will not occur in a clay that already is wet, which is not particularly reassuring because damaging shrinkage still can occur if and when the clay dries out.

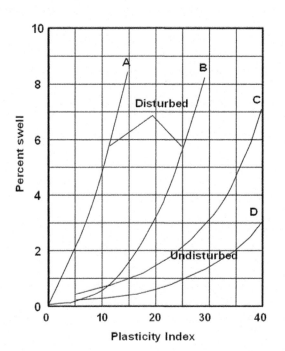

**Figure 12.8**

Data indicating that remolding and a loss of structure greatly increase swelling pressures. All but curve D, which has a surcharge pressure of 1000 lb/ft$^2$ (44 kPa) have a surcharge pressure of 1 lb/in.$^2$ (6.9 kPa). (Modified from Chen, 1988.)

**Figure 12.9**

Seasonal volume changes in Poona clay, India. Left graph shows that expandability is limited to the upper meter despite deeper variations in moisture content. (From Katti and Katti, 1994.)

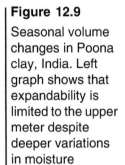

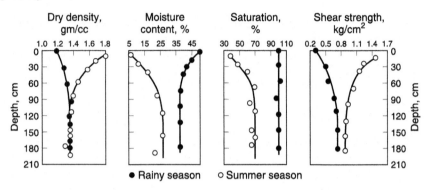

### 12.8.7 Summary of Factors Influencing Expandability

The amount of expansion that can be anticipated depends on at least seven variables: (1) clay mineralogy, and (2) clay content, both of which are reflected in the consistency limits; (3) existing field moisture content; (4) surcharge pressure; (5) whether or not the soil is remolded; (6) thickness of the expanding layer; and (7) availability of water.

### 12.8.8 Thickness of the Active Layer

One of the most extensive expansive clay areas in the world is in India, but detailed field investigations indicate that only about the upper 90 cm (3 ft) of the expansive soil actually experiences seasonal volume changes. Below that depth the clay is volumetrically stable, even though, as seen in the second graph of Fig. 12.9, the moisture content is not. As previously indicated, saturation is not required for full expansion of Ca-smectite, which is the most common expansive clay.

The surficial layer involved in seasonal volume change is called the *active layer*, and determines the depth of shrinkage cracking and vertical mixing, which by disrupting the soil structure tends to increase its expandability.

**Example 12.1**

Calculate the seasonal ground heave from data in Fig. 12.9.

*Answer:* If the soil is divided into three layers, 0–30, 30–60, and 60–90 cm, average increases in density from the left-hand graph are approximately $0.5/1.22 = 41\%$; $0.2/1.3 = 15\%$; and $0.1/1.3 = 8\%$, respectively. Multiplying these percentages by the layer thicknesses gives total volume changes from the dry to the wet seasons of $(0.41 + 0.15 + 0.08) \times 30 = 19$ cm (7.5 in.). However, part of this will go toward closing open ground cracks, in which case one-third of the volume change will be directed vertically, about 6 cm or 2.5 in. The answer therefore is between 6 and 19 cm, depending on filling of the shrinkage cracks and amount of lateral elastic compression of the soil.

## 12.8.9 Controlling Volume Change with a Nonexpansive Clay (n.e.c.) Layer

One of the most significant discoveries for controlling expansive clay was by Dr. R. K. Katti and his co-workers at the Indian Institute of Technology, Mumbai. Katti's group conducted extensive full-scale laboratory tests to confirm field measurements, such as shown in Fig. 12.9, and found that expansion can be controlled by a surficial layer of compacted non-expansive clay. A particularly severe test for the design was the canal shown in Fig. 12.10. The most common application of Katti's method is to stabilize the upper meter (3 ft) of expansive clay by mixing in hydrated lime, $Ca(OH)_2$. If only the upper one-third, 30 cm (1 ft), is stabilized, volume change will be $(0.15 + 0.08) \times 30 = 7$ cm (3 in.), a reduction of about 60 percent. If the upper 60 cm (2 ft) is stabilized, the volume change will be $0.08 \times 30 = 2.4$ cm (1 in.), a reduction of over 85 percent. Stabilization to the full depth has been shown to eliminate volume change altogether. The next question is, why?

An answer may be in the curves in Fig. 12.8, as a loss of clay structure greatly increases clay expandability. As a result of shrink-swell cycling and an increase in horizontal stress, expansive clays are visibly sheared, mixed, and remolded, so by destroying soil structure expansion probably begets more expansion. According to this hypothesis, substituting a layer of nonexpansive clay for the upper highly expansive layer may help to preserve the structure and integrity of the underlying layer. It was found that using a sand layer or a foundation load instead of densely

**Figure 12.10**

The Malaprabha Canal in India was successfully built on highly expansive clay using Katti's method, by replacing the upper 1 m of soil with compacted nonexpansive clay (n.e.c.) to replicate the conditions shown in Fig. 12.9 for the underlying soil.

compacted clay is less effective, perhaps because it interrupts the continuum (Katti and Katti, 2005).

## 12.9 PLASTICITY INDEX VS. LIQUID LIMIT

### 12.9.1 Concept

The relationship between PI and LL reflects clay mineralogy and has an advantage over the activity index because a particle size analysis is not required. Because the liquid limit appears on both sides of the relationship, data can plot only within a triangular area defined by the PL = 0 line shown in Fig. 12.11.

### 12.9.2 The A-Line

A line that approximately parallels the PI versus LL plot for particular soil groups is called the *A-line*, which was proposed by A. Casagrande and for the most part separates soils with and without smectite clay minerals. However, the separation is not always consistent, as can be seen in Fig. 12.11 where loess crosses the line. At low clay contents loessial soils also are more likely to show collapse behavior instead of expanding.

## 12.10 A SOIL CLASSIFICATION BASED ON THE A-LINE

### 12.10.1 Background

During World War II, Arthur Casagrande devised a simplified soil classification system for use by the armed forces. The objective was a system that could be used to classify soils from visual examination and liquid/plastic behavior. In 1952

**Figure 12.11**

Representative relationships between PI and LL. (Adapted from U.S. Dept. of Interior Bureau of Reclamation, 1974.)

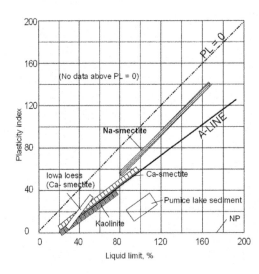

the system was adopted for civilian uses by the U.S. Bureau of Reclamation and the U.S. Army Corps of Engineers and became known as the *Unified Classification*. The ASTM Designation is D-2487. The system applies not only to fine-grained soils but also to sands and gravels, and is the most widely used system for soil investigations for building foundations and tunneling.

## 12.10.2 "S" Is for Sand

One advantage of the Unified Classification system is its simplicity, as it uses capital letters to represent particular soil properties: S stands for sand, G for gravel, and C for clay. Because S already is used for sand, another letter, M, was selected for silt, from the German word *Moh*.

A sand or gravel can either be well graded, W, or for poorly graded, designated by P, respectively indicating broad or narrow ranges of particle sizes. Thus, SP is a poorly graded sand, GW a well-graded gravel.

Fine-grained soils are characterized on the basis of liquid limit and the PI and LL relationships to the A-line. A silt or clay with a liquid limit higher than 50 percent is designated by H, meaning high liquid limit, and if the data plot above the A-line the soil is CH, clay with a high liquid limit. If the liquid limit is higher than 50 percent and the data plot below the A-line, the designation is MH, silt with a high liquid limit.

The Unified Classification system therefore distinguishes between silt and clay not on the basis of particle size, but on relationships to the liquid limit and plasticity index. In order to avoid confusion, clay and silt that are defined on the basis of size now usually are referred to as "clay-size" or "silt-size" material.

If the silt or clay liquid limit is lower than 50, the respective designations are ML and CL, silt with a low liquid limit or clay with a low liquid limit. However, if the plasticity index is less than 4, silt dominates the soil behavior and the soil is designated ML. This is shown in the graph in Table 12.2. Soils with a plasticity index between 4 and 7 show properties that are intermediate and are designated CL-ML.

## 12.10.3 Details of the Unified Classification System

Letter abbreviations for the various soil characteristics are as follows:

| | |
|---|---|
| G = Gravel | O = Organic |
| S = Sand | W = Well graded |
| M = Nonplastic or low plasticity | P = Poorly graded |
| C = Plastic fines | L = Low liquid limit |
| Pt = Peat, humus, swamp soils | H = High liquid limit |

## Table 12.2
Unified soil classification system

| Major Divisions | | | Group Symbols | Typical Names | Field Identification Procedures (Excluding particles larger than 3 in. and basing fractions on estimated weights) |
|---|---|---|---|---|---|
| 1 | 2 | | 3 | 4 | 5 |
| Coarse-grained Soils — More than half of material is *larger* than No. 200 sieve size. The smallest particle visible to the naked eye. | Gravels — More than half of coarse fraction is larger than No. 4 sieve size. (For visual classification, the 1/4-in. size may be used as equivalent to the No. 4 sieve size) | Clean Gravels (Little or no fines) | GW | Well-graded gravels, gravel-sand mixtures, little or no fines. | Wide range in grain sizes and substantial amounts of all intermediate particle sizes. |
| | | | GP | Poorly graded gravels or gravel-sand mixtures, little or no fines. | Predominantly one size or a range of sizes with some intermediate sizes missing. |
| | | Gravels with Fines (Appreciable amount of fines) | GM | Silty gravels, gravel-and-silt mixture. | Nonplastic fines or fines with low plasticity (for identification procedures see ML below). |
| | | | GC | Clayey gravels, gravel-sand-clay mixtures. | Plastic fines (for identification procedures see CL below). |
| | Sands — More than half of coarse fraction is smaller than No. 4 sieve size. | Clean Sands (Little or no fines) | SW | Well-graded sands, gravelly sands, little or no fines. | Wide range in grain size and substantial amount of all intermediate particle sizes. |
| | | | SP | Poorly graded sands or gravelly sands, little or no fines. | Predominantly one size or a range of sizes with some intermediate sizes missing. |
| | | Sands with Fines (Appreciable amount of fines) | SM | Silty sands, sand-silt mixtures. | Nonplastic fines or fines with low plasticity (for identification procedures see ML below). |
| | | | SC | Clayey sands, sand-clay mixtures. | Plastic fines (for identification procedures see CL below). |

|  |  |  |  |  | Identification Procedures on Fraction Smaller than No. 40 Sieve Size | | |
|---|---|---|---|---|---|---|---|
| | | | | | Dry Strength (Crushing characteristics) | Dilatancy (Reaction to shaking) | Toughness (Consistency near PL) |
| Fine-grained Soils — More than half of material is *smaller* than No. 200 sieve size. The No. 200 sieve size is about the smallest particle visible to the naked eye. | Silts and Clays Liquid limit is less than 50 | | ML | Inorganic silts and very fine sands, rock flour, silty or clayey fine sands or clayey silts with slight plasticity. | None to slight | Quick to slow | None |
| | | | CL | Inorganic clays of low to medium plasticity, gravelly clays, sandy clays, silty clays, lean clays. | Medium to high | None to very slow | Medium |
| | | | OL | Organic silts and organic silty clays of low plasticity. | Slight to medium | Slow | Slight |
| | Silts and Clays Liquid limit is greater than 50 | | MH | Inorganic silts, miscaceous or diatomaceous fine sandy or silty soils, elastic silts. | Slight to medium | Slow to none | Slight to medium |
| | | | CH | Inorganic clays of high plasticity, fat clays. | High to very high | None | High |
| | | | OH | Organic clays of medium to high plasticity, organic silts. | Medium to high | None to very slow | Slight to medium |
| Highly Organic Soils | | | Pt | Peat and other high organic soils. | Readily identified by color, odor, spongy feel and frequently by fibrous texture. | | |

(1) Boundary classification: Soils possessing characteristics of two groups are designated by combinations of group symbols. For example GW-GC, well-graded gravel-sand mixture with clay binder: See Col. 7.
(2) All sieve sizes on this chart are U.S. standard.

**FIELD IDENTIFICATION PROCEDURES FOR FINE-GRAINED SOILS OR FRACTIONS**
These procedures are to be performed on the minus No. 40 sieve size particles, approximately 1/64 in. For field classification purposes, screening is not intended, simply remove by hand the coarse particles that interfere with the tests.

Dilatancy (reaction to shaking)

After removing particles larger than No. 40 sieve size, prepare a pat of moist soil with a volume of about one-half cubic inch. Add enough water if necessary to make the soil soft but not sticky.

Place the pat in the open palm of one hand and shake horizontally, striking vigorously against the other hand several times. A positive reaction consists of the appearance of water on the surface of the pat which changes to a livery consistency and becomes glossy. When the sample is squeezed between the fingers, the water and gloss disappear from the surface, the pat stiffens, and finally it cracks or crumbles. The rapidity of appearance of water during shaking and of its disappearance during squeezing assist in identifying the character of the fines in a soil.

Very fine clean sands give the quickest and most distinct reaction whereas a plastic clay has no reaction. Inorganic silts, such as a typical rock flour, show a moderately quick reaction.

Dry Strength (crushing characteristics)

After removing particles larger than No. 40 sieve size,

Adopted by Corps of Engineers and Bureau of Reclamation, January 1952

| Information Required for Describing Soils | Laboratory Classification Criteria |
|---|---|
| 6 | 7 |

| | | |
|---|---|---|
| For undisturbed soils add information on stratification, degree of compactness, cementation, moisture conditions, and drainage characteristics. | $C_u = \dfrac{D_{60}}{D_{10}}$ Greater than 4 $\qquad$ $C_c = \dfrac{(D_{30})^2}{D_{10} \times D_{60}}$ Between 1 and 3 | |
| | Not meeting all gradation requirements for GW | |
| Give typical name; indicate approximate percentages of sand and gravel, maximum size; angularity, surface condition, and hardness of the coarse grains; local or geologic name and other pertinent descriptive information; and symbol in parentheses. | Atterberg limits below "A" line or PI less than 4 | Above "A" line with PI between 4 and 7 are *borderline* cases requiring use of dual symbols. |
| | Atterberg limits above "A" line with PI greater than 7 | |
| | $C_u = \dfrac{D_{60}}{D_{10}}$ Greater than 6 $\qquad$ $C_c = \dfrac{(D_{30})^2}{D_{10} \times D_{60}}$ Between 1 and 3 | |
| Example: *Silty sand*, gravelly; about 20% hard, angular gravel particles 1/2-in. maximum size; rounded and subangular sand grains, coarse to fine; about 15% nonplastic fines with low dry strength; well compacted and moist in place; alluvial sand; (SM). | Not meeting all gradation requirements for SW | |
| | Atterberg limits below "A" line or PI less than 4 | Limits plotting in hatched zone with PI between 4 and 7 are *borderline* cases requiring use of dual symbols. |
| | Atterberg limits above "A" line with PI greater than 7 | |

*Left vertical label:* Use grain-size curve in identifying the fractions as given under field identification.

*Column notes:* Determine percentages of gravel and sand from grain-size curve. Depending on percentage of fines (fraction smaller than No. 200 sieve size) coarse-grained soils are classified as follows:
Less than 5% — GW, GP, SW, SP
More than 12% — GM, GC, SM, SC
5% to 12% — *Borderline* cases requiring use of dual symbols.

For undisturbed soils add information on structure, stratification, consistency in undisturbed and remolded states, moisture and drainage conditions.

Give typical name; indicate degree and character of plasticity; amount and maximum size of coarse grains; color in wet condition; odor, if any; local or geologic name and other pertinent descriptive information; and symbol in parentheses.

Example: *Clayey silt*, brown; slightly plastic; small percentage of fine sand; numerous vertical root holes; firm and dry in place; loess; (ML).

Plasticity chart
For laboratory classification of fine-grained soils

Comparing soils at equal liquid limit toughness and dry strength increase with increasing plasticity index

mold a pat of soil to the consistency of putty, adding water if necessary. Allow the pat to dry completely by oven, sun, or air-drying, and then test its strength by breaking and crumbling between the fingers. This strength is a measure of the character and quantity of the colloidal fraction contained in the soil. The dry strength increases with increasing plasticity.

High dry strength is characteristic for clays of the CH group. A typical inorganic silt possesses only very slight dry strength. Silty fine sands and silts have about the same slight dry strength, but can be distinguished by the feel when powdering the dried specimen. Fine sand feels gritty whereas a typical silt has the smooth feel of flour.

Toughness (consistency near plastic limit)

After particles larger than the No. 40 sieve size are removed, a specimen of soil about one-half inch cube in size is molded to the consistency of putty. If too dry, water must be added and if sticky, the specimen should be spread out in a thin layer and allowed to lose some moisture by evaporation. Then the specimen is rolled out by hand on a smooth surface or between the palms into a thread about one-eighth inch in diameter. The thread is then folded and re-rolled repeatedly. During this manipulation the moisture content is gradually reduced and the specimen stiffens, finally loses its plasticity, and crumbles when the plastic limit is reached.

After the thread crumbles, the pieces should be lumped together and a slight kneading action continued until the lump crumbles.

The tougher the thread near the plastic limit and the stiffer the lump when it finally crumbles, the more potent the colloidal clay fraction in the soil. Weakness of the thread at the plastic limit and quick loss of coherence of the lump below the plastic limit indicate either inorganic clay of low plasticity, or materials such as kaolin-type clays and organic clays which occur below the "A" line.

Highly organic clays have a very weak and spongy feel at the plastic limit.

These letters are combined to define various groups as shown in Table 12.2. This table is used to classify a soil by going from left to right and satisfying the several levels of criteria.

## 12.10.4 Equivalent Names

Some equivalent names for various combined symbols are as follows. These names are appropriate and should be used in reports that may be read by people who are not familiar with the soil classification symbols. For example, describing a soil as an "SC clayey sand" will be more meaningful than to only refer to it as "SC," and illustrates the logic in the terminology. The meanings of the symbols for coarse-grained soils are fairly obvious. Names used for fine-grained soils are as follows:

CL = lean clay        CH = fat clay
ML = silt        MH = elastic silt
OL = organic silt        OH = organic clay

If a fine-grained soil contains over 15 percent sand or gravel, it is referred to as "with sand or with gravel;" if over 30 percent, it is "sandy" or "gravelly." If over 50 percent it goes into a coarse-grained classification. If a soil contains any cobbles or boulders it is referred to as "with cobbles" or "with boulders." Other more detailed descriptors will be found in ASTM D-2487.

## 12.10.5 Detailed Descriptions

Descriptions of the various groups that may be helpful in classification are as follows:

### GW and SW
Soils in these groups are well-graded gravelly and sandy soils that contain less than 5 percent nonplastic fines passing the No. 200 sieve. The fines that are present do not noticeably affect the strength characteristics of the coarse-grained fraction and must not interfere with its free-draining characteristic. In areas subject to frost action, GW and SW soils should not contain more than about 3 percent of soil grains smaller than 0.02 mm in size.

### GP and SP
GP and SP soils are poorly graded gravels and sands containing less than 5 percent of nonplastic fines. The soils may consist of uniform gravels, uniform sands, or nonuniform mixtures of very coarse material and very fine sand with intermediate sizes lacking, referred to as skip-graded, gap-graded, or step-graded.

### GM and SM
In general, GM and SM soils are gravels or sands that contain more than 12 percent fines having little or no plasticity. In order to qualify as M, the

plasticity index and liquid limit plot below the A-line on the plasticity chart. Because of separation of coarse particles gradation is less important, and both well-graded and poorly graded materials are included in these groups. Some sands and gravels in these groups may have a binder composed of natural cementing agents, so proportioned that the mixture shows negligible swelling or shrinkage. Thus, the dry strength is provided either by a small amount of soil binder or by cementation of calcareous materials or iron oxide. The fine fraction of noncemented materials may be composed of silts or rock-flour types having little or no plasticity, and the mixture will exhibit no dry strength.

## GC and SC
These groups consist of gravelly or sandy soils with more than 12 percent fines that can exhibit low to high plasticity. The plasticity index and liquid limit plot above the A-line on the plasticity chart. Gradation of these materials is not important, as the plasticity of the binder fraction has more influence on the behavior of the soils than does variation in gradation. The fine fraction is generally composed of clays.

## Borderline G and S Classifications
It will be seen that a gap exists between the GW, SW, GP, and SP groups, which have less than 5 percent passing the No. 200 sieve, and GM, SM, GC, and SC soils, which have more than 12 percent passing the No. 200 sieve. Soils containing between 5 and 12 percent fines are considered as borderline and are designated by a dual symbol such as GW-GM if the soil is a well-graded gravel with a silt component, or GW-GC if well-graded with a clay component. Many other dual symbols are possible, and the meaning should be evident from the symbol. For example, SP-SC is a poorly graded sand with a clay component, too much clay to be SP and not enough to be SC.

## ML and MH
ML and MH soils include soils that are predominantly silts, and also include micaceous or diatomaceous soils. An arbitrary division between ML and MH is established where the liquid limit is 50. Soils in these groups are sandy silts, clayey silts, or inorganic silts with relatively low plasticity. Also included are loessial soils and rock flours.

Micaceous and diatomaceous soils generally fall within the MH group but may extend into the ML group when their liquid limit is less than 50. The same is true for certain types of kaolin clays and some illitic clays having relatively low plasticity.

## CL and CH
The CL and CH groups embrace clays with low and high liquid limits, respectively. These are mainly inorganic clays. Low-plasticity clays are classified as CL and are usually lean clays, sandy clays, or silty clays. The medium-plasticity and high-plasticity clays are classified as CH. These include the fat clays, gumbo

clays, certain volcanic clays, and bentonite. The glacial clays of the northern U.S. cover a wide band in the CL and CH groups.

### ML-CL

Another type of borderline classification that already has been commented on is when the liquid limit of a fine-grained soil is less than 29 and the plasticity index lies in the range from 4 to 7. These limits are indicated by the shaded area on the plasticity chart in Fig. 12.11, in which case the double symbol, ML-CL, is used to describe the soil.

### OL and OH

OL and OH soils are characterized by the presence of organic matter and include organic silts and clays. They have plasticity ranges that correspond to those of the ML and MH groups.

### Pt

Highly organic soils that are very compressible and have very undesirable construction characteristics are classified in one group with the symbol Pt. Peat, humus, and swamp soils with a highly organic texture are typical of the group. Particles of leaves, grass, branches of bushes, or other fibrous vegetable matter are common components of these soils.

## 12.11  FIELD USE OF THE UNIFIED SOIL CLASSIFICATION SYSTEM

### 12.11.1  Importance of Field Identitication

A detailed classification such as indicated in Table 12.2 requires both a gradation and plasticity analysis. Even after this information is available from laboratory tests of soil samples, it is important to be able to identify the same soils in the field. For example, if a specification is written based on the assumption that a soil is an SC, and the borrow excavation proceeds to cut into ML, it can be very important that somebody serves notice and if necessary issues a stop order. This is a reason why all major construction jobs include on-site inspection. Suggestions for conduct of a field identification using the Unified Classification system are in ASTM D-2488.

### 12.11.2  Granular Soils

A dry sample of coarse-grained material is spread on a flat surface to determine gradation, grain size and shape, and mineral composition. Considerable skill is required to visually differentiate between a well-graded soil and a poorly graded soil, and is based on visual comparisons with results from laboratory tests.

The durability of coarse aggregate is determined from discoloration of weathered materials and the ease with which the grains can be crushed. Fragments of shale or other rock that readily breaks into layers may render a coarse-grained soil unsuitable for certain purposes, since alternate wetting and drying may cause it to disintegrate partially or completely. This characteristic can be determined by submerging thoroughly dried particles in water for at least 24 hours and observing slaking or testing to determine a loss of strength.

## 12.11.3   Fine-Grained vs. Coarse-Grained Soils

As shown in Table 12.2, fine-grained soils are defined as having over 50 percent passing the No. 200 sieve. The percentage finer can be estimated without the use of a sieve and weighing device, by repeatedly mixing a soil sample with water and decanting until the water is clear, and then estimating the proportion of material that has been removed. Another method is to place a sample of soil in a large test tube, fill the tube with water and shake the contents thoroughly, and then allow the material to settle. Particles retained on a No. 200 sieve will settle out of suspension in about 20 to 30 seconds, whereas finer particles will take a longer time. An estimate of the relative amounts of coarse and fine material can be made on the basis of the relative volumes of the coarse and fine portions of the sediment.

## 12.11.4   Fine-Grained Soils

Field identification procedures for fine-grained soils involve testing for dilatancy, or expansion on shaking, plasticity, and dry strength. These tests are performed on the fraction of soil finer than the No. 40 sieve. In addition, observations of color and odor can be important. If a No. 40 sieve is not available, removal of the fraction retained on this sieve may be partially accomplished by hand picking. Some particles larger than this sieve opening (0.425 mm, or nominally 0.5 mm) may remain in the soil after hand separation, but they probably will have only a minor effect on the field tests.

### Dilatancy

For the dilatancy test, enough water is added to about $2\,cm^3$ ($\frac{1}{2}$ in.$^3$) of the minus-40 fraction of soil to make it soft but not sticky. The pat of soil is shaken horizontally in the open palm of one hand, which is struck vigorously against the other hand several times. A fine-grained soil that is nonplastic or has very low plasticity will show free water on the surface while being shaken, and then squeezing the pat with the fingers will cause the soil structure to dilate or expand so that the soil appears to dry up. The soil then will stiffen and finally crumble under increasing pressure. Shaking the pat again will cause it to flow together and water to again appear on the surface.

A distinction should be made between a rapid reaction, a slow reaction, or no reaction to the shaking test, the rating depending on the speed with which the pat changes its consistency and the water on the surface appears or disappears. Rapid reaction is typical of nonplastic, uniform fine sand, of silty sand (SP or SM), of inorganic silt (ML), particularly the rock-flour type, and of diatomaceous earth (MH). The reaction becomes more sluggish as the uniformity of gradation decreases and the plasticity increases, up to a certain degree. Even a small amount of colloidal clay will impart some plasticity to the soil and will materially slow the reaction to the shaking test. Soils that react in this manner are somewhat plastic inorganic and organic silts (ML or OL), very lean clays (CL), and some kaolin-type clays (ML or MH). Extremely slow reaction or no reaction to the shaking test is characteristic of typical clays (CL or CH) and of highly plastic organic clays (OH).

## Field Estimate of Plasticity

The plasticity of a fine-grained soil or the binder fraction of a coarse-grained soil may be estimated by rolling a small sample of minus-40 material between the palms of the hand in a manner similar to the standard plastic limit test. The sample should be fairly wet, but not sticky. As it is rolled into $\frac{1}{8}$-inch threads, folded and re-rolled, the stiffness of the threads should be observed. The higher the soil above the A-line on the plasticity chart (CL or CH), the stiffer the threads. Then as the water content approaches the plastic limit, the tougher are the lumps after crumbling and remolding. Soils slightly above the A-line (CL or CH) form a medium-tough thread that can be rolled easily as the plastic limit is approached, but when the soil is kneaded below the plastic limit, it crumbles.

Soils below the A-line (ML, NH, OL, or OH) form a weak thread and with the exception of an OH soil, such a soil cannot be lumped into a coherent mass below the plastic limit. Plastic soils containing organic material or much mica form threads that are very soft and spongy near the plastic limit.

In general, the binder fraction of a coarse-grained soil with silty fines (GM or SM) will exhibit plasticity characteristics similar to those of ML soils. The binder fraction of a coarse-grained soil with clayey fines (GC or SC) will be similar to CL soils.

## Field Estimate of Dry Strength

Dry strength is determined from a pat of minus-40 soil that is moistened and molded to the consistency of putty, and allowed to dry in an oven or in the sun and air. When dry the pat should be crumbled between the fingers. ML or MH soils have a low dry strength and crumble with very little finger pressure. Also, organic silts and lean organic clays of low plasticity (OL) and very fine sandy soils (SM) also have low dry strength.

Most clays of the CL group and some OH soils, as well as the binder fraction of gravelly and sandy clays (GC or SC), have medium dry strength and require considerable finger pressure to crumble the sample. Most CH clays and some organic clays (OH) having high liquid limits and located near the A-line have high dry strength, and the test pat can be broken with the fingers but cannot be crumbled.

**Color and Odor**

Dark or drab shades of gray or brown to nearly black indicate fine-grained soils containing organic colloidal matter (OL or OH), whereas brighter colors, including medium and light gray, olive green, brown, red, yellow, and white, are generally associated with inorganic soils.

An organic soil (OL or OH) usually has a distinctive odor that can be helpful for field identification. This odor is most obvious in a fresh sample and diminishes on exposure to air, but can be revived by heating a wet sample.

The details of field identification are less imposing and more easily remembered if they are reviewed in relation to each particular requirement. For example, if a specification is for SC, the criteria for an SC soil should be reviewed and understood and compared with those for closely related soils, SM and SP and respective borderline classifications.

## 12.12 THE AASHTO SYSTEM OF SOIL CLASSIFICATION

### 12.12.1 History

A system of soil classification was devised by Terzaghi and Hogentogler for the U.S. Bureau of Public Roads in the late 1920s, predating the Unified Classification system by about 20 years. The Public Roads system was subsequently modified and adopted by the American Association of State Highway Officials (now Highway and Transportation Officials) and is known as the AASHTO system (AASHTO Method M14S; ASTM Designation D-3282).

As in the Unified Classification system, the number of physical properties of a soil upon which the classification is based is reduced to three—gradation, liquid limit, and plasticity index. Soil groups are identified as A-1 through A-8 for soils ranging from gravel to peat. Generally, the higher the number, the less desirable the soil for highway uses.

### 12.12.2 Using the AASHTO Chart

The process of determining the group or subgroup to which a soil belongs is simplified by use of the tabular chart shown in Table 12.3. The procedure is as follows. Begin at the left-hand column of the chart and see if all these known properties of the soil comply with the limiting values specified in the column. If they do not, move to the next column to the right, and continue across the chart until the proper column is reached. The first column in which the soil properties fit the specified limits indicates the group or subgroup to which the soil belongs. Group A-3 is placed before group A-2 in the table to permit its use in this manner even though A-3 soils normally are considered less desirable than A-2 soils.

**Figure 12.12**

Chart for
classifying
fine-grained soils
by the AASHTO
system.

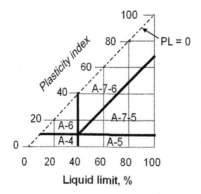

The ranges of the liquid limit and the plasticity index for fine-grained soils in groups A-4, A-5, A-6, and A-7 are shown in Fig. 12.12, which has been arranged to be comparable to the Unified chart.

**Example 12.2**
Classify a soil containing 65% of material passing a No. 200 sieve and having a liquid limit of 48 and a plasticity index of 17.

*Answer:* Since more than 35% of the soil material passes the No. 200 sieve, it is a silt-clay material and the process of determining its classification can begin by examining the specified limits for group A-4, where the maximum is 40. Since the liquid limit of the soil being classified is 48%, it cannot be an A-4 soil so we proceed to the columns to the right, where it will be seen that it meets the liquid limit requirement of A-6 and A-7, but meets the plasticity index requirements of A-7. The soil therefore is A-7. This procedure is simplified by reference to Fig. 12.12, which also is used to separate A-7 soils into two subgroups, A-7-5 and A-7-6.

## 12.12.3  Size Grade Definitions

AASHTO definitions of gravel, sand, and silt-clay are as follows:

**Gravel**
Material passing a sieve with 75 mm (3 in.) square openings and retained on a No. 10 (2 mm) sieve.

**Coarse Sand**
Material passing the No. 10 sieve and retained on the No. 40 (425 μm) sieve.

**Fine Sand**
Material passing the No. 40 sieve and retained on the No. 200 (75 μm) sieve.

**Table 12.3**
Soil classification by the AASHTO system

| General Classification | Granular Materials (35% or less passing No. 200) | | | | | | | | Silt-Clay Materials (More than 35% passing No. 200) | | | | |
|---|---|---|---|---|---|---|---|---|---|---|---|---|---|
| | A-1 | | A-3 | A-2 | | | | A-4 | A-5 | A-6 | A-7 | | |
| Group Classification | A-1-a | A-1-b | | A-2-4 | A-2-5 | A-2-6 | A-2-7 | | | | | A-7-5 | A-7-6 |
| Sieve analysis, percent passing: | | | | | | | | | | | | | |
| No. 10 | 50 max. | | | | | | | | | | | | |
| No. 40 | 30 max. | 50 max. | 51 min. | | | | | | | | | | |
| No. 200 | 15 max. | 25 max. | 10 max. | 35 max. | 35 max. | 35 max. | 35 max. | 36 min. | 36 min. | 36 min. | 36 min. | | 36 min. |
| Characteristics of fraction passing No. 40: | | | | | | | | | | | | | |
| Liquid limit | | | | 40 max. | 41 min. | 40 max. | 41 min. | 40 max. | 41 min. | 40 max. | 41 min.[b] | | 41 min. |
| Plasticity index | 6 max. | | NP | 10 max. | 10 max. | 11 min. | 11 min. | 10 max. | 10 max. | 11 min. | 11 min. | | 11 min. |
| Usual types of significant constituent materials | Stone fragments, gravel and sand | | Fine sand | Silty or clayey gravel and sand | | | | Silty soils | | Clayey soils | | | |
| General rating as subgrade | Excellent to good | | | | | | | Fair to poor | | | | | |

[a] Classification procedure: With required test data available, proceed from left to right on above chart and correct group will be found by the process of elimination. The first group from the left into which the test data will fit is the correct classification.
[b] Plasticity index of A-7-5 subgroup is equal to or less than LL minus 30. Plasticity index of A-7-6 subgroup is greater than LL minus 30 (see Fig. 12.12)

### Silt-Clay or Combined Silt + Clay
Material passing the No. 200 sieve.

### Boulders
Boulders retained on the 75 mm (3 in.) sieve are excluded from the portion of the sample being classified, but the percentage of such material is recorded.

The term "silty" is applied to fine material having a plasticity index of 10 or less, and the term "clayey" is applied to fine material having a plasticity index of 11 or more after rounding to the nearest whole percent.

## 12.12.4  Descriptions of Groups

The following generalized observations may be applied to the various AASHTO soil groups:

### A-1
Typical of this group are well-graded mixtures of stone fragments or gravel, volcanic cinders, or coarse sand. They do not contain a soil binder or have a nonplastic or feebly plastic binder. Subgroup A-1-a is mainly stone fragments or gravel, and A-1-b is mainly coarse sand.

### A-3
Typical of this group is fine beach sand or fine desert blow sand without silty or clayey fines, or with a very small amount of nonplastic silt. The group also includes stream-deposited mixtures of poorly graded fine sand with limited amounts of coarse sand and gravel.

### A-2
This group includes a wide variety of granular materials that are at the borderline between A-1 and A-3 and silt-clay materials of groups A-4 through A-7. A-2 includes materials with less than 35 percent passing a No. 200 sieve that do not classify as A-1 or A-3, because either the fines content or plasticity, or both, are in excess of the amounts allowed in those groups.

Subgroups A-2-4 and A-2-5 include various granular materials with not more than 35 percent passing a No. 200 sieve and containing a minus No. 40 portion that has characteristics of the A-4 and A-5 groups, respectively. These subgroups include such materials as gravel and coarse sand with silt content or plasticity index in excess of those allowed in A-1, and fine sand with nonplastic silt content in excess of the limitations of group A-3.

Subgroups A-2-6 and A-2-7 include materials similar to those described under subgroups A-2-4 and A-2-5, except that the fine portion contains plastic clay having the characteristics of the A-6 or the A-7 group. The group index, described below, is 0 to 4.

## A-4
The typical material of this group is a nonplastic or moderately plastic silty soil, 75 percent or more of which passes the No. 200 sieve. However, the group also can include mixtures of fine silty soil with up to 64 percent retained on the No. 200 sieve.

## A5
Typical of this group is soil that is similar to that described under group A-4, but has a diatomaceous or micaceous content that makes it highly elastic, indicated by a high liquid limit. These soils are "springy" and may be difficult to compact.

## A-6
The material of this group typically is plastic clay soil with 75 percent or more passing the No. 200 sieve, but can include fine clayey soil mixtures with up to 64 percent retained on the No. 200 sieve. Materials of this group usually have high volume change between wet and dry states.

## A-7
A-7 soils are similar to A-6 but have higher liquid limits. Subgroup A-7-5 materials have moderate plasticity indexes in relation to liquid limit, and which may be highly elastic as well as subject to considerable volume change on wetting or drying. Subgroup A-7-6 materials have high plasticity indexes in relation to liquid limit, and are subject to very high volume changes.

## A-8
A-8 soil is peat or muck soil in obviously unstable, swampy areas. A-8 soil is characterized by low density, high compressibility, high water content, and high organic matter content. Attention is directed to the fact that the classification of soils in this group is based largely upon the character and environment of their field occurrence, rather than upon laboratory tests of the material. As a matter of fact, A-8 soils usually show laboratory-determined properties of an A-7 soil, but are properly classified as group A-8 because of the manner of their occurrence.

## 12.12.5  Group Index

The group index gives a means for further rating a soil within its group or subgroup. The index depends on the percent passing the No. 200 sieve, the liquid limit, and the plasticity index. It is computed by the following empirical formula:

$$\text{Group index} = (F - 35)[0.2 + 0.005(LL - 40)] + 0.01(F - 15)(PI - 10) \quad (12.6)$$

in which $F$ is the percent passing the No. 200 sieve, expressed as a whole number and based only on the material passing the 75 mm (3 in.) sieve, $LL$ is the liquid limit, and *PI* is the plasticity index. When the calculated group index is negative it is reported as zero (0).

The group index is expressed to the nearest whole number and is written in parentheses after the group or subgroup designation. A group index should be given for each soil even if the numerical value is zero, in order to indicate that the classification has been determined by the AASHTO system instead of the original Public Roads system. A nomograph has been devised to solve eq. (12.6), but it now is more conveniently solved with a computer spreadsheet.

# 12.13 LIMITATIONS AND COMPARISONS OF SOIL CLASSIFICATION SYSTEMS

The classification systems described above use disturbed soil properties and therefore do not take into account factors such as geological origin, fabric, density, or position of a groundwater table. The classifications nevertheless do provide important information relative to soil behavior so long as the limitations are recognized. Classification is no substitute for measurements of important soil properties such as compressibility, shear strength, expandability, permeability, saturation, pore water pressure, etc.

Boundary lines for fine soils in the Unified and AASHTO classification systems do not precisely coincide, but the systems are close enough that there is considerable overlapping of designations, so a familiarity with one system will present at least a working acquaintance with the other.

Some approximate equivalents that will include most but not all soils are as follows:

### A-1-a or GW
Well-graded free-draining gravel suitable for road bases or foundation support.

### A-1-b or SW
Similar to A-1-a except that it is primarily sand.

### A-2 or SM or SC
Sand with appreciable fines content. May be moderately frost-susceptible.

### A-3 or SP
Sand that is mainly one size.

### A-4 or ML
Silt that combines capillarity and permeability so that it is susceptible to frost heave. Low-density eolian deposits often collapse when wet.

### A-5 or Low-Plasticity MH

Includes micaceous silts that are difficult to compact.

### A-6 or CL

Moderately plastic clay that has a moderate susceptibility to frost heave and is likely to be moderately expansive. All A-6 is CL but not all CL is A-6.

### A-7-5 or Most MH

Silty clay soils with a high liquid limit, often from a high mica content.

### A-7-6 or CH

Highly plastic clay that is likely to be expansive. Low permeability reduces frost heave. All A-7-6 soils also classify as CH.

### A-8 and Pt

Peat and muck.

## 12.14 OTHER DESCRIPTIVE LIMITS

Other tests and descriptive terms have been devised or defined that are not as widely used or have fallen into disuse. Some are as follows:

### 12.14.1 Toughness

Toughness is defined as the flow index from the liquid limit test, which is the change in moisture content required to change the blow count by a factor of 10, divided by the plasticity index.

### 12.14.2 Shrinkage Limit

The shrinkage limit test was suggested by Atterberg and has been used as a criterion for identifying expansive clay soils. However, the test involves complete destruction of the soil structure and drying from a wet mud, which makes correlations less reliable. The shrinkage limit generally is lower than the plastic limit, and the transition from the intermediate semisolid state to a solid is accompanied by a noticeably lighter shade of color due to the entry of air.

The shrinkage limit test also fell into disfavor because it used a mercury displacement method to measure the volume of the dried soil pat. An alternative method now coats the soil pat with wax for immersion in water (ASTM D-4943).

In order to perform a shrinkage limit test a soil-water mixture is prepared as for the liquid limit but with a moisture content that is considerably above the liquid limit, and the moisture content is measured. A sample is placed in a shallow dish that is lightly greased on the inside and struck off even with the top of the dish, which has a known weight and volume. The soil then is oven-dried at 110°C and the weight recorded. The soil pat then is removed and suspended by a thread in melted wax, drained and allowed to cool, and re-weighed.

During oven-drying the volumetric shrinkage equals the volume of water lost until the soil grains come into contact, which is defined as the shrinkage limit. Then,

$$SL = w - \left[ \frac{V - V_d}{m_s} \right] \times 100 \tag{12.7}$$

where $w$ and $V$ are the soil moisture content and volume prior to drying, $V_d$ is the volume of the pat after oven-drying, and $m_s$ is the mass of the dry soil in grams. The determination assumes that the density of water that is lost during drying is $1.0 \, \mathrm{g/cm^3}$.

A so-called "shrinkage ratio" equals the dry density of the soil at the shrinkage limit:

$$SR = m_s/V_d \tag{12.8}$$

where $SR$ is the shrinkage ratio and other symbols are as indicated above.

### 12.14.3  COLE

The "coefficient of linear extensibility" (COLE) test is used by soil scientists to characterize soil expandability, and has an advantage over the shrinkage limit test in that the original soil structure is retained, which as previously discussed can greatly reduce the amount of soil expandability. No external surcharge load is applied. A soil clod is coated with plastic that acts as a waterproof membrane but is permeable to water vapor. The clod then is subjected to a standardized moisture tension of 1/3 bar, and after equilibration its volume is determined by weighing when immersed in water. The volume measurement then is repeated after oven-drying, and the volume change is reduced to a linear measurement by taking the cube root:

$$COLE = \sqrt[3]{(V_m/V_d)} - 1 \tag{12.9}$$

Where $V_m$ is the volume moist and $V_d$ is the volume dry. Volumes are obtained from the reduction in weight when submerged in water, which equals the weight of the water displaced. For example, if the reduction in weight is 100 g (weight), the volume is 100 cm³. A COLE of <3 percent is considered low, 3 to 6 percent moderate, and >6 percent high for residential construction (Hallberg, 1977).

### 12.14.4  Slaking

Shale may be subjected to a slaking test that involves measuring the weight loss after wetting and tumbling in a rotating drum (ASTM D-4644). Dry clods of soil also may slake when immersed in water as the adsorptive power may be so great that air in the pores is trapped and compressed by water entering the capillaries, causing the soil clod literally to explode and disintegrate. The same soil will not slake when saturated. Slaking therefore can provide an immediate clue that a soil has been compacted too dry, discussed in the next chapter.

## 12.15 SUMMARY

This chapter describes laboratory tests relating the plastic behavior to moisture content, which form the basis for engineering classifications. Two classification methods are presented, one that is more commonly used in highway soil engineering and the other in foundation engineering. Soils may be classified by either or both methods as part of a laboratory testing program. Classification is useful for determining appropriate uses of soils for different applications, but is not a substitute for engineering behavioral tests.

Results of classification tests can be influenced by air-drying, so soil samples preferably are not air-dried prior to testing. If they are air-dried, considerable mixing and aging are required to ensure complete hydration of the clay minerals prior to testing. As soils used in classification tests are remolded, the results are not directly applicable to most field situations, exceptions being soils that are being remolded in the base of active landslides, in mudflows, and soils that have been liquefied by vibrations such as earthquakes. Classification therefore is more commonly a diagnostic than a performance tool.

### Problems

12.1. Define liquid limit, plastic limit, plasticity index, and activity index.

12.2. Four trials in a liquid limit test give the following data. Plot a flow curve and determine the liquid limit.

| Number of blows | Moisture content, % |
|---|---|
| 45 | 29 |
| 31 | 35 |
| 21 | 41 |
| 14 | 48 |

12.3. If the plastic limit of the soil in Problem 12.2 is 13%, what is the plasticity index?

12.4. If the soil in Problem 12.2 contains 30% $2\,\mu m$ clay, what is the activity index?

12.5. The liquid limit of a soil is 59%, the plastic limit is 23, and the natural moisture content is 46%. What is the liquidity index? What is its significance?

12.6. The liquid limit of a soil is 69% and its natural moisture content is 73%. Is this soil stable, metastable, or unstable? What is the dictionary definition of metastable?

12.7. Define shrinkage limit and shrinkage ratio.

12.8. The volume of the dish used in a shrinkage limit test is measured and found to be $20.0\,cm^3$, and the volume of the oven-dry soil pat is $14.4\,cm^3$. The weights of the wet and dry soil are 41.0 and 30.5 g, respectively. Calculate the shrinkage limit and shrinkage ratio.

12.9. Describe a nonplastic soil and explain how this characteristic is determined in the laboratory.

12.10. Distinguish clearly between a nonplastic soil and one that has a PI equal to zero.

12.11. Can you think of a reason why a fine-grained binder soil should be close to or below the plastic limit when it is added to a coarse-grained soil to form a stabilized soil mixture?

12.12. If the PI of a stabilized soil pavement is too high, what adverse characteristics are likely to develop under service conditions? What may happen if the PI is too low?

12.13. Is soil containing water in excess of the liquid limit necessarily a liquid? Explain.

12.14. A soil clod coated with a semipermeable plastic membrane and equilibrated at 1/3 bar moisture tension weighs 210 g in air and 48 g submerged in water. After oven-drying, the corresponding weights are 178 g and 47 g. (a) What is the COLE? (b) Rate the expansive potential of this soil. (c) If you have no choice but to put a light slab-in-grade structure on the soil, what precautions might be taken to prevent damage?

12.15. What is the significance of the group index in connection with the AASHTO system of classification?

12.16. State the broad general character of soils included in groups A-1, A-2, and A-3 of the AASHTO system and give approximate equivalents in the Unified Classification system. What are the specific differences?

12.17. What are the principal differences between two soils classified as A-4 and A-5 in the AASHTO system?

12.18. What are the approximate Unified Classification equivalents of AASHTO groups A-4, A-5, A-6, A-7, and A-8? Which pairs are most nearly identical?

12.19. Give the major characteristics of soils included in (a) the GW, GC, GP, and GF groups of the Unified Classification system; (b) the SW, SC, SP, and SF groups; (c) the ML, CL, and OL groups; (d) the MN, CH, and OH groups.

12.20. Give four examples of borderline classifications in the Unified Classification system and explain what each means.

*The following problems are with reference to data in Table 7.5 of Chapter 7.*

12.21. Classify soils No. 1, 2, and 3 in Table 7.5 according to the AASHTO and Unified systems.

12.22. Classify soils No. 4, 5, and 6 in Table 7.5 according to the AASHTO and Unified systems.

12.23. Classify soils No. 7, 8, and 9 in Table 7.5 according to the AASHTO and Unified systems.

12.24. Classify soils No. 10, 11, and 12 in Table 7.5 according to the AASHTO and Unified systems.

12.25. Which soil in each system is most susceptible to frost heave? What characteristics contribute to this susceptibility?

12.26. Which soil in each system is most expansive? Which is moderately expansive?

12.27. A loess soil changes from A-4 to A-6 to A-7-6 depending on distance from the source. Predict the volume change properties including expansion and collapsibility.

12.28. State the Denisov criterion for loess collapsibity. Does it take into account the increase in density with depth?

12.29. Seasonal changes in moisture content of an expansive clay deposit extend to a depth of 4 m (13 ft). Does that depth coincide with the thickness of the active layer? Why (not)?

12.30. Why classify soils?

## References and Further Reading

American Society for Testing and Materials. *Annual Book of Standards*. ASTM, Philadelphia.

Chen, F. K. (1988). *Foundations on Expansive Soils*, 2nd ed. Elsievier, Amsterdam.

Grim, R. E. (1968). *Clay Mineralogy*. McGraw-Hill, New York.

Hallberg, G. (1977). "The Use of COLE Values for Soil Engineering Evaluation." *J. Soil Sci. Soc. Amer.* 41(4), 775–777.

Handy, R. L. (1973). "Collapsible Loess in Iowa." *Soil Sci. Soc. Amer. Proc.* 37(2), 281–284.

Handy, R. L. (2002). "Geology, Soil Science, and the Other Expansive Clays." *Geotechnical News* 20(1), 40–45.

Katti, R. K., Katti, D. R., and Katti, A. R. (2005). *Primer on Construction in Expansive Black Cotton Soil Deposits with C.N.S.L. (1970 to 2005)*. Oxford & IBH Publishing Co., New Delhi.

Skempton, A. W. (1953). "The Colloidal Activity of Clays." *Proc. 3rd Int. Conf. on Soil Mech. and Fd. Engg.* 1, 57.

U.S. Department of Interior Bureau of Reclamation (1974). *Earth Manual*, 2nd ed. U.S. Government Printing Office, Washington, D.C.

# 13 Compaction

## 13.1 HISTORICAL BACKGROUND

### 13.1.1 Paths of Least Resistance

Compaction is the oldest and most common method for soil stabilization, at first being accomplished automatically as herd animals followed the same trails—some of which are followed by the routes of modern highways.

Compaction has some obvious benefits. A compacted soil is harder, and can support more weight and shed water better than the same soil that has not been compacted. The extensive road system of the Roman Empire consisted of stones laid flat on top of compacted soil. Roman road builders even attempted to compact soil with elephants, which is not very efficient because elephants prefer to step in the same tracks. The modern version of the elephant walk is a "sheepsfoot roller" that emulates tracking by feet or hoofs, but does so randomly.

### 13.1.2 The Twentieth Century

Compaction was largely intuitive and hit-or-miss until the 1930s when the Los Angeles county engineer, R. R. Proctor, devised a laboratory test that allowed systematizing and optimizing the compaction process. Proctor's procedure revolutionized the construction of roads and earth dams, respectively the longest and the largest human-made structures in the world. Without these control methods, costs would be higher and failures commonplace.

## 13.2 PROCTOR'S DISCOVERY

### 13.2.1 A Direct Approach

Proctor recognized that moisture content is a major variable influencing compaction, and therefore devised a test that isolates the role of moisture content by holding other variables constant. Compaction was accomplished in the laboratory by ramming a soil into a standardized steel mold using a standardized amount of energy, and it was discovered that there is an optimum moisture content that gives the highest density for a given compactive effort. If a soil is too dry, more energy is required to attain a particular density, and it is too wet, no amount of energy will compact it to the same density. The optimum moisture content is designated by *OMC*.

Proctor's test was designed to simulate the action of a sheepsfoot roller. An earlier field standard was if a roller "walked out" of a soil layer as it was compacted from the bottom up, and some contractors still use this as a criterion for adequate compaction. However, "walking out" measures bearing capacity, not density, and a roller will walk out of a dry soil much more readily than a wet one. In a properly compacted soil the air content is so low that the soil has very little capacity for additional water.

The hand tamper devised by Proctor is shown in Fig. 13.1. The original tamper, shown on the left, involved dropping a 5.5 lb (24.5 N) weight 12 in. (0.305 m) to compact soil in three layers in a steel mold, using 25 blows per layer. The volume of the mold is $1/30\,ft^3$ ($944\,cm^3$). The lifting height is controlled by housing the rammer inside a pipe; the bottom of the pipe is placed in contact with the soil and the rammer lifted by hand until it contacts the top, when it is dropped.

### 13.2.2 AASHTO T99 Procedure

Four or five tests are performed with soil at different moisture contents, and if the soil is not re-used (ASTM 698) about 5 lb (2.25 kg) will be required. *Some laboratories re-use soil after it has been thoroughly pulverized between tests, but this is not an ASTM-approved procedure and if done should be so stated in the test report.* Soils containing significant amounts of mica should not be re-used as the mica breaks down. Soils containing halloysite clay mineral are irreversibly changed by drying, and should *not* be air-dried prior to testing.

Air-dried soils are moistened and allowed to age overnight to allow clay expansion. The soil is compacted in three approximately equal layers using a collar temporarily attached to the top of the mold to hold the soil. The collar is removed, the top of the soil struck off with a knife or straight edge, and the net weight and moisture content of the soil determined. The test then is repeated with

**Figure 13.1**

Hand-operated Proctor and modified Proctor density rammers. The mold holding the soil is standardized, but the modified Proctor uses a heavier rammer that is dropped farther on thinner soil layers. The modified test was developed to meet the need of air-field construction.

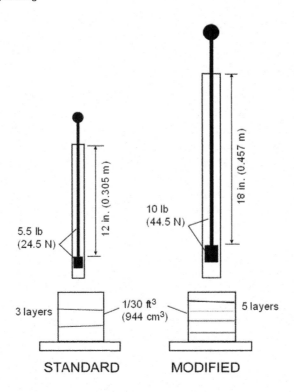

more or less added water. The AASHTO T180 (ASTM D-1557) modified Proctor procedure is similar except for a higher ramming energy and the use of five instead of three layers (Fig. 13.1).

## 13.2.3 Results

Figure 13.2 shows some typical results from standard compaction of a silty sand (SM) soil. A smooth curve is drawn through the data points, from which the optimum moisture content (OMC) and maximum compacted density are read from the graph.

An important feature of all such graphs is the theoretical "zero air void line," which is a curve that is calculated from the soil unit weight and specific gravity of the solid components. As can be seen, one leg of the density curve approximately parallels the zero air void line, indicating a persistent air content. It is physically impossible for the curve to cross the line, and the horizontal departure from the line indicates the amount of air that still is remaining in the soil after compaction. Measurements of air permeability indicate that as a soil is compacted and air voids are constricted, permeability decreases so that the last bit of air is retained in the fine pores.

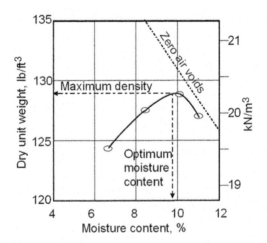

**Figure 13.2**

Compacted moisture-density relationship for an SM soil and a standard Proctor compactive effort. Modified Proctor compaction moves the curve upward and to the left to a higher density and lower OMC.

## 13.2.4   Not Lubrication

It once was assumed that the optimum moisture content was optimum because it provides just enough water to lubricate the soil grains. However, the effect of water on non-clay minerals is to increase instead of reduce sliding friction. As previously discussed, water weakens ionic bonding in a mineral surface so that it scratches more readily, increasing sliding friction.

The anti-lubricating effect of water also is shown by compaction curves such as Fig. 13.3, which is the same as in the previous figure but with additional data on the dry side of the OMC. Compacting the dry soil resulted in a higher density when the soil was dry than was obtained at the OMC.

The explanation for this behavior can be seen in Figs. 11.6 and 11.7 and the accompanying discussion in Chapter 11: annular rings of water linking soil grains

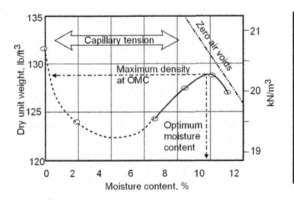

**Figure 13.3**

Additional test data for the soil in Fig. 13.2 demonstrates that friction and capillary tension define creates optimum moisture contents for compaction.

have negative pressure that draws the grains together, increasing sliding friction. Unless augmented by the presence of clay minerals, capillary tension and negative pressure disappear at zero moisture content. Capillary tension also disappears as the soil moisture content approaches saturation, giving the two highest compacted densities in a granular soil.

## 13.3  ZERO AIR VOIDS LINE

The zero air voids line normally is included on moisture-density (or moisture-unit weight) plots because it defines the edge of prohibited territory, which is the saturation moisture content for a particular density. If voids are completely filled with water there is no air left to be expelled. The saturated unit weight can be calculated from a block diagram or from the following formula:

$$\gamma_{max} = G_s \gamma_w / [1 + G_s(m/100)] \tag{13.1}$$

where $\gamma_{max}$ is the zero air voids unit weight, $\gamma_w$ is the unit weight of water, $G_s$ is specific gravity of the soil grains, and $m$ is the moisture content in percent. Solutions for several values of $G_s$ and $m$ are shown in Table 13.1 for plotting on compaction curves.

## 13.4  COMPACTING IN LAYERS

### 13.4.1  A Critical Requirement

Compaction requires time and energy, and it may seem a huge bother to have to spread and compact a soil in a series of layers. Occasionally an inexperienced operator will decide that it is not worth the bother, with grim and predictable consequences. The reason for this requirement is because pressure applied over a limited area at the ground surface spreads and dissipates with depth. This is illustrated in Fig. 13.4, where stress contours are drawn on the basis of elastic theory and decrease as the square of the distance from the contact area.

### 13.4.2  A Virtual Compactor

As a soil is compacted layer on layer, an important function of a previously compacted lower layer is to in effect cause stresses to "bounce up" and intensify. This is illustrated in Fig. 13.5. The argument is as follows: if the previously compacted soil layer is sufficiently rigid that no deflection occurs, its influence can be exactly duplicated by replacing the lower layer with vertical uplift pressures in the same pattern as the downward pressures caused by the weight of the compactor. This "virtual compactor" therefore doubles vertical pressure in the lower part of the soil layer that is being compacted.

| Moisture content, % | Specific gravity of solids | | | | |
|---|---|---|---|---|---|
| | 2.5 | 2.6 | 2.65 | 2.7 | 3 |
| 0 | 156.0 | 162.2 | 165.4 | 168.5 | 187.2 |
| 5 | 138.7 | 143.6 | 146.0 | 148.4 | 162.8 |
| 10 | 124.8 | 128.8 | 130.7 | 132.7 | 144.0 |
| 15 | 113.5 | 116.7 | 118.3 | 119.9 | 129.1 |
| 20 | 104.0 | 106.7 | 108.1 | 109.4 | 117.0 |
| 25 | 96.0 | 98.3 | 99.5 | 100.6 | 107.0 |
| 30 | 89.1 | 91.1 | 92.1 | 93.1 | 98.5 |
| 35 | 83.2 | 84.9 | 85.8 | 86.6 | 91.3 |
| 40 | 78.0 | 79.5 | 80.3 | 81.0 | 85.1 |
| 45 | 73.4 | 74.8 | 75.4 | 76.1 | 79.7 |
| 50 | 69.3 | 70.5 | 71.1 | 71.7 | 74.9 |
| 60 | 62.4 | 63.4 | 63.8 | 64.3 | 66.9 |
| 70 | 56.7 | 57.5 | 57.9 | 58.3 | 60.4 |
| 80 | 52.0 | 52.7 | 53.0 | 53.3 | 55.1 |
| 90 | 48.0 | 48.6 | 48.9 | 49.1 | 50.6 |
| 100 | 44.6 | 45.1 | 45.3 | 45.5 | 46.8 |

| Moisture content, % | Specific gravity of solids | | | | |
|---|---|---|---|---|---|
| | 2.5 | 2.6 | 2.65 | 2.7 | 3 |
| 0 | 24.51 | 25.49 | 25.98 | 26.47 | 29.41 |
| 5 | 21.78 | 22.56 | 22.94 | 23.32 | 25.57 |
| 10 | 19.61 | 20.23 | 20.54 | 20.84 | 22.62 |
| 15 | 17.82 | 18.34 | 18.59 | 18.84 | 20.28 |
| 20 | 16.34 | 16.77 | 16.98 | 17.19 | 18.38 |
| 25 | 15.08 | 15.45 | 15.63 | 15.80 | 16.81 |
| 30 | 14.00 | 14.32 | 14.47 | 14.62 | 15.48 |
| 35 | 13.07 | 13.34 | 13.48 | 13.61 | 14.35 |
| 40 | 12.25 | 12.49 | 12.61 | 12.73 | 13.37 |
| 45 | 11.53 | 11.75 | 11.85 | 11.95 | 12.51 |
| 50 | 10.89 | 11.08 | 11.17 | 11.26 | 11.76 |
| 60 | 9.80 | 9.96 | 10.03 | 10.10 | 10.50 |
| 70 | 8.91 | 9.04 | 9.10 | 9.16 | 9.49 |
| 80 | 8.17 | 8.28 | 8.33 | 8.38 | 8.65 |
| 90 | 7.54 | 7.63 | 7.67 | 7.72 | 7.95 |
| 100 | 7.00 | 7.08 | 7.12 | 7.15 | 7.35 |

**Table 13.1**

Densities and moisture contents that define zero air voids lines from eq. (13.1). The upper part of the table is in lb/ft$^3$, the lower part in kN/m$^3$

**Figure 13.4**

Compaction pressure dissipates as the square of the distance from the contact area, so soil must be compacted in thin layers or "lifts."

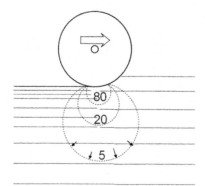

**Figure 13.5**

Concept of a "virtual compactor" that doubles the pressure in a compacting layer over a rigid previously compacted layer.

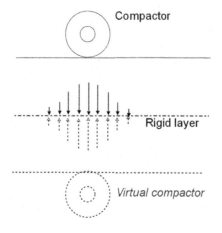

### 13.4.3 The Importance of Roughness

Vertical stress is not the only requirement for compaction, as soil also must be confined laterally or it will simply squeeze out. This effect can be demonstrated by pouring a small mound of dry sand on a glass plate and pressing it down with the palm of a hand; the sand squeezes out laterally. The experiment then is repeated with the sand supported on a sheet of sandpaper, and friction with the sandpaper reduces the amount of spreading so that more force can be applied to the sand and it is compacted.

Without a rough base, major principal stress vectors radiate outward from an applied surface load and tend to push the soil laterally, as shown at the right in Fig. 13.6. At the left in the figure, frictional restraint shown by the horizontal arrow tends to confine the soil and prevent spreading.

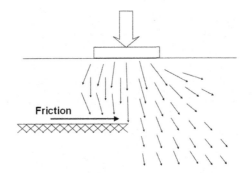

**Figure 13.6**
Frictional contact with a rough base layer helps to confine stress and aids compaction. Principal stress vectors on the right, without a firm base, are from Jürgenson (1934).

Frictional restraint also can be added on top of the compacting layer by substituting a grid or mesh pattern for a smooth-wheel roller, but it is at the bottom of the layer that restraint is most needed.

### 13.4.4  Tamping-Foot Rollers

Tamping-foot rollers have projections that intensify contact stress. The objective is to punch through and prevent compaction of the upper soil in each layer so that on repeated passes of the roller the soil compacts from the bottom up (Fig. 13.7). The wedge shape of the feet on tamping-foot rollers now is preferred over the knobby shape of the sheepsfoot as being less destructive to the compacted soil as the projections or feet walk out.

Design can be based on the same theory that is used for design of shallow foundations, presented later in this book, and in at least one case a roller weight was designed based on this theory because it is necessary for the bearing capacity to be exceeded by the individual projections for this type of roller to be effective. Bearing capacity, or support for a surface load, depends on several factors

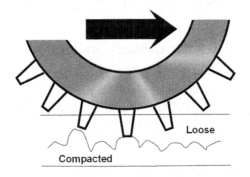

**Figure 13.7**
Foot pressure exceeds the bearing capacity of the uncompacted soil, penetrating through to compact the soil layer from the bottom up.

including the size of the contact area, the soil cohesion, and its angle of internal friction.

Successive roller passes on clay compact the soil and increase its cohesion, which increases the bearing capacity so that the roller walks out. Without the contribution of cohesion, as in cohesionless sands, there is very little walk-out so the projections only knead and stir the soil. Sands are compacted with smooth-wheel rollers or plates, and can be aided by vibration.

### 13.4.5   Proof Rolling

After a soil is compacted it may be subjected to "proof rolling" with a heavy roller or even a heavily loaded truck. Contrary to popular opinion, a track vehicle such as a bulldozer is *not* appropriate for proof rolling because it is designed to spread pressure and be supported on soft soils. The soil pressure during proof rolling equals the tire pressure in the vehicle.

Proof rolling can reveal pockets of soil that are insufficiently compacted or are compacted too wet, but the reliability of this procedure depends on the soil moisture content. If the moisture content is on the dry side of the OMC the soil will support a heavier load because of capillary tension or "apparent cohesion" that is destroyed when the soil becomes wet.

### 13.4.6   "Oreo Cookie Effect"

Even with a firm base and compaction in layers, stress distribution is not uniform through the depth of a single layer, which results in a nonuniformity that has been called the "Oreo cookie effect" (H. Allender, personal communication). Density measurements therefore should involve an entire layer or multiple layers in order to obtain reliable averages.

## 13.5   OUT GOES THE AIR, NOT THE WATER

### 13.5.1   A Critical Distinction

*Compaction* means to squeeze air out of the soil, which is rapid and is in contrast to *consolidation*, which means to squeeze out water. The viscosity of water is about $1 \times 10^{-2}$ poise, or about 50 times that of air. Compaction therefore is rapid and consolidation is at least 50 times slower. A saturated soil can be consolidated but not compacted.

As compaction proceeds and air is squeezed out, the soil eventually will reach the point where there is very little air left, and the soil is close to saturation as a result of compaction. The remaining air is trapped in tiny pockets, *and compaction stops regardless of the number of additional passes of a compactor.* If processing continues past that point, the result is called *overcompaction.*

### A Conflict over Nomenclature

"Consolidation" was first described by a geologist, Sir Charles Lyell, in a textbook published in 1851. Lyell noted that sediments on the sea bottom are compressed and consolidate under their own weight and not from the weight of hundreds or thousands of feet of sea water. Despite this prior definition by an eminent geologist, geologists in the petroleum industry call the process "compaction" and refer to shale as being "highly compacted." The *Oxford English Dictionary* favors the engineering usage, defining consolidation as bringing together and compaction as packing together. It is important that geotechnical engineers not confuse the two terms and their distinctive meanings.

## 13.5.2  Overcompaction

A continued application of compactive effort after soil reaches near-saturation point not only wastes energy, but the energy is redirected into shearing and remolding the compacted and nearly saturated soil. Shearing smears clay particles so that they are oriented parallel to the shear surfaces and permanently weaken the soil. This must be carefully guarded against, as the soil then must be dried and pulverized before it can be recompacted.

The effect of overcompaction on strength of a clay is shown in Fig. 13.8, which shows stress-strain curves for two samples of the same soil at the same moisture content and same density, the difference being that the weaker soil was compacted at a higher moisture content than the stronger one.

Clues to overcompaction are vibrations that can be felt under foot, and an occasional "thunky" sound when stomped on. A bulge of soil sometimes may be

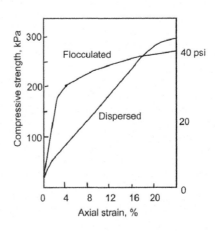

**Figure 13.8**

Stress-strain curves show the reduction in strength after a soil becomes dispersed through overcompaction and shearing. (After Seed and Chan 1959, with permission of the American Society of Civil Engineers.)

seen moving as a wave ahead of tires on heavy equipment, increasing energy consumption because the equipment always is moving up out of a hole of its own making. A relatively dry crust can develop that if broken through causes miring of the equipment. Overcompaction commonly occurs on construction haul roads on wet, fine-grained soils after repeated passes of heavy construction and earth-moving equipment.

The remedy for overcompaction obviously is not more compaction. Instead the soil must be ripped out and aerated or kiln-dried. If the soil contains active clay minerals, in particular smectite, an alternative procedure that may be less expensive and saves time is to add several percent hydrated lime or lime-rich fly ash, which raise the plastic limit and the optimum moisture content.

## 13.6  COMPACTION OF GRANULAR SOILS

### 13.6.1  Free-Draining Soils

Water content is not critical for compaction of free-draining soils such as gravel and crushed rock. Vibratory compaction is most effective with granular soils including sands, as vibrations create oscillations of intergranular contact stresses and friction.

### 13.6.2  Bulking and Flooding

Intergranular friction caused by capillary tension causes a granular soil to "bulk up" or decrease in density after it is disturbed. Thus, a bucket filled with dry sand will weigh more than a bucket filled with wet (not saturated) sand, even though the latter includes some additional weight from water. This effect is called *bulking*, and must be taken into account during batching operations for materials used in concrete.

Bulking also occurs when sand or gravel is spread as a base for a floor or foundation. Since the cause is capillary action, the density can be increased by saturating the soil with water. This has led to a somewhat haphazard practice of "compacting" granular soils by flooding with water, but it should be emphasized that the result is equilibrium with the self-weight of the soil, and not to any additional load. Flooding induces consolidation, not compaction. A soil that is consolidated to equilibrium with its self-weight is still compressible under an additional load. It therefore is said to be *normally consolidated*. The low densities achieved by flooding may go unnoticed with light structures as settlement is almost immediate, hopefully occurring before the concrete has set. Although flooding is better than no treatment at all, it is not a substitute for compaction.

### 13.6.3 Relative Density Test

The Proctor density test originally was intended to simulate the action of a sheepsfoot roller, but as previously mentioned, sheepsfoot rolling is not effective in sandy soils containing little clay. Furthermore, the coarser the soil, the less effective are capillary forces at preventing densification, so moisture content becomes less critical.

Because sands are most effectively compacted with a vibratory roller or plate vibrator, a different reference density is used based on densities achieved with a vibrating table (Fig. 13.9). Both maximum and minimum densities are determined, and the position of the compacted density in relation to these two limits is called the *relative density*, which is expressed as a percent.

The formula for relative density is

$$D_d = 100(e_{max} - e)/(e_{max} - e_{min}) \qquad (13.2)$$

where $e_{max}$ and $e_{min}$ are the maximum and minimum void ratios determined from the test, and $e$ is the void ratio of the soil.

A more convenient form of this equation that uses unit weights is

$$D_d = 100 \frac{\gamma_{max}(\gamma - \gamma_{min})}{\gamma(\gamma_{max} - \gamma_{min})} \qquad (13.3)$$

where $\gamma_{max}$ and $\gamma_{min}$ are, respectively, the maximum and minimum dry unit weights, and $\gamma$ is the dry unit weight of the soil.

### 13.6.4 Defining Minimum and Maximum Density

A minimum density, $\gamma_{min}$, is determined by oven-drying and pulverizing the soil to a single-grain structure, then slowly pouring it through a funnel into a container of known volume. The soil must be dry to prevent bulking.

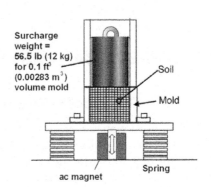

**Figure 13.9**

Vibrating table for obtaining a standardized maximum density of granular soils.

A maximum density, $\gamma_{max}$, is more difficult to define because it depends on the compaction mechanics and confinement. The determination has been standardized and involves putting loose soil into a cylindrical mold that is set on a vibrating table (Fig. 13.9). A surcharge weight is placed on top to confine the soil, and the table is actuated with electromagnetic vibrations at 50 or 60 Hz depending on availability. The vibration time is adjusted to give a uniform number of cycles.

Tests have shown that $\gamma_{max}$ also depends on the geometry of the equipment, the applied load, and the acceleration ratio, which is the ratio of acceleration from the vibrations to that of gravity, so the conditions under which the test is performed must be carefully controlled (ASTM Designation D-4253).

Minimum and maximum densities often are determined with the soil wet and after oven-drying, and the higher sets of values used for determining relative density. While capillary effects reduce the efficiency of compaction, they also tend to hold grains together after densification instead of letting them fly apart from the continuing vibrations.

### 13.6.5 Applications of Relative Density

The relative density determination is limited to soils with 15 percent or less noncohesive fines passing the No. 200 sieve. Soils containing 5 to 15 percent noncohesive fines also may be impact-compacted for comparative purposes.

**Example 13.1**
A field density determination gives $\gamma = 110 \, lb/ft^3$ (17.3 kN/m³). Laboratory tests give $\gamma_{max} = 127 \, lb/ft^3$ and $\gamma_{min} = 102 \, lb/ft^3$ (19.95 and 16.0 kN/m³). What is the relative density of the field soil?

*Answer:*

$$D_d = 100 \frac{127(110 - 102)}{110(127 - 102)} = 37\%$$

## 13.7 COMPACTION SPECIFICATIONS FOR FINE-GRAINED SOILS

### 13.7.1 The Proctor Relationship

Most compaction specifications are written around an optimum moisture content and maximum density from a Proctor density test (Fig. 13.2). The moisture content of a soil in the field is adjusted by adding water or aerating and drying until it is within a specified range of the optimum moisture content.

A 100 percent density is difficult to achieve unless the moisture content is on the dry side of the OMC, so the density requirement in the field normally is set at 95 percent of the maximum density obtained in the laboratory test. A target area may be defined by a moisture content within ± 2 percent of the OMC and a compacted density to equal or exceed 95 percent of the maximum is indicated by the specification area in Fig. 13.10. This area can be modified to meet special requirements.

Figure 13.10 also shows "compaction growth curves" for different percentages of the standard compactive effort. As the compactive effort increases, the OMC is reduced along a line that roughly parallels the zero air voids curve.

## 13.7.2 Too Dry for Comfort

Proctor discovered a problem that still sometimes is overlooked or ignored—that compacting a soil on the dry side of the OMC leaves too much air in the soil. The soil remains permeable to water, and when saturated may become so weakened that it densifies or collapses under its own weight—even though the compacted density can meet a 95 percent specification requirement. Some potential consequences are illustrated in Fig. 13.11. The problem is most common in trench backfill where the compaction moisture content often is not monitored or controlled. As a result a strip of pavement that is over a sewer running down the middle may be unsupported from one manhole to the next.

A particularly serious consequence of settlement of soil backfill in a trench is if the soil weight is carried by a pipe crossing through the trench so that the pipe breaks.

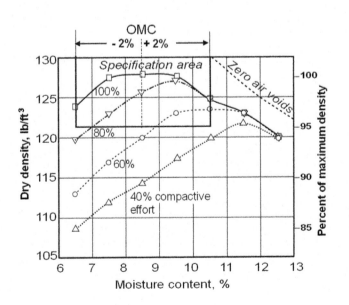

**Figure 13.10**

Compaction growth curves for a silty clay. The shaded area is for a specification requirement of 95 percent minimum density with the moisture content within 2 percent of the OMC. (Data courtesy of Prof. J. M. Hoover.)

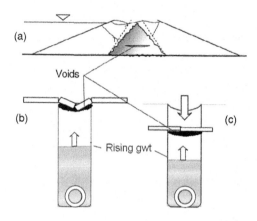

**Figure 13.11**

Serious problems caused by collapse of soil compacted on the dry side of the OMC: (a) earth dam piping; (b) pavement collapse; (c) fracture and leak of a gas or water pipe: gas flowing through loose trench backfill soil and into a building can cause fire or explosion.

Broken water mains can cause locally severe underground erosion, so the dramatic collapse of the street is sure to make the 6 o'clock news.

Gas line breaks can be more tragic, as escaping gas follows the path of least resistance through loose trench backfill into a building or basement, where it can cause a fire or explosion. Pipes crossing through soil in a trench should be supported by some type of beam.

Collapse of soil within an earth dam can lead to serious consequences by allowing "piping," or flow of water through the dam, leading to erosion and failure.

### 13.7.3   Preventing Collapse of Compacted Soil

Figure. 13.12 shows volume changes after compacting and wetting. The area of zero volume change is shown shaded, and for the most part falls within a box drawn for the specification limits of OMC ± 2 percent, and a compacted unit weight between 95 and 100 percent of the standard Proctor density. Thus, compaction within these specification limits normally will result in a soil that is relatively stable against future volume changes. An exception is expansive clays, which are discussed later in this chapter.

Where even a small amount of collapse cannot be tolerated, as in the case of the core material in an earth dam, the specification limits for moisture content often are increased to between the OMC and the OMC plus 2 percent to as high as 4 percent. However, compaction on the wet side weakens the soil and makes it more susceptible to overcompaction and development of excess pore water pressure.

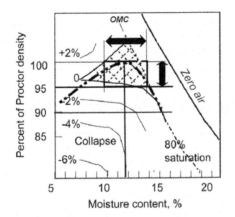

**Figure 13.12**

Contours show volume changes from wetting of a compacted silty clay. Compaction within usual specification limits (shown by the heavy arrows) results in negligible volume change if the soil is not highly expansive. (After Lawton et al., 1992.)

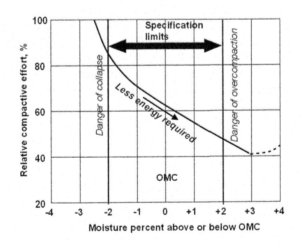

**Figure 13.13**

Relation of compaction energy to moisture content, determined from energy data along the 95 percent density line in Fig. 13.12.

## 13.7.4 Compacting for Low Permeability

In applications such as in the confined core of an earth dam where low permeability is more important than strength, the moisture content requirement usually is 0 to +2 percent or +4 percent above the OMC.

Compacting on the wet side of the OMC also can save time and fuel by reducing the required compactive effort. In Fig. 13.13 about 65 percent of the standard compactive effort is required for soil to reach the specification density at the optimum moisture content. This is reduced to about 50 percent at 2 percent above optimum. However, compacting at above the OMC narrows the target range and increases the possibility for overcompaction.

### 13.7.5 Relaxing Density Requirements

For applications where the compacted soil is not heavily relied on for strength and where some settlement is permissible, such as for landscape grading that does not involve steep slopes or high retaining walls, the density requirement may be reduced to 90 percent of the maximum. In the example in Fig. 13.13, 90 percent density could be achieved with about 40 percent effort at the OMC, or compared with 65 percent for 95 percent density. The same criterion applied to the soil in Fig. 13.12 would result in about a 3 percent reduction in volume upon saturation, so saturating the lower 1 m (3 ft) of a fill placed by this criterion would generate a settlement of 30 mm or a little over an inch.

### 13.7.6 Modified Proctor Density

The trend toward a higher density and lower optimum moisture content continues with modified Proctor compaction, which moves the density curve higher and farther to the left (Fig. 13.14). The energy for laboratory compaction is 56,00 ft-lb/ft$^3$ (2700 kN-m/m$^3$) for the modified test compared with 12,400 ft-lb/ft$^3$ (600 kN-m/m$^3$) for the standard test, an increase of over 450 percent. The modified density requirement therefore is energy-intensive and is used only in special circumstances.

**Figure 13.14**

Modified Proctor density is achieved with a higher compactive effort and a lower moisture content, and results in a denser, stronger, and less compressible soil. It is used in heavy, duty applications such as airport landing strips and runways.

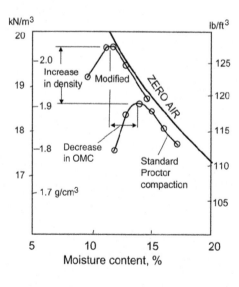

*Question*: Can a modified Proctor density be specified with a standard Proctor moisture content? What about the reverse situation?

*Answer:* No and no. Why?

## 13.8   SPECIAL PROBLEMS WITH COMPACTION OF EXPANSIVE CLAYS

### 13.8.1   So There's a Problem; Can We Cut a Deal?

Expansive clays do not give up easily. As discussed in the preceding chapter, Section 12.8.9, expansion may be reduced or prevented by burying under 1 m (3 ft) of compacted nonexpansive clay (n.e.c.) that can be manufactured by adding and mixing several percent hydrated lime. What other methods, if any, are available?

### 13.8.2   Adjusting the Compaction Moisture Content

Logically, if expansive clay is compacted wet enough it should be fully expanded, but unfortunately it then will shrink if it is dried out. On the other hand, compaction on the dry side of optimum will leave it permeable and subject to expansion on wetting. The latter is illustrated in Fig. 13.12, so the moisture content for compaction of an expansive clay often is specified to be between the OMC and OMC + 2 percent.

The two options, compacting on the dry side or on the wet side of the optimum moisture content, are further illustrated by data in Fig. 13.15 for an expansive sandy clay. From the upper graph it may be seen that expansion is over 2 percent if the soil is compacted to maximum density at the optimum moisture content, is reduced about one-half if the soil is compacted at 2 percent above optimum, and becomes zero if compacted at 4 percent above optimum. However, such a soil will be greatly weakened by dispersion and possible overconsolidation, and if it dries out it will shrink excessively. On the other hand, if the soil is compacted at 2 percent below the OMC, shrinkage is reduced one-half compared to that which will occur if the soil is compacted at the OMC, but shrinkage is not completely eliminated.

Most problems with expansive clays occur when they are under low pressure and subject to either an increase or a decrease in the moisture content, so the compaction moisture content can be selected in anticipation of future changes. For example, moisture tends to accumulate in soil that is sealed off from the atmosphere, such as under pavements or floor slabs, so to reduce the likelihood of volume change, expansive clay may be compacted on the wet side of the OMC.

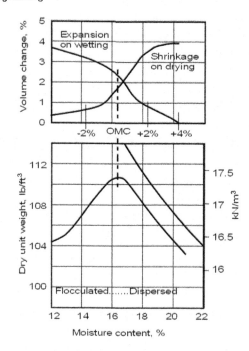

Seasonal moisture changes in expansive clay can be reduced by (a) periodically watering around the outer perimeter of a house or other structure, (b) extending a concrete slab or membrane horizontally outside of the structurally loaded areas, or (c) installing a vertical moisture barrier around the perimeter. Complete isolation is difficult because of the high suction forces exerted by dry expansive clay and the disruption caused by shrinkage cracking. In cooler climates such as in Canada, interior heating may drive moisture out of expansive clay during the winter, causing shrinkage that is very difficult to control.

Because some variations in moisture content are unavoidable, another option that may be economical for small structures is to support them on a structurally designed slab-on-grade that can bridge between support areas. Larger structures can be supported on deep foundations that extend below the active zone of the clay, in which case floor joists are supported on bails of straw or cardboard "crush boxes" so that the beams will not be lifted off the foundations, and floors are structurally supported high enough off the ground to allow for clay expansion.

A procedure that should not be overlooked is to replace the upper layer of expansive clay with a soil that is nonexpansive and treat it with hydrated lime. Treatment of the upper 1 ft (0.3 m) can reduce expansion, of the upper 2 ft (0.6 m) can reduce it even further, and of the upper 3 ft (1 m) may bring it under control.

# 13.9  STRENGTH AND MODULUS OF COMPACTED SOIL SPECIMENS

## 13.9.1  Another Emphasis

Compaction specifications not only are intended to control future volume changes of a soil, they also may be intended to increase the soil strength. It sometimes is assumed that higher density means higher strength, but this trend is trumped by the moisture content: compacting on the dry side of the OMC may leave the soil such that it can collapse under its own weight when wet with water, or too vigorous compaction on the wet side can shear and remold the soil.

Most compacted soil is used in embankments for roads, highways, or earth dams. In these applications, soil in the embankments not only should resist volume changes, it must have sufficient shearing strength that side slopes are stable and do not develop landslides. Railroad embankments, many constructed a century ago, usually were not compacted and, significantly, still require continual maintenance and repair. It therefore is important that the shear strength characteristics of the compacted soil should be determined.

Compacted soils must be strong enough to support structures and minimize settlement that not only can affect the integrity of the structure but also that of connecting utility lines; a broken gas or water line or sewer that slopes the wrong way is more than just an inconvenience.

## 13.9.2  Strength of Compacted Soils

Since compacted soil is a manufactured, quality-controlled product, its suitability for particular applications often is based on experience with the same or similar soils under similar applications. However, if a compacted soil is to be used for founding heavy structures that are outside of the range of experience, the soil strength and compressibility should be measured at the worst-case densities and moisture contents that are within specification limits, and those values used in design. After a soil is compacted it can be drilled, tested, and analyzed using methods presented in later chapters.

The simplest option is to subject compacted density specimens to an unconfined compression test, where ends are loaded until the specimen fails. This test is discussed in Chapter 18.

## 13.9.3  Laboratory "K Test"

Specimens compacted for laboratory density determinations usually are discarded, but they can easily and quickly be tested for strength and

**Figure 13.16**

The expanding steel mold of the "K test" enables rapid estimates of soil strength and compressibility parameters from Proctor density specimens, but drainage may be incomplete.

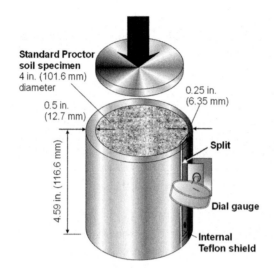

compressibility in an expandable steel mold (Fig. 13.16). A specimen from a Proctor density test is inserted in the mold and compressed at the ends, which causes the mold to expand and automatically exert a gradually increasing lateral confining stress on the sample.

In geotechnical engineering, $K$ denotes the ratio between horizontal and vertical stress. In a "K test," vertical stress is measured and the lateral stress is obtained from a suitable calibration from expansion of the steel mold. As both stresses simultaneously increase it is possible to obtain running estimates of both the soil cohesion and the angle of internal friction, as well as a compression modulus and soil-to-steel sliding friction. Although the soil is confined between porous stones to allow drainage, compacted specimens contain sufficient air that little drainage occurs. K-test results have been used with bearing capacity theory for design of tamping-foot rollers.

## 13.10   GRADATION AND COMPACTED DENSITY

Density after compaction is closely related to the range of particle sizes, as progressively smaller particles fill in the voids, whereas density also is influenced by the largest particles present, as a solid particle is more dense than a cluster of many smaller particles.

A variety of moisture-density curves representing different gradations of soil is shown in Fig. 13.17, where it will be seen that, in general, the coarser the soil, the higher the compacted density and lower the OMC. However, it also may be noted

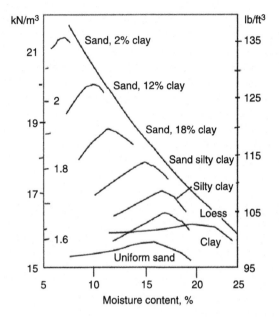

**Figure 13.17**

Some representative tests showing the influence of gradation on compacted density (Johnson and Sahlberg, 1962).

that the lowest density is obtained with a sand that is nearly all one size, which is the reason such a soil is said to be poorly graded and classified as SP or A-3.

## 13.11    FIELD DENSITY TESTS

### 13.11.1    Density as a Means for Compaction Control

Measurements of soil density and moisture content are the primary means for quality control of compaction. In-place density measurements also are used for determining cut-and-fill quantities. For example, if the density of soil in a borrow pit will be increased 15 percent as a result of compaction operations, this 15 percent reduction in volume (which is loosely termed "shrinkage") must be taken into account when calculating borrow quantities.

### 13.11.2    Direct Measurement Methods

The classic methods for measuring compacted density involve hand-excavating a hole and measuring the volume of the hole, and the weight and moisture content of the soil taken from the hole. The volume of the hole is determined by expanding a rubber balloon into the hole with water and measuring the volume of water required (a volumeter test), or by filling the hole with a standard sand and measuring the amount of sand required (a sand-cone test). Construction traffic must be halted during sand-cone tests because vibrations affect packing of the sand. Details of these tests will be found in laboratory testing manuals or standard methods of the ASTM.

Another method involves driving a thin-walled steel tube into the soil and measuring the weight and volume of soil extracted. This is considered less accurate than the methods described above because of sample disturbance and compaction during sampling.

A thin-walled steel tube called a Shelby tube can be pushed instead of driven, by use of a drilling machine. The sample then is extruded in the laboratory, and the weight, volume, and moisture content of representative samples measured. Strengths also can be determined. This is the same procedure that is used in exploration drilling for foundation soils, and has the advantage that it also is used for sampling and testing soil underneath a compacted layer. Routine samples for purposes of soil identification are obtained by driving a thick-walled Standard Penetration Test (SPT) device, but the soil is so disturbed and compacted that it is not acceptable for density determinations. Details of these tests are found in Chapter 26.

### 13.11.3  Nuclear Density Measurements

The most common method for control of field compaction operations involves measuring both the soil density and moisture using a nuclear density-moisture meter. Two modes of operation are available, *backscatter*, which is as the name implies, and *transmission*. For transmission measurements a probe containing a radioactive source is pushed down into a hole in the soil and radiation is transmitted upward at an angle to detectors in the base of the device. In the backscatter mode, both the source and counter are at the ground surface, which gives lower penetration and is less accurate. The backscatter mode is particularly useful for measuring densities of pavements.

For the transmission mode a probe is driven into the soil to the required depth and withdrawn, or a hole is made with an auger. The nuclear unit is placed over the hole and a rod that is slightly smaller than the probe hole is lowered from within its shield down into the hole, as shown in Fig. 13.18. The rod contains a radioactive source that emits gamma rays. Best accuracy is obtained if the source is against the side of the hole closest to the detectors. Gamma radiation transmitted through the soil up to the bottom of the device is counted for 15 seconds to 4 minutes, depending on the accuracy desired and the transmission characteristics of the soil.

Although nuclear devices measure density in $kg/m^3$, they also are calibrated to read unit weights in $lb/ft^3$ or $kN/m^3$. Test units require daily calibration, which can be done by testing in concrete blocks of known density or other materials such as limestone, granite, or aluminum (ASTM Designation D-2922). Operation requires trained and certified personnel.

**Figure 13.18**
Nuclear density gauge. For transmission measurements the probe is lowered the desired distance into a hole previously prepared in the soil. (Photo courtesy Dr. David White.)

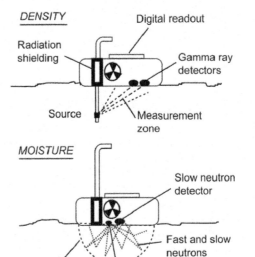

**Figure 13.19**
Principle of nuclear density-moisture content testing device for monitoring field compaction operations.

## 13.11.4   Nuclear Determinations of Moisture Content

Moisture contents are determined using a fast neutron source that is incorporated in the same instrument, shown diagrammatically in Fig. 13.19. As neutrons encounter hydrogen ions in the soil they are slowed, and those returning to the unit are counted. The count includes all hydrogen ions including those that are structurally held as $OH^-$ ions in mica and the clay minerals, plus hygroscopic water that is not driven off by normal oven-drying, plus hydrocarbons in organic matter. Calibration therefore is required by counting similar soils with moisture contents that are determined by oven-drying (ASTM Designation D-3017).

### 13.11.5   Accuracy of Nuclear Determinations

With modern instruments and accurate calibrations, density determinations are accurate to within about $\pm 1$ percent, and moisture contents to within about $\pm 2.5$ to $10 \, kg/m^3$ (0.2 to $0.6 \, lb/ft^3$), depending on the counting time. The upper part of the tested layer is emphasized, as about 50 percent of the counts come from the upper 25 percent of the soil layer tested by the transmission method.

### 13.11.6   Operating Nuclear Devices

The U.S. Nuclear Regulatory Commission requires that nuclear density testers have radiation warnings and be kept under lock and key when not in use. Operators must be specially trained and certified, and wear film badges to ensure that dosage does not exceed prescribed levels. Training normally is available from instrument manufacturers. Units are periodically checked for radioactivity leakage, and require periodic replacement of nuclear source materials. With proper safety precautions the radiation danger is negligible, particularly when compared to hazards from construction traffic.

## 13.12   FIELD COMPACTION EQUIPMENT

A broad catalog of compaction equipment is available and is continually being improved as manufacturers seek a competitive edge. Most common are heavy rollers, vibratory rollers, and tamping rollers. In close quarters such as trench backfills, hydraulic hammers can be mounted on backhoes. Hand-operated vibratory or "jumping jack" equipment may be used for moderate needs close to buildings. Applicability of different equipment for different soil types is indicated in Table 13.2.

## 13.13   CHEMICAL COMPACTION AIDS

Many different chemical compounds have been marketed as "compaction aids." These generally are detergents that are added in small mounts to water to decrease capillary tension, which reduces intergranular friction so that a specification density can be achieved with less effort. However, reducing capillary attraction

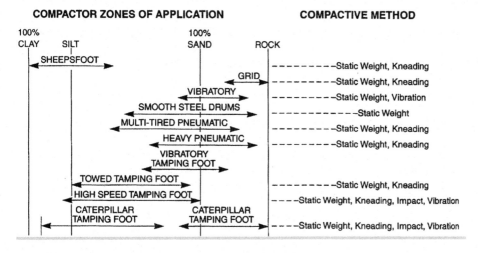

Adapted from *Caterpillar Performance Handbook*, © Caterpillar Inc

also reduces the strength of the compacted soil and can increase dispersion of clays. In other words, a small amount of chemical can make a soil easier to compact by making it weaker. *In recognition of this, chemical "compaction aids" are not permitted in most compaction specifications.*

Whereas hydrated lime flocculates an active clay soil, increasing the OMC and lowering the compacted density, chemical compaction aids do just the opposite. Field inspectors should be wary of water additions that have a soapy feel or tend to foam. Many aids are rendered ineffective by expansive clay minerals that remove organic chemicals by adsorption between clay layers.

$CaCl_2$, calcium chloride, also is the reverse of a compaction aid as it increases the surface tension of water, and densification occurs as a result of higher suction forces when the soil dries out. Calcium chloride occasionally is used to increase the surface density of granular unpaved roads and as a dust palliative, but has a low pH that accelerates rusting.

Another chemical additive, lignin, which is a byproduct from paper manufacturing, is used as a dust palliative and behaves as a glue until rain leaches it out.

## 13.14  STATISTICS OF COMPACTION

### 13.14.1  Variability and Accuracy of Testing

The more variable a material, the more tests are required to characterize the final product, and soil can be quite variable. The most common statistical parameter is the *arithmetic mean*, or *average*, but this says nothing about variability. If the average density meets a design specification, it means that one-half of the tests do not meet the specification, so the average must be higher than the specification requirement.

### 13.14.2 Scraper Mixing

Soil variability can be reduced by a simple procedure illustrated in Fig. 13.20, where scrapers load while moving downslope across soil layers instead of picking them up one by one. Additional mixing is obtained by disking and grading, or a specialized soil mixer can be used.

### 13.14.3 Maxima and Minima

Maximum and minimum values indicate the range or spread of the data. If minimum values fall below the design requirement, the engineer must make a choice: should the area where the deficiency occurred be reworked and recompacted, or can the deficiency be ignored? Usually the first option is selected because no construction person cares to approve a test that does not meet specifications.

The engineer also has other choices: if one determination does not meet specifications it may only reflect a natural variability of data, but if half of the tests do not pass, the entire project is suspect and should be tested and, if necessary, torn out and recompacted under more carefully controlled conditions.

If moisture contents are too high, as sometimes happens when compaction is attempted during rain, the soil must be scarified and aerated before being

**Figure 13.20**
Mixing reduces soil variability and energy is saved by cutting across soil layers. (Duwayne McAninch, personal communication.)

recompacted, which can introduce construction delays and cost overruns. However, the integrity and reputation of the engineer must remain above question.

### 13.14.4  The Control Chart

A useful tool used in manufacturing is a *control chart* (Fig. 13.21). The control chart is simply an as-measured day-to-day plot of critical data such as soil compacted density and moisture content. Adverse weather or other conditions should be noted on the chart and can help to explain anomalies and decide on corrective measures.

For example, the control chart may show a group of tests that gave marginal or unacceptable density data. Is the reason bad data, or is it perhaps coincident with an early-morning rain, a change in the soil, a change in operator or operating procedure, a change in equipment, or is it none of those causes and attributable to carelessness or inattention to details?

One of the first jobs for an engineer-in-training often is that of inspector. It is a very important job, one that requires and develops self-confidence. The inspector must not be intimidated or fooled into looking the other way, but should be discreet about offering suggestions for improvement because of liability issues. The duty of the inspector is to determine whether specifications are satisfied and to report and if necessary close down a job if they are not. The inspector should be wary of the contractor who knows in advance when tests will be conducted and then suggests to "test over here."

### 13.14.5  Measures of Variability

If all tests are required to meet specifications, from the point of view of the contractor one test that passes is better than 10 tests if one does not pass, and the more tests there are the more likely it is that one or more will not pass. The current practice of requiring additional compaction in areas that do not meet

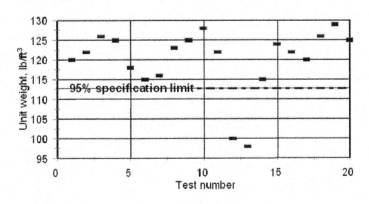

**Figure 13.21**

Control chart for compacted density. A similar chart can be prepared for moisture content measurements and can help pinpoint problem areas.

specifications gives no assurance that other similar or worse areas may not have been missed by the tests. Ideally, a large number of tests will be randomly performed and they all will meet specifications, but it is not a perfect world.

The most common statistical measure for variability is the *standard deviation*, which unfortunately is given the same symbol, $\sigma$, that engineers use for stress. The standard deviation is routinely calculated by computer programs that calculate averages.

An average usually is designated by $\bar{x}$, and the standard deviation by a number following a $\pm$ symbol; therefore 5.5 $\pm$ 1.2 means an average value of 5.5 and a standard deviation of 1.2, unless a different definition is stated.

The formula for the square of the standard deviation, called the *variance*, is

$$\sigma^2 = \sum (x_i - \bar{x})^2 / n \tag{13.4}$$

where $x_i$ represents individual data readings, $\bar{x}$ is the average, and $n$ is the number of data points. It will be noted that "variance" has a specific meaning and should not be confused with "variability."

### 13.14.6 The Meaning of a Standard Deviation

Mathematically a standard deviation is the second moment of the data about the mean. The mean is first calculated, and the departure of each data point from the mean is determined. Each departure then is squared so that negative departures become positive and additive. The sum of the squared departures is divided by the number of data points to obtain an average called a *variance*. In order to return to a linear scale, the square root of the variance is the *standard deviation*.

"Standard error" is the standard deviation divided by the square root of the number of observations. The term carries an unfortunate connotation and is not popular with engineers.

Manufacturing tolerances may have limits of acceptability as $\bar{x} \pm 3\sigma$, where $\sigma$ is the standard deviation. This is satisfactory when the standard deviation is small, but in soils it typically is much too large for this criterion to be valid or acceptable. In the case of compacted soil, which is a manufactured product, the limits are set by density and moisture content criteria obtained from laboratory density tests.

### 13.14.7 Coefficient of Variation

An important measure in engineering is the *coefficient of variation*, $C_v$, which is defined as the standard deviation divided by the mean and expressed as a percent:

$$C_v = 100\sigma / \bar{x} \tag{13.5}$$

The coefficient of variation allows comparing the variability of apples and that of oranges. $C_v$ usually is in the range 10 to 25 percent for most soil properties. The lower the coefficient, the more confidence one may have that a soil is uniform or a job is uniformly constructed

## 13.14.8 The Normal Distribution Curve

Most statistical theory is based on the concept of random error, which theoretically results in a "normal" or "Gaussian" distribution of data. This is shown by the symmetrical, "normal" curve in Fig. 13.22. However, this distribution is not strictly applicable to most engineering problems because the ends trail off to infinity in both directions, which may be physically impossible. For example, soil moisture contents cannot exceed limits imposed by saturation, densities cannot exceed those dictated by a zero air void condition, and neither can be negative. Furthermore it is the lower "tail" of a distribution, which is least well defined, that defines a probability of failure.

## 13.14.9 Is There a Better Theoretical Distribution?

The normal distribution is based on probability theory that says that if the number of measurements is very large, than the distribution is symmetrical around an average value and follows a bell-shaped or Gaussian normal distribution curve. This led to a concept of "probable error" that is based on the area under a normal curve. The term has fallen into disuse with limited measurements because they probably do not properly define the distribution. The "probable error" equals $0.6745\sigma$, and means that there is a 50 percent probability that a measurement will be outside of bounds determined by the mean and the standard deviation. For example, if a mean is 100 and the standard deviation is 10, 50 percent of the measurements should fall in a range $100 \pm 6.745$, or between 93 and 107.

The concept of probable error is used when there is a large number of tests that do follow a normal distribution, and is a basis for control chart limits of $\pm 3\sigma$ that

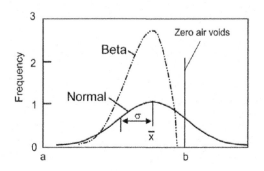

**Figure 13.22**

A beta distribution may be preferred over a normal distribution because it can be adjusted to stay within physical limits that are absolute.

theoretically include 99.73 percent of the data. Thus, in the manufacture of 1000 light bulbs the $3\sigma$ control chart limits should pass 997 and reject 3. In order to obtain the most efficient manufacture that is closest to the tolerance limits, the product should be consistent so that the standard deviation is low and $3\sigma$ is low.

### 13.14.10  Beta Distribution

The use of a *beta distribution* has been pioneered in geotechnical engineering by Harr at Purdue University. A beta distribution has defined upper and lower limits that customarily are adjusted to be 0 and 1. An example of a beta distribution is shown in Fig. 13.22, and has been drawn to stop before it reaches the zero air voids condition. The beta shape is variable, and in this case a lopsided shape is reasonable as measurements tend to be concentrated close to the zero air voids line.

A beta distribution may be narrower or broader than that which is shown, may be symmetrical or skewed one way or the other, may be higher at the ends than at the middle, or may even be a horizontal line indicating an even distribution. With that kind of versatility it may be difficult to assign the most appropriate descriptive parameters. As a first estimate a beta distribution may be selected that approximates a symmetrical normal distribution, but with discreet upper and lower boundary values based on physical limits such as not allowing a negative factor of safety.

### 13.14.11  Reliability and Factor of Safety

Engineers usually require that a design meet a certain minimum *factor of safety*, which is a ratio between the design value and a failure value. For example, if the calculated load on a beam to cause failure is 100, a factor of safety of 3 will mean that the design value must be 300. This not only allows for variability of the beam strength due to inconsistencies in manufacture, but also can to a limited extent compensate for future overloading.

In the case of compaction, it is inevitable that some tests will not meet specifications, and a beta distribution can help to translate a failure ratio into reliability. A formula for reliability is (Harr, 1987):

$$R = \frac{a + 1}{a + b + 2} \tag{13.6}$$

where $a$ is the number of successes and $b$ is the number of failures.

A beta distribution for eq. (13.6) is shown in Fig. 13.23, which in this case indicates a reliability of 0.71—that is, based on this limited number of tests, 71 percent of the product can be expected to meet specifications, instead of 80 percent indicated by the number of tests that actually passed.

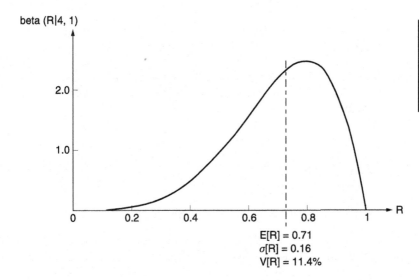

**Figure 13.23**
Beta distribution
calculated for 4
successes and 1
failure. (From Harr,
1987.)

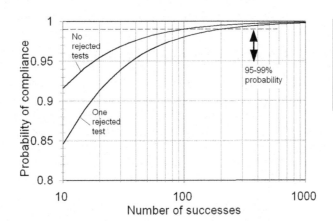

**Figure 13.24**
Reliability
prediction based
on number of
successes with or
without one failure.

## 13.14.12   Number of Consecutive Tests with No Failures

Equation (13.6) can be written for no failures or $b = 0$:

$$R = \frac{a+1}{a+2} \tag{13.7}$$

where $a$ is the number of tests. Some solutions for this and the preceding equation are shown in Fig. 13.24. For example, if 20 tests are performed and one test does not pass, $R = 0.91$, but if all tests pass, $R = 0.95$, which is more acceptable. This does not necessarily mean that 5 percent of the job will fail because failure also depends on other factors. According to this analysis, 10 tests without a failure indicate a reliability of $11/12 = 92$ percent. The higher the number of tests and lower the number of failures, the higher the reliability.

A common procedure is to repeat a test that does not pass, in which case it will have approximately the same probability of success as the first batch of tests. The repeated test therefore might be regarded as somewhat superfluous—unless it fails.

## 13.15    PARTICLE STACKING

### 13.15.1    Vertical Orientation of Voids

As discussed in Chapter 8, sedimentation leads to stacking of soil particles, as the contact angle must be high enough to cause slipping. A similar observation was made by Oda (1978) for sand held in a container that was tapped on one side (Fig. 13.25). The phenomenon later was modeled by steel rods by Oda et al. (1985). Such an arrangement is metastable and should be prevented by changing directions of the major principal stress, as occurs under a roller or tamping foot.

### 13.15.2    Dilatancy or Compression

Vaid and Thomas (1995) found that when specimens of loose sand were compressed vertically they tended to dilate, which would occur if there is bending and opening of a columnar structure. Compressed horizontally, the sand densified, which also is consistent with the concept of a columnar arrangement.

Pham (2005) vertically compressed a loessial silt, some specimens with and others without lateral strain. Vertical compression alone resulted in a much lower degree

**Figure 13.25**

Preferred vertical orientation of sand grains after self-weight densification. The dashed circles show random orientation. (Modified from Oda, 1978.)

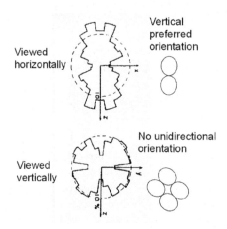

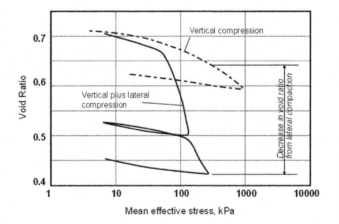

**Figure 13.26**

Comparison of void ratios obtained by vertical compression and vertical plus lateral compression of a silt initially compacted to 95 percent maximum density. The change in void ratio at 300 kPa mean stress is from 0.64 to 0.42, a decrease of 34 percent. (From Pham, 2005.)

of compression, shown by comparing the curves in Fig. 13.26. Specimens that were compacted horizontally as well as vertically also had a much higher modulus as a result of the additional densification.

While these observations may appear to be academic, they affect laboratory compaction of soil specimens in preparation for testing. Laboratory compaction should model the field process, which was the basis for the Proctor density test and other methods such as a kneading compactor. Static loading of soil in a steel mold does not duplicate these conditions.

## 13.16  SUMMARY

Compaction is the most frequently used method for improving soil properties such as strength, low hydraulic conductivity, and stability from volume changes, but requires careful tests and controls to achieve these purposes. Compaction of non-expansive soil to 95 percent density at or near the optimum moisture content should yield a product that is stable from future volume changes, whereas a soil that is compacted on the dry side of optimum may collapse when wet even though it meets the density requirement. Compaction on the wet side of optimum is useful for some purposes but increases the danger of overcompaction that will damage the soil.

Expansive clays can be approached by replacing or treating the clay, or to a limited extent by adjusting the compaction moisture content and preventing future changes in moisture content.

Compaction normally improves strength and reduces compressibility, but these properties must be measured or their adequacy is based on experience with particular soils. Statistical variability indicates that occasional tests will not meet specifications, in which case a reliability can be calculated based on the number of tests that pass and that do not pass. Compaction is most effective if it produces horizontal as well as vertical compressive stresses.

## Problems

13.1. Suggest compaction specifications for the following: (a) base course for a highway, (b) base course for an airport runway for large aircraft, (c) residential landscaping, (d) trench backfill that will support pavement, (e) clay core of an earth dam.

13.2. Calculate the compactive energy per unit volume of soil in the standard and modified Proctor tests.

13.3. Explain the differences and applications between compaction and consolidation.

13.4. (a) Why are soils compacted in layers? (b) Why is it important to have a firm, rough base for compaction?

13.5. Soil in a haul road develops a soft, bouncy feel and a rolling swell ahead of scraper wheels so that in effect the equipment is always going uphill. What is the reason and what is your recommendation?

13.6. A bucket filled with dry sand weighs more than a bucket filled with wet sand. Explain.

13.7. A nuclear gauge is used to test soil at the standard Proctor OMC and maximum density in Fig. 13.2. What percent accuracy can be expected for the moisture content?

13.8. Identify the soil in Fig. 13.2 by comparing the compacted density curve with those of Fig. 13.17. Does this identification appear to be correct?

13.9. Repeat Problem 13.8 with the soil in Fig. 13.15.

13.10. Give two reasons why a field density might give data above the zero air voids line. What should be done to mitigate this problem?

13.11. In Fig. 13.14, 90% modified Proctor density is higher than 95% standard density, and there is a shortage of water at the construction site. Which specification do you recommend? Is there any risk involved in specifying 90% maximum density?

13.12. (a) Relative density tests on a dry sandy gravel give a maximum and minimum of 20.1 and $14.5\,kN/m^3$ (128 and $92\,lb/ft^3$), respectively. (b) If $G = 2.65$, find $e_{max}$ and $e_{min}$. (c) Dry densities of this soil are found to be

19.2, 18.2, and 17.0 kN/m³ (122, 116, and 108 lb/ft³). Calculate the relative densities.

13.13. What is the percent saturation at the OMC and maximum density in Fig. 13.2? Assume $G = 2.70$.

13.14. List some of the hypotheses that have been proposed to explain the failure of Teton Dam in Idaho and comment on each. (Additional reading required.)

13.15. A proposed earth dam will contain 4,098,000 m³ (5,360,000 yd³) of earth. It will be compacted to a void ratio of 0.80. There are three available borrow pits, which are designated as A, B, and C. The void ratio of the soil in each pit and the estimated cost of moving the soil to the dam is as shown in the tabulation. What will be the least earth-moving cost, and which pit will it be most economical to use?

| Pit | Void ratio | Cost of moving per m³ | per yd³ |
|-----|-----------|-----------------------|---------|
| A | 0.9 | $1.67 | $1.28 |
| B | 2.0 | 1.19 | 0.91 |
| C | 1.6 | 1.56 | 1.19 |

13.16. (a) Calculate the mean, standard deviation, and coefficient of variation of the data shown in Fig. 13.21. (b) Two of the tests do not meet specifications. According to the subcontractor all of the soil received the same number of roller passes so he questions the validity of the tests. As the inspector, what are your options?

13.17. The data points shown in Fig. 13.27 are obtained in a Proctor density test of a silty sand. Which is more likely to be the correct interpretation of the data, linear regression as in (a), conventional curve fitting (b), or curve fitting as in (c)? Explain.

13.18. The following data were obtained during compaction of clay fill to support a shopping mall. Use a computer spreadsheet to prepare a control chart. The density is specified to at or above 95% of the maximum

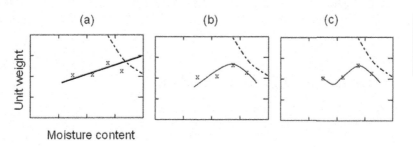

(a)          (b)          (c)

Unit weight

Moisture content

**Figure 13.27**
Data interpretations in Problem 13.17.

laboratory density, which is $98.6 \, \text{lb/ft}^3$ ($15.5 \, \text{kN/m}^3$) and the moisture content is specified to be between the optimum of 18.8% and 4% above optimum. (a) With reference to Fig. 13.17, is this a reasonable maximum density? (b) Prepare a control chart for density and for moisture content showing the specification limits. Are any of the data outside of these limits? (c) Calculate the reliability. If 500 tests were conducted, what number might be expected to pass?

| Test No. | Dry density lb/ft$^3$ | Moisture kN/m$^3$ | content,% |
|---|---|---|---|
| 1 | 95.4 | 15.0 | 19.8 |
| 2 | 99.3 | 15.6 | 18.0 |
| 3 | 99.0 | 15.55 | 17.8 |
| 4 | 97.6 | 15.3 | 18.0 |
| 5 | 98.9 | 15.5 | 18.0 |
| 6 | 99.3 | 15.6 | 17.0 |
| 7 | 97.8 | 15.4 | 18.8 |
| 8 | 95.8 | 15.1 | 19.0 |
| 9 | 96.7 | 15.2 | 17.4 |
| 10 | 96.6 | 15.2 | 18.9 |
| 11 | 95.7 | 15.0 | 18.5 |
| 12 | 98.9 | 15.5 | 20.0 |

13.19. Calculate the mean, standard deviation, and coefficient of variation of the density data in Problem 13.18. Does the uniformity of the data indicate good, average, or poor quality control on compaction? Would a control chart based on $\bar{x} \pm 3\sigma$ define acceptable limits? What are those limits? *Partial ans.:* $\bar{x} = 97.58 \bar{x} \pm 1.497$.

13.20. A statistical "t" test is based on a symmetrical distribution and is used to estimate confidence limits on the mean. With 12 observations, one can be 95% confident that the actual mean is $\bar{x} \pm 0.664\sigma$. Calculate the limits on the mean.

13.21. According to the area under a normal curve, 5% of the densities will exceed $\bar{x} + 1.645\sigma$ and 1% will exceed $\bar{x} \, 2.875\sigma$. Is this reasonable? Why (not)?

## References and Further Reading

Harr, M. R. (1987). *Reliability-Based Design in Civil Engineering*. McGraw-Hill, New York.

Johnson, A. W., and Sahlberg, J. R. (1962). "Factors Influencing Compaction Test Results." *Highway Research Board Bull. 319.*

Lawton, E. C., Fragaszy, R. J., and Hetherington, M. D. (1992). "Review of Wetting-Induced Collapse in Compacted Soils." *ASCE J. Geotech. Eng.* 118(9), 1376–1394.

Leonards, G. A., ed. (1987). *Dam Failures*. Elsevier, Amsterdam.

Oda, M. (1978). "Significance of Fabric in Granular Mechanics." *Proc. U.S.-Japan Seminar on Continuum Mechanics and Statistical Approaches in the Mechanics of Granular Materials* 7–26.

Oda, M., Nemat-Nasser, S., and Konishi, J. (1985). "Stress-Induced Anisotropy in Granular Masses." *Soils and Foundations* 25(3), 85–97.

Olson, R. E. (1963). "Effective Stress Theory of Soil Compaction." *ASCE J. Geotech. Eng.* 89(SM2), 27–45.

Proctor, R. R. (1993). "Fundamental Principles of Soil Compaction." *Engineering News Record* (Aug. 31 and Sept. 7, 21, and 28).

Seed, H. B., and Chan, C. K. (1959). "Structure and Strength Characteristics of Compacted Clays." *ASCE J.* 8(SM5, Pt. 1), 87–128.

Vaid, Y. P., and Thomas, J. (1995). "Liquefaction and Post-liquefaction Behavior of Sand." *ASCE J. Geotech. Geoenviron. Eng.* 89(SM2), 27–45.

Jürgenson, L. (1934). "The shearing resistance of soils." *Jour. Boston Soc. Civil Engrs.* July: 184–217. Reprinted in *Contributions to Soil Mechanics 1925–1940*. Boston Soc. of Civil Engrs., Boston, Mass.

Pham, Ha (2005). "Support Mechanisms of rammed aggregate piers." Unpublished Ph.D. thesis, Iowa State University Library, Ames, Iowa.

# 14

## Seepage

## 14.1  GRAVITATIONAL FLOW OF WATER IN SOIL

### 14.1.1  The Overall Direction Is Down

Unlike capillary water that can move in any direction in response to suction or negative pore water pressure, the net direction of gravitational flow is downward. Even uphill flow in a siphon requires that the exit must be lower than the level of the entrance.

The flow of water in soil in some ways is similar to flow in a pipe, but soil is a wild array of tiny channels consisting of extremely irregular, angular crevices between grains of the soil. Resisting the flow is viscous drag along these pores, and the smaller the effective diameter of the pores, the larger the drag and the slower the flow. The seepage rate in sand, other factors being the same, differs by a factor of a thousand or more from that in clay.

### 14.1.2  Significance of Groundwater Flow

Any excavation that penetrates below the groundwater table becomes a target for seepage, which unless controlled will flood the excavation up to the original groundwater level. Because seepage exerts a viscous drag on the soil, it can affect stability of the sides of the excavation. Groundwater elevations should be measured as part of an exploration drilling program.

The groundwater table can be lowered locally with wells, and perimeter wells and drains are commonly used for this purpose around a construction site. The flow of water into an excavation also can be restricted by sheet-pile walls, but pumping still is required to lower the water level and compensate for leaks.

The rate of groundwater flow affects the number and spacing of wells for site dewatering, the size and number of pumps, and the means for disposal of the extracted water. For this reason it is important that the flow rate be estimated prior to construction dewatering. Sandy and gravelly soils that allow relatively high flow rates are said to be aquifers, highly valued as sources for water but also posing an inconvenience or danger if they are encountered unexpectedly in an open excavation or tunnel.

Earth dams are designed to minimize seepage through, under, and around the dam. Nevertheless some leakage is inevitable, and seepage through the dam is directed into toe drains so that it will not exit and endanger the integrity of the downstream face of the dam. Seepage force in an upward direction, as can occur below an earth dam or near a levee, is the cause of quicksand.

### 14.1.3 Seepage Forces

As soil restrains water from quickly draining out, water exerts an equal and opposite drag on the soil in the direction of flow—for every action there is an equal and opposite reaction. This becomes a matter of considerable importance in the initiation of landslides, slope failures, and mudflows. It is no coincidence that most slope failures occur after prolonged periods of rain that increase downslope seepage of water within the soil.

### 14.1.4 Laminar Flow

Gravitational flow can be laminar or turbulent, depending on the fluid viscosity, flow velocity, and the size, shape, and smoothness of the conduit or channel through which the fluid is flowing. In soils the velocity of groundwater rarely, if ever, becomes high enough to produce turbulence, so flow normally is laminar, although exceptions can occur in underground caverns and possibly in rapidly draining road base course materials.

Water in laminar flow moves in essentially parallel paths. The quantity of water flowing past a fixed point in a stated period of time therefore equals the cross-sectional area of the water multiplied by the average velocity of flow. This relationship, referred to as a continuity condition, is expressed by the formula

$$Q = vA \qquad (14.1)$$

where $Q$ is the volume of flow per unit of time, $v$ is the average flow velocity, and $A$ is the cross-sectional area of flowing water.

## 14.2   MECHANICS OF GRAVITATIONAL FLOW

### 14.2.1   Gravitational Potential

In hydraulics, gravitational potential is conveniently expressed in units of head. Head includes three components: pressure head, elevation head, and velocity head, all of which are expressed in terms of equivalent height above a datum. Velocity head is negligible for seepage through soils where fluid velocities are low.

As shown in Fig. 14.1, in a static body of water there is a tradeoff between pressure head and elevation head: as elevation decreases, pressure increases and total head remains the same, so there is no head difference to drive gravitational flow.

Water head is expressed as a height or equivalent height of water above a datum. For example, if the datum is the ground surface, artesian water confined in a vertical pipe will rise to the level indicated by its combined pressure and elevation head.

### 14.2.2   Hydraulic Gradient

Hydraulic gradient is one of the factors that determines rate of flow, and is defined as the reduction in total head divided by the distance over which it occurs. By definition,

$$i = \frac{h}{d} \qquad (14.2)$$

where $i$ is the hydraulic gradient, $h$ is the change in head, and $d$ is the distance over which it occurs. For example, if an open channel is 1000 m long and declines 10 m in that distance, the hydraulic gradient is 10/1000, or 0.01. If the decline is 100 m,

**Figure 14.1**

In standing water such as in this standpipe total head everywhere is the same, so there is no gravitational flow.

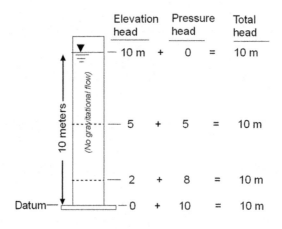

| | Elevation head | | Pressure head | | Total head |
|---|---|---|---|---|---|
| | 10 m | + | 0 | = | 10 m |
| | 5 | + | 5 | = | 10 m |
| | 2 | + | 8 | = | 10 m |
| | 0 | + | 10 | = | 10 m |

the hydraulic gradient is 0.1, and the water flow will be substantially faster. Streams descend rapidly from mountains because of a high hydraulic gradient, whereas rivers on floodplains lower their gradient by extending their channel lengths through meandering, and have a much slower rate of flow.

## 14.2.3 Darcy's Law

In 1856 a French engineer, Henry Darcy, measured the volume of water flowing through saturated sand columns and discovered the relationship now known as Darcy's Law:

$$Q = kiA \tag{14.3}$$

where $Q$ is the volume of water flowing through the soil in a unit of time, $k$ is a proportionality constant, $i$ is the hydraulic gradient, and $A$ is the cross-sectional area of the soil. This relationship is illustrated by flow through a container filled with sand in Fig. 14.2. As the reduction in head occurs entirely within the sand, it is possible to contour the loss with vertical lines called *equipotential lines*. This relationship is general and may be applied to any flow that is laminar in character.

The value of the proportionality constant $k$ depends on soil properties, in particular the size and number of soil pores. The equation simply means that

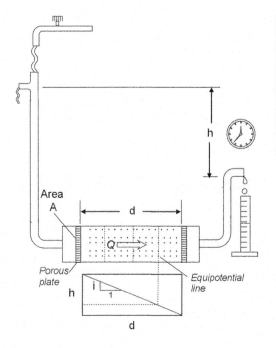

**Figure 14.2**

Diagram illustrating the loss of head, *h*, as water *Q* flows through sand. The hydraulic gradient, *i*, equals *h* divided by the flow distance *d*.

the quantity of water flowing through a given cross-sectional area of soil is equal to a constant multiplied by the hydraulic gradient.

### 14.2.4 Hydraulic Conductivity

The constant $k$ in eq. (14.3) commonly is called the coefficient of permeability, but since permeability also applies to gas flow, a more appropriate term for water flow is *coefficient of hydraulic conductivity*. As $k$ depends on the size and number of voids in a unit cross-sectional area, it is indirectly related to the size, shape, and packing of the soil grains. A clayey soil with very fine grains will have a much lower coefficient than will a sand with relatively coarse grains, even though the void ratio and the density of the two soils may be nearly the same. On the other hand, when we consider the same soil in two different states of density, as density increases due to compaction or consolidation, pore spaces are reduced in size, resistance to flow is increased, and $k$ decreases.

### 14.2.5 Approach Velocity

In the application of the Darcy Law and eq. (14.3), the cross-sectional area $A$ is the area of the soil including both solids and void spaces. Obviously, flow does not occur through the solid fraction of a soil so the velocity, $v$, in eq. (14.1), or the product $ki$ in eq. (14.3), is a factitious or "made-up" velocity. It is referred to as the "velocity of approach" or the "superficial velocity" of the water just before entering or after leaving a soil mass. The true velocity can be approximated by dividing $v$ by the soil porosity.

As $i$ in eq. (14.3) is dimensionless, $k$ has the dimensions of a velocity, that is, a distance divided by time. However, it should be emphasized that having units of velocity does not mean that $k$ is a velocity because the actual velocity also depends on the hydraulic gradient. The value of $k$ represents a "superficial velocity of water flowing through soil under unit hydraulic gradient." The actual flow velocity of water through soil must be higher than the superficial velocity $k$, because not only is the cross-section partly occupied by soil grains, but the flow is not straight-line and follows tortuous paths around the soil grains.

### 14.2.6 Hagen-Poiseuille Equation

The *Hagen-Poiseuille* derivation assumes laminar flow through a straight tube of uniform diameter and therefore does not take into account the irregular cross-section and tortuosity of flow channels in soil. Nevertheless the derivation is useful because it shows the relationship between flow velocity and an effective pore diameter.

In Fig. 14.3 an axially symmetrical element of radius $y$ is acted on at the ends by uniformly distributed pressure heads $h_1$ and $h_2$, and on the sides by viscous shear. Hydraulic head is expressed in terms of height of a water column, so to convert to

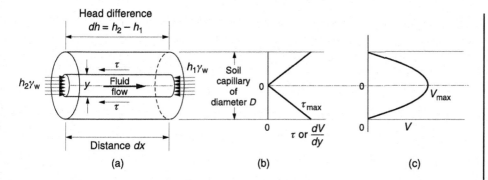

**Figure 14.3**
(a) Forces on a
fluid element in
laminar flow; (b)
distribution of
viscous shearing
resistance, which
is zero at the
center of the tube;
(c) distribution of
fluid velocity in the
tube.

pressure, $h$ is multiplied by the unit weight of water, $\gamma_w$. Equating forces in the $x$-direction gives

$$\pi y^2 \gamma_w \, dh = 2\pi y \tau \, dx$$

$$\tau = \frac{y}{2} \cdot \frac{\gamma_w dh}{dx} \tag{14.4}$$

where $\tau$ is the viscous shearing resistance. Therefore if $y = 0$ at the center of the tube, $\tau = 0$.

In an ideal Newtonian fluid, the shearing stress, $\tau$, is proportional to the rate of shearing, which can be represented as the change in velocity across a distance $y$. If $V$ is the velocity at any point within a capillary, the shearing rate is $dV/dy$. Along the center axis of a tube where shearing stress is zero, the rate of change of velocity with $y$ also is zero, shown in Fig. 14.3(b). Since, according to eq. (14.4), shearing stress is proportional to $y$, it increases linearly to a maximum at the tube boundaries, where the velocity is zero.

Integration with respect to $y$ indicates that the velocity distribution follows a parabola across a tube with laminar flow (Fig. 14.3(c)). In a unit of time the amount of fluid moved through a cylindrical tube is described by rotation of a parabola about its central axis, which is a paraboloid. The volume $q$ moved per unit of time then is

$$q = \frac{\pi D^4 i}{128} \frac{\gamma_w}{\eta} \tag{14.5}$$

where $D$ = pore diameter

$\eta$ = fluid viscosity

$i = dh/dx$, the hydraulic gradient or head loss per unit length

$\gamma_w$ = unit weight of water.

Equation (14.5) emphasizes the important role of pore diameter, which is to the fourth power. However, this is partly offset because a smaller pore diameter means that there are more pores per unit cross-sectional area of soil. As the number of pores $n$ is approximately inverse to $D^2$, and the flow per unit area then is

$$nq = \frac{Q}{A} = C_1 D^2 i \frac{\gamma_w}{\eta} = ki \tag{14.6}$$

$$k = C_1 D^2 \frac{\gamma_w}{\eta} \tag{14.6a}$$

where $C_1$ is a constant and $k$ is the coefficient of hydraulic conductivity of the soil. However, as the pore diameter $D$ is not known and is not readily measured, empirical investigations have for the most part attempted to relate fluid flow characteristics of soils to their particle size and void ratio.

A commonly used empirical equation developed by Hazen is

$$k = C_2 (D_{10})^2 \tag{14.7}$$

where $D_{10}$ is the soil *effective size*, $C_2 = 100$, and $k$ is in centimeters per second.

Hazen's equation does not account for variability in void ratio or influences from other grain sizes, and used alone is not considered reliable for prediction. Other studies have shown that, as an approximation, with particle size constant,

$$k = C_3 \left( \frac{e^3}{1+e} \right) \tag{14.8}$$

where $C_3$ is a constant and $e$ is the void ratio.

In summary, the coefficient of hydraulic conductivity, $k$, increases approximately as the square of both the particle size and the void ratio. Equation (14.8) is a simplified version of the *Kozeny-Carman equation* for flow through porous media.

Typical ranges of $k$ values are shown in Table 14.1. Traditional metric units are in centimeters per second, but because of the extended range of the values, the coefficient frequently is expressed in terms of those units and an exponent to base 10. Therefore a coefficient of "minus 1" normally will mean $10^{-1}$ cm/s.

In Table 14.1 it will be noted that $k$ less than "minus 5" is considered low, "minus 3 to minus 4" is medium, and larger than "minus 2" is high. It should be emphasized that the other major influence on $k$, density, is not included in this table, and it is possible to convert a medium to a low coefficient simply by compaction.

| Descriptive term | $k$, cm/s | Soils | **Table 14.1** |
|---|---|---|---|
| | $>10^0$ | Clean gravel | Representative |
| High | $10^0$ to $10^{-2}$ | Gravel, clean coarse sand | values of the |
| | $10^{-2}$ to $10^{-3}$ | Well-graded sand, fine sand | hydraulic |
| Medium | $10^{-3}$ to $10^{-4}$ | Silty sand | conductivity |
| | $10^{-4}$ to $10^{-5}$ | Dense silt, clayey silt | coefficient, $k$ |
| Low | $10^{-5}$ to $10^{-6}$ | Clay, silty clay | |
| | $<10^{-6}$ | Clay | |

# 14.3   MEASURING HYDRAULIC CONDUCTIVITY OF SOIL

## 14.3.1   The Need to Know

Seepage affects water losses through and under earth dams and levees, from irrigation ditches, and into open excavations. Seepage rates also affect pumping capacity for dewatering, and spacing and depths of drains for lowering the groundwater table. Seepage rates also influence settlement of structures as water is forced out of the foundation soil, which is the subject of a later chapter.

## 14.3.2   Different Measurement Methods

Hydraulic conductivity measurements can be conducted in the field with various pumping tests, or in a laboratory on disturbed or undisturbed samples. Each procedure offers advantages for particular types of problems, and the method that is most feasible and appropriate for a particular problem is the one that should be used. For example, in order to predict seepage through an anticipated earth dam, the most appropriate test would be conducted in the laboratory on samples of soil compacted to represent the range in densities and moisture contents specified for the dam. On the other hand, a field test of the soil in place is more appropriate for studies related to dewatering of an excavation. In every case, the objective should be to determine the conductivity of the soil either in its natural or an anticipated condition and to do so as accurately as possible.

## 14.3.3   Precision of Measurements

Values of $k$ vary by many orders of magnitude, depending on soil voids that are highly sensitive to soil structure, gradation, and density. These variations influence measurements that therefore are not precise and should not be reported to more than 2 significant figures. Other factors, including remolding during sampling, cracks, root holes, and perimeter leakage in the measuring chamber, also can profoundly affect the data, which therefore in all cases must be examined to determine if the numbers are reasonable. Anomalous tests should be repeated, and if results are not consistent, the measurement setup and procedure should be examined for possible sources of error.

For example, if a field permeability test gives $k$ that is many times higher than other tests in the vicinity, and the test is repeated and gives the same high value, it could mean proximity to a fissure or an abandoned mine tunnel or cavern.

Another common cause of high $k$ measurements is if the water pressure used for testing opens fissures or even fractures the soil, a process called *hydraulic fracturing*. If that happens, more realistic measurements will be obtained if the tests are conducted at lower pressures. However, a pressure that can lead to hydraulic fracturing also can be relevant. For example, it would be extremely unwise for fluid pressures near or under a dam to be allowed to exceed those required for hydraulic fracturing, and some major dam failures have been attributed to this cause (Leonards, 1987).

The sensitivity of hydraulic conductivity to small changes in void ratio emphasizes the importance of minimizing disturbance during sampling of soils for laboratory tests. In addition, samples must be tight against walls of the test containers to prevent seepage around the margins.

Another important factor influencing results is that natural soil deposits often are nonisotropic with respect to flow, that is, the conductivity coefficient in the vertical direction differs considerably from that in the horizontal direction, a situation that is most evident in layered alluvial or lacustrine soil deposits. If such a condition exists, horizontal and vertical conductivity measurements may be conducted, and the ratio between the two is used to quantify flow in the field in relation to the seepage direction.

Field measurements should simulate anticipated flow directions. For example, flow into the sides of an open excavation will be horizontal, whereas flow up into the bottom will be vertical. Flow rates in field tests should be reproducible within 5 percent, and repeated determinations of $k$ should be within 25 percent of one another.

# 14.4 LABORATORY MEASUREMENTS OF $k$

### 14.4.1 Constant-Head Permeameter

Coarse-grained soils can be tested in a laboratory *constant-head permeameter* that uses an overflow to maintain a constant hydraulic head on the sample. The hydraulic gradient then remains constant throughout the testing period. A schematic diagram is shown in Fig. 14.4. The amount of water passing through the soil sample during a period of time is measured, and appropriate values that

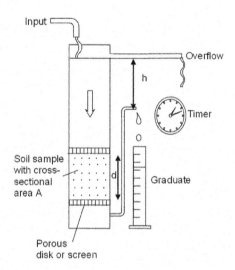

Input

Overflow

$h$

Timer

O

Soil sample with cross-sectional area A

$d$

Graduate

Porous disk or screen

**Figure 14.4**
Schematic diagram of a constant-head permeameter.

include dimensions of the apparatus and soil sample are substituted in eq. (14.3) to obtain a value for the conductivity coefficient.

In preparation for testing, *sand or silt soil* is poured into the permeameter in layers that are tamped in place to a predetermined relative density. Then to prevent air blockage within the sample, air is removed from the soil by use of a vacuum (ASTM Designation D-2434) or by purging with $CO_2$ gas, which dissolves and is removed by permeating water. Another potentially serious problem is if air that initially is dissolved in the permeating water comes out of solution and forms bubbles in the soil. This possibility is reduced by having both the inlet and outlet pressures high to keep the air in solution, as shown in Fig. 14.4.

The test may be performed on the same sample with several values of $h$ to determine linearity. If the flow rate is not linear with increasing $h$, this signals a nonuniform response such as turbulent flow, air in the system, segregation of fines, or development of channels, or *piping*. Piping also may be evident from muddying of the exit water.

## Example 14.1

A soil sample in a constant-head apparatus is 152 mm (6 in.) in diameter and 203 mm (8 in.) long. The vertical distance from headwater to tailwater is 279 mm (11 in.). In a test run, 347 kg (766 lb) of water passes through the sample in 4 hr 15 min. Determine the coefficient of hydraulic conductivity. Is this value considered high or low?

*Answer:* In this test, $h = 279$ mm and $d = 203$ mm, so $i = h/d = 1.37$. $A = \Pi(15.2/2)^2 = 181$ cm$^2$. $Q = 347$ kg/255 min $= 1.36$ kg/min $= 2.27$ g/s $= 2.27$ cm$^3$/s. From eq. (14.3),

$$Q = kiA$$
$$2.27 \text{ cm}^3/\text{s} = k \times 1.37 \times 181 \text{ cm}^2$$
$$k = 9.1 \times (10)^{-3} \text{cm/s}$$

which is rounded to $10^{-2}$ cm/s. This in the medium range and representative of sand.

In computing the value of the conductivity coefficient from data obtained in a test of this type, as in all fluid flow problems, it is important to carry dimensions through the calculations in order to ensure that they are dimensionally correct. A relatively easy and sure way to do this is to decide in advance the final units that are desired, and reduce the values of $Q$ and $A$ to those units before making the computation, as was done in the example.

### 14.4.2 Falling-Head Permeameter

Fine-grained soils that require long testing times are most efficiently tested by simply filling a permeameter and letting the head decrease as the water runs out. The decline in head is measured in a small tube to increase precision. This arrangement is shown in Fig. 14.5. In the conduct of the test, the water passing through the soil sample causes water in the standpipe to drop from $h_0$ to $h_1$ in time $t_1$. Because the head is constantly decreasing, an integration is necessary.

**Figure 14.5**

Schematic diagram of a falling-head permeameter.

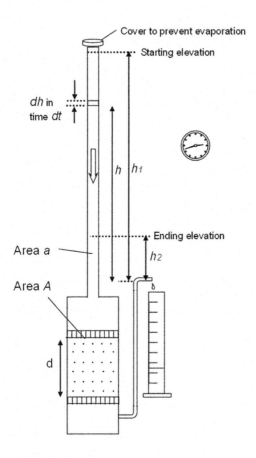

The head on the sample at any time $t$ between the start and finish of the test is $h$, and in any increment of time $dt$ there is a decrease in head $dh$. From these definitions the following relationships may be written. The minus sign is required as the head is decreasing with time.

$$k = \frac{h}{d}A = -a\frac{dh}{dt} \tag{14.9}$$

where $A$ is the cross-sectional area of the soil specimen and $a$ is that of the standpipe. Then

$$k\frac{A}{d}\int_{t_0}^{t_1} dt = -a\int_{h_0}^{h_1} \frac{dh}{h} \tag{14.10}$$

from which

$$k = \frac{ad}{At_1}\ln\frac{h_0}{h_1} \tag{14.11}$$

### Example 14.2

A sample of clay soil having a cross-sectional area of 78.5 cm² (12.15 in.²) and a height of 50 mm (1.97 in.) is placed in a falling-head permeameter in which the area of the standpipe is 0.53 cm (0.082 in.²). In a test run, the head on the sample drops from 800 mm to 380 mm (31.5 to 15.0 in.) in 1 hr 24 min 18 sec. What is the coefficient of hydraulic conductivity of this soil? Is this value reasonable?

*Answer:* Substituting $a = 0.53$ cm², $d = 5.0$ cm, $A = 78.5$ cm² and the appropriate values for $t_1$, $h_1$, and $h_2$ in eq. (14.11) gives

$$k = [(0.53 \text{ cm}^2)(5.0 \text{ cm}) \div (78.5 \text{ cm}^2)(5058 \text{ s})] \ln (800 \text{ mm}/380 \text{ mm})$$

$$= 6.7(10)^{-6} \times 0.744 = 5(10)^{-6} \text{ cm/s}$$

A similar procedure is used with English units, in which case the answer will be in inches/sec. Note that by writing the units they cancel out and give the answer in cm/s. The value for $k$ should be carefully inspected to determine if it is reasonable. This value for $k$ is very low and indicates clay.

## 14.4.3 Flexible-Wall Permeameter

Perimeter leakage is reduced if the soil is contained in a rubber membrane and placed inside a pressure cell, as shown in Fig. 14.6. Undisturbed samples of fine-grained soils are trimmed to remove remolded soil around the edges, and water pressure on the outside of the membrane pushes it into intimate contact with the soil. Ideally the cell pressure should simulate lateral pressure existing in the field. The sample can be de-aired with back-pressure on the water to dissolve all air. Simultaneous testing of several samples is facilitated by using regulated air pressure to drive the input.

**Figure 14.6**

Schematic diagram of a flexible-wall permeameter.

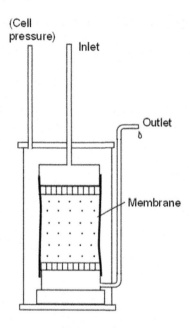

Either constant or falling-head methods may be used, and specimen length can be monitored to ensure that the soil does not consolidate as a result of seepage forces. This test method is described in ASTM Designation D-5084.

### 14.4.4 Selecting Appropriate Test Conditions

Hydraulic conductivity is so variable and is influenced by so many factors that it is important that soil samples have a structure, density, and lateral confining pressure representative of field conditions. For example, air is present in all compacted soils and eventually will dissolve in the seepage water, increasing conductivity. Therefore compacted soil samples should be evacuated and de-aired prior to testing, or time allowed for air under pressure in the sample to dissolve and be carried away in the seepage water. In that case several changes of water are necessary, and the test should be continued until the flow rate becomes constant. Water used in a test may be de-aired by evacuating or boiling, and it is important that water be kept under pressure in the sample to keep any air in solution.

Tap water ordinarily is used as a permeant, as distilled water becomes acidic from capture of $CO_2$ gas from air to create $H_2CO_3$. The surplus of $H^+$ ions can have a dispersing effect on soil clay that will influence the soil structure and hydraulic conductivity. Unsoftened the water also is preferred to avoid the dispersing effects of $Na^+$ ions in the water.

Another method for eliminating perimeter leakage is to test soil samples while they are still inside the special thin-walled steel tubes that are used to acquire the samples, cutting the tubes and the contained soil samples into lengths that fit into a permeameter. In this case, conductivity will be reduced by remolding of soil around the perimeter of the sample.

## 14.4.5 Temperature and Viscosity of Water

The coefficient of hydraulic conductivity is influenced by the viscosity of the permeating water, which depends on its temperature. Permeability data therefore are adjusted to a standard laboratory temperature of 20°C. However, it will be noted that that temperature does not represent most field conditions, as evidenced by the temperature of groundwater in a well or spring, where the temperature approximates that of the mean annual air temperature. Therefore a temperature correction is required to simulate most field conditions—the cooler the water, the higher the viscosity and the lower the hydraulic conductivity.

The coefficient of viscosity in SI is in Newton-seconds per square meter ($N s/m^2$). The coefficient of viscosity of water at 20.20°C (68.36°F) is $1 mN s/m^2$, which also equals 1 centipoise in the cgs metric system. Correction factors based on the viscosity of water are presented in Table 14.2.

**Example 14.3**
Correct a coefficient of hydraulic conductivity $k_{27} = 12(10)^{-4}$ cm/s measured in the laboratory at 27°C to a $k_{10}$ for a field temperature of 10°C.

*Answer:* The first step is to adjust the laboratory data to the standard temperature of 20°C, which, being lower than the measurement temperature, means that the water will have a higher viscosity so the corrected $k$ should be lower. The correction factor for 27°C is 0.85. Therefore

$$k_{20} = 0.85 \times 12(10)^{-4} = 10.2(10)^{-4}$$

The second step is to adjust $k_{20}$ to $k_{10}$, and again the corrected viscosity should be lower. The correction factor is 1.3, which is divided into $k_{20}$:

| °C | Correction |
|----|-----------|
| 10 | 1.30 |
| 12 | 1.23 |
| 14 | 1.16 |
| 16 | 1.11 |
| 18 | 1.05 |
| **20** | **1.00** |
| 22 | 0.953 |
| 24 | 0.910 |
| 26 | 0.869 |
| 28 | 0.832 |
| 30 | 0.797 |

**Table 14.2**

Correction factors to adjust $k$ to a standard reference temperature of 20°C

$$k_{10} = 10.2(10)^{-4} \div 1.3 = 7.8(10)^{-4}\,\text{cm/s}$$

As a result of the two-stage correction the final value for $k$ is 35% lower than that measured in the laboratory.

# 14.5 FIELD TESTS FOR HYDRAULIC CONDUCTIVITY

## 14.5.1 Advantages of Field Tests

Because most sampling is conducted vertically, laboratory tests normally use vertical flow even though hydraulic conductivity often is higher in a horizontal direction, sometimes by orders of magnitude. Field measurements therefore are more reliable for predicting lateral seepage into excavations or under dams, and in many other situations involving groundwater flow through natural soils. Field measurements also avoid problems from sampling disturbance and are conducted on a much larger scale that averages the effects of local variations in the soil.

## 14.5.2 Drawdown Pumping Test

A relatively simple method for measuring hydraulic conductivity in the field is to pump the water level down a test well and measure water levels in two or more observation wells spaced at different distances from the test well. The test well has a perforated casing and ideally is drilled through the water-bearing soil to an underlying impervious stratum, but if no impervious stratum is present, the well is drilled to a considerable depth below the water table so there will be little upward flow of water into the well. The perforations in the well casing must be sufficiently close to one another to permit the groundwater to flow into the well as fast as it reaches the casing.

Observation wells are put down at various radial distances from the test well, the number of such wells and the radial distances depending on site uniformity and extensiveness of the investigations. If casing is required, it must be perforated or equipped with a screen at the lower end. In some cases observation wells can be installed by pushing in casing with a "sand point."

Elevations of the groundwater table are measured and recorded at each of the wells before pumping begins. Pumping from the test well then proceeds and continued until a steady flow is established, which is indicated when water levels in the test well and observation wells become constant. The decrease in elevation of the water table at each of the various observation wells, the radial distances out to these wells, and the rate of discharge from the test well, provide the information necessary for computing the coefficient of permeability of the soil within the zone of influence of the test well.

## 14.5.3 Calculation of Hydraulic Conductivity

With steady flow and uniform soil conditions, the groundwater surface assumes the shape of an inverted hyperbolic cone. A radial section through the well and two observation wells is shown in Fig. 14.7. The flow toward the test well is assumed to be radial at the observation wells so the hydraulic gradient is $dh/dr$. Water flows through a cylindrical area $2\pi rh$ toward the test well where the cylinder has a radius $r$ and height $h$. Substituting these values into eq. (14.3) gives

$$Q = kIA = 2\pi rhk\frac{dh}{dr} \tag{14.12}$$

Integrating between $r$ and $h$ for the two observation wells gives

$$\int_{r_2}^{r_1} \frac{dr}{r} = \frac{2\pi k}{Q} \int_{h_2}^{h_1} h\,dh \tag{14.13}$$

from which

$$k = \frac{Q\ln(r_1/r_2)}{\pi(h_1^2 - h_2^2)} \tag{14.14}$$

### Example 14.4

A test well yields a steady discharge of 7.89 l/s (125 gal/min). Elevations of the water table above an impervious layer (or above the bottom of the test well in case no impervious layer exists) are 10.39 m (34.1 ft) and 10.58 m (34.7 ft), respectively, and these wells are located 18.3 m (60 ft) and 30.5 m (100 ft) from the center of the test well. Determine the coefficient of permeability of the soil.

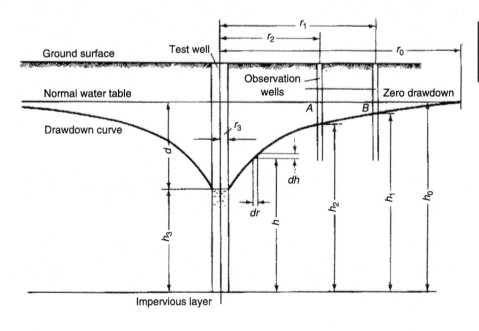

**Figure 14.7**
Determination of $k$ from a drawdown pumping test.

*Answer:* In SI, $Q$ is expressed in m³/s; in English units in ft³/day. Note that 1 liter $= (10)^{-3}$ m³ and 1 gal $= 0.1337$ ft³. Substitution of appropriate values in eq. (14.14) gives

$$k = \frac{7.89(10)^{-3} \ln(30.5/18.3)}{\pi(10.58^2 - 10.39^2)} = 320 \, \mu m/s = 91 \, ft/day = 3.2(10)^{-3} \, cm/s$$

Again it should be emphasized that $k$ does not represent a flow rate, which depends on the hydraulic gradient.

## 14.5.4 Approximate Drawdown Method

If point A in Fig. 14.7 is taken at the circumference of the test well and B is on the circle of zero drawdown of the water table, eq. (14.14) becomes

$$k = \frac{Q \ln(r_0/r_3)}{\pi(h_0^2 - h_3^2)} \tag{14.15}$$

where $r_3$ is the radius of the test well, $r_0$ is the maximum radius of influence or radius of the circle of zero drawdown, $h_0$ is the height of the groundwater table above an impervious layer or the bottom of the test well, and $h_3 = h_0 - d$ where $d$ is the drawdown in the test well at a steady flow condition.

Since the radius of the circle of zero drawdown will always be several hundred times the radius of the test well, the value of $\ln(r_0/r_3)$ varies over only a relatively narrow range, so a value may be assumed for $r_0$. It then is not necessary to observe the drawdown in observation wells.

## Example 14.5

The diameter of the test well of Fig. 14.7 is 0.61 m (24 in.) and the drawdown is 7.65 m (25.1 ft) with a steady pumping discharge of 7.89 l/s (125 gal/min). The distance from the impervious layer to the normal water table is 10.7 m (35.2 ft). Assume that the radius $r_0$ of zero drawdown is 150 m (500 ft). Determine the coefficient of permeability. (1 liter $= 1000$ cm³ $= 10^{-3}$ m³.)

*Answer:* Substitution of the appropriate values in eq. (14.15) gives

$$k = \frac{7.89(10)^{-3} \ln(150/0.305) \, m^3/s}{\pi[(10.7^2) - (10.7 - 7.65)^2] \, m^2} = 0.148 \, m/s = 41.9 \, ft/day = 15 \, cm/s$$

---

*Questions:* Is this answer reasonable for sand? What if $r_0$ is assumed to be twice as far, 300 m (1000 ft)?

*Answers:* The answer is high even for coarse sand. Recalculation with $r_0 = 300$ m increases $k$ by 11%, so the answers are within a range of acceptable accuracy for $k$.

---

## 14.5.5 Open-End Pumping Tests

Less accurate but faster and less costly is to simply pour water into an open-ended drill casing that is bottomed in a test hole that extends below the level of the groundwater table. In a procedure developed empirically by the U.S. Bureau of Reclamation the depth to the water table is measured, then casing is lowered and carefully cleaned out just to the bottom of the boring. The rate at which clear water must be added to maintain a constant head near the top of the casing is measured, and the hydraulic conductivity is calculated from

$$k = \frac{Q}{5.5rh} \tag{14.16}$$

in which $Q$ is the flow rate for a constant head, $r$ is the internal radius of the casing, and $h$ is the head of water above the water table. In this equation $r$ and $h$ have units of length that must be consistent with the units of $Q$ and $k$.

### Example 14.6
An NX, 76 mm (3 in.) i.d. casing is set at 9.14 m (30 ft) depth, 4.57 m (15 ft) below the water table. The casing is cleaned out, and it is found that by adding 0.0315 l/s (0.5 gal/min) of water, the average water level is at the ground surface. Find $k$.

*Answer:*

$$k = \frac{31.5 \, \text{cm}^3/\text{s}}{5.5(3.8 \, \text{cm})(9.14 - 4.57) \times 1000 \, \text{cm}} = 0.33(10)^{-3} \, \text{cm/s}$$

Is this answer reasonable for a silty sand?

A disadvantage of open-end pumping tests is that only a very limited volume of soil is tested, just below the end of the casing. The factor 5.5 in eq. (14.16) was arrived at experimentally by use of an electrical analog where electrical current, voltage, and resistance are analogous to $Q$, $h$, and the reciprocal of $k$, respectively.

## 14.5.6 Packer Pumping Tests

Packer tests are conducted by partitioning off sections of a boring with expandable packers (Fig. 14.8). These tests are commonly used to evaluate foundation soils and rocks for dams, and have an advantage that different depth intervals can be discretely tested. For best accuracy the depth interval should be 5 and preferably 10 times the hole diameter. The equation for $k$ is

$$k = \frac{Q}{2\pi Lh} \ln \frac{L}{r} \tag{14.17}$$

where $L$ is the length of hole tested, $r$ is the radius of the hole, and $h$ is the pumping pressure expressed as hydraulic head.

**Figure 14.8**

Packer test for
determining
hydraulic
conductivity.

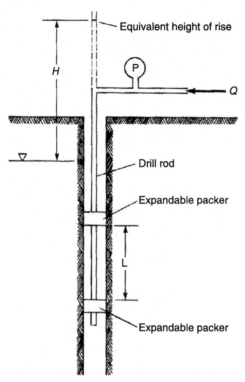

In eq. (14.17) it will be seen that if $k$ is constant, $Q$ is proportional to the pumping pressure head, $h$. The test therefore may be repeated with different values of $h$ to observe the effects of hydraulic pressure and flow on $k$. A sharp increase in $k$ usually means that cracks are being opened, indicative of hydraulic fracturing. This becomes likely if the water pressure exceeds the overburden pressure, as water injects horizontally and lifts the soil. The pressure to cause hydraulic fracturing also has been used to estimate lateral in-situ stress in the soil, but only if that stress is lower than the overburden pressure, and other methods are considered more reliable. Hydraulic fracturing can be deleterious if it occurs during pressure grouting, which is injection of a fluid that then hardens and stabilizes soil. This is discussed in Section 24.6.2.

### 14.5.7 Slug Tests

"Slug tests" refer to either pouring a measured amount of water or "slug" into a boring, or removing a measured amount of water from the boring, and observing the rate of change of elevation of water in the hole (Fig. 14.9). Boast and Kirkham (1971) showed that if the auger hole extends completely through a pervious stratum to an impervious layer, the flow is subject to exact mathematical analysis. In the absence of an impervious layer the results can present a good

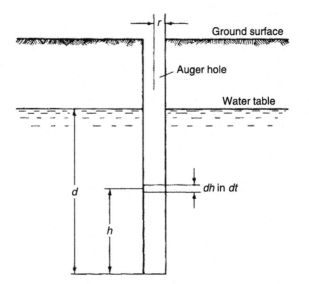

approximation if the ratio of depth to diameter of the auger hole is large. In this method, as in any method where water is taken from a boring, it is advisable to pump water from the hole several times and allow the hole to refill, in order to flush out soil particles from the remolded zone along the sides.

The formula for determining the $k$ by this procedure is

$$k = 0.617 \frac{r}{Sd} \frac{dh}{dt} \tag{14.18}$$

where $r$ is the radius of the hole, $d$ is the depth of the hole below the groundwater table, $dh/dt$ is the rate of rise of the water in the hole, and $S$ is a coefficient that depends on $h/d$ where $h$ is the depth of water in the hole at the time the rate of rise is measured. Values of the coefficient $S$ are given in Fig. 14.10.

In order to perform this test a hole is bored down to an impermeable layer or to a depth below the water table that is at least 10 times the diameter of the hole. Water in the hole is allowed to rise to a constant level, which is considered to be the elevation of the water table. Water then is pumped from the hole and the hole allowed to refill two or three times, more if the soil puddles readily. The water is then pumped out of the hole again and the rate of rise of the water level measured at several different elevations by measuring the rise in a short period of time.

As flow into the auger hole is almost entirely horizontal, this method is especially useful for determining horizontal permeability in nonisotropic soils. If the soil is stratified, the flow into the auger hole is from all strata penetrated by the boring, with the greatest inflow being through the most permeable layer.

**Figure 14.10**

Values of $S$ for use in eq. (14.18), after Boast and Kirkham (1971).

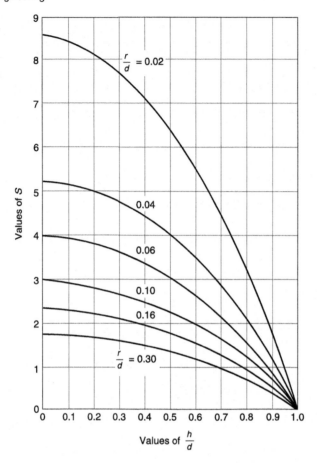

### Example 14.7

A 4 in. diameter hole is augered to extend 2.5 ft below the groundwater table and is baled several times to remove the water. When the water depth below the water table is 1.5 ft, the rate of rise is 1 inch in 12 minutes. What is $k$?

*Answer:* h = 1 ft and $d = 1.5$ ft so $h/d = 0.67$. $r = 2$ in. so $r/d = 2/18 = 0.11$. From Fig. 14.10, $S = 1.6$. Then

$$k = 0.617 \, (2\,\text{in.}/1.6 \times 18\,\text{in.}) \times 1\,\text{in.}/12\,\text{min} = 0.0036\,\text{in.}/\text{min} = 0.43\,\text{ft/day} = 1.5 \, (10)^{-4}\,\text{cm/s.}$$

Is this a reasonable value? Yes, if the soil is fine-grained.

## 14.5.8 Dye Injection Tests

A dye or salt solution can be injected into a well and the time determined for it to be detected at another well or seepage outlet that is downslope. This gives a direct measurement of the true flow velocity through the pores. This method is most applicable to granular soils because of the time required and because dyes and salts may interact with and be adsorbed by clays. The hydraulic gradient $i$ equals the decrease in elevation of the groundwater table divided by the distance downslope.

As discussed in Section 14.2.4, $k$ defines a superficial flow velocity under a unit hydraulic gradient in a soil that is all voids. The relation between flow velocity measured by dye injection and $k$ is

$$k = \frac{v_s n_e}{i} \tag{14.19}$$

where $v_s$ is the seepage velocity, $n_e$ the porosity, and $i$ the hydraulic gradient.

### Example 14.8

Dye injected into a boring in sand is detected entering an open excavation 100 m away after 85 minutes. The groundwater level in the boring is 4 m higher than in the excavation, which is being dewatered to the capacity of the pumps. Estimate $k$ and recommend how much more pump capacity will be needed to lower the water level an additional 3 m.

*Answer:* The seepage velocity $v_e = 100/85 = 1.18$ m/min $= 2.2$ cm/s. The hydraulic gradient $i = 4$ m$/100$ m $= 0.04$. Porosity is not known and must be estimated; it is assumed to be 0.3. Then

$$k = 2.2\,\text{cm/s} \times 0.3/0.04 = 16\,\text{cm/s}.$$

This value of $k$ is high but is reasonable for a sand aquifer. If $k$ is constant, eq. (14.19) indicates that the seepage velocity $v_s$ is proportional to $i$, so dewatering to a depth of 7 m instead of 4 m will increase the required pump capacity by that ratio, other factors being constant. However, the increased seepage also may cause slope instability, so that possibility should be investigated. This problem can be avoided by reversing the seepage direction with perimeter wells.

## 14.5.9 Tests in the Vadose Zone

"Vadose zone" means a zone of unsaturated soils. The hydraulic conductivity in such a zone normally is lower than if the soil were saturated because of partial blocking of water flow by air. The rate of infiltration of rainwater into soil is in part controlled by hydraulic conductivity in the vadose zone. Compacted soils contain air, so they also constitute a vadose zone unless they later become saturated with water.

Compacted clay layers used to seal the bottoms of ponds, lagoons, and solid waste containments can contain air even though the soil is in contact with water, so long as water drains out of the bottom as fast as or faster than it enters at the top. However, a saturated coefficient often is used for design of such containers to be on the safe side. The vadose zone also is a repository for insoluble low-density pollutants such as gasoline that float on the groundwater table, and geoenvironmental engineering procedures are available for sampling pore fluids from the vadose zone.

## 14.5.10  Measuring Infiltration in the Vadose Zone: Overview

An indication of hydraulic conductivity in the vadose zone is obtained from infiltration tests. The simplest of these is a borehole infiltration test or "perc test" that is widely used as an indicator of the suitability of a soil for a septic drain field. In this test, water is poured into an auger hole and the rate of decline of the water level is measured. Since the test is performed at shallow depth it is strongly influenced by capillary attractions, so the hole should be kept filled with water for 4 to 24 hours prior to measurement. Because infiltration is both horizontal and vertical, results are considered empirical, and the acceptability for drain fields is based on experience as reflected in local building codes.

Auger hole "perc" and other infiltration tests differ from slug tests in that they do not fully saturate the soil. Keeping an auger hole filled with water results in a "field-saturated" condition but the soil still contains air. The coefficient of hydraulic conductivity therefore is designated by $k_{fs}$ (fs for field saturated), and normally is only about 50 percent of the saturated $k$, or it may be as little as 25 percent of the saturated $k$ for fine-grained soils. A high conductivity is required for operation of septic drain fields, so these results are on the safe side for design.

## 14.5.11  Infiltrometers

More sophisticated infiltration tests are part of *soil physics*, which is a branch of agronomic soil science. A condensed discussion of various test methods is given in ASTM Designation D-5126. Only a brief treatment is included here.

The most common method for testing is with an *infiltrometer*, which in its simplest form is a metal ring that is driven a short distance into the ground surface and filled with water. This creates a downward-penetrating wetting front that also spreads laterally. A *double-ring infiltrometer* adds a concentric outer ring so that the space between the rings also can be filled with water, which directs infiltration from the inner ring downward, as shown in Fig. 14.11.

The driving force for infiltration includes both gravitational and matric or capillary potential. Gravity head at the wetting front is zero, so the gravitational head is measured from the surface of the ponded water down to the wetting front. Therefore, as the wetting front advances, if the ponded water is maintained at a constant elevation both the gravitational head and flow distance increase. If the level of the ponded water approaches the ground surface, the gravitational head $h$ equals the thickness of the

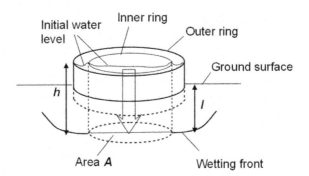

**Figure 14.11**

Schematic diagram of a double-ring infiltrometer. As the water level decreases, $h \rightarrow l$ and $i \rightarrow 1$.

partially saturated soil layer behind the wetting front. As this also equals the flow distance $l$, the hydraulic gradient $i = h/l = 1$. The Darcy equation (14.3) then becomes

$$Q/A = k_{fs} \qquad (14.20)$$

where $Q$ is the quantity of water infiltrating per unit of time, $A$ is the infiltrating area, and $k_{fs}$ is the coefficient of hydraulic conductivity in the field-saturated soil behind the wetting front. This relationship can be used for sands or silts where the wetting front has advanced sufficiently that $h \gg h_F$, where $h_F$ is the matric potential.

In clay, matric potential is the main driving force. A model developed by J. R. Philip indicates that the horizontal infiltration rate without any influence of gravity should go as the square root of time, whereas vertical infiltration that is in response to both gravitational and matric potential goes as the square root of time plus a time function (Jury and Horton, 2004). The constants in the required equations are evaluated experimentally.

## 14.6 MANAGING THE GROUNDWATER TABLE

### 14.6.1 Drain Tiles

Perimeter drains are installed to lower the groundwater table prior to excavating, and permanent drains are installed around building foundations to prevent seepage into basements. Many natural wetlands have been drained with tile lines to convert the land for agriculture, and soils at hazardous waste sites destined for burning in kilns can be dewatered with drains and the water treated separately instead of having to be burned off.

Drains draw down a groundwater table as shown in Fig. 14.12. A simplified approach called the Dupuit-Forchheimer, or DF theory, defines the groundwater surface between two tile lines as an ellipse with focal points located on the surface

**Figure 14.12**

Recharge $R$ from infiltration of rain, and elliptical drawdown curve to two parallel drainage tile lines.

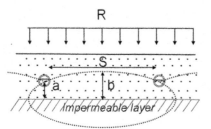

of an underlying impermeable stratum. The idealized "DF soil" is discussed by Kirkham and Powers (1971), who show that the error from the assumptions in most cases is relatively small. Then

$$\frac{S^2}{b^2 - a^2} = \frac{4k}{R} \tag{14.21}$$

where $S$ is the tile line spacing, $a$ and $b$ are the depths to an impermeable barrier under the tile and halfway between the tile, respectively, $k$ is the coefficient of hydraulic conductivity of the soil, and $R$ is the infiltrating rainfall expressed in the same units as $k$. A materials balance requires that the amount of water going in equals the amount going out, so the quantity per unit of length of tile is $R \times S$. However, because low areas where drains are most likely to be used collect runoff water from adjacent hillsides, $R$ may exceed the annual rainfall.

**Example 14.9**

A silty soil layer 5 m (16.4 ft) thick is to be drained so that the water table at its highest point is 1 m below the ground surface. Assume $k = 5(10)^{-5}$ cm/s (1.6 m/yr) and $R = 0.76$ m/yr (30 in./yr), and try two alternative tile depths and spacings.

*Trial 1.* Let the tile depth be 2 m (6.6 ft); $a = (5 - 2) = 3$ and $b$ is unchanged:

$$S^2 = (4^2 - 3^2) \times \frac{4 \times 1.6}{0.76} = 59, \text{ and } S = 7.7 \text{ m(22ft)}$$

*Trial 2.* Let the tile depth be 3 m (9.8 ft). Then $a = (5 - 3) = 2$ m and $b = (5 - 1) = 4$ m. Therefore

$$S^2 = (4^2 - 2^2) \times \frac{4 \times 1.6}{0.76} = 100, \text{ and } S = 10 \text{ m(33 ft)}$$

While the Trial 2 solution may appear more economical because it would reduce the length of tile lines by 50 %, every solution that looks good on paper also must be checked for feasibility or "buildability." A failure to do so will lend to droll observations concerning impractical solutions proposed by inexperienced engineers. In this case it should be noted that the deeper the trench required to install the drains, the less stable are the trench walls, and even the shallower trench might be unsafe with a seasonally high groundwater table. Laws governing worker safety usually require that a rigid trench box be used to protect workers inside such a trench.

## 14.6.2 Average Drawdown Between Tile Lines

Drain tile lines are widely used to stabilize landslides. Because of the difficulties and dangers of trenching in unstable, sliding soil, even with a trench box, the drains often are installed by horizontal probing or drilling.

In order to analyze the effect of drainage on slope stability it is necessary to approximate the percentage of the soil volume that will remain undrained, which includes all soil below the level of the drain tile plus an upper part that is defined by an ellipse extending between the drain tile, as shown in Fig. 14.12. The problem is simplified if that part of the ellipse is approximated by a parabola, in which case the cross-sectional area of soil above the level of the tile is

$$A = \frac{2}{3}S(b-a) \tag{14.22}$$

where $b$ and $a$ are as indicated in the figure and $S$ is the tile spacing. Dividing by $S$ gives the average distance of the groundwater table above the tile line:

$$(b-a)_{av} = \frac{2}{3}(b-a) \tag{14.23}$$

Adding $a$ to each side gives the average groundwater elevation above the impermeable layer:

$$b_{av} = \frac{2b+a}{3} \tag{14.24}$$

### Example 14.10
For each of the tile spacings indicated in the previous example calculate the percent of the soil that will remain undrained, assuming that it initially was saturated to the ground surface.

*Answer:* Trial 1: $b_{av} = (2 \times 4 + 3)/3 = 3.7\,\text{m}$; $100\,(3.7/5) = 73\%$.

Trial 2: $b_{av} = (2 \times 4 + 2)/3 = 3.3\,\text{m}$; $100\,(3.3/5) = 66\%$.

## 14.6.3 Emergency Procedures for Landslides

Landslides usually develop after periods of heavy rainfall that raise the groundwater table, so steps may be taken to reduce the infiltration rate $R$ so that it can be accommodated by the natural soil drainage. Unfortunately, the broken ground in landslides ponds water, and ground cracks funnel water down into the soil. Ponded water should be drained and cracks closed by spading in soil wedges. Slopes also can be temporarily covered with plastic film (Fig. 14.13). Such measures obviously are only temporary, but may delay sliding through a wet season to allow later use of more effective control measures.

**Figure 14.13**

Temporary landscaping with plastic to reduce infiltration $R$ and try and put some brakes on a landslide in southern California.

## 14.6.4 Groundwater Mounding under Lagoons

Unless steps are taken to prevent it, submerging soil under a lagoon will increase the infiltration rate so that the groundwater level mounds up underneath the lagoon, and eventually will come into contact with the polluted lagoon water. Lagoons therefore are lined with clay or other material to decrease infiltration. Even without mounding, seepage losses should be intercepted before they enter the groundwater supply. A sand or permeable geotextile layer may be placed underneath the clay liner and kept pumped out and the seepage recycled. This procedure also is frequently used underneath sanitary landfills.

Another alternative to keep lagoon water from entering the groundwater supply was suggested by a soil physicist, Robert Horton. Horton's tests showed that adding clean water to the permeable layer so that the head is slightly higher than that of the lagoon will reverse the flow direction through the clay liner so that it is upward instead of downward, even if the liner has been punctured and compromised.

**Example 14.11**

A sewage lagoon 55 ft deep is to be constructed with a clay liner 3 ft thick with $k = (10)^{-6}$ cm/s. Underneath the clay is unsaturated sand. The mean annual rainfall is 35 in. Should one anticipate groundwater mounding?

*Answer:* From eq. (14.3), $Q/A = ki$, where $i = 55/3 = 18.3$. $Q/A$ therefore equals $18.3(10)^{-6}$ cm/s $= 230$ in./year $\gg 35$ in./year, so there should be groundwater mounding.

*Question:* What if the lagoon is 5.5 ft deep?

*Answer:* This would reduce $i$ and $ki$ by a factor of 10, so $Q/A = 23$ in./year $< 35$ in./year, and there should be no mounding—so long as there are no leaks.

## 14.7 FLOW NETS

### 14.7.1 Solving Two-Dimensional Flow Problems

The tile drain problem presented above is a two-dimensional flow problem. A graphical scheme for solving such problems is called a *flow net*. An example of a flow net is shown in Fig. 14.14 for a constant-head permeameter. Horizontal lines represent *flow paths*, which are paths taken by water in laminar flow, and vertical lines show contours of head, the *equipotential lines*.

The space between adjacent flow lines is a *flow path*, which is analogous to a pipe and *must be continuous*, having both an inlet and an outlet. This is obvious in Fig. 14.14, but becomes less obvious when flow nets are drawn for more complicated geometries.

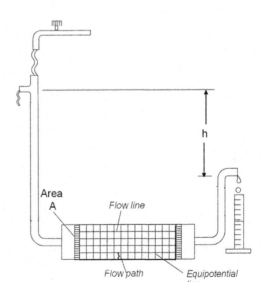

**Figure 14.14**

Two-dimensional flow net for the permeameter of Fig. 14.4, based on the soil being contained in a box having a rectangular cross-section.

Because head decreases in the direction of flow, the intersections of equipotential lines with flow lines must be at 90° even though the flow lines are curved. If the intersections are not at 90°, the flow net is incorrect.

### 14.7.2 Boundary Lines of a Flow Net

In Fig. 14.14 the boundary flow lines are along the outer walls of the permeameter, but in field situations the boundaries may be less obvious unless one remembers that they *must be flow lines*. For example, in the flow net of Fig. 14.15 the flow lines dip sharply downward under the lower ends of the sheet-pile cofferdam. The upper boundary is the flow path down *along the surface* of the sheet pile and up *along the surface* on the other side. The lower boundary is the contact between the soil and an impermeable layer. In other situations the upper boundary is at the level of the groundwater table because that is the upper limit of saturation by gravitational water.

### 14.7.3 Curved Flow Paths

In Fig. 14.15 the flow paths are sharply curved so the geometrical figures cannot be true squares. However, it will be seen that the flow lines and equipotential lines *all intersect at right angles*, and if the geometric figures were subdivided with more flow paths and equipotential lines they would approach squares, in which case *the median dimensions of each figure will be equal*.

**Figure 14.15**

Flow net for a sheet-pile cofferdam showing flow lines, flow paths, and equipotential lines. The surfaces of the sheet pile and of the impervious stratum are boundary flow lines.

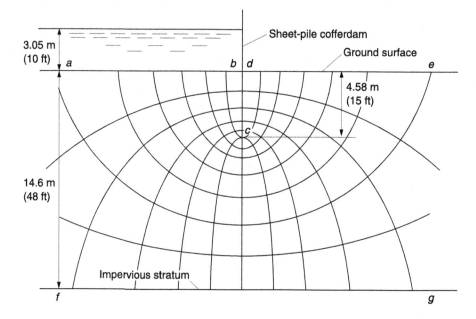

### 14.7.4 Variable Hydraulic Gradient

The curved flow paths that are typical of flow nets drawn for field situations mean that the hydraulic gradient, which is the head loss per unit of travel, is not the same throughout the flow region, which is a reason for drawing a flow net. For example, as shown in Fig. 14.15, the seepage path for water entering soil at $b$ and exiting at $d$ travels a much shorter distance and has a higher hydraulic gradient, and therefore a higher velocity, than that entering at $a$ and exiting near $e$. As will be shown later, this is important because a high emerging hydraulic gradient can create quicksand.

### 14.7.5 Laplace's Equation

While it is not necessary to comprehend the mathematical justification for flow nets in order to use them, the derivation is not complicated and the principles upon which it is based are important. From this it can be seen why flow-net theory also applies, for example, to the flow of electricity, but not to the flow of a gas, which is compressible, or of water in unsaturated soil, because of the content of compressible air and suction derived from the matric potential.

Pierre-Simon Laplace (pronounced with two ah's) was an eighteenth/nineteeth-century French mathematician and astronomer. The basis for Laplace's equation relating to flow nets is the principle of continuity, that the volume of water flowing into a unit volume of saturated soil must equal the volume of water flowing out. By dividing the flow into $x$ and $y$ components, this principle can be represented as shown in Fig. 14.16, where the flow direction changes within the element. The principle of continuity states

$$q = q_{xe} + q_{ye} \tag{14.25}$$

where $q$ is the total flow quantity and $q_{xe}$ and $q_{ye}$ are flow quantities in the $x$ and $y$ directions. This expression also applies to flow of electricity, which allows two-dimensional flow nets to be modeled with an electrical analog such as a resistance paper or even a layer of cardboard wet with salt water with electrodes contacting the paper at the entrance and exit faces.

Equation (14.25) may be expressed in terms of fluid velocity $V$ at the two entry faces, since $q_{xe} = V_x$ times area, and $q_{ye} = V_y$ times area. Then for a unit thickness, as shown in the left-hand segment of Fig.14.16,

$$q = V_x dy + V_y dx \tag{14.26}$$

The velocities at the two departure faces usually are not the same as at the corresponding entry faces because of a changing flow direction within the element. This change can be represented by a rate of change with respect to distance,

**Figure 14.16**

Principle of
continuity for
derivation of
Laplace's
equation:
entrance and exit
rates in a square
element must be
consistent
with volume
in = volume out.

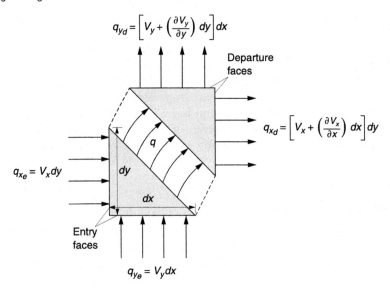

$$q_{yd} = \left[ V_y + \left( \frac{\partial V_y}{\partial y} \right) dy \right] dx$$

Departure
faces

$$q_{xd} = \left[ V_x + \left( \frac{\partial V_x}{\partial x} \right) dx \right] dy$$

$$q_{xe} = V_x dy$$

$q$

$dy$

$dx$

Entry
faces

$$q_{ye} = V_y dx$$

$\partial V_x/\partial x$ and $\partial V_y/\partial y$, times the corresponding distances $dx$ and $dy$. Then, as shown
by the right-hand segment of the figure, the departing velocities are

$$V_x + (\partial V_x/\partial x)dx \quad \text{and} \quad V_y + (\partial V_y/\partial y)dy \tag{14.27}$$

respectively. The departing volumes $q_{xd}$ and $q_{yd}$ are obtained by multiplying the
departure velocities times the corresponding areas. Then

$$q = V_x dy + (\partial V_x/\partial x)dxdy + V_y dx + (\partial V_y/\partial y)dydx \tag{14.28}$$

Equating (14.27) with (14.28) and canceling identities gives

$$(\partial V_x/\partial x)dxdy + (\partial V_y/\partial y)dxdy = 0$$
$$(\partial V_x/\partial x) = (\partial V_y/\partial y) \tag{14.29}$$

This is the equation of continuity for two-dimensional flow through a
homogeneous mass.

Darcy's Law states that $q = kia$ or $V = ki$, where $k$ is the coefficient of hydraulic
conductivity and $i$ is the hydraulic gradient. Hence,

$$V_x = ki_x = -k(\partial h/\partial x) \tag{14.30a}$$
$$V_x y = ki_y = -k(\partial h/\partial y) \tag{14.30b}$$

where the $(-)$ sign indicates that a flow velocity $V_x$ or $V_y$ corresponds to a
reduction in hydraulic head, $\partial h/\partial x$ and $\partial h/\partial y$, respectively. Equations (14.30)
differentiated with respect to $x$ and to $y$ are

$$\partial V_x/\partial x = -k(\partial h/\partial x) \quad \text{and}$$
$$\partial V_y/\partial y = -k(\partial h/\partial y) \tag{14.31}$$

Substituting in eq. (14.28) gives

$$\partial^2 h/\partial x^2 = -k\partial^2 h/\partial y^2 \tag{14.32}$$

This is Laplace's differential equation in two dimensions. It says that a rate of change in head loss with respect to distance in one coordinate direction requires an opposite rate of change in the other. A similar equation may be derived for three-dimensional flow.

Laplace's equation gives the mathematical justification for the existence of nonparallel, curvilinear flow lines and equipotential lines. The rate of change in head is a maximum in the direction of flow; hence equipotential lines cross flow lines at right angles, this being one of the binding requirements for a flow net. It also can be shown that the discharge quantity of water flowing between any two adjacent flow lines is the same.

Laplace also was the first to suggest that randomness of measurements follows a statistical normal or Gaussian distribution.

## 14.7.6 Determining Flow Quantities from a Flow Net

The total quantity of water $Q$ flowing through a unit thickness of a soil mass equals the sum of the quantities in all the flow paths of the flow net. Since a basic requirement of a flow net is that every flow path must transmit the same quantity of water,

$$Q = F\,dQ \tag{14.33}$$

where $F$ is the number of flow paths and $dQ$ is the quantity in each path. Similarly, the total head loss $h$ is the sum of the drops in head in all equipotential spaces of the flow net:

$$h = N\,dh \tag{14.34}$$

where $N$ is the number of equipotential spaces and $dh$ is the reduction in head across each space. In an arbitrarily selected square of length $l$, width $w$, and thickness $= 1.0$, the area normal to flow is $w \times 1$, and the hydraulic gradient is $i = dh \div l$. Substituting these quantities in the Darcy Law for a single square gives

$$Q = kiA$$
$$dQ = k\frac{dh}{l}w \tag{14.35}$$

where $dQ$ is the quantity in each flow path, $dh$ is the head loss between equipotential line, and $w$ and $l$ are dimensions of the square. Since $w/l = 1$,

$$dQ = k\,dh \tag{14.36}$$

Substituting from eqs (14.33) and (14.34),

$$\frac{Q}{F} = k\frac{h}{N}$$

$$Q = kh\frac{F}{N}$$

(14.37)

This means that the total quantity of water that will seep through a unit thickness (normal to the page) of a soil structure can be found by drawing a flow net on a cross-section of the structure, and multiplying the coefficient of permeability $k$ times the total head loss $h$, times the ratio of the number of flow paths $F$ to the number of equipotential spaces $N$. To find the total seepage through the entire structure, the flow through a unit dimension normal to the page is multiplied by the total corresponding dimension of the prototype structure. For a dam or similar structure this would be times the length of the dam.

The validity of the flow-net procedure for computing seepage quantities may be demonstrated by comparing the result obtained by use of the Darcy equation (14.3) with the result obtained by use of the flow-net equation (14.37) when both are applied to the same flow situation such as represented in Fig. 14.14.

### Example 14.12

Assume that the soil mass in Fig. 14.14 is 10 cm thick, 5 cm high, and 14 cm long; that the head loss is 12 cm; and that $k = 10^{-3}$ cm/s. Determine the total seepage (a) by eq. (14.3) and (b) by eq. (14.31).

*Answer:* (a) In the figure, there are 5 flow paths and 14 equipotential spaces, so according to eq. (14.37),

$$Q = (10)^{-3} \text{cm/s} \times 12 \text{cm} \times (5 \div 14) = 0.043 \text{cm}^2/\text{s}$$

which times the thickness of 10 cm gives 0.43 cm$^3$/s.

(b) By eq. (14.3)

$$Q = kiA = (10)^{-3} \times (12/14) \times (5 \times 1) = 0.043 \text{cm}^2/\text{s}$$

which gives a total flow of 0.43 cm$^3$/s, which is the same answer. A flow net obviously is not necessary to solve a simple rectilinear flow problem, but becomes invaluable when flow paths curve and change width.

## 14.8 HOW TO DRAW A FLOW NET

### 14.8.1 Why Engineers Draw Flow Nets

The Laplace solution means that there are many different ways to arrive at a flow net, including mathematical solutions, electrical analogies, finite element and finite difference solutions, and trial sketching. However, in soils the most significant source of error in seepage problems is not the flow net but measurement of $k$.

A useful exercise is to assign a seepage problem to a class in order to compare results from independently sketched flow nets. If the flow nets meet basic requirements of flow paths—that is, continuity, 90° intersections, and quasi-square elements—the variation in $F/N$ should not exceed about 10 percent, compared with variations in $k$ measurements that are many times that amount. The time required to sketch a rudimentary flow net is minimal and may be less than required to set up a problem on a computer where errors can be less obvious. One advantage of a flow net is that mistakes generally are obvious. Many engineering organizations therefore use trial sketching even though it may be supplemented by electrical analogy or other methods.

## 14.8.2    Steps in the Trial-Sketching Method

1.  The first step is to draw or trace a cross-section of the soil structure normal to the direction of seepage flow.
2.  Next, the two boundary flow lines should be established.
3.  The entrance and exit boundaries should be established. These may or may not be equipotential lines. In Fig. 14.15 both boundary lines are equipotential lines along the ground surfaces, so flow lines must intersect ground surfaces at right angles.

---

*Question:* What. if the ground surfaces on the two sides of the sheet pile in Fig. 14.15 were sloping?

*Answer:* Submergence on the left side means that the head is constant even though the elevation varies because of the compensating nature of pressure and elevation head. However, on the right side the emerging water is at atmospheric pressure, so the ground surface would not be an equipotential line, as the potential would depend on the elevation.

---

4.  In order to avoid discontinuous flow paths, the next step is to lightly sketch in two or three trial flow lines so that flow paths do not pinch off or come to a dead end. As a check, every flow path must link the entrance and exit boundaries.
5.  Now sketch trial equipotential lines with right-angle intersections and essentially square figures, and adjust the positions of flow lines as inconsistencies become apparent, such as squares that are rectangular or nonperpendicular intersections. This can be done with pencil and paper, or with a computer drawing program that allows lines to be shifted with a mouse. Each inconsistency will indicate the direction and magnitude of a necessary change.

**Figure 14.17**

Flow net for a masonry dam.

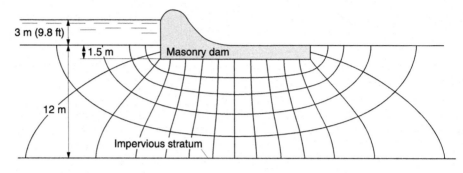

**Figure 14.18**

Flow net for an earth dam on an impervious stratum.

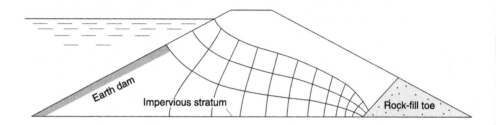

6. Adjustments are continued until the net appears to be a reasonable representation. More flow lines and equipotential lines then can be added during trial re-sketching for better precision, as $F$ and $N$ are whole numbers, but ordinarily three to five flow lines will be sufficient.

Flow nets made by experts should be studied, leading to a sense of fitness and intuition that will assist trial sketching. Some examples of flow nets are shown in the Figs. 14.17 and 14.18. It will be noted that every transition in a homogeneous soil is smooth and curvilinear, and flow paths taper gradually.

### 14.8.3   Flow Nets when *k* Varies Depending on the Direction of Flow

A common situation is where layering in soil causes horizontal permeability to be higher, in some cases many times higher, than vertical permeability. This could pose a difficult situation for analysis, but is readily taken into account in a flow net by compressing the horizontal scale. The scale is compressed according to the ratio of horizontal to vertical $k$, the flow net is drawn according to the conventional rules, and the scale then is re-expanded back to its normal

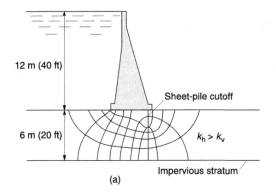

(a)

**Figure 14.19**
Flow net where
$k_h > k_v$. (a) Flow
net constructed on
a transformed
section. (b) Section
re-expanded to
normal scale.

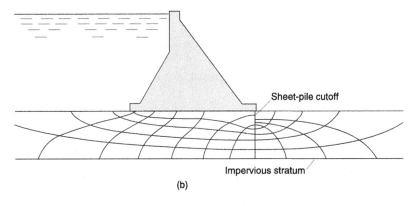

(b)

dimensions. This is easily and instantly accomplished with a computer drawing program.

When the horizontal permeability is the greater, this transformation is made as follows where $x'$ is the adjusted scale of $x$:

$$x' = \frac{x}{\sqrt{k_h/k_v}} \tag{14.38}$$

A transformed flow net with right-angle intersections and nominally square figures is shown in Fig. 14.19(a), and the flow net expanded to the original scale is shown in Fig. 14.19(b), where the flow lines and equipotential lines no longer intersect at right angles, nor are the figures squares. Because the elements of the transformed net are squares, it can be shown that the geometric mean of the two coefficients of permeability may be used to obtain an effective $k'$ to estimate the flow quantity:

$$k' = \sqrt{k_h k_v} \tag{14.39}$$

Derivations of eqs. (14.38) and (14.39) will be found in Harr (1962).

### 14.8.4 Weighted Average Permeabilities

The cause of high horizontal permeabilities usually is horizontal layers of coarser, more permeable soils. With normal vertical tube sampling it often is not practical to measure horizontal permeabilities, but vertical permeabilities of samples from individual layers can be measured and assumed to be the same as horizontal permeabilities for the same layers. An adjusted mean can by obtained by use of an electrical analogy.

A coefficient of hydraulic conductivity $k$ is the equivalent of electrical conductance, which is the reciprocal of resistance. The analog for parallel flow of water along layers with different permeabilities is electricity flowing through resistors with different resistances in parallel. The analog for water transmitted across a series of soil layers is electrical resistors arranged in series.

For an electrical circuit the total resistance of resistors in parallel is

$$\frac{1}{R} = \frac{1}{R_1} + \frac{1}{R_2} + \cdots \frac{1}{R_n} \tag{14.40}$$

where the $R's$ represent total and individual layer resistances in ohms. The analogous equation for permeabilities is

$$k_h = \frac{t_1 k_1 + t_2 k_2 + \cdots t_n k_n}{t} \tag{14.41}$$

where $k_h$ is the coefficient for flow along the layers, which in this case is in the horizontal direction, $t$ represents the total thickness, and $t_1, \ldots, t_n$ represent thicknesses of layers with permeabilities $k_1, \ldots, k_n$.

In the case of vertical flow through horizontal layers of differing permeabilities, the electrical analogy for resistance is

$$R = R_1 + R_2 + \cdots + R_n \tag{14.42}$$

The analogy for fluid flow across layers is

$$\frac{t}{k_v} = \frac{t_1}{k_1} + \frac{t_2}{k_2} + \cdots \frac{t_n}{k_n} \tag{14.43}$$

where $k_v$ is the coefficient for flow across the layers, which in this case is in the vertical direction, and the $t$'s are as above. Application of these equations is simplified if all layers having the same permeability are grouped together. It will be noted that the respective equations apply to flow along and across layers of varying permeability and regardless of the layer orientation.

### Example 14.13

A boring log in Fig. 14.26 shows 10 m of stratified soil that includes 6 silt layers with a total thickness of 8 m and $k = 10^{-4}$ cm/s, and 2 sand seams with a total thickness of about 2 m and $k = 10^{-2}$ cm/s. (a) Estimate the horizontal and vertical permeabilities on the basis of these test data, and (b) recommend a horizontal compression factor for the flow net and (c) an effective $k'$ for interpretation of flow net quantities.

*Answer:\rm* (a) From eq. (14.42) the horizontal permeability is

$$k_h = 8(10^{-4}) + 2(10^{-2}) = \underline{2.1(10^{-2})}\,cm/s$$

which is essentially the same $k$ as that of the sand. From eq. (14.43) the vertical permeability is obtained from

$$\frac{10}{k_v} = \frac{8}{10^{-4}} + \frac{2}{10^{-2}} = 8.02(10)^4$$

$$k_v = \underline{1.2(10^{-4})}\,cm/s$$

which is approximately the same $k$ as the silt.

(b) The horizontal compression factor for the flow net is $\sqrt{2.1(10^{-2}/1.2(10^{-4})} = 13$

(c) The effective $k$ value is $k' = \sqrt{[(2.1(10^{-2}) \times 1.2(10)^{-4}]} = 1.6(10^{-3})\,cm/s$.

## 14.9  HYDROSTATIC UPLIFT PRESSURE ON A STRUCTURE

Any structure that extends below a water surface, whether it is a free water surface or the groundwater table, is subjected to uplift pressure from the water. The uplift pressure below a free water surface is simply the depth times the unit weight of water, which is the principle of Archimedes.

Uplift pressures from water seeping through soil are easily calculated from equipotential lines in a flow net. Uplift pressure is important because it in effect partially floats a structure, which reduces sliding friction between the structure and the soil. For example, a dam that is subjected to water pressure on the upstream face and uplift pressure on the base must be evaluated for stability against sliding. Secondly, since uplift pressure acts within the footprint of a structure, it also can contribute to overturning.

In the masonry dam and flow net in Fig. 14.17, the total head loss $h$ is 3 m. There are 14 equipotential spaces in the net, so the head loss represented by each space is $3 \div 14 = 0.214$ m of water. The second equipotential line contacts the basal upstream corner so the total head at this point is $3 - 2(0.214) = 2.57$ m, which equals pressure head plus elevation head. If the tailwater elevation is taken as the datum, the uplift pressure at this point is $2.57 + 1.5 = 4.07$ m of water.

A similar calculation for the basal downstream corner gives a total head of $1.5 + 2(0.214) = 1.93$ m of water. As the equipotential lines are evenly spaced,

**Figure 14.20**

Uplift pressures at the base of a dam determined by intersections with equipotential lines.

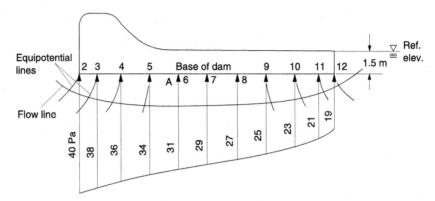

the average uplift pressure across the base of the dam is 2.25 m of water with the center of pressure at the middle of the dam. The distribution of pressure across the base of the structure is shown in Fig. 14.20.

Sliding friction is reduced because of a reduction in effective stress contact pressure from the weight of the dam. *Effective stress* is an important principle in geotechnical engineering: total stress equals the weight of the dam divided by the contact area, and effective stress is the same total stress minus the uplift pressure. It is the effective stress that determines sliding friction. The part of the stress carried by water sometimes is called the "neutral stress" since it acts in all directions.

## 14.10   SEEPAGE FORCE AND QUICKSAND

### 14.10.1   Overview

Water flowing in a stream or river has the power to move solid material ranging from sand-size to boulders and pickup trucks, depending on the velocity of the stream, which in turn depends on the hydraulic gradient. Similarly, water flowing through soil exerts a force on the soil mass that acts in the direction of flow. As this also is a viscous force acting on all soil grains, it is proportional to the fluid velocity and therefore to the hydraulic gradient.

Seepage force is a distributed force acting throughout an affected soil mass, so strictly speaking it is not a force or a pressure; it is a *force per unit volume*, with dimensions $FL^{-3}$.

Perhaps the most sudden and dramatic display of seepage force for one not accustomed to surprises is quicksand, where vertical seepage forces are sufficient to lift and separate soil grains so that there is no friction at grain contacts. Contrary to popular conceptions, stepping on quicksand is not like stepping on ball bearings, because assorted ball bearings interlock and have friction. Stepping on quicksand is like stepping onto dense liquid and offers virtually no resistance to sinking, until opposed by a buoyant force as a result of partial submergence.

Quicksand is a dense liquid, so any object with a lower density will float, and according to the principle of Archimedes the weight of a floating object equals the weight of the fluid it displaces. Without saturation and an upward flow of water there is no quicksand. There is no quicksand without water, and the density of quicksand is such that it is not possible to sink below the armpits unless one happens to be wearing heavy jewelry.

## 14.10.2 Calculating Seepage Force

A laboratory experiment to demonstrate seepage force and the development of quicksand is shown in Fig. 14.21. The net upward force on the bottom of the soil column is the pressure times the area, or

$$F = A\gamma_w h \tag{14.44}$$

where $A$ is the cross-sectional area, $h$ is the head loss in the soil, and $\gamma_w$ is the unit weight of water. The volume of the soil is $AL$ where $L$ is the length of the soil column, so the seepage force per unit volume is

$$S = \frac{F}{AL} = \frac{\cancel{A}\gamma_w h}{\cancel{A}L}$$

Since $h/L = i$,

$$S = i\gamma_w \tag{14.45}$$

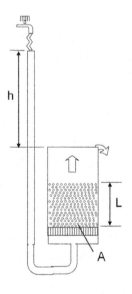

**Figure 14.21**
Seepage force and quicksand.

where $S$ is the seepage force per unit volume, $i$ is the hydraulic gradient, and $\gamma_w$ is the unit weight of water. This is the basic equation for seepage force that acts in the direction of seepage. Therefore if seepage is horizontal or at some angle other than vertical, the seepage force can be combined vectorially with the weight of the soil in order to obtain a resultant body force acting on the soil mass. For example, horizontal seepage into an excavation adds a horizontal component to the weight of the soil, which affects stability of the excavation walls.

### 14.10.3 Quicksand

Opposing the seepage force in Fig. 14.21 is the submerged weight of the soil column. The submerged weight of solids in a unit volume is $V_s(G-1)\,\gamma_w$:

$$\frac{W}{AL} = V_s(G-1)\gamma_w \tag{14.46}$$

where $V_s$ is the volume of solids in a unit volume of soil and $G$ is the specific gravity of the solids. From the definition of void ratio for a saturated soil, $e = V_w/V_s$, and appropriate substitutions it can be shown that $V_s = 1/(e+1)$. Therefore

$$\frac{W}{AL} = \frac{G-1}{e+1}\gamma_w \tag{14.47}$$

A quicksand condition develops when the seepage force $S$ equals the weight per unit volume, or

$$S = i_c\gamma_w = \frac{G-1}{e+1}\gamma_w$$

where $i_c$ is the critical hydraulic gradient. Then

$$i_c = \frac{G-1}{e+1} \tag{14.48}$$

This is the basic equation for quicksand. As an approximation, $G$ for quartz is 2.65, so if $e$ is 0.65,

$$i_c \approx 1.0$$

Therefore as a general guide quicksand can be expected to develop if the hydraulic gradient in an upward direction exceeds 1.0. However, as $i_c$ is sensitive to void ratio, a more accurate determination can be made from Fig. 14.22, which is based on $G = 2.70$.

It can be seen that the critical gradient does not depend on size of the soil particles except insofar as they influence permeability and the quantity of flow necessary to achieve $i_c$. That is, if $k$ in the expression $Q = kiA$ is large, the flow quantity $Q$ must have a proportionate increase to create the same hydraulic gradient. Quick conditions therefore are much more likely to develop in sand than in gravel, although a quick gravel theoretically is possible. A quick condition with relatively

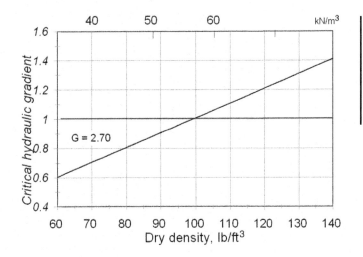

**Figure 14.22**
Effect of soil density on the hydraulic gradient required to generate a quick condition.

little seepage can develop in cohesionless silt, and sometimes occurs where a slip zone subjected to upward seepage is exposed in the toe area of a landslide.

## 14.10.4 Occurrences of Quicksand

Quicksand commonly occurs in the bed of a lake or river where water seeps upward as a result of artesian pressure. The points of emergence are underwater springs. Quicksand conditions therefore may be seasonal when artesian pressures are highest, and absent during drier seasons.

Excavations that are pumped dry in preparation of placing foundations for buildings or bridges can be an invitation for quicksand or "boiling" of soil as water seeps upward into the excavation. Sand boils also can occur on the downstream side of an earth dam as the reservoir is filled and develops high seepage pressures. Sand boils are common on the landward side of temporary levees built during flood stages of major rivers.

Unpredicted quick conditions can result if an undiscovered aquifer, buried tile line, gravel layer, or fracture zone in rock conveys water without diminishing the head. This is illustrated in Fig. 14.23.

## 14.10.5 Liquefaction

Temporary quicksand conditions can be induced by ground vibrations, particularly from earthquakes, if they densify saturated, loose sand so that stress is transferred to the water and the effective stress is reduced or becomes zero. Then, as the sand settles, the water they displace rises. A vibration-induced quick condition is called *liquefaction*. Evidence for liquefaction includes conical "sand volcanoes" where sand is carried up to the ground surface by the ejected

**Figure 14.23**

A fugitive tile line or stringer of gravel can cause quick conditions. Back-pressure is applied by encircling the seepage area with sandbags.

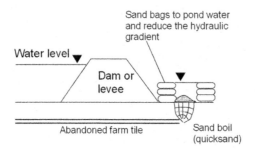

water. The occurrence of sand volcanoes buried in ancient sediment layers is used to date the recurrence interval of earthquakes.

### 14.10.6  Managing Quicksand

A simple and effective way to stop a quick condition is to reduce the hydraulic gradient by building a wall of sand bags to pond water, as shown in Fig. 14.23. A more permanent measure that is more suitable for foundation excavations is to use interceptor wells to reduce or reverse the gradient.

Another method is to cover a potential quicksand area with a layer of heavy, coarse-grained material that adds additional weight that must be lifted, but which also is sufficiently permeable that it allows seepage water to escape without appreciable resistance. Such a blanket of material is called a protective filter. A layer of filter fabric or intermediate particle size material can be used to prevent soil material from being carried upward into the filter. The design of filters is discussed later in Section 14.11.3.

## 14.11  FLOW NETS IN EARTH DAMS

### 14.11.1  Dam Arguments

Earth dams are the largest man-made structures and provoke some of the biggest arguments. On the positive side, dams are cost-effective and save energy for production of electrical power, and reservoirs created by dams are essential for survival of places like Las Vegas. (As grandmother would say, more's the pity.) Dams allow irrigation that turns deserts into gardens, and reservoirs are used for recreation and for flood control. Dams can allow river navigation that is the most economical means for shipping large quantities of grain, coal, etc.

Negative aspects are that dams drown valuable floodplain soils, displace towns and people, corrupt nature, and have a limited useful life because of sedimentation in the reservoir. Sediment trapped behind a dam is not available for delta-building at the mouth of the river, which starts a domino effect, the next domino being the reduction in sediment to maintain beaches, and encroaching shorelines as a result of wave erosion.

Of these problems, sedimentation may be the most serious as it will gradually diminish and eventually will destroy the usefulness of the dam. A bypass canal can be used to divert sediment-laden flood water, but this may only borrow time, and the amount of sediment that eventually may be deposited often will be hundreds of times larger than the volume of the dam itself. Optional procedures include lowering the reservoir level to let the river entrench and carve out a new floodplain, or fluidizing and pumping sediment around the dam (Jenkins et al., 1992). An option that does not yet appear to have been tried is to let sediment-laden water form a density current along the bottom of a reservoir to an outlet at the base of the dam.

Engineers should not be saddled with the decision of whether or not to build a dam, but can provide an objective assessment of the data. Engineers should develop an awareness of both the positive and negative aspects of dams, and if a decision is made to proceed, it is the engineer's obligation to ensure that the dam and reservoir are safe, functional, and long-lasting. Large dam construction is particularly active in developing countries of the world.

## 14.11.2  Flow-Net Boundary Lines in Earth Dams

The lower flow boundary in an earth dam normally is an impermeable layer that may or may not be at the base elevation of the dam. The upper flow boundary was investigated by A. Casagrande using dyes introduced at the upstream face of model dams. He found that the upper flow line can be approximated with a parabola that is modified at the upstream side to make a 90° angle with the dam face, which is an equipotential line. Casagrande's method, while empirical, has been compared with an analytical method and shown to be adequate for general use (Jepson, 1968).

Figure 14.24 shows a cross-section of a trapezoidal dam of homogeneous soil. The parabola begins at the water surface of the reservoir at point A, which is a

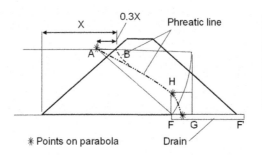

**Figure 14.24**

Casagrande method for drawing an upper boundary flow line through an earth dam.

horizontal distance 0.3 times the horizontal width of the upstream face that is under water.

Next, the focal point of the parabola, F, is assigned at the boundary of the drainage layer. The directrix is found by swinging an arc centered at A from F to a horizontal line and dropping a vertical line to G. One point on the parabola is halfway between F and G. Another point, H, is directly over F such that F, G, and H are corners of a square. These points should be sufficient to sketch in the parabola that then is modified with a smooth curve at B.

Without a drainage gallery the focal point of the parabola would move to F′, and extending a parabola to this point indicates that seepage would emerge on the upstream face, which should be avoided.

## 14.11.3  Drain and Filters

A layer of a more permeable granular soil often is incorporated into an earth dam as an underdrain, *EF* in Fig. 14.24. Another alternative is a toe drain that is triangular in cross-section, Fig. 14.27. In both instances soil is separated from the drain by a suitable filter to prevent soil fines from penetrating into voids in the coarser material. As design involves particle sizes in one medium and void sizes in the other, it is empirical based on laboratory tests. Representative criteria are as follows:

$$\frac{d_{15}}{D_{85}} < 5 \tag{14.49a}$$

$$\frac{d_{15}}{D_{15}} \text{ between 5 and 40} \tag{14.49b}$$

$$\frac{D_{85}}{\text{hole diameter in pipe drain } D_{50}} \geq 2 \tag{14.49c}$$

where *d* and *D* refer to particle sizes in the finer and coarser materials, respectively, and the subscripts denote percentages finer than the respective sizes. These requirements apply to both boundaries of a filter layer, so if a single layer does not satisfy all criteria two or more layers can be used. The thickness of each layer depends on the hand that must be resisted; a rule-of-thumb being that it should be at least 5 percent of the pressure head.

Another option is the use of a synthetic filter fabric, which is discussed by Koerner (1990). Whereas filter layers of soils have progressively larger void spaces and permeability, the capability of a geotextile to transmit water varies, and must be high enough not to impede drainage. The design therefore involves two criteria, one for flow of water and the other for retention of the soil.

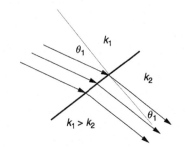

**Figure 14.25**
Diffraction of flow lines resulting from a change in *k*.

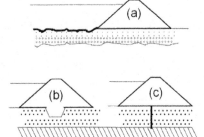

**Figure 14.26**
Some methods for reducing underdam seepage.

A geotextile having a sufficient thickness and transmissivity (sometimes called permitivity) can substitute for the drain itself. In this case it must resist crushing under pressure from the soil. Some applications include drains along the backsides of retaining walls under pavements or earth embankments.

## 14.11.4 Diffraction of Flow Lines

Earth dams often incorporate a clay core to reduce seepage, and in addition employ a granular drainage area in the toe. Seepage crossing a boundary between two permeabilities is diffracted, much in the same way that light is diffracted when crossing a boundary between two refractive indices, and the same equation applies. An analogy is soldiers marching in a parade and entering an area of mucky soil at an angle: as the marchers reach the muck they slow down, which causes the direction of the column to change. If the velocity decreases, the angle increases, and vice versa. The equation is

$$\frac{\tan \theta_1}{\tan \theta_2} = \frac{k_1}{k_2} \tag{14.50}$$

The derivation for this equation can be found in a physics textbook. It means that flow lines entering the clay core of an earth dam will be diffracted downward, lowering the phreatic line, and upon exiting the clay core or entering a toe drain will be diffracted upward. The nature of the diffraction is indicated in Fig. 14.25.

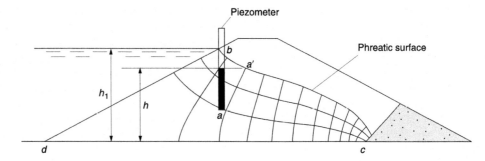

### 14.11.5   Controlling Underdam Seepage

A typical alluvial section on which a dam may be founded consists of clay
overlying sand over gravel. Upstream from the dam the river most likely will be
running in sand that connects to the gravel and can become a major path for
leakage. Several methods may be used to reduce such leakage. As shown in
Fig. 14.26(a), an area upstream, in particular the river channel, may be covered
with a compacted clay layer called an "upstream blanket." In Fig. 14.26(b)
relatively impermeable soil in the dam is extended down into the foundation soils,
which is called a "key trench." A trench that extends all of the way through the
permeable layers is a "cutoff trench." A third possibility is a "curtain wall" such
as shown in Fig. 14.26(c), which can be made with sheet pile or with a
concrete-filled trench called a "slurry trench."

### 14.11.6   Piping

Bad things still can happen after a dam is constructed and in service, and walk-
over inspections should be routinely performed. For example, floodplain soils
supporting earth dams normally are soft and slowly compressible, and over a
period of time a dam may settle unevenly and develop internal tension cracks
running transverse across the axis of the dam. Such cracks form a direct route
through the dam and if undetected and not corrected by grouting can be a prelude
to a major leak and failure.

Any leakage observed through an earth dam must be carefully examined to see
if the water is clouded by soil particles, which would indicate internal erosion
or "piping." If piping is observed, remedial treatments should immediately be
initiated.

### 14.11.7   Pinhole Test

Some soils are more susceptible to piping than other soils. Most vulnerable
are silts that are easily erodible, and soils containing clays that tend to
disperse as water reduces the content of flocculating ions. Normally these

are sodium clays, and are referred to as "dispersive clays." Such clays are regarded as unsatisfactory for use in earth dams. However, their detection can be complicated, and for this purpose a "pinhole test" was devised by Sherard et al. (1976). Details of the test are given in ASTM Designation D-4647.

The pinhole test essentially consists of drilling a 1 mm diameter hole through a compacted soil specimen and passing water through the hole. A constant head of 50 mm (2 in.) is used initially and the test continued for 5 minutes, but these values may be modified for additional testing. Erosion is determined from cloudiness of the water, increasing flow rate as the pinhole enlarges, and final dimensions of the pinhole. If after 5 minutes the hole has enlarged 50 percent, the soil is considered dispersive. The test is performed with distilled water, which will tend to increase dispersion.

## 14.11.8  Measuring Seepage Conditions within an Earth Dam

Cracks that do not go all of the way through or under a dam still can be dangerous, but are difficult to detect except as they affect the equipotential lines in the flow net. Anomalous behavior therefore can be detected by measuring the pore water pressure at various points within a dam. This is done with tubes called *piezometers* that have a porous tip that allows water to enter up into the tube. The height of the water in the tube is a measure of the total head at that point in the dam, and can be compared with that predicted from the flow net. An example is shown in Fig. 14.27. The level of water in a piezometer head is *not* at the phreatic surface, but is as determined by the contact with an equipotential line.

Let us now assume that the head measured by the piezometer is at the $h_1$ level of the water surface behind the dam instead of at $h$. That would mean that there is a direct connection between the location of the piezometer tip and the reservoir behind the dam, which should be thoroughly investigated as it eventually could develop into seepage and possible quick conditions in the face of the dam, and breaching of the dam. Piezometers are a valuable tool for assessing seepage conditions in and under a dam, and often are permanent installations. Some major dam failures might have been prevented had the seepage conditions been monitored with piezometers and corrective measures taken before it was too late.

## Problems

14.1. The change in elevation of a stream is 947 μm/m (5 ft/mile). What is the hydraulic gradient?

14.2. State the Darcy Law for flow of water through soil. Define the coefficient of hydraulic conductivity. Is it the same as the coefficient of permeability?

14.3. Explain the term "velocity of approach," as used in connection with the Darcy Law.

14.4. Some soils are said to be nonisotropic with respect to flow of gravitational water. What does this mean? What are some examples?

14.5. What is the difference between a constant-head and a falling-head permeameter? Explain why one type may be preferred for particular soils.

14.6. A constant-head permeameter is set up with a soil specimen 203 mm (8 in.) long and 76 mm (3.0 in.) in diameter. The vertical distance between the headwater and tailwater surfaces is 305 mm (12 in.). In a test run, 163.9 kg (361.5 lb) of water passes through the sample in a period of 18 hr 20 min. What is the coefficient of hydraulic conductivity of this soil?

14.7. A falling-head permeameter is set up with a soil sample 152 mm (6 in.) long and 51 mm (2 in.) in diameter. The area of the standpipe is 213 mm$^2$ (0.33 in.$^2$). In a test run, the water in the standpipe falls from an elevation of 1.22 m (48 in.) above the tail water to 0.538 m (21.2 in.) in 7 hr 36 min 32 s. What is the coefficient of hydraulic conductivity of this soil?

14.8. The conductivity coefficient of a soil determined in the laboratory at a temperature of 24.4°C (76°F) is 3.5 mm/min. What is the coefficient of the same soil when the permeating water is at 4.4°C (40°F)?

14.9. A test well is installed in permeable sand to a depth of 9.1 m (30 ft) below the groundwater table, and observation wells are put down at distances of (a) 15.2 m (50 ft) and (b) 30.5 m (100 ft) from the test well. When pumping reaches a steady state at 965 l/min (255 gal/min), the water table at (a) is lowered 1.158 m (3.8 ft) and at (b) 0.945 m (3.1 ft). Compute the coefficient of permeability of the sand.

14.10. A test well 762 mm (30 in.) in diameter is drilled through a water-bearing sand down to an impervious layer that is 7.32 m (24 ft) below the normal water table. Steady pumping at the rate of 757 l/min (200 gal/min) causes the water in the test well to be lowered 2.84 m (9 ft 4 in.). What is the approximate value of the coefficient of hydraulic conductivity of the sand?

14.11. A 114 mm (4.5 in.) diameter casing is set at 6.7 m (22 ft) depth in a hole with the water table at 1.04 m (3.4 ft) depth. Water added at the rate of 2.08 l/min (0.55 gal/min) maintains the water level at 51 mm (2 in.) depth. Find $k$.

14.12. A test head with packers spaced 1.52 m (5 ft) apart is lowered to a depth of 13.7 m (45 ft) in a 102 mm (4 in.) diameter hole. After sealing, a pumping pressure of 240 kPa (35 lb/in.$^2$) resulted in a pumping rate of 2.78 l/s (44 gal/min), increasing to 4 l/s (63 gal/min) after 12 min. Find the initial and final $k$ and give possible explanations for the difference.

14.13. A hole 152 mm (6 in.) in diameter is bored in the soil to a depth of 1.90 m (75 in.) below the water table. After the hole has been pumped out several times and allowed to refill, it is pumped out again and the rate of rise of water at an elevation 1.02 m (40 in.) below the water table is observed to be

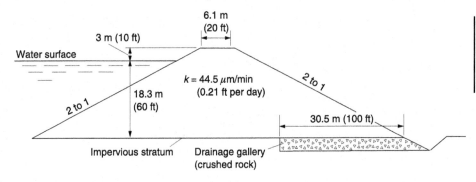

**Figure 14.28**
Earth dam for flow-net construction in Problem 14.21.

19.0 mm (0.75 in.) in 1 min. What is the coefficient of hydraulic conductivity of the soil?

14.14. What is the basis for the flow net?

14.15. Define flow line; equipotential line; flow path; equipotential space; three kinds of head.

14.16. State five fundamental properties of a flow net.

14.17. Identify the boundary flow lines and equipotential lines in Fig. 14.15.

14.18. Assume that the coefficient of hydraulic conductivity of the soil in Fig. 14.15 is 0.0042 mm/min. Determine the seepage per 30 m (100 ft) of length of the sheet-pile cofferdam.

14.19. The pervious foundation soil beneath the masonry dam in Fig. 14.17 has a coefficient of hydraulic conductivity of 2.96 mm/min (0.14 ft/day). Determine the seepage per 30 m (100 ft) length of the dam.

14.20. The pervious foundation soil in Fig. 14.19 has a coefficient of hydraulic conductivity of 1.22 mm/min in the vertical direction and 6.1 mm/min (0.02 ft/min) in the horizontal direction. Compute the seepage per linear meter (foot) of dam.

14.21. Construct a flow net and estimate the seepage per linear meter (foot) of the dam in Fig. 14.28.

14.22. What is the critical hydraulic gradient of a soil having a specific gravity of 2.70 and a void ratio of 0.85?

14.23. Determine the hydraulic gradient of seepage water at point $d$ in Fig. 14.15. If the specific gravity of the soil inside the cofferdam is 2.62 and the void ratio is 1.14, will sand boils develop? What is the factor of safety against boiling at this point?

14.24. What is the hydraulic gradient at point $e$ in Fig. 14.15?

14.25. Calculate the effective stress on a horizontal plane 4.9 m (16 ft) below the surface of a soil mass in which the water table is 1.5 m (5 ft) below the surface. Assume that the dry density of the soil is 14.1 kN/m$^2$ (90 lb/ft$^2$), the moisture content above the water table is 15%, and the specific gravity of the soil is 2.70.

**Figure 14.29**

How not to draw a flow net.

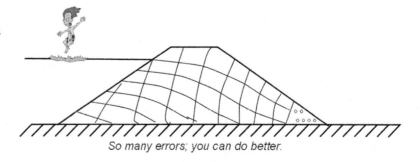

*So many errors; you can do better.*

14.26. Permeability tests on samples from several horizontal soil layers under a proposed dam give the following:

| Material | k (cm/s) | % of total section |
|----------|----------|--------------------|
| Clay | $8 \times 10^{-5}$ | 45 |
| Silt | $1 \times 10^{-4}$ | 35 |
| Sand | $2 \times 10^{-2}$ | 18 |
| Gravel | $10^{-1}$(est.) | 2 |

(a) Calculate the horizontal and vertical coefficients of permeability.
(b) Suggest methods to reduce the horizontal permeability.

14.27. Identify errors in the flow net in Fig. 14.29.

## References and Further Reading

Boast, C. W., and Kirkham, D. (1971). "Field Measurement of Soil Permeability Using Auger Holes." *Proc. Soil Sci. Soc. Amer.* 35, 365.

Casagrande, A. (1937). "Seepage Through Dams." *J. New Engl. Waterworks Assoc.* Reprinted in *Contributions to Soil Mechanics 1925–1940*, pp. 295–336. Boston Soc. Of Civil Engrs., Boston, Mass.

Cedergren, H. R. (1989). *Seepage, Drainage, and Flow Nets*, 3rd ed. John Wiley & Sons, New York.

Harr, M. E. (1962). *Groundwater and Seepage*. McGraw-Hill, New York.

Jenkins, S. A., Wasyl, J., and Skelly, D. W. (1992). "Tackling Trapped Sediments." *Civil Engineering* (Feb.), pp. 61–63.

Jepson, R. W. (1968). "Seepage Through Dams in the Complex Potential Plane." *ASCE J. Irrigation and Drainage Div.* 94(IR1), 23–39.

Jury, W. A., and Horton, R. (2004). *Soil Physics*, 6th ed. John Wiley & Sons, Hoboken, N. J.

Kirkham, D., and Powers, W. L. (1971). *Advanced Soil Physics*. Wiley Interscience, New York.

Koerner, R. M. (1990). *Designing with Geosynthetics* 2nd ed. Prentice-Hall, Englewood N.J.

Leonards, G. A., ed. (1987). *Dam Failures*. Elsevier, Amsterdam. Reprinted from *Engineering Geology* 24,1–4.

Sherard, J. L., Dunnigan, L. P., Becker, R. S., and Steele, E. F. (1976). "Pinhole Test for Identifying Dispersive Soils." *J. Geotech. Eng. Div.* ASCE 102(GT1), 69–85.

# 15

## Stress Distribution in Soil

## 15.1 STRESS IS WHERE YOU FIND IT

### 15.1.1 A Halfway Point

The first part of this book emphasizes how the geological and pedological history of soils affects their composition and engineering properties, and how to manipulate those properties with drainage and compaction. Some would say that the best is yet to come, using all of this valuable information to build on.

### 15.1.2 Approximations

The variability and complexity of soils are such that simplifications must be made, or analysis would come to a halt from being tied into an incomprehensible knot. For example, an approximation was used to measure particle diameter as individual particles are assumed to be spheres. Even though this is not appropriate for mica particles, relatively few analyses involve abundant mica particles, so the simplification is useful as long as its limitations are recognized. Even the most sophisticated computer modeling by finite elements requires simplifications. The artistry comes in making simplifications that have only a minimal effect on the outcome, and preferably with the effect on the safe side for design.

It also is possible to be overzealous in making simplifications, so that results not only miss the mark but they are on the unsafe side. For example, local statutes governing trench excavation safety may indicate that soil can be cut back to its "angle of repose." Ditch digging leaves little time for introspection, and some contractors observed that since clay can stand in a vertical bank, vertical is its angle of repose. The problem is that "angle of repose" applies only to a cohesionless soil such as sand, not to clay, where, as the depth of a vertical cut increases, the cut becomes less stable and eventually will fail, often taking lives with it. It took decades and many bad experiences to defeat this oversimplification.

This chapter deals with stress (pressure) that is induced in soils by a surface load, which is a factor determining the settlement of structures. An assumption is made that the pressure is uniformly applied to a soil surface by a flat, rigid plate. This is needed for a mathematical treatment that does not go to pages of equations and qualifiers, because in the real world there is either (a) an increase in pressure from resistance to vertical shearing in soil under plate edges, or (b) a decrease in shearing resistance near the edges if the soil squeezes out. The key question is, to what extent does the assumption of uniform pressure affect the results? Compression is cumulative with depth and these departures affect only near-surface stresses, so the simplification is not particularly damaging—but it nevertheless should be recognized.

## 15.2   TAKING THE MEASURE OF STRESS

### 15.2.1   Stress and Pressure

Much of geotechnical engineering deals with forces and pressures from the weight of buildings or other structures, and from the weight of the soil itself. Unlike structural loads that are held within confines of slabs, beams, columns, and trusses, stresses in soils radiate downward and outward, spreading like rays that decrease in intensity with distance. Fortunately, this distribution is predictable and subject to an approximate analysis by the theory of elasticity.

There is some confusion between the use of "pressure" and "stress," and the terms sometimes are used interchangeably. Pressure is appropriate to describe the force per unit area such as that exerted on the inside of a balloon or a vessel of water, and is oriented perpendicular to the contact surface. Stress is similar to pressure but exists within a material and therefore requires no surface contact. Furthermore, stress is directional. A foundation can simultaneously exert vertical pressure and lateral stress in the underlying soil if it tends to squeeze out.

Stresses existing in a material can be resolved into three principal stresses that are at right angles to one another and are purely tensile or compressive. However, the stressed element must be in perfect alignment for this to occur. Generally, soil weight acting vertically over a large area imposes a major or maximum principal stress that is oriented vertically, and the two other principal stresses caused by the soil tending to bulge laterally are horizontal. The mutually perpendicular planes on which principal stresses act are *principal planes*.

If the alignment of the stressed element is not parallel, as shown in Fig. 15.1, its surfaces no longer are principal planes and will support shearing stresses. In soils it is shearing stress that causes grains to slide over one another and can lead to a shear failure. In a landslide the shearing stress in the basal soil zone equals the maximum resistance to shearing. If the maximum shearing resistance decreases as a result of remolding, look out below because the landslide speeds up.

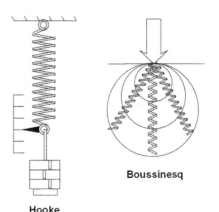

**Figure 15.1**

Two applications of elasticity, one for weighing and the other for determining stresses in soils that can cause settlement of foundations or may be transferred to underground structures.

Boussinesq

Hooke

The stress perpendicular to a shearing stress is called a *normal stress*, and it should be remembered that a normal stress is not a principal stress. Geotechnical engineering revolves around considerations of principal stresses on principal planes, and shearing and normal stresses on surfaces that are not principal planes.

## 15.2.2  Body Stresses

"Body stresses" are those contributed by the weight of the soil itself, which in many instances exceed the external loads that are imposed on the soil. Although the density of soil typically is only about 60 to 90 percent of that of Portland cement concrete, it still is roughly two times that of water. The calculation of body stresses was discussed in Chapter 9.

Both applied stresses and body stresses must be taken into account in settlement predictions for foundations, which are discussed in the next chapter. Both are involved in calculations of the bearing capacity of foundations against soil shear failure, also discussed in a later chapter. Retaining walls, landslides, and other slope stability problems involve mainly body stresses that in turn are affected by differences in soil density and moisture content, and buoyant forces from a groundwater table.

## 15.3  THEORY OF ELASTICITY

### 15.3.1  Ideal Elastic Behavior

In engineering mechanics, *strain* is the ratio of deformation to length and has nothing to do with working out. In an elastic material such as steel, strain is proportional to stress, which is why spring scales work.

Soil is not an ideal elastic material, but a nearly linear stress-strain relationship exists with limited loading conditions. A simplification therefore is made that under these conditions soil can be treated mathematically during vertical compression as an elastic material. (The same assumption frequently is made in finite element analyses.) However, an ideal elastic material rebounds like a spring when a load is removed, and rebound of soil is incomplete, indicating that some soil grains have moved relative to one another. For this reason soil is considered "quasi-elastic," or is described as exhibiting "near-linear elastic behavior." In this book these inferences will be implied instead of being repeatedly stated when the term "elastic" is used for soils.

There is a limit to near-linear elastic behavior of soils as loading increases and shearing or slipping between individual soil particles increases. Every tiny slip causes a transfer of stress to adjacent particles, and as this continues slipping will become coordinated along some most critical surface or zone of slippage. When that happens any semblance to an elastic response is lost as shearing more closely simulates *plastic behavior*. This is the behavioral mode of soils in landslides, bearing capacity failures, and behind most retaining walls. Only stress distributions that can be characterized as elastic are discussed in this chapter.

### 15.3.2 Elastic Theory Applied to Soils

Elastic behavior has been a favorite of mathematicians since the mid-seventeenth century, when an English physicist, Robert Hooke, dangled some weights on a spring, measured deflections, and discovered what now is known as Hooke's Law.

A landmark contribution to the study of stress distribution in soils was published in 1885 by a French mathematician, Joseph Valentin Boussinesq (pronounced Boo-sin-esk). Boussinesq solved for stresses in a "semi-infinite elastic medium." The semi-infinite medium also is called a half space, which is everything on one side of an infinite plane that cuts space in half. In many applications that plane represents the ground surface, and Boussinesq in effect solved for stresses radiating from a point load applied vertically at the ground surface.

As can be seen in Fig. 15.1, vertical compression under a point load induces vertical stress underneath the load and inclined stress farther away, so the amount of lateral confinement affects the amount of compression. Take away the inclined springs on one side by removing lateral restraint, for example next to a vertical cut, and the stress distribution changes and may allow shear failure, in which case elastic theory no longer is applicable.

### 15.3.3 Boussinesq's Approach

Boussinesq defined a "half space" that is made up of an *ideal elastic, homogeneous, isotropic material*. "Half space" means everything on one side of a plane. Soil is not ideally elastic, homogeneous, or isotropic (meaning that it is the same in all directions), but stress measurements indicate that the classical Boussinesq solution gives reasonably good approximations. However, when conditions are substantially different from the assumed conditions, for example in layered soils, modifications to the formulas become necessary.

Geotechnical engineers should understand the conditions upon which the formulas are based, even though they are not necessarily skilled in the advanced mathematical procedures by which the solutions are derived. While the formulas appear complicated, results are presented in simplified forms that are relatively easy to use.

Figure 15.2 shows the geometry of the Boussinesq solution, with a point load at $P$ and orthogonal axes $X$–$X'$ and $Y$–$Y'$ in the horizontal $XY$ plane that is the upper boundary of the semi-infinite mass. The axis $Z$–$Z'$ extends vertically downward into the mass, and passes through the origin of the coordinates. The location of any point $A$ down within the soil mass is indicated by the coordinate distances $x$, $y$, and $z$.

Stresses acting at point $A$ on planes normal to the coordinate axes are shown in Fig. 15.2(b). They consist of normal (compressive) stresses, denoted by $\sigma$ (sigma), and shearing stresses denoted by $\tau$ (tau). Subscripts on $\sigma$ indicate

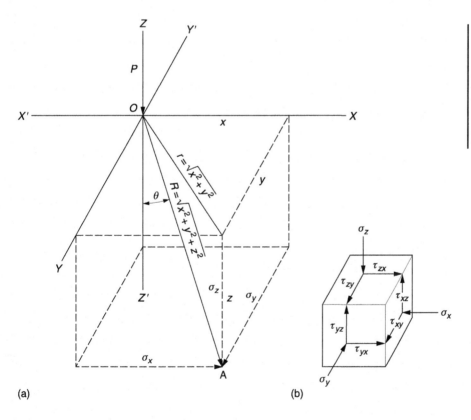

(a)  (b)

**Figure 15.2**

Symbols used in the Boussinesq solution for stresses transferred from a point load at $O$ to an element at $A$.

the orientation of the lines of action of the three orthogonal normal stresses. Thus, $\sigma_z$ is a normal stress parallel to the Z-axis through point $A$.

Shearing stresses have two subscripts. The first letter indicates the plane in which the shearing stress acts, and the second letter indicates the orientation of its line of action. Thus, $\tau_{zx}$ is a shearing stress acting in a plane normal to the Z-axis and on a line of action parallel to the X-axis. $\tau_{zy}$ is a shearing stress acting in the same plane but on a line of action parallel to the Y-axis. The stresses on only three faces of a cube of soil at point $A$ are shown in Fig. 15.2(b). Since this cube is in equilibrium, the stresses on the other three faces will have the same magnitudes as those shown, but they must act in the opposite directions.

The Boussinesq equations for normal stresses are as follows.

$$\sigma_z = \frac{3P}{2\pi} \frac{z^3}{R^5} \tag{15.1}$$

or

$$\sigma_z = \frac{3P \cos^5 \theta}{2\pi z^2} \tag{15.1a}$$

A special case for maximum vertical stress that is directly underneath a point load is $R = z$, and

$$\sigma_z = \frac{3P}{2\pi z^2} \tag{15.1b}$$

Stresses in the $X$ and $Y$ directions are

$$\sigma_x = \frac{P}{2\pi} \left[ \frac{3x^2 z}{R^5} - (1 - 2v) \left( \frac{x^2 - y^2}{Rr^2(R+z)} + \frac{y^2 z}{R^3 r^2} \right) \right] \tag{15.2}$$

$$\sigma_y = \frac{P}{2\pi} \left[ \frac{3y^2 z}{R^5} - (1 - 2v) \left( \frac{y^2 - x^2}{Rr^2(R+z)} + \frac{x^2 z}{R^3 r^2} \right) \right] \tag{15.3}$$

in which $v$ is Poisson's ratio, discussed in the next section, and other symbols have the meanings indicated in Fig. 15.2(a).

The equation for $\sigma_z$ is the most used of the above three because the distribution of vertical stress is required for prediction of foundation settlement. It also affects the transmission of surface loads to underground structures or strata. The horizontal stress equations are used to obtain pressures transmitted from surface loads to retaining walls.

An additional set of equations describes shearing stresses and occasionally is used for retaining walls, or to test whether the shearing strength of a soil locally may be exceeded because of a surface load:

$$\tau_{zx} = -\tau_{xz} = \frac{3P}{2\pi} \frac{z^2 x}{R^5} \tag{15.4}$$

$$\tau_{zy} = -\tau_{yz} = \frac{3P}{2\pi} \frac{z^2 y}{R^5} \tag{15.5}$$

$$\tau_{xy} = -\tau_{yx} = \frac{P}{2\pi} \left[ \frac{3xyz}{R^5} - (1 - 2v) \left\{ \frac{(2R + 2)xy}{(R + z^2)R^3} \right\} \right] \tag{15.6}$$

Note that immediately underneath the load point, where $x = y = 0$, all $\tau$ shearing stresses are zero.

## 15.4  POISSON'S RATIO

### 15.4.1  Strain and the Mechanics of Bulging

"Strain" in engineering mechanics has a specific meaning that must not be confused with stress, which is force per unit area. Strain has to do with deflection, and is defined as the *linear change in a dimension divided by that dimension*. It is analogous to percent but normally is not expressed as a percentage. For example, if a cylinder 100 units long is compressed 5 units, the strain is $5/100 = 0.05$. Strain sometimes is called "unit strain," but that term is redundant because strain is always a deflection per unit length in the strain direction.

A more formal definition of strain in the $X$, $Y$, and $Z$ directions is

$$\varepsilon_x = \frac{\Delta x}{x} \tag{15.7}$$

$$\varepsilon_y = \frac{\Delta y}{y} \tag{15.8}$$

$$\varepsilon_z = \frac{\Delta z}{z} \tag{15.9}$$

Siméon-Denis Poisson was a French mathematician who worked in the nineteenth century on a wide range of topics, ranging from electricity to astronomy and mechanics. The ratio that bears his name is defined as the ratio between strain normal to the direction of an applied stress, to strain parallel to that stress. In other words, Poisson's ratio is a measure of elastic bulging or squashing out sideways in response to an application of a major principal stress, and is an inherent property of elastic materials.

$$v = \frac{\varepsilon_x}{\varepsilon_z} \tag{15.10}$$

where $\varepsilon_x$ and $\varepsilon_z$ are strains in the $X$ and $Z$ directions. It may be seen from eqs. (15.1)–(15.3) that, according to the Boussinesq analysis, directly under a surface load Poisson's ratio does not affect the vertical compressive stress but it does influence the horizontal responses, which appears reasonable.

## 15.4.2 Values of Poisson's Ratio

As illustrated in Fig. 15.3, if a cubic element of a material is loaded along a $Z$-axis and Poisson's ratio is 0.5, expansion in both the $X$ and $Y$ directions is one-half that in the $Z$ direction. For this value of Poisson's ratio the net volume change is zero, and the material is said to be incompressible. On the other hand, if Poisson's ratio is 0, there is no expansion in the $X$ and $Y$ directions, so the change in volume occurs only in the $Z$-axis direction. An example of the latter is Styrofoam, which has $v = 0$.

The normal range of values of Poisson's ratio for all kinds of engineering materials is between 0 and 0.5. Poisson's ratio for soils is closer to the upper limit of 0.5 than it is to the lower limit of zero. According to the theory of elasticity Poisson's ratio cannot exceed 0.5 as this would imply an increase in volume from compression, which would create energy. There are situations where this can occur in soils as they dilate during shearing, but that is not ideal elastic behavior.

Examination of eqs. (15.2), (15.3) and (15.6) indicates that, if $v = 0.5$, the equations are very much simplified. They therefore are frequently written in the simplified form

$$\sigma_x = \frac{3P}{2\pi}\left[\frac{x^2 z}{R^5}\right] \tag{15.2a}$$

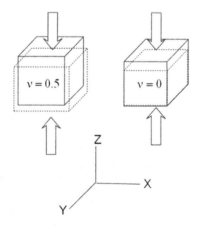

**Figure 15.3**

Relation of volume change to Poisson's ratio. On the left if $v = 0.5$, which is representative of some soils, compressive strain in the $Z$ direction equals the sum of expansive strains in the $X$ and $Y$ directions so the net volume change is zero. On the right, $v = 0$ in soft foam plastics.

$$\sigma_y = \frac{3P}{2\pi}\left[\frac{y^2 z}{R^5}\right] \tag{15.3a}$$

$$\tau_{xy} = -\tau_{yx} = \frac{3P}{\pi}\left[\frac{xyz}{R^5}\right] \tag{15.6a}$$

Note that the equations for $\sigma_z$, $\tau_{zx}$, and $\tau_{zy}$ do not contain Poisson's ratio and are not affected by it. The use of the simplified forms of the Boussinesq equations is an approximation because it assumes that the soil is incompressible, which it obviously is not. If Poisson's ratio is less than 0.5, it will be seen from eq. (15.1) that $\sigma_z$ will not be affected, but according to eqs. (15.2) ff., $\sigma_x$, $\sigma_y$, and $\tau_{yx}$ all will decrease as a result of the reduced tendency for bulging. An exception is directly underneath the load where $x = y = 0$, so lateral stresses and shearing stresses under a point load theoretically are zero. However, it also should be noted that there is no such thing as a point load; the assumption of a point load is a simplification.

## 15.5  ELASTIC STRESS-STRAIN RELATIONS

### 15.5.1  The Basis for Elastic Theory

Applications of the theory of elasticity are based on the premise that under uniaxial loading conditions, strain is proportional to applied stress:

$$\varepsilon_z = \frac{1}{E}\sigma_z \tag{15.11}$$

where $\varepsilon_z$ is strain in the $z$ direction, $\sigma_z$ is a compressive or tensile stress in the $z$ direction, and $E$ is the *modulus of elasticity*. In geotechnical engineering, compressive stress is positive, tension negative. Because strain is dimensionless, $E$ has units of stress, or force per unit area, designated dimensionally by $FL^{-2}$.

From the definition of Poisson's ratio, strain in the lateral $X$ and $Y$ directions is

$$\varepsilon_x = \varepsilon_y = \nu\varepsilon_z = \nu\frac{\sigma_z}{E} \tag{15.12}$$

where $\varepsilon_x$ and $\varepsilon_y$ are strain in the $x$ and $y$ directions and $\nu$ is Poisson's ratio. In most situations stress is applied from all sides, which is referred to as a triaxial loading condition. The total strain in any direction is the sum of strains from the three applied stresses, so by writing three equations for each stress direction one obtains the following equations for vertical and lateral strain, respectively:

$$\varepsilon_z = \frac{1}{E}\left[\sigma_z + \nu(\sigma_x + \sigma_y)\right] \tag{15.13}$$

$$\varepsilon_x = \frac{1}{E}\left[\sigma_x + \nu(\sigma_y + \sigma_z)\right] \tag{15.14}$$

$$\varepsilon_y = \frac{1}{E}\left[\sigma_y + \nu(\sigma_x + \sigma_z)\right] \tag{15.15}$$

These equations describe interrelationships between vertical and lateral strain from vertical elastic compression, based on the modulus of elasticity and Poisson's ratio.

# 15.6   CRITERIA FOR ELASTIC BEHAVIOR

## 15.6.1   Requirements

Several conditions must be met for the Boussinesq solution to be considered appropriate:

- The stresses must be in equilibrium, that is, the total vertical stress on a horizontal plane at any depth in the soil must equal the applied load at the ground surface. This criterion is always complied with and can be considered axiomatic in any static stress distribution problem.

- The soil must be homogeneous and isotropic, which ideally means no layering. Nor should there be a vertical alignment of soil particles shown in Fig. 8.5. As part of this requirement strains must be compatible; for example, there must not be any separations or cracks in the soil. The last criterion is out of compliance because the Boussinesq stress solution shows a shallow tensile zone near the surface and adjacent to the load, and soil has little or no strength in tension. This effect is minor except at shallow depths and close to a point load.

- Boundary conditions obviously are not infinite, but must be a reasonable representation or boundary conditions will influence the stress distribution. The most common situation is the presence of a rigid rock stratum or retaining wall, and these conditions can be accounted for by a modification of elastic theory. However, laboratory modeling of soil behavior with boundary conditions imposed by a bin or a box can be a major source of error that unfortunately sometimes is overlooked or ignored.

## 15.6.2   Layering

Soils frequently are layered in nature, and layering is utilized in road construction, for example by a flexible-type pavement resting on a soil base course and subgrade. The stiffness of the upper layers usually is very large compared with the soil subgrade, and an increase in modulus means a decrease in strain under the same loads.

Theoretical analyses of layering were developed by Burmister at Cornell University in 1943, and extended by Burmister and others to various two- and three-layer systems. The equations are quite lengthy, so tables or charts are usually used to estimate stresses under specific circumstances. Some graphical solutions are given by Poulos and Davis (1974). The problem can be approached with finite element modeling, which is case-specific.

Depth-related changes in soil modulus also occur as a result of compression under past and present overburden pressures, and in general the larger the depth in an otherwise uniform soil, the higher the modulus. These changes in modulus are taken into account in settlement calculations, but their effect normally is ignored in the calculation of vertical stress in a soil. As the modulus increases with depth, this is on the safe side for design.

Soil stress-strain curves become nonlinear as the soil approaches failure. This and other nonlinear elastic behaviors come under the aegis of "constitutive equations." These usually are empirically defined and can be case-specific. They are discussed by Christian and Desai (1977) in relation to finite element modeling.

## 15.7  INTEGRATION OF THE BOUSSINESQ EQUATION

### 15.7.1  Applications

The most important use of the Boussinesq solution is to predict vertical stresses in soils under foundations. Foundations impose area loads, not point loads, so it is necessary to integrate the Boussinesq relationships across the foundation contact area. As previously mentioned, the integration is based on the assumption that the foundation pressure is evenly distributed at the ground surface.

An integration for rectangular loaded areas was accomplished in the 1930s by Professor Nathan Newmark at the University of Illinois (Newmark, 1935). A similar integration with a different objective was made a decade later by Holl at Iowa State College (Holl, 1941). As illustrated in Fig. 15.4, Newmark solved for point stress from an area surface load, and Holl reversed the geometry and solved for load on a subsurface area such as a buried culvert pipe resulting from a concentrated surface load. Although details of the equations differ, there is some assurance in the fact that the results are identical.

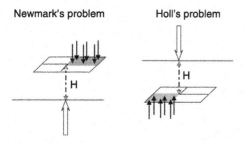

**Figure 15.4**

Two problems that give the same elastic solution. Note that both integrations solve for stresses at corners of loaded areas.

### 15.7.2 Nature of the Integration

An integration needs a starting point, and in order that the horizontal coordinates will be consistently positive, the starting point is at the corner of a rectangular area that is in line with the load. This is shown in Figs. 15.4 and 15.5.

Newmark's equation appears complicated but is readily programmed, and solutions can be obtained using tables and graphs. The equation is as follows, where symbols are as shown in Fig. 15.5:

$$\frac{\sigma_z}{p} = \frac{1}{4\pi}\left[\frac{2ABH\sqrt{(A^2 + B^2 + H^2)}}{H^2(A^2 + B^2 + H^2) + A^2B^2} \cdot \frac{A^2 + B^2 + 2H^2}{A^2 + B^2 + H^2} + \left(\sin^{-1}\frac{2ABH\sqrt{(A^2 + B^2 + H^2)}}{H^2(A^2 + B^2 + H^2) + A^2B^2}\right)\right] \tag{15.16}$$

A uniform load $p$ is applied to a rectangular area with one corner at the origin. The area has linear dimensions $A$ and $B$, and $H$ is the depth at which the vertical stress $\sigma_z$ is evaluated.

The procedure to obtain pressures at positions other than under one corner is discussed in later sections. Before proceeding with exact solutions we will examine simplifications that are frequently used for everyday applications.

**Figure 15.5**

Dimensional notations in the Newmark equation (15.16).

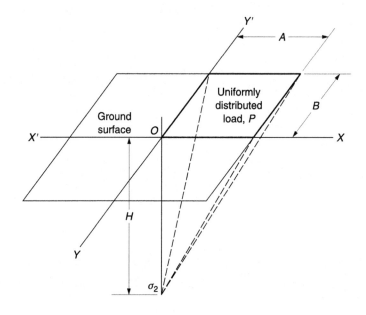

### 15.7.3 The Pressure Bulb

Vertical stresses under loaded areas can be contoured as shown in Fig. 15.6, which illustrates the decrease in stress with depth and with lateral distance from the center of a foundation. The shape of the contours is the "pressure bulb." Two solutions are shown, one for a long foundation and the other for a square foundation. The graphs are normalized to the foundation width $B$, that is, depths and horizontal distances are in terms of $B$, so the graphs are easily applied to actual foundations, and are sufficiently accurate for many purposes.

The horizontal spread of the pressure bulb also illustrates how pressure from one foundation can apply additional stress to soil under an adjacent foundation. According to elastic theory stresses are additive, which means that constructing a heavy building close to an existing building can contribute to adverse settlement of a nearby edge or corner of the older building.

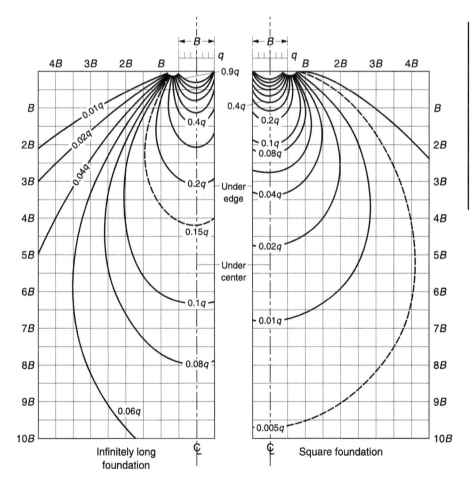

**Figure 15.6**

Pressure bulbs obtained from the Newmark integration of the Boussinesq solution of vertical stress under uniform area loads. (Modified from Sowers and Sowers, 1961.)

### 15.7.4 Influence Coefficients

Numbers on the contours are *influence coefficients*, designated by $I_c$. Influence coefficients are multiplied times the applied foundation pressure, $q$, to obtain vertical soil stress induced by the foundation load. That is, if $q$ is the applied foundation stess and $I_c = 0.5$, the added stress at the point in question is $0.5q$. The isostress contours are for vertical stresses, not principal stresses, except directly under the load axis. Contours constructed for the major principal stresses are circles that intersect the ground surface at the outer edges of the foundations.

### 15.7.5 Depth of the Pressure Bulb

Theoretically, the stress only becomes zero at an infinite depth or distance, but for practical purposes it is considered to reach a negligible value when the added stress is reduced 90 percent, that is, where $I_c = 0.1$. As shown in Fig. 15.6, this occurs at a depth of between $6B$ and $7B$. This depth therefore is relevant for determining depths of exploration borings. While soils normally increase in stiffness with depth, there is no guarantee that this will occur, and selected borings usually are extended deeper to determine if there are any softer or more compressible layers.

It should be emphasized that a very wide structure such as an earth dam, extensive fill area, or artificial island will cause the influence depth to extend proportionately deeper.

### Example 15.1
The compressible soil layer in Fig. 15.7 occurs between 1.0 m and 1.6 m depth. A 1 m square foundation exerts a pressure of approximately 100 kPa ($1\,\text{T/ft}^2$) at the ground surface. (a) What is the increase in vertical stress in the middle of the soil layer directly underneath

**Figure 15.7**

Drawing for Example 15.1.

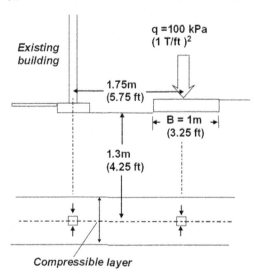

*Compressible layer*

the foundation? (b) What influence will this have on a foundation that is 1.75 m from the center of the new foundation?

*Answer:*

(a) The depth to the center of the compressible layer is $1.3B$. The graph at the right in Fig. 15.6 indicates that $I_c$ is approximately 0.24 under the center, so the additional vertical stress on the middle of the compressible layer under the center of the new foundation is approximately $100 \times 0.24 = 24\,\text{kPa}$ ($500\,\text{lb/ft}^2$).

(b) The center of the existing wall is $1.75B$ from the center of the new foundation. At a depth of $1.3B$, Fig. 15.6 indicates that $I_c$ is approximately 0.02, so additional stress under the existing foundation will be about $2\,\text{kPa}$ ($42\,\text{lb/ft}^2$), which should have only a moderate effect on the adjacent foundation.

## 15.7.6 Linear Approximations

A procedure that is used for routine settlement analyses of small structures is to assume that the pressure bulb extends from the edges of a foundation downward along a linear path having a slope of 2 vertical to 1 horizontal. This is shown at the left in Fig. 15.8 for a long foundation, with dashed lines showing the resulting $I_c$ values of 0.5, 0.25, and (at the bottom) 0.10, for comparison with the Boussinesq distribution. It will be seen that this construct in general is conservative.

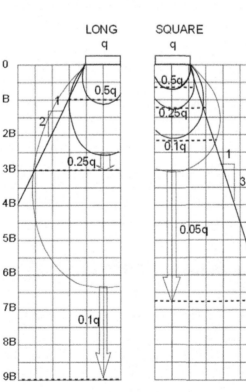

**Figure 15.8**
Linear slopes to approximate the Boussinesq distribution, 1:2 for long and 1:3 for square or round surface area loads.

In the case of a square foundation, shown at the right in Fig. 15.8, a 2:1 slope is unconservative, but if the slope is adjusted to 3:1, as shown in the figure, the stress is more concentrated and the approximation is reasonably accurate. *For square or round foundations a slope of 3 vertical to 1 horizontal therefore is suggested.*

An objection to this procedure is that it gives a false impression that the induced stress is constant across an area, then suddenly drops to zero at the edges of the area. For example, at depth $B$ and a distance from the edge $0.5B$ from the edge of a long foundation, $I_c$ according to the linear approximation is either 0.5 or 0, and the Bousinesq distribution gives 0.3. *It therefore is recommended that linear approximations be used with caution or not used at all when estimating stresses under nearby foundations.*

### Example 15.2

Re-work the previous example using the linear distribution shown at the right in Fig. 15.8.

*Answer:*

(a) At a depth of 1.3 m the expanded side dimension of the contact area is $2 \times 1.3(1/3) + 1 = 1.87$ m, so the expanded area is $(1.87)^2 = 3.5$ m$^2$. The stress on this area therefore is indicated to be reduced by a factor $1/3.5 = 0.29$, so the linear approximation gives 29 kPa under the foundation, compared with 24 kPa obtained from Fig. 15.4, so the linear approximation is on the conservative side.

*Question:* What if the 2:1 slope is used?

*Answer:* The expanded side dimension then would be $2 \times 1.3(1/2) + 1 = 2.3$ m, giving an area of 5.3 m$^2$ and a indicated stress of $100/5.3 = 19$ kPa instead of 24 kPa, which is on the unsafe side.

(b) The expanded side dimension indicates a spread of $1.87/2 = 0.94$ m from the center of the new foundation, so according to the linear approximation method the existing foundation is outside of the influence area and will not be affected. This is approximately correct but nevertheless could instill false confidence because it is on the unsafe side.

*Question:* If the 2:1 slope is used, would the adjacent foundation be affected?

*Answer:* One-half of the expanded side dimension is $2.3/2 = 1.15$ m, which still does not extend to soil underneath the existing foundation.

## 15.8 USING THE NEWMARK COEFFICIENTS

### 15.8.1 Stresses under Foundations

A more accurate means for calculating vertical stress is to use influence coefficients determined from eq. (15.16), either by use of a computer or by interpolation of solutions in Table 15.1. Usually it is not the stress under one corner that is desired, but stresses at different locations, including at the middle, where it should be a maximum. In order to accomplish this the foundation area is divided into quadrants, as shown in Fig. 15.9, the stress calculated for one corner and multiplied times 4. If the stress under one side is desired, the area is divided along a line intersecting that side, and the corner stresses evaluated and added. It would be incorrect to simply multiply a corner stress times 2 because the contributing areas change.

Other examples of this procedure are shown in Fig. 15.9. By an appropriate selection of quadrants, stress may be determined at any depth and distance that is either inside or outside the contact area, by adding or subtracting the contributions of each to the total influence coefficient.

### Example 15.3

Calculate the vertical stress at a depth of 6.1 m (20 ft) under the center of the short side of a 6.1 m (20 ft) by 12.2 m (40 ft) rectangular area that supports 191.4 kPa (2 tons/ft$^2$).

*Answer:* This is Case V in Fig. 15.9. Place the origin of coordinates at the center of the short side and divide the loaded area into two equal halves. The values of $A$, $B$, and $H$ for each half are 6.1 m $(12.2/2) = 6.1$ m, and 6.1 m, respectively, which gives $m = 1.0$, and $n = 1.0$. From Table 15.1, $I_c = 0.175$, so the answer is

$$\Delta\sigma_z = 2 \times 0.175 \times 191.4 = 67 \, \text{kPa} \, (0.7 \, \text{T/ft}^2)$$

## 15.8.2 Long Foundations

It will be noted from Table 15.1 that influence coefficients become asymptotic to constant values as one foundation dimension becomes large, being essentially constant after the length:width ratio equals 10.

## 15.8.3 Circular Foundations

Circular foundations are commonly used for water towers, fuel storage tanks, stand-alone grain elevators, etc. Grain elevators pose an additional problem because down-drag of grain on the inside walls concentrates load on the walls, and a doughnut-shaped foundation is often used.

Influence coefficients at various depths under the center of a uniformly loaded circular area are given by

$$\frac{\sigma_z}{P} = 1 - \left[1 - \frac{1}{1 + (R/H)^2}\right]^{3/2} \tag{15.17}$$

where, as shown in Fig. 15.10, $H$ is the depth under the center and $R$ is the radius of the loaded area. For points at a horizontal distance $A$ from under the center,

**Table 15.1**

Vertical influence coefficients for a corner of a loaded rectangular area, after Newmark (1935)

| $m = A/H$ or $n = B/H$ | $n = B/H$ or $m = A/H$ | | | | | | | | |
|---|---|---|---|---|---|---|---|---|---|
| | 0.1 | 0.2 | 0.3 | 0.4 | 0.5 | 0.6 | 0.7 | 0.8 | 0.9 |
| 0.1 | 0.005 | 0.009 | 0.013 | 0.017 | 0.020 | 0.022 | 0.024 | 0.026 | 0.027 |
| 0.2 | 0.009 | 0.018 | 0.026 | 0.033 | 0.039 | 0.043 | 0.047 | 0.050 | 0.053 |
| 0.3 | 0.013 | 0.026 | 0.037 | 0.047 | 0.056 | 0.063 | 0.069 | 0.073 | 0.077 |
| 0.4 | 0.017 | 0.033 | 0.047 | 0.060 | 0.071 | 0.080 | 0.087 | 0.093 | 0.098 |
| 0.5 | 0.020 | 0.039 | 0.056 | 0.071 | 0.084 | 0.095 | 0.103 | 0.110 | 0.116 |
| 0.6 | 0.022 | 0.043 | 0.063 | 0.080 | 0.095 | 0.107 | 0.117 | 0.125 | 0.131 |
| 0.7 | 0.024 | 0.047 | 0.069 | 0.087 | 0.103 | 0.117 | 0.128 | 0.137 | 0.144 |
| 0.8 | 0.026 | 0.050 | 0.073 | 0.093 | 0.110 | 0.125 | 0.137 | 0.146 | 0.154 |
| 0.9 | 0.027 | 0.053 | 0.077 | 0.098 | 0.116 | 0.131 | 0.144 | 0.154 | 0.162 |
| 1.0 | 0.028 | 0.055 | 0.079 | 0.101 | 0.120 | 0.136 | 0.149 | 0.160 | 0.168 |
| 1.2 | 0.029 | 0.057 | 0.083 | 0.106 | 0.126 | 0.143 | 0.157 | 0.168 | 0.178 |
| 1.5 | 0.030 | 0.059 | 0.086 | 0.110 | 0.131 | 0.149 | 0.164 | 0.176 | 0.186 |
| 2.0 | 0.031 | 0.061 | 0.089 | 0.113 | 0.135 | 0.153 | 0.169 | 0.181 | 0.192 |
| 2.5 | 0.031 | 0.062 | 0.090 | 0.115 | 0.137 | 0.155 | 0.170 | 0.183 | 0.194 |
| 3.0 | 0.032 | 0.062 | 0.090 | 0.115 | 0.137 | 0.156 | 0.171 | 0.184 | 0.195 |
| 5.0 | 0.032 | 0.062 | 0.090 | 0.115 | 0.137 | 0.156 | 0.172 | 0.185 | 0.196 |
| 10.0 | 0.032 | 0.062 | 0.090 | 0.115 | 0.137 | 0.156 | 0.172 | 0.185 | 0.196 |
| $x$ | 0.032 | 0.062 | 0.090 | 0.115 | 0.137 | 0.156 | 0.172 | 0.185 | 0.196 |

| | 1.0 | 1.2 | 1.5 | 2.0 | 2.5 | 3.0 | 5.0 | 10.0 | $x$ |
|---|---|---|---|---|---|---|---|---|---|
| 0.1 | 0.028 | 0.029 | 0.030 | 0.031 | 0.031 | 0.032 | 0.032 | 0.032 | 0.032 |
| 0.2 | 0.055 | 0.057 | 0.059 | 0.061 | 0.062 | 0.062 | 0.062 | 0.062 | 0.062 |
| 0.3 | 0.079 | 0.083 | 0.086 | 0.089 | 0.090 | 0.090 | 0.090 | 0.090 | 0.090 |
| 0.4 | 0.101 | 0.106 | 0.110 | 0.113 | 0.115 | 0.115 | 0.115 | 0.115 | 0.115 |
| 0.5 | 0.120 | 0.126 | 0.131 | 0.135 | 0.137 | 0.137 | 0.137 | 0.137 | 0.137 |
| 0.6 | 0.136 | 0.143 | 0.149 | 0.153 | 0.155 | 0.156 | 0.156 | 0.156 | 0.156 |
| 0.7 | 0.149 | 0.157 | 0.164 | 0.169 | 0.170 | 0.171 | 0.172 | 0.172 | 0.172 |
| 0.8 | 0.160 | 0.168 | 0.176 | 0.181 | 0.183 | 0.184 | 0.185 | 0.185 | 0.185 |
| 0.9 | 0.168 | 0.178 | 0.186 | 0.192 | 0.194 | 0.195 | 0.196 | 0.196 | 0.196 |
| 1.0 | 0.175 | 0.185 | 0.193 | 0.200 | 0.202 | 0.203 | 0.204 | 0.205 | 0.205 |
| 1.2 | 0.185 | 0.196 | 0.205 | 0.212 | 0.215 | 0.216 | 0.217 | 0.218 | 0.218 |
| 1.5 | 0.193 | 0.205 | 0.215 | 0.223 | 0.226 | 0.228 | 0.229 | 0.230 | 0.230 |
| 2.0 | 0.200 | 0.212 | 0.223 | 0.232 | 0.236 | 0.238 | 0.239 | 0.240 | 0.240 |
| 2.5 | 0.202 | 0.215 | 0.226 | 0.236 | 0.240 | 0.242 | 0.244 | 0.244 | 0.244 |
| 3.0 | 0.203 | 0.216 | 0.228 | 0.238 | 0.242 | 0.244 | 0.246 | 0.247 | 0.247 |
| 5.0 | 0.204 | 0.217 | 0.229 | 0.239 | 0.244 | 0.246 | 0.249 | 0.249 | 0.249 |
| 10.0 | 0.205 | 0.218 | 0.230 | 0.240 | 0.244 | 0.247 | 0.249 | 0.250 | 0.250 |
| $x$ | 0.205 | 0.218 | 0.230 | 0.240 | 0.244 | 0.247 | 0.249 | 0.250 | 0.250 |

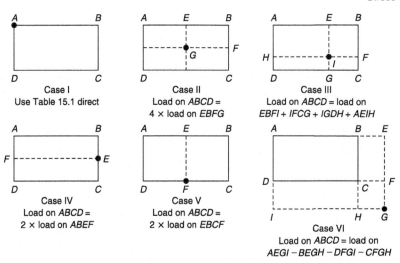

**Figure 15.9**

Application of the Newmark integration to different locations relative to a rectangular foundation.

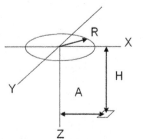

**Figure 15.10**

Identification of variables used in eq. (15.17) and Table 15.2.

the solution involves elliptical integrals that have been evaluated in Table 15.2. The influence coefficient obtained from Table 15.2 should be multiplied by the uniform load on the circular area to obtain the vertical stress at any point in the undersoil.

### Example 15.4

A circular foundation 10 m in diameter will support a fuel tank that will create a maximum foundation pressure of 93 kN/m². Compute the vertical stress in soil at a depth of 10 m under the center and under the perimeter of the foundation.

*Answer:* $R = 5$ m and $H = 10$ m, so under the center $A/R = 0$ and $H/R = 10/10 = 1$. From Table 15.2, $I_c = 0.646$. Under the edge, $A = 5$ m so $A/R = 1$ while $H/R$ remains 1, so $I_c = 0.332$. The respective soil pressures therefore are $93 \times 0.646 = 60$ kN/m² under the center and $93 \times 0.332 = 31$ kN/m² under the edge.

## 15.8.4 Average Stresses under a Foundation

As most foundations are rigid and therefore do not adapt to differences in settlement between corner, edge, and center positions, average soil stresses may be

**Table 15.2**

Vertical influence coefficients at different radial distances and depths under a circular loaded area, after Jumikis (1971)

| | | | | | A/R | | | | | |
|---|---|---|---|---|---|---|---|---|---|---|
| H/R | 0 | 0.25 | 0.50 | 1.0 | 1.5 | 2.0 | 2.5 | 3.0 | 3.5 | 4.0 |
| 0.25 | 0.986 | 0.982 | 0.966 | 0.460 | 0.014 | 0.002 | 0.000 | 0.000 | 0.000 | 0.000 |
| 0.50 | 0.911 | 0.896 | 0.840 | 0.417 | 0.060 | 0.010 | 0.003 | 0.000 | 0.000 | 0.000 |
| 0.75 | 0.784 | 0.763 | 0.692 | 0.374 | 0.102 | 0.026 | 0.007 | 0.003 | 0.001 | 0.001 |
| 1.00 | 0.646 | 0.626 | 0.562 | 0.332 | 0.127 | 0.042 | 0.015 | 0.006 | 0.003 | 0.001 |
| 1.25 | 0.524 | 0.507 | 0.458 | 0.292 | 0.137 | 0.055 | 0.023 | 0.010 | 0.005 | 0.003 |
| 1.50 | 0.424 | 0.411 | 0375 | 0.256 | 0.138 | 0.065 | 0.030 | 0.015 | 0.007 | 0.004 |
| 1.75 | 0.345 | 0.336 | 0.310 | 0.224 | 0.134 | 0.071 | 0.036 | 0.019 | 0.010 | 0.006 |
| 2.00 | 0.284 | 0.278 | 0.259 | 0.196 | 0.126 | 0.073 | 0.041 | 0.022 | 0.013 | 0.008 |
| 2.5 | 0.200 | 0.196 | 0.186 | 0.151 | 0.109 | 0.072 | 0.045 | 0.028 | 0.017 | 0.011 |
| 3.0 | 0.146 | 0.144 | 0.138 | 0.118 | 0.092 | 0.067 | 0.046 | 0.031 | 0.021 | 0.014 |
| 4.0 | 0.087 | 0.086 | 0.084 | 0.076 | 0.065 | 0.053 | 0.041 | 0.031 | 0.023 | 0.017 |
| 5.0 | 0.057 | 0.057 | 0.056 | 0.052 | 0.047 | 0.040 | 0.034 | 0.028 | 0.022 | 0.018 |
| 7.0 | 0.030 | 0.030 | 0.029 | 0.028 | 0.027 | 0.025 | 0.022 | 0.020 | 0.017 | 0.015 |
| 10.0 | 0.015 | 0.015 | 0.015 | 0.014 | 0.014 | 0.013 | 0.013 | 0.012 | 0.011 | 0.010 |

more meaningful. Average stresses can be calculated based on the following assumptions:

(a) If the loaded area is infinitely long and the low part of the pressure bulb is approximated by a parabola, the average stress is two-thirds of the center stress plus one-third of the edge stress.
(b) If the loaded area is circular and the low part of the pressure bulb is approximated by a paraboloid of revolution, the average is one-half of the center stress plus one-half of the edge stress.

For a square loaded area a circular equivalent can be obtained from equal areas:

$$\pi R_e^2 = X^2$$
$$R_e = X/\pi$$
(15.10a)

where $R_e$ is the radius of the equivalent circle and $X$ is one side of the square. The same approach might be used for a rectangular area, but as it is transitional to a long loaded area, the weighting factor changes.

In summary, based on the above assumptions the average stresses can be approximated as follows:

Long foundation : $\Delta\sigma_{zav} = 0.67\Delta\sigma_{zcenter} + 0.33\Delta\sigma_{zedge}$ (15.10b)

Circular foundation : $\Delta\sigma_{zav} = 0.5\Delta\sigma_{zcenter} + 0.5\Delta\sigma_{zedge}$ (15.10c)

Square foundation: use eq. (15.10c) with an equivalent radius from (15.10a).

Rectangular foundation: because $\Delta\sigma_{zav}$ is transitional to the long foundation case, a conservative estimate will use the formula for a long foundation.

### Example 15.5
Estimate the average stress for the preceding example.

*Answer:* Vertical stress under the center was determined to be 60 kPa and under the edge 31 kPa, so based on a paraboloid distribution the average stress is the actual average, 46 kPa. Note that no more than two significant figures are reported in the final answer because of the assumptions made in the analysis.

## 15.8.5   Stress from a Surface Point Load to a Buried Structure

The Holl solution is applied similarly to Newmark's and uses the same influence coefficients. A common use is to determine pressures transmitted by wheel loads to buried structures such as culverts, and shows how stresses are greatly magnified by a shallow burial depth. As in the case of foundation loading, the maximum stress is obtained by dividing the stressed area into four quadrants, solving for a corner stress, and multiplying times 4. A solution normally is required for only one depth that represents the depth to the top of a culvert.

The incorporation of a rigid body such as a concrete culvert, or a flexible body such as a flexible conduit, into a soil mass violates the Boussinesq criterion of a uniform, homogeneous, isotropic material. However, measurements indicate that the results are within acceptable limits.

## 15.9   NEWMARK INFLUENCE CHART

## 15.9.1   Introduction

Foundations may have irregular shapes that can become quite cumbersome to solve, involving overlapping areas where one must be subtracted to prevent a double exposure. In addition, buildings often are supported by separate foundations to support each column and wall, leading to a confused network of areas as pressure from one will influence pressure and settlement of another. Newmark therefore devised a graphical procedure for determining influence coefficients for irregular contact areas.

The graphical procedure involves drawing a scaled tracing of the footing plan and laying it on top of a special scaled chart, as shown in Fig. 15.11, so that the location to be evaluated is at the center of the chart. The influence coefficient then is simply the number of square-like areas of the chart that are covered by the foundation plan. Fractional areas are estimated and included in the total. The most critical stage is selection of a proper scale, which is different for each depth.

**Figure 15.11**

A Newmark chart, each small area representing $0.005I_c$.

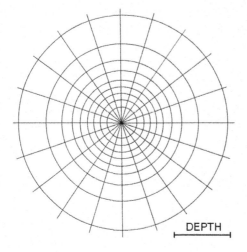

DEPTH

The use of Newmark charts was somewhat labor-intensive prior to the introduction of personal computers, but now a drawing can be repositioned and the scale changed with a click and drag of a mouse.

## 15.9.2 Using the Newmark Chart

The following are for use of computer graphics. A similar procedure can be used by drawing a transparency to each required depth scale.

1. The Newmark chart is scanned, inserted, and locked in a computer drawing.
2. The foundation plan then is scanned and transferred to a different page of the same drawing.
3. A depth is selected and marked at a convenient point on the foundation plan at the same scale as the plan.
4. The plan and depth scale then are moved by holding down the left mouse button so that the depth scale is superimposed on the depth scale of the Newmark chart.
5. The foundation plan is expanded or reduced until the two scales match. This is done by clicking and holding the left mouse button on a corner marker, not a side marker as that will distort the drawing.
6. The foundation plan is moved so that the location to be evaluated is over the center of the chart, and the number of squares covered by the plan is counted and multiplied by the $I_c$ scale increment of the chart. The sum of the areas is the total influence coefficient, which then is multiplied by the contact pressure at the base of the footing to give vertical stress at the location and depth being examined.
7. The chart is saved to disk and printed in hard copy for future reference and possible incorporation into a geotechnical report.

It is a simple matter to determine the influence from an adjacent footing, as all that is necessary is to count the squares covered by the footing and add that number to the total.

When the stress at a different point is desired, for example under a corner instead of the center of the footing, the foundation drawing is shifted so that point is over the center of the chart and the counting process repeated.

When stresses at different depths are desired, the scale of the footing plan is changed according to step 3 above and the process repeated.

The same chart is used to determine stresses from a point surface load to any point on a buried object such as a culvert. Higher-resolution Newmark charts for vertical, horizontal, and shear stresses are presented by Poulos and Davis (1974).

### Example 15.6
Solve the second part of Example 15.1 using the Newmark chart.

*Answer:* The footing and distance to the adjacent footing are drawn to scale in Fig. 15.12. As the value of each square is 0.005, the influence coefficient is estimated at $5(0.005) = 0.025$, which may be compared with the previous estimate of 0.02. The plan is shifted relative to the chart to obtain stresses as any location at this depth.

## 15.10   OTHER SOLUTIONS

### 15.10.1   Embankments

The contact pressure from a trapezoidal cross-section such as a road embankment or earth dam is not uniform but decreases to zero at the outer edges. A general

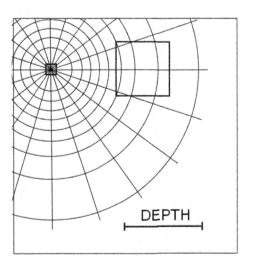

**Figure 15.12**
Newmark chart to determine $I_c$ in Example 15.6.

DEPTH

solution was obtained by Holl (1941). By simplifying the cross-section to a triangle, influence coefficients are obtained as shown in Table 15.3 for soil underneath the centerline and under the edges. Net pressures then can be obtained by removing stresses from an upper triangle where there is no soil—that is, of a triangular section having a base width equal to that of the upper surface of the embankment.

## 15.10.2 Horizontal Stress

Elastic solutions included in a later chapter define pressures transmitted to various depths on retaining walls from concentrated surface loads.

## 15.10.3 A Particulate Model

Harr (1977) suggested a particulate analog that, as shown in Fig. 15.13, approximates the ideal elastic distribution of vertical stress. This should not be surprising, as an elastic solution involves stress transfer between atoms or

| Table 15.3 | $z/a$ | Centerline, $I_c$ | Edge $I_c$ |
|---|---|---|---|
| Vertical influence coefficients at depth $z$ under a triangular embankment of width $2a$ | 0 | 1.00 | 0.00 |
| | 0.1 | 0.937 | 0.032 |
| | 0.2 | 0.874 | 0.062 |
| | 0.3 | 0.814 | 0.091 |
| | 0.4 | 0.758 | 0.117 |
| | 0.5 | 0.705 | 0.139 |
| | 0.6 | 0.656 | 0.158 |
| | 0.8 | 0.570 | 0.187 |
| | 1.0 | 0.500 | 0.205 |
| | 1.2 | 0.442 | 0.214 |
| | 1.4 | 0.395 | 0.216 |
| | 1.6 | 0.356 | 0.215 |
| | 1.8 | 0.323 | 0.211 |
| | 2.0 | 0.295 | 0.205 |
| | 2.5 | 0.242 | 0.187 |
| | 3.0 | 0.195 | 0.170 |
| | 3.5 | 0.177 | 0.153 |
| | 4.0 | 0.156 | 0.139 |
| | 5.0 | 0.126 | 0.177 |
| | 10.0 | 0.063 | 0.062 |
| | 20 | 0.032 | 0.032 |
| | 50 | 0.013 | 0.013 |

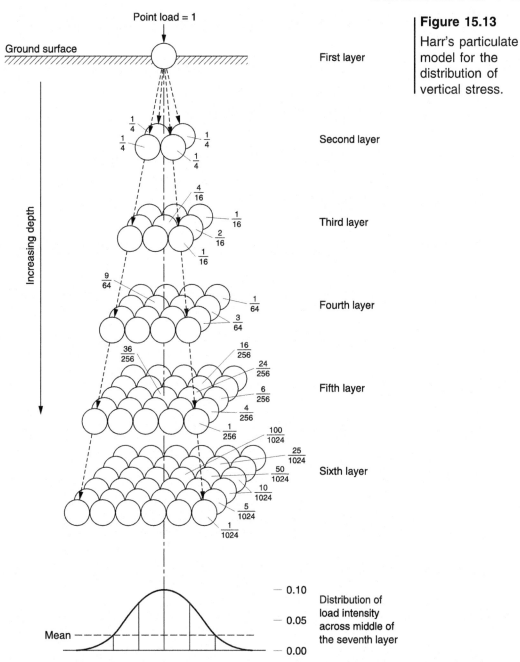

**Figure 15.13**
Harr's particulate
model for the
distribution of
vertical stress.

molecules that also are particles, even though tightly bound into a continuous mass.

### 15.10.4 Finite Element Modeling

Complicated stress fields are readily modeled by finite element solutions that also can be programmed to include nonlinear behavior. However, finite element solutions are case-specific and are difficult to check other than by repeating the analysis, so if erroneous data are inadvertently included a problem may remain undiscovered. Solutions based on elastic theory can provide a valuable check for more sophisticated computer solutions.

### 15.10.5 Soil as a Composite: Westergaard's Solution

A composite is made up of layers having different strength properties: automobile tires are composites of rubber plus layers of steel and Nylon fabrics. In 1938 H. M. Westergaard, a professor at the University of Illinois, modified Boussinesq's analysis by assuming that alternating stiff layers of soil prevent lateral strain. In addition Poisson's ratio may be assumed to be zero, in which case vertical stresses directly under a point load are reduced exactly one-third. For area loads with $\nu = 0$, a formula given by Taylor (1948) is

$$\sigma_z = \frac{p}{2\pi} \cot^{-1} \sqrt{\frac{H^2}{2A^2} + \frac{H^2}{2B^2} + \frac{H^2}{4A^2B^2}} \tag{15.18}$$

where the $\cot^{-1}$ term is in radians. A graph showing pressure bulbs from this relationship is shown in Fig. 15.14.

Most layered systems such as alluvial sequences probably do exhibit some lateral compression, in which case the developed stresses should be between those predicted from the Boussinesq and Westergaard relationships. Other factors also may restrict lateral compression, such as incorporating layers of horizontal tensile reinforcement in MSE (mechanically stabilized earth) walls and from lateral compaction by Rammed Aggregate Piers. Both topics are discussed in later chapters.

### Problems

*In the following problems use the Boussinesq and Newmark relationships:*

15.1. For a concentrated surface load of 44.5 kN (10,000 lb), compute and plot the vertical stresses along a diametric line in a horizontal plane at each of the following depths: 0.6, 12, 1.8, and 3.0 m (2, 4, 6, and 10 ft). Use the Boussinesq analysis.

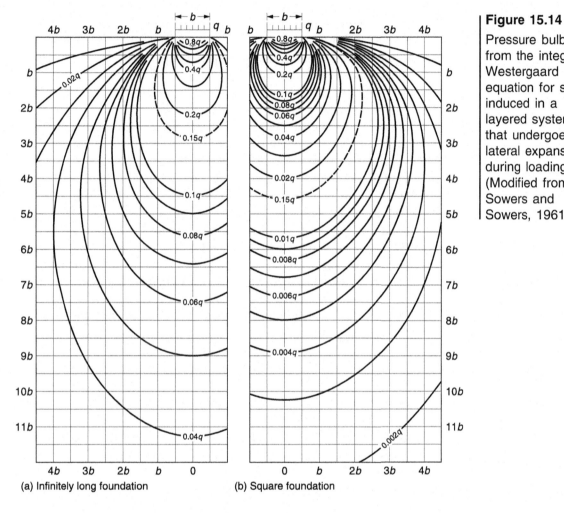

**Figure 15.14**
Pressure bulbs from the integrated Westergaard equation for stress induced in a layered system that undergoes no lateral expansion during loading. (Modified from Sowers and Sowers, 1961.)

(a) Infinitely long foundation

(b) Square foundation

15.2. A concentrated load of 66.7 kN (15,000 lb) is applied at the soil surface. Compute the vertical stress at the following points:

| Point | Coordinates, m (ft) | | |
|---|---|---|---|
| | $x$ | $y$ | $z$ |
| A | 12 (4) | 1.8 (6) | 0.9 (3) |
| B | 30 (10) | 0 | 1.5 (5) |
| C | 0 | 1.2 (4) | 30 (10) |
| D | 0.9 (3) | 0.9 (3) | 1.8 (6) |
| E | 2.1 (7) | 2.1 (7) | 2.1 (7) |
| F | 0 | 0 | 1.8 (6) |

15.3. Compute and plot the vertical stresses at several points along the vertical axis under a concentrated load of 1000 units applied at the surface.

15.4. Compute the horizontal stress in the direction parallel to the $x$-axis, due to a concentrated surface load of 35.6 kN (8000 lb), at a point whose coordinates are: $x = 1.2$ m (4 ft), $y = 1.5$ m (5 ft), and $z = 1.8$ m (6 ft). Assume the Poisson's ratio of the soil is 0, 0.25, and 0.5.

15.5. Compute and plot the vertical stresses along a diametric line in a horizontal plane at a depth of 1.5 m (5 ft) due to a concentrated surface load of 17.8 kN (4000 lb).

15.6. Determine the total load on an area 2 by 2.5 m (6.56 by 8.30 ft) whose center is 1.5 m (4.92 ft) under a truck wheel load of 45 kN (10,100 lb) applied at the ground surface.

15.7. Using the data of Problem 15.6, determine the total load on the area if the wheel load is over the center of the long side.

15.8. Determine the total load on an area 0.61 by 0.91 m (2 by 3 ft) at a depth of 1.22 m (4 ft), due to an axle load of 80.1 kN (18,000 lb) applied at the soil surface. One wheel is directly over the center of the area. The wheels are spaced 1.83 m (6 ft), center to center, and the axle is oriented parallel to the long sides of the area.

15.9. Determine the vertical stress at a point 6.1 m (20 ft) under the center of an area 6.1 by 12.2 m (20 by 40 ft), on which a load of 213.5 MN (2400 tons) is uniformly distributed.

15.10. Using the data of Problem 15.9, determine the vertical stress at a point under one corner of the area.

15.11. Using the data of Problem 15.9, determine the vertical stress under a point on the surface located outside the loaded area at a distance of 3.05 m (10 ft) each way from one corner.

15.12. A system of spread footings for a building is shown in Fig. 15.15. All footings exert a pressure of 287 kPa (6 kips/ft$^2$) on the underlying soil. Using a Newmark chart, estimate the vertical stress at a depth of 4.88 m (16 ft) under footings A-1 A-3, and B-3.

| Footing No. | Size | |
|---|---|---|
| | (ft) | (m) |
| A-1, A-5, D-1, D-5 | 5 × 5 | 1.52 × 1.52 |
| A-2, A-3, A-4, B-1, B-5, C-l, C-5, D-2, D-3, D-4 | 6 × 6 | 1.83 × 1.83 |
| B-2, B-3, B-4, C-2, C-3, C-4 | 8 × 8 | 2.44 × 2.44 |

15.13. A pressure of 239 kPa (5000 lb/ft$^2$) is applied over a circular area 6.1 m (20 ft) in diameter. Determine the vertical stress at a depth of 9.14 m (30 ft) (a) under the center of the area, (b) under the perimeter of the loaded area, and (c) under a point 3.05 m (10 ft) outside of the loaded area.

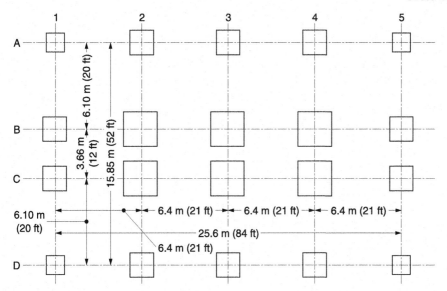

**Figure 15.15**
Foundation plan
for Problem 15.12.

15.14. Compare the vertical stress on the center ball of the 7th layer in Fig. 15.13 with that obtained from elastic theory with $z = 32$.

15.15. A long temporary surcharge load is triangular and symmetrical in cross-section, 45 ft (13.7 m) wide at the base, with slope angles of $25°$. The soil in the surcharge has a unit weight of $110 \, lb/ft^3$ ($17.3 \, kN/m^3$). Calculate the added vertical stress at a depth of 4 and 8 ft (1.2 and 2.4 m) under the center and under the edges of the embankment.

15.16. After time is allowed for settlement, the top of the embankment in the preceding problem is cut down to make room for a roadway 22 ft (6.7 m) wide. (a) Calculate the final stress at 4 and 8 ft (1.2 and 1.4 m) depths under the center of the embankment using an equivalent rectangular cross-section. (b) Calculate the final stress at the center by subtracting the weight of the triangular mass removed.

*In the following problems use the Westergaard analysis:*

15.17. What would prompt you to use the Westergaard analysis?

15.18. Re-work Problem 15.1 using the Westergaard analysis.

15.19. Re-work Problem 15.9 using the Westergaard analysis.

15.20. Determine stresses in Problem 15.9 from averaging the results of Boussinesq and Westergaard analysis. What would be the justification for this procedure?

## References

Burmister, D. M. (1943). "The Theory of Stresses in Layered Systems and Applications to Design of Airport Runways." *Proc. Highway. Res. Bd.* 23, 126.

Christian, J. T., and Desai, C. S. (1977). "Constitutive Laws for Geological Materials." In Desai and Christian, ed., *Numerical Methods in Geotechnical Engineering.* McGraw-Hill, New York.

Harr, M. E. (1977). *Mechanics of Particulate Media.* McGraw-Hill, New York.

Holl, D. L. (1941). "Plane-Strain Distribution of Stress in Elastic media." Iowa Engineering Experiment Station Bull. 48, Iowa State College, Ames, Iowa.

Jumikis, A. R. (1971). "Vertical Stress Tables for Uniformly Distributed Loads on Soil." Rutgers Univ. Engr. Res. Publ. No. 52.

Newmark, N. M. (1935). "Simplified Computation of Vertical Pressures in Elastic Foundations." Engineering Experiment Station, University of Illinois, Circular No. 24.

Poulos, H. G., and Davis, E. H. (1974). *Elastic Solutions for Soil and Rock Mechanics.* John Wiley & Sons, New York.

Sowers, G. B., and Sowers, G. F. (1961). *Introductory Soil Mechanics and Foundations,* 2nd ed. The Macmillan Co., New York.

Taylor, D. W. (1948). *Fundamentals of Soil Mechanics.* John Wiley & Sons, New York.

Westergaard, H. M. (1938). "A Problem in Elasticity Suggested by a Problem in Soil Mechanics: A Soft Material Reinforced by Many Strong Horizontal Sheets." In *Mechanics of Solids, Timoshenko 60th Anniversary Volume.* The Macmillan Co., New York.

# 16

## Settlement

## 16.1  SLOW BUT SURE

### 16.1.1  Overview

Generally, the heavier and more concentrated the load from a structure the more it will settle, but much also depends on the soil. The most extreme case of a bad-soil day is from founding a structure on peat, which compresses like caramel on a car seat. Even structures founded on rock settle as the rock undergoes elastic compression. Soils compress elastically under light loads, but then as loads become heavier than anything they previously have experienced, particle shifts and rearrange to allow increasing amounts of settlement. This stage of settlement is attributed to *primary consolidation*, where the rate of settlement is controlled by the time required to squeeze the pore water out. Quantifying this time factor is considered to be one of Terzaghi's greatest contributions.

Settlement of a building relative to its surroundings can shear off water, gas, or other utility lines where they enter the building. Settlement can reverse the flow in sewers if the amount is not anticipated and included in design. Centuries-old buildings sometimes have settled to the extent that the ground floor is a step or two below the street or sidewalk level, and sidewalks can develop a dangerous tilt as adjacent buildings settle. In a few instances settlement has reduced a ground floor to a basement level.

In addition to elastic and consolidation settlement, a third category of settlement can occur if a load is heavy enough to cause a shear failure, in which case a structure suddenly sinks into the ground or tips over. This is not good. It is treated in Chapter 22 on bearing capacity. Foundation bearing pressures often are decided on the basis of allowable consolidation settlement, which should be safe from a bearing capacity failure.

### 16.1.2 Declining Settlement Rate with Time

As a soil consolidates and becomes denser it becomes stiffer and more resistant to further compression. Settlement therefore decreases with time, and eventually in theory will stop. Settlement that continues after all excess pore water pressure has dissipated is called *secondary consolidation*, and relates to a time dependency of breaking and remaking chemical bonds. This is much slower and can continue for many years. It is a manifestation of *soil creep.*

### 16.1.3 Diminishing Compressibility with Depth

Generally the deeper a soil is buried the more it is compressed under the pressure of overburden, and the lower is its compressibility. This trend is not inevitable because it can be trumped by a change in the soil strata, for example from sand to clay. Nevertheless the increase in soil stiffness with depth is a justification for using piles or piers to transfer loads to a deeper, less compressible soil layer, which is the most common method for controlling and reducing building settlement.

The same effect can be accomplished by use of a temporary *surcharge load* to simulate an overburden pressure. For example, if a building or other structure is to be constructed on a compressible soil, the site can be prepared by applying a temporary load and waiting for settlement to occur. This typically will involve a few weeks but may continue for a year or more, depending on the amount of the surcharge in relation to the final bearing pressure, and the consolidation rate of the soil involved. If the surcharge load equals or exceeds the weight of the future structure, after the settlement has been accomplished the surcharge can be removed and replaced by the structure, which then should experience little or no settlement.

On a large development with many construction sites the same soil surcharge can be moved about from one site to the next, in which case the timing will depend on the settlement rate. This is predicted from laboratory tests and the actual rate then is monitored in order to make a final prediction. The amount of time required can be shortened by using a surcharge weight that is higher than the anticipated foundation pressure.

### 16.1.4 Consolidation Is Better than a Bearing Capacity Failure

Foundation pressure on a saturated, compressible soil increases the pore water pressure, which causes the hydraulic gradient that pushes water out of the soil. Now let us consider what will occur if the water is prevented from escaping. In this case all of the added pressure will be carried by the pore water so there can be no increase in the effective stress, or in friction between the soil grains. As the load increases while the effective stress remains the same, the system eventually will reach a point where the soil shear strength no longer is adequate to support the load, and the soil will shear and allow a bearing capacity failure.

Prevention of a bearing capacity failure therefore is a primary concern during consolidation of soft clay soils, and drains may be incorporated into the soil if necessary to help relieve excess pore water pressure and ensure stability. Where it is anticipated that this will be a problem, pore water pressures are monitored with piezometers to ensure that drainage keeps up with the rate of loading, and if it does not, consolidation may be halted until it does.

## 16.2  DIFFERENTIAL SETTLEMENT

### 16.2.1  Damage to Structures

Differential settlement, where one part of a structure settles more than another part, causes floors to tilt and walls to crack. Crack patterns in exterior masonry walls usually pass through weak areas such as window and door openings, and indicate parts of a structure that have gone down relative to other parts. Some representative crack patterns are illustrated in Fig. 16.1, and the student should use a practiced eye to survey existing structures for damages.

Differential movements also can be obvious from cracking of interior walls of buildings, separations and shearing at ends of walls, and distortion of door and

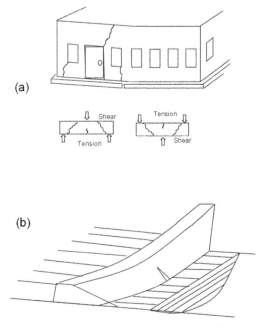

(a)

(b)

**Figure 16.1**
(a) Crack patterns resulting from differential settlement of buildings can be interpreted by analogy to shear and tension cracks in a beam.
(b) Sagging of an earth dam from a varying thickness of compressible foundation soil can lead to overtopping at the top and/or transverse tension cracks and leaks through the bottom.

window frames into parallelograms so that windows don't open and doors don't close. This also is not good.

Another cause of damage that can induce similar crack patterns is differential lifting of a structure by expansive clay. This possibility always should be checked or the remedy may not address the right cause. Elevation surveys can be made with reference to a stable benchmark, and it should be noted that in an expansive clay area, a street or pavement is not a stable benchmark. A stable benchmark must be founded below the zone of clay expansion that normally extends 2 m or more (over 6 ft) below the ground surface, and if there is any question, more than one benchmark should be used.

As settlement usually is slow and can continue for years, patching often is cosmetic and only temporary. Corrective measures include pressure grouting or underpinning, which means jacking against piles placed underneath a structure. Underpinning is the surest remedy but is labor-intensive and expensive. In extreme cases uneven settlement can impair structural integrity to the extent that a building will be condemned and must be torn down.

### 16.2.2 Settlement of Building Additions

As previously mentioned, the rate of consolidation settlement gradually decreases and slows to a stop with time. It therefore is impossible to synchronize settlement rates of a building addition with that of the existing building unless its settlement has stopped and the addition is supported on a deep foundation such as piles. A separation should be made and brick lines not be carried across, and hallways crossing from one structure to the other may have ramps that can tilt as the structures continue to settle at their respective rates. Supporting an addition on piles can greatly reduce its settlement but does not influence continuing settlement of the existing building.

### 16.2.3 Towers

Uneven settlement is exaggerated by leaning of tall structures such as smokestacks, monuments, or church spires, and as tilting occurs, the center of gravity shifts toward the low side, further aggravating the situation. This is illustrated in Fig. 16.2(a). Settlement of grain elevators can cause unloading mechanisms to become inoperative. Often several storage silos will be supported on a single concrete pad foundation and settlement is monitored during first filling, so that tilting can be prevented by adding more weight to the high side.

The famous Leaning Tower is perhaps the best-known example of differential settlement. Level measurements of successive tiers indicate that the tower leaned one way and then another as it was being built. Many medieval towers developed a similar zigzag mentality because the corrections did not take into account the

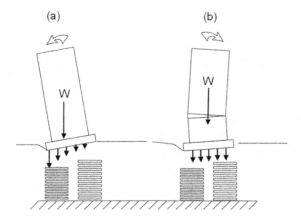

(a)                              (b)

**Figure 16.2**

(a) Tilting increases the foundation contact pressure on the low side, which can make matters worse, or (b) a correction can be made during construction. Then tilting can pursue in a different direction because soil that already is compressed becomes less compressible.

changes in compressibility as settlement occurred. This is indicated by the springs in Fig. 16.2. Lead weights added on the high side of the Leaning Tower arrested the tilt but detracted from its appearance, so soil underneath the foundation was removed by auger holes, sufficient to at least temporarily arrest tilting while preserving the famous lean. What makes the Leaning Tower unique among medieval towers is that it is still standing.

## 16.2.4   Settlement of Earth Dams

Earth dams obviously are located in river valleys, and therefore are usually supported on soft alluvial soils that vary in compressibility and are thickest near the center of the valley. The center part of the dam therefore normally settles more than the abutment areas, with differential settlements that sometimes can be measured in feet or fractions of a meter. If this is not predicted and taken into account in the design, the sag can cause transverse tension cracking in the base of the dam and/or allow overtopping, with disastrous consequences.

Transverse cracking, illustrated in Fig. 16.1(b), is perhaps the more insidious of the two dangers because it is out of sight and underdam seepage may be carried away in granular soil layers that are common in alluvial sediments. Rapid seepage can erode a channel through the process called *piping* and can lead to failure of the dam. Earth dams that may be subject to such differential settlement will have a relatively weak clay core soil that can distort without cracking, and the clay is supported on the outer flanks by stronger soils. One indication of transverse cracking is the effect on piezometric levels within and under the dam, and exposed slopes of earth dams are regularly inspected for signs of piping. Piping can be stopped with pressure grouting.

### 16.2.5 Can Differential Settlement Be Avoided?

As illustrated in Fig. 16.3(a), the more heavily loaded the foundation, the more it will tend to settle. However, even if foundation contact pressures are the same, a large footing will settle more than a small one because the pressure bulb extends deeper (Fig. 16.3(b)). The size effect is taken into account in predictions using stress distributions discussed in the preceding chapter.

Unforeseen changes in soil compressibility (Fig. 16.3(c)), are an all too common cause of differential settlement. For example, the boundary between floodplain point bar sand and its associated oxbow clay may be obvious on airphotos but

**Figure 16.3**

(a)–(d) Four causes for differential settlement and (e), (f) methods for reducing or eliminating differential settlement.

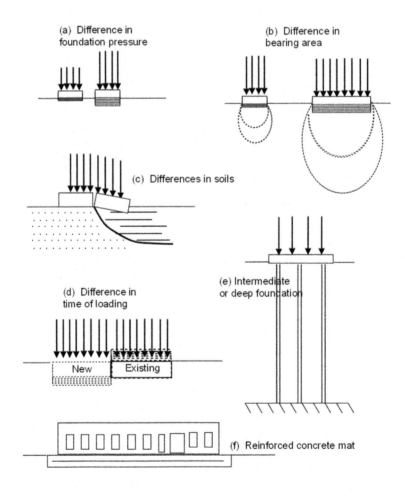

(a) Difference in foundation pressure

(b) Difference in bearing area

(c) Differences in soils

(d) Difference in time of loading

New    Existing

(e) Intermediate or deep foundation

(f) Reinforced concrete mat

obscured up close by later deposits of silt and clay or by construction activities. Exploration borings normally are conducted at outer corners of a building and near the middle, but still may miss an important geological feature, and everything between the borings is inferred by interpolation.

A common source of difficulty is loose fill that has been pushed into trenches, ravines, or old excavations. Concrete rubble can be particularly difficult, especially if in a tangle of reinforcing steel, as it obstructs exploration boring and can make it difficult to install deep foundations. Adverse soil conditions discovered during excavations for foundations can lead to an adjustment in the location, or support of part or all of a building on deep foundations.

Finally, and as previously mentioned, because settlement is nonlinear with time, foundations loaded at different times will be out of synch (Fig. 16.3(d)).

A way to avoid differential settlement is to use deep or intermediate foundations (Fig. 16.3(e)), or to place a structure on a reinforced concrete mat foundation (Fig. 16.3(f)), both of which are discussed in later chapters.

## 16.3  SOIL MECHANICS OF COMPRESSION

### 16.3.1  Elastic Compression

Elastic compression of soil is rapid and normally occurs as a structure is being built. Although settlement is approximately proportional to the applied load, soil does not conform to ideal elastic criteria because of incomplete rebound upon removal of the load. Elastic compression of soil is more accurately described as *"near-linear elastic behavior."*

### 16.3.2  Primary Consolidation

Primary consolidation initiates when pressure on the soil is high enough to break down the soil structure, and involves rearrangement and movement of soil particles into adjacent voids as the soil compresses and densifies. Consolidation settlement is larger and slower than elastic settlement, being held back by the time required to squeeze water out, which can require many months or years.

The concept of primary consolidation is illustrated in Fig. 16.4, where (a) shortly after the load is applied it is almost entirely carried by water pressure, shown on the pressure gauge and by the height of the jet of the outflow. Then (b) as water leaks out, pore pressure decreases and more of the load is transferred to the soil grains, indicated by compression of the spring. Finally at (c) water pressure is zero and the load is entirely carried by the soil grains, signaling an end to primary consolidation.

**Figure 16.4**

Principle of the time delay in primary consolidation.

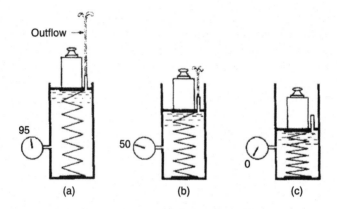

Outflow →

95

50

0

(a)      (b)      (c)

In primary consolidation the rate of settlement is controlled by the rate at which water can escape, which in turn is a function of permeability and drainage geometry of the compressing soil. The theory applies only to saturated soils; if a soil is not saturated, negative pore water pressure (suction) and clay-water bonding normally will prevent consolidation until such time as the soil becomes saturated, introducing an element of surprise.

### 16.3.3 Secondary Consolidation Settlement

After all excess pore water pressure has been dissipated, consolidation can continue as *secondary consolidation*, which is time-related plastic restructuring of particle-to-particle bonds. A similar phenomenon in soil on hillsides is called *soil creep*, evidenced by tilted trees and tippy-dippy retaining walls. However, soil creep on hillsides does not increase the density or resistance of the soil, where during secondary consolidation soil increases in density and shearing resistance so that the rate gradually slows down and eventually stops.

A major distinction between primary and secondary consolidation is that, while the rate of primary consolidation depends on drainage, the rate of secondary consolidation does not. Drainage accelerates primary consolidation but not secondary consolidation unless the soil dries out.

The focus of most geotechnical investigations is on primary consolidation, but in some instances in old structures the settlement attributed to secondary consolidation has exceeded that from primary consolidation. Fortunately, both primary and secondary consolidation are evaluated by the same laboratory test, as primary consolidation occurs first and is followed by secondary consolidation.

Because until recently there has been no rational method for defining the rate of secondary consolidation, it for the most part has been evaluated based on observations and experience with particular soils. This is discussed in the next chapter.

## 16.4 TERZAGHI'S THEORY FOR PRIMARY CONSOLIDATION

### 16.4.1 A Case for Versatility

Terzaghi's academic training was in mechanical engineering and geology and illustrates a transfer of technology from one field to another, in this case from heat transfer to soil drainage, which certainly is a step. Koestler (1964) considered such transfers as a key to creativity, describing it as "a juxtaposition of conflicting matrices" in science and engineering, art and humor. The more skillful the juxtaposition, the keener the invention or the better the joke.

Terzaghi observed that sand compresses more rapidly than saturated clay, and sensed an analogy between the thermodynamics of heat flow and escape of pore water, both of which involve a gradually declining gradient. He wondered if the thermal conductivity of a metal might play the same role as hydraulic conductivity of a soil, and applied essentially the same equations to develop a time relationship.

The most critical step was not the speculation, but in thoroughly testing the hypothesis in the laboratory. To this end he invented and built what now is called a consolidometer, and is standard equipment in every geotechnical laboratory.

This chapter discusses the amounts of consolidation measured in the laboratory and how these translate into amounts in the field. Consolidation rates, which are the focal point of Terzaghi's theory, are discussed in the next chapter.

### 16.4.2 Teraghi's Odometer

The basic test apparatus was devised and used by Terzaghi in the early 1920s at Robert College in Turkey. As shown in Fig. 16.5, a short cylinder of soil is compressed between two porous plates. The soil initially is saturated with water, and the porous end plates permit water to escape freely at the top and bottom of the soil specimen. After an increment of load is applied, the amount of compression after prescribed time intervals is measured and recorded. The soil specimen intentionally is thin in order to decrease the distance that the water must flow, which speeds up the test. Even so, small amounts of compression often could be observed continuing even after a load had been applied for 24 hours. Changes in the height of the soil sample are measured to the nearest 0.0025 mm (0.0001 in.), and the apparatus must be guarded against vibrations or accidental bumping.

In order to develop and test theoretical relationships, the early test apparatus incorporated a vertical glass tube connected with the base to serve as a standpipe for a falling-head permeameter.

**Figure 16.5**

Schematic
diagram of a
floating-ring
consolidometer,
also called an
odometer.

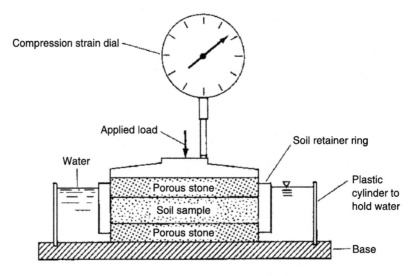

### 16.4.3 Test Procedure

An undisturbed sample of soil is carefully trimmed using a cutting ring that has
the same internal diameter as the confining ring of the consolidometer. The
sample is struck off flat at the ends using the cutting ring as a guide, the height is
measured, and the sample carefully pushed into the consolidometer. Any holes
resulting from plucking out of large particles at the ends are carefully packed with
some of the cuttings.

Because the soil is confined in a rigid steel ring for the test, it models loading
saturated soil under a broad area, such that lateral strain is negligible.

An initial reading is made on the compression-strain dial with no load on the
sample, and then a suitable increment of load, say 10–20 kPa (0.1–0.2 T/ft²) is
applied using a lever arrangement or an air cylinder. The compression dial is read
at various time intervals, more frequently at first, and then less frequently with
time as compression slows down. When compression has nearly reached a
maximum under the applied load, another increment of load is applied, which
initiates a new cycle of consolidation so that the time-rate of compression is
measured. Partly as a matter of convenience, each load normally is left on for
24 hours, so a complete test requires about a week.

Variations on the traditional approach include using different load increments or
applying a continuously increasing load, but these procedures are outside the
scope of this text.

Loads often are increased geometrically because, as the soil consolidates, it
becomes stiffer, so a larger load increment is required to obtain the same amount
of compression. This cycle of loading and measuring the time-rate of strain is

repeated until the total applied load exceeds that to which the prototype soil will be subjected by a proposed structure plus the overburden pressure. A procedure for the one-dimensional consolidation test is described in ASTM Designation D-2435.

# 16.5   ANALYSIS OF CONSOLIDATION TEST DATA

## 16.5.1   Test Data

Some actual test results are shown in Table 16.1. It will be noted that in most cases each load was selected to be two times the previous load. After a load was applied, the specimen height, shown by vertical columns of data, decreased rapidly at first and then slowed down and almost came to a stop after 24 hours. The near-equilibrium 24-hour readings will be examined in this chapter, and the timed reading will be analyzed in the next chapter.

The soil specimen initially was cut to 1.25 in. (31.8 mm) thickness, placed in the consolidometer, and the dial gauge read 0.3147 in. The specimen then was saturated with water, and the dial reading increased to 0.4896 under a nominal pressure of 0.1 tons/ft$^2$. An increase in the dial reading signifies an increase in specimen thickness. How is that possible?

*The answer is that this is an expansive clay*, the amount of expansion upon wetting being $0.4896 - 0.3147 = 0.1749$ in. (4.44 mm), or 14 percent. As the soil was obtained from relatively shallow depth above a groundwater table, it was naturally partly desiccated. A graph of the 24-hour readings in Fig. 16.6 indicates a load equivalence from desiccation of 3.3 tons/ft$^2$.

## 16.5.2   Decrease in Thickness with Pressure

The initial straight-line relationship shown by the dashed line in Fig. 16.6 suggests elastic behavior. Then at A the thickness begins to decrease more rapidly, indicating that consolidation has begun following a breakdown of the soil structure. As pressure further increased and the soil compressed, it became progressively stiffer as a result of consolidation.

## 16.5.3   Change in Modulus with Pressure

The modulus, or pressure divided by strain, is a measure of soil stiffness. As strain is dimensionless, for example in./in. or mm/mm, the units of modulus are the same as the units of pressure. Figure 16.7 shows the modulus calculated from 24-hour readings with increasing pressure. The first part of the curve may reflect changes during seating of the porous plates, discussed below. Elastic compression halts at 3 tons/ft$^2$ (0.3 MPa) as consolidation begins and results in a

# Table 16.1

Consolidation test data for an expansive clay

Load increments, kPa (tons/ft$^2$), and compression dial readings, 0.0025 mm (0.0001 in.)

| Elapsed time $t$ (min) | $\sqrt{t}$ | 9.6–19.1 (0.1–0.2) Dial | Diff. | 19.1–4.8 (0.2–0.5) Dial | Diff. | 48–96 (0.5–1) Dial | Diff. | 96–192 (1–2) Dial | Diff. | 192–383 (2–4) Dial | Diff. | 383–766 (4–8) Dial | Diff. | 766–1530 (8–16) Dial | Diff. |
|---|---|---|---|---|---|---|---|---|---|---|---|---|---|---|---|
| 0 | 0 | 4896 | 0 | 4850 | 0 | 4761 | 0 | 4478 | 0 | 3793 | 0 | 2954 | 0 | 2332 | 0 |
| 0.25 | 0.5 | 4892 | 4 | 4842 | 8 | 4742 | 19 | 4400 | 78 | 3723 | 70 | 2897 | 57 | 2280 | 52 |
| 1 | 1 | 4889 | 7 | 4836 | 14 | 4728 | 33 | 4343 | 135 | 3672 | 121 | 2855 | 99 | 2241 | 91 |
| 2.25 | 1.5 | 4886 | 10 | 4830 | 20 | 4714 | 47 | 4296 | 182 | 3621 | 172 | 2814 | 140 | 2203 | 129 |
| 4 | 2 | 4883 | 13 | 4825 | 25 | 4700 | 61 | 4230 | 248 | 3570 | 223 | 2773 | 181 | 2165 | 167 |
| 6.25 | 2.5 | 4880 | 16 | 4819 | 31 | 4686 | 75 | 4173 | 305 | 3518 | 275 | 2732 | 222 | 2127 | 205 |
| 9 | 3 | 4877 | 19 | 4813 | 37 | 4673 | 88 | 4116 | 362 | 3468 | 325 | 2692 | 262 | 2090 | 242 |
| 12.25 | 3.5 | 4874 | 22 | 4807 | 43 | 4659 | 102 | 4059 | 419 | 3417 | 376 | 2651 | 303 | 2052 | 280 |
| 16 | 4 | 4871 | 25 | 4801 | 49 | 4645 | 116 | 4011 | 467 | 3374 | 419 | 2610 | 344 | 2014 | 318 |
| 25 | 5 | 4867 | 29 | 4794 | 56 | 4627 | 134 | 3946 | 532 | 3308 | 485 | 2562 | 392 | 1967 | 365 |
| 36 | 6 | 4864 | 32 | 4788 | 62 | 4614 | 147 | 3900 | 578 | 3259 | 534 | 2525 | 429 | 1936 | 396 |
| 49 | 7 | 4863 | 33 | 4785 | 65 | 4607 | 154 | 3880 | 598 | 3228 | 565 | 2503 | 451 | 1916 | 416 |
| 64 | 8 | 4862 | 34 | 4783 | 67 | 4603 | 158 | 3872 | 606 | 3207 | 586 | 2494 | 460 | 1908 | 424 |
| 100 | 10 | 4861 | 35 | 4781 | 69 | 4597 | 164 | 3860 | 618 | 3171 | 622 | 2473 | 481 | 1889 | 443 |
| 225 | – | 4858 | 38 | 4777 | 73 | 4587 | 174 | 3838 | 630 | 3124 | 669 | 2443 | 511 | 1861 | 471 |
| 400 | – | 4853 | 43 | 4767 | 83 | 4548 | 213 | 3821 | 657 | 3034 | 759 | 2366 | 588 | 1820 | 512 |
| 1440 | – | 4850 | 46 | 4761 | 89 | 4478 | 283 | 3793 | 685 | 2954 | 839 | 2332 | 622 | 1762 | 570 |

Note: True specific gravity = 2.68; dry weight = 410.5 g (0.905 lb); area of sample = 81.076 cm$^2$ (12.566 in.$^2$); thickness of sample = 31.75 mm (1.2500 in.) with dial reading at 0.3147. Temperature, 27°C (80°F)

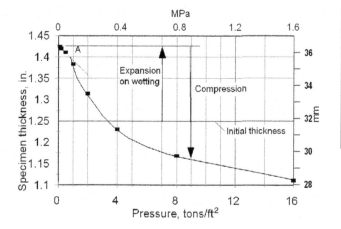

**Figure 16.6**
24-hour test data from Table 16.1, showing initial expansion from wetting and subsequent compression under load.

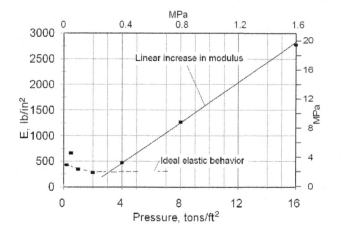

**Figure 16.7**
24-hour compression data for the example data indicates a linear increase in constrained modulus with increasing pressure.

linear increase in modulus with pressure. This is called a "constrained modulus" because lateral expansion is prevented.

The change from elastic to consolidating behavior probably relates to the soil stress history, by representing a former level of consolidating pressure. The indicated preconsolidation pressure in this case would be $3.1\,\mathrm{T/ft^2}$ (297 kPa), which coincides with the equivalent pressure developed by clay expansion.

### 16.5.4 Specimen Height vs. Void Ratio

From the definition of void ratio, $e$, a change in void ratio is

$$\Delta e = \frac{\Delta V_\mathrm{v}}{V_\mathrm{s}} \qquad (16.1)$$

If no lateral strain is allowed, as in a consolidation test, the corresponding change in specimen height is

$$\Delta h = \frac{\Delta V_v}{(V_s + V_v)} \qquad (16.2)$$

from which

$$\Delta e = \Delta h(1 + e) \qquad (16.3)$$

### Example 16.1

As a review, calculate the void ratio prior to and after expansion from the soil information in Table 16.1.

*Answer:* Prior to expansion the soil specimen was 1.25 in. $= 3.18$ cm thick, and the volume was $3.18 \times 81.076 = 257.4\,\text{cm}^3$. This value, the total mass including water, and the specific gravity of the soil grains can be written into a block diagram. The next step is to calculate volumes that are shown in parentheses: $410.5\,\text{g} \div 2.68 = 153.2\,\text{cm}^3$, and $257.4 - 153.2 = 104.2\,\text{cm}^3$.

(104.2)
(153.2) | 2.68 g/cm³ | 410.5 g
25.7 cm³

Then the answer almost writes itself: Prior to expansion, $e = 104.2/153.2 = 0.680$. Afterwards the height is $1.25 + (0.4896 - 0.3147) = 1.425$ in. $= 3.619$ cm, so the expanded volume is $81.076 \times 3.619 = 293.4\,\text{cm}^3$. The void ratio should be determined by the student, and is $e = 0.915$. If this block diagram and calculation no longer are familiar, the student needs to re-familiarize.

## 16.5.5   The Linear e-log P Relationship

For convenience, Casagrande (1936) plotted void ratio versus the logarithm of pressure. This has become the standard method for presentation of consolidation test data, and is referred to as an *e-log P graph*, shown in Fig. 16.8. As can be seen, at higher pressures the relationship becomes nearly linear.

On the basis of the linear increase in modulus such as shown in Fig. 16.7, Janbu (1969), of the Norwegian University of Science and Technology, wrote the following expression, which means that the higher the pressure $P$, the larger must be the increment in pressure $dP$ for a given change in void ratio $-de$:

$$\frac{dP}{de} = -kP \qquad (16.4)$$

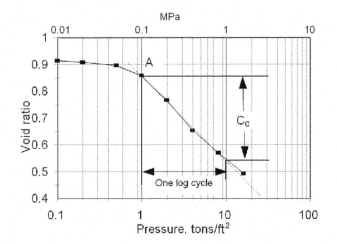

**Figure 16.8**
An *e*-log *P* plot
calculated from the
24-hour
consolidation data
in Table 16.1, and
showing definition
of the compression
index, $C_c$.

where the negative sign indicates that an increase in pressure brings about a
decrease in the void ratio. Rearranging,

$$\frac{dP}{P} = -k\, de \tag{16.5}$$

from which

$$\ln P = -ke + C \tag{16.6}$$

where $C$ is a constant of integration. Therefore, if the soil modulus increases
linearly with pressure, as it does in Fig. 16.7, *the void ratio should decrease linearly
with the logarithm of pressure*, as it does in Fig. 16.8.

The above statement is valid regardless of the base of the logarithms. The slope of
a plot with logarithms to the base 10 is defined as the *compression index*, $C_c$. Thus,
as shown in Fig. 16.8, the compression index is the change in void ratio with one
log cycle of pressure to the base 10:

$$C_c = \frac{\Delta e}{\Delta \log P} \tag{16.7}$$

The equation for the linear portion of the *e*-log *P* graph is obtained by
substituting

$$\Delta e = e_0 - e_1$$

and

$$\Delta \log P = \log P_1 - \log P_0 = \log(P_1/P_0)$$

which gives

$$e_1 = e_0 - C_c \log(P_1/P_0) \tag{16.8}$$

where $e_0$ is the void ratio at pressure $P_0$, and $e_1$ is the void ratio at pressure $P_1$.
This is the basic equation used to calculate void ratio under any pressure for a

normally consolidated soil, based on the compression index and a value of the void ratio corresponding to a particular pressure.

Some curvature of the linear portion of the $e$-log $P$ plot typically occurs at high pressures, which is reasonable because there must exist some minimum void ratio that corresponds to the densest packing arrangement of soil grains without breaking the grains.

### 16.5.6   Sources of Experimental Error

No soil sample is truly "undisturbed," and X-ray photographs of soil samples that were obtained by pushing a thin-walled steel sampling tube show a zone of disturbance around the perimeter. Sampling preferably is oversize so that the disturbed perimeter zone can be trimmed off. A procedure is described later in this chapter to adjust test results to compensate for the influence of unavoidable sample disturbance.

Another source of error is because the lateral confining stress that exists in the field is not reinstated prior to compression testing in the laboratory. Instead, lateral stress must develop during compression, which tends to increase compressibility. This may allow a premature transition from elastic to consolidating behavior in early stages of the test.

Whereas the above factors can affect the determination of soil compressibility, particularly at low applied pressures, other factors affect the rate of consolidation. Decompression of soft soil that is a considerable distance below the groundwater table may create air bubbles as a result of dissolution, causing the sample to expand and be more compressible. To counteract this, a consolidation test can be performed in a sealed chamber so that back-pressure can be applied to put the air back in solution. Water that is used for saturating consolidation test samples should have recently been de-aired with a vacuum or by boiling.

Finally, because the rate of consolidation depends on hydraulic conductivity, it also depends on the viscosity and temperature of the water. Consolidation usually pursues more rapidly in the laboratory than in the field where ground temperatures are lower. Times therefore are adjusted based on the viscosity of water.

### 16.5.7   Preliminary Estimate of Compression Index from the Liquid Limit

The linear portion of an $e$-log $P$ plot represents soil void ratios that are in equilibrium with the applied pressure. This is the *normally consolidated* stage of consolidation, and relates to the soil composition. A relationship suggested by Skempton (1944) is

$$C_c = 0.007(LL - 10) \tag{16.9}$$

where $LL$ is the liquid limit in percent. Terzaghi and Peck (1967) then made a correction for remolding during the liquid limit test and proposed

$$C_c = 0.009(LL - 10) \tag{16.9a}$$

The latter equation is more likely to be on the safe side for design. As $C_c$ cannot be negative, the implication is that a soil with a liquid limit less than 10, such as sand or relatively clean silt, will not consolidate but still can compress elastically. While useful for making preliminary estimates, *eq. (16.9a) is not a substitute for consolidation testing*.

**Example 16.2**
A soil sampled from a depth of 2 m has a liquid limit of 55% and a void ratio of 0.80. What minimum void ratio might be expected if the pressure is increased by a factor of 5, and what is the change in void ratio?

*Answer:* The largest change in void ratio will occur if the soil is normally consolidating, in which case eqs. (16.7) and (16.8) apply. From eq. (16.9a) $C_c$ is estimated to be 0.315. Equation (16.7) then becomes

$$e = e_1 - C_c \log(P/P_1) = 0.80 - 0.315 \log(5) = 0.58$$
$$\Delta e = 0.80 - 0.58 = 0.22$$

## 16.5.8 Does Transferring Load Deeper Really Decrease Settlement?

Obviously transferring a foundation load down to rock will reduce settlement; however, many pile or other deep foundations do not extend to rock, but simply transfer the load downward where it is supported by friction along the perimeter of the piles. The two respective types are called "end-bearing" and "friction" pile.

The answer to this question is shown in Fig. 16.9, where the increase in pressure is the same at A and at B, but the corresponding change in void ratio is much lower at B because of the semilogarithmic relationship.

# 16.6 RELATION BETWEEN VOID RATIO AND SETTLEMENT

## 16.6.1 Settlement Equation

A relationship between the change in void ratio and amount of settlement can be determined with the aid of block diagrams (Fig. 16.10), where the initial thickness

**Figure 16.9**

Transferring a load deeper decreases settlement: a unit increase in pressure at A results in a larger change in void ratio than the same increase in pressure at B.

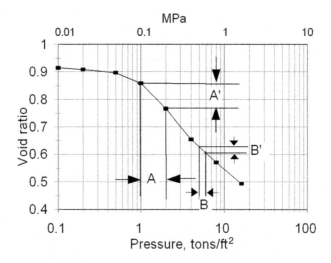

**Figure 16.10**

Block diagrams for calculation of settlement from a change in void ratio.

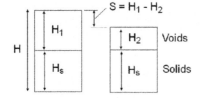

of a consolidating layer is designated by $H$. Because only the voids are affected by a consolidating pressure, the respective contributions to thickness before and after consolidation are designated by $H_1$ and $H_2$. Settlement attributable to compression of this layer therefore is

$$S = H_1 - H_2 \tag{16.10}$$

The initial and final void ratios by definition are

$$e_1 = \frac{H_1}{H_S} \quad \text{and} \quad e_2 = \frac{H_2}{H_S}$$

where $H_S$ is the thickness of the solids. Substituting in eq. (16.10) gives

$$S = H_S(e_1 - e_2) \tag{16.11}$$

Because $H_S = H - H_1 = H - H_S e_1$, solving for $H_S$ gives

$$H_S = \frac{H}{1 + e_1} \tag{16.12}$$

Substituting in eq. (16.11) gives

$$S = H\left[\frac{e_1 - e_2}{e_1 + 1}\right] \tag{16.13}$$

where $S$ is the amount of settlement, $H$ is the thickness of the compressible layer, $e_1$ is the initial void ratio, and $e_2$ is the final void ratio. *This is the fundamental equation relating the amount of compression to soil layer thickness and change in the void ratio.* It also confirms the obvious, that the amount of settlement is directly proportional to the thickness of the compressible layer.

### Example 16.3

A large foundation exerting a pressure of 3 tons/ft$^2$ is to be founded at a depth of 4 ft in the soil represented by the data in Fig. 16.8. If the compressible soil layer extends 4 ft below the foundation level, estimate the amount of settlement. Assume that a saturated condition exists at the ground surface.

*Answer:* The remainder of the block diagram of Example 16.1 can be solved to obtain a unit weight for the saturated expanded soil:

$$\begin{array}{lll}
104.2 & \boxed{1.0\,\text{g/cm}^3} & (104.2\,\text{g}) \\
153.2 & \boxed{2.68\,\text{g/cm}^3} & 410.5\,\text{g} \\
257.4\,\text{cm}^3 & & (514.7\,\text{g})
\end{array}$$

The saturated density of the soil therefore is $514.7 \div 257.4 = 2.00\,\text{g/cm}^3$, and the saturated unit weight is $\gamma_{\text{sat}} = 2.00 \times 62.4 = 125\,\text{lb/ft}^3$. The initial stress at the middle of the compressible layer, which is at a depth of $4+2 = 6\,\text{ft}$, therefore is $6 \times 125 = 625$ lbft/$^2 = 0.312$ tons/ft$^2$. From Fig. 16.8 the initial void ratio corresponding to this pressure is approximately $e_1 = 0.91$.

For purposes of estimation the unit weight of concrete in the foundation, which is nominally 150 lb/ft$^3$, is assumed to be the same as the unit weight of the soil removed from the excavation, so the final stress at the middle of the compressible layer is $3+0.312 = 3.3$ tons/ft$^2$. From Fig. 16.8 the corresponding void ratio is $e_2 = 0.68$. Substituting these values in eq. (16.13) gives

$$S = 4\,\text{ft} \times \frac{0.91 - 0.68}{1.91} = 0.5\,\text{ft} = 6\,\text{in.}$$

This is an unacceptable amount of settlement. What are some options?

## 16.6.2  Settlement as a Sum of Compressions of Individual Soil Layers

Because of the nonlinear relationship between added stress and depth described in the preceding chapter, a thick stratum of compressible soil is divided into arbitrary sublayers that are treated individually. Total settlement then is the sum of the compressions of the individual sublayers. Because of the repetitive nature of the calculations they are most conveniently accomplished using a computer spreadsheet.

If different layers of soil occur at different depths, these are tested and treated separately, and if necessary divided into arbitrary sublayers. As in Example 16.3, initial pressures are calculated at the midplanes of each layer or sublayer, based on overburden thickness and unit weight, and position of the groundwater table. Final pressures include initial stress plus the distributed vertical stress from the weight of the structure.

Because both the soil compressibility and the amount of distributed vertical stress normally decrease with depth, calculations ordinarily proceed from the top down and are continued until the incremental amount of settlement becomes negligible.

### Example 16.4

Make a preliminary estimate of settlement of the water standpipe in Fig. 16.11, based on the weight of the water plus 15% for the weight of the structure. The unit weight of the soil is $125\,\text{lb/ft}^3$, the submerged unit weight $125 - 62.4 = 63\,\text{lb/ft}^3$. Use the soil compression information in Fig. 16.8. (The same procedure is followed with SI units.)

*Answer: Step 1.* The initial in-situ vertical stresses at the center of each layer are:

Layer A : $P_1 = 6.5 \times 125 = 812\,\text{lb/ft}^2$

B : $P_1 = 812 + 5 \times 125 + 5 \times 63 = 1750\,\text{lb/ft}^2$

C : $P_1 = 1750 + 10 \times 63 = 2380\,\text{lb/ft}^2$

D : $P_1 = 2380 + 10 \times 63 = 3010\,\text{lb/ft}^2$

E : $P_1 = 3010 + 10 \times 63 = 3640\,\text{lb/ft}^2$

The calculations carry three significant figures, but the answer will be valid for no more than two and should be reported as such. Note that the summation can be continued for deeper layers if necessary.

*Step 2.* The vertical stress imposed by the filled water tower is calculated at the center of each layer. Since the tower foundation is round, the influence coefficients in Table 15.2

**Figure 16.11**

Standpipe for the problem of Example 16.4.

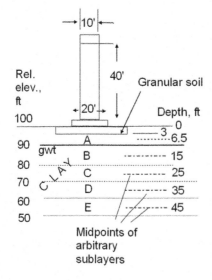

are appropriate. In order to obtain a maximum estimate of settlement, coefficients will be determined along the centerline of the tower.

The weight of the tower is $1.15 \times \pi(5)^2 \times 40 \times 62.4 = 22{,}540\,\text{lb}$, so the foundation contact pressure is $22{,}540 \div \pi(10)^2 = 718\,\text{lb/ft}^2$. A spreadsheet can be set up as shown below, where the influence coefficient, $I_c$, is obtained (by interpolation if necessary) from Table 15.2 and is multiplied times the foundation contact pressure to obtain $\triangle P$. This is added to the initial overburden pressures at each depth to obtain $P_2$. The values of $e_1$ and $e_2$ are obtained from the graph in Fig. 16.9, which for ease of reading can be replotted to a linear pressure scale. Let $Z$ represent depths to the middle of each sublayer and $H$ be the corresponding sublayer thicknesses, shown in Table 16.2.

The indicated amount of settlement is excessive. From the right-hand column in the table it might be assumed that transferring the foundation load to the bottom of the B layer with pile would reduce settlement by 2.6 inches, but *that is not correct* because the pressure bulb also would extend deeper, increasing the $I_c$ coefficients for the deeper layers.

### Example 16.5

(a) What depth of surcharge soil with a unit weight of $100\,\text{lb/ft}^3$ would be required to cause future settlement to be zero? (b) What amount would reduce it to 1 inch?

*Answer:* (a) For zero settlement the surcharge pressure should equal or exceed the foundation contact pressure, or $718 \div 100 = 7.2\,\text{ft}$ of surcharge, which therefore can be expected to reduce settlement by the predicted amount of 3.6 inches. One inch of settlement would mean a reduction of 2.6 inches. If a linear relationship is assumed, $7.2\,\text{ft} \times 2.6/3.6 = 5.2\,\text{ft}$ of surcharge.

An excessive settlement problem usually is approached by suggesting different options to bring it within acceptable limits, and the geotechnical engineer will point out the advantages and disadvantages of each so that the owner may make an intelligent decision.

## 16.6.3 Influence of Excavation on Settlement

Excavation and removal of soil, whether for a basement or to reshape the terrain, removes overburden pressure that can be looked upon as the equivalent of

**Table 16.2**
Results from Example 16.4

| Layer | H (ft) | T (ft) | R (ft) | H/R | $I_c$ | $\triangle P + P1 = P_2$ (lb/ft$^2$) | | | $e_1$ | $e_2$ | $\triangle H$ (ft) | (in.) |
|-------|--------|--------|--------|------|-------|------|------|------|-------|-------|------|-------|
| A | 6.5 | 7 | 10 | 0.65 | 0.834 | 599 + | 812 = | 1410 | 0.91 | 0.88 | 0.11 | 1.3 |
| B | 15 | 10 | 10 | 1.5 | 0.424 | 304 | 1750 | 2050 | 0.87 | 0.85 | 0.11 | 1.3 |
| C | 25 | 10 | 10 | 2.5 | 0.20 | 144 | 2380 | 2520 | 0.84 | 0.83 | 0.05 | 0.7 |
| D | 35 | 10 | 10 | 3.5 | 0.116 | 83 | 3010 | 3090 | 0.81 | 0.805 | 0.03 | 0.3 |
| E | 45 | 10 | 10 | 4.5 | 0.072 | 52 | 3640 | 3690 | 0.78 | 0.78 | 0 | 0 |

Total $= 3.6$ inches

removing a surcharge load. The net consolidating pressure is the foundation contact pressure minus the pressure from the weight of the excavated soil, which can be considerable.

For example, the weight of residential housing often is less than the weight of soil that is removed for a basement, which is reasonable because buildings are mostly air. However, the weight of the building normally is concentrated on load-bearing walls and columns that do settle, whereas the basement floor, having virtually no load, is more likely to come up as a result of soil expansion.

In a "floating foundation," the weight of the structure equals the weight of the soil removed so the structure essentially becomes a boat. True floating foundations are rare because of the high cost and requirement of continuous reinforced concrete mat under the structure. On the other hand, a partially floating foundation may be used to reduce foundation pressures to a manageable level on very weak soils. Another alternative will be the use of deep foundations such as pile.

## 16.7   DEFINING A PRECONSOLIDATION PRESSURE

### 16.7.1   Unloading Curve

A preconsolidated soil is defined as one that previously has been consolidated under a higher stress than exists today, a situation that readily can be simulated in a laboratory consolidation test. Unloading is shown by *de* in Fig. 16.12, which in general follows a straight line on a semilogarithmic plot. That means that re-expansion is not linearly elastic, but involves a gradual decrease in modulus as the

**Figure 16.12**

Load-unload-reload curves and the Casagrande method for defining preconsolidation pressure, $P_c$.

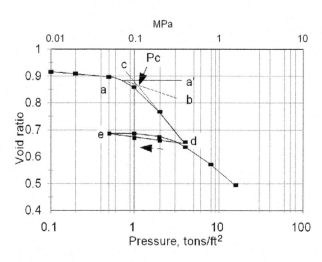

soil structure partly recovers. An equation analogous to eq. (16.8) therefore can be written to describe expansion, with the sign changed to indicate an increasing void ratio, and substituting an expansion modulus, $C_e$, for the compression modulus, $C_c$:

$$e_1 = e_c + C_e \log(P_1/P_c) \tag{16.14}$$

where $e_1$ is the void ratio at pressure $P_1$ and $e_c$ is the void ratio at the overconsolidation pressure $P_c$. Since most problems involve loading instead of unloading and $C_e$ is small compared with $C_c$, this equation is infrequently used, but it nevertheless can be useful, for example to predict the amount of rebound in the bottom of an open foundation excavation.

## 16.7.2   Reloading Curve

As shown by d-e-d in Fig. 16.12, when the laboratory-preconsolidated soil is reloaded, there is a hysteresis loop until the recompression curve reaches and merges with the normally consolidated leg of the curve. Casagrande suggested a method for determining preconsolidation pressure based on the shape of the reloading curve. The Casagrande procedure is as follows. With reference to Fig. 16.12:

1. Select a point of *maximum curvature* on the graph, shown at a.
2. Draw a horizontal line, aa'.
3. Draw an extension of the straight-line portion of the graph to c.
4. Bisect the angle between the extension and aa', shown as ab.
5. The intersection of ab with the extension line c defines a preconsolidation pressure, $P_c$.

The student should draw an arbitrary e-log P graph and practice this procedure until the steps become obvious.

### Example 16.6
Compare the preconsolidation pressure in Fig. 16.12 with that obtained from Figs. 16.7 and 16.8, which are from the same data.

*Answer:* $P_c$ in Fig. 16.12 is approximately 1.1 tons/ft$^2$, close to the value at A in Fig. 16.7. The Janbu construction based on modulus in Fig. 16.8 gives a much higher value. Disturbance decreases the preconsolidation pressure determined from a semilogarithmic plot, and a correction method is discussed later in this chapter.

## 16.7.3   Dominance of Overconsolidated Soils and Importance of Overconsolidation Pressure

Most soils in nature are at least to some degree overconsolidated because of a reduction in overburden pressure from erosion or melting of glacial ice.

Even recently deposited alluvial soils will be overconsolidated after a single lowering and recovery of the groundwater table.

Expansive clays exhibit characteristics of overconsolidation as a result of drying shrinkage, which pulls soil grains together by suction of negative pore water pressure instead of by vertical external pressure.

Preconsolidation pressure is important for foundation design because it defines the transition from low to a higher compressibility, and limiting foundation pressures to less than the overconsolidation pressure limits settlement to the near-linear elastic stage of compression. For example, in Fig. 16.12 it is only when pressures exceed about 1 ton/ft$^2$ that there is a substantial decrease in the soil void ratio, as the soil in effect "remembers" an earlier consolidating pressure.

### 16.7.4  Overconsolidation Ratio

The ratio of preconsolidation pressure to existing pressure is called the "overconsolidation ratio," or $OCR$:

$$OCR = \frac{P_c}{\gamma' z} \qquad (16.15)$$

where $P_c$ is the preconsolidation pressure, $z$ is the depth, and $\gamma'$ is the unit weight on an effective stress basis—that is, corrected for buoyant influences from the groundwater table. $OCR$ sometimes is used to characterize a soil and may be estimated from field testing procedures, but the term also is ambiguous because as depth $z$ decreases, $OCR$ increases and becomes infinite the ground surface. If $OCR$ is used as a soil descriptor, the depth or vertical stress should be noted.

## 16.8  SOME WAYS TO REDUCE ERROR

### 16.8.1  Putting the Soil Back Together

Extrusion of soil samples from a sampling tube can produce "wafering" or fracturing across the sample as a result of slight rebound as the soil emerges from the tube. A simple and obvious way to help alleviate this problem is to preload the soil in the consolidometer to the calculated overburden pressure, then remove the load and start the test.

### 16.8.2  Reinstating Lateral Stress

As previously mentioned, lateral field stresses are not reinstated prior to testing in a consolidation test, which can allow the soil structure to yield prematurely and lower the measured preconsolidation pressure. Unfortunately, it is not possible to reinstate lateral field stress in a conventional consolidometer.

The "stress path method," developed by Lambe (1967) at MIT, encases a cylindrical soil sample in a thin rubber membrane that then is confined in a pressure chamber so that lateral pressure can be regulated as the sample is loaded at the ends. This is the "triaxial test" apparatus, which normally is used to measure the soil shearing strength, and can allow a reinstatement of the stress environment existing in the field prior to sampling. A disadvantage is that the longer soil sample takes longer to drain, and a difficulty is that the field in-situ lateral stress usually is not known and seldom is measured. The general test procedure is as follows:

1. Obtain several samples from different depths under a proposed structure even though the soil is the same, since the lateral stress regimes are different.
2. Calculate the effective overburden stress for each sample depth, and estimate or measure the lateral effective stresses for each depth. As discussed later in this text, fairly reliable estimate of the "$K_0$ stress" can be made if the soil is normally consolidated.
3. Insert a soil sample in a triaxial cell that includes a provision for saturating the soil prior to testing, and for drainage of excess pore water expelled during consolidation.
4. Apply lateral and vertical effective stresses that simulate the stress environment existing in the field, and allow the sample to equilibrate.
5. Use elastic theory to predict changes in both the lateral and vertical stresses as a result of foundation loading.
6. Adjust vertical and lateral stresses to the new values and measure the vertical compression. The test requires considerable time and therefore is expensive.

## 16.8.3 Reducing Sample Disturbance

The ideal undisturbed soil sample is one that has been hand-cut from an exposed face or block of soil, but that seldom is feasible. Instead, "undisturbed" samples are obtained by pushing a thin-walled steel tube into the soil. Even though the bottom of the tube is chamfered inward so that it has a slightly smaller diameter, side friction still may occur between the soil and the internal walls of the tube. Disturbance can be reduced by use of a "piston sampler," which uses suction to help draw the sample into the tube, or by encasing the sample in a sock as it enters the sampling tube, but these procedures are not in common use.

A second cycle of damage can occur as samples are extruded in the laboratory by applying pressure on one end of the sample. A long residence time in the sampling tube can increase frictional resistance to extrusion, particularly if the tube has rusted. If the resistance is high, the tube should be sacrificed by milling a slot down one side, after which the tube will spring open and the sample may be removed.

### 16.8.4    Compensating for Sample Disturbance

Schmertmann (1955) while at Northwestern University conducted consolidation tests on samples having varying amounts of disturbance up to and including complete remolding. Examples for a Chicago glacial clay are shown in Fig. 16.13, where it will be seen that as disturbance increases, the bend in the $e$-log $P$ curve flattens out and almost disappears and decreases the apparent preconsolidation pressure by as much as a factor of 10. As shown at the bottom of the figure, the amount of reduction in void ratio, $\triangle e$, is fairly symmetrical around the preconsolidation stress determined by the Casagrande method for relatively undisturbed soil.

With increasing consolidating pressures the effects of sample disturbance diminish and become negligible when the void ratio is about one-half of the original void ratio. Tests of different soils give an average ratio of $0.42e_0$, which is taken as an anchor point for a corrected curve.

It also can be seen that the unloading curves are essentially parallel and nearly linear, a geometry that is assumed to apply to the changes in void ratio that occurred during geological unloading.

On the basis of these evidences, Schmertmann suggested reconstructing the consolidation curve to correct for sample disturbance as follows. With reference to Fig. 16.14:

1. Determine an initial void ratio $e_0$ prior to the sample being loaded. This establishes the initial part of a virgin curve because elastic compression does

**Figure 16.13**

Influence of sample disturbance on $e$-log $P$ curves, after Schmertmann (1955).

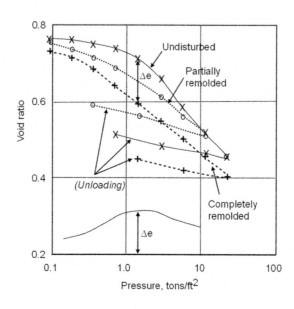

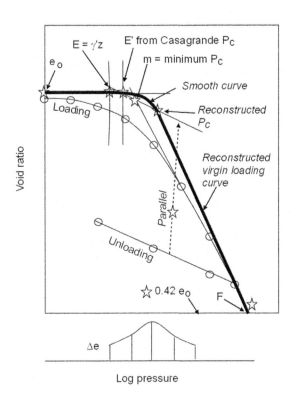

**Figure 16.14**
Schmertmann's
method to correct
for sample
disturbance.
Important points in
the reconstruction
are labeled with
stars.

not substantially change the void ratio. The line extends along $e_0$ to point $E$, which is defined by the effective overburden pressure, and beyond point $E$ if the soil is overconsolidated.

2. A preliminary estimate of $P_c$ is obtained using the Casagrande procedure so that the $e_0$-$E$ line extends to $E'$. In the case of an expansive clay, $e_0$ should reflect the soil density prior to expansion.

3. Draw a line from $E'$ parallel to an experimentally determined *unloading* curve. This is assumed to be the path taken by geological unloading. The intersection at $m$ represents a minimum preconsolidation pressure.

4. Establish a trial $P_c$ on an extension of the unloading line from $E'$.

5. Draw a line from the trial $P_c$ to $F$, where the test data extrapolate to $0.42e_0$.

6. Sketch a smooth curve through $E'$ and the trial $P_c$ to represent hysteresis between loading and unloading.

7. Test the trial $P_c$ by plotting values of $\Delta e$ as shown at the bottom of Fig. 16.14, and if necessary relocate $P_c$ to obtain a more symmetrical distribution. The graph of $\Delta e$ also is useful because it delineates the zone of major correction.

8. After the above criteria have been met, the virgin curve is along $e_0$ through $E'$, $P$, and $F$.

9. Curves may be reconstructed for other normally consolidated depths in the same stratum by recalculating $e_0$ and $\gamma'z$ to establish $E$, drawing the geological unloading curves as shown above to the reconstructed virgin compression curve.

### 16.8.5 Summary: Are Corrections Necessary?

It generally is recognized that except for very soft clays, conventional consolidation tests tend to overpredict settlement, sometimes by a significant margin. Sample disturbance reduces the apparent preconsolidation pressure and therfore increases predicted settlement. If the soil is not underconsolidated, it can safely be preloaded to the overburden pressure prior to testing. This tends to mend obvious disturbance such as transverse cracking that sometimes develops as soil specimens are being extruded from a thin-walled tube sampler. The Schmertmann procedure then may be used based on experience with particular soils.

## 16.9   SOME SPECIAL e-log P RELATIONSHIPS

### 16.9.1   Unsaturated and Cemented Soils

Apparent cohesion in soil that is not saturated will tend to carry the elastic response to a pressure that is higher than the preconsolidation pressure, when the soil structure abruptly collapses when the soil strength is exceeded. This tendency becomes even more pronounced in the case of soil that has been cemented, for example with calcium carbonate or gypsum. This is shown by the stair-stepped curve in the middle of Fig. 16.15.

### 16.9.2   Underconsolidated Saturated Soils

A soil is *underconsolidated* if its density is lower than the equilibrium saturated density. Clays in geologically youthful deposits such as deltas may be underconsolidated because excess pore water has not completely drained away. A clue is obtained in the field if pore water pressure measurements are higher than indicated by the groundwater table, but this also can reflect an artesian condition.

Sands deposited in slack water such as in deltas often are underconsolidated because of a tendency for grains to remain loosely stacked where they fall, as discussed in Chapter 8. This condition is metastable and potentially hazardous if the soil is disturbed by vibrations from machinery or earthquakes, as sudden collapse will transfer stress from grain-to-grain contacts to pore pressure, weakening the soil. In extreme cases all strength is lost and the sand becomes liquid, losing all resistance for supporting structures that then sink or tip over. Liquefaction is a major cause of damage by earthquakes to structures on loose sand deposits.

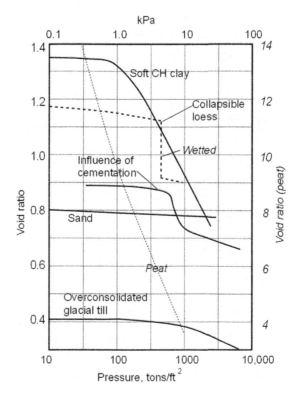

**Figure 16.15**
*e*-log *P* curves for
selected soils.

Undisturbed sampling of underconsolidated sand is difficult and can require freezing in situ and then using rock coring procedures. Identification usually is by means of field tests that include the *Standard Penetration Test* and *Cone Penetration Test* described in Chapter 26.

### 16.9.3 Collapsible Loess

Low-density loess is permeable and does not retain a high groundwater table, and therefore may be underconsolidated. Structures founded on such soil can be stable so long as the soil does not become wet or saturated. If excessive wetting does occur, as from a misdirected drain or a broken water pipe, a low-density loess can suddenly densify or collapse under its own weight. Even leveling off a hilltop, by interrupting natural drainage, can lead to collapse.

Eolian deposition of silt results in low density and an open structure that is held together by negative pore water pressure associated with the clay. Significantly, collapsible loess usually contains smectite as the clay mineral. A simple test for collapsibility is to place an undisturbed sample of the soil at its natural moisture content into a consolidometer, load it to the field overburden pressure, and then introduce water. The is illustrated by the heavy dashed line in Fig. 16.15. A sudden decrease in height of 5 percent or more is not uncommon. After collapse, the loess will go through a liquid stage, indicated by lateral and vertical stresses being equal (Handy, 1994).

In a "double odometer test" two samples are tested, one at its natural moisture content and the other in a saturated condition. The difference in void ratio at any applied pressure shows the extent of collapsibility.

A simple criterion for collapsibility based on the saturation moisture content and liquid limit is discussed in Chapter 12. Based on this criterion, there is over a 90 percent probability that loess deposited close to a source and remaining unsaturated will be collapsible.

The density of collapsible loess gradually increases with depth, so collapsibility is most pronounced at shallower depths. More significant is a gradual increase in clay content and decrease in thickness with increasing distance from an eolian source, which greatly reduces the prospects for collapsibility as the soil changes from ML or A-4 silt to CL or A-6 clay.

Where collapse is unavoidable, as in a foundation soil for a dam or a canal, loess may be presaturated by ponding. After collapse, the groundwater table then is lowered to initiate normal consolidation.

## 16.9.4  Sensitive Clays

Sensitive clays are those that lose a substantial part of their strength upon remolding. All clays show some degree of sensitivity, which is a reason for applying a correction to the consolidation curves.

Saturated, highly sensitive *quick clays* pose special problems because they are difficult to sample, losing all of their strength when agitated and changing to a viscous liquid. Consolidation tests of quick clays sometimes show a delay similar to that caused by apparent cohesion as they retain a weak structure, followed by collapse.

## 16.9.5  Overconsolidated Soils

The majority of soils are overconsolidated from removal of overburden by excavation or geological erosion, or from melting of glacial ice.

Secondary consolidation that slowly proceeds after excess pore water pressure has dissipated gradually increases soil density and the apparent preconsolidation pressure. Aging therefore increases the *overconsolidation ratio*, or *OCR*, defined in eq. (16.15).

## 16.10 SETTLEMENT OF STRUCTURES ON SAND

### 16.10.1 Advantages and Disadvantages of Sand

The high permeability of sand allows rapid drainage, so primary consolidation normally occurs during construction. Secondary consolidation may occur but is relatively small and depends on the content of fines.

Sand is difficult or impossible to sample without causing significant disturbance. Consolidation testing therefore requires that the sand be reconstituted to a density that existed in the field prior to sampling. Sand therefore usually is tested in situ. Most common are penetration tests that measure the resistance of sand to pushing a cone-tipped rod or driving of a standard sampler. Details of these tests are discussed in Chapter 26.

### 16.10.2 Using Results from the Standard Penetration Test (SPT)

In this test, a special thick-walled sampling tube is driven with a standardized hammer, and the number of blows is counted for each foot (0.3 m) of penetration. Because the blow count, $N$, is higher for dense sands than for loose sands, a correlation is obtained between $N$, foundation load, and settlement.

Building codes often use $N$ values with recommendations for bearing capacity based on experience. A summary chart was devised by Hough (1969) for noncohesive soils from silt to sand to gravel and is the basis for Fig. 16.16, with gravel having the highest bearing capacity for a given blow count, and silt and clay the lowest. Bearing capacities are reduced 50 percent if there is submergence because of buoyant reduction in intergranular friction. This graph may be

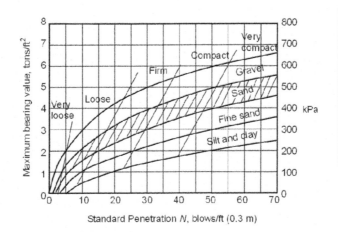

**Figure 16.16**

Chart based on Hough's (1969) presumptive bearing value for 1 inch (25 mm) settlement, based on experience and information in building codes.

useful for small structures that do not justify a detailed foundation investigation, and for the most part is on the conservative side. The chart was devised without corrections to $N$ based on overburden pressure and described in Chapter 26. A similar chart has been devised for clayey soils but is considered less reliable.

### Example 16.7

A wall footing will carry 3 tons per linear foot (87.5 kN/m). SPT $N$ values are 16, 24, and 26 blows per 6 in. (0.1 m) interval. Estimate an appropriate footing width.

*Answer:* In the Standard Penetration Test the last two incremental blow counts are added, which gives 25 blows per foot. In Fig. 16.16 the intersection of a line drawn up from 25 blows to sand indicates that the soil is considered a "compact sand" that has an allowable bearing pressure of 3 tons/ft$^2$ (300 kPa). The width requirement is 3 tons/ft $\div$ 3 tons/ft$^2 = 1.0$ ft (0.3 m), which is a minimum that is buildable. As there would be little added expense if a larger width were selected, the recommendation probably would be for 1.5 ft (0.5 m).

## 16.10.3  Elastic Compression of Sand

Settlement of foundations on sand might be predicted from an elastic modulus and the stress distributions discussed in Chapter 15 if the modulus remained constant, but it increases with increasing confinement and overburden pressure. This can be taken into account by numerical modeling using finite element analysis. This approach was used by Schmertmann et al. (1970, 1978) at the University of Florida, based in part on modulus measurements made with a "screwplate test." The screwplate is a circular steel plate that has a radial cut and has been bent so that it can be screwed into the ground.

Test results then were simplified as shown in Fig. 16.17, where the analysis indicated that square or round footings cause compression of the sand roughly to a depth $2B$, where $B$ is the footing width. Penetration test data therefore should represent soil to a depth of $2B$ below the foundation level. For long footings the influence extends about twice as deep, through a depth range 0 to $4B$. Settlement is calculated by summing the areas under the respective distribution diagrams, using a modulus derived from in-situ tests such as the SPT or cone penetrometer.

The graphs in Fig. 16.17 are based on cone penetrometer values that have units of pressure. Various conversion factors from SPT blow counts have been suggested based on empirical studies; a simple conversion is $q_c = 4N$ for sands and $q_c = 2N$ for silts, where $q_c$ is the cone penetration value in tons/ft$^2$ or kg/m$^2$, and $N$ is the SPT blow count. However, this conversion introduces an additional source of experimental error.

### Example 16.8

Predict settlement from the preceding example using Schmertmann's method.

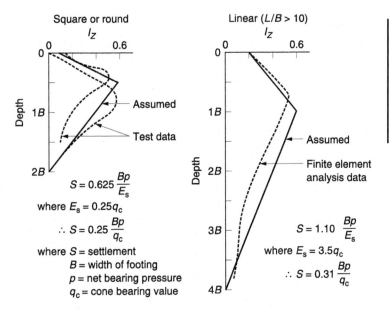

**Figure 16.17**
Schmertmann
et al. (1978)
method for
predicting
settlement of
foundations on
sand.

*Answer:* First $N$ must be converted to an equivalent cone value—which implies that the result will carry only *one* significant figure: $q_c = 4 \times 25 = 100$ tons/ft$^2$. From the information at the lower right in Fig. 16.17,

$$S = 0.31 \times 1\,\text{ft} \times 3\,\text{tons/ft}^2 \div 100\,\text{tons/ft}^2 = 0.009\,\text{ft} = 0.1\,\text{in.}$$

This suggests that the presumptive bearing value obtained in the preceding example is overly conservative.

## 16.10.4   Corrections for Shallow Footings and for Creep

Two empirical corrections are suggested by Schmertmann et al. to take into account the lower lateral restraint of sand underneath shallow footing depths, and for continuing slow settlement in time, or creep. The first correction factor is

$$C_1 = 0.83 + 0.17(p/\sigma_B) \tag{16.16}$$

where $C_1$ is a settlement correction factor, $p$ is the net increase in bearing pressure at the footing elevation, and $\sigma_B$ is the initial pressure at a depth $B$ under a square footing, or $2B$ under a continuous footing. Because the pressure is expressed as a ratio the same equation applies regardless of the system of units. This factor also takes the level of the groundwater table into account.

The second factor considers creep that occurs even in sand:

$$C_2 = 1 + 0.2\log(10t) \tag{16.17}$$

where $t$ is in years.

### Example 16.9

Assume that the sand in the previous examples has a unit weight of $125\,lb/ft^3$ ($19.6\,kN/m^3$) and the footing is at a depth of 3 ft (1 m).

*Question:* What amount of settlement can be anticipated after 1 year and after 100 years?

*Answer:* The net increase in bearing pressure at the footing level is $p = 3\,tons/ft^2 - (3 \times 125)/2000 = 2.81\,tons/ft^2$. The initial pressure at a depth of $2B$ is $\sigma_B = 2 \times 1 \times 125/2000 = 0.125\,ton/ft^2$. The correction factors are

$C_1 = 0.83 + 0.17(p/\sigma_B) = 0.83 + 0.17(2.81/0.125) = 3.4$

After 1 year, $C_2 = 1 + 0.2\ \log(10) = 1.2$

After 100 years, $C_2 = 1 + 0.2\ \log(1000) = 1.6$

Settlement after 1 year: $S = 0.1 \times 3.4 \times 1.2 = 0.4\,in.$

Settlement after 100 years: $S = 0.1 \times 3.4 \times 1.6 = 0.5\,in.$

*Question:* What would be the effect of submergence?

*Answer:* With submergence $C_1 = 0.83 + 0.17(2.81/0.0625) = 8.5$, which will increase settlement by a factor of $8.5/3.4 = 2.5$ times, to about 1 inch.

The final predictions still are about one-half of those obtained from Fig. 16.16.

## 16.10.5  Simplified Method Based on a Constant Modulus and Stress Distribution

By assuming a constant modulus and a 2 vertical to 1 horizontal spread of stress, an integration may be performed to predict settlement (Calhoon, 1972). This is illustrated at the left in Fig. 16.18, where the vertical stress from load $P$ is distributed across an increasing area with depth according to:

$$\text{Linear footing} : \sigma_v = \frac{P_1}{z} \tag{16.18a}$$

$$\text{Square footing} : \sigma_v = \frac{P}{z^2} \tag{16.18b}$$

Where $\sigma_v$ is vertical stress, and $P_1$ and $P$ are vertical loads per unit length and unit area, respectively. The total compression of soil between depths $z_0$ to $z_1$ is

$$S = \int_{z_0}^{z_1} \!\!\cdot\cdot\ dz = \int_{z_0}^{z_1} \frac{\sigma_v}{E} dz \tag{16.19}$$

where $\cdot\cdot$ is strain in the vertical direction and $E$ is the average modulus. From Fig. 16.18 it will be seen that $z_0 = B$ and $z_1 = z_0 + h = B + h$. The compression through depth $h$ is obtained by subtracting that for $z_0$ from that for $z_1$. Solutions are:

$$\text{Linear footing} : S = \frac{P_1}{E}[\ln(B + h) - \ln(B)] \tag{16.20a}$$

$$\text{Square footing} : S = \frac{P}{E}\left[\frac{1}{B+h} - \frac{1}{B}\right] \tag{16.20b}$$

The width of a rectangular or circular footing can be based on the area equivalent for a square footing, so for a footing having dimensions $2 \times 4$, $B = \sqrt{(2 \times 4)} = 2.8$. For a thick soil layer the summation can be terminated at a depth where the average added pressure is reduced to 10 percent of the applied pressure, which from Fig. 15.6 is about 5 or 6 times $B$. As the logarithmic term in Eq. 16.20 a must be dimensionless, length dimensions are relative to a unit dimension.

Care must be used that units are consistent; if $E$ is in tons/ft$^2$, $P$ is in tons or tons/ft and $S$ is in feet, or if $E$ is in MPa, $P$ is in MN or MN/m and $S$ is in meters.

The most difficult part of the procedure is a reliable determination of the modulus, $E$. $E$ can readily be measured with laboratory compression tests of undisturbed samples, but such sampling is not possible for sands, so testing must be conducted *in situ*. Various *in-situ* tests such as the pressuremeter and Dilatometer that can be used to estimate $E$ are discussed in the Chapter 26. Approximate relationships between $E$ and cone penetration data are shown by formulas in Fig. 16.17. Results from the standard penetration test (SPT) can be converted to approximate equivalent cone values by $q_c = 2N$ for silty sand to $4N$ for sand, where $N$ is the number of standardized blows required to drive a sampler one foot (0.3 m). Combining the ratios gives $E \approx 5$ to 10 times $N$, with $E$ in tons/ft$^2$ ($1N$ in MPa). It will be noted that this adds another layer of empiricism to tests that are strongly influenced by factors besides the soil modulus.

### Example 16.10
Calculate settlement for Example 16.7 using the approximate method.

*Answer.* With $P_1 = 3$ tons/ft, $E = 10(25) = 250$ tons/ft$^2$, and $B = 1$ ft, the compression to a depth $h = 5B = 5$ ft, is from Equation 16.30a.

$S = (3/250)[\ln(1 + 5) - \ln(1)] = 0.022$ ft $= 0.3$ in. This is approximately the same as obtained by the Schmertmann method in Example 16.9 after corrections for shallow foundations and for creep.

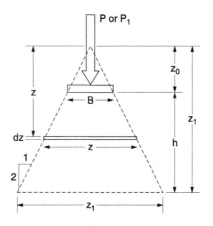

**Figure 16.18**

Simplified method for predicting settlement, modified from Calhoon (1972).

**Table 16.3**

Representative values of limiting settlements

| Limiting factor or type of structure | Maximum allowable settlement | |
|---|---|---|
| | Differential | Total |
| Drainage of floors | 0.01–0.02L | 6–12 in. (150–300 mm) |
| Stacking, warehouse lift trucks | 0.01L | 6 in. (150 mm) |
| Tilting of smokestacks, silos | 0.004B | 3–12 in. (75–300 mm) |
| Framed structure, simple | 0.005L | 2–4 in. (50–100 mm) |
| Framed structure, continuous | 0.002L | 1–2 in. (25–50 mm) |
| Framed structure with diagonals | 0.0015L | 1–2 in. (25–50 mm) |
| Reinforced concrete structure | 0.002–0.004L | 1–3 in. (25–75 mm) |
| Brick walls, one-story | 0.001–0.002L | 1–2 in. (25–50 mm) |
| Brick walls, high | 0.0005–0.001L | 1 in. (25 mm) |
| Cracking of panel walls | 0.003L | 1–2 in. (25–50 mm) |
| Cracking of plaster | 0.001L | 1 in. (25 mm) |
| Machine operation, noncritical | 0.003L | 1–2 in. (25–50 mm) |
| Crane rails | 0.003L | [a] |
| Machines, critical | 0.0002L | [a] |

Note: $L$ is the distance between adjacent columns, $B$ is the width of base. [a]Consult manufacturer

## 16.11   ALLOWABLE SETTLEMENT

### 16.11.1   Allowable Total Settlement

The maximum permissible settlement is rather arbitrary, and should consider structural criteria, safety, and appearance, matters that should be discussed with the responsible structural engineer and architect. Differential settlement obviously is more critical than total settlement, but as an extreme case the two can be considered equal if part of a structure does not settle. Some general guidelines for allowable settlement are presented in Table 16.3. A common arbitrary criterion is that settlement should not exceed 1 inch (25 mm).

**Example 16.11**

Suggest a maximum allowable settlement for the standpipe in Fig. 16.11.

*Answer:* The base width is 20 ft, so the maximum allowable settlement for the tower is $0.004 \times 20 = 0.08$ ft $= 1$ inch.

### 16.11.2   Influence of the Superstructure

Settlement predictions for individual footings are based on the assumption that they are free-standing and are not affected by rigidity of the superstructure. This may be referred to as a *design settlement* that may or may not be attained depending on the flexibility of the structure. The more rigid the structure, the more the loads will become distributed according to the resistance of the soil, the stiffer soil carrying the larger loads.

This acts to decrease differential settlement but adds stress in the structure. The designer ordinarily assumes that the full predicted design settlements will occur, and examines the structure to determine if the resulting stresses and differential movements can be tolerated.

## Problems

16.1. What are some differences between primary consolidation and secondary consolidation?

16.2. What is meant by the term differential settlement? Give three examples of damages that can result from differential settlement.

16.3. A building is designed with uniform pressures on all footings, yet there is damaging differential settlement. Give two possible reasons and suggest an investigation into the cause.

16.4. A lawsuit revolves around whether damages to a house were caused by differential settlement or by expansive clay. Differential settlement would be a design or construction problem, but site conditions such as expansive clay often are legally regarded as an owner responsibility. Explain how both conditions are time-related and what clues, including weather factors, should be examined.

16.5. The void ratio of a soil is 0.798 at a pressure of 95.8 kPa and the compression index is 0.063. Draw a void ratio-pressure diagram for this soil on both semilogarithmic and natural scales.

16.6. Plot the void ratio-pressure data in Table 16.4 on a semilogarithmic scale and estimate the preconsolidation load of the soil by the Casagrande method.

16.7. Repeat Problem 16.6 with the Schmertmann correction.

16.8. Determine the preconsolidation pressure for the data in Table 16.4 using Janbu's procedure.

| Load increment, kPa (tons/ft$^2$) | Void ratio | $\varepsilon$ |
|---|---|---|
| 9.6 (0.1) | 1.072 | |
| 19.2 (0.2) | 1.063 | 0.0036 |
| 47.9 (0.5) | 1.044 | 0.0127 |
| 95.8 (1.0) | 1.013 | 0.0277 |
| 191.5 (2.0) | 0.941 | 0.0625 |
| 383 (4.0) | 0.834 | 0.1142 |
| 766 (8.0) | 0.725 | 0.1668 |
| 1530 (16.0) | 0.616 | 0.2195 |

**Table 16.4**

Data for Problem 16.6 ff

16.9. The data in Table 16.4 are for a soil sample from a depth 6.1 m (20 ft) below the ground surface and 4.57 m (15 ft) below the water table. Assume that soil above the water table weighs $17.3\,kN/m^3$ ($110\,lb/ft^3$) and the buoyant unit weight below water is $10.4\,kN/m^3$ ($66\,lb/ft^3$). Is this soil normally consolidated or overconsolidated?

16.10. Determine the compression index of the soil in Problem 16.9 (a) without, (b) with the Schmertmann correction.

16.11. A clayey soil has a liquid limit of 28. Estimate the compression index and explain its meaning and how it is used to predict settlement.

16.12. Is there any limitation to the use of a compression index with an over-consolidated soil?

16.13. A compressible soil layer is 8.5 m (28 ft) thick and its initial void ratio is 1.046. Tests and computations indicate that the final void ratio after construction of a building will be 0.981. What will be the settlement of the building over a long period of time?

16.14. A foundation 6.1 by 12.2 m (20 by 40 ft) in plan is to be constructed at a site where the geological profile is as shown in Fig. 16.19. The foundation load is 287 kPa (3 tons/ft$^2$) and the unit weights of various soils are as shown in the figure. Consolidation tests of the compressible clay indicate that it has a void ratio of 1.680 at 95.8 kPa (1 ton/ft$^2$) and a compression index of 0.69. Calculate the total settlement of the center of the foundation.

16.15. Give two examples of underconsolidated soils and potential consequences for each.

16.16. A loess soil has a liquid limit of 32 and a moisture content of 18%. What minimum unit weight is required to prevent collapse?

16.17. An earth dam is to be built with one end on alluvial soil and the other on a loess-covered gravel terrace. What investigations should be conducted to ensure safety and functionality of the dam?

**Figure 16.19**

Foundation conditions for Problem 16.14.

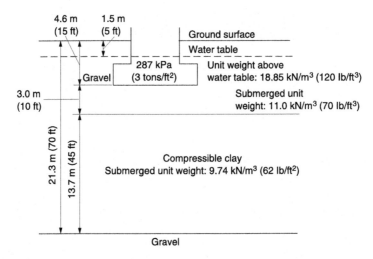

16.18. By use of a block diagram derive an equation relating loess collapsibility to unit weight, specific gravity, and liquid limit.

16.19. What measures can be taken to prevent loess collapse? Is there any situation where collapse should be induced by wetting prior to construction?

16.20. A square column footing, founded on sand at a depth of 1.2 m (4 ft) below ground level, will carry a load of 890 kN (100 tons). The average standard penetration value of the sand is 20 blows/0.3 m (4 ft). Make a preliminary determination of the size of footing to the nearest 1 ft (0.3 m) that will be required to limit settlement to 1 inch (25 mm), (a) if the water table is 3 m (10 ft) below ground level, (b) if the water table is at ground level.

16.21. In Problem 16.20, make a more detailed determination of the required footing size to the nearest 1 ft based on 50-year settlement.

16.22. A trapezoidal combined footing will be 3 m (10 ft) long with widths of 0.5 m (1.6 ft) at one end, 1 m (3.3 ft) at the other. It will rest at a depth of 1 m (3.3 ft) in sand with a static cone value $q_c = 300$ kg/cm$^2$. Estimate the long-term design settlement.

16.23. Use the Calhoon approximation to predict settlement and prepare a graph of settlement versus footing width for the conditions in Problem 16.20.

## References and Further Reading

Calhoon, M.L. (1972). "Simplified Method to Calculate Settlement beneath Footings." *ASCE Civil Engineering*, 42: 2: 73–74.

Casagrande, A. (1936). "The Determination of the Pre-consolidation Load and its Practical Significance." *Proc. 1st Int. Conf. on Soil Mechanics and Foundation Eng.*, Harvard University, Cambridge, Mass., 3, 60.

Gibbs, H. J., and Bara, J. P. (1967). "Stability Problems of Collapsible Soils." *ASCE J. Soil Mechanics and Foundation Eng., DW.* 93(SM4), 577–594.

Handy, R. L. (1994). "A Stress Path Model for Collapsible Loess." In E. Derbyshire et al., ed., *Genesis and Properties of Collapsible Soils*. NATO ASI Series, Kluwer Academic Publ., Dordrecht, The Netherlands.

Hough, B. K. (1969). *Soils Engineering*, 2nd ed. Ronald Press, New York.

Janbu, N. (1969). "The Resistance Concept Applied to Deformations of Soils." *Proc. 7th Int. Conf. on Soil Mechanics and Foundation Eng.*, I, 191–196.

Koestler, A. (1964). *The Act of Creation*. The Macmillan Co., New York.

Lambe, T. W. (1967). "Stress Path Method." *ASCE J. Soil Mechanics and Foundation Eng. DW.* 93(SM6), 309–31.

Schmertmann, J. (1955). "The Undisturbed Consolidation of Clay." *ASCE Trans.* 120, 1201–1227.

Schmertmann, J. H. (1970). "Static Cone to Compute Static Settlements over Sand." *ASCE J. Soil Mechanics and Foundation Eng. Div.* 96(SM3), 1011–43.

Schmertmann, J. H., Hartman, J. P., and Brown, P. R. (1978). "Improved Strain Influence Factor Diagrams." *ASCE J. Geotech. Eng. Div.* 104(GT8), 1131–35.

Skempton, A. W. (1944). "Notes on the Compressibility of Clays." *Quart. J. Geol. Soc. London* 100, 119–135.

Terzaghi, K., and Peck, R. B. (1967). *Soil Mechanics in Engineering Practice*, 2nd ed. John Wiley & Sons, New York.

# 17

# Time-Rate of Settlement

## 17.1 OVERVIEW

### 17.1.1 Time Is Timely

A settlement analysis not only should indicate the amount of anticipated settlement, but also should give an indication of the rate at which it will occur. Measurements in laboratory consolidation tests indicate that after each increment of pressure is applied, consolidation is nearly complete after 24 hours, whereas the time for foundations to reach equilibrium is measured in months or years. Some type of scaling factor is necessary.

### 17.1.2 Terzaghi's Research

Terzaghi's consolidation apparatus included small standpipes to measure excess pore water pressure developed during compression of soil, and pore water pressures were observed to quickly increase after a compressive load was applied, and then gradually decrease as water drained out and allowed the soil to compress. Sand was found to compress much more rapidly than clay, indicating that the soil permeability might be a controlling factor—but why should so much more time be required in the field compared with in the laboratory?

## 17.2 TERZAGHI'S THEORY

### 17.2.1 Development of a Theory

Terzaghi's theory of consolidation is considered one of the cornerstones of geotechnical engineering. His theory is based on (1) continuity of fluid flow and (2) an application of Darcy's Law. The derivation is highly mathematical, but as is the case with many classical engineering formulas, results have been simplified to make them easier to use. It is suggested that the student follow through the theoretical development in order to gain a better understanding of the basis for the theory.

The principle of continuity is illustrated in Fig. 17.1, where the flow rate of water upward into the bottom face of the differential element, $q_z$, equals the area, $dx\,dy$, times velocity, $v$:

$$q_z = v\,dx\,dy \tag{17.1}$$

The amount of water departing from the differential element is increased as a result of vertical consolidation of the element:

$$q_{zd} = \left[v + \frac{\partial v}{\partial z}dz\right]dx\,dy \tag{17.2}$$

In eq. (17.2) the partial differential term, $\partial v/\partial z$, represents the increase in flow rate in the $z$ direction due to consolidation of the element. Subtracting the volume of water departing from the volume entering gives the volume of pores lost, or the amount of consolidation, per unit of time:

$$q_{zd} - q_z = \frac{\partial v}{\partial z}dx\,dy\,dz = \frac{\partial n}{\partial z}dx\,dy\,dz \tag{17.3}$$

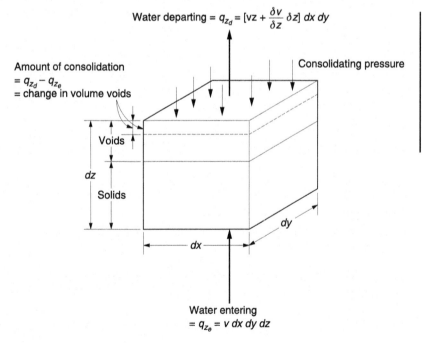

Water departing $= q_{z_d} = [vz + \frac{\delta v}{\delta z}\,\delta z]\,dx\,dy$

Amount of consolidation
$= q_{z_d} - q_{z_e}$
= change in volume voids

Consolidating pressure

Voids

dz

Solids

dx

dy

Water entering
$= q_{z_e} = v\,dx\,dy\,dz$

**Figure 17.1**
Differential element of compressed soil. The volume change of saturated soil equals the volume of water ejected.

$$\frac{\partial v}{\partial z} = \frac{\partial n}{\partial z} \tag{17.4}$$

In eq. (17.4), $\partial v/\partial z$ is the change in velocity across the element and $\partial n/\partial z$ is the change in porosity, $n$ (expressed as a fraction) with time. This is the continuity equation for one-dimensional consolidation. It also may be expressed in terms of the void ratio, since $n = e \div (1+e)$:

$$\frac{\partial e}{\partial z} \left[ \frac{1}{1+e} \right] \frac{\partial v}{\partial z} \tag{17.5}$$

That is, the higher the rate of change of fluid velocity with respect to vertical distance, the higher the rate of change of the void ratio.

## 17.2.2  Consolidation and Darcy's Law

The next step is to introduce Darcy's Law, $q = kia$, or

$$v = ki$$

where $k$ is the coefficient of hydraulic conductivity and $i$ is the hydraulic gradient. This may be written

$$v = k \frac{\partial n}{\partial z} \tag{17.6}$$

where $\partial h/\partial z$ is the rate of change in head with respect to $z$ within the element. Changes in elevation head are compensated by equal and opposite changes in static pressure head, so $h$ signifies excess pressure head. Multiplying by the unit weight of water converts this to excess pore water pressure, $u$:

$$u = h\gamma_w$$

where $h$ is the excess pore pressure expressed as head and $\gamma_w$ is the unit weight of water. Then

$$v = \frac{k}{\gamma_w} \frac{\partial u}{\partial z} \tag{17.7}$$

Differentiating eq. (17.7) with respect to $z$ gives

$$\frac{\partial v}{\partial z} = \frac{k}{\gamma_w} \frac{\partial u^2}{\partial z^2} \tag{17.8}$$

Combining eqs. (17.7) and (17.5) gives

$$\frac{\partial e}{\partial t} \left[ \frac{1}{1+e} \right] = \frac{k}{\gamma_w} \frac{\partial u^2}{\partial z^2} \tag{17.9}$$

For small strains the change in void ratio $\partial e$ is proportional to the change in pore pressure $\partial u$:

$$a_v = \frac{\partial u}{\partial e} \tag{17.10}$$

where $a_v$ is a coefficient of compressibility. Substitution for $\partial e$ in eq. (17.9) gives

$$\frac{\partial u}{\partial t} = \frac{k(1+e)}{a_v \gamma_w} \frac{\partial u^2}{\partial z^2} \qquad (17.11)$$

If we define a *coefficient of consolidation*, $C_v$:

$$C_v = \frac{k(1+e)}{a_v \gamma_w} \qquad (17.12)$$

then

$$\frac{\partial u}{\partial t} = C_v \frac{\partial u^2}{\partial z^2} \qquad (17.13)$$

This is Terzaghi's consolidation equation, and indicates that the reduction in excess pore pressure with time is a function of the *rate of change of pore pressure with distance*. The longer the distance, the slower the rate of drainage. This is the scaling equation that enables laboratory time to be converted into field time. By definition, compression that depends on the rate of ejection of pore water is *primary consolidation*. It is important to note that the coefficient of consolidation, $C_v$, is a time function, not a pressure function as in the case of the compression index, $C_c$.

## 17.2.3   Solving for Primary Consolidation

Drainage time is instantaneous at the boundaries of a consolidation layer and increases with distance from a boundary. Solution of eq. (17.13) therefore requires evaluating and summing excess pore pressures at different levels within a consolidating layer. The distribution of pore water pressure in a consolidating layer between two drainage faces is illustrated in Fig. 17.2, which is the drainage geometry in a laboratory consolidometer. This same arrangement occurs in the field where a consolidating clay layer is sandwiched between two sand layers and the drainage directions are vertical, both upward and downward.

The solution of eq. (17.13) is obtained by defining boundary conditions and integrating between those boundaries with Fourier series. This solution is presented by Taylor (1948), Jumikis (1962), Holtz and Kovacs (1981), and others. Jumikis develops a useful identity to define the percent total pore pressure dissipation $U$ in a consolidating layer, and hence the percent of total consolidation. In Fig. 17.2, $U$ is the enclosed area labeled "pore water pressure," and can be seen to decrease as consolidation goes from 50 to 90 percent completion. The series is

$$U = 1 - \frac{8}{\pi^2}\left(e^{-Nt} + \frac{1}{9}e^{-9Nt} + \frac{1}{25}e^{-25Nt} + \ldots\right) \qquad (17.14)$$

where $t$ equals time and

$$N = \frac{\pi^2}{4}\frac{C_v}{H^2} \qquad (17.15)$$

**Figure 17.2**

Distribution of
pore water
pressure and
effective stress in
a consolidating
clay layer
between two sand
layers.

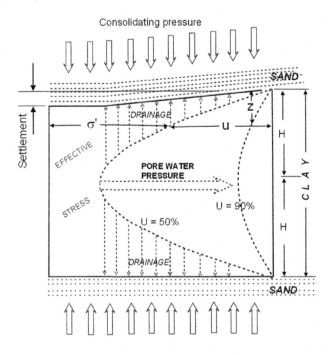

where $H$ is the maximum vertical distance to a drainage face. *Thus, in a clay layer that drains in two directions, upward and downward, H equals one-half of the layer thickness.* Equation (17.14) quickly converges toward a final value but in theory never reaches that value. The relationship can be simplified by defining a time factor, $T$, as

$$T = \frac{C_v t}{H^2} \qquad (17.16)$$

Substituting in eq. (17.15) and solving for $N$,

$$N = \frac{\pi^2}{4}\frac{T}{t}$$

$$Nt = \frac{\pi^2}{4}T \qquad (17.17)$$

which is the negative exponent in eq. (17.14).

### 17.2.4 Importance of the Time Factor *T*

Transposing eq. (17.16) gives

$$t = \frac{H^2}{C_v}T \qquad (17.18)$$

Consolidation time therefore is proportional to $T$ and inverse to the coefficient of consolidation, $C_v$. The coefficient of consolidation in turn is proportional to $k$, the coefficient of hydraulic conductivity, in eq. (17.12), so the higher the permeability, the shorter the time required for consolidation. Terzaghi's analysis demonstrates why consolidation of sand is many times faster than that of clay, other factors being the same.

## 17.2.5 Importance of $H^2$

The maximum drainage distance, $H$, is particularly significant because it is squared, so doubling the drainage distance increases consolidation time by a factor of 4. The drainage distance is doubled if drainage can only occur from one face instead of two, which is the reason why a large foundation on a consolidating layer will be placed on a bed of sand, so that consolidation can occur more rapidly.

Because the drainage distance is squared, it can explain the huge difference between consolidation times in the field and those that are measured in a laboratory consolidometer. For example, if a soil layer 2.5 m (8 ft) thick in the field is represented by a laboratory consolidometer sample 2.5 cm (1 in.) thick, the thickness ratio is $250/2.5 = 100$, and the ratio of consolidation times is $(250/2.5)^2 = 10,000$. It is this relationship that gives practicality to the consolidation test and the use of thin samples to reduce testing time.

Figure 17.2 is a schematic representation of drainage from a clay layer between two sand layers. The pore pressure at any depth $z$ is $u$, and the percent conversion from pore water pressure to effective stress is indicated by the area $U$. Two solutions are shown, one for an area ratio indicating 50 percent completion of consolidation, and the other for 90 percent consolidation. It will be noted that because of the double drainage geometry $H$ is one-half of the clay layer thickness.

## 17.2.6 Values of $T$

Table 17.1 shows values of $T$. Rearranging eq. (17.18) gives

$$C_v = \frac{H^2}{t_p} T_p \qquad (17.19)$$

where the subscript p designates a particular percent completion of consolidation. Equation (17.19) is used to evaluate $C_v$ for a particular soil and load increment from time-compression data in the consolidation test. Once $C_v$ is known, eq. (17.18) is used to solve for time $t$ for a given percentage of consolidation in the field. Results typically are plotted as a *settlement versus time curve*. This is used to predict the time that will be necessary for primary consolidation settlement to reduce to an amount that is manageable.

**Example 17.1**
A laboratory consolidation test gives $C_v = 2.5\,m^2/year$. How much time will be required for 80% consolidation of a 1.5 m (4.9 ft) thick layer of this clay that is underlain by shale and overlain by sand?

| | Percent of total settlement, $U \times 100$ | Percent remaining | Time factor $T$ |
|---|---|---|---|
| **Table 17.1** Calculated time factors $T$ for various values of $U$ expressed as percentages | 0 | 100 | 0.000 |
| | 5 | 95 | 0.002 |
| | 10 | 90 | 0.008 |
| | 15 | 85 | 0.018 |
| | 20 | 80 | 0.031 |
| | 25 | 75 | 0.049 |
| | 30 | 70 | 0.071 |
| | 35 | 65 | 0.096 |
| | 40 | 60 | 0.126 |
| | 45 | 55 | 0.159 |
| | 50 | 50 | 0.197 |
| | 55 | 45 | 0.239 |
| | 60 | 40 | 0.286 |
| | 65 | 35 | 0.340 |
| | 70 | 30 | 0.403 |
| | 75 | 25 | 0.477 |
| | 80 | 20 | 0.567 |
| | 85 | 15 | 0.684 |
| | 90 | 10 | 0.848 |
| | 95 | 5 | 1.13 |
| | 98 | 2 | 1.50 |
| | 99 | 1 | 1.78 |
| | 99.9 | 0.1 | 2.71 |
| | 100.0 | <0.05 | 3.00 |

*Answer:*

*Step 1.* With single drainage $H$ equals the layer thickness, or 1.5 m.

*Step 2.* $T$ from Table 17.1 for 80% completion is $T_{80} = 0.567$.

*Step 3.* Substituting these values into eq. (17.18) gives

$$t_{80} = \frac{(1.5\,\text{m})^2}{(15.26)} \times 0.567 = 0.51 \text{ year} = 190 \text{ days}$$

## 17.3 DETERMINING $C_V$

### 17.3.1 Taylor's Square-Root-Of-Time Method

In order to determine $C_v$ for a particular soil and loading condition, it is necessary to fit experimental consolidation data for a particular soil to eq. (17.19). Taylor (1948), at MIT, noted that if the percent completion of primary consolidation determined by Terzaghi's theory is plotted versus the square root of the time factor $T$, the graph initially is a straight line, as shown in Fig. 17.3. Since time $t$ is proportional to the time factor $T$, the same relationship may be expected to hold for consolidation data, which has been confirmed experimentally.

In Fig. 17.3 the linear portion of the theoretical relationship between percent consolidation and $\sqrt{T}$ is extended from $O$ to $B$. A point then is established on the theoretical curve that represents 90 percent consolidation, and a line is drawn through that point, from $O$ to $C$. The horizontal distance measured from the ordinate is $AC = 1.15AB$.

The same construction can be applied to experimental data, as in Fig. 17.4, to define the $\sqrt{t}$ for 90 percent primary consolidation. This graph uses measured deflections and does not require calculating void ratios. The value of $C_v$ then is determined from eq. (17.19) by substituting appropriate values for drainage distance $H$, the measured $\sqrt{t}$, and $T_{90} = 0.848$.

The Taylor or "square root of time" procedure for estimating $C_v$ is as follows:

1. Plot timed consolidometer compression data on the $y$-axis, versus corresponding values of $\sqrt{t}$ on the $x$-axis.
2. Draw a straight line through the linear portion of the plot, omitting the first point(s) if necessary as they may be influenced by elastic compression.
3. Draw a horizontal line from the $y$-axis to intersect the linear plot, measure the distance to the $y$-axis, and mark off that distance times 1.15. Then draw another line from that point to the origin at the upper left of the graph. The intersection of this line with the data plot defines $\sqrt{t_{90}}$ for 90 percent primary consolidation, and is read from the graph. Square to obtain $t_{90}$.
4. Substitute $t_{90}$ in eq. (17.19) and calculate $C_v$.

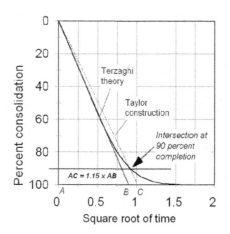

**Figure 17.3**

Graph of theoretical percent consolidation versus $\sqrt{T}$ that is the basis for Taylor's construction to determine $C_v$.

### Example 17.2

The Taylor construction in Fig. 17.4 indicates that $\sqrt{t_{90}} = 5.3\,\text{min.}^{0.5}$, so $t_{90} = 5.3^2 = 28.1\,\text{min}$. The sample thickness at the beginning of this episode of consolidation was 1.383 in. (35.13 mm) and at the end was 1.315 in. (33.40 mm), giving an average of 1.349 in. (34.26 mm). (1) Calculate $C_v$, and (2) predict the time for 95% consolidation of a clay stratum 3 m thick with double drainage.

*Answer:* (1) The determination of $C_v$ is very sensitive to the sample thickness so the average thickness before and after consolidation is used. The next step (and the one that is most often overlooked) is to divide the thickness by 2 to obtain $H$ for a double-drainage situation: $H = 34.27/2 = 17.13\,\text{mm}$. Then from eq. (17.19),

$$C_v = \frac{H^2}{t_{90}} T_{90} = \frac{(17.13\,\text{mm})^2}{28.1\,\text{min}} \times 0.848 = \underline{8.86\,\text{mm}^2/\text{minute}}$$

The preferred units for application to a field situation are $\text{m}^2/\text{year}$ or $\text{ft}^2/\text{year}$. To avoid mistakes, the series of conversion factors should be written down so that units can be cancelled. Thus in the first two terms, minutes cancel minutes, which converts the time to hours, and so on, until the remaining units are $\text{m}^2/\text{year}$:

$$c_v = 8.86\,\frac{\text{mm}^2}{\text{min.}} \times \frac{60\,\cancel{\text{min.}}}{1\,\cancel{\text{hour}}} \times \frac{24\,\cancel{\text{hours}}}{1\,\cancel{\text{day}}} \times \frac{365\,\cancel{\text{days}}}{1\,\text{year}} \times \left(\frac{1\,\text{mm}}{1000\,\text{mm}}\right)^2 = \underline{4.66\,\text{m}^2/\text{year}}$$

(2) The value of $C_v$ is substituted into eq. (17.18) with the appropriate values for $H$ and $T$ for the field situation, where the maximum drainage distance $H$ again is one-half of the layer thickness:

$$t_{95} = \frac{H^2}{C_v} T_{95} = \frac{(1.5\,\text{m})^2}{4.66\,\text{m}^2/\text{year}} \times 1.13 = 0.55\,\text{years} = 7\,\text{months}$$

Note that the final answer is given to no more than two significant figures because that is the limit of accuracy of reading the graph, so to include more would be to claim an unwarranted precision in the results.

### Figure 17.4

Application of the Taylor method to data for the pressure interval 1 to 2 tons/ft$^2$ in Table 16.1.

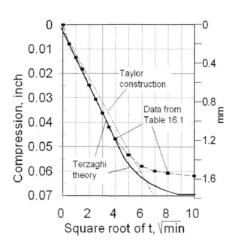

*Question:* Upon careful inspection of the soil boring logs and samples, a 2-inch layer of sand is discovered that divides the clay layer into two equal layers. How might that influence settlement?

*Answer:* The total clay thickness is about the same but the sand divides it into two layers, thereby reducing the drainage distance of each by a factor of 2. As the drainage distance is squared, time will be divided by a factor of 4, reducing the time requirement from 7 months to 7 weeks.

### 17.3.2 Accuracy of the Taylor Method

The data in Fig. 17.4 deviate from the ideal theoretical curve as consolidation progresses toward completion. This is reasonable because $C_v$ depends on the coefficient of hydraulic conductivity, which decreases as soil compresses. This can be confirmed by determining $C_v$ for different load increments. However, will this affect the determination of $\sqrt{t_{90}}$?

In Fig. 17.4 the total amount of compression is 0.061 in. Ninety percent of that is 0.055 in., which closely agrees with the amount predicted by the intersection of the data plot with the 1.15-sloping line.

### 17.3.3 Logarithm of Time Method

An empirical procedure suggested by A. Casagrande at Harvard University involves plotting compression dial readings versus the logarithm of time. This is shown in Fig. 17.5 for the same data as in Fig. 17.4. The inter-section of the two straight lines through the data, at 40 minutes, was proposed to define the end of primary consolidation, although this is more accu-rately defined from measurements of pore water pressure. One-half of the compression at this point is 0.0300 in., which as shown in Fig. 17.5 corresponds to a time of 6 minutes. This value and $T_{50}$ are substituted in eq. (17.19) to obtain $C_v$.

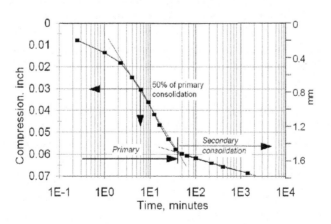

**Figure 17.5**

The log-time method applied to the same data as in Fig. 17.4.

### Example 17.3

Evaluate $C_v$ for the previous example using the log-time method. Note that in this case time $t = 6$ min is read directly from the graph and is not squared.

*Answer:*

$$C_v = \frac{H^2}{t_p} T_p = \frac{(17.13\,\text{mm})^2}{6\,\text{min}} \times 0.197 = \underline{9.6\,\text{mm}^2/\text{minute}}$$

which is very close to the estimate of $8.9\,\text{mm}^2/\text{minute}$ from the Taylor method. The initial data points that are not on the line occur during the first 3 minutes and indicate a quasi-elastic response, discussed in the next section.

## 17.4  FORE

### 17.4.1  Overview

In many natural phenomena the rate of approach to an end condition is proportional to the departure from that condition, similar to an automobile gradually slowing to a stop. This concept of a steadily diminishing rate is formalized in a "first-order rate equation" of chemical kinetics. Its application in geotechnical engineering recently was given the acronym "FORE" (Li, 2003).

A requirement of FORE is that the primary thesis must be met, that a rate of change is proportional to the departure from some end condition. Often conditions will change and require definition of a new end point and a new rate equation, such as the shift from primary to secondary consolidation.

In some cases a change may be continuous and closely coupled with the primary rate, in which case it will not be discerned by the rate equation (Handy, 2002).

### 17.4.2  FORE for an End Value

Let $D$ equal an unknown end value of $y$ that in theory will never be fully achieved. The amount of departure from the end value at any time is $(y - D)$.

A first-order rate equation states that the rate of change $d(y - D)$ with respect to an independent variable $x$ is proportional to $(y - D)$. This can be stated mathematically by

$$\frac{d(y - D)}{dx} = -k(y - D) \tag{17.20}$$

where $k$ is a constant and the minus sign indicates a declining rate. Rearranging,

$$\frac{d(y - D)}{(y - D)} = -k\,dx \tag{17.21}$$

from which

$$\ln (y - D) = -kx + C \tag{17.22}$$

where $C$ is a constant of integration. This is the basic rate equation expressed in the form $y = ax + b$, which means that if the rate equation method is applicable, data inserted into eq. (17.22) must plot as a straight line. Converting to logarithms to the base 10, and in recognition that there is no logarithm of a negative number,

$$\log_{10} |y - D| = -k_{10}x + C_{10} \tag{17.22a}$$

The final value, $D$, is not known so it is determined by trial and error, by substituting trial values of $D$ into the equation until it yields a straight line. If $D$ is not correct, the line will curve and a new value should be tested. If a single $D$ value does not meet this criterion it may be necessary to departmentalize the data and define two separate rate equations.

As $D$ is varied, convergence to develop a straight line is rapid and is easily accomplished in a minute or two with a computer spreadsheet. Both $y$ and $D$ must have the same sign convention: if compression $y$ is negative, the maximum amount of compression $D$ also must be negative.

## 17.4.3  When Conditions Change

The rate equation changes if the system changes. Such a change can be expected as primary consolidation that depends on the rate of water being expelled from soil yields ground to secondary consolidation that is much slower and related to soil creep. Test data should include some overlap because drainage and the initiation of secondary consolidation should begin immediately in soil next to a drainage face.

The procedure to find $D$ is as follows and is illustrated in Table 17.2:

1. Set up a spreadsheet with $x$ and $y$ data, for example time and compression or settlement amount data in the first two columns.
2. Reserve the third column for trial values of $D$, which are entered at the top and automatically copied to other cells down the column.
3. In the fourth column calculate $\log_{10}|y - D|$ where $|y - D|$ is the absolute or positive number.
4. Plot $\log_{10}|y - D|$ from column 4 versus $x$ from column 1.
5. Visually observe linearity of the plot and vary $D$ to obtain a straight line.

For convenience the graph can be displayed on the spreadsheet to instantly display effects of changes in $D$. A simultaneous calculation of the statistical $R^2$ allows fine-tuning of $D$ to obtain the best linearity and highest $R^2$, shown at the lower right in Table 17.2. Regression coefficients $a$ and $b$ also are calculated and are used to write an equation for settlement versus time.

Figure 17.6 shows a solution for $D$ from the data in Table 17.2, and the effect of varying $D$ by $\pm 0.01$, or in this case $\pm 1.7$ percent, illustrating the excellent sensitivity of this procedure.

Error messages in Table 17.2 result from attempting to calculate the logarithm of a negative number, because the best-fit $D = 0.061$ is less than the actual final

| Table 17.2 | Time in minutes | Dial, inch | Compression $y$ in inches | Trial $D$ in inches | $|y - D|$ |
|---|---|---|---|---|---|
| Spreadsheet for application of FORE to primary consolidaton data for the 1 to 2 tons/ft² load increment in Table 16.1. For convinience a graph such as shown in Fig. 17.6 can be plotted on the spreadsheet to show changes in linearity with different ultimate trial values | 0 | 4478 | 0.0000 | 0.061 | — |
| | 0.25 | 4400 | 0.0078 | 0.061 | −1.274 |
| | 1 | 4343 | 0.0135 | 0.061 | −1.323 |
| | 2.25 | 4296 | 0.0182 | 0.061 | −1.369 |
| | 4 | 4230 | 0.0248 | 0.061 | −1.441 |
| | 6.25 | 4173 | 0.0305 | 0.061 | −1.516 |
| | 9 | 4116 | 0.0362 | 0.061 | −1.606 |
| | 12.25 | 4059 | 0.0419 | 0.061 | −1.719 |
| | 16 | 4011 | 0.0467 | 0.061 | −1.845 |
| | 25 | 3946 | 0.0532 | 0.061 | −2.108 |
| | 36 | 3900 | 0.0578 | 0.061 | −2.495 |
| | 49 | 3880 | 0.0598 | 0.061 | −2.921 |
| | 64 | 3872 | 0.0606 | 0.061 | −3.398 |
| | 100 | 3860 | 0.0618 | 0.061 | ERR |
| | 225 | 3838 | 0.0640 | 0.061 | ERR |
| | 400 | 3821 | 0.0657 | 0.061 | ERR |
| | 1440 | 3793 | 0.0685 | 0.061 | ERR |
| | | | | $R^2 =$ | 0.9992 |
| | | | | $a =$ | −0.0333 |
| | | | | $b =$ | −1.3090 |

**Figure 17.6**

$D$ values are varied to obtain a straight-line plot. Sensitivity is to <0.001 in. (<0.025 mm). For convenience in testing $D$ values this graph can be plotted directly on a spreadsheet.

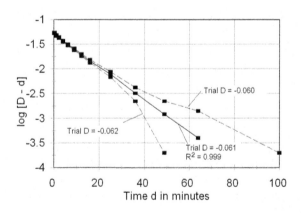

measurement of 0.0685. This indicates that there is a discontinuity in the data, which is the anticipated shift to secondary consolidation as time exceeded 64 minutes. Longer times therefore were excluded from the determination of the first trial $D$ and $R^2$, and then treated separately to obtain another $D$ end value for secondary consolidation.

### 17.4.4 Prediction Equations

After a value for $D$ has been evaluated and the regression coefficients obtained, the values are entered into eq. (17.22a) solved for $y$:

$$\log (D - y) = ax + b$$
$$y = D - 10\exp(ax + b)$$

(17.23)

The equation for primary consolidation for the data in Table 17.2 is

$$y = 0.061 - 10\exp(-0.033t - 1.309) \text{ (inches)}$$

where $y$ is the amount of compression in inches and $t$ is time in minutes. The same procedure is used with SI units. Applying the procedure to the remaining data that represents longer times gives

$$y_s = 0.069 - 10\exp(-0.0014t - 0.199) \text{ (inches)}$$

for secondary consolidation. The two respective end amounts are 0.061 and 0.069 in. (1.55 and 1.75 mm), indicating that approximately 89 percent of the final compression is attributed to primary and 11 percent to secondary consolidation. These two equations, one for primary and the other for secondary consolidation, are shown by heavy solid lines in Fig. 17.7, and show convergence and the anticipated overlap.

The time at which primary gives way to secondary consolidation is indicated to be between 50 and 60 minutes, compared with 40 minutes in Fig. 17.5.

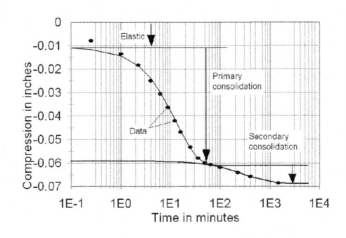

**Figure 17.7**

The solid lines show results from a FORE analysis for primary and secondary consolidation, and show close agreement with measured values.

### 17.4.5 Initial Compression

As in the other methods, there are initial deviations from the ideal equation during the first 3 minutes. This rapid response suggests elastic compression. Substituting $t = 0$ in the FORE equation gives a starting value for primary consolidation, which then may be compared with the initial zero-load specimen thickness to obtain a modulus. The most accurate determination will be if the soil has been preloaded to the overburden pressure and then unloaded, as discussed in the preceding chapter.

It should be noted that the modulus measured during compression in a consolidometer is a *confined modulus* because lateral expansion is prevented. The confined modulus generally is higher than the Young's elastic modulus where lateral expansion is allowed, but is applicable to a field compression situation under a broad loaded area.

**Example 17.4**
Calculate the confined modulus from the initial response to loading.

*Answer:* When $t = 0$, $y = 0.061 - 10$ exp $1.309 = 0.0119$ in. (0.48 mm). The specimen thickness at the beginning of this load increment (see Example 17.2) was 1.383 in. (35.1 mm), so elastic strain was $\varepsilon = 0.0119/1.383 = 0.00859$.

The load increment is $1 \, \text{ton/ft}^2$, or $13.9 \, \text{lb/in.}^2$ (95.8 kPa). Therefore $E_c = 1/0.00859 = 120 \, \text{tons/ft}^2$ or $13.9/0.00859 = 1600 \, \text{lb/in.}^2$ (11 MPa). Note that calculations carry three significant figures, but the answers are reported only to two.

### 17.4.6 Transforming Thickness and Time Scales to the Field

In order to apply laboratory findings to a field situation, the settlement scale is a simple ratio of the soil layer thickness in the field to the sample thickness in the consolidometer, $H/h$.

According to Terzaghi's theory and as in eq. (17.18), time is proportional to the square of the drainage distance, or

$$t_f = \left[\frac{H}{h}\right]^2 t \tag{17.24}$$

where $t_f$ is time in the field, $t$ is the time for a comparable amount of primary consolidation in the laboratory, and $H$ and $h$ are the corresponding field and laboratory maximum drainage distances. Thus, for a double drainage situation the multiplier for time is the *square of the ratio of field and laboratory thicknesses*, but for single drainage situation in the field *it is four times that ratio*.

**Example 17.5**

Prepare a graph showing percent primary consolidation settlement versus time for a 3 m (10 ft) thick layer of the clay represented in Table 17.2. Assume double drainage.

*Answer:* The ratio of field to laboratory thickness is $3000 \, \text{mm}/34.27 \, \text{mm} = 87.5$, so the scale factor for thickness is 87.5, and for time is $(87.5)^2 = 7660$. Both the thickness and time conversions are accomplished by adding two additional columns on the spreadsheet. A transformed graph for primary consolidation is shown in Fig. 17.8.

---

*Question:* How does this result compare with that from the Taylor method?

*Answer:* The Taylor method (Example 17.2) indicated that 95% consolidation would be completed in 7 months, which agrees with the FORE analysis. FORE indicates that 2 months will be required for 60% primary consolidation, compared with the Taylor method determination of

$$t_{60} = \frac{(1.5 \, \text{m})^2}{4.6 \, \text{m}^2/\text{yr}} \times 0.286 = 0.138 \, \text{years} = 1.7 \, \text{months}$$

---

## 17.4.7 Consolidation Test of Quick Clay

Consolidation tests and simultaneous pore pressure measurements of a quick clay by Crawford (1964), of the National Research Council of Canada, gave a puzzling result that appeared to conflict with Terzaghi's theory, as pore water pressures declined to zero before primary consolidation was complete. Might an explanation be hidden away somewhere within the clay?

A FORE analysis of the primary consolidation data showed that pore water pressures reached zero; exactly as predicted by Terzaghi theory (Fig. 17.9). A FORE analysis applied to long-term data then left a large gap shown in the figure, and a speculation that consolidation involved three instead of two stages: (1) primary consolidation during which pore water pressure, which gradually increased pressure on the structure, caused (2) a time-dependent breakdown of the structure, followed by (3) secondary consolidation as soil grains came into more intimate contact.

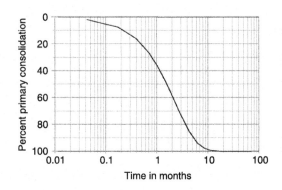

**Figure 17.8**

Transformed thickness and time scales from Fig. 17.7 for primary consolidation settlement in the field for Example 17.5.

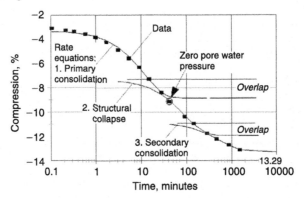

**Figure 17.9**

FORE predicts the end time for primary consolidation of a quick clay, and suggests two other stages of compression. Data are from Crawford (1964). (From Handy, 2002, with permission of the American Society of Civil Engineers.)

## 17.5 CORRECTING SYSTEMATIC ERRORS AFFECTING CONSOLIDATION TIME

### 17.5.1 Temperature Correction

Because $C_v$ is directly proportional to the coefficient of hydraulic conductivity, which in turn depends on viscosity of the pore fluid, temperature differences between laboratory and field can introduce systemic error. The groundwater temperature in the field generally reflects the mean annual temperature, so in temperate climates $C_v$ may be overestimated and time underestimated from laboratory tests. Times therefore should be corrected by multiplying by the ratio of water viscosities at the field and laboratory temperatures:

$$k_{temp} = \frac{\eta}{\eta_f} \tag{17.25}$$

where $k_{temp}$ is the temperature correction factor for primary consolidation time and $\eta$ and $\eta_f$ are respective viscosities of water in the laboratory and in the field. Viscosities are shown in Fig. 17.10.

**Example 17.6**

The mean annual temperature obtained from weather records is 58°F (14.4°C) and the laboratory temperature during a consolidation test is 72°F (22.2°C). What correction should be made to primary consolidation times?

*Answer:* Inserting respective values of $\eta$ from Fig. 17.10 gives

$$k_{temp} = 1.22/0.95 = 1.3$$

That is, primary consolidation times in the field should be multiplied times 1.3 to account for the lower viscosity of water compared with the laboratory. This correction should be

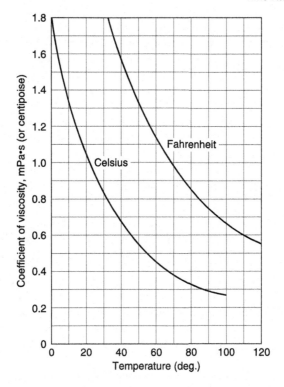

**Figure 17.10**
Viscosity of water in relation to temperature.

made regardless of the analysis method. In this case, not making the temperature correction would introduce 30% error which, as it would underestimate consolidation time, would be on the unsafe side. The correction only applies to primary consolidation where viscosity is a controlling factor, and not to secondary consolidation where chemical bonding controls. A correction to secondary consolidation rate would be on the basis of the absolute temperature, that is, the temperature range calculated relative to $-273°C$, so there is little percent change.

## 17.5.2   Drainage Rate, Varved Clays, and Vertical Drains

The consolidometer models compression of a layer of soil in the field sandwiched between two drainage layers. Some soils such as varved clays are composed of regular interlayers of clay and silt in relatively thin layers that greatly reduce consolidation time provided that there is lateral drainage—some place for the water to go.

Vertical drains frequently are installed to hasten primary consolidation settlement. These may be vertical borings filled with loose sand, called *sand drains*, or a more modern equivalent is constructed by stitching in vertical channels of plastic, called *wick drains*, with an apparatus that operates like a massive sewing machine.

Where vertical drains are used it is imperative that they not be covered up without some provision for a drainage outlet at the ground surface. If rapid drainage is anticipated, vertical drains will be connected to drain pipes that run laterally offsite. Slower drainage can be through a layer of sand that may contain perforated drain tiles for draining large areas.

The drainage rate of soil between the vertical drains is governed by horizontal instead of vertical permeability, and cannot reliably be predicted from a conventional one-dimensional consolidation test. The design of vertical drains is specialized and is beyond the scope of this text. After drains are installed, the progress of settlement can be predicted from settlement measurements and FORE. An example is given later in this chapter.

### 17.5.3   Imperfect Modeling by the Consolidation Test

A consolidometer models compression of a broad layer of soil in the field, but also incorporates side friction that does not exist in the field. Side friction is minimized by smooth sides on the inside of the consolidometer ring, which may include a Teflon liner. Side friction is eliminated in the triaxial compression test discussed in the next chapter, but specimens are much thicker, drainage times many times longer, and air pressure trapped in the pores can become significant.

A second difference between field and laboratory conditions is a much higher hydraulic gradient in the laboratory test, which could be troublesome if it results in movement of clay particles. If that occurs, clay will intrude into the porous stone disks used to confine the soil layer. The hydraulic gradient is reduced by using smaller load increments, which increases the time required for the test.

Another aspect of the modeling is lateral confining pressure, which will be reduced by lateral bulging of soil under a small or narrow foundation area, which tends to increase settlement. The effect of bulging can be duplicated in triaxial stress-path testing, which was discussed in the previous chapter. Allowable foundation pressures are based both on acceptable settlement and on bearing capacity failure, which employs a generous factor of safety to minimize lateral bulging and prevent shear failure.

### 17.5.4   Comparison of Measured and Predicted Settlements

Settlement of buildings normally is limited to less than 1 inch (25 mm), so comparing measured with actual settlement requires accurate elevation measurements to stable benchmarks. A much larger amount of settlement can be tolerated in earth dams, and one such comparison is shown in Fig. 17.11, which shows best and worst settlement predictions for points in the Ft. Randall Dam, an earth dam constructed in South Dakota on the Missouri River floodplain.

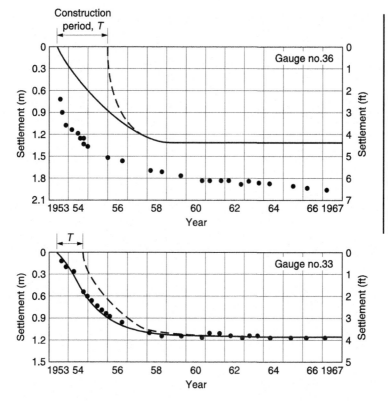

**Figure 17.11**

Best and worst predictions of settlement at Ft. Randall Dam. The main discrepancy at Gauge 36 is from immediate settlement that should not be considered as primary consolidation.

In both cases in Fig. 17.11 settlement initiated almost immediately when construction began, suggesting an initial elastic response that was not considered in calculations of primary consolidation settlement. This was not included in the settlement prediction and can explain the large discrepancy at Gauge 36, where the immediate settlement was approximately 2.3 ft (0.7 m). If that amount is added to the predicted primary consolidation settlement, the result closely agrees with the measured total settlement. Settlement at this gauge had not stopped after 14 years, indicating a contribution from secondary consolidation. Therefore, despite simplifications in the development and applications of the theory of primary consolidation, if care is used in sampling and testing, predictions can be reliable.

## 17.6 SECONDARY CONSOLIDATION

### 17.6.1 Overview

Secondary consolidation used to be like the crazy aunt in the attic: everybody knows she is up there but don't quite know how to deal with it. Unlike primary consolidation, there has been no guiding theory except to recognize that the behavior is time-related and in some ways is similar to soil creep.

## 17.6.2 Empirical Procedures

The conventional method for analyzing secondary consolidation assumes that a linear second part of a void ratio-log time curve extends indefinitely. The *coefficient of secondary consolidation* is

$$C_\alpha = \frac{\Delta e}{\Delta \log t} \tag{17.26}$$

where $\Delta e$ is the change in void ratio over the time interval represented by $\Delta \log t$. In order to calculate $C_\alpha$, strain measurements must be reduced to equivalent void ratios. $C_\alpha$ is readily evaluated from the change in void ratio during one log-time interval. A more complete discussion of this method is presented by Holtz and Kovacs (1981).

A more direct procedure is to define a "modified compression index," $C_{\alpha\varepsilon}$, which is expressed in terms of strain instead of void ratio. The modified compression index is determined by reading the change in strain over one log-time interval:

$$C_{\alpha\varepsilon} = \frac{\Delta \varepsilon}{\Delta \log t} \tag{17.27}$$

In eq. (17.27), $\Delta \varepsilon$ is the increment of strain over the time interval represented by $\Delta \log t$.

### Example 17.7
According to eq. 17.27 and the data in Fig. 17.5, (a) what is the modified compression index, and what will be the percent increase in strain (b) 1 year and (c) 10 years after primary consolidation is completed?

*Answer:* (a) The change in strain over one log time interval in Fig. 17.5 gives $C_{\alpha\varepsilon} = 0.007$.

(b) Equation (17.26) is rearranged to solve for $\Delta \varepsilon$. If compression is taken as positive, after 60 minutes or 1 hour and at the end of primary consolidation $\varepsilon_0 = 0.06$. One year = 8160 hours, so the change in $\Delta \log t = (\log 8160 - \log 1) = 3.94$. Multiplying times $C_{\alpha\varepsilon}$ gives $\Delta \varepsilon = 0.028$, which added to $\varepsilon_0 = 0.06$ gives strain after one year, $\varepsilon = 0.088$.

(c) After 10 years, $\Delta \log t = (\log 81,600 - \log 1) = 4.91$, giving $\Delta \varepsilon = 0.034$. Total strain is $\varepsilon = 0.094$.

Are these answers reasonable? They indicate that after 1 year 32% of the strain is from secondary consolidation, and after 10 years, 35%. These estimates appear to be excessive, and do not take into account a probable decline in $C_{\alpha\varepsilon}$ with time.

### 17.6.3 Characteristics of Secondary Consolidation

An arguable "top 10" features that may help to explain secondary consolidation are as follows:

1. There is no relation between rate of secondary consolidation and distance to a drainage face. However, this does not preclude a relationship to layer thickness.

2. According to the model shown in Fig. 17.2, primary and secondary consolidation should begin simultaneously at a drainage face, but primary consolidation will dominate until nearly all excess pore water pressure has dissipated.

3. It usually is assumed that $C_\alpha$ remains constant with time, but field and laboratory data indicate that it decreases with time (Mesri and Godlewski, 1977).

4. Data on different soils indicate that the ratio of $C_\alpha/C_c$ is approximately constant, in the range 0.03–0.06. The more compressible the soil, the higher the rate of secondary consolidation.

5. Secondary consolidation is not ideally plastic behavior because it involves volume change.

6. Secondary consolidation ejects water and therefore must induce a hydraulic gradient, however small, or the water would not move.

7. The slow rate of secondary consolidation may reflect a higher viscosity of water loosely held in the electrical double layer. That being the case, eqs. (17.6) ff. may still apply but with a much higher value of $k$.

8. The logarithmic relationship for $C_\alpha$ indicates that the modulus increases linearly with time. This may be one aspect of "aging."

9. Secondary consolidation increases the apparent preconsolidation pressure by shifing the $e$-log $P$ curve to the right. From studies of Pleistocene clays, Bjerrum(1954), of the Norwegian Geotechnical Institute, showed a relation to the logarithm of time. This aspect of aging may help explain the scarcity of normally consolidated soils.

10. Secondary consolidation settlement of structures still can be troublesome decades or even centuries after construction, and in thick clay strata eventually may exceed settlement from primary consolidation. A saving feature is that many structures are built with only a 50- to 100-year life expectancy.

### 17.6.4 Application of the FORE to Secondary Consolidation

Some of the difficulties of predicting secondary consolidation may be managed with FORE. Data after the end of primary consolidation are included in Table 17.3, and the corresponding linear relationship for $D_s = 0.0686$ in. (1.74 mm) is shown in Fig. 17.12.

**Table 17.3**

Spreadsheet for secondary consolidation for data in Table 17.2

| Time in minutes | Dial, inch | Compression in inches | Trial $D$ in inches | Log $(D-d)$ |
|---|---|---|---|---|
| 64 | 3872 | 0.0606 | 0.0686 | −2.097 |
| 100 | 3860 | 0.0618 | 0.0686 | −2.167 |
| 225 | 3848 | 0.0630 | 0.0686 | −2.252 |
| 400 | 3821 | 0.0657 | 0.0686 | −2.538 |
| 1440 | 3793 | 0.0685 | 0.0686 | −4.000 |
| | | | $R^2 =$ | 0.9984 |
| | | | $a =$ | −0.0014 |
| | | | $b =$ | −1.9902 |

**Table 17.4**

Some representative values for soil modulus $E$, after Kulhawy (1991)

| Soil | $E$, tons/ft$^2$ | MN/m$^2$ |
|---|---|---|
| Sand | | |
|   Loose | 50–200 | 5–20 |
|   Medium | 200–500 | 20–50 |
|   Dense | 500–1000 | 50–100 |
| Clay | | |
|   Soft | 25–150 | 2.5–15 |
|   Medium to stiff | 150–500 | 15–50 |
|   Very stiff to hard | 500–2000 | 50–200 |

**Figure 17.12**

FORE prediction of secondary consolidation end value for data of Fig. 17.7.

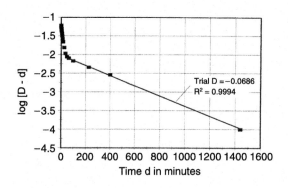

In Table 17.2 the 24-hour reading of 0.0685 in. is comparable with the projected final predicted $D = 0.0686$ in.

### 17.6.5   Transforming the Secondary Consolidation Time Scale

As in the case of primary consolidation, the time involved in secondary consolidation in the field is larger by orders of magnitude than in the laboratory, so a time transformation is necessary in which drainage rate is not the governing factor. A consolidometer test is a scale model, so the problem may be examined using principles of similitude.

The scale factor is

$$n = \frac{H}{h} \tag{17.28}$$

where $H$ is the soil layer thickness in the field and $h$ is the initial thickness of the sample tested in the laboratory. The soil itself is not scale modeled because the grain size is the same in the laboratory test as in the field, but it is more important that other physical properties, including clay mineralogy, particle interference, and amount, viscosity, density, and surface tension of the pore water, should be the same in the laboratory as in the field. As the model is scaled downward the hydraulic gradient is scaled upward, but that should influence only primary consolidation.

A model can be developed on the basis of dimensionless "pi terms" that are equal in the model and in the prototype. The Reynolds number is a pi term that is used to differentiate between lamellar and turbulent fluid flow. As the Reynolds number is defined on the basis of viscosity and, as shown in the next chapter, viscous behavior appears to be involved in soil creep and secondary consolidation, the number can be written for the field prototype (on the left) and for the model (on the right):

$$\frac{\rho v r}{\eta} = \frac{\rho_m v_m r_m}{\eta_m} \tag{17.29}$$

In this expression $\rho$ and $\rho_m$ represent fluid density, $v$ and $v_m$ velocity, $r$ and $r_m$ a corresponding length term, and $\eta$ and $\eta_m$ fluid viscosity. Because fluid density $\rho$ and viscosity $\eta$ are the same in the model and in the prototype, velocity and length terms are the remaining variables:

$$v r = v_m r_m$$

The length term can be any linear measure that distinguishes between the model and the prototype, and therefore may be represented by the respective layer thicknesses, $h$ and $H$. Substituting $H/t$ for $v$ and $H$ for $r$ and their equivalents in the model gives

$$\frac{H^2}{t} = \frac{h^2}{t_m}$$

where $t$ is time. Transposing,

$$\frac{t}{t_m} = \frac{H^2}{h^2} = n^2 \tag{17.30}$$

**Figure 17.13**

Laboratory consolidation test data of Fig. 17.7 transformed to the field prototype scale.

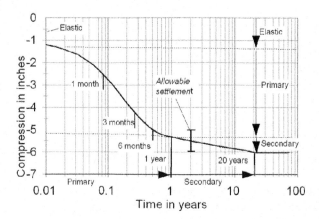

This is the same result as that obtained by Terzaghi based on drainage distance. It suggests that field time for secondary consolidation equals model time multiplied by the scale factor squared, which is the thickness ratio squared.

The assumption of equal Reynolds numbers implies that viscosity is a controlling factor—not the viscosity of water but of the soil-water complex. If surface tension were the controlling factor, which might be the case in an unsaturated soil, the Weber number would be more appropriate and would give a time ratio of $n^3$. If compressibility were the control, as in the case of gases, the Cauchy number (which is the square of the Mach number) would gives a time ratio of $n$. If no fluid behavior were involved, $n$ would equal 1, which observation shows is not the case. Squaring the scale factor therefore appears to be the most likely scenario, even though this will distort the scaling of surface tension and compressibility if the soil contains any undisolved air.

Use of the same time transformation for both primary and secondary consolidation simplifies the prediction procedure, and is the basis for Fig. 17.13, with primary consolidation settlement predicted to be essentially complete after 1 year, and secondary after 20 years.

## 17.6.6  A Rational Basis for the Time Transformation

Results from a dimensional analysis, like Einstein's time-space relationship, sometimes appear to come from another world. The basis for Terzaghi's time transformation is that it takes longer for fluid to flow through a long pipe than through a short one, although without the mathematical derivation one might intuitively (and incorrectly) assume a linear relationship. Might there be a rational explanation for the time transformation for secondary consolidation?

One possibility might relate to probability and a "weakest link first" hypothesis. This is a concept that the weakest grain contacts will slip first, thereby transferring stress to adjacent grain contacts so that they will be more disposed to slip next. However, secondary consolidation is a viscous phenomenon, so each slip eats up time. In a thin layer of soil, slipping may more readily become a joint effort than in a thicker layer.

### 17.6.7 Applications of FORE for Surcharge Loading

Figure 17.13 predicts a total final settlement of 6 inches (150 mm), which is unacceptable for most buildings where total settlement typically is limited to a maximum of 1 inch (25 mm). The usual options are to go to an intermediate or deep foundation system, discussed in later chapters, or to use a temporary surcharge to induce settlement prior to construction. The compelling question then is how long should a surcharge be left in place?

If a full surcharge is applied, the answer is as shown in Fig. 17.13. For example, if 1 inch (25 mm) is allowable, this can be subtracted from the total and the time read from the graph. In Fig. 17.13 only 6 months will be required and will mainly involve primary consolidation, which means that it could be speeded up by use of vertical drains.

The effect of a surcharge load that is less than the anticipated foundation load can be assessed from the $e$-log $P$ curve. Even a small reduction in load can result in an unsatisfactory performance because the desired void ratio may not be attainable, or if it is, time-consuming secondary consolidation may be required to accomplish the same settlement as will occur during primary consolidation under the foundation load.

### 17.6.8 Adjustment for a Long Construction Time

A long construction time, as for an earth dam, can be represented by compressing the time-settlement prediction curve vertically to represent one-half of the final settlement. The starting time is when construction is estimated to be at 50 percent completion. The curve then is duplicated and moved to a second starting time at 100 percent completion. The curves are plotted to a linear time scale and connected by a common tangent, and a tangent is drawn from the first curve to the origin. The tangents then simulate additions of load that are linear with time during the construction period. An example is shown in Fig. 17.14.

## 17.7 MONITORING SETTLEMENT

### 17.7.1 Overview

As indicated in Fig. 17.14, the field performance has the last word, and as settlement is monitored it may be necessary to revise expectations. A prediction method by Asaoka (1998) involves settlement measurements made at regular

**Figure 17.14**

Settlement predictions and measurements for Ft. Randall Dam adjusted for the construction schedule.

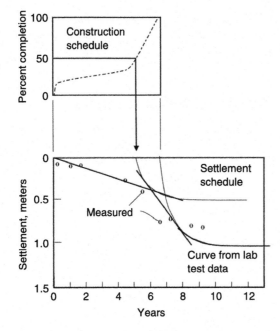

intervals. The amount of settlement at the end of each interval is plotted versus the amount at the end of each *preceding* interval. The point at which settlement will stop is defined by a line at 45°, so an intersection with the extrapolated data line defines the time and amount of expected total settlement. A difficulty with the application of this method is because the two lines are nearly parallel, an intersection is not precisely defined.

## 17.7.2  FORE in the Field

The Kansai artificial island airport in Osaka Bay, Japan, afforded an opportunity to test FORE, particularly as the upper supporting strata were drained with vertical sand drains while the lower strata were too deep for economical installation.

The FORE treatment to predict final settlement, $D$, at the location of maximum settlement is shown in Fig. 17.15. A steel plate was located at the contact between the two soil layers and attached to a pipe extending above the ground surface to differentiate between compression to the two layers. The FORE settlement predictions for the two layers is shown in Fig. 17.16, and shows rapid compression of the upper layer and a much slower compression of the lower layer, confirming the benefit from the drains. The FORE analysis indicated that the upper layer that contained the drains quickly went into secondary consolidation. Of the predicted final total settlement, 40 percent will come from compression of the upper layer and 60 percent by compression of the lower layer. As the lower layer is a stiffer

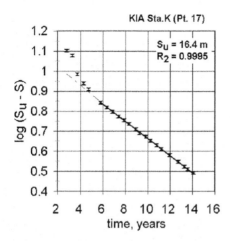

## Figure 17.15

FORE fit to settlement measurements at the artificial island supporting the Kansai International Airport, Osaka Bay, Japan (Handy, 2002). Progresssive loading during the first 5 years was not included in the analysis. Settlement data are courtesy of Professor Koichi Akai, University of Kyoto, Japan. (Reproduced with permission of the American Society of Civil Engineers.)

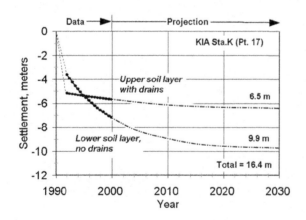

## Figure 17.16

Data and projected total settlement from compression of two soil layers at the Kansai International Airport, Osaka Bay, Japan (Handy, 2002). (Reproduced with permission of the American Society of Civil Engineers.)

Pleistocene deposit, the amount of its compression was not fully anticipated in the original design.

Additional fill will be required in order to maintain freeboard, but an additional load will reinitiate settlement. The amount of fill necessary can be estimated from the ratio of the existing fill thickness to the total predicted settlement. This ratio is almost exactly 2:1, indicating that 10 m (30 ft) of fill will be required to add 4 m (15 ft) of freeboard to the low spots.

The Kansai airport is considered an outstanding feat of engineering at a very difficult construction site, and a second airport is under construction.

## Problems

17.1. The time factor $T$ in Table 17.1 is 0.848 for 90% consolidation. Where does this come from? Show the calculation.

17.2. Explain why $H$ sometimes equals the thickness of a consolidating layer and sometimes is one-half of the layer thickness.

17.3. A large mat foundation will support a supermarket. Explain how incorporating a layer of sand between the foundation slab and the soil might influence the rate of settlement and the amount of differential settlement. What precautions are necessary to ensure that the sand layer will be effective?

17.4. Determine $C_v$ by the Taylor method from time-settlement data for the load increment 0.5 to 1 ton/ft$^2$ in Table 16.1.

17.5. (a) Repeat Problem 17.4 for load increments 2 to 4 and 4 to 8 tons/ft$^2$, and (b) plot $C_v$ versus the average consolidating pressure. Explain any observed trends.

17.6. (a) Make a sketch showing a method to accelerate primary consolidation of clay soil under a highway embankment. (b) What two kinds of measurements can be made to indicate the progress of primary consolidation and show when it is completed?

17.7. (a) What can be done to accelerate secondary consolidation? (b) What measurements should be made to show progress and indicate when settlement is completed?

17.8. Repeat Problem 17.4 using the log-time method. How does the answer compare with that from the square root of time method? Which method do you recommend?

17.9. In a consolidation test, the void ratio of the soil decreases from 1.239 to 1.110 when the pressure is increased from 192 to 383 kPa (2 to 4 tons/ft$^2$). If the coefficient of permeability of the soil at this increment of pressure is $8.4 \times 10$ mm/s, determine the coefficient of consolidation, in square meters (feet) per year.

17.10. A foundation 6.1 by 12.2 m (20 by 40 ft) in plan is to be constructed at a site where the geological profile is as shown in Fig. 17.17. The foundation load is 287 kPa (3 tons/ft$^2$) and the unit weights of various soils are as shown in the diagram. Consolidation tests of the compressible clay indicate that it has a void ratio of 1.680 at 95.8 kPa (1 ton/ft$^2$) and a compression index of 0.69. The coefficient of consolidation is $8.3 \times 10^{-4}$ cm$^2$/s.

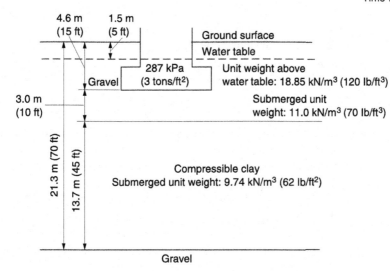

**Figure 17.17**
Geological profile
for problem 17.10.

Calculate the total settlement, and plot the rate of settlement at the center of the foundation, in millimeters (inches) versus years.

17.11. Assume that the material beneath the compressible clay in Problem 17.10 is impervious rock instead of gravel. Calculate and plot the rate of settlement.

17.12. The foundation described in Problem 17.10 is to be investigated by the stress path method, with a soil sample from the middle of the compressible layer directly under the center of the loaded area subjected to a loading sequence simulating that which will occur in the field. Assume that $K_0 = 0.4$, that is, horizontal stress is 0.4 times the vertical stress before and after the foundation load is imposed. Define vertical and lateral stress conditions for the test.

17.13. (a) Set up a spreadsheet such as Excel or QuatroPro and use FORE to predict final primary and secondary consolidation amounts from the 16 to 32 ton/ft$^2$ load increment in Table 16.1. (b) Determine the regression coefficients and plot a graph of both contributors to settlement versus time using a logarithmic scale for time. Is there any apparent overlapping between primary and secondary consolidation? (c) What percentage of final consolidation can be attributed to elastic compression, and to primary and secondary consolidation?

17.14. (a) Plot compression amounts versus time for data in the 8 to 16 tons/ft$^2$ load increment in Table 16.1. (b) Select a representative average sample thickness during this time interval. (c) Apply a temperature correction, 16°C in the field and 21°C in the laboratory. (d) Calculate thickness and time scale factors for a soil layer 2.5 m thick with drainage in one and in two directions. (e) Replot the data to predict settlement in the field with both of these drainage conditions.

17.15. Using 24-hour compression data in Table 16.1, predict a minimum void ratio attainable by increasing pressure to a maximum that can be sustained by the soil grains without breaking. Is this amount reasonable? How may this compare with typical void ratios in shale?

17.16. Sketch a line through the data points representing secondary consolidation in Fig. 17.10, select four or five representative amounts of settlement, and perform a FORE analysis to predict final settlement.

17.17. A warehouse building on linear footings on clay is stable for 20 years after construction, and then appears to settle sufficiently to affect the operation of forklifts. (a) Can this be explained by consolidation theory, and if so, how? (b) What other reasons might contribute to a sudden increase in settlement? (c) Suggest another explanation for cracking, related to soil mineralogy and moisture content, and suggest how the true cause can be identified.

17.18. The plan for Kansai International Airport was to use 18 m of fill plus an extra 12 m to allow for settlement, making 30 m total fill. It appears that this will result in as much as 16.4 m of settlement. Draw a sketch showing how much additional fill will be required in the low areas to bring the elevation up to the originally planned grade.

17.19. In the problem of Example 17.9 the owner requests that the surcharge load be reduced by one-half. How will this affect the time-rate of settlement?

17.20. As an illustration of the unlimited scope of geotechnical engineering, apply FORE to men's world track records for 100 m and project an ultimate minimum time.

| 1912 | 10.6 seconds | Donald Lippincott, U.S. |
|------|--------------|--------------------------|
| 1921 | 10.4 | Charley "World's Fastest Human" Paddock, U.S. |
| 1930 | 10.3 | Eddie Tolan, U.S. |
| 1936 | 10.2 | Jesse Owens, U.S. (Berlin Olympics. Hitler left early.) |
| 1956 | 10.1 | Willie Williams, U.S. |
| 1960 | 10.0 | Armin Hary, W. Ger. |
| 1968 | 9.9 | James Hines, U.S. |
| 1991 | 9.86 | Carl Lewis, U.S. |
| 1994 | 9.85 | Leroy Burrell, U.S. |
| 1996 | 9.84 | Donovan Bailey, U.S. |
| 1999 | 9.79 | Maurice Green, U.S. |
| 2001 | 9.78 | Tim Montgomery, U.S. |
| 2005 | 9.77 | Asafa Powell, Jamaica |

# References and Furthur Reading

Akai, K. and Tanaka, Y. (1999). "Settlement Behavior of an Off-shore Airport KIA." In Barends et al., ed., *Geotechnical Engineering for Transportation Infrastructure.* Balkema, Rotterdam.

Bjerrum, L.(1954). "Geotechnical properties of Norwegian marine clays." *Geotechnique,* 4. 49–69.

Casagrande, A. (1936) "The Determination of the Pre-consolidation Load and its Practical Significance." *Proc. 1st Int. Conf. on Soil Mechanics and Foundation Engineering* 3, 60.

Crawford, C. B. (1964). "Interpretation of the Consolidation Test." *ASCE J. Soil Mech. and Foundation Eng. Div.* 90(5), 85–102.

Gibbs, H. J. and Bara, J. P. (1967). "Stability Problems of Collapsible Soils." *ASCE J. Soil Mech. and Foundation Eng.Div.* 93(SM4), 577–594.

Handy, R. L. (2002). "First-Order Rate Equations in Geotechnical Engineering." *ASCE J. Geotech. Geoenviron. Eng.* 128(5), 416–425.

Holtz, R. D. and Kovacs, W. D. (1981). *Introduction to Geotechnical Engineering.* Prentice-Hall, Englewood Cliffs, N. J.

Janbu, N. (1969). "The Resistance Concept Applied to Deformations of Soils." *Proc. 7th Int. Conf. on Soil Mechanics and Foundation Engineering* 1, 191–196.

Jumikis, A. R. (1962). *Soil Mechanics.* D. Van Nostrand Co., Inc., Princeton, N. J.

Kansai International Airport Co., Ltd. (2001). "Kansai International Airport (KIX) Brief on Land Settlement," http://www.kiac.co.jp/english/default.htm

Kulhawy, F. H. (1991). "Drilled Shaft Foundations." In H. Y. Fang, *Foundation Engineering Handbook,* 2nd ed. Van Nostrand Reinhold, New York.

Lambe, T. W. (1967). "Stress Path Method." *ASCE J. Soil Mech. and Foundation Eng. Div.* 93(SM6), 309–331.

Lawson, G. (1997). *World Record Breakers in Track and Field Athletics.* Human Kinetics, Champaign, Ill.

Li, K. S. (2003). "Discussion of 'First-Order Rate Equations in Geotechnical Engineering.'" *ASCE J. Geotech. and Geoenviron. Eng.* 128(8), 778–780.

Mesri, G. and Godlewski, P. M. (1977). "Time and Stress-Compressibility Interrelationship." *ASCE J. Geotech. Eng. Div.* 103(Gt5), 417–430.

Mitchell, J. K. (1993). *Fundamentals of Soil Behavior,* 2nd ed. John Wiley & Sons, New York.

Spangler, M. G. (1968). "Further Measurements of Foundation Settlements in the Ft. Randall Dam Embankment." *Highway Research Record* 223, 60–62.

Taylor, D. W. (1948). *Fundamentals of Soil Mechanics.* John Wiley & Sons, New York.

Terzaghi, K. (1943). *Theoretical Soil Mechanics.* John Wiley & Sons, New York.

# 18

## Soil Shear Strength

## 18.1 OVERVIEW

### 18.1.1 Letting Go

Soil strength is largely a matter of resistance to shearing, whether the soil is bulging out from under one side of a foundation or sliding off down a hill. Soil shears as individual particles slide or roll past one another, so an important objective of design is to limit shearing stresses so that slipping is discontinuous and involves only densification. Some examples of shear failures are shown in Fig. 18.1, where it will be seen that the shear surfaces are continuous and are inclined at an angle from the axis of compression.

The load tending to cause shearing may be external, as in Fig. 18.1(b) and (c), or it may come from the weight of the soil itself. As a load increases, sliding at the weakest contacts transfers part of the stress to adjacent particles so that they also slide, until a continuity is achieved that defines a slip surface. Shearing automatically seeks out the most critical combination of stress direction and sliding resistance, and therefore finds the most critical surface.

Because it must support its own weight, shearing stresses exist everywhere within soil even though the soil is stable. The maximum available shearing resistance is the shearing strength, which enables soil to remain in place on a hillside or in an embankment, levee, or earth dam. Shearing strength also reduces soil pressure against retaining walls, and is responsible for the bearing capacity of foundations and piles.

### 18.1.2 Shearing and Normal Stresses

A major component of soil shearing strength is friction between individual soil grains, which in turn depends on the normal forces pushing the soil grains

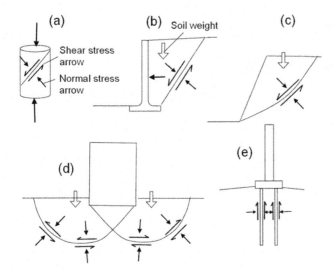

**Figure 18.1**

Some applications of soil shearing strength: (a) measured in an unconfined compression test; (b) reducing soil pressure against a retaining wall; (c) holding up a hill; (d) holding up a building; (e) holding up a pile foundation. Open arrows indicate soil weight, half arrows friction, full arrows normal stress on the shear surface.

together. Both kinds of forces are shown in Fig. 18.1. Both act over discrete areas. Stress is a force per unit area, so the larger the force or the smaller the area on which it acts, the higher the stress.

### 18.1.3   Slip Zones

A chain breaks at the weakest link, but soil grains spread the bad news by passing off part of the load to the neighbors. Even after the neighborhood concedes and a continuous slip surface develops, as sliding continues it picks up and involves more soil grains and a slip surface becomes a *slip zone*. Generally, the farther the distance of sliding, the thicker the slip zone. Landslides develop slip zones that are millimeters to a meter or more in thickness.

### 18.1.4   Friction

Much of the resistance to sliding comes from friction between soil grains. Because the friction is developed internally in a soil mass, it is called *internal friction*. Internal friction in soil not only depends on sliding friction but also includes other sources of resistance such as soil grains getting in the way of one another. Before addressing those complications, let us review the nature of sliding friction.

## 18.2 SLIDING FRICTION

### 18.2.1 Amontons' Law

In the seventeeth century a French physicist, Guillaume Amontons, discovered a basic law of sliding friction:

$$F = Nf \tag{18.1}$$

where $F$ is a resisting force, $N$ is the normal force perpendicular to the slip plane, and $f$ is the coefficient of sliding friction. Dividing by a contact area $A$ gives the equivalent expression in terms of stresses:

$$\frac{F}{A} = \frac{N}{A}f$$

In soil mechanics this formula is expressed in terms of stresses as

$$\tau_f = \sigma_n f \tag{18.2}$$

where $\tau_f$ (tau-sub f) is the conventional notation for shearing stress at failure, $\sigma_n$ (sigma-sub n) is the normal stress on the slip plane, and $f$ is the coefficient of sliding friction. These two stresses are shown in Fig. 18.2(c).

For reasons that will become apparent later, the coefficient of friction, $f$, usually is expressed in terms of a *friction angle*, $\phi$ (phi, pronounced "fee") as follows:

$$f = \tan \phi \tag{18.3}$$

Combining the two formulas gives

$$\tau_f = \sigma_n \tan \phi \tag{18.4}$$

where $\tau_f$ is the shearing stress at failure, or the shearing strength, $\sigma_n$ is the stress normal to the slip surface, and $\phi$ is the friction angle. This is the basic formula for the shearing strength of soils such as sands that derive their strength from friction.

### 18.2.2 Obliquity

Obliquity comes from the word "oblique," which means at an angle. In engineering mechanics, obliquity refers to the angle between the directions of a normal and a shearing stress along a particular surface. The *angle of obliquity* is designated by $\alpha$ (alpha), and is shown in Fig. 18.2(b). If the obliquity angle is zero, all stress is normal to the potential slip surface, and shearing stress is zero. As shearing stress increases, the obliquity angle also increases until sliding occurs, at which point the maximum obliquity angle equals the friction angle. The stress polygon in Fig. 18.2(d) shows the relationship between normal and shearing stresses, the coefficient of friction, and the friction angle, $\phi$.

The term "friction angle" is preferred over "coefficient of friction" for soil because, as a particulate material, the resistance to internal slippage involves other factors besides sliding friction. One of these is interlocking, which occurs between individual grains and depends on the soil density and gradation.

---

*Question:* If the block in Fig. 18.2 is placed on end, how will that affect sliding friction? *Hint:* Note that eq. (18.1) defines a ratio of *forces*, not stresses. Dividing both by a contact area $A$ does not change the ratio regardless of the size of $A$. It was this mind-bending observation that led to Amontons' Law.

---

### 18.2.3    Sliding Friction Angle along an Inclined Surface

If a block is placed on a board and the board tilted as shown in Fig. 18.3(a), an angle will be reached where the block slides off. From common perpendiculars it will be seen that the angle at which sliding initiates is the same as the friction angle, $\phi$.

### 18.2.4    Angle of Repose

Another common example of friction angle is the sloping surface of a pile of sand, gravel, coal, grain, or other particulate material that has been dropped on top of

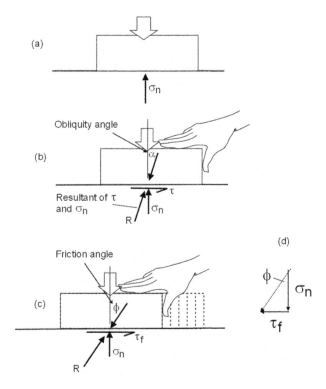

**Figure 18.2**

As a pushing force increases, the angle of obliquity $\alpha$ increases to a maximum and the block slides. That maximum angle is the *friction angle,* $\phi$. The equivalent angle in soils is called the *angle of internal friction.*

**Figure 18.3**

(a) A table tilted sufficiently for the block to slide will be inclined at the friction angle, $\phi$.
(b) Particulate materials dropped on a pile tend to flow out to the *angle of repose*, which equals the angle of sliding friction of the surface material.

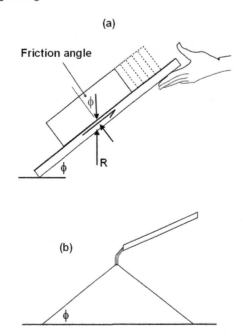

the pile. In this case the maximum tilt angle of Fig. 18.3 becomes the *angle of repose* for individual particles. The sand trickling down in an hourglass comes to rest at the angle of repose. If one disregards the possibility for interlocking, the angle of repose equals the angle of sliding friction for the materials involved.

---

*Question:* What is the friction angle of a liquid such as water?

---

### 18.2.5 Adhesion Theory of Sliding Friction

Friction is such a universal commodity that few people ask why it happens so long as they can trust that it happens. Friction keeps the car on the road and, when called upon, brings it to a stop. What may be the explanation for Amontons' Law? Why is friction proportional to normal stress? Why is friction reduced by a lubricant, whether it is an oil or a banana peel?

A theory of friction proposed for soils by Terzaghi in 1925 has since been demonstrated to apply to many other materials including metals (Bowden and Tabor, 1950, 1964). This theory suggests that, as illustrated in Fig. 18.4(a), apparently smooth surfaces in contact actually touch each other at only a few tiny areas. These areas, called asperities, temporarily bond chemically by van der Waals forces.

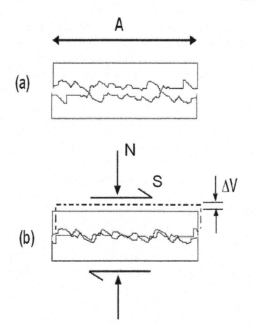

**Figure 18.4**
(a) Actual areas of contact are very small and are not known, so normal and shearing stresses are expressed on the basis of total area, *A*.
(b) Interlocking of rough surfaces leads to a dilatant increase in volume, $\Delta V$, during shearing.

As surfaces are pushed together by a normal force, the contact areas grow proportionately, which causes a proportionate increase in frictional resistance. This simple theory is supported by indirect evidence such as electrical conductivity through the contacts. In elastic materials when the normal stress is relieved, the asperities spring back, bonding is relieved, and friction decreases. However, in soft materials such as clays, rebound is incomplete so part of the resistance to sliding is more permanent.

This theory also can explain how a lubricant such as an oil works, as it prevents surface contacts and bonding. If a lubricant fails at this function, for example in a metal bearing, the bearing "seizes" as asperities grab and weld to one another. Non-roller bearings incorporate dissimilar metals such as steel against bronze, or against a lead alloy or Nylon, in order to discourage this bonding.

### Is Water a Lubricant?
The tendency for landslides to develop during wet seasons can lead to the assumption that water must be an effective lubricant, but that is not the whole story. Water can decrease the shearing resistance of clays by increasing the thickness of the electrical double layer surrounding each particle. However, in non-clay minerals water is an *antilubricant*, bonding to mineral surfaces. The same forces draw water up into a capillary tube, or into soil. The solids themselves also have a very high surface tension, meaning that a large amount of energy is required to create more surface area, and by weakly bonding to the mineral surface, water decreases the surface tension of the solid. Therefore wetting glass

with water makes it easier for the surface to be scratched. A glass "cutter" actually is a glass scratcher, and is more effective if the glass surface is wet.

Water does not lubricate landslides any more than they are lubricated already, because in most cases the slip surface is permanently wet. It is the water *pressure* that acts as a lubricant by carrying part of the normal stress. It is a rise in groundwater level that buoys up the soil, reduces normal stress and friction, and can trigger a landslide.

### 18.2.6  A Simplification

According to the adhesion theory of friction, both normal and shearing forces are concentrated on tiny asperities, as shown in Fig. 18.4(a). However, because the true contact area is virtually impossible to evaluate, stress in soils is expressed on a *total area basis*. Thus the stresses in Fig. 18.4(a) are the normal force $N$ and shearing force $S$ divided by area $A$. The actual contact stresses between hard materials may be many thousands of times higher than the normal stress expressed on a total area basis. This convention therefore is "factitious," or made up, and will be discussed further relative to the influence of pore water pressure acting on shear surfaces.

## 18.3  DILATANCY: A SECOND COMPONENT OF INTERNAL FRICTION

### 18.3.1  Glass vs. Sandpaper

Sliding friction is not the only component of internal friction in soils, because if one places two sheets of sandpaper in contact with sand against sand, the resistance to sliding is markedly increased. In fact, the only way that sliding can occur is if grains lift up and slide over one another, as shown in Fig. 18.4(b). This is described as *interlocking* of the grains. This same phenomenon occurs internally in dense sands: grains that interlock must be lifted over one another in order to shear. The result is an increase in volume during shearing, which was named *dilatancy* in 1885 by an English engineer and physicist, Osborn Reynolds—the same Reynolds who also used dimensional analysis to define the Reynolds number used in fluid mechanics.

### 18.3.2  Dilatant Contribution to Friction

Dilatancy requires expansion against an applied normal stress, so the dilatant response to sliding mimics sliding friction because it also is proportional to normal stress. Thus a dilatant component automatically adds to a measured friction angle. Then as sliding continues, the shear zone eventually achieves a reduced density that is stable at what is known as the *critical void ratio*. Without a dilatant contribution, the friction angle reduces and represents sliding friction.

Dilatant behavior is a goal of soil compaction, which pushes soil grains into voids and increases the expansion required for shearing. Dilatancy also is enhanced by a uniform gradation that decreases the volume of voids, as coarse particles form a skeleton and voids are filled by progressively finer particles.

As density and gradation improve, the friction angle increases. For example, the friction angle of sand without a dilatant component is approximately 25°, which is the angle of sliding friction for quartz. With densification and a broad gradation it can become 35° to 40° or even higher if the grains are angular. Shearing resistance increases as the tangent of $\phi$, so as a friction angle increases from 25° to 35°, an increase of 40 percent, internal friction increases 50 percent at the same normal stress.

Taylor (1948, p. 346) suggested that the dilatant contribution equals the work done against the normal stress as shearing progresses. This later was elaborated by Rowe et al. (1964):

$$\phi = \phi_s + \frac{\Delta V/V}{\Delta S/S} \tag{18.5}$$

where $\phi$ is the angle of internal friction, $\phi_s$ is the angle of sliding friction, and $\Delta V/V$ and $\Delta S/S$ respectively are the changes in volume per unit volume, and a unit shearing deformation.

The ratio $\Delta V/V \div \Delta S/S$ equals the slope of a sample thickness-shear displacement curve, and means that, as long as the volume is increasing during shearing, there is a dilatant contribution to the friction angle. The reverse also can be true—if during initial shearing of loose sand there is a *decrease* in volume, negative dilatancy *subtracts* from the friction angle. These relationships are demonstrated in the next section, where the term "dilatancy" is preferred as being more definitive than "interlocking."

## 18.4 DIRECT SHEAR TESTS

### 18.4.1 A Direct Approach to Soil Shear Strength

A direct shear test provides a simple means for measuring soil internal friction, and as can be seen in Fig. 18.5, is not unlike the sliding block experiment of Fig. 18.2. In fact, in 1846 in a classic study of landslides, Alexandre Collin reported experiments in which a vertical stress was varied on blocks of clay that then were slid relative to one another, and he presented graphs that showed friction angles.

The earliest shear box tests appear to have been conducted by Marston and Anderson (1913) in connection with their studies of loads on buried pipe. In 1919 Arthur Bell described a direct shear type test using a segmented tube to hold a soil

**Figure 18.5**

Diagram of a direct shear test apparatus. The vertical dial gauge measures dilatant expansion, which affects the shearing strength.

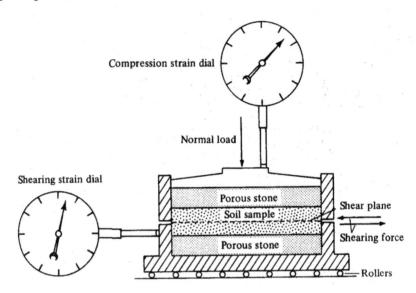

sample that was confined at the ends (Skempton, 1985). In 1921 Terzaghi independently devised a direct shear apparatus for his seminal studies at Robert College in Turkey (Altay Birand, personal communication).

In a direct shear test a constant normal stress is applied vertically to soil held in a shear box and the soil allowed to consolidate; then a shearing stress is applied by pulling or pushing one-half of the shear box relative to the other half. The normal stress is held constant and the shearing stress increased until the soil fails.

## 18.4.2 Stress-Deflection Curves

In a direct shear test one part of the shear box is pushed at a constant rate relative to the other with a screw mechanism, and the shearing resistance is monitored. The shearing force therefore gradually increases, reaches a peak value, and then decreases to a stable value that represents an *ultimate* or *residual* strength. Dividing the normal load and shear force by the cross-sectional area of the specimen at the shear plane gives the normal stress and shearing stresses at any stage of the test. These values may be plotted versus displacement on a shear diagram. Since the normal stress is held constant during a test, another option is to plot the stress ratio, $\tau_f/\sigma_n$, as shown in the upper graph of Fig. 18.6.

In Fig. 18.6 the dashed upper line is for dense sand. With increasing shearing displacement there is a high peak strength indicating a contribution from dilatancy, and then as shearing continues the strength decreases to a residual strength. This interpretation is confirmed in the lower graph, which shows volume

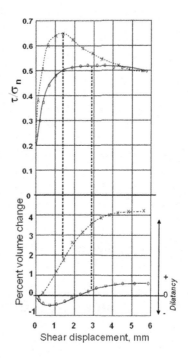

**Figure 18.6**
Two direct shear tests on Ottawa sand, one loose and other compacted. Upper graph shows shearing resistance, lower graph the volume change. Peak strengths on the upper graph occur where the slope of the volume change curve is steepest. (After Taylor, 1948.)

change. As long as the volume was increasing, shown by the dashed line, there was an additional component of shearing strength, but when the volume change curve leveled off, the shearing strength declined to a residual value representing sliding friction.

The peak stress at 0.05 in. (1.27 mm) displacement coincides with the steepest slope of the volume change line, as predicted in eq. (18.5).

In contrast, the solid lines in Fig. 18.6 show shearing of an initially loose sand that decreased in volume upon shearing; initial dilatancy was negative, reducing the shearing resistance. Then as sliding continues this strength is recovered when the soil densifies until the residual strength is the same as that for the initially dense sand. The critical void ratio was the same regardless of whether the sand initially was dense or loose, as dilatancy made up the difference.

### 18.4.3 Importance of the Critical Void Ratio

A void ratio that is larger than a critical void ratio can occur in loose alluvial or deltaic sands and indicates that the soil is *metastable*—it is held together by sliding friction, but if disturbed, negative dilatancy will subtract from the frictional resistance and the soil structure will collapse. If the soil is saturated with water, individual grains will become temporarily suspended in water and duplicate the stress condition in quicksand. This process of *liquefaction* is a major source of damage during earthquakes. A condition is defined as metastable if energy going in can trigger a larger amount coming out.

### 18.4.4 A Shear Failure Envelope

Direct shear tests are conducted with different applied normal stresses in order to establish a friction angle. This requires identical soil specimens, as any variations between test specimens will contribute random error. Generally a minimum of three separate tests is conducted, and if the friction angle is constant, a graph of shearing strength versus normal stress should be a straight line. A statistical linear regression analysis may be performed to obtain a best-fit straight line through test data points.

An important feature of the direct shear test is that the soil specimens are allowed to drain during the test so pore pressures normally are negligible, but drainage also is affected by permeability of the soil and the rate of shearing.

### 18.4.5 Can a Direct Shear Test Measure Modulus?

Shearing can define a shear modulus, $G$, that is, the ratio of shearing stress to shearing strain. The shear modulus in turn relates to the modulus of elasticity, $E$, through $E = 2G(1 + \mu)$, where $\mu$ is Poisson's ratio. However, in order to define shear strain the shear zone thickness must be known, and in a conventional direct shear test the shear zone occupies only a portion of the total specimen thickness. Also, a shear modulus imples elastic behavior.

Several approaches have been used in an attempt to remedy this deficiency without compromising the influence from dilatancy. A "shear box" is hinged at the corners. It is loaded at opposing corners so that it deforms from a square shape into a parallelogram and the entire soil sample shears. However, the change in shape also forces a decrease in volume, and results often reveal a diagonal tension crack that interrupts shear surfaces.

The "simple shear test" was developed by Bjerrum and Landva (1966) in Norway to create a more uniform shearing displacement by confining a cylindrical soil sample in a rubber membrane with an embedded spiraling steel wire (ASTM Designation D-6528).

### 18.4.6 Shear Plane Orientation and Soil Fabric

As shown in Fig. 18.1, most field applications for shear test data involve shearing in an inclined direction that crosses horizontal layering that may be present in the soil. However, soil sampling normally is accomplished in vertical boreholes, so the orientation of the shear plane in a laboratory direct shear test normally is horizontal. This limits the applicability of laboratory direct shear testing, which with conventional sampling is most useful for homogeneous soils.

### 18.4.7 Borehole Shear Test

The BST is a direct shear test that is performed in the sides of a vertical borehole in the field, and measures average shear strengths across stratification or bedding planes. Horizontal or inclined tests may be performed in a cut face in the soil, and the test can be conducted in the laboratory. Details of the test are discussed in Chapter 26, and a diagram of the test apparatus and photograph are shown in Figs. 26.13 and 26.15. Like the laboratory direct shear test, the BST is a "drained test," which means that excess pore water pressures are dissipated during the test. Drainage is encouraged because the BST is a "stage test" in which shearing is repeated at successively higher normal stresses so that drainage times are cumulative. The test succeeds because the normal stress is more concentrated next to the shear plates and acts to recompress the previously sheared soil to a higher shearing strength, moving the shear surface outward.

## 18.5  COHESION

### 18.5.1  Coulomb's Experiments

Charles-Augustin de Coulomb was a French engineer most famous for discovery of the inverse square law for electrostatic forces. The discovery was made possible by his invention of a sensitive balance called a torsion balance, which he invented in response to a contest for a more accurate marine compass. Although he did not win the contest, his discovery ironically laid out the map for orbiting satellites and for space travel.

As a military engineer Coulomb was concerned with the stability of soil in trenches, and in 1773 he presented a classic paper that modified Amontons' Law by hypothesizing that the shearing resistance of soil is made up of two components, friction that is proportional to normal stress, and cohesion that is constant regardless of the normal stress. This adds a cohesion term to eq. (18.4) to give what is known as Coulomb's equation:

$$\tau_f = c + \sigma_n \tan \phi \qquad (18.6)$$

where $c$ is soil cohesion and the other variables are as in the previous equation: $\tau_f$ is shearing stress at the instant of failure, $\sigma_n$ is the normal stress on the shearing surface, and $\tan \phi$ is the coefficient of internal friction. Although $c$ actually is a cohesive component of shearing strength, it usually is designated simply as "cohesion."

Equation (18.6) is a linear equation that defines a *failure envelope*. In mathematics an envelope is a threshold that cannot be crossed, and that also is the case here: no data points can plot above a failure envelope. Equation (18.6) is graphed in Fig. 18.7, which shows cohesion, $c$, and the friction angle, $\phi$.

### 18.5.2 The Nature of Cohesive Shear Strength

If the failure envelope is extended to the left past the ordinate it is in a zone of tension instead of compression. The intersection with the $x$-axis is a negative stress called the *intrinsic pressure*. This seldom is used in calculations, but indicates a relationship between cohesive shear strength and true cohesion, or the attractive forces between clay particles. These forces in turn relate to double-layer attractions that depend in part on the moisture content.

The cohesion of dry sand is zero, but temporary "apparent cohesion" is caused by capillary attractions and negative pore water pressures discussed in Chapter 11. "Apparent cohesion" was introduced by Terzaghi to indicate that part of cohesive shear strength that is lost upon saturation.

### 18.5.3 Cohesion and Preconsolidation Pressure

Figure. 18.8 shows results from a series of laboratory direct shear tests reported in 1937 in a textbook by C. A. Hogentogler, who was with the U.S. Bureau of Public Roads. The tests were performed on a clayey soil that was preconsolidated at various normal stresses prior to shearing. That is, each test involved first loading the soil to a predetermined pressure, removing part of the load, and then shearing the soil . The first loading therefore becomes a preconsolidation pressure, and as can be seen from the figure, the higher the preconsolidation pressure the higher the cohesion. Furthermore, the data points obtained without the application of a preconsolidation pressure all are on a line that extrapolates through the origin.

**Figure 18.7**

A graph of Coulomb's equation (18.6), indicating that soil shearing strength includes both a friction and a cohesion component.

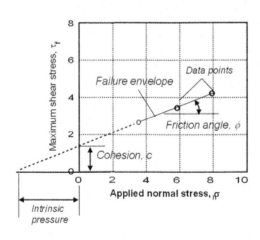

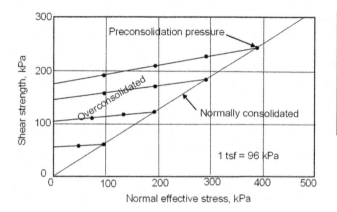

**Figure 18.8**

Shear envelopes
showing relation
of cohesion
to preconsolidating
pressure, after
Hogentogler
(1937).

These tests indicate that without some type of preconsolidation, cohesion is zero. Soils in nature are preloaded by the weight of overlying soil, including that which has been removed by erosion, or preconsolidation can be simulated by drying shrinkage of expansive clay. An example of a very soft fine-grained soil that has had little or no preconsolidation pressure is at the bottom of a lake.

Each of the failure envelopes in Fig. 18.8 is bilinear, with a break at the preconsolidation pressure. This break usually is not defined by laboratory shear tests because of the limited number of tests. A similar break frequently is obtained with the BST and is shown in Fig. 26.16. As indicated in the previous chapter, settlement is minimized if pressures are lower than the preconsolidation pressure.

### 18.5.4  The "$\phi = 0$" Simplification

In some dense clays the overconsolidated portion of the shear envelope is nearly flat. This also is shown by line $DB$ in Fig. 18.8, as cohesive strength that develops as clay grains are pushed together tends to be retained after the consolidating pressure has been removed. The assumption of a friction angle of zero, while an approximation, simplifies analytical solutions for slope stability and bearing capacity because the soil weight then does not affect shearing strength.

A $\phi = 0$ condition also occurs during rapid loading that does not allow time for drainage, so that all added normal stress is taken by pore water pressure. An example is a car driving on a muddy road. This is discussed later in more detail in relation to the development of pore water pressure.

### 18.5.5  Linear Regression of Direct Shear Data

In direct shear type tests the normal stress is an independent or controlled variable, and shearing stress is a measured or dependent variable. The data therefore are suitable for statistical linear regression analysis.

Best-fit straight lines are defined by the statistical *method of least squares*, such that the sum of the squares of the vertical offsets of data from the line is minimized. This procedure, which at one time required extensive hand calculations, now is performed almost instantly with a digital computer or programmable calculator.

In addition to defining an equation that best fits the data, a regression analysis generates a correlation coefficient, $R$, that is a measure of the goodness of fit. Perfect agreement between the data and the regression line gives $R = 1.00$ or $-1.00$, the latter if the slope is negative. In order to eliminate the sign convention most programs report $R^2$, which is a "coefficient of determination." As $R^2$ is smaller than $R$, this creates a larger spread of values if $R$ is close to 1.0. As the value of $R^2$ depends on the number of data points, $n$, this also should be reported.

### Example 18.1

Determine the friction angle, cohesion, and correlation coefficient from the following direct shear test data:

| Test specimen | Normal stress | | Maximum shear stress | |
|---|---|---|---|---|
| | lb/in.$^2$ | kPa | lb/in.$^2$ | kPa |
| 1 | 1.77 | 12.2 | 2.39 | 16.5 |
| 2 | 3.51 | 24.2 | 3.81 | 26.3 |
| 3 | 7.02 | 48.4 | 5.70 | 39.3 |
| 4 | 14.4 | 99.6 | 6.32 | 69.5 |

*Answer:* A regression analysis gives $y = a + bx$, where $a = 1.52$ lb/in.$^2$ (10.5 kPa), slope $b = 0.592$, and $R^2 = 0.998$. The $a$ intercept equals the cohesion. The slope $b = \tan \phi$, so $\phi = 30.6°$.

Regression analysis gives a more accurate interpretation of the results than can be obtained by fitting a straight line by eye. However, every data set should be plotted and examined for data that for one reason or another may not be linear.

## 18.5.6 Shear Tests of Fractured Rocks and Soils

Some soils, particularly expansive clays, are highly fractured as a result of desiccation, and at low normal stresses behave more like granular soils with a high friction angle than like clays. Then when normal stresses are high enough to cause shearing through the blocks, the behavior becomes that of a cohesive clay. This develops a bilinear shear envelope, but with the first segment at a high angle that extrapolates through the origin, and the second representing shearing strength of the clay.

An example of this behavior is in Canadian quick clays, where back-calculation of stresses initiating landsides indicates that the soil was restrained by a granular behavior until movement started, and then it became liquefied (Eden and Mitchell, 1970).

## 18.5.7 Approximate Representations for *c* and $\phi$ Used in Design

Ideally shear strength data should be compartmentalized to represent over-consolidated and normally consolidated portions of a shear envelope, as shown in Fig. 18.9. However, laboratory determinations usually rely on three or at most four tests that do not define shear behavior boundaries.

Three failure modes are possible in a particular soil (Fig. 18.9): (1) at low normal stresses a highly fractured clay can exhibit granular behavior with a high friction angle and near-zero cohesion, shown at *A*; (2) as normal stress increases but is less than the preconsolidation pressure, shearing through the soil matrix generates both cohesion and a friction angle; and (3) at normal stresses exceeding the preconsoliation pressure, the failure envelope will follow that of a normally consolidating soil with a higher friction angle and zero cohesion.

In Fig. 18.9 a best-fit line through unassigned data points generally is on the safe side and is adequate for design. However, the failure to resolve a preconsolidation pressure shown by $P_c$ can lead to consolidation that in turn will increase pore water pressure and subtract from the shear strength. This can be determined from measurements of pore water pressure during a test.

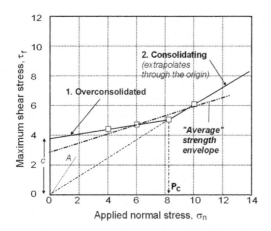

## Figure 18.9

Two-stage shear envelope for an overconsolidated clay and an approximate "average" common in design. A high-angle segment such as at *A* can occur in gravel or in highly fractured hard clay.

**Figure 18.10**

Creep rates with constant normal and shear stresses measured after different times all point to a minimum shearing stress for creep to occur. (From Borehole Shear Test data of Lohnes et al., (1972), with permission of the American Society of Civil Engineers.)

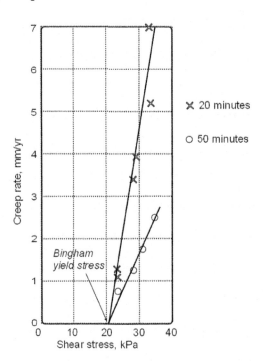

### 18.5.8 Residual Shearing Strength

As shearing of a dense soil progresses, the soil dilates and expands, but the amount of expansion is limited, and eventually the density in the shear zone reaches a stable condition at the critical void ratio. The shear strength at this void ratio has no dilatant component and is called the "ultimate" or "residual" shear strength. It in effect is a measure of an average coefficient of sliding friction of minerals in the soil. However, it also is affected by changing orientations of soil particles as a result of shearing. For example, shearing of clay can create a smearing effect that will reduce the residual shear strength.

Residual shearing strength is measured in a direct shear test by continuing shearing past the point of maximum shearing stress, as shown in Fig. 18.6. Not only is the friction angle reduced by the removal of a dilatant component, but the continuous breaking of bonds causes cohesion to be at or near zero (Skempton and Petley, 1967).

In the BST the shear zone moves outward, so residual strength measurements require reducing the normal stress before continuing shearing to ensure that it will occur along the same surface.

A conservative view is that residual strength instead of the peak strength should be used in design. However, this ignores dilatancy, which is a major benefit from soil compaction. In practice the design that is based on peak shear strength *with a factor of safety* often will result in field stresses that are lower than the residual strength.

## 18.6   SOIL CREEP

### 18.6.1   Influence of Shear Rate

In soft clays, the faster the rate of shearing, the higher the shear strength. This rate dependence is the equivalent of a *viscosity* and contributes to a phenomenon observed in the field known as *soil creep*. Creep can explain slow, imperceptible movement of soil on a steep slope, and tilting of retaining walls and tombstones over the years because of eccentric loading or being placed on slopes. (In areas of ground freezing similar movements can be attributed to frost heave.)

### 18.6.2   Measuring Soil Creep

The minimum shearing stress to cause creep is the *Bingham yield stress* (Fig. 18.10). The yield stress was determined with the BST by applying different shearing stresses and a constant normal stress and measuring the rate of deformation. After an initial consolidation phase the rate slowed down but continued as long as the shearing stress exceeded the yield stress. Although the Bingham yield stress is an important consideration in soils subject to creep, it normally is not measured. However, by knowing the characteristics of creep behavior its influence can be incorporated in design.

### 18.6.3   Design for Soil Creep

As creep is a constant-volume phenomenon, it is prevented by a tendency for dilatancy, and therefore is prevented by compaction unless, as in the case of an expansive clay, the soil can "de compact" upon wetting.

In Fig. 18.11, the yield stress at different normal stresses essentially parallels the shear envelope and is separated from it by the soil cohesion. This indicates that cohesion in creeping soil is rate-dependent—in other words, it is a viscosity, and for design should be assumed to be zero. Also since a creeping soil is not dilating, design should be based on the residual and not the peak friction angle. That angle is further lowered if the soil becomes saturated so that water seeping downslope adds a viscosity component called a *seepage force*.

---

*Question:* Have you ever observed the effects of soil creep on trees on a slope or on a retaining wall? How can creep be prevented?

---

**Figure 18.11**

The graph of Bingham yield stress versus normal stress in an actively creeping soil shows a rate dependence of soil cohesion, which in a creeping soil should be assumed to be zero. The physical chemistry of creep resembles that of secondary consolidation.

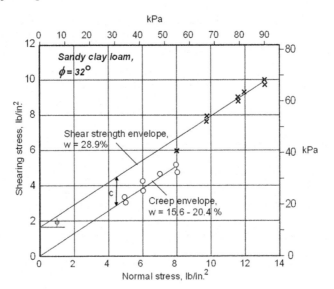

### 18.6.4  Creep and Secondary Consolidation

Secondary consolidation involves sliding of soil grains over one another, much as in soil creep. However, in secondary consolidation there is no continuity of a shearing surface. Furthermore, instead of proceeding at a constant void ratio as in the case of creep behavior, secondary consolidation results in a gradual reduction in void ratio and increase in density. The separation of the two lines in Fig. 18.10 probably can be attributed to secondary consolidation.

## 18.7  PORE WATER PRESSURE AND SHEAR STRENGTH

### 18.7.1  Critical Role of Pore Water Pressure

Landslides usually occur after prolonged periods of rain, when soil is saturated and there is a substantial head of pressure on water in the shear zone. Water pressure acts to reduce the contact pressure between soil grains by tending to push them apart. Although water pressure acts only on areas of grains that are not in actual contact, these areas are so small that they conventionally are ignored, which means that pore water pressure, like normal and shearing stress, can be expressed on an overall area basis.

The influence of pore water pressure is illustrated in Fig. 18.12, where it subtracts from the applied normal stress but does not affect the shearing stress as the water itself has no shearing resistance. One of Terzaghi's most important contributions was his modification of the Coulomb equation for shearing strength by

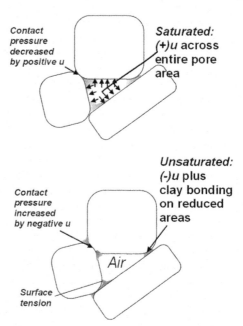

**Figure 18.12**

Whereas positive pore pressure is simply subtracted from normal stress, negative pore pressure acts over limited areas near grain contacts, and is enhanced by double-layer attractions that vary with moisture content.

incorporating pore water pressure. The result is the *Coulomb-Terzaghi effective stress equation,* or simply the *effective stress equation*:

$$\tau_f' = c' + (\sigma_n - u) \tan \phi' \qquad (18.7)$$

where the symbols are as described in relation to eq. (18.6), and $u$ is the pore water pressure. In shear tests where the pore water pressure is measured, it therefore is subtracted from the normal stress to obtain a shear envelope on an effective stress basis. Prime symbols indicate effective stresses. Therefore,

$$\tau_f' = c' + \sigma_n' \tan \phi' \qquad (18.7a)$$

Equation (18.7) is one of the most widely used relationships in geotechnical engineering. In a drained test such as a slowly conducted direct shear or Borehole Shear test $u = 0$, so the soil strength parameters are $c'$ and $\phi'$.

### Example 18.2

Assume static groundwater conditions and calculate the shearing resistance of soil at the base of a landslide (a) without and (b) with the influence of pore water pressure from a groundwater table at the ground surface. The depth of the landslide is 12 ft (3.66 m), the unit weight of the soil is 120 lb/ft³ (18.9 kN/m³), and drained shear tests give $\phi' = 22°$ and $c' = 500$ lb/ft² (23.9 kPa).

*Answer:* From eq. (18.7),

(a) $\tau_f' = 500 + (12 \times 120) \tan 22° = 500 + 582 = 1080$ lb/ft² without pore water pressure.

(b)  $\tau_f' = 500 + (12 \times 120 - 12 \times 62.4)\tan 22° = 500 + 279 = 779\,\mathrm{lb/ft^2}$  with pore water pressure.

This is a strength reduction of 28%. In SI,

(a)  $\tau_f' = 23.9 + (3.66 \times 18.9)\tan 22° = 23.9 + 27.9 = 51.8\,\mathrm{kPa\,lb/ft^2}$  without pore water pressure.

(b)  $\tau_f' = 23.9 + (3.66 \times 18.9 - 3.66 \times 9.80)\tan 22° = 23.9 + 13.5 = 37.4\,\mathrm{kPa}$  with pore water pressure.

Note that a more accurate answer to part (b) can be obtained by reading the pore water pressure from a flow net, as discussed in Chapter 14. As the pressures measured from a flow net are lower than the hydrostatic pressure, the simplification is on the safe side.

## 18.7.2  Using Buoyant Unit Weight

Another procedure that is based on the same principle and gives the same answer as eq. (18.7) is to use the *buoyant unit weight* to calculate resisting stress for soil below a groundwater table. Then

(a)  $\tau_f' = 500 + 12 \times (125 - 62.4)\tan 22° = 500 + 304 = 804\,\mathrm{lb/ft^2}$  or

(b)  $\tau_f' = 23.9 + 3.66 \times (18.9 - 9.80)\tan 22° = 23.9 + 13.5 = 37.4\,\mathrm{kPa}.$

This simplification can be used only for that part of the soil that is below the groundwater table.

## 18.7.3  True Contact Areas in Saturated Clays

According to eq. (18.7) the influence of excess pore water pressure becomes nil if $\phi = 0$, which can be explained if pore pressure from liquid water does not invade between and separate clay particles, but is stopped by interconnections and overlapping of double layers. The contact area hypothesis for friction does not apply if there is no friction.

## 18.7.4  Pore Water Pressure if $\phi = 0$

According to Equation 18.6, positive pore water pressure will have no influence on the shearing strength if $\phi = 0$, as discussed in Section 18.5.4. This at least in part can be attributed to a reduction in pore area and increase in true contact area as clay particles are jammed together under pressure.

The other process by which $\phi = 0$, a soft, saturated clay that does not drain under loading, occurs on muddy roads and also in sediments in the ocean bottom where the weight of the overlying water column has no influence on the shearing strength.

For this reason pore water pressure also is called "neutral stress." It was this observation that led a Scottish geologist, Sir Charles Lyell, to suggest that effective stress, or total stress minus neutral stress, must be the driving force for consolidation of seafloor sediments. This observation later was formalized by Terzaghi.

## 18.7.5  Negative Pore Pressure

In unsaturated soils the pore pressure term $u$ in eq. (18.7) is negative as a result of capillary forces and clay double-layer attractions. In eq. (18.7), a negative of a negative becomes positive, so negative pore pressure adds to instead of subtracts from soil shear strength. However, since the soil is not saturated, the negative pore pressure does not act across an entire cross-section but only on the area occupied by water (Fig. 18.12).

A. W. Bishop, of Imperial College in England, suggested a term $X$ to account for the area correction and also to include the influence of any pore air pressure. Pore air pressure is relatively small in a system open to the atmosphere, but can be influenced by changes in atmospheric pressure that in turn affect pore water pressure.

Negative pore pressure increases upon drying while the cross-sectional area occupied by water decreases. If active clay minerals are present, the increase in pore water pressure is more important than the decrease in area, so as a clay soil dries it becomes harder. On the other hand, as drying occurs in sand, negative pore pressure will reach the vapor pressure of water and the sand will spontaneously dry out. In clay the concentration of the double-layer attractions and increased negativity of pore pressure are the main cause of shrinkage, as the negative pressure can exceed many atmospheres. The soil mechanics of unsaturated soils is a complex topic that is the subject of a book by Fredlund and Rahardjo (1993).

## 18.7.6  Apparent Cohesion in Sand

Because negative pore pressure shifts a shear envelope to the left but can disappear upon saturation, Terzaghi referred to its effect on cohesion as *apparent cohesion*. Laboratory shear tests are routinely conducted with soil in a saturated state in order to minimize this effect. An exception can be made for soils such as loess, based on the premise that their field condition will never reach saturation.

## 18.7.7  Negative Pore Pressure and Elastic Rebound in Shale

Shale that has been compressed under a rock column for millions of years often experiences elastic rebound when unloaded. As voids tend to expand they draw in water that can permanently soften the shale. When such a shale is exposed in an excavation it should immediately be sealed with bitumen to prevent entry of water and preserve negative pore pressure, as otherwise any expansion will add to subsequent settlement under a foundation load. The horizontal layered structure of shale tends to direct expansion vertically, which can aggravate the problem.

Another procedure that can reduce later damages from shale rebound expansion is to scarify it while wet and recompact it under controlled conditions as a soil. This relieves suction, randomizes the shale structure, and provides a larger void content should there still be any later latent expansion. Geologically old (Pennsylvanian age and older) shales normally do not contain expansive clay minerals so expansion is physical and relatively moderate.

Nonexpansive shale in embankments can degrade in time and return to clay, resulting in slope failures years or decades after construction.

Geologically younger shale and clay deposits that contain expansive clay minerals are processed as expansive soils by compacting on the wet side of the optimum moisture content and preventing future drying, or by addition of a chemical stabilizer such as lime.

### 18.7.8 Changes in Pore Water Pressure During Shearing

Dilatancy increases the volume of voids in a soil and therefore decreases pore water pressure. This can be illustrated by a walk on a beach sand at the water's edge. If one should care to look down, it will be seen that the sand around the boundary of each footprint appears to dry up as pressure is applied, and if a foot pressure is maintained, the sand becomes wet again as water is drawn in to relieve the negative pore water pressure.

Bishop measured changes in pore water pressure during triaxial compression and proposed some descriptive *pore pressure parameters*. The effects of pore water pressure are most elegantly revealed by laboratory triaxial compression tests where cylinders of soil are encased in a rubber membrane that can prevent changes in moisture content during shearing, which is discussed in the next section.

## 18.8  COMPRESSIVE STRENGTH

### 18.8.1  Overview

One of the oldest and simplest methods for measuring the strength of soil is the unconfined compressive strength test, in which a cylinder of soil is compressed from the ends until it breaks. This test obviously is not satisfactory for sands that do not hold together. The test later was modified by covering the soil with a membrane and simultaneously applying lateral stress in a *triaxial compression test*. Even though in both cases a soil specimen is compressed from the ends, failure is by internal shearing, as shown in Fig. 18.13. The most critical inclination angle for shearing is found automatically, and as will be shown relates to the internal friction angle of the material being tested.

## 18.8.2 Triaxial Test

In the triaxial test a cylindrical soil specimen is enclosed in a rubber membrane and placed in a pressure chamber (Fig. 18.14). In the conventional test, a constant all-around stress is applied with fluid or air pressure. This is followed by gradually increasing vertical stress so that the specimen is tested in compression. This is a

**Figure 18.13**

Soil specimens after triaxial compression testing. Both specimens show bulging characteristic of plastic behavior, but also reveal inclined surfaces indicating internal shearing or sliding of particles over one another.

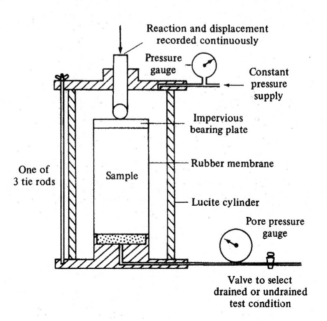

Reaction and displacement recorded continuously

Pressure gauge

Constant pressure supply

Impervious bearing plate

One of 3 tie rods

Sample

Rubber membrane

Lucite cylinder

Pore pressure gauge

Valve to select drained or undrained test condition

**Figure 18.14**

Schematic diagram of a triaxial test apparatus. Here lateral stress is applied using regulated air pressure. Water may be used in order that volume changes can be monitored.

"triaxial test" because pressure is applied from the three principal stress directions, one vertical and the other two equal and horizontal.

Triaxial tests give more freedom for location of the shear plane than do direct shear tests, where the location is fixed by the horizontal separation of the shear box. Also, shear directions in vertically oriented samples are inclined and therefore cut across any horizontal layering, thereby giving an averaging effect and being more in keeping with slip surface orientations in the field. A disadvantage of triaxial testing is that the normal and shear stresses are not measured on the shear plane, but are evaluated indirectly from measuring the major and minor principal stresses. This is accomplished using Mohr's failure theory.

As in the case of a direct shear test, several tests are required on the same soil in order to evaluate $\phi$ and $c$. A direct shear test involves application of different normal stresses; in a triaxial test different lateral confining stresses are used.

Drainage is optional in triaxial testing: if drainage is not allowed, the test is referred to as *undrained*, and pore water pressures normally are monitored. Pore water pressures then are subtracted from principal stresses to give effective stresses that are used to determine effective stress shear strength parameters, $\phi'$ and $c'$.

## 18.8.3  Mohr Circle Representation of Stresses

In 1882 a German civil engineer and professor, C. Otto Mohr, suggested a geometric construction called a *Mohr circle* that greatly simplifies the conversion of principal stresses on principal planes to shearing and normal stresses on a plane having any given orientation. He also proposed a failure theory that will be discussed in more detail.

In Fig. 18.15, assume a unit thickness vertical to the page, which allows a conversion of stresses to forces by multiplying by respective line lengths. Letting $A = 1$ and summing forces in the $x$ and $y$ directions leads to

$$\sigma_n = \sigma_1 \cos^2 \theta + \sigma_3 \sin^2 \theta \tag{18.8}$$

$$\tau = (\sigma_1 - \sigma_3) \sin \theta \cos \theta \tag{18.9}$$

where stress symbols are as indicated in the figure and $\theta$ is the angle with the major principal plane. This development will be found in most mechanics of materials textbooks. By means of another series of identities it is possible to eliminate the angle $\theta$, which results in

$$\tau^2 + \left[ \sigma - \left( \frac{\sigma_1 + \sigma_3}{2} \right) \right]^2 = \left( \frac{\sigma_1 - \sigma_3}{2} \right)^2 \tag{18.10}$$

This is the identity developed by Mohr, who noted that it is the equation of a circle, with the form

$$y^2 + (x - x_c)^2 = R^2 \tag{18.10a}$$

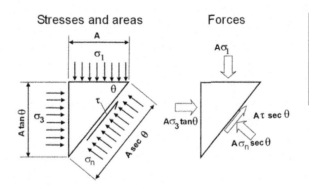

where $R$ is the radius of a circle with its center located on the $x$-axis at a distance $x_c$ from the origin. The following equivalencies may be noted between eq. (18.10) and (18.10$a$):

$$y = \tau$$
$$x_c = \tfrac{1}{2}(\sigma_1 + \sigma_3)$$
$$R = \tfrac{1}{2}(\sigma_1 - \sigma_3)$$

The diameter of the circle is

$$2R = \tfrac{1}{2}(\sigma_1 - \sigma_3)$$

Therefore,

(1)  the diameter of a Mohr circle equals the difference between the major and minor principal stresses, and

(2)  the center is offset to the right by a distance equal to the average of the two principal stresses.

**Example 18.3**
Draw a Mohr circle to represent a vertical major principal stress of 4 units and a horizontal minor principal stress of 2 units in a triaxial test.

*Answer:* The diameter of the circle is $2R = 4 - 2 = 2$. The center is located on the $x$-axis at a distance $x_c = \tfrac{1}{2}(4 + 2) = 3$. This Mohr circle is shown in Fig. 18.16(a).

### 18.8.4  Mohr Circle and Shear Failure

Let us suppose that the Mohr circle in Fig. 18.16(a) represents the stress condition at shear failure in sand. It has been found experimentally that, if other compression tests are conducted at different values of lateral stress $\sigma_3$, the circles will have a common tangent as shown in Fig. 18.16(b). As this line defines the principal stress ratio for failure, it is identical to the Coulomb failure envelope from direct shear tests. The point of tangency therefore presents a method for

**Figure 18.16**

(a) A Mohr circle allows a conversion of major and minor principal stresses to normal and shearing stresses on any plane.
(b) The tangent to several Mohr circles defines a *Mohr-Coulomb failure envelope.*

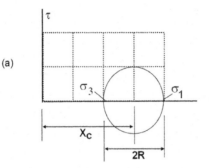

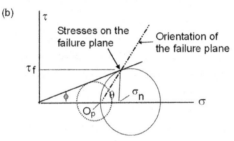

converting principal stresses at failure to the corresponding normal and shear stresses in the Coulomb equation, $\tau_f$ and $\sigma_n$.

By simply connecting the circles, Mohr developed a theory that failure does not occur at the maximum shearing stress at the top (or bottom) of the Mohr circle, but at a point of tangency to a failure envelope that is tangent to the circles. Mohr's theory ignores any influence from the intermediate principal stress, and other failure theories have been suggested. However, Mohr's theory is considered adequate for most geotechnical engineering purposes.

Mohr's theory does not require that the failure envelope be a straight line, but again this approximation appears adequate for most purposes, particularly if the failure envelope is segmented to correspond to different failure modes. If a straight line is assumed, it is referred to as the *Mohr-Coulomb failure envelope.*

### 18.8.5 Mohr Circle and Inclination of the Shear Plane

Another important feature of the Mohr circle is that with a linear failure envelope the failure angle, designated by $\theta$ in Fig. 18.17, is constant regardless of the size of the circle. The $\theta$ shear angle also is not affected if a soil has cohesion, as shown in Fig. 18.17.

The relationship between $\theta$ and $\phi$ is developed from Fig. 18.17:

- Through similar triangles $\angle FDC = 90 - \phi$.
- Since $\angle BFE = 90°$, it can be shown that $\beta = 90 - (90 - \theta) - \phi$, so $\beta = \theta - \phi$.

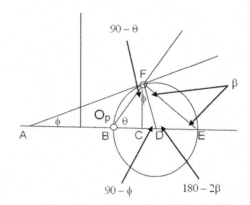

**Figure 18.17**
Diagram for
development of
eq. (18.11).

- $\angle ACF = 90 - \phi$. As an exterior angle of triangle $DEF$ it also equals $2\beta = 2\theta - 2\phi$.
- Equating $90 - \phi = 2\theta - 2\phi$ and solving for $\theta$ gives

$$\theta = 45 + \phi/2 \tag{18.11}$$

This is a defining equation that describes the failure angle in a soil sample that is compressed vertically, as in unconfined or triaxial compression. By use of this equation, $\theta$ angles measured on failed specimens can be used to estimate the friction angle. The equation also gives the slip angle for soil behind a yielding retaining wall, where the major principal stress is vertical from weight of the soil and the minor principal stress is horizontal, representing restraint of the wall.

**Example 18.4**
Estimate $\phi$ from the failure angles in Fig. 18.13.

*Answer:* In both cases $\theta$ is approximately 60°. Therefore $60 = 45 - \phi/2$; $\phi/2 = (60 - 45) = 15$; $\phi = 30°$.

Equation (18.11) relating shear angle and principal stress directions enables a comparison of triaxial and direct shear test failure geometries, as in Fig. 18.18.

## 18.8.6 Origin of Planes

The point designated $O_p$ in Figs. 18.16 and 18.17 is the *origin of planes*, because a line drawn in any direction through $O_p$ will intersect the Mohr circle to give values of $\tau$ and $\sigma_n$ on that plane. Thus when a line is drawn from $O_p$ to the failure point, the values of $\tau_f$ and $\sigma_n$ correspond to those obtained from a direct shear test. By convention, if the shear direction is counterclockwise, as in Fig. 18.15, $\tau$ is positive; or if clockwise, it is negative.

The location of the origin of planes depends on the directions of the principal stresses. For example, if the major principal stress is horizontal,

**Figure 18.18**

The shear direction at $45 + \phi/2$ shows the relationship to the direct shear test except for the minor principal stress that is normal to the page.

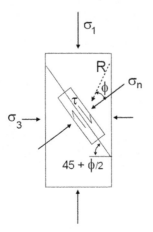

$O_p$ moves to the top of the Mohr circle. The procedure to locate an origin of planes is as follows:

1. Draw a Mohr circle for any value of $\sigma_1$ and $\sigma_3$.
2. Draw a line through the $(\sigma_n, \tau)$ stress point *parallel to the plane on which it acts*.
3. $O_p$ is at the intersection of this line with the Mohr circle.

In the example, and in most cases, the major principal stress is vertical and therefore acts on a horizontal plane, so drawing a horizontal line through $\sigma_1$ in Fig. 18.16 gives the $O_p$ as shown. The same point is indicated by $\sigma_3$, which acts on a vertical plane: a vertical line through $\sigma_3$ contacts the Mohr circle at only one point, which is $O_p$.

### 18.8.7  Unconfined Compressive Strength

As previously mentioned, the unconfined compressive strength test is a special case of a triaxial test in which the lateral confining stress is zero. A Mohr circle for an unconfined compressive strength is shown in Fig. 18.19.

It sometimes is assumed that this test is undrained and that $\phi = 0$, indicated by the horizontal dashed line in the figure. This is applicable for either of two conditions: for overconsolidated stiff clays where anticipated field stresses are less than the preconsolidation pressure, or for a soft, saturated clay with rapid loading that does not allow drainage of excess pore water pressure. Then according to the Mohr circle the shearing strength is entirely represented by cohesion, which equals one-half of the unconfined compressive strength:

$$c = \frac{q_u}{2} \tag{18.12}$$

where $q_u$ is the unconfined compressive strength. Also according to eq. (18.11) the failure angle $\theta = 45°$. If the failure angle is not at $45°$, the $\phi = 0$ condition is not appropriate.

**Example 18.5**

Soil at the bottom of a vertically cut open trench has no confinement on one side. The unit weight of the soil is $125\,\text{lb/ft}^3$ $(19.6\,\text{kN/m}^3)$. The unconfined compressive strength is $800\,\text{lb/ft}^2$ $(38\,\text{MPa})$. How deep will the trench be when it caves in?

*Answer:* $125 \times h = 800$; $h = 6.4\,\text{ft}$; or, $19.6 \times h = 38$; $h = 1.9\,\text{m}$, *with no margin for safety.*

## 18.8.8  Unconfined Compressive Strength in Relation to *c* and $\phi$

The following general equation relates $q_u$, $c$, and $\phi$:

$$q_u = 2c \tan(45 + \phi/2) \tag{18.13}$$

The student can develop this equation as follows: in Fig. 18.20,

$$c = \frac{q_u}{2} \frac{1 - \sin\phi}{\cos\phi} \tag{18.14}$$

A trigonometric identity is required to obtain eq. (18.14): $\tan \frac{1}{2}\,(90 - \phi = (\sin 90 - \sin \phi)/(\cos 90 + \cos \phi)$.

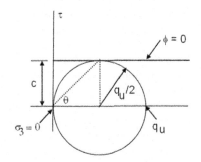

**Figure 18.19**

Mohr circle representing unconfined compressive strength $q_u$ if $\phi = 0$.

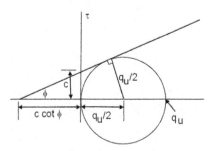

**Figure 18.20**

Development of a relationship between $q_u$, *c*, and $\phi$.

### 18.8.9  Enhancing Shear Strength with Lateral Stress

The magnifying effect of lateral confining pressure, $\sigma_3$, on the compressive strength, $\sigma_1$, can be demonstrated by a derivation similar to that for eq. (18.14), substituting $(q_u + c \cot \phi + \sigma_3)$ for $(q_u + c \cot \phi)$, which gives

$$\sigma_1 = \frac{2c \cos \phi}{1 - \sin \phi} + \frac{\sigma_3}{1 - \sin \phi} \qquad (18.15)$$

The second term in this equation shows the effect of lateral stress on strength for soils having a friction angle that is larger than zero.

**Example 18.6**

A retaining wall is tied back with soil anchors that are post-tensioned to create an average horizontal stress of $100\,\text{lb/ft}^2$ (4.8 MPa) in soil against the wall. How will that affect the load-bearing capacity of soil behind the wall (a) if $\phi = 0$, (b) if $\phi = 30°$?

*Answer:* (a) With zero friction angle it will increase the load-bearing capacity by the amount of the lateral stress. (b) With a 30° friction angle it will increase the load-bearing capacity by $(100)/(1 - \sin 30°) = 200\,\text{lb/ft}^2$, or $(4.8)/(1 - \sin 30°) = 9.6\,\text{MPa}$, or two times the amount of the applied lateral stress.

### 18.8.10  Role of the Intermediate Principal Stress, $\sigma_2$

One of the shortcomings of the triaxial test is that it does not address the influence from an intermediate principal stress that is larger than the minor principal stress, as often occurs in a field condition. In the conventional laboratory test, $\sigma_2 = \sigma_3$, which is valid for soil that is under the center of a square or circular foundation. In most cases in the field the intermediate principal stress truly is intermediate.

According to Mohr failure theory, failure occurs in the plane of the major and minor principal stresses regardless of the intermediate principal stress. There nevertheless is a confining effect from an intermediate stress that will tend to restrict the shear direction and therefore can affect the results. Because this is a matter of probability, it includes a size effect that is difficult to evaluate. One option is a "plane strain" test in which soil is confined between two smooth walls so that $\sigma_1 > \sigma_2 > \sigma_3$. This tends to give slightly higher shear strengths than a triaxial test but introduces wall friction, and will overcorrect a field condition that allows some lateral yielding. The triaxial test, by allowing failure at slightly lower stresses, is generally considered to be on the safe side and is widely used for design. The intermediate principal stress is higher in direct shear type tests, which can contribute to a discrepancy of a degree or two in measured friction angles between the two kinds of tests.

### 18.8.11  End Restraint in the Triaxial Test

Samples that are compressed vertically in a triaxial test initially tend to expand laterally as a consequence of Poisson's ratio. This mobilizes radial restraining friction on ends of the sample that adds to the applied $\sigma_3$, with the result that, as shown in Fig. 18.13, bulging is restricted close to the ends of the test specimen. The influence of end friction is reduced by using smooth platens in the testing machine, and by increasing the sample height. It is for this reason that the height-to-diameter ratio has been standardized at 2. Testing shorter samples increases end restraint and is on the unsafe side for design.

## 18.9  STRESS PATHS

### 18.9.1  Overview

Stress paths afford a simple way to graph soil stress history, beginning with the geological origin and compression under the weight of overburden. Stress paths were developed as an engineering tool at MIT by T. W. Lambe, who showed that the stress path for a laboratory triaxial shear test can be designed to simulate an anticipated progression of loading conditions in the field and used to predict settlement. Stress paths also are useful to show stress changes without the clutter of a series of Mohr circles. This is shown in Fig. 18.21.

### 18.9.2  Stress Path During Consolidation

Computer-controlled triaxial tests that allow no lateral strain show that as vertical stress is increased there is a proportional increase in horizontal stress. The proportionality constant is

$$K_0 = \frac{\sigma_h'}{\sigma_v'} \tag{18.16}$$

which is a ratio of effective stresses. This simulates a stress sequence during consolidation either in the field or in a laboratory consolidation test, as shown by

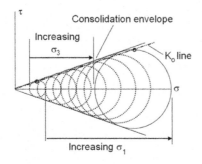

**Figure 18.21**
During consolidation both the major (vertical) and minor (horizontal) principal stresses increase according to a constant ratio, $K_0$. Connecting peaks of the Mohr circles generates a $K_0$ line.

the series of Mohr circles moving to the right in Fig. 18.21. This progression also can be simply represented by connecting the points of maximum $\tau$ values at the tops of the circles. This is called the $K_0$ *line*, and shows the stress path during consolidation.

### 18.9.3 Stress Path During Vertical Compression

In a conventional triaxial test, lateral confining stress is held constant so that as vertical stress increases, lateral bulging occurs that is partly restrained by the applied lateral pressure and partly by the soil internal friction. As loading continues, internal friction becomes more fully mobilized until it reaches a maximum level and the soil fails. The stress path is defined by a series of circles having an increasing diameter, and is oriented at 45° to the $x$-axis. This is shown in Fig. 18.22(a). A graph showing a stress path is called a *p-q diagram* where $q$ is the radius of the Mohr circle plotted on the $y$-axis and $p$ is the location of the center plotted on the $x$-axis. The relationship between Mohr circles and a *p-q* diagram is shown in Fig. 18.22.

### 18.9.4 $K_f$ Line

Mohr circles for tests conducted at different lateral stresses define a failure envelope as shown in Fig. 18.16, or the envelope also can be represented by a line connecting maximum $\tau$ values. This is the $K_f$ *line*. The slope angle and intercept of a $K_f$ line are designated by $\alpha$ and $a$ respectively, and relate to the corresponding failure envelope parameters $\phi$ and $c$ as follows:

$$\sin \phi = \tan \alpha \tag{18.17}$$

$$c = a \sec \phi \tag{18.18}$$

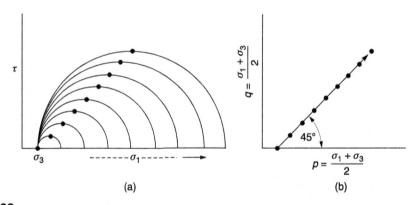

(a)                    (b)

**Figure 18.22**

(a) Increasing vertical stress while holding lateral stress constant results in a family of Mohr circles and a stress path oriented at 45° to the $x$-axis. (b) Peak shear stresses plotted as a "*p-q* diagram."

**Example 18.7**

A triaxial test of a clean sand is performed with a lateral stress of $10 \, \text{lb/in.}^2$ (7.0 kPa) (Fig. 18.23). The sample shears when the vertical stress reaches $30 \, \text{lb/in.}^2$ (21 kPa). What is the slope of the $K_f$ line? Of the failure envelope?

*Answer:* Cohesion of a clean sand may be assumed to be zero. The maximum shearing stress is the radius of the Mohr circle, which is $\frac{1}{2}(30-10)=10 \, \text{lb/in.}^2$ (7.0 kPa). The center of the circle is at $\frac{1}{2}(30+10)=20 \, \text{lb/in.}^2$ (14 kPa), so the slope of the $K_f$ line is $\alpha = \tan^{-1}(10/20)=26.6°$. From eq. (18.11), the slope of the failure envelope is $\phi = \sin^{-1}(10/20)=30°$.

### 18.9.5 Pore Water Pressure in a Triaxial Test

A stress path representation is particularly useful for depicting the effects of pore water pressure, since the stress path point is simply moved horizontally for a distance represented by $u$. An example is given in Fig. 18.24, which is for a CU or consolidated-undrained triaxial test. In this test drainage is allowed during the application of all-around pressure, so initially the total and effective stresses are equal and are represented by a Mohr circle that is a single point at A. Then the drain cock is closed and a vertical stress is applied so that the Mohr circle grows to the right and the stress path begins its ascent at 45°. However, as positive pore pressure develops it subtracts from the major and minor principal stresses, which moves the stress point to the left from B to C. The effective stress path therefore is from A to C. The path is curved because as vertical stress is applied the soil further consolidates, decreasing the voids.

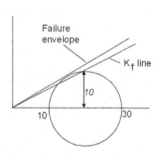

**Figure 18.23**
Diagram for Example 18.7.

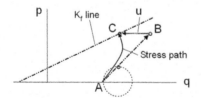

**Figure 18.24**
Influence of positive pore water pressure on a stress path, which is diverted from AB to AC.

### 18.9.6 Drainage Options During a Triaxial Test

There are several options for drainage during a triaxial test. If no drainage is allowed, the test is referred to as *UU*, unconsolidated-undrained, which is most useful for defining the strength of soft, saturated clay under rapid loading such as wheel loads. This is a forced "$\phi = 0$" condition wherein parameter *A* in the next section equals 1.0.

At the other extreme from a *UU* test is *CD*, a consolidated-drained test, in which excess pore water is allowed to drain from the specimen during both consolidating and shearing stages. This represents slow loading conditions that often occur in the field, but easily can require a week or more of laboratory time for a single specimen and therefore is rarely used.

Most common triaxial test is *CU*, consolidated-undrained, in which the drain cock is left open and time is allowed for consolidation during application of the confining stress, and then the drain is closed and pore water pressures are measured during vertical loading. Data from this test are used to predict drained conditions by subtracting pore pressures from the principal stresses or from the stress path points, as indicated in the preceding discussion. The simulation is not perfect, however, particularly in normally consolidated soils such as recently deposited subaqueous sandy and silty sediments, or in very sensitive or quick clays. When drainage is permitted, these soils gain strength as a vertical load is applied as a result of additional consolidation prior to shearing, whereas with undrained shearing there is no opportunity for further consolidation so strength is lower. While this is on the safe side for design, the most realistic procedure is to select test conditions that most nearly simulate those anticipated in the field.

### 18.9.7 Pore Pressure Parameters

Pore water pressures measured during triaxial tests define stress paths that are valuable indicators of soil behavior. For example, a dense soil that dilates during shearing will show a reduction in pore water pressure, whereas a consolidating soil will show an increase. Although Bishop proposed several parameters to describe changes in pore water pressure during consolidation and shearing, the most relevant for triaxial tests is the *A* parameter, which is defined as

$$A = \frac{\Delta u}{\Delta \sigma_1} \tag{18.19}$$

where $\sigma_2 = \sigma_3$, both being held constant. Maintenance of constant lateral stress is critical to this expression as it allows the soil to either compress or dilate, which will decrease or increase the volume of the voids, respectively. In a consolidometer test where there lateral strain is not allowed, the increment of pore pressure momentarily equals the increment of applied vertical stress.

A more general expression for $A$ that takes into account changes in the two minor principal stresses is

$$A = \frac{\Delta u - \Delta \sigma_3}{\Delta \sigma_1 - \Delta \sigma_3} \qquad (18.20)$$

where $\sigma_2 = \sigma_3$, both allowed to vary. For this derivation see Lambe and Whitman (1979). Stress path directions showing different values of $A$ are shown in Fig. 18.25.

### Example 18.8

Characterize the soil behavior in Fig. 18.24 from the stress path directions.

*Answer:* The stress path initially rises to the right, and a comparison with the slopes in Fig. 18.25 indicates that the soil is overconsolidated. As the test proceeds, the change in direction and trend toward a more positive pore water pressure suggests that the soil may be entering a consolidation stage. This will be confirmed if there is a break in the shear envelope.

## 18.9.8 Predicting Pore Pressures in the Field

To be more useful for field prediction of pore water pressures eq. (18.9) can be solved for $\Delta u$:

$$\Delta u = A(\Delta \sigma_1 - \Delta \sigma_3) + \Delta \sigma_3 \qquad (18.21)$$

### Example 18.9

Predict the undrained pore water pressure and vertical effective stress in a soft, saturated clay under a tire having an inflation pressure of $30 \, \text{lb/in.}^2$ (210 kPa).

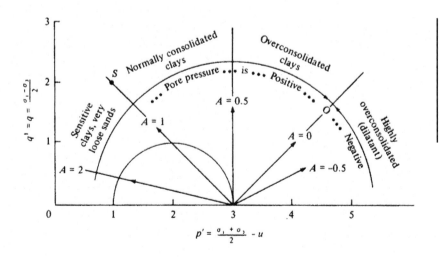

**Figure 18.25**

Stress path directions for different values of pore pressure parameter $A$, and relations to the soil stress history.

*Answer:* In a soft clay assume $A = 1$: $\Delta u = \Delta \sigma_1$ and all applied stress will go to pore water pressure, there will be no increase in effective stress regardless of the horizontal stress, and you are stuck in the mud.

## 18.9.9   Uses and Limitations of Pore Pressure Parameters

It should be emphasized that the use of pore pressure parameters is appropriate *only if the soil is saturated and the soil changes in volume.* Pore pressure parameters that are incorporated into computer programs for slope stability are relevant only if the soil volume changes, for example by imposing an additional load such as fill soil at the top of the hill.

A pore pressure parameter also is applicable if the soil dilates during initial shearing, but in that case the parameter is negative and gives only a temporary increase in shearing resistance. This will delay but not prevent a landslide as the negative pore becomes relieved by sucking in water.

Positive pore pressures developed during foundation loading have a deleterious effect on the foundation bearing capacity by subtracting from the normal stress and therefore the shearing strength. The procedure to calculate excess pore water pressure is as follows:

- Measure parameter $A$ from a triaxial test of the soil.
- Determine the change in vertical stress, $\Delta \sigma_1$, based on an elastic stress distribution.
- Determine the change in horizontal stress, $\Delta \sigma_3$, based on an elastic stress distribution.
- Determine $\Delta u$ from eq. (18.21).

This will give a worst-case scenario based on undrained conditions such as might occur during rapid construction with poor drainage conditions. Where pore water pressures are anticipated to pose a problem for bearing capacity, as in the case of an earth dam or road embankment constructed on soft clay, they should be monitored during construction and if they approach a dangerous level, construction should be halted to allow drainage to occur.

### Example 18.10

Assume a point load of 5 kN (1.12 kip) is applied at the surface of a saturated soil. If $A = 0.5$, calculate (a) the excess pore water pressure, (b) the vertical effective stress, and (c) the percent change in horizontal shear strength at a depth of 1 m immediately after application of the load and later after excess pore pressure dissipates. The soil unit weight is $20 \, \text{kN/m}^3$ ($125 \, \text{lb/ft}^3$).

*Answer:* Prior to application of the load, $\sigma_z = 1 \times 20$ and $\sigma_z' = 1 \times (20 - 9.8)$ $= 10.2 \, \text{kPa}$. The increase in vertical stress from theory of elasticity is

$\Delta\sigma_z = 3P/2\pi z^2 = (3 \times 5\,\text{kN})/(2\pi \times 1^2) = 2.39\,\text{kPa}$, part of which is carried by excess pore water pressure. Then

   (a) From eq. (18.10), $\Delta u = \Delta\sigma_1 A = 2.39 \times 0.5 = 1.19\,\text{kPa}$.

   (b) After application of the load, $\sigma_z' = 10.2 + \Delta\sigma_z - \Delta u$

$$= 10.2 + 2.39 - 1.19 = \underline{11.4\,\text{kPa}}$$

   (c) The percent change in vertical effective stress and hence in horizontal shear strength is from 10.2 to 11.4, or +12%, so soil at a depth of 1 m will be stronger after application of the load. On the other hand, the additional strength may be required to support the load. Then as the excess pore pressure dissipates, the situation will become less critical as the vertical effective stress changes to $10.2 + 2.39 = 12.59\,\text{kPa}$, which is a net increase of 23% over the original condition.

## 18.9.10 Pore Air Pressure

Soils that are not saturated prior to testing add a complication to undrained stages of triaxial testing, as air and water trapped inside the membrane cannot escape, although under pressure the air can go into solution. Soils compacted to simulate optimum field compaction conditions contain at least 1or 2 percent air, in which case positive pore pressure measurements mainly reflect pressure in the pore air, because pore water pressure in an unsaturated soil is negative relative to atmospheric pressure. The compressibility of air compared to pore water provides a better cushion against changes in pore pressure, and a removal of air by solution will be reflected by a change in parameter $A$.

Pore air and pore water pressures may be measured separately during triaxial testing, air through the top platen and water through the bottom. Because a positive pore air pressure tends to hold grains apart, it should be subtracted from the principal stresses to obtain an effective stress that includes the influence of suction in the capillary contact areas.

Significant pore air pressures are less likely to develop in an open system in the field than in a sealed triaxial test specimen. Pore air pressure in the field nevertheless is subject to changes in atmospheric pressure. A low-pressure weather system therefore may be conducive to landslides—particularly if it brings rain. A hurricane, by reducing atmospheric pressure, therefore can influence the occurrence of submarine landslides in soils on the ocean bottom.

## 18.9.11 Stress Path Analysis of Compaction

Chapter 13 described the need to compact soil in layers in order to take advantage of lateral confining stress provided by friction with a firm underlying layer. A more quantitative determination can be made if vertical and horizontal stresses with and without a frictional underlayer are calculated from Burmister and

Boussinesq elastic theories, respectively. Representative results are shown at the left in Fig. 18.26, where it will be seen that the $\sigma_1$ vertical stress is relatively unaffected, whereas the $\sigma_3$ horizontal stress is greatly influenced by the presence of a rough underlayer.

The middle graph in Fig. 18.26 shows $q/p$ ratios calculated from the data on the left. In order to prevent shearing and promote compaction, the maximum allowable $q/p$ ratio therefore is defined by the soil internal friction, as cohesion is low or essentially zero in an initially loose soil. The compacting zones from a first roller pass are shown with and without a rough base, on soil having an intial friction angle of 25° that corresponds to sliding friction. As the friction angle is larger, the $q/p$ ratio becomes less critical, indicating that a granular soil having a higher friction angle, such as crushed stone, can be compacted in thicker layers.

As each pass of a compactor builds up the underlayer, the loose soil layer compacts from the bottom up a well as from the top down. Top-down compaction can be a disadvantage by creating a stiff upper crust that then will spread and diminish the pressure from the compactor, so projections often are used that concentrate stress and punch through the upper layer. The roller projections therefore can be designed to cause a localized bearing capacity failure, discussed in the next chapter.

**Figure 18.26**

$p/q$ ratios calculated from elastic analyses show the importance of having a firm, rough layer to compact against. Boussinesq stresses are without a firm layer; Burmister with a firm layer.

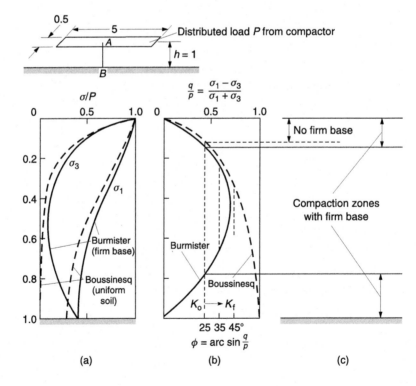

## 18.10 CONDUCT OF A TRIAXIAL TEST

### 18.10.1 Preserving Saturation

Samples intended for triaxial testing in a saturated condition should be prevented from drying, as resaturation is difficult and time-consuming. A partial vacuum can pull water into the sample, and back-pressuring can facilitate solution of pore air. Sampling of ocean or lake bottom soils can result in dissolution of gases that appear as tiny bubbles throughout the sample so that the soil literally grows out of the ends of a sampling tube. Sampling tubes should be sealed and kept refrigerated to reduce this source of error.

### 18.10.2 Reduction of Triaxial Test Data

Unlike a direct shear type test where the shear area is governed by the apparatus, triaxial tests usually involve bulging that changes the cross-sectional area. Since lateral stress, $\sigma_3$, is an applied pressure it does not change, but the vertical stress is the axial load divided by the cross-sectional area that changes. Because shearing occurs through a considerable length of the specimen, the average area may be used and can be obtained from height and volume change measurements (Bishop and Henckel, 1962).

Let $A_0$, $V_0$, and $h_0$ be the initial cross-sectional area, volume, and height of the specimen. Then

$$A_0 = \frac{V_0}{h_0} \tag{18.22}$$

Changes in these values are represented by $\Delta$:

$$A_1 = \frac{V_0 + \Delta V}{h_0 + \Delta h} \tag{18.23}$$

where $A_1$ is the cross-sectional area adjusted for bulging Dividing eq. (18.23) by (18.22) gives

$$A_1 = A_0 \frac{1 + \Delta V/V_0}{1 - \epsilon} \tag{18.24}$$

where strain is expressed as a positive number when decreases in volume and length are negative. In a UU (unconsolidated, undrained) test of saturated soil, if $\Delta V = 0$,

$$A_1 = A_0 \frac{1}{1 - \epsilon} \tag{18.24a}$$

### Example 18.11

A CU (consolidated, undrained) triaxial test of saturated, overconsolidated glacial till gives the data in Table 18.1. (a) Plot the stress-strain curve; (b) determine the elastic modulus; (c) determine the strength; (d) determine the pore pressure parameters at 80% of the

**Table 18.1**

Data from a triaxial test on saturated glacial till, sample depth 4 m (13 ft), OCR $= 8.2$. Test $= 138$ kPa (20 lb/in.$^2$).
Courtesy Dr. A. J. Lutenegger

| 1 | 2 | 3 | 4 | 5 | 6 | 7 | 8 | 9 | 10 | 11 | 12 |
|---|---|---|---|---|---|---|---|---|----|----|----|
| Vert. dial. 0.025 mm (0.001 in.) | $\epsilon$ | $\Delta V$ (cm$^3$) | $\Delta V/V_0$ | $a$ (cm$^2$) | $Q$ (N) | $\sigma_1 - \sigma_3$ (kPa) | $\sigma_1$ (kPa) | $u$ (kPa) | $q$ (kPa) | $P$ (kPa) | $p'$ (kPa) |
| 0 | 0 | 0 | 0 | 37.06 | 0 | 0 | 138 | 15 | 0 | 138 | 123 |
| 100 | 0.018 | −0.52 | −0.00099 | 37.7 | 400 | 106 | 244 | 22 | 53 | 191 | 169 |
| 200 | 0.036 | −0.90 | −0.00017 | 38.4 | 667 | 174 | 312 | 34 | 87 | 225 | 191 |
| 300 | 0.054 | −1.25 | −0.00024 | 39.1 | 810 | 207 | 345 | 41 | 104 | 242 | 201 |
| 400 | 0.071 | −1.59 | −0.00030 | 39.8 | 890 | 223 | 361 | 43 | 112 | 250 | 207 |
| 500 | 0.089 | −1.92 | −0.00036 | 40.5 | 943 | 233 | 371 | 43 | 117 | 255 | 212 |
| 600 | 0.107 | −2.16 | −0.00041 | 41.4 | 979 | 236 | 374 | 42 | 118 | 256 | 214 |
| 700 | 0.125 | −2.39 | −0.00045 | 42.2 | 996 | 236 | 374 | 41 | 118 | 256 | 215 |

*Note*: $h_0 = 142.3$ mm (1.503 in.); $d_0 = 68.71$ mm (2.705 in.).
$V_0 = 527.6$ cm$^3$ (32.2 in.$^3$); $a_0 = 37.06$ cm$^2$ (5.747 in.$^2$).

failure stress. The initial height $h_0 = 142.3$ mm and the initial diameter is 68.7 mm.

*Step 1: Data reduction.* $A_0 = \pi (68.7/2)^2 = 37.07$ cm$^2$; $V_0 = 37.07 \times 14.23 = 527.5$ cm$^3$.

In the data row of a dial reading 100 in the table,

Column 1, $\Delta h$ is from a dial reading, $100 \times 0.025$ mm

Column 2, $= 100 \times 0.025/142.3 = 0.0176$

Column 3, $\Delta V$ measured from displacement of the liquid filling the triaxial chamber, −0.52 cm$^3$

Column 4, $\Delta V/V = -0.52/527.5 = 0.00099$

Column 5, eq. (18.24) $A_1 = 37.07 \times (1 - 0.00099)/(1 - 0.0176) = 37.7$ cm$^3$

Column 6, measured axial load, $400 \text{ N} = 0.4$ kN

Column 7, axial loading equals $\sigma_1 - \sigma_3 = Q/A_0 = 0.4 \text{ kN}/37.7 \text{ cm}^2 \times (10^4 \text{ cm}^2/1 \text{ m}^2) = 106$ kPa

Column 8, $\sigma_1 = 106 + 138 = 244$ kPa

Column 9, pore water pressure measured with manometer, $u = 22$ kPa

Column 10, $q = \sigma_1 - \sigma_3/2 = 106/2 = 53$ kPa

Column 11, $p = (\sigma_1 + \sigma_3)/2 = (244 + 138)/2 = 191$ kPa

Column 12, $p' = p - u = 191 - 22 = 169$ kPa.

*Step 2: Solutions.*

(a) The stress that is plotted versus strain usually is the *deviator stress*, which is $\sigma_1 - \sigma_3$. That is because $\sigma_3$ is applied all around the specimen at the start of the test. Also, the deviator stress, being a shear stress, is the same on a total and on an effective stress basis. The stress-strain plot is shown in Fig. 18.27.

(b) A true elastic modulus does not exist because of the curvilinear stress-strain plot. Two estimates of the elastic modulus are the *tangent modulus*, shown by *OA* in Fig. 18.27, and the *secant modulus*, shown at *OB* for 5% strain. The tangent modulus is $E = 200$ kPa$/0.019 = 10,500$ kPa (1520 lb/in.$^2$). The secant modulus at $= 5\%$ is $E = 200$ kPa$/0.05 = 4000$ kPa (580 lb/in.$^2$).

(c) Failure usually is defined on the basis of the peak strength, but it also may be the ultimate (final) strength, strength at a particular strain, or strength at minimum volume. In Fig. 18.27 the peak and ultimate strengths are the same, at $\sigma_1 - \sigma_3 = 236$ kPa (34.2 lb/in.$^2$). A minimum volume is not defined by the data.

(d) 80% of the peak failure stress is $\sigma_1 - \sigma_3 = 0.8(236) = 189$ kPa. From Table 18.1, at a deviator stress of 174 kPa, parameter $A = \Delta u/\Delta \sigma_1 = (34 - 15)/(312 - 138) = 0.11$.

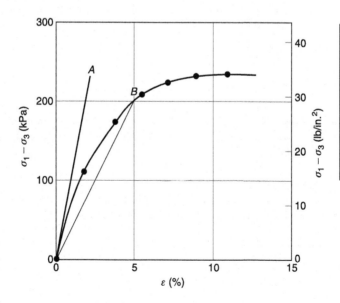

**Figure 18.27**

Deviator stress vs. strain in a triaxial test. The tangent modulus is represented by the line from the origin to A, the secant modulus from the origin to B.

*Questions:*

(1) Why is $\Delta V$ so small?

(2) What consolidation state is indicated by pore pressure parameter $A$?

(3) What can you infer from $u$ being positive at the beginning of the test?

## 18.10.3 Best-Fit Failure Envelope from Triaxial Test Data

A major annoyance in processing of data from triaxial tests is difficulty in fitting a common tangent to a series of Mohr circles whose diameter is affected by random error. The larger the degree of randomness, the larger the degree of difficulty. If the soil is quite variable it is even possible for Mohr circles to be concentric so that there can be no common tangent, and direct shear type tests such as the BST are used to sort things out (Handy et al., 1985).

A simple linear regression of $p$ on $q$ *is not correct* and will give results that are on the unsafe side. The reason is because a linear regression analysis *requires* that the dependent variable be in the $y$ direction so that random variations are vertical, in the direction of the least squares determination. In a triaxial test $\sigma_3$ is the independent variable that is held constant, but as shown in Fig. 18.22(a), with $\sigma_3$ constant and random variations in strength, $\sigma_{3max}$, variability will be at $45°$.

One option is to rotate the axes before and after the regression (Handy, 1981). Another that was suggested by the U.S. Bureau of Reclamation is to regress $\sigma_{1max}$ versus $\sigma_3$ and obtain a linear equation having the form $\sigma_1 = a(\sigma_3) + b$. It then can be shown that

$$\phi = \tan^{-1}(a-1)/2\sqrt{a} \tag{18.25}$$

$$c = \frac{b}{2\sqrt{a}} \tag{18.26}$$

The regression also will yield a value of $R^2$ as an indication of reliability, and a more detailed statistical analysis can define confidence limits for mean values of $c$ and $\tan \phi$.

### Example 18.12
A regression of failure stress to lateral stress in triaxial tests yields $a = 2.4$ and $b = 90\,\text{kPa}$. Determine the best-fit friction angle and cohesion.

*Answer:*

$$\phi = \tan^{-1}(2.4 - 1)/2\sqrt{2.4} = 24°$$
$$c = 90/2\sqrt{2.4} = 29\,\text{kPa}$$

### 18.10.4 Failure Criteria

Engineers usually think of failure in terms of maximum stress, but other criteria may be preferable for certain purposes. For example, in a fissured clay or in weathered soils susceptible to creep, cohesion may be omitted and only the residual strength friction angle used in stability calculations (Bjerrum, 1967).

In loose sand the ultimate strength may not be attainable within strain tolerances, in which case a maximum allowable strain may be specified. For example, if the sand were 3 m thick and the allowable settlement is 0.03 m, one could evaluate the stress giving 0.01 axial unit strain in the triaxial test.

If a potential shear surface cuts through two soils that do not have the same relationship of stress to strain, the peak strength for one may be used in conjunction with the residual strength from the other.

Figure 18.6 also shows that most soils coming under load will first decrease in volume owing to densification or elastic deformation or both, and then increase because of reduction in interlocking. Holtz, of the U.S. Bureau of Reclamation, therefore suggested that the initial increase in volume signals initiation of failure, even though the soil has not yet reached a maximum shearing resistance. The minimum volume point also is the point where pore pressure reaches a maximum, although there may be some lag in tests due to permeability effects. The strength at minimum volume (or maximum pore pressure) may be used where initiation of failure is not allowable, as in road bases subjected to repeated loads. Repeated loading to stresses above the minimum volume strength can result in rapid disintegration and failure.

## 18.11  IMPORTANCE OF FIELD OBSERVATIONS

Sometimes soil is so variable that it is practically impossible to obtain representative samples. A shear strength test may be meaningless in a mass of clay that is highly fissured and slickensided unless strength is measured along these partings and discontinuities. In this case the slope of a steep failing bank may give a better measure of the shear angle and hence the friction angle.

Another indication of shear strength is the angle of repose of a cohesionless material that is impossible to sample and test, for example on a talus slope or the slip face on the lee side of a dune.

As will be discussed in Chapter 21 on slope stability, the maximum height of a vertical cut gives a measure of unconfined compressive strength.

A steeply sloping hillside that shows evidence of soil creep (such as curved tree trunks) gives a measure of the creep friction angle with zero cohesion.

Natural hillslopes are products of rockfalls, landslides, creep, and slope wash. Any changes that decrease stability must be compensated by changes that increase it. Measurements of natural slope angles have been used to assign minimum setbacks for housing developments and were used to define slopes of cuts in various soils in connection with planning for a Panama sea-level canal.

## Problems

18.1. A 220 N (50 lb) weight resting on a concrete floor is acted on by a horizontal force of 43 N (9.7 lb). What is the obliquity angle? What is the developed coefficient of friction?

18.2. If a horizontal force of 133 N (30 lb) is required to move the weight in Problem 18.1, what are the coefficient of friction and the friction angle?

18.3. Suppose that the weight in Problems 18.1 and 18.2 rests on a concrete ramp that makes an angle of 30° with the horizontal. Will the weight slide down the ramp?

18.4. Describe internal friction and cohesion in a dense sand as compared with a loose sand.

18.5. Three trials in a direct shear test yield the following results. Draw the shear diagram, and determine the cohesion and the angle of friction of the soil. Also determine the intrinsic stress in the soil.

| | Stresses at failure, kPa (lb/ft$^2$) | |
| --- | --- | --- |
| Specimen No. | Normal | Shearing |
| 1 | 14.4 (300) | 29.0 (605) |
| 2 | 28.7 (600) | 35.7 (745) |
| 3 | 50.3 (1050) | 45.5 (950) |

18.6. The maximum and minimum principal stresses at a point in a stressed body are 1000 and 300 stress units, respectively. Determine the normal and shearing stresses on a plane that passes through the point and makes an angle of 25° with the maximum principal plane. Also determine the resultant of the normal and shearing stresses and find the obliquity of the resultant.

18.7. What is the maximum shearing stress at the point under consideration in Problem 18.6? What normal stress, resultant stress, and obliquity angle are associated with this maximum shearing stress?

18.8. An unconfined compression test is conducted on a cylinder of stiff, fine-grained cohesive soil. The stress at failure is 30.6 kPa (640 lb/in.$^2$). What is the shearing strength of this soil? Draw the Mohr diagram and the $\phi = 0$ failure envelope.

18.9. Triaxial compression tests on three identical cylinders of soil give the following results. Draw the Mohr diagram, determine the cohesion and the angle of friction of the soil, and estimate the failure angle and the normal stress, shearing stress, and the resultant stress on the failure plane of each specimen at failure.

| | Stresses at failure, kPa (lb/in.$^2$) | |
|---|---|---|
| Specimen No. | Lateral | Axial |
| 1 | 47.9 (1000) | 230 (4800) |
| 2 | 95.8 (2000) | 364 (7600) |
| 3 | 144 (3000) | 469 (9800) |

18.10. Perform a regression analysis on the data from the preceding example and define $\phi$, $c$, and $R^2$.

18.11. When an unconfined compression test is conducted on a cylinder of soil, it fails under an axial stress of 105 kPa (2200 lb/in.$^2$). The failure plane makes an angle of 50° with the horizontal. Draw the Mohr diagram, and determine the cohesion and the angle of friction of this soil sample.

18.12. (a) Give two examples of $\phi = 0$ and discuss the role of pore water pressure in each example. (b) What is the $A$ pore pressure parameter for each of the examples? How well do the $A$ parameters describe the properties of these soils?

18.13. A UU triaxial test with $\sigma_3 = 188$ kPa (20 lb/in.$^2$) on an unsaturated loess soil sample with $h_0 = 142$ mm (5.6 in.) and $h_0/d = 2.0$ gives the following data:

| | | Axial load | | $u$ | |
|---|---|---|---|---|---|
| $_{,1}$ (%) | $\triangle V$ (cm$^3$) | (N) | (lb) | (kPa) | (lb/in.$^2$) |
| 0 | 0 | 0 | | 32 | (4.7) |
| 0.34 | −103 | 29 | (6.6) | 34 | (4.9) |
| 0.69 | −2.33 | 223 | (50) | 34 | (5.0) |
| 1.37 | −4.50 | 497 | (112) | 37 | (5.4) |
| 2.06 | −6.56 | 618 | (139) | 39 | (5.6) |
| 3.43 | −9.76 | 695 | (156) | 43 | (6.2) |
| 5.15 | −12.74 | 703 | (158) | 45 | (6.5) |

Note: (a) Plot the deviator stress versus strain, and determine (b) the tangent and secant moduli, (c) the failure stress, (d) parameter $A$ at failure, (e) an optional failure stress.

18.14. The test of Problem 18.9 gave the following measured pore pressures at the respective failure stresses. Draw the p-q diagram on an effective stress basis, draw and label the $K_f$ line, and find $\alpha$, $c'$, and $\phi'$.

| Specimen No. | $u$, kPa (lb/in.$^2$) |
|---|---|
| 1 | 21.4 (3.1) |
| 2 | 34.5 (5.0) |
| 3 | 66.2 (9.6) |

18.15. In Fig. 18.25 explain why the $A = 0.5$ line is vertical.

18.16. A CU triaxial test on clay gives the following data with $\sigma_3$ maintained constant at 68.9 kPa (10 lb/in.$^2$). Plot the total and effective stress paths. Draw the $A = 0$ line and evaluate parameter $A$ at failure.

Pore pressures under various axial stresses, kPa (lb/in.$^2$)

| $\sigma_1$: 69 (10) | 97 (14) | 124 (18) | 152 (22) | 169 (24.5) (failure) |
|---|---|---|---|---|
| $u$: 14 (2) | 31 (4.5) | 48 (7) | 69 (10) | 83 (12) |

Qualitatively characterize the amount of preconsolidation relative to the applied $\sigma_3$.

18.17. In consolidometer tests of a saturated clay, increments of pore pressure developed immediately after loading approximately equal the increments of load. Assuming that $\Delta\sigma_3 = K_0\Delta\sigma_1$, evaluate pore pressure parameter $A$ from eq. (18.20). Does this value appear reasonable?

18.18. (a) Discuss the role of pore pressure in compaction. (b) Give reasons for scarifying the surface of a compacted clay layer prior to the addition of more material.

18.19. A single-cylinder water tower resting on sand for 20 years suddenly develops a severe tilt and is in danger of collapse. Prepare for a class discussion by developing at least three working hypotheses to explain the difficulty, and indicate what evidence should be gathered to test each hypothesis. (Do not overlook possible contributions from weather.) Why not just propose one hypothesis and stick to it?

18.20. Discuss relationships between preconsolidation pressure, settlement, and shear strength.

18.21. An expert witness for the other side in a lawsuit states that ocean-bottom soils are hard because they have been compacted by the weight of the water. Can you refute that statement in terms that will be understood by a lay person?

18.22. A building on a floodplain is partly founded on point bar sand and partly on clay that has been overconsolidated by desiccation. During a brief period of high water either one part of the building settles or another part is lifted up. Explain both possibilities.

18.23. Can there be pore water pressure where there is a neutral stress?

18.24. Is there any problem applying eq. (18.7) to negative pore water pressures?

# References and Further Reading

Bishop, A. W., and Henkel, D. J. (1962). *The Measurement of Soil Properties in the Triaxial Test,* 2nd ed. Edward Arnold, London.

Bjerrum, L. (1967) "Progressive Failure in Slopes of Overconsolidated Plastic Clay and Clay Shales." *ASCE J. Soil Mech. and Foundation Eng. Div.* 93(SM5), 3–49.

Bjerrum, L., and Landva, W. (1966). "Direct Simple Shear Tests on a Norwegian Quick Clay." Norwegian Geotech. Inst. Publ. No. 70.

Bowden, F. P., and Tabor, D. (Part 1, 1950 and Part 2, 1964). *The Friction and Lubrication of Solids.* Clarendon Press, Oxford.

Eden, W. J., and Mitchell, R. J. (1970). "The Mechanics of Landslides in Leda Clay." *Canadian Geotech. J.* 11(3), 285–296.

Fredlund, D. G., and Rahardjo, H. (1993). *Soil Mechanics for Unsaturated Soils.* John Wiley & Sons, New York.

Gibbs, H. J., and Coffey, C. T. (1969). "Techniques for Pore Pressure Measurements and Shear Testing of Soil." *Proc. 7th Int. Conf. on Soil Mechanics and Foundation Engineering* I, 151–157.

Handy, R. L. (1981). "Linearizing Triaxial Test Failure Envelopes." ASTM *Geotech. Testing J.* 4(4), 188–191.

Handy, R. L., Schmertmann, J. H., and Lutenegger, A. J. (1985). "Borehole Shear Tests in a Shallow Marine Environment." *ASTM Spec. Tech. Publ.* 883, 140–153.

Hogentogler, C. A., et al. (1937). *Engineering Properties of Soil.* McGraw-Hill, New York.

Lambe, T. W., and Whitman, R. V. (1979). *Soil Mechnanics.* John Wiley & Sons, New York.

Lohnes, R. A., Millan, A., and Handy, R. L. (1972). "Measurement of Soil Creep." *ASCE J. Soil Mech. and Foundation Eng. Div.* 98(SM1), 143–147.

Lutenegger, A. J., and Timian, D. J. (1987). "Reproducibility of Borehole Shear Test Results in Marine Clay." *ASTM Geotech. Testing J.* 10(1), 13–18.

Marston, A., and Anderson, A. O. (1913). "The Theory of Loads on Pipes in Ditches" Bull. 31, Iowa Engineering Experiment Station, Ames, Iowa.

Mitchell, J. K. (1993). *Fundamentals of Soil Behavior.* John Wiley & Sons, New York.

Rowe, P. W., Borden, L., and Lee, I. K. (1964). "Energy Components During the Triaxial Cell and Direct Shear Tests." *Geotechnique* 14, 247–261.

Skempton, A. W. (1985). "A History of Soil Properties." *Proc. 11th Int. Conf. on Soil Mechanics and Foundation Engineering,* Golden Jubilee Volume, 95–121.

Skempton, A. W., and Petley, D. J. (1967). "The Strength along Structural Discontinuities in Stiff Clays." *Proc. Geotech. Conf. Oslo* 2, 29–46.

Taylor, D. W. (1948). *Fundamentals of Soil Mechanics.* John Wiley & Sons, New York.

Terzaghi, K. (1925). *Erdbaumechanik.* Franz Deuticke, Vienna.

# 19

## Lateral Stress and Retaining Walls

## 19.1 OVERVIEW

### 19.1.1 Uses of Retaining Walls

Earth-retaining structures hold hillsides from crashing into houses or dumping out onto roadways and bike trails. Bridge abutments are retaining walls that support ends of bridges, and basement walls are retaining walls that serve a similar dual purpose, holding soil back while providing support for a building.

The design of retaining walls was first approached analytically by military engineers. An analysis made by Coulomb in the eighteenth century still is the basis for most computer programs.

### 19.1.2 Kinds of Retaining Walls

Short retaining walls can be stacked timbers or railroad ties but, as shown in Fig. 19.1, if such a wall is over about a meter (3–4 ft) high it almost inevitably will bulge or tilt and fail if the timbers are not anchored into soil behind the wall.

Another approach for landscaping walls is to use patented concrete blocks with extended edges so that when the blocks are stacked they automatically lean back into the soil, which is referred to as "batter." Such walls are flexible and are called "segmented walls."

Higher walls are either heavy enough that they do not slide or tilt excessively, or they are anchored with tension elements or "tiebacks" that extend back into the soil. High walls pose a higher risk if they should suddenly tilt and fall over.

**Figure 19.1**
This dog-gone wall was not built right! Too many ties and no tiebacks. According to the classical triangular distribution of earth pressure the wall should have kicked out at the bottom. It didn't. More on this later.

Engineering expertise is required for their design, and skill and care are required in their construction. Death by wall can be extremely fatal.

"Gravity" retaining walls are massive structures that are held in place by their own weight. Soil is excavated to make room for the wall, the wall is built, and finally soil is backfilled into the space behind the wall. This is a "bottom-up" construction sequence.

Instead of depending on gravity alone, a tension member can be attached to the wall and anchored in the soil in a kind of "boot-strap" approach. One type is mechanically stabilized earth, or MSE, invented by a French engineer, H. Vidal. MSE is constructed from the bottom up as facing panels are held in place by horizontal steel or plastic strips that extend back into the soil. The design of MSE walls is discussed in the next chapter.

Retaining walls also can be built from the top down, for example by driving steel sheet pile or filling a trench with concrete, and then excavating the soil away from one side. As these walls are thin they are not gravity walls, but are held in place by tiebacks that are rods or cables anchored in the soil.

A somewhat similar procedure called "soil nailing" essentially consists of the rod reinforcement without a wall, as steel rods hold the ground in place by friction. The exposed surface is protected from erosion by steel mesh and shotcrete, which is sprayed-on concrete. Soil nailing also is discussed in the next chapter.

### 19.1.3  Plastic vs. Elastic Soil Behavior

Soil weight is converted to lateral pressure against a wall by either of two different mechanisms: (1) internal slipping as the soil tends to shear, or (2) elastic deformation if the soil expands or bulges laterally according to Poisson's ratio and does not shear. Walls are designed to yield slightly in order to mobilize shearing resistance in the soil, so design is based on the soil internal friction. However, after a wall and soil have been in place for a period of time the soil gains sufficient strength that a concentrated load on soil behind the wall causes an elastic response that is solved using the theory of elasticity. This distinction is important because stress transferred elastically is concentrated higher on the wall.

### 19.1.4  Yielding of a Retaining Wall

In order for shearing resistance of a soil to partially restrain it from pushing against a wall, the wall must be allowed to move slightly so that it acts to partly retain itself. This is referred to as the *active case* of lateral earth pressure. Walls are designed based on active-case soil pressures but include a factor of safety so that the restraining friction is not fully mobilized.

The ratio of horizontal to vertical stress for soil in active-case shearing is $K_a$. If a wall is prevented from moving, as in a bridge abutment or basement wall, soil pressures are higher, and the ratio of horizontal to vertical stress is designated $K_0$, for *earth pressure at rest*. Both $K_a$ and $K_0$ involve internal friction in the soil, but $K_0$ is larger than $K_a$.

### 19.1.5  Influence of Surface Loads

Additional pressure is transferred to a retaining wall if a load is placed on top of the soil behind the wall, whether the load is from a building, a road, or a parking lot. After soil has been in place for a few months or years it gains cohesive strength from aging. This even occurs in sandy soils. Walls also are built with a factor of safety so that some additional load can be applied without stressing the soil to its maximum strength.

As a result, a load imposed on the surface of soil behind a wall ordinarily is not sufficient to overcome the shearing resistance of the soil, so the soil response is elastic. While this may sound like an advantage, an elastic response tends to concentrate the additional pressure higher on the wall where it is more likely to cause tilting. However, as will be shown, elastically transferred pressure also decreases rapidly with increasing distance from a wall. It is for this reason that trucks should not be allowed to park close to the top of a retaining wall.

The first part of this chapter deals with wall design that is based on soil shear strength, and the second part discusses elastic response to surface loads.

## 19.2 DESIGN REQUIREMENTS

### 19.2.1 Lateral Pressure and Friction Angle

Soil pressure like water pressure increases with depth, but soil is heavy, about twice as dense as water. The higher the wall, the higher the soil pressure against the wall. However, unlike water, soil has internal friction that if allowed to mobilize can greatly reduce pressure on the wall. The amount of the reduction depends on the internal friction, which can be evaluated from shear strength testing. For low or inexpensive walls a shortcut is to use an empirical relationship such as shown in Fig. 19.2. Such approximations may err from the true values of friction angle by as much as 5°, so the designer must incorporate a larger factor of safety that in turn increases the cost. One of the more challenging aspects of geotechnical engineering is convincing a client that not only is design confidence increased but money can be saved by quality soil testing.

Most soils also have cohesion, but that is not included in design, particularly as cohesion is zero in loose soil backfill.

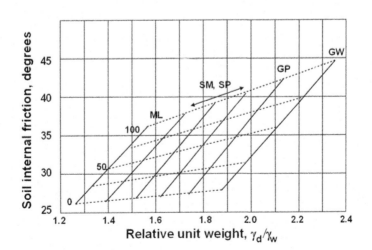

**Figure 19.2**

Estimates for friction angle in relation to soil classification and density. Dotted lines are for relative densities in percent. (Modified from Lazarte et al., 2003.)

### Example 19.1

Estimate the angle of internal friction for an ML soil having a dry unit weight of 90 lb/ft³.

*Answer:* The unit weights relative to that of water are $90/62.4 = 1.44$. The friction angle from the chart is $30 \pm 2°$, which is higher than some measured values.

## 19.2.2 Failure Modes

Retaining walls must be designed to resist several different failure modes that include:

- overturning;
- sliding along the base;
- bulging in the center area that may be preliminary to rupture;
- sinking and tilting as a result of eccentric loading and consolidation of the foundation soil;
- sinking caused by a foundation bearing capacity failure; and
- being part of a landslide, referred to as "global stability."

Some of these possibilities are illustrated in Fig. 19.3. Settlement calculations can proceed as outlined in the preceding chapter; design for global stability is discussed in Chapter 21 on slope stability, and design to resist a foundation failure is addressed in Chapter 22 on foundation bearing capacity.

**Figure 19.3**

Failure modes for retaining walls. Ignore one possibility and still lose it all.

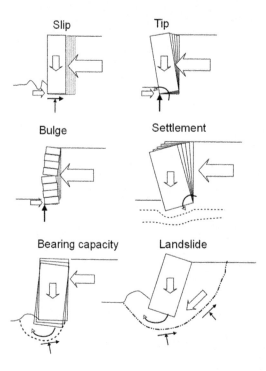

Slip  Tip

Bulge  Settlement

Bearing capacity  Landslide

### 19.2.3  Stability and Height

Vertical stress in a soil, or overburden pressure, is a product of depth and the soil unit weight, and lateral stress is proportional to vertical stress times the $K$ factor. Thus lateral stress is proportional to height, ideally varying from zero at the top of the wall to a maximum value at the bottom. The factors influencing stability then pile up like Chinese acrobats: force equals stress times area, and therefore is proportional to the height squared. Overturning moment equals force times a lever arm, and therefore is proportional to the height cubed. Costs go up likewise.

#### Short Walls

Generally retaining walls shorter than about 1.2 m (4 ft) that are commonly used for landscaping can be safely built without a sophisticated design, particularly if they use commercial interlocking concrete blocks available from home improvement stores. Failure of a low wall is not so much a tragedy as an inconvenience. Low walls built with timbers or railroad ties require tiebacks because wood is less dense than soil and cannot be held in place by the weight of the ties alone. A simple procedure is to attach and extend short sections of timber back into the soil to act as tiebacks.

#### Basement Walls

Stand-alone walls built with hollow concrete blocks also have a low survival rate. They survive as basement walls because of vertical pressure from floor joists spanning the structure. A failure to recognize this requirement can allow a heavy load such as a bulldozer or ready-mixed concrete truck that is too close to an unsupported wall to slip off into a basement.

### 19.2.4  Options

The constantly increasing variety of retaining wall designs is testimony to the ingenuity of engineers and builders. Some common types are illustrated in Fig. 19.4.

Many retaining walls are "gravity walls" that are held in place by their own weight. These may be massive blocks of concrete or can involve a cantilever arrangement that uses weight of soil to bear down on a projecting heel of the wall to prevent tipping or sliding, as shown at the center of Fig. 19.4. (The heel and toe of a retaining wall are designated on the basis that the wall is facing outward.) Gravity walls normally are built from the bottom up.

Driven steel sheet piles have interlocking edges that create a continuous wall and have great versatility. They often are used for waterfront situations where dewatering is impossible, and for temporary walls, for example to hold an excavation open and facilitate dewatering. This is a type of top-down wall. Tiebacks are installed as necessary to ensure stability.

**Figure 19.4**

Some kinds of retaining structures. Resistance comes from several sources including weight of the wall, passive resistance from soil in front of the wall, tiebacks, and in the case of a cantilever wall, weight of soil behind the wall.

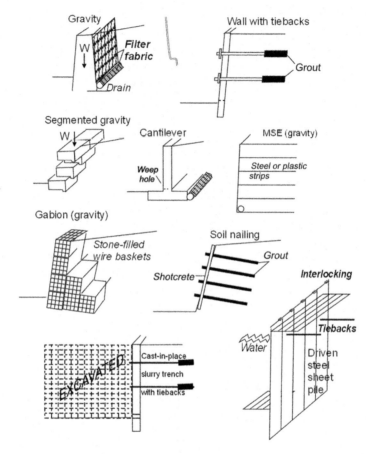

A slurry trench uses pressure from a bentonite-water slurry to temporarily support the sides of a trench until the wall is built. Steel reinforcing cages are lowered into the slurry, which then is displaced from the bottom by pumping relatively dense fluid concrete through a canvas chute or "tremie" that reaches to the bottom of the trench and is raised as the trench is filled. After the concrete has set and gained strength, excavation proceeds on one side and tiebacks are installed to help support the wall. The walls sometimes are called "diaphragm walls."

Yet another type of top-down wall is made by filling large vertical borings with soil mixed with a stabilizing chemical such as Portland cement, hydrated lime, or reactive fly ash, which is the dust collected from burning powdered coal in power plants. Adjacent borings overlap to create a continuous wall.

Tiebacks may be steel rods or cables with the ends grouted in place, or can be screwed-in soil anchors. Usually each tieback is pulled with a hydraulic jack, not only to test its anchoring capacity, but also to apply tension that helps to hold the wall in place, a process called "post-tensioning."

In mechanically stabilized earth, or MSE, the wall functions as a decorative facing and is held in place by steel or plastic strips extending back into the soil, but is not post-tensioned.

It can be seen that each new type of wall uses lessons from the past, and also can provide valuable new insights that will be helpful in the future.

### 19.2.5   Quality Control During Construction

The internal friction of soil depends in part on the degree of compaction, so the compacted soil density and moisture content are part of the specification, particularly for a high wall. The usual procedure is to specify a required density and moisture content, which are measured with nuclear gauges. A more recent trend is to directly measure the soil internal friction with a rapid in-situ testing method, such as the Borehole Shear Test, as a quality control measure. This can eliminate the requirement of empirically correlating strength with density, and is particularly advantageous if the fill soils are variable.

## 19.3   ACTIVE, PASSIVE, AND AT-REST EARTH PRESSURES

### 19.3.1   Internal Friction

Soil has characteristics of both a solid and a liquid, in that it exerts pressure against a vertical or inclined surface that increases approximately linearly with depth. However, unlike liquid, soil has an internal restraint from intergranular friction and cohesion. This is obvious because soil can be piled up, whereas a true liquid such as water flows out flat.

The ratio of lateral to vertical stress or pressure is designated by the ever-popular coefficient $K$. One factor affecting soil $K$ is whether the soil is pushing or is being pushed. Soil pressure against a retaining wall is many times lower than that on the blade of a bulldozer having the same height, and if the bulldozer were designed on the basis of active retaining wall pressures it could only back up.

### 19.3.2   Active Earth and Passive Pressures

The upper limit of soil pressure is *passive earth pressure*, which develops as soil passively resists being pushed. The lower limit is called *active earth pressure* as soil acts to retain itself. $K$ for the active case is designated $K_a$, and for the passive case, $K_p$. These are two of the most important parameters in geotechnical engineering, affecting not only pressures on retaining walls but also foundation bearing capacity, supporting strength of piles, and landslides. $K_a$ always is smaller than 1.0 and $K_p$ always is larger than 1.0. $K$ for a liquid is of course 1.0.

A simple model was devised by Terzaghi to demonstrate the relationship between active and passive earth pressures and is illustrated in Fig. 19.5. The model wall is hinged at the bottom and held by a spring scale at the top, and the box filled on one side with sand. First the wall can be pulled toward the soils. A considerable force is required to move the wall into the soil, and when the wall moves the scale measures passive resistance. Then the pulling force is relaxed so that the soil does the pushing, and the scale reading when the wall moves back is much lower and represents active resistance.

**Figure 19.5**

Model to illustrate soil in active ($K_a$) and passive ($K_1$) states. The force with zero movement is a measure of $K_0$, the earth pressure at rest.

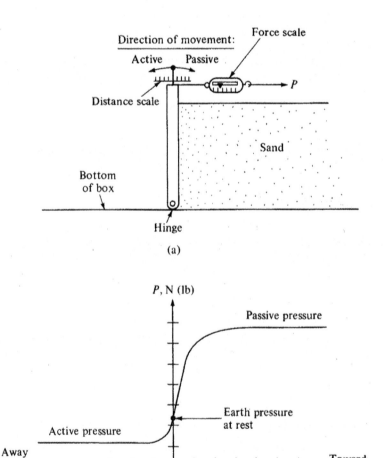

Considerably more movement is required to mobilize full passive pressure than when it moves the other way to mobilize active pressure. This is fortunate because both pressures can exist simultaneously on opposite sides of a retaining wall, active pressure as the wall moves away from the high side, and passive pressure as the wall moves into soil on the low side. Therefore as the wall moves, active restraint develops first, and then passive restraint develops only to the extent necessary to prevent further wall movement. The lower, active pressure acts on the full height of the wall and the higher, passive pressure acts only on that part that is dug into the soil. Both require that the wall must move.

### 19.3.3 Amount of Wall Movement

The amount of wall movement to mobilize active soil resistance to moving is of the order of 0.4–0.5 percent of the height of the wall. Because walls are designed with a factor of safety, they generally move less than the amount required to fully mobilize active case pressures. However, outward tilting of a vertical wall can cause an adverse psychological reaction in those who are in the shadow of the wall. Retaining walls that are not locked in place to prevent movement usually are built to lean into the soil so that they tilt toward vertical. The initial lean is called *batter*.

Batter also has an advantage for design because it moves the center of gravity of the wall into the soil and away from the tipping point, increasing the resisting moment from the weight of the wall.

### 19.3.4 Designing for Earth Pressure at Rest

If a wall does not move, as illustrated in Fig. 19.5, the intermediate stress state is *earth pressure at rest*. In this case the ratio of horizontal to vertical stress designated by $K_0$.

A $K_0$ condition occurs in nature during consolidation of sedimentary deposits, and occurs in a laboratory consolidation test because lateral strain is not allowed. Basement walls, bridge abutments, and walls that are keyed in to adjacent buildings and are not allowed to move are designed to resist earth pressure at rest, which is higher than active pressure and lower than passive pressure.

Removal of overburden by erosion or excavation causes a proportionate reduction in vertical stress but does not significantly relieve lateral stress. Some reduction of lateral stress will occur from elastic rebound, but for the most part it remains locked in. The inherited lateral stress in an overconsolided soil therefore increases $K_0$ so that it often exceeds 1.0, which would be impossible from application of vertical stress alone. A high value of $K_0$ therefore indicates that a soil is overconsolidated, and as discussed in Chapter 16 on settlement, a foundation pressure that is less than a soil overconsolidation pressure will allow relatively little settlement.

# 19.4   RANKINE'S THEORY OF ACTIVE AND PASSIVE EARTH PRESSURES

### 19.4.1   Overview

The problem of determining the lateral pressure against retaining walls is one of the oldest in engineering literature. In 1687 a French military engineer, Marshal Vauban, set forth certain rules for the design of revetments to withstand the lateral pressure of soil. Since then many theories of earth pressure have been proposed by various investigators, and numerous experiments have been conducted in this field. The theories presented by Coulomb in 1773 and Rankine in 1860 both involve the angle of internal friction in soils, and are often referred to as the "classical earth pressure theories." They were developed to apply only to cohesionless soil backfill, which is the condition that is most appropriate for initially loose backfill placed against a wall.

Although the Rankine theory of lateral earth pressure was proposed nearly a century later than the Coulomb theory, the principals advocated by Rankine will be presented first, as they are based on the concept that a conjugate relationship exists between vertical pressure and lateral pressure at every point in a yielding soil mass. Coulomb's approach is one of static equilibrium of the entire soil mass behind a wall, and has the advantage that it incorporates variable friction between soil and a wall. Without wall friction, both theories give the same answers.

### 19.4.2   Simplified Derivation of Rankine's Coefficients

Rankine's analysis preceded that of Mohr by over 20 years, but the use of Mohr's circle greatly simplifies the derivation of the Rankine coefficients. From the right triangle in Fig. 19.6 it will be seen that

**Figure 19.6**

Mohr circle for derivation of the Rankine active and passive coefficients.

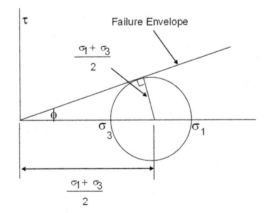

$$\sin \phi = \frac{(\sigma_1 - \sigma_3)/2}{(\sigma_1 + \sigma_3)/2} \tag{19.1}$$

Cross-multiplying gives

$$\sigma_1 \sin \phi + \sigma_3 \sin \phi = \sigma_1 - \sigma_3$$

$$\sigma_3(1 + \sin \phi) = \sigma_1(1 - \sin \phi) \tag{19.2}$$

$$\frac{\sigma_3}{\sigma_1} = \frac{1 - \sin \phi}{1 + \sin \phi}$$

This is the equation for a principal stress ratio, and applies to both active and passive cases. In the active case the major principal stress is vertical, and in the passive case it is horizontal.

**Active Case**

In the active case the major and minor principal stresses are vertical and horizontal, respectively, so $\sigma_1 = \sigma_v$, and $\sigma_3 = \sigma_h$. Making these substitutions,

$$K_a = \frac{\sigma_h}{\sigma_v} = \frac{1 - \sin \phi}{1 + \sin \phi} \tag{19.3}$$

Through a trigonometric identity it can be shown that

$$K_a = \tan^2(45 - \phi/2) \tag{19.3a}$$

The two equations may be used interchangeably.

**Passive Case**

For the passive state the stresses are simple so $\sigma_1 = \sigma_h$ and $\sigma_3 = \sigma_v$. Then

$$K_p = \frac{1 + \sin \phi}{1 - \sin \phi} \tag{19.4}$$

Through a trigonometric identity it can be shown that

$$K_p = \tan^2(45 + \phi/2) \tag{19.4a}$$

Obviously, $K_p = 1/K_a$.

### 19.4.3 Orientation of Slip Lines

A laboratory unconfined compressive strength test shown in Fig. 19.7 illustrates the active case. It will be recalled from Mohr's analysis that the origin of planes is at the left edge of the Mohr circle, and the orientation of slip lines is $45 + \phi/2$ from horizontal. As there are actually two failure envelopes, this orientation becomes $\pm(45 + \phi/2)$, as shown in Fig. 19.7. While both orientations are active in a compression test, behind a retaining wall one set of parallel slip surfaces is activated and the other is not because it slopes into the soil.

**Figure 19.7**

Mohr analysis to show orientation of slip lines in active and passive cases with a horizontal round surface.

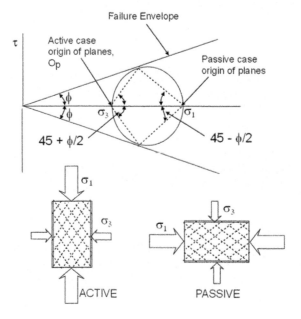

In the passive case the origin of planes is at the right edge of the Mohr circle, and the failure plane orientations are $\pm(45 - \phi/2)$ from horizontal. As in the active case, because of geometric constraints only one set of slip lines is activated in soil in front of a retaining wall.

**Example 19.2**

Soil has $\phi = 30°$. What are the active and passive Rankine coefficients, and what is the ratio of active to passive earth pressures?

*Answer:* $K_a = (1 - \sin 30)/(1 + \sin 30) = 0.33$, and $K_p = 1/K_a = 3.0$. The ratio of stress to passive pressures at any depth therefore is $3/0.33 = 9$. The higher the friction angle, the higher the ratio. It also may be noted that if $\phi = 0$, $K_a = K_p$ and the ratio is 1.0.

Active and passive states also play critical roles in foundation bearing capacity and landslides, so it is important that this concept be thoroughly understood.

## 19.4.4 Influence of Sloping Backfill

Rankine's original derivation assumed that with a sloping backfill the line of force is parallel to the slope, and produces a relationship that is shown in Fig. 19.8, which indicates a sharp increase in lateral stress ratio as the slope approaches the soil friction angle. When the two angles are equal, $K = 1.0$ (the slope angle cannot exceed the friction angle, which is the angle of repose). This requires that the right

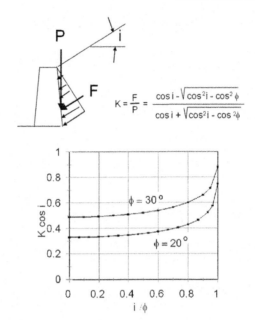

**Figure 19.8**

Rankine's analysis assumes that the line of action *F* is parallel with the slope, which requires a corresponding development of wall friction. Graph shows the horizontal pressure in relation to slope angle *i* as a fraction of $\phi$.

amount of friction be mobilized between the soil and the wall, and the Coulomb analysis is more general. However, Rankine's concept of internal shearing within an entire soil mass is widely applicable and is the basis for bearing capacity theory discussed in the next chapter.

---

*Note:* Most walls now are designed using off-the-shelf computer programs that involve simplifying assumptions that are covered by incorporating a "factor of safety." For example, if a worst-case failure condition can be precisely evaluated, increasing the design resisting forces by 5 percent should give an adequate margin of safety. In reality, in order to be on the safe side the safety margin, which is arbitrary, normally will equal or exceed 50 percent. In addition every computer program that is used for design must be tested with problems for which there are known solutions. It is not the programmer who bears the ultimate responsibility for an error; it is the engineer.

---

### Example 19.3
Sand is being dumped from a conveyor belt and sliding down the surface of a pile against a wall. The angle of repose is 25°. By what ratio does the slope of the pile increase soil pressure against the wall?

*Answer:* The $K_a$ values from the graph in Fig. 19.8 give a ratio of approximately $0.8/0.4 = 2$; pressure is doubled by the slope angle.

### 19.4.5 Horizontal Stress at Any Depth

Equations (19.3) can be transposed to solve for the Rankine active stress at any depth:

Active case: $\sigma_h = K_a \gamma h$           (19.5)

A similar equation can be written for the passive case:

Passive case: $\sigma_{hp} = K_p \gamma h_p$         (19.6)

where $\gamma$ is the soil unit weight, $h$ is depth below a level surface, and the subscript p designates the passive case.

### 19.4.6 "Equivalent Hydrostatic Pressure"

Structural engineers sometimes prefer that Rankine active state soil pressures be expressed in terms of hydrostatic equivalents, or the product $K_a \gamma$.

**Example 19.4**
The unit weight of a soil is 100 and $K_a = 0.4$. Calculate the Rankine active stress at a depth of 5 (a) using eq. (19.5), and (b) using the hydrostatic equivalent pressure.

*Answer:* (a) From eq. (19.5), $\sigma_h = K_a \gamma h = 0.4(100)(5) = 200$. (b) The equivalent hydrostatic pressure is $0.4(100) = 40$; multiplying by the depth gives $40(5) = 200$.

### 19.4.7 Real Hydrostatic Pressure

Far more important than the quasi-hydrostatic pressure discussed in the preceding paragraph is real hydrostatic pressure from water behind a retaining wall. A major cause of retaining wall failures is poor drainage so that soil behind the wall becomes saturated. Then it is not just the saturated unit weight of the soil that is acting against the wall, it is the pressure of the water plus that from the submerged unit weight of the soil. The reason for the larger influence from water is because $K$ for water is 1.0, as water pressure acts across the area of the voids between the relatively small true grain contacts.

**Example 19.5**
A 20 ft (6.1 m) high wall is designed to retain soil having a unit weight of 120 lb/ft³ (18.9 kN/m³) and a friction angle of 30°. Estimate the average pressure on the wall with and without the soil becoming saturated with water.

*Answer:* $K_a = 0.5$ for the soil, 1.0 for the water. The average pressure is at one-half the wall height, which is 10 ft (3.05 m). Soil pressure at this depth without hydrostatic pressure is

$\sigma_{hav} = K_a \gamma h = 0.5(120)(10) = 600$ lb/ft², or
$\sigma_{hav} = 0.5(18.9)(3.05) = 28.8$ kPa

With the soil fully saturated let us assume that its submerged unit weight is approximately that of the soil minus the unit weight of water. (This assumes that air remains trapped in the soil. For a more exact answer the saturated unit weight should first be calculated using a block diagram.) Then

$$\sigma_{hav} = K_a\gamma_{sub}h + (1.0)\gamma_w h = (0.5)(120 - 62.4)(10) + (1.0)(62.4)(10)$$
$$= 288 + 624 = 912\,lb/ft^3,\ or$$
$$\sigma_{hav} = 0.5(18.9 - 9.81)(3.05) + (1.0)(9.81)(3.05) = 13.9 + 29.9 = 43.8\,kPa$$

which is slightly over a 50% increase in pressure on the wall.

### 19.4.8　Drain that Wall!

Some common methods for drainage are shown in Fig. 19.4. If the backfill is relatively permeable, drainage can be accomplished with drain tile along the inner toe of the wall and/or with weep holes through the base of the wall. Drain tile often is covered with a layer of geofabric to prevent soil from entering the tile. The inner face of a wall also can be covered with a conducting geofabric mat to funnel water into the drains.

Segmented masonry or stone walls either are not mortared or vertical joints are left open to allow drainage. Gabion walls are interconnected wire mesh baskets filled with stones and are free-draining.

## 19.5　PREDICTING STABILITY

### 19.5.1　Total Force from Soil Against a Wall

Rankine pressure at the base of the wall is

$$(\sigma_h)_{max} = K\gamma H \tag{19.7}$$

where $H$ is the height of the wall and $K$ is the active or passive coefficient of earth pressure. If pressure at the top of the wall is zero, the total force equals the average pressure times the wall height, or

$$P = \tfrac{1}{2}K\gamma H^2 \tag{19.8}$$

where $P =$ the resultant force from soil pressure per unit length of the wall;

$\gamma =$ unit weight of the soil;

$H =$ height of the wall;

$K =$ coefficient of lateral earth pressure: $K_a$ for the active state, $K_p$ for the passive state, or $K_0$ for earth pressure at rest.

Equation (19.8) is the most widely used relationship for predicting the force of soil against a retaining wall. It applies to active, passive, and at-rest states, so long as $K$ is constant with depth, and with a level backfill no friction is mobilized between the soil and the wall. This assumption is questionable since backfill normally settles and creates wall friction. Nevertheless this equation is an important guide and is useful for routine calculations.

## 19.5.2 Resistance to Sliding

The first question concerning stability of a wall is will the wall slide? The answer is obtained by summing horizontal forces on the wall and determining if the force from active soil pressure is exceeded by potential friction on the base plus passive resistance from soil in front of the wall. In many walls an additional tensile restraint is obtained from tiebacks, and a factor of safety is used for design.

A high soil friction angle is advantageous because it reduces active soil pressure on the back of the wall and increases passive resistance from soil in front of the wall. A high friction angle also can increase basal sliding friction.

### Example 19.6

Calculate the factor of safety ($FS_s$) against sliding of the retaining wall in Fig. 19.9. Assume that the coefficient of sliding friction on the base is one-half the internal friction of the soil and the unit weight of concrete is $150 \, lb/ft^3$ ($23.6 \, kN/m^3$).

*Answer:*

1. Force from active soil pressure,

   $P_a = \frac{1}{2}(0.333)(120)(10)^2 = 2000 \, lb \leftarrow$, or
   $= \frac{1}{2}(0.333)(18.9)(3.05)^2 = 29.3 \, kN$

2. Force from passive soil pressure,

   $P_p = \frac{1}{2}(3)(120)(3)^2 = 1620 \, lb \rightarrow$, or
   $= \frac{1}{2}(3)(18.9)(0.91)^2 = 23.5 \, kN$

3. Base friction: the weight of the wall is

   $W = 2(10)(150) = 3000 \, lb$, or
   $= 0.61(3.05)(23.6) = 43.9 \, kN$

Coefficient of friction $= 0.5 \tan 30° = 0.289$

$S = 0.289(3000) = 866 \, lb \rightarrow$, or
$= 0.289(43.9) = 12.7 \, kN$

The factor of safety against sliding is $(1620 + 866)/2000 = 1.2$, or

$(23.5 + 12.7)/29.3 = 1.2$

---

*Question:* As a designer how would you increase this factor of safety?

*Answer:* The most economical way would be to cut a shallow trench under the base of the wall to key it into the soil.

---

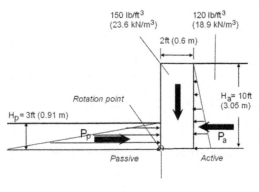

**Figure 19.9**

Walls in Examples 19.6 through 19.8.

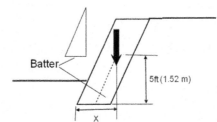

### 19.5.3 Safety from Overturning

The overturning moment equals $P$ times the height of the line of action of $P$ on the wall. With a triangular distribution of pressure, $P$ acts at one-third of the height of the wall. Hence

$$M = 0.333PH \tag{19.9}$$

where $M$ is the overturning moment. The same equation applies to both the active and passive sides by use of the appropriate $K$ and depth $H$ below the ground surface.

As in the case of sliding resistance, the resistance to overturning derives from (1) passive resistance of soil in front of the wall, (2) the weight of the wall, and (3) tiebacks if present.

**Example 19.7**
Calculate the factor of safety against overturning ($FS_o$) of the wall in the upper part of Fig. 19.9.

*Answer:*

1. Active moment:

$$M_a = 0.333(10)(2000) = 6670 \, \text{ft-lb, or}$$
$$= 0.333(3.05)(29.3) = 29.8 \, \text{kN-m}$$

2. Passive moment:

$$M_p = 0.333(3)(1620) = 1620\,\text{ft-lb, or}$$
$$= 0.333(0.91)(23.5) = 7.1\,\text{kN-m}$$

3. Wall weight moment:

$$M_w = 1(3000) = 3000\,\text{ft-lb, or}$$
$$= 0.3(43.9) = 13.2\,\text{kN-m}$$

$$FS_o - (1620 + 3000)/6670 = 0.7,\text{ or}$$
$$= (7.1 + 13.2)/29.8 = 0.7$$

According to this design the wall will fall over, which is not good.

### 19.5.4 Importance of Passive Resistance

As shown in the upper part of Fig. 19.9, creating passive resistance at the toe of a wall normally is only a matter of extending the base of the wall below the ground surface. This contributes to design efficiency, the larger the friction angle the larger the difference. It therefore can be of critical importance that the passive pressure be maintained throughout the life of the wall, as removing soil from the toe can lead to consequences such as shown in Fig. 19.11.

### 19.5.5 Batter

A simple method for increasing resistance to overturning is to increase the lever arm with batter, which means to lean the wall back into the soil. This is accomplished with batter, or leaning the wall back into the soil. It is illustrated in the lower diagram in Fig. 19.9. This is very common practice, with batter usually being in a range from 1:12 to 1:4 (horizontal to vertical), or 4° to 15° from vertical.

#### Example 19.8
What batter would be required for the wall in the previous example to have a FS against overturning of 1.3?

*Answer:*

1. The required resisting moment is $1.3(6720) = 8740\,\text{ft-lb}$, or $1.3(29.8) = 38.7\,\text{kN-m}$.
2. The moment required from weight of the wall is $8740 - 1620 = 7120\,\text{ft-lb}$, or $38.7 - 7.1 = 31.6\,\text{kN-m}$.
3. Equating the required moment and the moment from weight of the wall gives $7120 = 3000X$, where $X$ is the moment arm. $X = 2.37\,\text{ft}$, or $31.6 = 43.9X$; $X = 0.72\,\text{m}$.

The batter must be such that the centroid of the wall is 2.37 ft (0.72 m) from the corner of the base. The required angle is $\tan^{-1}[5/(2.37-1)] = 75°$, or $\tan^{-1}[1.52/(0.72-0.30)] = 75°$ or 1:3.6.

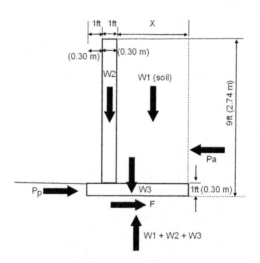

**Figure 19.10**
Cantilever wall in
Example 19.9 (not
to scale).

## 19.5.6 Cantilever Walls

Another approach to effectively increase resistance to overturning is to extend the heel of the wall back into the soil so that it supports the weight of the overlying soil. The design of the reinforced concrete in cantilever walls requires a knowledge of the moments and is part of structural engineering. An example of a cantilever wall is shown in Fig. 19.10. Solving a wall problem is mainly through the application of statics and bookkeeping.

### Example 19.9

Adjust the length of the toe section of the wall in Fig. 19.10 to give factors of safety equal to or greater than 1.5 for overturning and for sliding. Assume unit weights of $120\,\text{lb/ft}^3$ $(18.9\,\text{kN/m}^3)$ for the soil and $150\,\text{lb/ft}^3$ $(23.6\,\text{kN/m}^3)$ for the wall material, and $\phi = \delta = 30°$.

*Answer:* With $X$ as the toe length of the wall the weights of the various sections are as follows:

$$W1 \text{ (soil)} = 120(8X) = 960X \text{ lb}$$
$$W2 \text{ (concrete)} = 150(8)(1) = 1200 \text{ lb}$$
$$\underline{W3 \text{ (concrete)} = 150 \times 1(X + 2) = 150X + 300 \text{ lb}}$$
$$\Sigma W = 960X + 1200 + 150X + 300 = 1110X + 1500 \text{ lb}$$
$$F = (1110X + 1500)\tan 30° = (1110X + 1500) \times 0.577 \text{ lb}$$

From soil properties

$$K_a = 0.333; \ K_p = 3.0$$
$$P_a = K_a\gamma H^2 = 0.333(120)(9)^2 = 3240 \text{ lb}$$
$$P_p = K_p\gamma H_p^2 = 3(120)(1)^2 = 360 \text{ lb}$$

**Figure 19.11**

This 55 ft (16.8 m) wall suddenly collapsed after soil was improperly removed from the toe. The failure broke a water main (dark dot) and took half of the adjacent street; fortunately there was no traffic at the time. The soil is loess.

From $\Sigma X$

Let $F + P_p = 1.5 P_a$

$640 X + 865 = 1.5(3240)$

$X = 6.25\,\text{ft}\,(1.91\,\text{m})$

From $\Sigma$ moments

Resisting $\geq 1.5 \times$ Acting

$W1(X/2 + 2) + W2(1.5) + W3(X+2)/2 + P_p(0.5) \geq 1.5 P_a$

$960X(0.5X + 2) + 1200(1.5) + (150X + 300)(0.5X + 1) + 0.5(360) \geq 1.5(3240)$

A solution can be obtained by the quadratic equation, but a first step should be to substitute $X = 6.25$ ft as a trial value, in which case

$37,835 \gg 4680$

Sliding therefore is the critical requirement, and the FS against overturning is excessively high.

---

*Question:* How might the design be improved?

*Answer:* The distance $X$ can be reduced and the reduced resistance to sliding compensated by cutting a trench to create a keyway under the wall.

*Exercise:* Rework the problem using SI. Are the answers the same?

---

# 19.6 COULOMB'S ANALYSIS FOR ACTIVE EARTH PRESSURE

## 19.6.1 Concept

While Rankine's and Coulomb's theories for soil pressure on a wall give identical answers, the Coulomb analysis includes a variable wall friction and is more suited for computer analyses. However, as will be shown, the analysis still is a simplified version of what actually occurs in soil behind a retaining wall and can be on the unsafe side.

Coulomb reduced the retaining wall problem to a consideration of static equilibrium of a triangular single block sliding down a plane that extends down through the backfill to the heel of the wall, shown in Fig. 19.12. The triangular mass of soil sometimes is referred to as the sliding wedge. An analysis of the forces acting on the wedge at incipient failure reveals the amount of horizontal thrust that is necessary for the wall to remain in place. Coulomb's analysis is widely used even though experimental results indicate that the failure surface is curved instead of linear, and

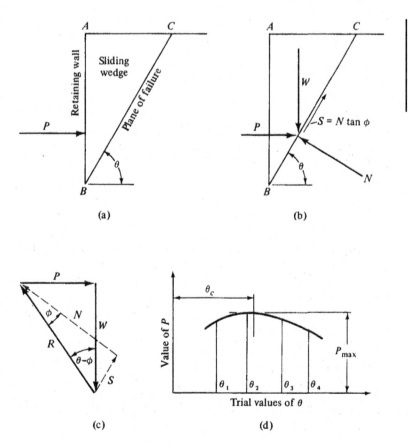

**Figure 19.12**

Development of Coulomb's equation for active pressure on retaining walls.

wall friction introduces arching action in the soil that is not considered in the analysis.

A plane of failure is arbitrarily located at an angle $\theta$ with horizontal, $BC$ in Fig. 19.12(a). If we disregard friction between the soil and the wall, the forces acting on the sliding wedge are shown in Fig. 19.12(b) and consist of the weight $W$ of the soil within the wedge, which acts through the centroid of the triangle $ABC$; a thrust $N$ normal to the plane of failure; and a shearing force $S$, which acts upward along $BC$ and which, at the limit of equilibrium, is equal to $N \tan \phi$.

If the unit weight of the soil and the friction angle are known, the force $W$ acting through the centroid of the sliding wedge can be calculated. The magnitudes of $N$ and $S$ are not known, but the line of action of their resultant makes an angle $\theta$ with a normal to the plane of sliding. It is therefore possible to construct a polygon of forces (Fig. 19.12(c)). Then, by choosing a series of failure planes having different values of $\theta$ and obtaining the thrust on the retaining wall corresponding to each value, the maximum thrust can be determined graphically, as shown in Fig. 19.12(d). While this illustrates the concept of a critical angle, in practice the slip angle $\theta_c$ is more easily determined analytically.

## 19.6.2  The Critical Slip Angle $\theta_c$

The critical value of $\theta_c$ can be obtained by writing an equation for $P$ in terms of $\theta$ and setting the differential equal to zero. Figure 19.12(c) shows a triangle of forces $P$, $W$, and $R$, which is the resultant of forces $N$ and $S$. $R$ therefore makes an angle $\phi$ with a vector that is normal to the failure plane. It also makes the angle $(\phi - \theta)$ with vertical. In this triangle,

$$P = W \tan(\theta - \phi) \tag{19.10}$$

$$W = \tfrac{1}{2}\gamma H^2 \cot\theta \tag{19.11}$$

Hence,

$$P = \tfrac{1}{2}\gamma H^2 \cot\theta \tan(\theta - \phi) \tag{19.12}$$

Since $\gamma$, $H$, and $\phi$ are constant for any particular wall and soil backfill, this equation shows that $P$ varies with the angle $\theta$, as was indicated in the graphical procedure of Fig. 19.12(d). Setting $dP/d\theta$ to zero and solving for $\theta$ gives

$$\theta_c = 45 + \phi/2 \tag{19.13}$$

which is the same as obtained from the Rankine analysis with a level backfill. Substituting this value of $\theta$ into eq. (19.13) gives

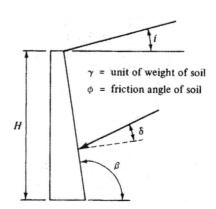

**Figure 19.13**
Variables in Coulomb's retaining wall equation.

$\gamma$ = unit of weight of soil
$\phi$ = friction angle of soil

$$P = \tfrac{1}{2}\gamma H^2 \cot(45 + \phi/2)\tan(45 - \phi/2)$$
$$= \tfrac{1}{2}\gamma H^2 \tan^2(45 - \phi/2) \tag{19.14}$$

which is the same as the Rankine equation for the active case with a level backfill.

A general solution involving a sloping backfill, a battered wall, and wall friction is given by

$$P_R = \frac{\gamma H^2}{2} \times \left[ \frac{\sin^2(\beta - \phi)}{\sin^2 \beta \sin(\beta + \delta)\left(1 + \sqrt{\dfrac{\sin(\phi + \delta)\sin(\phi - i)}{\sin(\beta - i)\sin(\beta + \delta)}}\right)^2} \right] \tag{19.15}$$

where symbols are as shown in Fig. 19.13. For the derivation of eq. (19.15) see, for example, Jumikis (1962). For a vertical wall and level backfill, eq. (19.15) reduces to

$$P = \frac{\gamma H^2}{2} \left[ \frac{\cos^2 \phi}{\cos \delta \left[1 + \sqrt{\dfrac{\sin(\phi + \delta)\sin \phi}{\cos \delta}}\right]^2} \right] \tag{19.16}$$

It is important to note that $P$ in eqs. (19.15) and (19.16) is not horizontal, but is sloping at an angle $\delta$ to the back surface of the wall. The $\delta$ angle represents the wall-to-soil friction and is equal to or less than $\phi$.

### 19.6.3 Accuracy of $K_a$ and $K_p$

Both the Rankine and Coulomb analyses assume linear failure surfaces, but experimental measurements indicate that the surfaces are curved. This means that the curved surfaces are more critical, so assumption of linear surfaces does not define the most critical condition and therefore can underestimate active pressures and overestimate passive pressures. The error, which is of the order of 10 percent, therefore must be included in a factor of safety.

An error that can be several times larger comes from the assumption of a linear distribution of pressure with depth. Friction between soil and a wall introduces arching action tht raises the center of pressure on a wall, adversely affecting the overturning moment. This is discussed in section 19.8.

## 19.7  COHESION

### 19.7.1  Soils that Stand Alone

Whereas cohesionless sand cannot hold a vertical slope but slides down to the angle of repose, soils with cohesive strength do have the ability to stand in a steep or vertical slope to a height that is designated as the *critical height, $H_c$*. As discussed in Chapter 21, the value of $H_c$ is the depth at which the overburden pressure equals the unconfined compressive strength, $q_u$:

$$H_c = \frac{\gamma}{q_u} \tag{19.17}$$

where $\gamma$ is the soil unit weight and $q_u$ is the unconfined compressive strength. The tendency for a soil to stand alone will reduce active stress on a wall to zero at depth $H_c$. From a consideration of the Mohr circle a substitution can be made for $q_u$:

$$H_c = \frac{\gamma}{2c}(45 - \phi/2) \tag{19.18}$$

where $c$ is the soil cohesion and $\phi$ is the angle of internal friction.

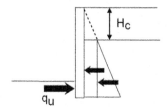

**Figure 19.14**

Cohesion allows the upper soil to stand alone to a height $H_c$ and therefore modifies the stress diagram. Even though the soil may stand to $H_c$ without a wall, the cohesion is not dependable because with saturation and time it can give way to soil creep.

The excavated "top-down" method of construction allows soil to retain its cohesion down to a depth $H_c$. The stress diagram therefore is simply truncated at the top as shown in Fig. 19.14. Forces on the active side are represented by the two arrows at the right and are evaluated from the respective areas, whereas passive resistance approximately equals the soil unconfined compressive strength times the wall height.

### 19.7.2  Why Not Include Cohesion?

(1) Loose backfill against a wall has no cohesion. (2) Later, even with internal drainage, a wall prevents evaporation of water from an exposed soil surface, and as moisture content increases, cohesion decreases. (3) Cohesion of moist clayey soils tends to be rate-dependent, which means that over time the soil may creep. The effects are readily seen in older retaining walls that were not designed for creep. Most engineered retaining walls are designed without including a cohesion component.

## 19.8  WALL FRICTION

### 19.8.1  Wall Friction and Rankine's Theory

Rankine's theory suggests two sets of parallel slip surfaces at an angle $90 - \phi$ to one another to create diamond patterns such as shown in Fig. 19.7. The short dimension of each diamond represents the direction of the minor principal stress and the long dimension that of the major principal stress. If there is no wall friction, the diamonds lay as shown in Fig. 19.15(a) with a principal stress axis normal to the wall, with the diamonds upright for the active case and laying flat for the passive.

With fully developed wall friction the wall becomes a shear path, causing the array to be tilted as shown in Fig. 19.15(b). Shear paths then must curve upward if no shearing stress exists at the ground surface. The transition curve is assumed to be a log spiral, for reasons that are discussed in Chapter 21 on slope stability.

### 19.8.2  Effect of Wall Friction on Sizes of the Active and Passive Wedges

A visual comparison of the straight and curved slip boundaries in Fig. 19.15 suggests that the size of the active wedge is not greatly affected by wall friction, but the passive wedge is considerably larger and extends deeper, which should cause a significant increase in the passive resistance. A bulldozer is not as effective if it has a rusty blade.

**Figure 19.15**

(a) Idealized Rankine stress distribution with zero wall friction; (b) distortion of Rankine active and passive wedges and tilting of principal stress directions by wall friction.

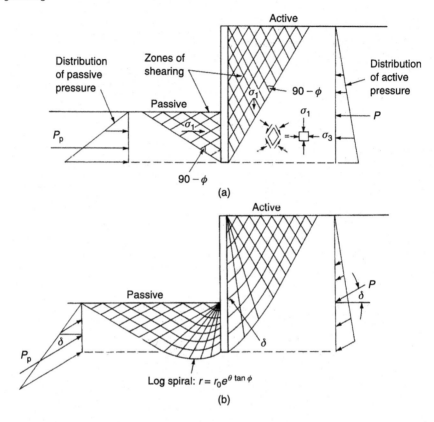

### 19.8.3 The Twentieth Century Discovery of Arching Action

Marginally designed segmented walls typically bulge out near the middle instead of kicking out at the bottom, which is where both Rankine and Coulomb theories indicate that lateral pressure is highest. In the 1930s, Terzaghi conducted full-scale wall experiments for the U.S. Bureau of Public Roads that showed that the maximum pressure exists at about 40 percent of the wall height instead of at the base. Terzaghi attributed this departure from classical theory to *arching action*. It is readily explained by considering the soil wedge behind a retaining wall as being partially supported by friction on both sides; then, as the wedge is deeper and becomes narrower, the increment of weight is reduced until at the bottom it is fully supported, shown in Fig. 19.16. Terzaghi did not hold back, and even wrote a paper describing the "fundamental fallacy of earth pressure calculations," but with little effect and the fallacy lives on.

### 19.8.4 Arching with a Difference

Geotechnical engineering is sometimes viewed as a maverick because of its tendency to rattle basic concepts, and arching is no different. The prevailing concept of an

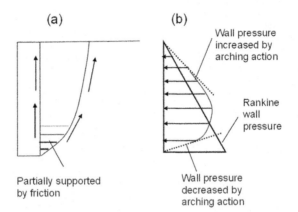

**Figure 19.16**

An additional influence of wall friction is to (a) decrease active pressure at the bottom of a wall, after Terzaghi (1936), and (b) increase it above the midpoint.

arch is that it is a stand-alone structure that can support a wide variety of structures including bridges, roofs, and aqueducts, some of which have remained standing for many hundreds of years. The free-standing arch often is assumed to be a Roman invention, but dates to prehistoric times, for example in stone structures and in the igloo of the native American Inuit. Arching occurs naturally in roofs of caverns.

The ideal free-standing arch mathematically is a catenary, which is Latin for chain, and is the shape taken by a chain held at the ends. A chain represents pure tension with no bending moment, so when the shape is flipped over it represents pure compression with no bending moment. Igloos are catenaries, which offers a huge advantage because as the snow creeps in compression the structure gradually shortens but does not cave in, and when it gets too short for comfort they build another. In contrast, the more symmetrical arches of classical architecture are circular and require iron hoops, chains, tie bars, or flying buttresses to hold the sides from bulging out and allowing the roofs to cave in. The Gothic arch is circular with the center cut out, which does not alleviate the problem, and it is for this reason that classical cathedrals often were built first as models, to see what bracing would be required to keep them up.

Arching action in soils gives only partial support, so the major principal stress is not continuous as it is in a stand-alone arch. Instead, the *minor* principal stress forms a continuous curve that is shown in the lower part of Fig. 19.17. This discovery led to a mathematical solution. The minor principal stress arch is a trajectory of reduced compression, and it is helpful to think of it as similar to a tensile reinforcement. Ironically, many retaining walls now incorporate horizontal tensile reinforcing strips of steel or plastic in the soil that when deformed tend to follow the sagging trajectory of the minor principal stress.

## 19.8.5 Increased Pressure Due to Arching

The reduction in wall pressure near the bottom of a wall friction is illustrated in Fig. 19.16(a). However, this does not explain an increase in pressure that has been

**Figure 19.17**

Soil partially supported by wall friction develops an arch-like trajectory of the *minor* principal stress. The arch has the approximate shape of a catenary, or a chain held at the ends.

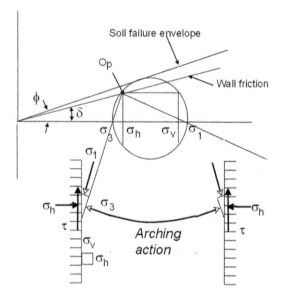

observed higher on the wall—where it also does the most potential harm by increasing the overturning moment.

The cause of the increase in pressure was first proposed by D. P. Krynine (1945) in a brief discussion of a paper by Terzaghi. Krynine noted that rotation of principal stress directions near the wall should have the effect shown in the Mohr diagram in Fig. 19.17: friction along a rough vertical wall moves the origin of planes $O_p$ from the abscissa to a point higher on the Mohr circle. Dropping a vertical line from the origin of planes to the opposite position on the Mohr circle defines a shearing stress and a horizontal stress $\sigma_h$ that is larger than the minor principal stress $\sigma_1$. This mobilization of wall friction can contribute to bursting of grain bins as they start to be emptied from the bottom.

Krynine's formula is

$$K_w = \frac{1 - \sin^2 \phi}{1 + \sin^2 \phi} \tag{19.19}$$

where $K_w$ is the ratio of horizontal to vertical soil stress at the wall. Note the similarity to the Rankine active state equation, except that the sine functions are squared.

## 19.8.6 A Solution for Arching Action

The distribution of horizontal stress on a rough wall is given by the following equation (Handy, 1985). Solutions are shown in Figs. 19.21 and 19.22.

$$\sigma_h = \frac{\gamma}{\mu}(H-h)\tan\left(45-\frac{\phi}{2}\right)\left[1-\exp\left(\frac{K_\omega\mu}{\tan\left(45-\frac{\phi}{2}\right)}\frac{h}{H-h}\right)\right] \qquad (19.20)$$

where $H$ is the height of the wall;

$h$ is depth measured from the top;

$\mu$ is the coefficient of friction between the soil and the wall; and

$K_w$ is the ratio $\sigma_h/\sigma_v$ at the wall. An adjustment was made to the Krynine $K_w$ to incorporate averaging of horizontal stresses across the span of the failure wedge. An assumption that is on the safe side is to use $K_0 = 1 - \sin\phi$.

At the base of the wall where $h = H$, the first term drops out and horizontal stress is zero, which agrees with measurements by many investigators. A stress distribution based on this equation is shown in Fig. 19.18, and comparisons with measured wall pressures are shown in Fig. 19.19. These comparisons indicate that the higher the wall, the less significant is arching action, probably because of reduced wall friction from slipping of soil on the wall.

## 19.8.7 Easy Arching

A simplified solution for arching pressures can be made by constructing three straight lines, one for the upper part of a wall based on $K_w$, a maximum pressure line down the middle part, and a lower pressure line that goes to zero at the bottom. This is illustrated in Fig. 19.22. A curve then can be sketched tangent to the three lines, or the intersections may be used to define three pressure zones for calculating total force and moment.

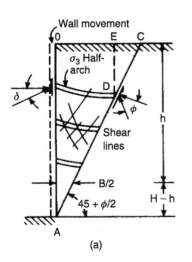

(a)

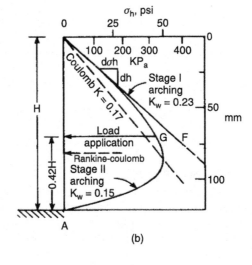

(b)

**Figure 19.18**

(a) Geometry used to analyze partial support from arching action in soil and

(b) nontriangular distribution that raises the center of stress on a vertical wall. (After Handy, 1985, with permission of the American Society of Civil Engineers.)

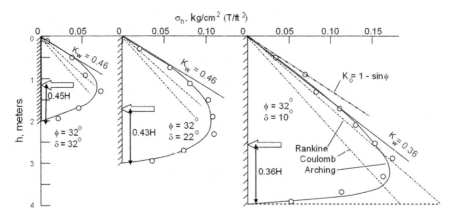

### Figure 19.19

Arching predictions of wall pressures compared with measurements by Tsagareli (1965). Total forces are about the same, but arching raises the center of pressure and increases the overturning moment. As wall friction is reduced the pressure distribution becomes transitional to Rankine active pressures. (After Handy, 1985, with permission of the American Society of Civil Engineers.)

### Figure 19.20

Numbers on the curved lines show arching $K_w$ values to be used for the upper part of a wall.

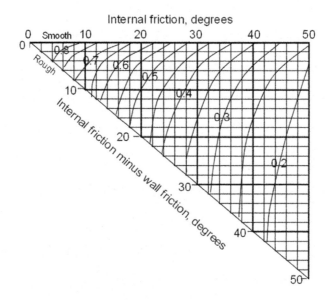

### Upper Wall Section

An iterative solution is required for $K_w$, or it may be obtained from Fig. 19.20. $K_w$ defines the slope of maximum pressure versus depth on the upper part of a wall. As shown in Fig. 19.21, an alternative that is more conservative is to use $K_0 = 1 - \sin \phi$. The Rankine and Coulomb coefficients should not be used as they are unconservative.

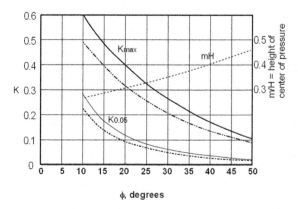

**Figure 19.21**

Solutions of eq. (19.23) used to define maximum pressure ($K_{max}$) and that existing at 5 percent of the wall height ($K_{0.05}$). $K$ values must be multiplied times the *total wall height*. Solid lines are for full wall friction with *mH* as the height of the center of pressure. Dashed lines are for 80 percent development of wall friction.

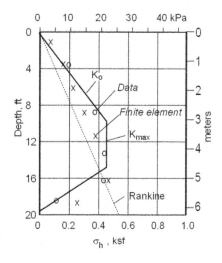

**Figure 19.22**

"Easy arching" pressure profile with $n = 0.8$. Circles show data and crosses show finite element solutions for this MSE wall (discussed in the next chapter) from Adib et al. (1990). The classic Rankine distribution is shown by a dotted line and places pressures too low on the wall.

### Middle Wall Section

In eq. (19.21), let $m$ = fractional height on a wall measured up from the bottom, and $n$ = fractional development of soil-to-wall friction compared with the soil internal friction. For a perfectly rough wall $n = 1.0$; for an ordinary wall $n$ is assumed to be 0.8. Actual measurements indicate that it may be somewhat lower than that, but assumption of a higher value is on the conservative side.

Making these substitutions in eq. (19.20) gives

$$K_H = \frac{\sigma_h}{\gamma H} = \frac{m}{n} \frac{\tan(45 - \phi/2)}{\tan \phi} \times \left[ 1 - \exp \left[ \frac{n \tan \phi (1 - \sin \phi)}{\tan(45 - \phi/2)} \times \frac{1 - m}{m} \right] \right] \quad (19.21)$$

The symbol $K_H$ is used to designate a $K$ value that relates to the *total wall height*, $H$, and not to the height $h$ measured on a wall. The maximum horizontal pressure is designated $K_{max}$, and is obtained numerically by varying $m$ for each value of friction angle, or graphically from Fig. 19.22. $K_{max}$ is multiplied times $\gamma H$ to give the maximum horizontal pressure for the middle-zone vertical line in Fig. 19.23.

### Lower Wall Section
The lower line is drawn through the zero point at the bottom and a point at 5 percent of the wall height, defined by $K_{0.05}$ from Fig. 19.21.

## 19.8.8 "Easy Arching" Calculation

The three straight lines approximating an arching pressure distribution define a trapezoid that may be divided into upper and lower triangular areas and a central rectangle. The force and its position can be calculated for each area, much the same as for the triangular distributions from classical earth pressure theory, and the respective forces and moments from each added to give a total force and total moment.

### Example 19.10
Define the three linear segments in Fig. 19.22 with $\gamma = 130 \, \text{lb/ft}^3$ (19.1 kN/m$^3$), $\phi = 40°$, $\delta = 32°$, and $H = 20 \, \text{ft}$ (6.1 m).

*Answer:*

1. In Fig. 19.20 $\phi = 40°$ and $(\phi - \delta) = 8°$, which gives $K_w = 0.28$ and defines the slope of the pressure distribution on the upper part of the wall. In order to draw a pressure line an arbitrary depth is selected and the pressure at that depth is defined by $\sigma_h = K_w \gamma h$; for example, at a depth of 10 ft, $\sigma_h = 10 \times 130 \times 0.28 = 364 \, \text{lb/ft}^2$ (or at 1 m depth $\sigma_h = 1 \times 19.1 \times 0.28 = 5.3 \, \text{kPa}$). This construct shows only the inclination of the line and not its terminal point, which is defined by the intersection with the vertical pressure line for the middle zone.
2. From Fig. 19.21 the maximum pressure is defined by $K_{max} = 0.17$. Multiplying this times the *total* wall height and soil unit weight gives $\sigma_{hmax} = 0.17 \times 20 \times 130 = 440 \, \text{lb/ft}^2$ ($0.17 \times 6.1 \times 19.1 = 20 \, \text{kPa}$). A vertical line is drawn through this pressure and is extended to intersect lines defined for the upper and lower zones. *It again must be emphasized that the maximum pressure depends on the total height of the wall.*
3. The pressure at 5% of the wall height is determined by $K_{0.05} = 0.03$ in Fig. 19.21. A line therefore is drawn from the base of the wall extending upward through a point defined by $\sigma_h = 0.03 \times 20 \times 130 = 78 \, \text{lb/ft}^2$ (3.7 kPa) and $h = 1 \, \text{ft}$ (0.3 m).

Average pressures on each of the three sections of the wall then can be considered separately to calculate overturning moment. As can be seen in Fig. 19.22, total force on the wall is not greatly affected by arching, but the overturning moment is influenced by the higher center of pressure on the wall. The following calculations can be made in the above example:

| | Force | Overturning moment | Ratio of FS |
|---|---|---|---|
| Arching analysis | 5400 lb (24 kN) | 52,000 ft-lb (71 kN-m) | 1.0 |
| Rankine | 5700 lb (25 kN) | 38,000 ft-lb (52 kN-m) | 0.7 |
| Coulomb | 5500 lb (24 kN) | 37,000 ft-lb (50 kN-m) | 0.7 |

The classical earth pressure theories therefore are on the unsafe side for walls that have a developed wall friction. The influence from a higher center of pressure is readily observed in marginally designed segmented walls that bulge in the midsection instead of kicking out at the bottom.

## 19.9  FASCIA WALLS

### 19.9.1  A Special Case of Arching Action

Fascia walls may be defined as walls that are close to and approximately parallel to exposed rock or resistant soil faces. The purpose of a fascia wall is to improve appearance, prevent erosion, and prevent potentially damaging rockfalls. Arching action dominates the pressure distribution on fascia walls because the column of soil is narrow and the soil weight is greatly reduced by supporting friction on both sides. In Fig. 19.23 the weight of the soil column at depth $h$ is

$$W = \gamma B h \tag{19.22}$$

Assume there is a depth at which this is supported by wall friction, $F$, which acts on both sides:

$$\gamma B h = 2F \tag{19.23}$$

Friction in turn equals the horizontal stress times a coefficient of friction, or

$$F = \sigma_h \tan \delta \tag{19.24}$$

where

$$\sigma_h = K_a \sigma_v \tag{19.25}$$

The relationship is complicated because frictional support is cumulative with increasing depth, which decreases $\sigma_v$ and therefore $\sigma_h$. This may be solved by integrating stress with depth, which was first applied to soils by Anson Marston, and is known as "Marston theory." This integration is presented in Chapter 25 on underground conduits. At any depth $h$, the lateral stress is

$$\sigma_h = \gamma B \frac{1 - e^{-2K(\tan \delta)(h/B)}}{2 \tan \delta} \tag{19.26}$$

At very large values of $h$ the exponential term becomes small and

$$\sigma_{h\,max} = \frac{\gamma B}{2 \tan \delta} \tag{19.27}$$

where $\gamma$ is the unit weight of the soil, $B$ is the horizontal distance between the back of the wall and the outcrop, and $\tan \delta$ represents wall friction on both sides. The maximum wall pressure therefore is *not* a function of the height of the wall as in the case of ordinary retaining walls, but depends on the separation distance between the wall and the exposed outcrop. A graphic illustration of wall pressures calculated by eqs. (19.26) and (19.27) and by the conventional Rankine analysis is shown in Fig. 19.23.

The simplified design method in Fig. 19.23 defines wall pressures in the upper zone according to $K_0$, below which the wall pressure is constant, defined by eq. (19.27). The substitution of $K_0$ is offset by conservatism from assuming straight-line relationships. The fascia wall design method was first presented in an earlier edition of this text and has been tested in practice.

### Example 19.11
Calculate the maximum active state pressure on the wall of Fig. 19.23 using soil data shown in the figure, and compare with the Rankine maximum pressure.

**Figure 19.23**

Simplified analysis for fascia walls showing the large reduction in wall pressures from arching action.

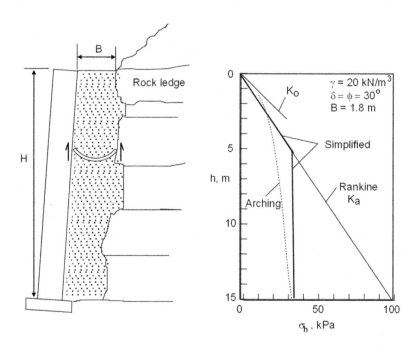

**Figure 19.24**
A standing section of this 30 ft (9.1 m) high wall is at the upper center of the photo. The design was was not engineered but was based on experience of building other walls in the vicinity. The collapse took four lives and led to development of the fascia wall theory.

*Answer:* From eq. (19.27),

$$\sigma_{hmax} = \frac{\gamma B}{2\tan\delta} = \frac{20\ kN/m^3 \times 1.8\ m}{2\tan30°} = \underline{31\ kN/m^2}$$

Rankine: $\sigma_{hmax} = 15\ m \times 20\ kN/m^3 \times \tan^2(45 - \phi/2) = \underline{100\ kN/m^2}$.

In this case both the total force and the overturning moment are reduced by arching action.

## 19.10 $K_0$, THE COEFFICIENT OF EARTH PRESSURE AT REST

### 19.10.1 Another Example of Arching Action?

A Hungarian engineer, J. Jáky (1944), derived a widely used equation that defines a ratio designated $K_0$ to define lateral stress in laterally confined consolidating soil. $K_0$ is intermediate between the limiting values designated by the Rankine active and passive $K_a$ and $K_p$.

Jáky's equation often is described as empirical but actually was derived, but on the basis of conditions that appear to bear little relationship to those existing in the field. Nevertheless the equation has repeatedly been confirmed by measurements in normally consolidated soil—that is, in soils with a density that is in equilibrium with the existing overburden pressure, as occurs in a laboratory consolidation test. Most soils in nature are overconsolidated from past pressures or drying shrinkage, which increases $K_0$. In order to distinguish the $K_0$ from normal consolidation from that resulting from overconsolidation, it sometimes is referred to as $K_{0nc}$.

Jáky assumed a sand pile with flanks lying at the angle of repose, as shown in Fig. 19.25. If the slope angle is $\phi$, according to Rankine theory a second set of shear surfaces is separated by an angle $90 - \phi$, which as shown in the figure makes the second set vertical. Jáky defined a horizontal element in the core area. If the core is settling relative to the flanks, and the horizontal element is continued laterally into the flanking zones, it will be seen that it is partially supported by friction at the ends, much as in a fascia wall.

Arching theory may be used to obtain a simplified derivation of the Jáky relationship; for the original derivation and a more detailed discussion see Michalowski (2005). A consideration of forces on a soil element adjacent to vertical frictional support (Handy, 1985) gives

$$\frac{\sigma_h}{\sigma_1} = \cos^2 \theta + K_a \sin^2 \theta \tag{19.28}$$

where $\sigma_h$ is horizontal stress, $\sigma_1$ is the major principal stress, and $\theta$ is the inclination of the major principal stress from horizontal. With fully developed wall friction, $\theta = 45 + \phi/2$, and

$$\frac{\sigma_h}{\sigma_1} = \cos^2(45 + \phi/2) + K_a \sin^2(45 + \phi/2) \tag{19.29}$$

By the use of half-angles this reduces to

$$\frac{\sigma_h}{\sigma_1} = 1 - \sin \phi \tag{19.30}$$

As can be seen by comparing $\sigma_v$ with $\sigma_1$ in Fig. 19.26, vertical stress is marginally lower than the major principal stress. If as an approximation they are assumed to be equal,

$$K_0 = \frac{\sigma_h}{\sigma_v} = 1 - \sin \phi \tag{19.31}$$

**Figure 19.25**

Jáky's sand pile model used to relate $K_0$ to $\phi$. The model incorporates vertical friction acting in partial support of a center element, as in arching action except that here the element is flat.

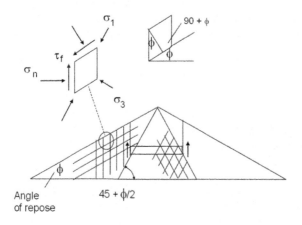

which is the Jáky approximate expression for the coefficient of earth pressure at rest. (Jáky also included an additional term that he later omitted.) As the vertical stress actually is less than the major principal stress, $K_0$ should be slightly larger than indicated by eq. (19.31), but this effect may be offset by incomplete development of the vertical friction.

### 19.10.2  Rationale for the Jáky Equation

On theoretical grounds Jáky's representation of $K_0$ is considered coincidental (Michalowski, 2005). Why should the vertical shearing stresses exist in homogeneous soil under a level ground surface? Such stresses have not been measured and may be impossible to measure, although they could be inferred from differences in settlement. Theoretical treatments assume that soil is homogeneous, but might localized variations in modulus and compressibility contribute to vertical shearing stresses that in turn may help explain Jáky's $K_0$? This question has not yet been answered, and even stresses existing in a uniform pile of sand are not fully understood (Watson, 1996).

## 19.11  SHOULD SOIL BE COMPACTED NEXT TO A RETAINING WALL?

### 19.11.1  Overview

Unquestionably compaction of soil next to a retaining wall puts additional pressure on the wall. The pattern of the pressure depends on whether the soil mass behaves plastically or elastically, and the actual behavior probably is transitional between the two as the soil compacts. Compaction of backfill mobilizes wall friction, and slight yielding of the wall allows soil to settle and assume an active state. A common evidence of settlement is the "bump at the end of the bridge," as a pavement supported on backfill settles and the bridge and its abutment do not.

Soils that are intended to support pavements or parking lots usually are at least lightly compacted, often with vibratory plates or rollers. Soils behind MSE walls are compacted to develop friction on the embedded horizontal tension strips. How much pressure from compaction equipment will be transmitted as horizontal stress on a retaining wall? Intuitively the heavier the equipment and the closer it is to the wall, the higher will be the added pressure.

### 19.11.2  Predicting Compaction Wall Pressures

As compaction involves internal shearing in the soil, induced wall pressures initially may be expected to be more in keeping with active or at-rest state pressures than with an elastic response. Duncan et al. (1991) assumed that lateral stress in a compacted soil is the mean from the Jáky and Rankine relationships, and then applied elastic criteria in a computer program called EPCOMP2. Figure 19.26 is an adaptation of their analysis showing lateral stresses induced by different roller contact pressures 0.5 ft (150 mm) from a wall, with soil $\phi = 35°$ and $\gamma = 117\,lb/ft^3$ (18.4 kN/m³). These pressures must be added to those contributed by weight of the soil.

Results for cohesive soils are higher at shallow depths, and pressures are almost doubled if the roller is immediately adjacent to the wall. On the other hand, pressures are reduced as much as one-third by doubling the distance to the wall and increasing the lift thickness. Maximum induced pressures usually occur between 4 and 8 ft (1.2 and 2.4 m) depth, where the following adjustments can be made for variations in $\phi$:

| $\phi$ | Multiplier at 4–8 ft depth |
|------|------|
| 25° | 0.7 |
| 30° | 0.8 |
| 35° | 1.00 |
| 40° | 1.1 |

Comparisons with pressures measured after compaction indicate agreement to within about ± 30 percent in sand and in silty clay. Duncan et al. (1991) indicate

**Figure 19.26**

Chart for estimating pressure transmitted to a wall from rollers having various contact pressures 0.5 ft (0.15 m) from the wall. (Adapted from data of Duncan et al., 1991.)

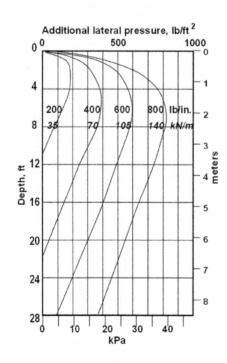

that lateral pressures inherited from compaction tend to gradually relax to at-rest values. They are even more relaxed if the wall moves.

### Example 19.12

A vibratory roller 60 in. (1.52 m) wide exerts a maximum total force from the roller weight plus impact load of 21.5 kips (95.6 kN) on soil with $\phi = 30°$ and $\gamma = 120\,\text{lb/ft}^3$ (18.8 kN/m³). (a) What is the approximate maximum induced pressure on a nearby retaining wall? (b) The approximate total force? (c) The effect on overturning moment for a wall 15 ft (4.57 m) high?

*Answer:*

(a)  The contact force is $21,500/60 = 360\,\text{lb/in.}$ (62.9 kN/m). From Fig. 19.26 the maximum induced wall pressure will be about $350\,\text{lb/ft}^2$ (17 kPa) at a depth of about 6 ft (2 m). For a friction angle of 30° the pressure is multiplied times 0.8 to give $\Delta\sigma_h = 280\,\text{lb/ft}^2$ (13 kPa).

(b)  Total force can be estimated from the area under the curve. This can be approximated by a trapezoid with an average pressure of $140\,\text{lb/ft}^2$ from 0 to 4 ft, 280 from 4 to 8 ft, and 140 from 8 to 14 ft. Then $\Delta P = (4 \times 140) + (4 \times 280) + (6 \times 140) = \underline{2500\,\text{lb}\ (11\,\text{kN})}$.

(c)  Multiplying the respective forces times lever arms gives $\Delta M = 13(4 \times 140) + 9(4 \times 280) + 4(6 \times 140) = \underline{21,000\,\text{ft-lb}\ (28\,\text{kN-m})}$.

(d)  Rankine $P = 0.5(0.33)(120)(15)^2 = \underline{4500\,\text{lb}\ (20\,\text{kPa})}$. Roller weight therefore would nullify a factor of safety of 1.5 against sliding.

(e)  Rankine $M = P \times (15/3) = 4500 \times 5 = \underline{22,500\,\text{ft-lb}}$, so the roller weight would nullify a factor of safety of 2 against overturning.

The roller might be viewed as a "proof load" if one is willing to accept that the wall could fail.

## 19.12  SURCHARGE LOADS

### 19.12.1  Two Choices

A load on the ground surface behind a retaining wall can either (a) reinstate active state internal shearing or (b) can leave the soil structure intact so that the soil behaves elastically. The two possibilities have distinctly different consequences.

A return to active shearing has the same effect as increasing the height of the wall by the equivalent amount of retained soil, as shown at the left in Fig. 19.27. From Rankine theory a constant stress equal to the surcharge pressure times $K_a$ is added throughout the height of the wall. The additional force against the wall is shown by the solid arrow in the figure, and equals the added stress times $H$, the height of the wall. The additional overturning moment equals that force times $H/2$, one-half of the wall height. This behavior is most likely to occur if the surcharge is broadly distributed and added soon after backfilling so that the soil has not had time to recover any cohesive shearing strength.

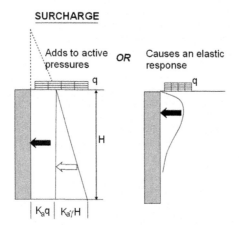

**Figure 19.27**

Two different soil reactions to a surcharge load. At the left, if the soil behaves elastically, added pressure is concentrated high on the wall. At the right, if a load is added immediately while the soil still is in an active state, a constant pressure is added to the full height of the wall and depends on $K_a$.

Most soils, including sands, gain substantial strength upon aging (Schmertmann, 1991), which can prevent returning to an active state. Furthermore, retaining walls are designed with a factor of safety so *the soil shearing resistance is not fully mobilized*. This can explain why experimental evidence supports the use of elastic theory to determine the influence of surcharge loads on existing retaining walls, first reported by Spangler (1938).

## 19.12.2   Elastic Response

A bulb of horizontal pressure developed as a consequence of an elastic soil response is shown at the right in Fig. 19.27. The total force is not the same as that developed from shearing because the additional wall pressure does not depend on the soil unit weight or internal friction, but only on the amount and location of the imposed stress relative to the wall and on Poisson's ratio.

Experimental evidence indicates that pressures induced on a wall are almost exactly two times the horizontal pressures predicted from the Boussinesq relationship, as shown in Fig. 19.28. The big question is why?

It will be recalled that the Boussinesq derivation assumes a "uniform, homogeneous half-space," which is the space on one side of a plane that extends to infinity. Soil behind a retaining wall is not in a half-space; it is half-space that is cut in half by the wall. A rigid wall prevents soil from expanding in the direction of the wall. The effect is most easily visualized by imagining the equivalent amount of horizontal pressure that would have to be applied to the soil to prevent

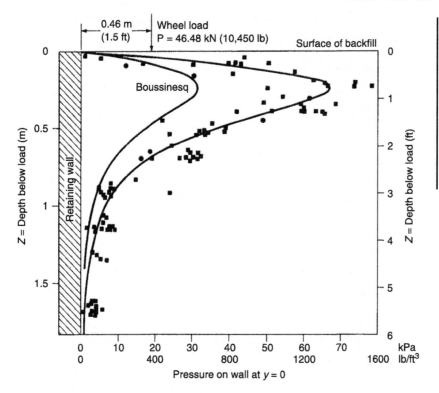

**Figure 19.28**
Measured wall
pressures from a
surface point load
are two times
those indicated by
the Boussinesq
solution. (From
Spangler, 1938,
and other
sources.)

its expansion if the wall were taken away; the pressure would be equal to that imposed on the wall, and acting in the opposite direction. This is illustrated in Fig. 19.29, and should double the pressure on the wall. This mirror-like behavior should occur regardless of the position of the surcharge load relative to the wall.

Because of the concentration of elastically induced pressures high on a wall, the effect on overturning moment can be more severe than if the soil were to shear and return to a Rankine active state.

### 19.12.3 Evaluating Elastically Induced Pressure

The Boussinesq expression for horizontal stress from a point load on the surface of the soil is doubled to depict horizontal pressure on an unyielding wall. Hence,

$$\sigma_h = \frac{3P}{\pi} \frac{x^2 z}{R^5} \qquad (19.32)$$

where $P$ is the point load and $R = \sqrt{x^2 + y^2 + z^2}$. The $x$, $y$, $z$, and $R$ dimensions are shown in Fig. 19.30.

**Figure 19.29**

An unyielding wall can mirror horizontal stresses and double soil pressure against the wall.

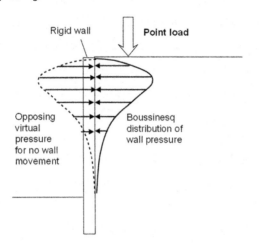

Rigid wall    Point load

Opposing virtual pressure for no wall movement

Boussinesq distribution of wall pressure

**Figure 19.30**

Dimensions for use in eq. (19.32).

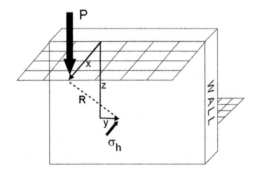

P

x

z

R

y

WALL

$\sigma_h$

### Example 19.13

A truck carrying ready-mixed concrete is parked with a back wheel 3 ft (1 m) from an 8 ft (2.4 m) basement wall. As you recently have completed a course in geotechnical engineering you recognize that this is not such a good idea, and you consider requesting that the concrete company instruct its drivers to park no closer than 10 ft (3 m) from a basement retaining wall, even though that means extending the chute in order to deliver the concrete. The axle load of the truck is 22,000 lb (100 kN). Are you right in your decision?

*Answer:* $P$ from one wheel load is 11,000 lb (50 kN). The maximum induced wall pressure will occur where the load is directly opposite to the wall, such that $y = 0$. Then, according to eq. (19.32) solved for different depths $z$ and distances $x$ normal to the wall,

$$\sigma_h = 0.95P \frac{x^2 z}{(x^2 + z^2)^{2.5}}$$

where $P = 11,000$ lb (50 kN). Some solutions are as follows. It will be noted that pressures will be appreciably higher for two wheel loads.

| z, ft (m) | x = 3 ft (0.9 m) | | x = 10 ft (3 m) | |
|---|---|---|---|---|
| z = 1 (0.3) | 299 lb/ft² | (14.3 kPa) | 10 lb/ft² | (0.48 kPa) |
| 2 (0.6) | 310 | (14.8) | 19 | (1.06) |
| 4 (1.2) | 121 | (5.8) | 29 | (1.62) |
| 6 (1.8) | 42 | (2.0) | 29 | (1.62) |

The highest pressure at 2 ft (0.6 m) depth may be compared with the Rankine pressure at that depth, which using average soil properties will be in the neighborhood of 40 lb/ft² (1.92 kPa) per foot (0.6 m) of depth. Wall pressure at that depth therefore is increased 310/80 (or 14.8/3.8), or over 400%. At 10 ft (3 m) a similar calculation indicates that the increase will be about 25%. This also is confirmed by an occasional concrete truck that ends up in a basement. A more precise evaluation could be made by integrating to obtain the area under the response curve.

## 19.12.4  Integrating Boussinesq

Equation (19.32) has been integrated from a point load to represent line loads and strip loads that are parallel or perpendicular to a wall, and to the general case of area loads. Equations expressing these relationships are presented in the Appendix. The equation is most complicated for area loads, and an approximation that may be used for small area loads is to perform a numerical integration, dividing the area into quadrants and using an equivalent point load at the center of each quadrant.

Long area loads parallel with a wall can be represented by one or more line loads, for which the equation is

$$\sigma_h = \frac{4p}{\pi} \frac{x^2 z}{R^4} \tag{19.33}$$

where $p$ is the load per unit length and $R = \sqrt{x^2 + z^2}$.

### Example 19.14
A building foundation is 4.5 ft (1.4 m) deep, located with the center parallel to and 21 ft (6.4 m) from a retaining wall. The foundation load is the equivalent of 2.5 kips/ft (36 kN/m). What is the approximate effect on the wall?

*Answer:* In eq. (19.33), $z$ is the depth below the foundation level, so the total depth below the top of the wall is $z + 4.5$ft ($z + 1.4$ m). Some solutions are as follows:

| Depth, ft (m) | | z, ft (m) | | $\sigma_h$, lb/ft² (kPa) | |
|---|---|---|---|---|---|
| 4.5 | (1.4) | 0 | (0) | 0 | (0) |
| 10 | (3.0) | 5.5 | (1.6) | 35 | (1.7) |
| 15 | (4.6) | 10.5 | (3.2) | 48 | (2.3) |
| 20 | (6.1) | 15.5 | (4.7) | 47 | (2.3) |

### 19.12.5 Effects on Forces and Moments

In order to determine the effects of lateral pressures on a wall they are multiplied times the areas on which they act. A numerical integration can be performed by considering the wall height in segments. This is illustrated in the following example, where it will be seen that with the visual area balancing in Fig. 19.31 it is not necessary to divide into many layers in order to obtain acceptable accuracy.

**Example 19.15**
Determine the influence of the wheel load in Fig. 19.28 on the force to cause sliding and on the overturning moment.

*Answer:* The first step is to select several depths, determine the horizontal pressure at those depths using eq. (19.32), and draw a pressure vs. depth curve as shown in Fig. 19.31. This is most conveniently done with a computer spreadsheet. The curve then is compartmentalized into arbitrary layers, and the average horizontal wheel-load-induced stress in each layer is obtained by drawing vertical lines such that the added and subtracted areas are equal, shown by the shaded areas in layer 2. Overturning moments are obtained by multiplying force times mid-layer heights. Results are as follows:

| Zone | Average pressure | Zone thickness | Force | Height to center | **Moment** |
|------|---------|---------|-------|---------|---------|
| 1 | 1050 lb/ft$^2$ | 1.5 ft | 1580 lb | 5.25 ft | 8270 ft-lb |
|   | 16 kPa | 0.46 m | 7.36 kN | 1.60 m | 11.8 kN-m |
| 2 | 500 lb/ft$^2$ | 1.5 ft | 750 lb | 3.75 ft | 2810 ft-lb |
|   | 6.1 kPa | 0.46 m | 2.81 kN | 1.14 m | 3.2 kN-m |
| 3 | 100 lb/ft$^2$ | 1.5 ft | 150 lb | 2.25 ft | 340 ft-lb |
|   | 1.9 kPa | 0.46 m | 0.87 kN | 0.68 m | 0.6 kN-m |
| **Totals** | | | **2500 lb** | | **10,000 ft-lb** |
|   | | | 11 kN | | 16 kN-m |

# 19.13   ALONG THE WATERFRONT: ANCHORED BULKHEADS

### 19.13.1   Overview

An earth-retaining structure that is widely used in waterfront construction is the sheet pile anchored bulkhead, shown in Fig. 19.32. An important difference from dry-land retaining wall systems is the inevitable presence of water pressures that can fluctuate with tides and affect the water level in soil behind the wall. The wall is embedded at the base and must have an adequate cross-section and tiebacks to resist bending. Tiebacks can be steel rods or cables that are made fast to anchor piles or "deadmen."

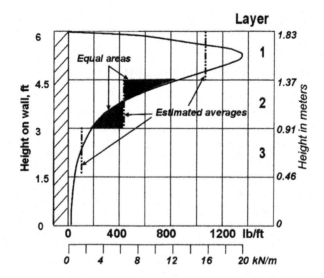

**Figure 19.31**
Drawing for
Example 19.15.

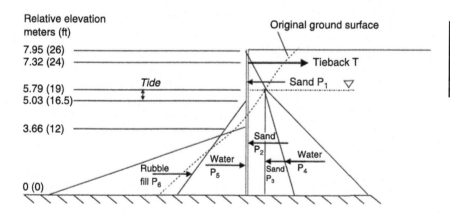

**Figure 19.32**
Pressure diagram
for a free-earth
support steel sheet
pile wall.

Active soil pressure on one side is opposed by the anchors or tiebacks near the top and by passive soil pressure at the bottom. Where tides are present, design should be for low tide on the passive side and high tide on the active side. Weep holes prevent retention of water in the soil above the level of the high tide.

## 19.13.2  Two Choices

Steel sheet pile surfaces are relatively smooth so arching action is negligible and the Rankine distributions are used. "Free-earth support" means that the pile acts like a simple beam set on end, with a load in the middle and support near the ends. "Fixed-earth support" is obtained when the sheet pile is driven deep enough to obtain "fixity" at the lower end so that the bottom does a reverse bend, as shown in Fig. 19.33. Fixed-earth support often is more economical but may be prevented by the presence of bedrock.

### 19.13.3 Free-Earth Support

Free-earth support is simplest to design but requires some careful bookkeeping. The soil active pressure, the maximum available passive pressure, and the maximum differential water pressures are calculated and plotted cumulatively from the top down. The respective forces then are obtained from the pressures times areas. The situation is complicated because tension in the tieback is not known, and the amount of the developed passive resistance is not known. However, the developed passive resistance can be determined by summing moments around the tieback point and setting them equal to zero. The developed passive resistance must be less than the maximum available passive resistance with a suitable factor of safety.

If the passive resistance is insufficient, the pile can be driven deeper, or if that is not possible because of a rock stratum, soil on the passive side can be replaced with coarse stone riprap that has a higher friction angle. Another advantage of riprap is that it protects against wave or current erosion.

**Figure 19.33**

Cumulative pressure diagram for a fixed-earth support steel sheet pile wall. The $P$ designations are the same as in Fig. 19.32.

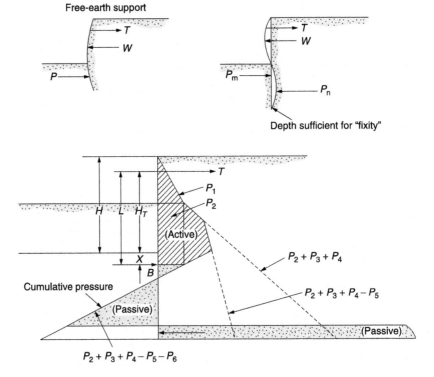

Finally, the required tension in the tiebacks is obtained by subtracting the developed passive resistance from the sum of the active forces. Tiebacks are designed with a suitable factor of safety. This ensures static equilibrium but does not address the bending moment that must be resisted by the pile. According to principles of engineering mechanics, the maximum moment occurs where the cumulative shear forces from one end are zero. This will occur at a depth below the tieback point where $T$ is balanced by cumulative active forces. It will be close to the middle of the pile. The design procedure is most easily understood by following an example calculation.

**Example 19.16**

Evaluate the factor of safety of the trial design in Fig. 19.32. On the active side $\phi = 30°$ and on the passive side a rubble backfill has $\phi = 40°$.

**Answer:** A spreadsheet can be arranged as follows. Note that submerged soil unit weights are used below water and $K_p$ is used for passive resistance of the soil.

| Force | Matl. | $\gamma$ kN/m$^3$ | $K$ | $H$ m | Force kN | Elev. m | Moment around T y, m | kN-m | Around base |
|---|---|---|---|---|---|---|---|---|---|
| Active: | | | | | | | | | |
| $P_1$ | Soil | 18.85 | 0.33 | 2.13 | 14.2[1] | 6.50 | 0.82 | 12 | 92 |
| $P_2$ | Soil | 9.04 | 0.33 | 5.79 | 101[2] | 2.90 | 4.42 | 446 | 293 |
| $P_3$ | Soil | 9.04 | 0.33 | 5.79 | 50.5[1] | 1.93 | 5.39 | 272 | 97 |
| $P_4$ | Water | 9.81 | 1.0 | 5.79 | 164[1] | 1.93 | 5.39 | 884 | 317 |
| Passive: | | | | | 330 | | | 1614 | 799 |
| $P_5$ | Water | 9.81 | 1.0 | 5.03 | 129[1] | 1.68 | 5.64 | −728 | 217 |
| $P_6$ | Soil | 9.04 | 4.6 | 3.66 | 278[1,3] | 1.22 | 6.10 | −1696[3] | 339[3] |
| | | | | | 407[3] | | | −2424[3] | 556[3] + 7.32T |

Note: [1]$P = 0.5K\gamma H^2$.
[2]$P = K\gamma H^2$.
[3]Maximum with fully developed passive resistance.

*Moments around T*: $6.10 \times P_6 + 728 = 1614$, from which $P_6 = 145\,kN$ and the FS for passive earth pressure is FS $= 278/145 = \underline{\mathbf{1.9}}$.

*Solution for T*: $T + 145 + 129 = 330$, from which $T = \underline{56\,kN/m\ of\ wall}$; tiebacks capable of sustaining 22.2 kN (50,000 lb) will be spaced every $56/22.2 = 2.5\,m$ along the wall.

*FS against sliding*: FS $= 407/330 = \underline{\mathbf{1.23}}$.

*FS against overturning*: FS $= (56 \times 7.32 + 556)/799 = \underline{\mathbf{1.21}}$.

A moment diagram is then prepared to determine the maximum bending moment that must be resisted by the sheet pile.

### 19.13.4 Fixed-Earth Support

Sometimes a more efficient design can be made if sheet pile is driven deep enough that the lower end remains fixed in the soil. This introduces a reverse bend, as shown in Fig. 19.33, which decreases the maximum bending moment and allows the use of lighter pile cross-sections. Fixed-earth support is analogous to a simple beam that is simply supported near one end and buried in a wall at the other.

The most commonly used method for fixed-earth support design is Tschebotarioff's (1951) "equivalent beam method." In this method the equivalent beam is horizontal with the soil load on top of a beam that is simply supported at the tieback point. The other end is supported by soil with the center of force located at a distance $X$ below the dredge line. A guide to estimation of $X$ is as follows:

| $\phi$ | $X$ |
|---|---|
| 30° | $0.08H_e$ |
| 35° | $0.03H_e$ |
| 40° | 0 |

where $H_e$ is the height of the exposed face of the sheet pile wall. The span of the equivalent beam therefore is $L = H_e + X$. The maximum bending moment then is approximated by $M = WL/8$, where $W$ is the soil load above the position of $X$ and is equal to the area under the pressure diagram, shown cross-hatched in Fig. 19.33.

**Figure 19.34**

Terzaghi and Peck (1967) pressure diagram for braced excavations.

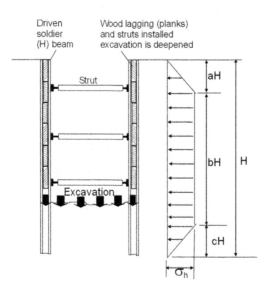

The next step is to calculate $B$, which represents the soil support for the embedded end of the beam. As in the case of the free-earth support method, this is obtained by taking moments around the tieback level. The embedment length, $t$, is obtained from

$$t = X + 1.2\sqrt{\frac{6B}{\gamma'(K_p - K_a)}} \qquad (19.34)$$

where $\gamma'$ is the submerged unit weight of the soil and $K_p$ and $K_a$ are the passive and active earth pressure coefficients. This value is added to $H_e$ to obtain the total required length of the pile.

The last step is to determine the tieback force by subtracting passive from active forces, as in the case of the free-earth support method.

## 19.14 DOWN IN THE TRENCHES: BRACED EXCAVATIONS

### 19.14.1 Trench Safety

Despite strict regulations and laws concerning safe practices in trenching, fatalities still occur from trench walls caving in on unprotected workers. A person need not be buried to prevent breathing, because the impact from collapsing soil forces air out of the lungs, and it is the passive resistance of soil that must be overcome in order for the worker to inhale. Time can be bought by supplying oxygen, but meanwhile the loss of circulation in the extremities can inflict lasting damage. Most such tragedies involve unskilled workers who often are college students in summer jobs, and involve contractors who can't get insurance and lack technical training.

The most common protection for workers in shallow trenches is a steel "trench box" that must be designed to withstand active soil plus impact pressures. The box is pulled along as workers install utilities in the open trenches, after which the trench is filled.

An engineer who observes dangerous violations of code is morally and professionally obligated to inform the contractor and, if necessary, the owner and the authorities—before it is too late.

**Example 19.17**

A worker is buried to mid-chest height in soil rubble lying at the angle of repose and having a unit weight of 80 lb/ft$^3$ (13 kN/m$^3$) and a friction angle of 35°. (a) What is the pressure against his body at a depth of 0.5 ft (0.15 m)? (b) What pressure must he exerted in order to breathe?

*Answer:* (a) According to the Rankine analysis of section. 19.44, $K$ for soil at the angle of repose is 1.0, so the soil pressure at the indicated depth is $0.5(80) = 40$ lb/ft$^2$, or $0.15(13) = 2 \text{kPa}$. Breathing will require overcoming the shearing resistance of the soil and pushing it upslope, which will be impossible. The only way to allow breathing will be to scoop away and hold back the soil so that it is level and not sloping into his body. Even then the pressure required will be impossibly high, because for this friction angle $K_p = 3.7$, so a pressure of $3.7 \times 40 = 148 \text{lb/ft}^2$ ($3.7 \times 2 = 7$ kPa) will be required.

## 19.14.2 Piling It On or Cutting Back

The soil excavated from a trench commonly is piled far enough back from the edge that loose "rollers" will be stopped before they fall into the trench. This does not take into account the effect of a surcharge load on trench safety, which has the same effect as deepening the trench. This problem can be minimized if instead of cutting a vertical trench the sides are cut back to a safe angle. This can easily involve doubling or tripling the amount of excavated soil, but this is easily accomplished with modern excavating equipment, and a wider installation increases compaction efficiency. However, space is required. The short-term stability of wide trenches may be based on experience with a particular soil. The design of stable slopes is discussed in a later chapter.

## 19.14.3 Bracing of Vertical Trench Excavations

Wide, sloping excavations may not be possible with deep trenches, and a trench box is difficult to manage and can become wedged in if the sides should yield. Deep trenches are protected with temporary cross-bracing that is installed as the trench is being deepened. This procedure influences the soil pressure that must be resisted by the bracing, because soil still in place below the trench bottom acts as a temporary brace that moves downward as the trench is deepened. Bracing is recycled so that emphasis is on safety instead of minimal design.

Strain measurements of bracing indicate that soil pressures are distributed fairly evenly throughout most of the depth of a trench, decreasing near the bottom, and that bracing normally does not yield sufficiently to develop active state pressures Peck (1969). The total force exerted by the soil therefore is about 30 to 50 percent higher than that calculated from Rankine active state analysis. Empirical guidelines for design are indicated in Table 19.1, with reference to Fig. 19.34.

**Table 19.1**

Empirical constants for design of bracing. Letters *a, b, c* refer to depth increments in Fig. 19.34. After Terzaghi and Peck (1967)

| Soil type | a | b | c | $\sigma_h$ |
|---|---|---|---|---|
| Sand | 0 | 1 | 0 | $0.65 K_a \gamma H$ |
| Soft to medium clay | 0.25 | 0.75 | 0 | $K_a \gamma H$ |
| Fissured clay | 0.25 | 0.5 | 0.25 | 0.2 to $0.4 \gamma H^*$ |

Note: *Lower value if movements are minimal with a short construction period.

## Example 19.18

What bracing will be required for an excavation 25 ft (7.6 m) deep in sand having a friction angle of 25° and a moist unit weight of 135 lb/ft³ (21.2 kN/m³)?

*Answer:* From Table 19.1 the pressure from the sand will be $0.65 \times 0.41 \times 135 \times 25$ = 900 lb/ft² throughout the entire depth of the excavation, or $900 \times 25 = 11.2$ tons per running foot or, with a factor of safety of 1.3, 14.6 tons/ft (1.4 MPa). As a trial design assume 15 ton jacks spaced every 5 ft along the trench, which will require a resistance of $14.6 \times 5 = 73$ tons per running foot, or $72/15 = 5$ jacks per installation.

## 19.14.4    Summary Statement

There are few responsibilities in geotechnical engineering that carry the weight of human life so much as the design and construction of retaining walls. This is a serious business, and every design should be independently checked by an experienced geotechnical engineer.

## Problems

19.1. Define active, passive, and at-rest pressures and explain in detail the differences between them.

19.2. Which cause(s) for failure of retaining walls are not specifically addressed in this chapter?

19.3. What lateral stress can be expected at a depth of 10 ft (3.0 m) in a normally consolidated soil with a unit weight of 120 lb/ft³ (18.9 kN/m³) and a friction angle of 30°?

19.4. The lateral stress measured at a depth of 3.5 ft (1.07 m) in the soil in Problem 19.3 is 1950 lb/ft² (93 kPa). Is this soil overconsolidated? If so, list three possible causes for overconsolidation.

19.5. Assume that the soil in the preceding problem is overconsolidated as a result of excavation of overburden. What amount of overburden has been removed?

19.6. What is a slurry-trench wall? Gabion? Cantilever wall? Fascia wall? MSE?

19.7. Under certain conditions the Rankine and the Coulomb earth pressure theories give the same answers. (a) When will the answers be different? (b) If the answers are the same, explain any conceptual differences.

19.8. A concrete wall 7.3 m (24 ft) high retains a backfill of gravel with a ground surface level with the top of the wall. The soil unit weight is 20.4 kN/m³ (130 lb/ft³) and the angle of internal friction is 35°. (a) From Rankine theory what is the pressure at the base of the wall and at its midheight? (b) What is the total force per unit length of wall? (c) What is the overturning moment per unit length?

19.9. Repeat Problem 19.8 using the Coulomb method and assuming that the wall friction angle is 0.8 times the internal friction. Which determination do you believe is more accurate? Why?

19.10. Repeat Problem 19.8 using "easy arching." Which determination do you believe is more accurate? Why?

19.11. A wall 4.27 m (14 ft) high is backfilled with soil weighing 18.85 kN/m$^3$ (120 lb/ft$^3$) and has an angle of friction of 25°. The back face of the wall is battered 1 in 12, and the backfill slopes upward from the top of the wall on a 6:1 slope. Assume that the angle of friction between the soil and the wall is 20°. Calculate the resultant force and overturning moment per unit length of wall.

19.12. A temporary cut is to be made through a cohesive soil that weighs 15.7 kN/m$^3$ (100 lb/ft$^3$) and has an unconfined compressive strength of 15.6 kPa (325 lb/ft$^2$). If the cut is 6.1 m (20 ft) deep and the sides are vertical, estimate the total force per unit length on timber sheeting used to restrain the cut faces. What will be the approximate location of the resultant pressure?

19.13. The space behind a retaining wall 3 m (10 ft) high is backfilled with cohesionless sand that will be saturated for the lower 1.8 m (6 ft) of the wall. Assume that the friction angle of the sand is 22° and the unit weight above the line of saturation is 19.6 kN/m$^3$ (125 lb/ft$^4$), and the buoyant unit weight is 11 kN/m$^4$ (70 lb/ft$^3$). Using Rankine theory, determine lateral pressure on the wall and draw a diagram showing the distribution of pressure from top to bottom of the wall.

19.14. A wall 7.9 m (26 ft) high retains a level backfill soil that weighs 18.1 kN/m$^3$ (115 lb/ft$^3$) and has a friction angle of 20°. A continuous building footing 1.8 m (6 ft) wide is to be constructed on the surface of the backfill at a clear distance of 3 m (10 ft) from the back of the wall. The load on this footing will be 192 kPa (2 tons/ft$^2$). Determine the effect this will have on a unit length of the wall and the position of the resultant force.

19.15. A concentrated load of 89 kN (20,000 lb) is applied to the backfill surface at a distance of 1.8 m (6 ft) back from a 3 m (10 ft) high retaining wall. Calculate the magnitude and distribution of lateral pressure caused by the superimposed load on a vertical element on the wall, which is 0.3 m (1 ft) long and directly opposite the load.

19.16. A sheet pile bulkhead is to be constructed as shown in Fig. 19.35. Using the free-earth support method, estimate the factor of safety of the structure and the stress in the anchorage per unit length of bulkhead. (*Note:* Consider water on both sides of bulkhead balanced without reference to the factor of safety.)

19.17. Rework Problem 19.16 using fixed-earth support and the equivalent beam method. What sheet pile depth is required for fixity?

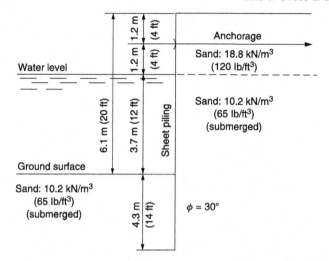

**Figure 19.35**
Steel sheet pile for
Problems 19.16
and 19.17.

19.18. What factors affect the distribution of horizontal pressure on braced excavations compared with retaining walls?

19.19. A wall is to be constructed 6 ft (2 m) in front of a limestone face, with a backfill consisting of limestone screenings having a unit weight of 130 lb/ft³ (20 kN/m³) and a friction angle of 30°. The wall will vary in height from 12 to 22 ft (3.7 to 6.7 m). Calculate (a) maximum and minimum pressures on the wall, (b) maximum and minimum total forces, and (c) the corresponding overturning moments.

19.20. A segmented wall is observed to be bulging near the middle, and after a heavy rain part of the wall topples. The designer blames the contractor, the contractor blames the designer, and the owner blames and sues them both. You are retained as an independent expert to help determine whose insurance should pay for the repair. You also are aware that most wall failures are caused by inadequate foundations. What tests and methods of analysis would you use to determine if the wall was improperly designed and/or poorly constructed?

## References and Further Reading

Adib, M., Mitchell, J. K., and Christopher, B. (1990). "Finite Element Modeling of Reinforced Soil Walls and Embankments." *Design and Performance of Earth Retaining Structures,* ASCE Spec. Geotech. Publ. No. 25, 409–423.

Duncan, J. M., Williams, G. W., Sehn, A. L., and Seend, R. B. (1991). "Estimation of Earth Pressures due to Compaction." *ASCE J. Geotech. Eng.* 117(12), 1833–1847.

Handy, R. L. (1985). "The Arch in Soil Arching." *ASCE J. Geotech. Eng.* 111(3), 302–318.

Jáky, J. (1944). "The Coefficient of Earth Pressure at Rest," *J. Soc. Hungarian Architects and Engineers* (October), pp. 355–358.

Jumikis, A. R. (1962). *Soil Mechanics.* Van Nostrand, Princeton, N.J.

Krynine, D. P. (1945). Discussion of "Stability and stiffness of cellular cofferesams," by Karl Terzaghi. *Trans. ASCE* 110, 1175–1178.

Lazarte, C. A., Elias, V., Espinoza, R. D., and Sabatini, P. J. (2003). *Geotechnical Engineering Circular No. 7, Soil Nail Walls.* Report No. FHWA0-1F-03-017, U.S. DOT FHWA.

Michalowski, R. L. (2005). "Coefficient of Earth Pressure at Rest." *ASCE J. Geotech. and Geoenviron. Eng.* 131(11), 1429–1433.

Peck, R. B. (1969). "Deep excavations and tunneling in soft ground." *State-of-the-Art Volume, 7th Intern. Conf. Soil Mech. and Foundation Eng.* 225–290.

Schmertmann, J. H. (1991). "The mechanical aging of soils." *ASCE J Geotech. Eng* 117, 9, 1288–330.

Spangler, M. G. (1938). "Horizontal Pressures on Retaining Walls Due to Concentrated Surface Loads." Bull. 140, Iowa Engineering Experiment Station, Ames, Iowa.

Spangler, M. G., and Mickle, J. L. (1956). "Lateral Pressures on Retaining Walls Due to Backfill Surface Loads." *Highway Res. Bd. Bull.* 141.

Terzaghi, K. (1936). "A Fundamental Fallacy in Earth Pressure Computations." *J. Boston Soc. Civil Engrs*, reprinted in *Contributions to Soil Mechanics 1924–1940* by the Society, Boston.

Terzaghi, K. (1943). *Theoretical Soil Mechannics.* McGraw-Hill, New York.

Terzaghi, K., and Peck, R. B. (1967). *Soil Mechanics in Engineering Practice*, 2nd ed. John Wiley & Sons, New York.

Tschebotarioff, G. P. (1951). *Soil Mechanics, Foundations, and Earth Structures.* McGraw-Hill, New York.

Watson, A. (1996). "Searching for the Sand-Pile Pressure Dip." *Science* 273(5275), 579–580.

# 20 Mechanically Stabilized Earth (MSE) Walls and an Introduction to Soil Nailing

## 20.1 OVERVIEW

### 20.1.1 Composites

Birds' nests, beaver dams, and straw in the bricks all are composites. Automobile tires are a composite, and it is not difficult to imagine what would happen if a tire were inflated without the tensile restraint from internal fibers. Plywood is a composite, as are layered pavement systems. The most common composite used in construction combines the compressive strength of Portland cement concrete with the tensile strength of steel.

Two composites involving soils were developed in France in the last half of the twentieth century, and now are used throughout the world. They are mechanically stabilized earth (MSE), and soil nailing. They also can be viewed as modified tieback walls.

### 20.1.2 Tres Caballeros

Three types of retaining structures employ tension elements embedded in soil. Tieback walls, discussed in the preceding chapter, are constructed from the top down, as tiebacks are put into place as soil is removed from one side of the wall. Soil nailing is similar in being constructed from the top down, and different because instead of functioning to hold a wall structure, the nails function to hold the soil to itself. Whereas tension in tiebacks results either from post-tensioning or the ends being pulled outward by movement of the wall, the tendency for soil to move pulls nails both ways from the middle. The nailed face is covered with steel mesh and sprayed-on concrete ("shotcrete") to protect it from erosion, and may have decorative panels attached, but the cover does not serve structurally as a wall.

Unlike these two construction methods, MSE walls are built from the bottom up instead of from the top down, by alternating horizontal tension members with layers of compacted soil. The outer ends of the tension strips are attached to a wall, but experimental evidence shows that, like soil nails, the strips are pulled both ways from near the middle.

The two construction procedures, top down or bottom up, have their own particular uses and sometimes may be combined, as when an MSE wall is built on top of a nailed wall.

# 20.2  MECHANICALLY STABILIZED EARTH (MSE)

## 20.2.1  Background

The story goes that in the late 1960s French engineer Henri Vidal was vacationing at the beach, and, like a true engineer, became more engrossed with the sand than with the scenery. Vidal observed that putting horizontal layers of pine needles in a pile of sand made it more tolerant to vertical loading. Thoughtful observation is a key to invention, and he designed a system consisting of layers of sand separated by flat, horizontal strips of steel. The strips are attached to facing panels to build a retaining wall. The resulting walls are flexible, drain readily, and cost considerably less than walls built by conventional methods. Reinforced Earth$^{TM}$ soon became one of the most popular options for retaining walls. The facing initially was folds of sheet metal that later were substituted by more decorative concrete facing panels (see Fig. 20.3, below).

After the Reinforced Earth patent expired, other kinds of stabilized earth walls were developed, leading to an all-inclusive generic term "mechanically stabilized earth," or MSE. Reinforcing elements can be steel rods or strips, or layers or grids of plastic having a suitable tensile strength and drag coefficient with the soil.

A simple model illustrating the principle of MSE can be made with sheets of paper and layers of sand. As shown in Fig. 20.1, lay a sheet of paper on the floor, cover the central area with a thin layer of dry sand, fold over the edges of the paper on top of the sand, and tape the paper where it overlaps at the corners. Then lay another sheet on top and repeat the process, a sheet of paper and a layer of sand, and keep repeating to make a stack that looks like a square Michelin Man. Cover the top and try stepping on it. We are not sure that this works but it sounds reasonable. Then throw away the paper and sweep up the sand.

## 20.2.2  Construction of MSE

Mechanically stabilized earth (MSE) consists of flat, horizontal strips or grids of steel or plastic mesh that are separated by layers of compacted granular soil. As soil layers are added, lateral expansion from pressure of the overburden is

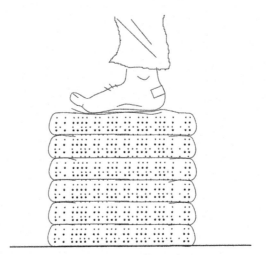

**Figure 20.1**
Tricky experiment to demonstrate the principle of mechanically stabilized earth. Keep a broom handy just in case.

opposed by tension in the strips, which has the same effect on the sand as lateral pressure applied externally in a triaxial test; the strips restrict or prevent further expansion and increase the frictional strength of the soil. The cohesive component of soil strength is not affected by lateral confinement, so the process is less effective with cohesive soils.

According to Rankine theory, lateral stress increases with depth, and it is important that sufficient friction be developed on the strips to prevent their pulling out. However, friction also increases with depth by adding more vertical pressure to the strips, so the two requirements in theory should balance out. Therefore, as conventionally constructed, all strips are the same length.

### 20.2.3 A Very Thick Wall

MSE walls have a facing, but the facing would be of no value without the tension strips. Therefore an MSE wall consists of the *entire structure* that extends from the facing back to the far ends of the strips. The MSE wall is a very thick gravity wall that must be large enough and heavy enough to prevent sliding or overturning. Also, it must have adequate foundation support so that it does not sink or tilt because of the eccentric loading as its weight acts downward and the retained soil pushes against the back of the block of reinforced soil.

The mass of the mechanically stabilized earth is designed to prevent internal failure from tension elements breaking or pulling out. Experiments have shown that the reinforcing elements are tensioned by friction that pulls both ways from near the middle. As shown in Fig. 20.2, this is in contrast to the stress in tiebacks, which are grouted at one end and fastened to the wall at the other.

**Figure 20.2**

Tiebacks are pulled at the ends as excavation proceeds downward. In MSE, reinforcing strips are installed as a wall is being built up from the bottom, and tension varies depending on friction.

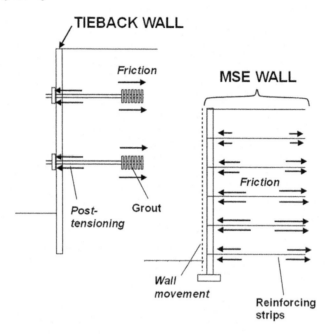

**Figure 20.3**

The facing on MSE wall often has the appearance of being an ordinary block or stone wall. Facing units or panels are held in place by connections to reinforcing elements that extend horizontally and are held in place by friction with soil.

Tieback tension therefore is constant throughout the ungrouted length of the tieback.

## 20.3 DESIGN OF MSE WALLS

### 20.3.1 Selection of Soil

MSE is best adapted for granular soils having high friction and low clay content. The soil often is brought in from off-site, and spread and compacted under controlled conditions.

Soil in an MSE wall provides friction that holds the tension strips, and MSE walls originally were intended to use only compacted granular gravel or sand (GW, GM, SW, or SM) soil backfill. Clayey sands (SC) and coarse silty (ML) soils also can be used if there is adequate drainage. These criteria apply to soil within the wall itself and not to soil retained by the wall structure.

Some MSE walls have been built using cohesive soils, *but that at best is risky and at worst irresponsible*. The lower hydraulic conductivity of cohesive soils leads to water retention, increasing the soil weight and contributing to excess pore water pressure and creep. Improper soil is a leading cause of failure of MSE walls.

A soil acceptability criterion proposed by the U.S. FHWA (Elias et al., 2001) and used by many state highway departments is (a) no more than 15 percent of the soil should pass a No. 200 sieve and (b) the plasticity index should be no higher than 6. The friction angle after compaction can be expected to be in the range 34° to 31°, depending on the gradation and plasticity.

Less rigorous requirements have been advanced by manufacturers of geogrids because of passive resistance from soil enclosed in the grid openings. These specifications may allow up to 35 percent passing the No. 200 sieve if there is internal drainage, in which case the drained friction angle is lowered to about 28°.

MSE also is used to stabilize slopes, in which case the wall is laid back to near the slope angle and the maximum allowable PI is 20, with up to 50 percent passing the No. 200 sieve. The strength requirement is reduced because of the lower slope angle compared with the relatively steep face of a high-angle retaining wall. For example, if a natural slope angle of 15° is stable, relatively less effort will be required to stabilize it at an angle of 20° or 25° than to make it near vertical. Factors influencing slope stability and landslides are discussed in the next chapter.

Compaction is specified to a minimum of 95 percent of the standard Proctor density, and is subject to field inspection and testing. The above criteria apply to soil used in the wall, and not to soil that is behind the wall, although its strength still has a direct bearing on wall stability.

### 20.3.2 External Stability

Design criteria for external stability are the same as for other gravity retaining walls, and the various possibilities shown in Fig. 19.3 should be reviewed. Pressure from the retained soil is assumed to act on a vertical plane through the ends of the reinforcing strips, as shown in Fig. 20.4. As the walls are segmented and therefore flexible, the lower part may move more than the upper part, resulting in the bowing shown in the figure. The direction of bowing can be a clue to the cause of distress. The other potential external failure modes include sliding, toppling, and involvement in a landslide.

**Example 20.1**

A 20 ft (6.1 m) high MSE wall is constructed with sandy soil having a friction angle of $20°$ and a total unit weight of $125 \, lb/ft^3$ ($19.6 \, kN/m^3$). The wall rests on and retains soil having a friction angle of $20°$ and a unit weight of $110 \, lb/ft^3$ ($17.3 \, kN/m^3$). The minimum length of the reinforcement is 70% of the height of the wall. (a) What is the factor of safety against sliding? (b) Against overturning? (c) If these factors are not adequate, what do you recommend? (d) What is the eccentricity of the foundation pressure?

**Figure 20.4**

MSE walls are manufactured with quality control while the foundation soils remain an uncontrolled variable, and must be evaluated for (a) differential settlement and (b) bearing capacity failure. Foundation problems are a main cause of distress in MSE walls.

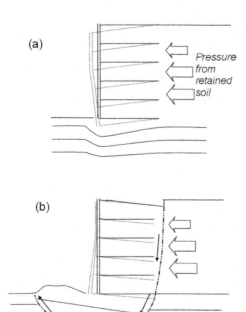

*Answer:*

(a) $K_a$ from the Rankine formula for a friction angle of $20°$ is 0.49. The force from the retained soil is

$P = \frac{1}{2}K_a\gamma H^2 = 0.5(0.49)(110\,\text{lb/ft}^3)(20\,\text{ft})^2 = 10,800\,\text{lb}$ per linear foot of wall, or

$\qquad = 0.5(0.49)(17.3\,\text{kN/m}^3)(6.1\text{m})^2 = 158\,\text{kN/m}$

The strip length represents the thickness of the wall: $0.7 \times 20 = 14\,\text{ft}$ $(0.7 \times 6.1 = 4.3\,\text{m})$, and the weight is

$W = 20\,\text{ft} \times 14\,\text{ft} \times 125\,\text{lb/ft}^3 = 35,000\,\text{lb}$ per linear foot or

$\qquad = 6.1\text{ m} \times 4.3\,\text{m} \times 19.6\,\text{kN/m}^3 = 514\,\text{kN/m}$

Frictional resistance is

$F = 35,000\,\text{lb/ft}\tan 20° = 12,740\,\text{lb/ft}$ or

$\qquad = 514\,\text{kN/m}\tan 20° = 187\,\text{kN/m}$

The factor of safety against sliding is $12,740/10,800 = 1.18$, or $187/158 = 1.18$.

(b) The overturning moment about the toe assuming a triangular distribution of active pressure is

$M_o = P\,(\text{lb/ft}) \times H\,(\text{ft}) \div 3 = 10,800 \times 20 \div 3 = 72,000\,\text{ft-lb}$ per foot of wall, or

$\qquad = 158\,(\text{kN/m}) \times 6.1\,(\text{m}) \div 3 = 320\,\text{kN-m per meter}$

The resisting moment from the weight of the reinforced soil wall is

$M_r = 35,000\,\text{lb} \times 7\,\text{ft} = 245,000\,\text{ft-lb}$ or

$514\,\text{kN} \times (4.3 \div 2)\,\text{m} = 1105\,\text{kN-m}$

The factor of safety against overturning is $245,000/72,000 = 3.4$, or $1105/320 = 3.4$.

(c) The factor of safety against sliding can be increased by seating the wall deeper into the soil to provide passive resistance at the toe.

(d) Eccentricity is defined as the distance the weight acts off center in order to maintain equilibrium, and is obtained by equating the weight couple and the net overturning moment about the toe of the wall:

$W \times e = P \times H \div 3$

$\qquad e = PH \div 3W = 10,800\,\text{lb/ft} \times 20\,\text{ft} \div 3 \times 35,000\,\text{lb/ft} = 2.1\,\text{ft}$ or

$\qquad\qquad = 158\,\text{kN/m} \times 6/1\,\text{m} \div 3 \times 514\,\text{kN/m} = 0.63\,\text{m}$

Eccentricity is used in foundation bearing capacity calculations.

## 20.3.3 Internal Stability

Whereas internal stability of a reinforced concrete is dealt with by structural engineers, the internal stability of MSE walls is in the realm of geotechnical engineering in order to establish design criteria for length and the strength of reinforcing elements, and for their spacing vertically and horizontally.

A sliding wedge usually is the assumed failure mode for the interior of the wall—that is, with the wedge within the wall. The wedge is restrained by the horizontal friction strips that anchor into soil behind the wedge. Tensions measured with strain gauges attached to the strips (Fig. 20.5) appear to reflect the distribution of horizontal pressure instead of locating a hypothetical slip surface. The data indicate a concentration of pressure higher on the wall than at the base, as indicated by classical soil mechanics, which can explain distress by bulging near the midriff.

Arching theory was applied to experimental data by Schlosser (1990), which was for an MSE wall (Fig. 19.22) in the preceding chapter.

### 20.3.4 Strip Length

According to the Rankine criteria, without wall friction the failure wedge should extend upward and back from the heel of the wall facing elements at an angle $45 + \phi/2$ from horizontal, as shown in Fig. 20.5. This assumption is on the safe side for determining an anchorage for the strips, particularly as the strip tension diminishes from a maximum near the middle to zero at the end. With a vertical

**Figure 20.5**

Tension measured in steel reinforcing strips in an MSE wall indicate that the center of pressure is near the middle of the wall. This wall was surcharged to failure, which increased tension in the strips but did not alter the line of action. (From data of Al-Hussaini and Perry, 1978.)

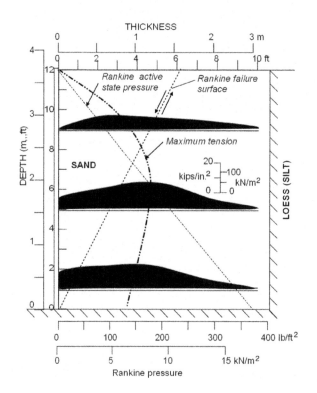

wall and a $30°$ friction angle the wedge extends into the wall a distance of $0.6H$. As the strips must embed in soil beyond the wedge, a minimum strip length of 0.6 to $0.7H$ is widely used. It will be noted that an MSE wall cannot be used if there is not enough room behind the facing for the required length of tension strips. MSE therefore is not appropriate, nor is it necessary, for a fascia wall.

## 20.3.5 Control of Compaction

Quality control for the backfill soil usually is on the basis of gradation and compacted density that is most readily measured with a nuclear gauge. The Borehole Shear test is adapted for direct measurement of friction angles in the compacted soil, and in sand the test can be performed very rapidly in hand-augered holes.

## 20.3.6 Pullout Resistance

Friction on the reinforcing strips depends on the contact pressure, which increases linearly with depth. The pullout resistance is obtained by multiplying the vertical soil pressure times a friction coefficient times the area on both sides of the strip. This is divided by a factor of safety that normally equals or exceeds 1.5. The friction coefficient is evaluated by laboratory pullout tests.

For steel strips the ratio of sliding friction to internal friction of the soil often is taken as 1.0. Geogrids have openings that interlock with fill so measured value can be higher than 1.0, but an allowance is made for stretching and creep of the plastic, and for possible installation damage. Suggested reduction factors for pullout friction are as follows for steel and for high-density polyethylene (HDP):

| Material | Soil | Reduction factor |
|---|---|---|
| High-density polyethylene (HDP) | Gravelly sand | 0.8 |
| | Sand | 0.75 |
| | CL, ML (clayey sand, silt) | 0.58 |
| Steel | Sand | 1.0 |

Other information is available in various manufacturers' trade literature. The pullout force increases with depth, so a calculation is made for each depth of grid.

**Example 20.2**
Calculate the pullout resistance per unit width of a geogrid needed to stabilize a 20 ft (6.1 m) high wall with compacted SC soil backfill having a dry unit weight of $102\,\text{lb/ft}^3$ ($16.0\,\text{kN/m}^3$). The factor of safety is to be 1.5.

*Answer:* The grid length is assumed to be 0.7 times the wall height, or 14 ft (4.3 m). From Fig. 19.2 and the ratio $\gamma_d/\gamma_w$ the soil friction angle is assumed to be $28°$. The total unit weight will be somewhat higher because of the retention of water, but the discrepancy will be small and on the safe side. The design pullout resistance per unit width per unit of depth per unit length of wall is as follows. Note that this does not depend on Rankine pressures, which apply only to horizontal stress.

$$T_{po} = (1\,\text{ft} \times 102\,\text{lb/ft}^3) \times (14\,\text{ft} \times 2) \times (0.58 \tan 28°) \div 1.5 = 590\,\text{lb/ft/ft} \text{ or}$$
$$= (1\,\text{m} \times 16.0\,\text{kN/m}^3) \times (4.3 \times 2) \times (0.58 \tan 28°) \div 1.5 = 28\,\text{kN/m/m}$$

The largest pullout resistance will be required for the deepest strip. Assuming this to be 2 ft (0.6 m) above the base of the wall, the pullout resistance including a factor of safety of 1.5 is

$$T_{po} = 18\,\text{ft} \times 590\,\text{lb/ft/ft} = 17,\!000\,\text{lb/ft length of wall or}$$
$$= 3.7\,\text{m} \times 28\,\text{kN/m/m} = 100\,\text{kN/m}$$

## 20.3.7  Tensile Strength of the Reinforcing Strips

As one requirement is that the pullout resistance must increase with depth, so must the tensile strength of the reinforcement. Unlike the pullout resistance, which depends only on vertical pressure, tensile strength depends on lateral stress, which conventionally is calculated from Rankine or Coulomb theories. The factor of safety is increased to 2.0 to account for the vagaries of lateral soil pressure including arching action.

A simple approach is to assume a unit length of wall and calculate the pressure times vertical height of each segment, then multiply by the horizontal spacing between tension members. If the reinforcement is continuous, as in the case of a grid or mat, the answer is left in terms of a unit width of the grid or mat.

**Example 20.3**
Calculate the required strip tensile strength per unit length of wall for panels extending between depths 16 to 20 ft (4.9 to 6.1 m) in the wall of the preceding example.

*Answer:* The Rankine active coefficient is $K_a = \tan^2(45 - 28/2) = 0.36$. The average depth for the bottom panel is 18 ft (5.5 m).

$$\sigma_h = 18\,\text{ft} \times 102\,\text{lb/ft}^3 \times 0.36 = 660\,\text{lb/ft/ft} \text{ or}$$
$$= 5.5\,\text{m} \times 16.0\,\text{kN/m}^3 \times 0.36 = 32\,\text{kN/m/m}$$

With a panel height of 4 ft (1.2 m) and a factor of safety of 2, the required tensile strength per unit length is

$$T = 660\,\text{lb/ft}^2 \times 4\,\text{ft} \times 2 = 5300\,\text{lb/ft of wall or}$$
$$= 32\,\text{kN/m}^3 \times 1.2\,\text{m} \times 2 = 77\,\text{kN/m}$$

Note that, to avoid confusion in calculations, units are carried on all quantities and canceled as appropriate. Also, the above examples incorporate only two significant figures so they may be easier to follow; in practice three significant figures would be carried through the calculations and the answers rounded to two.

The required tension must be equaled or exceeded by a grid that is commercially available or the spacing must be changed. The same calculations are made for higher panels where a lighter reinforcement may be used.

Arching action in soil behind conventional walls requires the development of wall friction. In MSE walls this load transfer may be aided by vertical support for the soil from the strips that are attached to the wall. Some trial calculations indicate that arching action will account for about a 15 percent friction reduction in the factor of safety for tensile strength of reinforcement near the midsection, and a 30 percent reduction in factor of safety against overturning.

## 20.4  SOIL NAILING

### 20.4.1  A Wall Without a Wall

As the name implies, soil nailing involves insertion of long steel rods into soil so that the soil holds the nails and the nails hold the soil. Soil nailing was first devised in France in the early 1970s, and offers advantages of low cost, versatility, and easily negotiating curves and corners. Furthermore, nailing does not require large equipment.

Nails can be shot into soft soil using compressed air, but the most common method is to insert them into prebored holes. Shooting in with air pressure can be used for soft soils, and the use of perforated steel tubes for nails provides drainage. Because the depth of embedment depends in part on resistance of the soil, after installation the projecting ends of the tubes are cut off.

Prebored soil nails usually have heads and may have bearing plates that act like large washers to distribute stress on the soil surface. After insertion, prebored nails are grouted for their full length with concrete mortar. Borings and nails are inclined downward at about 15° to keep the grout from running out.

### 20.4.2  Similarities and Differences

The concept of soil nailing in some ways is similar to that of MSE walls, as both employ regularly spaced tension members, but there also are important differences. A major distinction is that nailing works from the top down and treats soils in situ, whereas MSE is constructed from the bottom up with soil that is specially selected, spread, and compacted layer on layer. Soil nailing is performed by incrementally excavating to a depth of a meter or two, installing nails, and repeating the process for the entire excavated depth.

The holding power of soil nails often is tested by pulling, but they are not routinely post-tensioned as in the case of tiebacks.

### 20.4.3  Kinds of Soil

Soil nailing is less restrictive than MSE with regard to suitable soils, in part because the soils are not disturbed when they are nailed. In fact, granular soils that are required for MSE walls usually will be unsuitable for nailing because they will not hold a vertical face long enough for nails to be installed. Soil nailing requires that the soil can stand unsupported while an excavation is made a meter or two deep for the nails to be installed and grouted. If the soil is not self-supporting, nailing cannot be used.

The simplest method for evaluating soil suitability for nailing also is the direct method, to dig a trial excavation. However, a trial excavation normally will not be feasible for the full height of a wall, in which case soil borings and tests are needed. For example, a common alluvial sequence is a layer of clay over sand, and it would be a bad surprise to encounter sand after nailing has begun.

### 20.4.4  Protecting the Nailed Surface

After nails are installed the exposed soil surface is covered with steel mesh and shotcreted, or covered with sprayed-on, low-slump concrete. The jagged appearance of the shotcrete then can be concealed by attaching decorative panels.

Because soil nailing is a fairly recent development with few failures, design methods are not well established and probably lean toward overdesign. While the lack of failures is meritorious, it also means that only limited information is available to define failure mechanisms. Two nailed walls have intentionally been failed in order to gain this information. The Clouterre Test Wall in France was failed by saturating the soil, and a National Experimental Geotechnical Test Site wall at the University of Massachusetts was failed by overexcavation at the toe. As in the case of MSE walls, the slip surface is curved, as shown in Fig. 20.6.

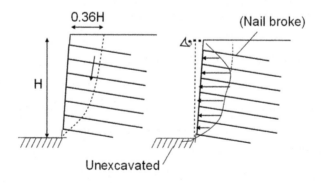

**Figure 20.6**

Results from some experimental nailed walls. The pressure diagram at the right is similar to that of a braced excavation. (Based on experiments reported by Plumelle et al., 1990.)

Nail stresses approach zero at the base of the wall and are higher than those predicted from Rankine theory higher on the wall, but as discussed below, this may be attributable to the construction sequence instead of to soil arching.

## 20.4.5 Soil Nailing Construction

The first step in soil nailing is to excavate a vertical or near-vertical cut to a depth of 1 to 2 m (3 to 6 ft). The usual procedure is to bore a row of holes for the length of the excavation and inclined at about 15° (1 on 3.5 to 4) from horizontal. Nails then are inserted into the holes. Since the nail tensile strength normally is not a governing criterion, the nails should be soft steel to allow bending without breaking, and they can vary from about 15 to 50 mm (0.5 to 1.5 in.) in diameter. As in MSE, the length is gauged to the height of the wall.

Because of the installation method there must be working space behind the drill for it to be fully retracted, and sufficient headspace above the drilling machine to allow inclined drilling. Where headspace is limited, the boring angle can be reduced, but should not be lower than about 5°. Nails then are grouted from the bottom of the borings.

The nailed soil face is covered with a layer of shotcrete, then a layer of wire mesh, then another layer of shotcrete for a total thickness of 150–250 mm (6–10 ins.). Although the mesh-reinforced concrete is not considered to be the same as a structural wall, it is designed to withstand up to 60 percent of the horizontal soil pressure in case a nail slips. After the concrete has set and/or the wall completed, panels may be installed to improve the rough appearance.

After a lift is completed the next soil layer is excavated and the process repeated, and heights as high as 75 ft have been stabilized. In a continuous cut, the nail length is about 80 percent of the height of the cut, which is about the same as the length of reinforcing strips in MSE walls.

## 20.4.6 Drainage

The shotcrete layer will seal off normal soil drainage, so weepholes are drilled through the concrete in a regular pattern. To prevent water from running down the outside face, perforated pipes or plastic drainage strips may be installed vertically prior to shotcreting each lift. Drains are connected vertically and can feed seepage water into a drain tile running along the base of the wall.

## 20.4.7 Principles of Design

The design of soil nailing is complex and sometimes seems to be more intuitive than by design. A rational design depends on whether the product performs like wall, in which case the guiding criteria relate to soil pressures, or as a stabilized slope, in which case the criteria relate to position and stresses along a slip surface.

Observed failures of nailed walls indicate that instead of failing by mass toppling they fail by internal slipping, so a slip-surface design is most frequently used. The distribution of pressure presents a major departure from an ideal triangular distribution, being zero at the bottom and then becoming nearly constant at shallower depths. This is close to the pressure pattern for braced excavations that also are constructed by top-down excavation, but with support from horizontal struts instead of from nailing. In both situations soil still in place below the bottom of the excavation resists yielding, in effect acting like a strut. Then as the excavation is deepened, the soil rebounds elastically and transfers nearly uniform stress to the struts or the nails. As shown in Fig. 20.6, movement tends to be by tilting outward instead of by developing a middle bulge, as in MSE walls.

## 20.4.8  Nail Length and Strength

As in the case of MSE walls, the incipient slip surface emerges at a distance of about $0.3-0.4H$ behind the edge of the wall, so tensile stresses are highest near the center of a nail. Typically nail lengths are about the same as for MSE walls, 60 to 80 percent of the wall height, the longer dimensions being for low friction angle soils.

Nails and their surrounding casing of grout obviously must be designed to resist slipping, and must have adequate tensile strength to prevent breaking. Furthermore the heads must not pull off. Resistance to bending is a consideration in some design methods, but is indicated to make a relatively minor contribution to stability.

The larger the spacing between nails, the lower the cost. A spacing of 1.5 m (5 ft) generally is close enough to prevent sloughing off of soil and has been adopted for uniformity in construction. Rows of nails can either be aligned vertically or staggered; vertical is preferred for attaching facing elements, but staggered is less likely to allow sloughing off of soil between the nails. For easier constructability, nail rows can be a constant distance above each step in the excavation, or rows can be laid out horizontally and stepped up or down as required to conform to a changing wall height. The lowest row should be no more than about 0.75 m above the final ground surface.

The pattern of deflection of the facing is similar to that of a mattress as nails hold and dimple the surface. However, because the facing shotcrete is rigid, if yielding occurs it will tend to develop concentric cracks around the nail heads. The facing must resist punching shear, and design is based on a premise that if one nail slips, its stress can be carried by the surrounding nails.

## 20.4.9  Design Methods

Soil-nailed walls may be battered or tilted back for better stability, and if room is available high walls often are stepped so that nail lengths relate to the height of each step. Stepping of an excavation also can help to localize and limit the extent of a failure. Design of soil nailing normally is based on a slip line approach. The German method and the CALTRANS method (for California DOT) approximate the failure geometry with straight lines. The Davis method (for University of California at Davis) assumes a parabolic failure surface, and the French method a circular failure surface. However, Sheahan and Ho (2003) report near-linear slip surfaces that enable a simplified solution. A relatively complex kinematic or energy approach assumes a log spiral (see Byrne et al., 1998, p. 94). The U.S. Federal Highway Administration has developed a detailed design procedure based on a circular failure surface.

Most design methods have been computer programmed, but the programs are expensive and may not be readily available for student use. Sheahan and Ho (2003) suggest a spreadsheet computational method that adapts Coulomb's analysis by assuming a planar slip but also includes a cohesion component of shear strength. Although according to the Coulomb/Rankine solutions the most critical slip surface with a vertical wall and no nails should be inclined at an angle of $\theta_c = 45 + \phi/2$, the spreadsheet solutions indicate that other angles are more critical, so trials are conducted with a varying slope angle.

A soil mass that has been stabilized by nailing also must be safe from global failure (a landslide), which is discussed in the next chapter.

## 20.5  SUMMARY

Despite similarities of purpose, different design philosophies and procedures are used for mechanically stabilized earth (MSE) walls, soil nailing, and conventional retaining walls.

Design of MSE and conventional walls is based on soil pressures determined from the soil internal friction. The MSE wall is constructed with cohesionless soil, and conventional walls are backfilled with disturbed soil that has little cohesion, so it is not incorporated into the design. Arching action raises the center of pressure and can significantly increase the overturning moment in conventional and MSE walls, but design factors of safety normally are sufficiently high that the wall is not seriously compromised if arching is ignored. If a design factor of safety is marginal, the system should be recalculated including arching action.

Current soil nailing design is shear-strength oriented, based on shearing resistance along a potential slip surface and pullout resistance of the nails. The location of the surface remains speculative. Part of the difficulty in nailing design is that the distribution of soil pressure appears to depend more on elastic rebound properties of the soil than on its internal friction, but this is not a consideration in a design that involves pressures that derive from the soil shearing strength. A similar situation exists with strutted excavations, where design remains empirical and based on measurements of strut stresses in different kinds of soil.

## Problems

20.1. List the conditions that must be tested to ensure external stability of an MSE wall. Which of these is the most common cause of distress or failure? The second most likely cause?

20.2. Prepare a table showing MSE soil specifications for walls with steel strips or plastic grids, and for stabilizing slopes. What problems can be anticipated in each case if soils do not meet these specifications?

20.3. Part of a parking lot on top of an MSE wall is settling, and a section of wall has failed by tipping outward. Investigation revealed that the soil was compacted by running over it with a bulldozer, and density tests gave about 90% of the standard density. The builder claims that overloading at the surface caused the problem, and that the reduced weight of the soil should result in lower, not higher, pressures on the wall facing. (a) Is that a valid argument? Why (not)? (b) How will a lower compacted density influence wall pressure, sliding, overturning, and pullout resistance? State your explanations in terms that can be understood by a jury.

20.4. Water is observed seeping from a level at about one-third of the wall height of an MSE wall. What can be concluded with regard to the soil conditions? How might this have an adverse effect on stability of the wall? What can be done to improve the situation without rebuilding the wall?

20.5. A reinforced earth wall 20 m (65.6 ft) high will utilize strips spaced every 0.3 m (1 ft) vertically and 1 m (3.3 ft) horizontally. The backfill is compacted sand, $\phi = 35°$, and $\gamma = 17.3 \, kM/m$ ($110 \, lb/ft^3$). The design factor of safety is 1.5. (a) What maximum tensile strength is needed in the strips? (b) If the steel in the strips has a yield stress of 345 MPa ($50,000 \, lb/in.^2$) and the strips are 102 mm (4 in.) wide, what is the maximum thickness?

20.6. Enter "Retaining wall failure" in a computer search engine. Select one that may or may not be an MSE wall and prepare a two-page summary indicating the kind and height of the wall, the kind of soil, and the nature of the failure. As an expert witness you will indicate what in your opinion are the possible cause(s), which in addition to design factors can include any unusual weather, flooding, surcharge, lawn watering, earthquakes, faulty construction, or other factors that either are based on solid evidence or can be a target for investigation.

20.7. Is lateral stress increased, decreased, or does it remain the same (a) in soil that is nailed, (b) in soil in an MSE wall, and (c) in a tieback wall with post-tensioning? Explain the significance.

# References and Further Reading

Al-Hussaini, M., and Perry, E. B. (1978). "Field Experiment of reinforced Earth wall." In *ASCE Symposium of Soil Reinforcement*, pp. 127–156.

Byrne, R. J., Cotton, D., Porterfield, J., Wolschlag, C., and Uelblacker, G. (1998). *Manual for Design and Construction Monitoring of Soil Nail Walls*. U.S. DOT-FHWA Publ. No. FHWA-SA-96-069R. Available through NTIS, National Technical Information Service.

Desai, C. S., and Al-Hosseini, K. E. (2005) "Prediction of Field Behavior of Reinforced Soil Wall Using Advanced Constitutive model." *J. Geotech. and Geoenviron. Eng.* 131(5), 729–739.

Elias, V., Christopher, B. R., and Berg, R. R. (2001). *Mechanically Stabilized Earth Walls and Reinforced Soil Slopes Design and Construction Guidelines*. Report No. FHWA-NHI-00-043. U.S. DOT FHWA.

Juran, I., and Elias, V. (1991). "Ground Anchors and Soil Nails in Retaining Structures." In H.-Y. Fang, ed., *Foundation Engineering Handbook*, pp.868–905. Van Nostrand Reinhold, New York.

Lazarte, C. A., Elias, V., Espinoza, R. D., and Sabatini, P. J. (2003). *Geotechnical Engineering Circular No. 7, Soil Nail Walls*. Report No. FHWA0-1F-03-017, U.S. DOT FHWA.

Plumelle, C., Schlosser, F., Delage, P., and Knochenmus, G. (1990). "French National Research Project for Soil Nailing: Clouterre." *Design and Performance of Earth Retaining Structures*, ASCE Geotech. Spec. Publ. 25, 600–675.

Schlosser, F. (1990). "Mechanically Stabilized Earth Retaining Structures in Europe." *Design and Performance of Earth Retaining Structures*, ASCE Geotech. Spec. Publ. 25, 347–378.

Sheahan, T. C., and Ho, C. (2003). "Simplified Trial Wedge Method for Soil Nailed Wall Analysis." *J. Geotech. and Geoenviron. Eng.* 129(2), 117–124.

# 21

# Slope Stability and Landslides

## 21.1 CUT SLOPES VS. BUILT-UP SLOPES

### 21.1.1 The Slippery Slope

Flat ground may be widespread, but sloping ground can be wider and deeper, where the land has been incised by streams and rivers. If the slope is too high or too steep, or if the soil is too weak, it will find ways to correct the deficiency. A common corrective measure is the landslide.

Landslides go by many different names from rockfalls to mudlows, but the point is that they all slide and they all involve shearing resistance. In terms of lives and property lost, landslides are among the most tragic and devastating events that involve soils. Landslides occur naturally as streams cut valleys deeper and wider, and as lakes and oceans erode their shores. Landslides are triggered by shaking from an earthquake, or by willful indiscretions of man.

The most common naturally occurring slopes are stream valley walls and eroding shorelines, but slopes also occur as exposed sides of faults and uplifted mountain ranges. Depositional slopes include sand dunes, deltas, and volcanic cinder cones. The most common depositional slopes are man-made—road embankments, earth dams, piles of sand, grain, or industrial or mining waste. Other man-made slopes include cuts made by trenching and by excavations for foundations, basements, or roadcuts. All slopes have one thing in common: if conditions are right, they have it in them to host a landslide.

Geotechnical engineers frequently are called on to analyze the mechanics of landslides and propose fixes. Engineering geologists also are closely involved with landslides, but their emphasis is more on recognition and defining landslide hazard areas. Landslide hazard maps of many critical areas now are available from geological surveys and on the Internet.

590

This chapter will discuss the soil mechanics of naturally occurring slopes and landslides, followed by design considerations for cut slopes and embankments to prevent slope failures.

### 21.1.2 When Push Exceeds Drag

Landslides occur when the downslope component of soil or rock weights exceeds the maximum resistance to sliding along a particular surface. Landslides can be shallow or deep. They differ from erosion by wind or running water because the main driving force is gravity acting on the weight of the soil.

The best time to stop a landslide is before it starts, because shearing and remolding of soil in the slip zone causes it to lose a substantial part of its shearing strength. Therefore, after sliding starts it may speed up, depending on geometric factors and how much strength has been lost. Sliding then continues until the shearing resistance improves or the ground reaches a configuration that compensates for the loss of soil strength. Sensitive soils move farther and much more rapidly that those that are less sensitive, sometimes becoming devastating mud flows that move so fast that they are virtually inescapable.

Normally for a landslide to stop it must move to a flatter surface. This often takes place gradually, as the sliding soil moves along a concave surface that is nearly vertical at the top and nearly horizontal at the bottom. Passive resistance of soil building up at the bottom stops the slide. If that soil is removed, either naturally or as a result of man's activities, the slide continues.

One of the worst things that can be done to encourage a landslide to continue also seems to be the most intuitive, which is to put the soil back where it belongs. Curiously it sometimes takes two or even three attempts before owners are willing to admit defeat. Putting the soil back not only removes passive resistance, it puts weight where it will be most effective at renewing sliding.

### 21.1.3 Reaching for More

A landslide releases lateral pressure on soil at the top of the slope, so soil that has not yet moved becomes less stable and more susceptible to sliding. Thus, as shown in Fig. 21.1, a landslide area will enlarge headward by biting off more ground until the geometry changes or the landslide runs out of soil.

The uphill sequence sometimes stops at a basement wall because soil removed to make a basement usually weighs more than the structure. However, the weight of the structure is carried by the foundation that is exposed by the landslide, which can lead to a bearing capacity failure. Putting soil back will reactivate the slide, but if necessary the structure often can be saved by underpinning.

**Figure 21.1**
Causes of this landslide were excavation at the toe of the slope and doing too much laundry at the top (waste water went into a septic drainfield). The slide enlarged uphill in steps as sliding removed lateral support from soil exposed in the slide scarp. This landslide was stabilized with drilled lime.

Man-made slopes obviously must be designed to be safe from landslides. An active landslide automatically selects the most critical slip surface, but in a stable slope that surface is not known. Because of the infinite number of possible slip geometries the problem is statically indeterminate, so a slip surface either is assumed or is arrived at by computational trial and error.

Active landslides are more readily analyzed because the slip surface can be located by inspection and by drilling. The analysis can determine the causes, or if more than one causal factor is involved, prorate responsibility to the several contributors. Analysis also is required to compare the effectiveness of different repair methods. Homeowner insurance policies almost inevitably have a waiver in fine print that excludes damages from ground movements, placing them in the category with nuclear wars and volcanic eruptions.

### 21.1.4 Infinite Slopes

The simplest example of an unstable slope is sand dumped off the end of a conveyor belt, or any other grains dribbling down a surface that is appropriately named the slip face. As discussed in Chapter 18, the material is deposited at the "angle of repose."

The forces involved in an angle of repose are shown in Fig. 21.2, where $W$ is the weight of an element. As shown in the figure, the weight may be resolved into $N$,

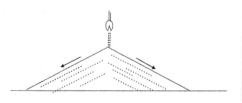

**Figure 21.2**
A simple case of borderline slope stability: sand at the angle of repose.

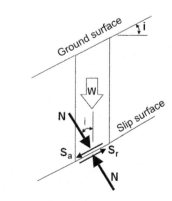

a force normal to the slip surface, and $S_a$, a force acting parallel with the surface and tending to cause sliding. The acting force is

$$S_a = W \sin i \qquad (21.1)$$

and the force normal to the shear surface is

$$N = W \cos i \qquad (21.2)$$

From the Coulomb equation, if cohesion is zero, the resisting force is only a function of $N$:

$$\begin{aligned} S_r &= N \tan \phi \\ &= W \cos i \tan \phi \end{aligned} \qquad (21.3)$$

At the instant of sliding $S_a = S_r$, or

$$\begin{aligned} W \sin i &= W \cos i \tan \phi \\ \tan i &= \tan \phi \\ i &= \phi \end{aligned} \qquad (21.4)$$

The angle of repose of a cohesionless granular material, $i$, therefore equals the angle of internal friction, $\phi$. Because $\phi$ tends to increase with an increasing thickness of overburden, shearing is concentrated near the surface.

### 21.1.5  Factor of Safety for a Slope in Cohesionless Material

The factor of safety against sliding may be defined as the resisting force divided by the acting force, or

$$FS = \frac{S_r}{S_a} = \frac{N\tan\phi}{W\sin i} = \frac{W\cos i \tan\phi}{W\sin i}$$

$$FS = \frac{\tan\phi}{\tan i} \tag{21.5}$$

If the factor of safety is 1.0, then $i = \phi$ (eq. 21.4).

### Example 21.1
A highway slope is to be cut through a stable sand dune with a factor of safety of 1.2 against slipping. The sand is predominately quartz having an angle of internal friction of 25°. What is the design slope angle assuming no influence from water?

*Answer:* $\tan i = \tan 25°/1.2 = 0.389$; $i = 21°$.

### 21.1.6  Seepage Force

As viscosity holds back the flow of water through soil pores, there is an equal and opposite push exerted by the water on the soil. This was discussed in relation to the formation of quicksand in Section 14.10, where the following formula is derived:

$$S = i_w \gamma_w \tag{14.44}$$

where $S$ is the seepage force which, as in the case of a unit weight, is expressed in force per unit volume of soil, $i_w$ is the hydraulic gradient, or head loss per unit length that the water travels, and $\gamma_w$ is the unit weight of water.

Equation (21.1) can be written for a unit volume of soil oriented in the direction of seepage (Fig. 21.3), so the soil unit weight can be substituted for $W$. Because the soil is submerged, the submerged unit weight is used to calculate the acting force. Then

$$\text{Soil: } S_a = (\gamma - \gamma_w)\sin i \tag{21.6}$$

**Figure 21.3**

Seepage force $S$ acting within a unit volume of soil having weight $W$.

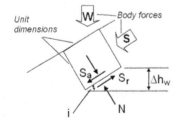

where $S_a$ is parallel with the slope. With seepage parallel to the slope the hydraulic gradient is

$$i_W = \Delta h_W = \sin i \tag{21.7}$$

Seepage force $S = \gamma_w \sin i$. Adding this to the soil acting force gives

$$S_a + S = \gamma \sin i \tag{21.8}$$

The submerged unit weight also is used for the calculation of normal force:

$$N = (\gamma - \gamma_w)\cos i \tag{21.9}$$

The factor of safety is

$$FS = \frac{S_r}{S_a} = \frac{N \tan \phi}{\gamma \sin i} = \frac{(\gamma - \gamma_w)\cos i \tan \phi}{\gamma \sin i}$$

$$FS = \frac{(\gamma - \gamma_w)}{\gamma} \frac{\tan \phi}{\tan i} \tag{21.10}$$

If the factor of safety is 1.0,

$$\tan i = \frac{(\gamma - \gamma_w)}{\gamma} \tan \phi \tag{21.11}$$

It will be noted that the acting force, in the denominator, depends on the total unit weight of the soil, whereas the resisting force in the numerator depends on $(\gamma - \gamma_w)$, the submerged unit weight.

As an approximation $\gamma_w = 0.5\gamma$. Making this substitution,

$$FS \approx 0.5 \frac{\tan \phi}{\tan i} \tag{21.11a}$$

Seepage forces are real forces. They also can be analyzed from a consideration of effective stresses, as discussed later in this chapter.

### Example 21.2
Assume a worst moisture condition for the highway cut in the previous example and determine the corresponding slope angle. Is this value appropriate for design?

*Answer:* $1.2 = 0.5 \tan 25°/\tan i$; solving for $i$ gives $11°$. This would be conservative because fully saturated conditions are not likely to occur in a sand dune.

## 21.1.7  Creep Angle

Equation (21.11a) also may be used to estimate a creep angle where cohesion is a viscosity function that approaches zero with a very slow shear rate. At incipient creep with $FS = 1$,

$$\tan i = 0.5 \tan \phi \tag{21.12}$$

**Example 21.3**

Estimate the creep angle for a saturated cohesive soil with $\phi = 20°$.

*Answer:* $\tan i = 0.5 \tan 20°$; $i = 10°$, or approximately 1 vertical to 5.5 horizontal.

# 21.2   SLOPES IN COHESIVE SOILS

### 21.2.1   Complications from Cohesion

The infinite slope analysis does not apply when a cohesion variable is added to the resisting force in eq. (21.3). Shearing no longer is along a shallow surface parallel with the slope, but takes a path that minimizes the combined influences of cohesion, which acts on the length of a slip surface, and internal friction, which depends on the normal force on that surface. In most landslides the failure path is curved and the problem is statically indeterminate, so solutions are obtained through a series of approximations and trial and error. This following discussion begins with a simple case of a planar slip that follows along the path of a weak stratum or fault, and then proceeds to more complicated geometries.

### 21.2.2   Slip Surface Fixed by an Inclined Weak Layer

A simple case that illustrates the method of analysis by separating soil weight into acting and resisting force components is shown in Fig. 21.4, where the slip surface has been fixed by a weak layer of soil inclined at an angle $\theta$ with horizontal. The acting force tending to cause slipping is the component of soil weight that is parallel with the slip surface:

$$S_a = W \sin \theta \tag{21.13}$$

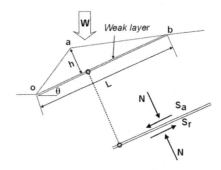

**Figure 21.4**

Sliding of unsaturated soil on an inclined weaker layer, where $W$ is the weight of the sliding soil. This was the case for a tragic failure at Vaiont Dam in northern Italy, where the slip surface was an old fault plane.

Thus, the steeper the $\theta$ angle, the higher the acting force. Friction depends on the weight component acting normal to the slip surface:

$$N = W \cos \theta \tag{21.14}$$

In this case the steeper the $\theta$ angle, the lower the normal force and the lower the frictional resistance to sliding. The Coulomb equation for maximum resisting force, which is the same as eq. (18.6) but on a total force basis, is

$$S_r = cL + N \tan \phi \tag{21.15}$$

where $c$ is the soil cohesion, $L$ is the length of the slip surface, and $\phi$ is the friction angle. Hence

$$S_r = cL + W \cos \theta \tan \phi \tag{21.16}$$

If $S_r$ exceeds $S_s$ and therefore is not fully mobilized, the factor of safety is

$$FS = \frac{S_r}{S_r} = \frac{cL + W \cos \theta \tan \phi}{W \sin \theta} \tag{21.17}$$

If $S_r$ is fully mobilized and equals $S_a$, the factor of safety is 1.0 and the soil mass will slide. In this form eq. (21.17) does not include pore pressure, which normally plays a critical role in landslides, so the equation is applicable only for unsaturated conditions.

## Example 21.4

The size of the triangular mass of soil in Fig. 21.4 is scaled from the drawing: $L = 35$ ft (10.7 m) and $\theta = 24°$. The height of the triangle normal to $L$ is $h = 7$ ft (2.13 m). Strength parameters for the inclined weak layer are determined to be approximately $c = 50 \, lb/ft^2$ (2.4 kPa) and $\phi = 20°$. The unit weight is $\gamma = 100 \, lb/ft^3$ (15.7 kN/m³). What is the factor of safety against sliding?

*Answer:* The area of the triangle is

$$A = \frac{1}{2}(35)(7) = 122.5 \, ft^2, \text{ or } \frac{1}{2}(10.7)(2.13) = 11.4 \, m^2.$$

The soil weight is

$$W = (122.5 \, ft^2)(100 \, lb/ft^3) = 12{,}250 \, lb/ft \text{ of length,}$$
$$\text{or } (11.4 \, m^2)(15.7 \, kN/m^3) = 179 \, kN/m.$$

Substituting the appropriate values into eq. (21.17) gives

$$FS = \frac{50(35) + 12{,}250 \cos 24° \tan 20°}{12{,}250 \sin 24°} = \frac{1750 + 4073}{4862} = 1.2, \text{ or}$$
$$= \frac{2.4(10.7) + 179 \cos 24° \tan 20°}{179 \sin 24°} = \frac{25.7 + 59.5}{72.8} = 1.2$$

---

*Question:* Which is the more important component of the resisting force, friction or cohesion?

---

### 21.2.3 Planar Slip in Homogeneous Soil

Steep slopes in homogeneous soils may fail along a slip surface that is approximately planar. The analytical solution usually is attributed to Culmann, who published it in 1866, but it appears to have been first developed by Française in 1820 (see *Geotechnique* 1(1), 66, 1948). In shallower slopes the restricted geometry of the Culmann method causes it to overestimate the factor of safety and is on the unsafe side for design.

The notation in Fig. 21.5 is the same as in Fig. 21.4, and eqs. (21.13) to (21.17) still apply, except that $\theta$ is not known. It therefore is a simple matter to apply eq. (21.17) to a series of trial values of $\theta$, plot the factor of safety versus the trial value, and pick off the most critical value. An analytical solution is more precise:

From trigonometric considerations the soil weight is

$$W = \tfrac{1}{2}\gamma HL \csc i \sin (i - \theta) \tag{21.18}$$

Equation (21.18) may be rewritten in terms of a *developed* cohesion, $c_d$, and a *developed* friction angle, $\theta_d$, for which the factor of safety would be 1.0. Then

$$W \sin \theta = c_d L + W \cos \theta \tan \phi_d \tag{21.19}$$

Substituting the value of $W$ from eq. (21.18) into eq. (21.19) and solving for $c_d$ gives

$$c_d = \tfrac{1}{2}\gamma H \csc i \sin (i - \theta)(\sin \theta - \cos \theta \tan \phi_d)$$

An additional substitution is

$$\tan \phi_d = \sin \phi_d / \cos \phi_d$$

**Figure 21.5**

Culmann's method to define the most critical slip angle for planar failure in homogeneous soil.

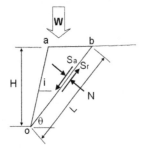

By use of a trigonometric identity, $\sin(\theta - \phi_d) = \sin\theta\cos\phi_d - \cos\theta\sin\phi_d$

$$c_d = \tfrac{1}{2}[\gamma H \csc i \sin(i - \theta)\sin(\theta - \phi_d)]/\cos\phi_d \qquad (21.20)$$

Setting the derivative of $c_d$ relative to $\theta$ equal to zero establishes the most critical value of $\theta$, which is

$$\theta_c = \tfrac{1}{2}(i + \phi_d) \qquad (21.21)$$

That is, the most critical failure angle for a planar failure is the *average of the slope angle and the angle of internal friction*. The steeper the slope, the higher the failure angle. This applies only to a planar failure surface.

### Example 21.5

According to eq. (21.21), what is the most critical failure angle for a vertical slope? How does this compare with the angle for Rankine active failure?

*Answer:* $\theta_c = (90 + \phi_d) = 45 + \phi_d/2$, which is identical to the Rankine active case that assumes parallel planar failures.

## 21.2.4 Critical Height of a Vertical Cut

Vertical cuts (Fig. 21.6), have a considerable practical significance as they are common during construction, for example for utility trenches and basements. Unsupported cuts still contribute to construction fatalities from trench cave-ins despite many precautions and regulations. It is critically important that contractors and construction workers be aware of the dangers and how to avoid them.

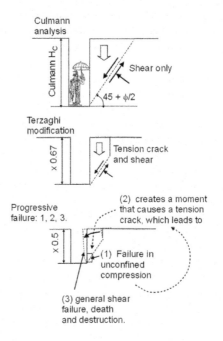

**Figure 21.6**

Three versions of critical height for a vertical cut. When the cliff lets go, better toss the umbrella and get out of there.

One method for predicting the maximum height of a vertical cut can be obtained by substituting eq. (21.21) into eq. (21.18) and setting the factor of safety equal to 1.0, but it must be emphasized that this is not an appropriate model, as discussed later in this section. This substitution gives

$$H_{cul} = \frac{4c \sin i \cos \phi}{\gamma[1 - \cos(i - \phi)]}$$

where $H_{cul}$ is the Culmann critical height. Setting $i = 90°$ gives

$$H_{cul} = \frac{4c}{\gamma} \frac{\cos \phi}{1 - \sin \phi}$$

$$H_{cul} = \frac{4c}{\gamma} \tan\left(45 - \frac{\phi}{2}\right)$$

(21.22—*mathematically correct but unsafe!*)

Equation (21.22) is unsafe for a vertical cut *because it does not address the most likely failure mechanism.*

### Terzaghi's Correction
As shown in Fig. 21.6, Terzaghi (1943) indicated that the Culmann $H_{cul}$ is unsafe because vertical tension cracks running parallel to the faces of an open vertical cut reduce the length and the amount of restraint along the slip plane. He therefore reduced the critical height by one-third. While that is a substantial improvement, it is empirical and inadequate. An analytical solution should be more accurate if the model is successfully drawn.

### Modification for Unconfined Compressive Strength
An earlier edition of this book pointed out that, because of the lack of lateral confinement, soil at the bottom of a vertical cut should yield if the overburden pressure exceeds the soil unconfined compressive strength, $q_u$. As shown at the bottom of Fig. 21.6, the tendency for a block of soil to settle and rotate will create a mechanism whereby tension cracks will develop, as observed by Terzaghi.

Let us therefore redefine the critical height, $H_c$, in terms of the unconfined compressive strength, $q_u$. At depth $H_c$ in the face of a vertical cut,

$$q_u = H_c \gamma$$
$$H_c = q_u/\gamma$$

(21.23)

The equation for unconfined compressive strength is

$$q_u = 2c \tan(45 - \phi/2)$$

(18.13)

which gives

$$H_c = \frac{2c}{\gamma} \tan(45 - \phi/2)$$

(21.24—*Safer but still requires a factor of safety*)

It will be noted that this is the critical height at failure and does not include a factor of safety, yet is only one-half of that obtained with eq. (21.22), emphasizing the importance of using an appropriate model for an analytical treatment.

The analysis based on unconfined compressive strength incorporates the concept of *progressive failure*, since shear failure is not assumed to be simultaneous all along a failure surface as in the case of the Culmann method, but starts with compression failure at the bottom that creates a bending moment that must be resisted by tension, which is low or nonexistent in soil. Unconfined compressive strength tests are routinely included in geotechnical reports.

Equation (21.24) still can be on the unsafe side if a trench cuts below a groundwater table because seepage forces will be directed toward the the open cut. This situation also is unsatisfactory from a construction standpoint. Restrictive regulations now require that workers in a trench deeper than a specific height must be protected by a steel "trench box," or the sides of the trench must be cut back to a stable angle.

### Example 21.6
A trench is to be excavated in soil that is reported to have an unconfined compressive strength of 500 lb/ft$^2$ (23.9 kN/m$^2$) and a unit weight of 120 lb/ft$^3$ (18.8 kN/m$^3$) What is $H_c$ with a factor of safety of 1.2?

*Answer:* According to the $q_u$ criterion $H_c = 500/120 = 4.2$ ft, or $23.9/18.8 = 1.27$ m. The allowable height is $H_a = 4.2 \div 1.2 = 3.5$ ft (1.1 m).

## 21.3  LANDSLIDES

### 21.3.1  A Slip Is Showing

A landslide does a credible job of selecting the most critical slip surface and establishing that, at the instant of sliding, the factor of safety was very close to 1.0. The factor of safety also is 1.0 when the landslide stops. It is a serious mistake to ignore an existing slip surface and impose some idealized shape such as a circle for the analysis, which can lead to some extremely misleading results. In one case a self-styled expert reported that the factor of safety was 0.2, which would provide a surplus of energy sufficient to carry the slide up and over the next hill.

### 21.3.2  Field Investigation

The first step in analysis of a landslide is a field investigation that involves (1) an elevation survey, usually along a traverse down the central axis of the landslide, and (2) exploration borings at selected points along that traverse. In natural landslides more than one soil type usually is involved, and it is important to identify weak strata encountered in borings that may reveal the location of the slip zone, and impermeable strata that can contribute to a perched groundwater table.

Particularly important is (3) to measure the position of the groundwater table in each boring, and (4) to extend borings deep enough that they encounter and penetrate through the slip zone.

An active landslide should be checked for sounds of running water and for any other evidence of broken water or sewer pipes indicated by welling-up of the groundwater table and local wet spots. Nearby residents should be asked if they have had any problems with sewer backups. Leakage is stopped by shutting the water off.

Landslides often are spoon-shaped with the long dimension of the spoon oriented across the slope. A three-dimensional analysis is possible but adds complexity to an already complicated problem. If the landslide is wide, two or more traverses may be investigated.

Borings, cone probings, and/or in-situ shear tests, discussed in Chapter 26, are made at selected locations along a traverse. Difficult terrain requires the use of a track vehicle. Shallow landslides may most conveniently be investigated by hand augering. Undisturbed soil samples also may be taken for laboratory tests, but it may be difficult or impossible to obtain duplicate samples, in which case tests can be performed in situ.

Ground elevations and soil boring information are plotted to scale to develop a cross-section, as shown in Fig. 21.7. There should be no vertical exaggeration because slip angles may be measured from the graph. The cross-section should show soil strata, the slip surface, and the position of the groundwater table. Average unit weight of the sliding mass of soil is needed, and cohesions and friction angles of the slip zone soil are required for an analysis.

Landslides generally are analyzed on an effective stress basis. Pore water pressures are determined from groundwater levels in the borings with the aid of flow nets. Pore pressures also can be monitored in the field with piezometers, although active sliding can destroy a piezometer. Anomalies in groundwater elevations can be critically important clues to broken sewers or water pipes, or to aquifers and artesian effects, all of which can be causal factors.

The factor of safety is 1.0 when a landslide starts, and immediately decreases as soil in the slip zone is remolded and loses strength. By the time a field investigation is performed the landslide probably will temporarily have stopped so the factor of safety still is 1.0, or it may be slightly higher if the pause has allowed water to drain out.

The analysis will provide valuable benchmark values that will enable predictions of the effectiveness of different repair methods. The design factor of safety will be influenced by drainage conditions. The worst-case scenario will involve full saturation of the sliding mass.

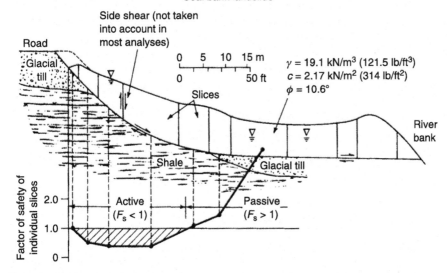

**Figure 21.7**
Landslide cross-section divided into arbitrary vertical slices for analysis. The factors of safety for individual slices shown at the bottom define active and passive parts of the landslide—the part doing the pushing and the part being pushed. This is an important consideration for stabilization. The analysis did not include vertical shearing stress, which adds to the factor of safety where the slip surface curves.

### 21.3.3 From Head to Toe

The landslide in Fig. 21.7 is typical in that the soil mass has moved down onto a more nearly level surface so soil in the toe area can buttress against further sliding. An unusual feature that should be noted in a field investigation is the river aggressively removing soil from the toe and thereby acting to promote further sliding.

The exposed part of a landslide slip surface at the top of a landslide is the landslide "scarp." The field investigation should include a walk along and above the scarp to examine for tension cracks running parallel to the scarp, as these normally will delineate the next part of the hill to go. The situation is aggravated by surface water running down into the cracks, which therefore should be sealed by cutting in a soil wedge with a spade.

### 21.3.4 Progressive Failure

Obviously the most favorable time to stop a landslide is before it starts. Most landslides involve progressive failure, like tearing a sheet of paper out of a notebook, so there are some worrisome clues such as development of a tension crack prior to failure. Another evidence for progressive failure is that back-calculation of the resisting force favors the residual, remolded shear strength instead of the peak shear strength (Duncan and Wright, 2005). Thus, as one part of the slide moves sufficiently to reduce the soil strength to a remolded state, additional stress is transferred to the remaining areas until the slip zone becomes continuous and the entire mass slides. The strength reduction also is confirmed by Borehole Shear tests that inevitably show lower strength in the shear zone compared with the same soil outside of the zone. Conventional analysis methods nevertheless assume simultaneous shear all along a slip surface, and progressive failure is taken into account by using the residual shear strength in calculations. If residual strength data are not available, cohesion may be assumed to be zero.

## 21.3.5 Development of a Shear Zone

As sliding continues, the slip zone gradually incorporates more soil until it constitutes a shear zone. One theory to explain this is that the first slipping involves the weakest set of grain contacts, so when movement occurs the contacts attain more nearly average strength so that slipping stops there and shifts to the next weakest set of contacts. This has been successfully simulated in laboratory models (Logani, 1973).

The development of a remolded zone at the base of a landslide has a parallel in grinding of rocks in a fault zone, where it has been found that the thickness of ground rock, or gouge, is between 0.1 and 10 percent of the distance moved, averaging about 1 percent (Scholtz, 1967). In a landslide, substantial additional thickening occurs in the toe area if the slip zone soil is close to its liquid limit, so higher overburden pressures upslope and confinement by overlying soil cause it to squeeze out at the bottom.

### Example 21.7
What average thickness of slip zone can be expected in a landslide that has moved 30 ft (10 m)?

*Answer:* 0.3 ft = 4 in. (0.1 m = 100 mm), not enough for replicate samples. Adequate for a Borehole Shear test.

## 21.3.6 After Sliding Stops

Nature's remedy for a landslide is to build up enough passive resistance from soil at the toe that the landslide stops. This of course can be overruled by nature or by man, by a river or by a bulldozer. If the landslide is left alone, in time it will partly repair itself through drainage and redevelopment of cohesive strength of soil in the shear zone. However, drainage can only return the slip zone soil to a normally consolidated state that is permanently weaker than when it was overconsolidated prior to sliding. An old landslide therefore remains a silent foe hiding under the trees and ready to harvest new victims. Often all that is necessary to reinvigorate an old landslide is to put a structure such as a house or apartment building on top. This was the case in Fig. 1.3 where affected structures were a total loss.

## 21.3.7 Earthquakes and Liquefaction

An earthquake shakes the ground back and forth like noodles in a frying pan. The movement is a vivid demonstration of Newton's Second Law, $f = ma$, where $f$ is force, $m$ is mass, and $a$ is acceleration provided by the earthquake. The inertia of the soil mass contributes to horizontal body forces that are in addition to horizontal components of gravitational and seepage forces, and therefore can trigger many landslides, particularly in marginally stable ground.

Soil liquefaction is a highly destructive phenomenon when shaking of a saturated, low-density granular soil breaks down its structure so that it collapses in on the soil water. Intergranular contact stress is transferred to pore water pressure so the soil becomes a dense liquid the same as quicksand. After a granular soil has liquefied and shaking stops, soil grains suspended in water settle out so excess pore water rises to the ground surface, carrying along sand to make small "sand volcanoes." The existence of ejected sand in layered sediments allows dating of earlier earthquakes in order to determine their periodicity. The influence of earthquakes and liquefaction is discussed in more detail in Chapter 27; see also Seed (1968).

## 21.3.8 Emergency Measures

Landslides can require immediate action to save properties from complete destruction:

1. If a structure is in imminent danger, local authorities must be informed and endangered structures evacuated.
2. Utility companies must be informed so that they can disconnect services or install flexible loops in the connections.
3. Lawn watering must be stopped on and above the slide area.
4. Transverse tension cracks in the soil are plugged with soil to prevent entry of surface runoff water.
5. Surface runoff water is diverted away from the slide area with trenches, and downspouts are extended away from the slide area.
6. Backward tilting of the individual segments ponds water that infiltrates into the soil, so channels should be cut to drain the water.
7. As a temporary emergency measure a slope may be given a plastic raincoat that must be securely staked down to prevent blowing.

8. Consider moving structures to a different site before it is too late and they become involved in the landslide.

More permanent repair measures then can be applied, such as removing load from the top, adding load at the toe, installing drains or structural measures, or chemical stabilization in responsive soils.

# 21.4  SLOPES WHERE THE MOST CRITICAL SLIP SURFACE IS NOT KNOWN

## 21.4.1  Slipping into the Modern Era

In 1918 a commission was established in Sweden to investigate a series of soil failures that had resulted in a number of deaths. The upper limit of each landslide was defined by an exposed near-vertical surface, the slide scarp, and downslope the slip surface was found to approximate a circle. The "Swedish method" or "Ordinary Method of Slices" was devised by Fellenius to find the most critical circle and predict the stability of existing slopes.

Because the soil mass is assumed to rotate about a common center, solution is based on moment equilibrium—that is, the sum of the acting forces along the failure arc times the radius to the center equals the sum of the resisting forces along the same arc times the same radius. Since the radius is constant, in polar coordinates this is the equivalent of a force equilibrium of tangential forces. Moment equilibrium still remains the basis for modern computer solutions, including those for which the slip surface is not a circle and the radius is not constant. However, there also are limitations to a moment equilibrium. For example, it cannot be used when part of the slip surface is flat because the radius is infinite. Then a force equilibrium is used, as in the preceding discussions.

## 21.4.2  Adaptation of the Ordinary Method of Slices to Noncircular Surfaces

Although the ordinary method was devised for a circular slip, the method of analysis also can be used with noncircular surfaces using a force equilibrium on orthogonal $xy$ axes instead of a moment equilibrium. In this text this will be referred to as the "simple method of slices."

The first step is to draw a cross-section of a slope (or landslide) to scale, and include the actual slip surface or, if the slip surface is not known, a trial surface. The mass of soil above the slip surface is then divided into vertical slices as in Fig. 21.7, and forces on each slice are calculated using the same principles and formulas as previously presented for a triangular mass of soil slipping along a planar surface. Analysis involves summing resisting and acting forces on all slices and dividing to obtain a factor of safety.

The relationships in eqs. (21.13) to (21.17) may be used to define acting and resisting forces along the base of each slice. A factor of safety then may be calculated that does not include side forces and can define active and passive zones, as in Fig. 21.7. Horizontal side forces are required for equilibrium, and it is these forces and resulting friction between slices that are ignored in the simplified method of slices. The result is to underpredict the overall factor of safety, particularly for deep slip surfaces and long slices.

### 21.4.3 Pore Water Pressure

Saturation and the resulting pore water pressure often become critical factors affecting slope stability. Pore water pressures can be measured in the field with piezometers, and can be estimated from a flow net. A flow net then can be modified to predict the influence of drainage on stability.

Whereas in static equilibrium the pore water pressure is simply the depth below a groundwater table times the unit weight of water, in a dynamic situation involving seepage, the resistance to seepage causes a reduction in pore pressure. In Fig. 21.8 this is indicated by sloping equipotential lines. Therefore the pore water pressure at point 8 on the slip surface does not equal the depth below the groundwater table, but a depth $H_w$ below the intersection of the equipotential line with the groundwater table.

### 21.4.4 Calculation Procedure for Each Slice

The following steps are required to evaluate the stability of each slice and of all slices, and may be used as a model for a spreadsheet. *Slices are vertical (Figs. 21.7 and 21.9).* The same procedure is used whether the location of the

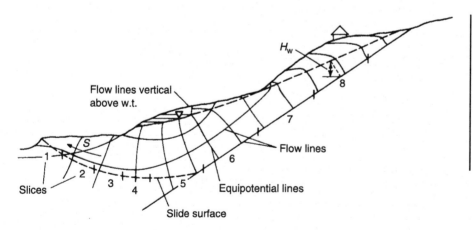

**Figure 21.8**

Flow net in a landslide. Upward seepage force in the toe area further destabilizes that area, which may be too weak to support the weight of a person.

**Figure 21.9**

Dimensions and forces in the simplified method of slices.

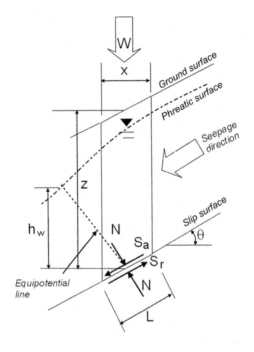

slip surface is known or is represented by a trial surface. Variables are identified in Fig. 21.9. Distances and slip angles are measured from a plotted cross-section, and soil properties are determined from density and shear strength tests.

**Soil Data:**

1. Slice number.
2. $\gamma$ = soil unit weight.
3. $\phi'$ = soil friction angle measured on an effective stress basis.
4. $c'$ = soil cohesion measured on an effective stress basis.

**Slice Data (Measured on the Cross-Section):**

5. $x$ = slice width.
6. $z$ = slice depth.
7. $\theta$ = slip angle from horizontal.
8. $h_w$ = pore pressure in units of water head.

**Calculations:**

9. Weight $W = xz\gamma$.
10. Acting force $S_a = W \sin \theta$.

11. Normal force $N = W \cos \theta$.

12. Length $L = x \cos \theta$.

13. Pore water pressure $u = h_w \gamma_w$, where $\gamma_w =$ unit weight of water.

14. Effective normal force $N' = N - uL$.

15. Resisting force $S_r = c'L + N' \tan \phi'$.

16. Slice factor of safety $FS_i = S_r/S_a$ for the individual slice.

## Summations:

17. Sum of acting forces is $\sum S_a$.

18. Sum of resisting forces is $\sum S_r$.

19. Global factor of safety $= FS_t = \sum S_r / \sum S_a$ for the slope.

## Notes:

Each of the steps can be set up as a column in a spreadsheet, and appropriate numbers and formulas copied vertically to enable a solution for each slice.

In Step 2 the soil unit weight is the total unit weight including water.

In Steps 3 and 4 the soil cohesion and internal friction are on an effective stress basis, which means that they either are measured with the soil in a fully drained condition, or stresses have been corrected by subtracting pore water pressures measured in a test specimen at the moment of failure.

In Step 8 the pore water pressure is measured from the equipotential line as shown in Fig. 21.9, which is not the same as the depth below the phreatic line because of seepage restraints of the water.

In Step 14 pore water pressure acts across the bottom of each slice, so it must be multiplied by length $L$ to give the force acting in opposition to $N$.

Steps 1–17 are calculated for each slice, and Steps 18–20 are summations that are performed only once per analysis.

After a spreadsheet has been set up it is a simple matter to change selected variables and instantly determine the effects on the factor of safety. For example, $h_w$ values can be decreased by installing drains, or shearing strengths can be increased by incorporating aggregate columns or by chemical stabilization. Slices can be changed by adding soil to increase passive resistance at the bottom of the landslide, or tension members can be introduced to transfer part of the acting forces by anchoring into stable soil outside of the shear surface.

### Example 21.8

Sum forces and calculate a factor of safety for the following slice:

1. Slice number $= 5$.
2. $\gamma =$ soil unit weight $= 125 \, \text{lb/ft}^3$ or $19.6 \, \text{kN/m}^3$.
3. $\phi' = 20°$.
4. $c' = 500 \, \text{lb/ft}^2$ or $23.9 \, \text{kN/m}^2$.
5. $X = 10 \, \text{ft}$ or $3.05 \, \text{m}$.
6. $Z = 16 \, \text{ft}$ or $4.88 \, \text{m}$.
7. $\theta = 32°$.
8. $h_w = 8 \, \text{ft}$ or $2.44 \, \text{m}$.

*Calculations*:

9. $W = XZ\gamma = 10(16)(125) = 20{,}000 \, \text{lb/ft}$ length, or $3.05(4.88)(19.6) = 292 \, \text{kN/m}$.
10. $S_a = W \sin \theta = 20{,}000 \sin 32° = 10{,}600 \, \text{lb/ft}$, or $292 \sin 32° = 155 \, \text{kN/m}$.
11. $N = W \cos \theta = 20{,}000 \cos 32° = 16{,}960 \, \text{lb/ft}$, or $292 \cos 32° = 248 \, \text{kN/m}$.
12. $L = X \sec \theta = 10 \sec 32° = 11.8 \, \text{ft}$, or $3.05 \sec 32° = 3.60 \, \text{m}$.
13. $N' = N - h_w \gamma_w L = 16{,}960 - 8(62.4)(11.8) = 16{,}960 - 5891 = 11{,}070 \, \text{lb/ft}$, or $= 248 - 2.44(9.81)(3.60) = 162 \, \text{kN/m}$.
14. $S_r = c'L + N' \tan \phi' = (500)(11.8) + 11{,}070 \tan 20° = 5900 + 4029$
    $= 9930 \, \text{lb/ft}$, or $= (23.9)(3.60) + 162 \tan 20° = 86 + 59 = 145 \, \text{kN/m}$.
15. $\text{FS}_i = 9930/10{,}600 = 0.94$, or $145/155 = 0.94$.

---

*Question:* Is this slice in the active or the passive zone?

---

## 21.4.5  Additional Horizontal Forces

Horizontal acting and resisting forces can be added to each slide to account for reinforcing members such as soil nails, or inertial forces developed from earthquakes. The latter are calculated from Newton's Law, $f = ma$, where $f$ is force, $m$ is mass, and $a$ is maximum acceleration from the earthquake, which is discussed in more detail in Chapter 27. The mass of each slice is obtained from the same formula, $m = W/g$, where $W$ is the weight and $g$ is the acceleration of gravity.

## 21.4.6  Seepage Force and Effective Stress

Seepage forces are automatically incorporated into an analysis made on an effective stress basis. This can be demonstrated from a consideration of forces on a single slice as shown in Fig. 21.10. Assume a unit volume of fully saturated cohesionless soil with weight $W = \gamma$. Then

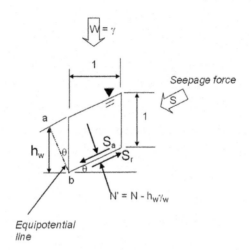

**Figure 21.10**
Diagram to illustrate equivalence of seepage force and effective stress analysis.

$$S_a = \gamma \sin \theta$$
$$N = \gamma \cos \theta$$

The normal force on an effective stress basis is

$$N' = N - U, \text{ where } U = uL, \text{ where } L = 1/\cos \theta$$

From the equipotential line $u = \gamma_w h_w$, where $h_w = ab \cos \theta$ and $ab = \cos \theta$. Then

$$U = \gamma_w (\cos^2 \theta / \cos \theta) = \gamma_w \cos \theta$$
$$N' = \gamma \cos \theta - \gamma_w \cos \theta = (\gamma - \gamma_w) \cos \theta$$
$$S_r = N' \tan \phi' = (\gamma - \gamma_w) \cos \theta \tan \phi'$$

$$\text{FS} = \frac{(\gamma - \gamma_w) \cos \theta \tan \phi'}{\gamma \sin \theta} = \frac{(\gamma - \gamma_w)}{\gamma} \frac{\tan \phi'}{\tan \theta} \tag{21.10}$$

which is the equation that was developed on the basis of seepage forces, with the notation on $\phi'$ indicating an effective stress basis. This illustration does not include soil cohesion as cohesion is not affected by seepage forces or buoyancy.

Generally the effective stress method is preferred for spreadsheet or computer programmed solutions as being easier to manipulate. Where seepage is upward, as at the toe of the slope in Fig. 21.8, the hydraulic gradient is negative and the slope angle is negative, so seepage still adds to the acting force. Seepage force remains an important consideration that should be separately evaluated and controlled in critical areas such as at the toe of a levee or earth dam, where vertical seepage can cause quicksand (cf. Chapter 14) or excessive seepage can cut a path through the dam and lead to eventual failure through piping.

### 21.4.7 Solution for a Worst-Case Scenario

A worst-case saturation condition with full mobilization of seepage forces is not unusual in an active landslide. A simple way to estimate this condition is to use the submerged soil unit weight to calculate resisting force and the total unit weight for acting force, with soil shear strength parameters measured on an effective stress basis. This errs on the safe side because it assumes a static condition for the pore water, such that the pore pressure is a function of depth below the ground surface instead of being determined from a flow net. For best accuracy the saturated and buoyant unit weights can be calculated using a block diagram, as discussed in Section 9.2.

**Example 21.9**

Use the buoyant unit weight to estimate a fully saturated factor of safety for the slice in the preceding example, using $125\,\mathrm{lb/ft^3}$ ($19.6\,\mathrm{kN/m^3}$) as the saturated unit weight.

*Answer:*

$$S_a = W\sin\theta = 10{,}600\,\mathrm{lb/ft}\ \text{or}\ 155\,\mathrm{kN/m\ (unchanged)}$$

$$W' = XZ\gamma' = (10)(16)(125 - 62.4) = 10{,}020\,\mathrm{lb/ft\ (buoyant\ weight)\ or}$$
$$= (3.05)(4.89)(19.6 - 9.81) = 146\,\mathrm{kN/ft}$$

$$N' = 10{,}020\cos 32° = 8490\,\mathrm{lb/ft}\ \text{or}\ 146\cos 32° = 124\,\mathrm{kN/m\ (based\ on\ buoyant}$$
$$\text{weight)}$$

$$cL = 5900\,\mathrm{lb/ft}\ \text{or}\ 86\,\mathrm{kN/m\ (unchanged)}$$

$$S_r = 5900 + 8490\tan 20° = 8990\,\mathrm{lb/ft}\ \text{or}\ 89 + 124\tan 20° = 134\,\mathrm{kN/m}$$

$$FS_i = 8990/10{,}600 = 0.85\ \text{ or }\ 134/155 = 0.86$$

compared with the previous answer of 0.94 for a partially saturated condition.

**Example 21.10**

Calculate a fully saturated factor of safety for the slice in the preceding example using effective stresses.

*Answer:*

$$S_a = 10{,}600\,\mathrm{lb/ft}\ \text{or}\ 155\,\mathrm{kN/m\ (unchanged)}$$

$$h_w = 16\cos^2 32° = 11.5\,\mathrm{ft}\ \text{or}\ 4.89\cos 232° = 3.52\,\mathrm{m}$$

$$u = 11.5(62.4) = 718\,\mathrm{lb/ft^2}\ \text{or}\ 3.52(9.81) = 34.5\,\mathrm{kN/m^2}$$

$$N' = N - uL = 16{,}960 - (718)(11.8) = 8187\,\mathrm{lb/ft\ or}$$
$$= 248 - (34.5)(3.60) = 124\,\mathrm{kN/m}$$

$$S_r = c'L + N'\tan\phi' = (500)(11.8) + 8187\tan 20° = 5900 + 3089 = 8989\,\mathrm{lb/ft\ or}$$
$$= (23.9)(3.60) + 124\tan 20° = 86.0 + 45.1 = 131\,\mathrm{kN/m}$$

$$FS_i = 8989/10{,}600 = 0.85\ \text{or}\ 131/155 = 0.85$$

In this case the solution involving submerged unit weight is simpler because of possible ambiguity in evaluating $h_w$.

# 21.5 SOLUTIONS FOR CIRCULAR AND LOG SPIRAL SHEAR FAILURES

## 21.5.1 Stability Numbers and Friction Circles

Dimensionless numbers such as the Reynolds number are used to reduce the number of experimental variables. A dimensionless number called the stability number can be defined for slopes:

$$s = \frac{c}{\gamma H} \tag{21.25}$$

where $s$ is Taylor's stability number, $c$ is the soil cohesion, $\gamma$ the soil unit weight, and $H$ is the height of the slope from the base to the crest. Solving for $H$ gives

$$H = \frac{c}{\gamma s} \tag{21.26}$$

Which shows that the higher the stability number, the lower the height at failure.

It is instructive to compare results from various methods for a homogeneous soil without any influence from a groundwater table. In Table 21.1, $s$ is the Taylor stability number. Equation (21.26) may be compared with eq. (21.24) for progressive failure of a vertical cut, in which case it may be shown that for this special case $s = 0.5(\tan 45° - \phi/2)$, which is included in the table. Also included in the table are results from a graphical solution called a friction circle, which has largely been replaced by computer programs employing methods of slices. From these data a planar slip is not the most critical surface and should not be

**Table 21.1**

Taylor stability numbers for various failure surfaces for soil having a friction angle $\phi = 25°$. The higher the number, the lower the indicated stability. Answers that are in close agreement are bracketed—but they still only apply to simultaneous failure along the entire slip surface

| Slope $i$, degrees | Surface | Method | $s$ | $\theta$, degrees |
|---|---|---|---|---|
| 90 (vertical) | Plane | Culmann | 0.159 | 57.5 |
| | Circle | Ord. slices | 0.165 | Variable |
| | Circle | $\phi$ circle | 0.166 | Variable |
| | Log spiral | Taylor | 0.165 | Variable |
| | Progressive | Unconfined | 0.319 | Variable |
| 30 | Plane | Culmann | 0.002 | 27.5 |
| | Circle | Ord. slices | 0.012 | Variable |
| | Circle | $\phi$ circle | 0.009 | Variable |
| | Log spiral | Taylor | 0.008 | Variable |

Note: Partly after Taylor (1948).

used unless it is directed by a weak sloping stratum. There is little to distinguish between a circular and a logarithmic spiral failure surface, but progressive failure must be simulated by using residual shear strength. With the lower slope angle with a deep circular failure, the Ordinary Method of Slices is overly conservative as it ignores frictional restraint from side forces on the slices, but this effect will diminish as the sliding mass becomes shallower.

## 21.5.2 Partially Mobilized Resisting Force

If the resisting force $S_r$ is fully mobilized, the resultant makes an angle $\phi$ (which equals the obliquity angle) with a normal to the slip surface. However, Bishop (1955) pointed out that in a stable slope $S_r$ is only partially mobilized, and the acting and normal force components of a slice weight depend on the obliquity angle, not on the friction angle, which is larger. A smaller angle increases the portion of weight $W$ that is converted to $N$, increasing friction, while a smaller portion is allocated to the acting force. This affects the calculated factor of safety for a stable slope but not for one that is actively sliding.

Because the obliquity and the factor of safety affect both the resisting force, the solution is by trial and error. Bishop proposed a simplified method that rapidly converges. The method was developed specifically for circular slip surfaces.

Bishop's concept is expressed mathematically as follows: in Fig. 21.11, vertical side forces are assumed to be equal and opposite so they do not affect the summation. Then

$$W = N \cos\theta + S_{rd} \sin\theta \qquad (21.27)$$

where $S_{rd}$ is the developed resisting force, which is the maximum resisting force divided by the slice factor of safety:

$$W = N \cos\theta + \frac{S_r \sin\theta}{FS_i}, \text{ or} \qquad (21.28)$$

**Figure 21.11**

The simplified Bishop method assumes a circular slip surface and adjusts the factor of safety for a partially developed resisting force $S_r$.

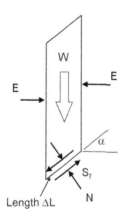

$$N \cos \theta = W - \frac{S_r \sin \theta}{FS_i} \qquad (21.29)$$

Thus as the factor of safety increases, $N$ decreases. By assuming that the factor of safety is constant along the entire slip plane, Bishop obtained the following expression:

$$FS = \sum \left[ \frac{c' \Delta l \cos \theta + (W - u \Delta l \cos \theta) \tan \phi'}{\cos \theta + (\sin \theta \tan \phi')/FS} \right] \div \sum W \sin \theta \qquad (21.30)$$

This may be solved with a computer program, or by iteration on a spreadsheet by substituting different values of FS.

If $\phi = 0$, eq. (21.30) reduces to

$$FS = \frac{\sum c' \Delta 1}{\sum W \sin \theta}$$

This is the same as the Ordinary Method of Slices. Bishop's method does not specify the radius or position of the center of the most critical circle, which are determined by repeating the calculations for different trial surfaces and selecting the one having the lowest factor of safety. This is quickly done with a computer that can be programmed to find the 10 most critical circles. An example solution is shown in Fig. 21.12, and it will be seen that a diversity of circular shear paths gives almost the same factor of safety. The actual path therefore can be expected to be sensitive to small variations in soil strength but can deviate from the most critical circle without greatly influencing the factor of safety.

The Bishop simplified method is an improvement over the ordinary method but generally has been replaced by other more rigorous and/or more general solutions. Nevertheless many newer methods still assume a circular slip surface.

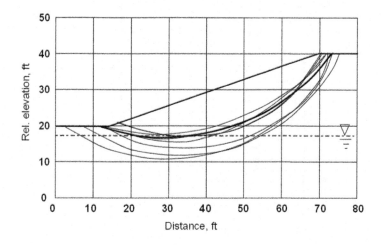

**Figure 21.12**

Trial circular surfaces from a STABL computer run using the simplified Bishop method of slices. All factors of safety are in a range ±1.5 percent, the most critical being shown with a heavy line.

Some methods incorporate a force as well as a moment equilibrium, and solutions are obtained using charts or computers.

### 21.5.3 Spencer's Method

Spencer (1967) approached the problem of too many unknowns by assuming that all forces acting on the sides of slices are parallel and inclined at a constant angle $\alpha$ with horizontal. Two equations result, one representing a force equilibrium and the other a moment equilibrium, and there are two variables, the $\alpha$ angle and the factor of safety. Solution is by trial and error, so Spencer presented a graphical summary shown in Fig. 21.13. The analysis now is more accurately accomplished with a computer.

**Figure 21.13**

Spencer's (1967) charts for estimating the factor of safety of a slope in homogeneous soil with selected pore pressure ratios.

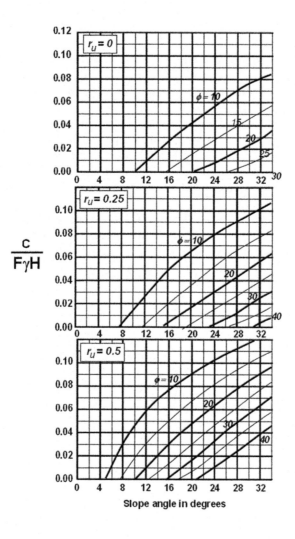

$$\frac{c}{F\gamma H}$$

Slope angle in degrees

Incorporation of pore water pressure into the analysis was simplified by defining a *pore pressure ratio*, defined as the ratio of pore water pressure at the base of each slice to the weight of the slice:

$$r_u = \frac{u}{\gamma h} \tag{21.31}$$

where $u$ is the pore water pressure at the slip surface, and $\gamma h$ represents the overburden pressure.

As a convenient approximation the pore pressure ratio is assumed to be constant throughout the potentially sliding mass. Spencer's charts in Fig. 21.13 show stability factors for pore pressure ratios of 0, 0.25, and 0.5. The first is for unsaturated conditions, and the last approximates a full saturation condition since the unit weight of water is roughly one-half that of soil.

Spencer's method was derived for a circular slip surface but also can be extended to noncircular surfaces. Duncan and Wright (2005) indicate that this is the simplest complete solution for estimating a factor of safety.

**Example 21.11**
Estimate the factor of safety for a 35 ft (10.7 m) high slope at 1 vertical to 3 horizontal, in soil having a friction angle of $25°$, cohesion of $300 \, lb/ft^2$ ($14.4 \, kN/m^3$), and unit weight of $120 \, lb/ft^3$ ($18.8 \, kN/m^3$). Assume a worst saturation condition.

*Answer:* The slope angle is $i = \tan^{-1} 0.333 = 18.4°$. For full saturation, the lower graph in Fig. 21.11 gives

$$\frac{c}{F\gamma H} = \frac{300}{F(120)(35)} = 0.028, \quad \text{or} \quad = \frac{14.4}{F(18.8)(10.7)}$$

where $F$ is the factor of safety. Solving for $F$ gives 2.5.

---

*Question:* What is the factor of safety with $c = 0$?

*Answer:* If $c = 0$ the failure is parallel to the soil surface and Spencer's charts cannot be used. The factor of safety for a continuous slope in cohesionless soil is discussed at the beginning of this chapter. From eq. (21.5), $FS = \tan \phi / \tan i = 1.4$.

---

## 21.6  SOLUTIONS FOR NONCIRCULAR SURFACES

### 21.6.1  Internal Restraint from Slice Side Forces

Some vertical repositioning of slices is necessary if they move downslope on a surface having a changing curvature. An analogy is a stack of books held vertically and allowed to slide along a curved surface: if the surface is circular and the books are allowed to rotate, there is no book-to-book friction, but if the curvature changes, individual books must slide up or down relative to one another. This vertical resistance to slipping adds to the factor of safety.

### 21.6.2 Janbu's Procedure

Janbu's (1973) simplified model, eq. 21.11, assumes that the line of thrust of side forces runs at an angle through lower parts of sides of the slices, analogous to pressures on a retaining wall. In addition the side forces are allowed to vary between slices, which is important because it recognizes active and passive zones, but it also increases the number of variables to be evaluated by trial and error.

Janbu's solution became increasingly popular after being incorporated into a computer program STABL developed by Siegle at Purdue University. The original program is in the public domain, and recently has been modified commercially to enable use with newer computer operating systems. Janbu's method includes both force and moment equilibria and can be applied to any shape slip surface.

### 21.6.3 Simplified Janbu

Janbu's simplified procedure assumes that interslice forces are horizontal, i.e., with zero friction. An empirical correction then is applied based on comparisons of results from both the complete and simplified procedures. As might be expected, the correction is largest for deep slip surfaces where slices are longer and have larger side forces. Correction factors are shown in relation to the thickness of the slide compared with its length, $d/L$, in Fig. 21.14. The correction is empirical and also can be applied to solutions by the ordinary method of slices.

**Figure 21.14**

Janbu's empirical correction to factors of safety to account for mobilization of side forces.

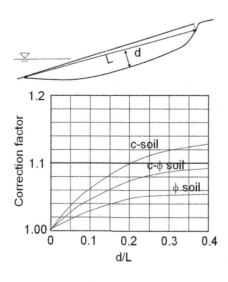

**Example 21.12**
An analysis of the landslide in Fig. 21.7 by the ordinary method of slices gave a factor of safety of 1.0 but the slide temporarily was stable. What is the Janbu correction?

*Answer:* $d/L$ is approximately 0.1, so the factor of safety should be multiplied by 1.05.

## 21.6.4   Other Methods

A comprehensive survey of various analytical procedures is beyond the scope of this book and is presented by Huang (1983) and by Duncan and Wright (2005). Many are included in commercially available computer programs. Assumptions made in some of the most commonly used are as follows: Lowe and Karafiath performed a force equilibrium assuming an inclination of side forces that is the average of the slope and slip angle for each slice, which appears to be a fairly realistic assumption. Morgenstern and Price and Chen and Morgenstern present rigorous solutions with both force and moment equilibria. Horizontal seismic forces and horizontal restraints from tensile reinforcement, nailing, etc., can be applied to slices in the various methods. Sarma's solution readily incorporates seismic influences.

## 21.6.5   Importance of Having Reliable and Accurate Soil Data

The variability of factors of safety from the various methods and procedures fades in comparison with the variability of the soil strength properties and accuracy in measurements. Soil unit weight is less variable, but its variability is magnified by the location of the groundwater table.

Slope stability analyses require *averaged* cohesions and angles of internal friction that can be obtained only with large numbers of tests that are specific for these values. The problem is both simplified and made more difficult in the case of an active landslide—simplified because remolding in the shear zone tends to move shearing strengths toward an average so that fewer tests are required, and more difficult because the shear zone usually is too thin to obtain multiple identical samples for laboratory testing. In that case a stage test, either a triaxial stage test or the in-situ Borehole Shear test, may be the only practical solution.

Where adequate soil data are lacking, as often is the case, analyses sometimes are performed using *assumed* values for cohesion and friction angle, or values obtained on the basis of empirical correlations such as shown in Fig. 19.2. The degree of sophistication of a computer program then has little to do with the reliability of the results, as any analysis based on a guess remains a guess. Chapter 27 presents somewhat more reliable estimates of friction angle based on penetration test data.

It is helpful to note that in an active landslide, soil cohesion is viscous or rate-dependent and therefore can be assumed to be zero. Cohesion is regained through thixotropic recovery if the landslide stops. Still, it is on the safe side to assume that it is zero.

### 21.6.6 Appropriate Factors of Safety

Engineers appreciate that a factor of safety is partly a "factor of ignorance" because it must take into account both the imprecision of the data and assumptions that may be made for analysis. An active landslide narrows the target area because the factor of safety is 1.00. Therefore a factor of safety that is only 1.05 based on a worst-case scenario with fully saturated conditions can give a reasonable assurance of stability, so long as nothing is done to aggravate the situation. On the other hand, if the factor of safety is known to only two significant figures, which often is the case, that variability must be incorporated into a design factor of safety.

In a stable slope, additional unknowns include location and shape of the most critical slip surface, location and anticipated changes in the groundwater table, and indeterminacy of the analysis. Because of these additional unknowns the design factor of safety usually is of the order of 1.2 or more.

Because of the potential for damages, an earth dam reasonably should be designed with a higher factor of safety than a roadcut. However, flattening the slopes of a large embankment increases costs geometrically so greater care is used in construction and control testing, and a moderate safety factor is tolerated. Design often is based on strengths measured on laboratory specimens shortly after compaction and therefore does not include long-term thixotropic strength gain, which can easily double the shear strength. Nevertheless dams do fail. Because of the additional care used in design and construction of large earth dams, the frequency of failures diminishes with increasing size.

## 21.7 RUMMAGING AROUND FOR THE BEST SLIP SURFACE

### 21.7.1 Computer-Drawn Trial Surfaces

Because of the ease and speed of electronic computers, it is a simple matter to let the computer step off trial surfaces and determine those having the lowest factors of safety. This procedure can be used with different computational methods, with nonhomogeneous soils, and with complicated slope geometries. A range in starting and ending positions is specified so that the search will not become infinite, and a number of steps are selected. With modern computers the number of trials readily is increased to over a hundred, compared with three or four using hand calculations.

As the number of computer runs increases, the computer can produce some unrealistic quirks in the slip surface. These probably relate to the assumptions made in the analysis, so even a "perfect" solution to an indeterminate problem is not perfect. It is doubtful that any model, whether analytical, numerical or physical, can duplicate the random perfection of reality. Nevertheless some idealized shapes will be considered, in part because of their application to foundation bearing capacity, discussed in the next chapter.

## 21.7.2  Log Spirals

A logarithmic spiral is shown in Fig. 21.15, and is most commonly used to predict foundation bearing capacity. Spirals also are applied to soil deformations on active and passive sides of retaining walls, as shown in Fig. 19.15.

A unique property of the log spiral is that the angle between the radius and the spiral is constant. If that angle is assigned to the angle of internal friction, any radius for the spiral defines the direction of the resultant force at failure. The corresponding equation is

$$r = r_0 e^{\prime -\omega \tan \phi} \tag{21.32}$$

where $r$ = any radius;

  $r_0$ = a reference radius;

  $\omega$ = the angle between $r$ and $r_0$;

  $e$ = the base of natural logarithms; and

  $\phi$ = the angle of internal friction.

The log spiral is most appropriate where pressure is externally applied to a soil mass, as in the case of foundations and passive pressure on retaining walls. The spiral is less suitable where driving forces change with depth because of

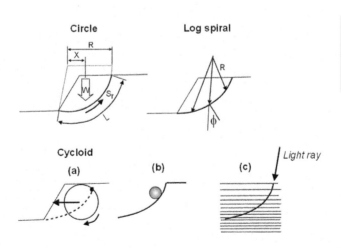

**Figure 21.15**

Ideal slip surfaces and (a–c) some developments involving cycloids.

weight of a soil mass, such as in slopes or the active side of retaining walls. From eq. (21.32) and Fig. 21.15 will be seen that a circle is a special case of a log spiral with $\phi = 0$.

### 21.7.3 Cycloids

A cycloid is generated by a point on the rim of a wheel that is moving along a horizontal surface (Fig. 21.15(a)). This does not appear particularly relevant, but the cycloid has some other features that have interesting parallels for slope stability.

Classic studies of landslides by a French engineer, Alexandre Collin (1846), led to a conclusion that the natural shape of the slip surface is most closely represented by a cycloid. This work was overlooked for many years until rediscovered and championed by Frontard (1948). Terzaghi objected that as the cycloid is unrelated to the friction angle, it can present unrealistic stress distributions. Ellis (1973) applied the cycloid to vertical cuts in trench excavations.

As pointed out by Fox (1981), the cycloid was shown by Bernoulli to be the solution to the "brachistochrone" problem that had intrigued Galileo: what is the shape surface that will provide the fastest and most efficient path for a ball rolling down a slope? The problem is illustrated in Fig. 21.15(b). Bernoulli used an analogy to solve the problem by noting a similarity to the path of a beam of light penetrating at an angle through a medium with a gradually increasing index of refraction (Fig. 21.15(c)). The solution was obtained by integrating Snell's Law of refraction. There also is a striking analogy in soil mechanics to a shear path that encounters increasing resistance with depth because of the increasing overburden pressure and friction. Although analogy can lead to new innovations and insights (Koestler, 1964), it is important to remember that, regardless of the depth of philosophical conjecture, it does not constitute a proof. However, the analogy that can be reinforced by reasoning can come closer.

Conceivably the energy-efficient cycloidal path for a rolling ball might also describe the most energy-efficient path for the slip surface in a landslide. This would be consistent with a concept of progressive failure, as each new segment of a slip surface searches for the most energy-efficient path. Fox (1981), who pointed out many of the features in this discussion, also pointed out that catenary curves, which according to arching theory are trajectories of major principal stress, for practical purposes are orthogonal to cycloids.

The $x$ and $y$ parameters of a cycloid with a starting point at the lower end and verticality at the upper end are

$$x = a(\theta + \cos\theta) \quad and \quad y = a(1 - \cos\theta) \tag{21.33}$$

where $\theta$ is in radians.

## 21.8  STOPPING LANDSLIDES

### 21.8.1  How to Be a Slide Stopper

The most obvious way to stop a landslide is to mitigate one or more of the contributing factors by (a) removing load from the top, which seldom is feasible, (b) adding load at the toe, (c) draining, or (d) increasing the average shearing strength in the slip zone. The effectiveness of different options is predicted using a method of slices. Some methods used to stop landslides are depicted in Fig. 21.16.

### 21.8.2  Drainage

Drainage is one of the oldest methods for stopping a landslide, and is often accomplished with corrugated plastic pipe that has been perforated with narrow slots to let water in. However, the slots also can let water out, and a tile line that crosses an active slip zone will be stretched, bent, compressed, or sheared off,

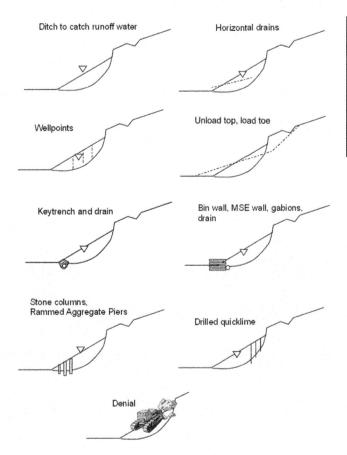

**Figure 21.16**

Some repair methods for landslides. The one at the bottom is too frequently used and can only make matters worse.

so it may drain directly into the slip zone. Trenching to install the tile is a risky adventure because it can reactivate the slide. Trenching can be done only during dry periods when the water table is low, and should be done so that only short sections are open at any one time.

Directional drilling such as shown in Fig. 21.17 is expensive, but can extend much deeper than is possible with trenching, and avoids the dangers associated with open trenches. Directional drilling is accomplished in soils by pushing a beveled mandrel using water jetting. The mandrel carries electronic direction sensors so that directions can be changed by changing the orientation of the bevel. A small-diameter perforated plastic pipe is attached to the mandrel at the exit point, and then pulled back and left in the boring.

Gravity drainage requires an exit point, and the groundwater table also can be lowered by the use of well points, vacuum well points, or electro-osmosis. The latter processes are relatively expensive but may be the only effective ways to drain low-permeable clay soils. Vertical wells can be effective but require intermittent pumping.

A key trench filled with granular material at the toe ideally will cross the slip zone and act as a drain. Another advantage is that part of the soil at the toe is replaced with soil having a higher friction angle, a factor that is readily included in stability calculations. However, a key trench is not necessarily a perfect solution because the landslide still can go over the top.

Drainage of a toe area can be improved by installation of gabions, which are interconnected woven wire baskets filled with rocks. They have the advantage of flexibility should the ground continue to move.

**Figure 21.17**
Directional drilling can install perforated plastic drain lines without trenching.

## 21.8.3 Soil Reinforcement

Construction operations on a landslide preferably are conducted during dry weather while the landslide has stopped. The area may be impassable for all but track-mounted equipment, and toe areas may be a quagmire. However, this lower, passive part of a landslide will not move without being pushed. Therefore the most effective means for stabilization will address the upper, active part of a landslide, which also can aid in stopping headward expansion. Headward expansion also can be stopped by soil nailing into the exposed scarp. Stabilizing the passive part of a landslide does not prevent it from being overrun.

One method for stabilization is to install vertical "stone columns" on a grid pattern and extending through the landslide slip zone. Design is based on an area replacement basis, substituting crushed stone with a friction angle of about 35° for soil that may have a friction angle of only one-half or one-third of that amount. "Rammed Aggregate Piers" are similar but smaller in diameter and have a higher friction angle, about 48–50°. These are discussed in Chapter 24 on intermediate foundations. Piles are more expensive and less effective unless used in large numbers because the sliding soil readily can go around. A formula for area replacement assuming that cohesion is zero in the slip zone is

$$m \tan R + (1 - m)\tan S + \; = F \tan S \qquad (21.34)$$

where $m$ is the fractional replacement of soil $S$ by $R$, and $F$ is the desired factor of safety.

### Example 21.13

The factor of safety of an active landslide is assumed to 1.0 and with full saturation the friction angle in the slip zone is back-calculated to be 12°. What percent replacement will be needed to increase the factor of safety to 1.2, (a) using stone columns having a friction angle of 35°, (b) using Rammed Aggregate Piers having a friction angle of 48°?

*Answer:*

(a) Stone columns: $m \tan 35° + (1 - m)\tan 12° + \; = 1.2 \tan 12°$

$$0.70\, m + 0.21 - 0.21\, m = 0.26$$

$$m = 0.10 \text{ or } 10\%$$

(b) Rammed Aggregate Piers: $m \tan 48° + (1 - m)\tan 12° = 0.26$

$$1.1 m + 0.21 - 0.21 m = 0.26$$

$$m = 0.06 \text{ or } 6\%$$

## 21.8.4 Drilled Lime

Seeding a landslide slip zone with quicklime was first tried in the early 1960s and has since been widely used where soils contain active clay minerals, smectites or montmorillonites. Pebble quicklime is introduced into borings made on a grid pattern in the active part of a landslide. Skill and experience are required in handling the chemical to prevent personal injuries. If the soil is too soft to hold an open boring, reverse augering is used to push lime down into the soil, and immediately creates a sheath of stabilized soil. In extreme cases lime can be introduced in holes created with a concrete vibrator, which has been used to save houses that were audibly creaking from pressures from landslides.

The drilled lime process works by immediately drawing water from the soil to hydrate the lime, and sliding immediately stops. However, this reduction in water content of the soil is only temporary. The lime column approximately doubles in diameter as the lime hydrates, drying and creating radial tension cracks in the surrounding soil. The cracks then are filled by a paste of hydrated lime, greatly enhancing its distribution into the soil (Handy and Williams, 1967). For economical reasons borings may only cover the upper part of the active zone so long as no other changes are made that would reduce the factor of safety. The boring diameters and spacing are designed to introduce from 1 to 3 percent lime, depending on the clay content of the soil. Excess lime is used in a pozzolanic reaction with soil clay minerals that creates hydrated silicates and aluminates that are the same products as in hydrating Portland cement.

A simple field test to determine applicability of the drilled lime method for a particular soil is to place small quantities of soft, saturated soil *from the slip zone* into two sealable plastic bags. A few percent *hydrated* lime is added to soil in one of the bags, and both are sealed and the samples remolded by hand. Within a few minutes a reactive soil will become crumbly and appear to dry out even though no moisture has been lost, as the plastic limit increases and exceeds the soil natural moisture content. If that does not occur, either the moisture content is too high or the soil does not contain the necessary active clay minerals.

# 21.9  SPECIAL PROBLEMS WITH EARTH DAMS

## 21.9.1  Relevance

Earth dams vary from some of the smallest to some of the largest man-made structures in the world. Design of large dams takes on additional importance because of safety issues. Many topics such as compaction, settlement, slope failure, dispersive soils, piping, etc., are covered elsewhere in this text. Because of the severity of failures and the continuing construction of new earth dams, particularly in developing countries, the topic merits some additional consideration.

## 21.9.2  Causes and Frequency of Dam Failures

A study by ICOLD (1995) indicates that of 5268 dams built between 1951 and 1986, 2.2 percent failed. Most failures involved small dams, as the probability of dam failure decreases with increasing dam height according to an approximate logarithmic relationship. Small dams are more likely to fail because

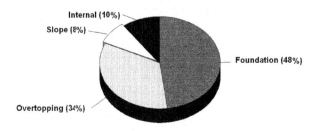

**Figure 21.18**

Causes of failures of earth dams. Slope failures include those occurring both in the dam and around the reservoir; internal failures include piping and improper compaction. (From data of Milligan, 1999.)

the consequences are not nearly so severe, and there typically is a lower level of care in design and construction. Small dams also are much easier to repair and restore to operating condition.

Most failures occurred before the dams were 10 years old, perhaps in part because soil aging results in a gradual gain in shear strength. A statistical study indicates that the odds of failure of a dam are about 1 in 7000 or 0.014 percent for any particular year, but the odds are influenced by seismicity as well as the size of the dam. In the typical 100-year useful life of a dam the probability is 1.4 percent, but the odds are improved after the dam has survived its first decade.

The pie chart in Fig. 21.18 indicates that slope failures are not the major cause of distress in earth dams, being eclipsed by foundation failures and overtopping. Nevertheless the basic tenets are followed regardless of size: using properly selected soil, having an adequate site preparation, and compaction of the embankment at a moisture content that is at or above the optimum. Some problems specific to earth dams and levees are discussed below.

## 21.9.3 Overtopping and Wave Action

Overtopping of an earth dam is a short ticket to disaster because of the large hydraulic gradient from the reservoir level to the base of the dam. The effect of the hydraulic gradient on erosion is effectively illustrated in Fig. 8.8 where a large section of the dam washed away. Overtopping usually is the result of an insufficient spillway capacity, and most dams have an emergency spillway during periods of unusually high rainfall. However, wave action also can be a factor.

Upstream faces of earth dams are covered with coarse stone, broken concrete, stabilized soil, or grout-filled plastic mattresses to prevent wave erosion. The dam obviously must be high enough above the reservoir level to prevent overtopping by waves, whose height depends on wind velocity and the distance the wind blows across water, or "fetch." The influence of fetch is illustrated by a guide to freeboard for small dams (Fig. 21.19).

**Figure 21.19**

Suggested freeboards for earth dams. (Adapted from U.S. Bureau of Reclamation, 1960.)

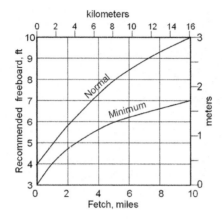

**Figure 21.20**

Rate of movement of the ground surface along a traverse running downslope on the Vaiont landslide. (Modified after Leonards, 1987, p. 608.)

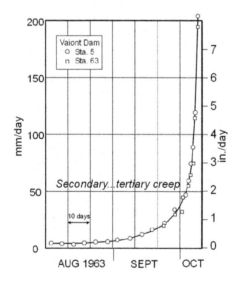

### 21.9.4  Hydraulic Gradient

Hydrological factors influencing the internal stability of earth dams were previously discussed in relation to dispersive soils and a critical hydraulic gradient that can cause quick conditions. It should be emphasized that the design hydraulic gradient may not be that which exists in the field, because if water finds a less restrictive path, pressure will be transferred that will increase the hydraulic gradient for the remainder of the path. For example, an uninterrupted gravel layer that can be expected to occur under an earth dam will conduct water with very little decrease in head, and the gradient may become critical for the remainder of the flow path.

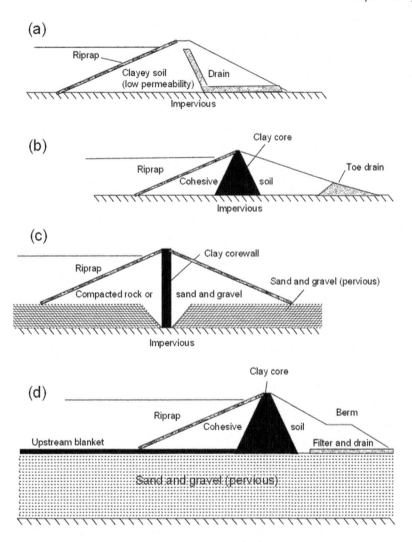

**Figure 21.21**
Some representative earth dam designs to direct and control seepage. Riprap protects the faces from erosion; berm adds to stability of the downstream slope. (Adapted from Creager et al., 1948.)

The resulting erosion and piping can have much more serious consequences than simply a leak.

## 21.9.5 Hydraulic Fracturing

High groundwater pressure near a filled reservoir can induce hydraulic fracturing, particularly in abutment areas that are not confined under the weight of the dam. Hydraulic fracturing is intentionally used in the petroleum industry to increase production from oil wells, and occurs in soils during pressure grouting, indicated by a sudden decrease in pumping pressure and increase in pumping rate. Hydraulic fracturing can create a bypass channel that in turn affects leakage and the hydraulic gradient.

The probability of fracturing is reduced by performing detailed "dental work" on foundation rock before a dam is built. This involves digging out and filling open seams and weathered zones that might yield and erode out under pressure.

The tip of a hydraulic fracture becomes a focal point for tension that can be calculated from a *stress concentration factor* from elastic theory and mechanics of materials. The smaller the radius at the tip of the fracture, the higher the tensile stress. Hydraulic fracturing is prevented by equalizing seepage pressure with pore water pressure, which means slow first-filling of a reservoir. Piezometers monitor pore water pressures within the dam and also preferably within the abutments to determine if they are as predicted from the flow net—a low pressure being conducive to hydraulic fracturing, and a high pressure indicating that fracturing already may have occurred. Piezometers are monitored throughout the life of a large dam to show any deviations from expected behavior.

### 21.9.6  Rapid Drawdown

An important factor contributing to slope instability of an earth dam is the possibility for a rapid drawdown, where the reservoir level is lowered sufficiently rapidly that soil in the dam and in slopes around the reservoir slope remains fully saturated, creating destructive downslope seepage forces. Reservoir levels therefore are lowered at a rate that will allow drainage of water from soil in the slopes. As pointed out earlier in this chapter, seepage forces can reduce a factor of safety against sliding by as much as 50 percent.

### 21.9.7  Leaky Reservoirs

While reservoir leakage does not constitute a failure on the same scale as the other factors discussed in this section, a reservoir leakage can cause the dam to be ineffective for its intended purpose, which insofar as the investors are concerned constitutes a failure. The likelihood of developing leaks is magnified because the reservoir area in contact with impounded water is orders of magnitude larger than the contact area with the dam itself. Reservoir leakage is most critical in karst areas where limestone is invaded by having underground caverns and sinks, and can be troublesome where a reservoir is in contact with areas of fractured rocks or gravel deposits.

This potential for leakage through soils and rocks underneath a dam is evaluated with pressure tests performed in borings, and these are an important part of site exploration. Underdam leakage is most effectively prevented by use of a cutoff wall that can be concrete, clay, or driven sheet pile. Grouting is frequently used but is difficult to direct to the desired area, and in severe cases of underdam leakage the grout is washed away as fast as it is injected, particularly after a reservoir is filled with water. In some cases the volume of grout pumped has exceeded the volume of the dam itself, resulting in a huge cost overrun. Leaky reservoirs also impact an entire neighborhood by raising the groundwater table and flooding nearby fields and basements.

In recognition that a river often flows on a bed of sand that grades downward into gravel, an upstream clay blanket may be used to seal off the river channel prior to reservoir filling. If the hydraulic conductivity of the potential seepage layer is 10 times that of the average soil underneath the dam, the clay blanket can extend a distance upstream equal to 10 times the width of the dam to achieve the same resistance to flow. Some common methods for controlling through-dam and under-dam seepage are illustrated in Fig. 21.21.

## 21.9.8 Case History of a Slope Failure

Vaiont Dam is a concrete arch dam that was subjected to a catastrophic landslide in a slope above the dam. The dam itself survived but the reservoir did not, as the tremendous splash overtopped the dam and swept away over 2000 lives in the valley below.

Vaiont Dam occupies a deep valley in the Dolomite Alps in northern Italy, and is one of the highest concrete arch dams in the world, standing tall at 860 ft (262 m). The landslide area was about 2 km wide by 1.5 km long (1.2 × 0.9 miles). The speed of sliding was back-calculated to be about 110 km/hr (70 mph), and a combination of wind and water from the slide blew out windows in a town 100 m above the reservoir on the opposite hillside. The failure in 1963 remains one of the worst dam tragedies to date.

The landslide at least in part developed along a prehistoric slip zone, and as the lake was being filled a smaller landslide developed in the same area. During the first filling of the reservoir livestock on the hillside were observed to be very nervous, probably as a result of rock noises that can be a prelude to shear failure but are inaudible to the human ear. Ground surveys were conducted and showed that the hillside was moving at a very slow rate, as shown at the left in Fig. 21.20. Attempts were made to reduce the rate of movement by reducing the lake level, but the lake level ominously kept rising as the hill squeezed in on the reservoir.

The engineers then were faced with a serious dilemma: should they increase the rate of drainage, which would create a rapid drawdown condition, or should they try and maintain the reservoir at a constant elevation, and/or should they require an evacuation? There were no easy answers, particularly as attempts to lower or maintain the reservoir level were frustrated by movement of the landslide.

The data in Fig. 21.20 present a classic example of different stages of soil creep, although this may not have been fully appreciated at the time. During primary creep, which is the mechanism of secondary consolidation, the rate is decreasing and may stop. During secondary creep the rate is constant, and during tertiary creep the rate is increasing and is preliminary to rapid shear failure. From the graph it appears that tertiary creep initiated about 2 months prior to failure and was assured by a month prior to failure, which is when warnings would have been appropriate.

The sudden and rapid rate of sliding was not expected, and it later was conjectured that heat generated from friction might have contributed by decreasing the viscosity. Steam pressure also was suggested but is unlikely under the huge confining pressure. A more likely explanation for the speed of the sliding comes from an examination of measurements at different points on the sliding mass. These indicated that the upper part of the slide had moved 4.03 m while the lower point moved only 2.87 m. This indicates that the rock mass between those two points had compressed 1.16 m, confirming the thesis that a landslide has an active, pushing part and a passive part that is being pushed. The separation distance between the points was about 700 m, so the compressive strain was about 0.17 percent as stored strain energy. Being elastic, that energy would be subject to instant release and conversion to acceleration.

Several approaches might have helped avert this tragedy, including a better appreciation of the geological factors contributing to the slide. Modern instrumentation, for example the use of inclinometers and acoustic microphones, could have provided a more stringent warning. The cows had the right idea.

## Problems

21.1. Describe two situations when the factor of safety of a landslide is exactly 1.0, and explain how this observation can be used to estimate soil sensitivity.

21.2. Explain the mobilization of shearing stresses within a sliding mass of soil, and two situations where they theoretically will not be mobilized.

21.3. Analysis of an active landslide yields a factor of safety of 2.8. Is that possible? In your opinion is this more likely to be due to erroneous data or to the method of analysis? What procedures should be used to check the results?

21.4. The friction angle of a soil is 25°. Calculate the angle of repose on the Moon where gravity is about one-fifth that on Earth.

21.5. A deep railroad cut intersects the plane of contact between a loess soil and an impermeable clay paleosol. The plane of contact makes an angle of 12° with horizontal. Soil information is given in Fig. 21.22. Assume that cohesion is zero and calculate the factor of safety with no contribution from seepage.

21.6. Repeat Problem 21.5 assuming a worst-case scenario for seepage.

21.7. In the preceding problem even though there is a plane along which sliding may occur, there is no assurance that a more critical curved slip surface

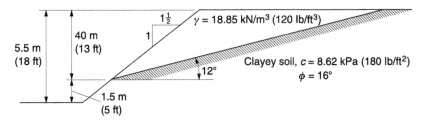

**Figure 21.22**

Conditions for Problems 21.5 and 21.6.

| | Shear component | | Normal component | |
|---|---|---|---|---|
| Slice No. | kN | (lb) | kN | (lb) |
| 1 | 14.31 | (980) | 8.17 | (560) |
| 2 | 14.90 | (1020) | 15.91 | (1090) |
| 3 | 12.57 | (860) | 29.76 | (2040) |
| 4 | 7.74 | (530) | 45.54 | (3120) |
| 5 | 5.97 | (410) | 49.06 | (3360) |
| 6 | 4.66 | (320) | 47.44 | (3250) |
| 7 | 3.94 | (270) | 40.00 | (2740) |
| 8 | 2.92 | (200) | 23.36 | (1600) |
| 9 | 2.20 | (150) | 18.37 | (1260) |
| 10 | 0.59 | (40) | 12.11 | (830) |

**Table 21.2**

Data for Problem 21.8

may not exist in the soil above that plane. Use Spencer's method with three different degrees of saturation to estimate factors of safety. Which do you feel is most appropriate for design of a roadcut along a county road? Of a cut bank behind a house?

21.8. The stability of a slope is being analyzed by the method of slices. On a particular curved surface through the soil mass, the shearing component of the weight of each slice and the normal component of the reaction at the base of each slice are presented in Table 21.2. The length of the curved surface is 29.9 m (98 ft). If the friction angle of the soil is 12° and the cohesion is 2.39 kPa (50 lb/ft²), what is the factor of safety along this particular curved surface?

21.9. A 21 m (70 ft) roadcut has side slopes of 2 to 1. The base of the cut is soil in limestone. The soil has the following properties: $\gamma = 19.6\,\text{kN/m}^3$ (125 lb/ft²); $c' = 26.8\,\text{kPa}$ (560 lb/ft²); $\phi' = 20°$. Draw a trial circular failure with the center at the crest of the cut and intersecting the cut at the toe. Use the simplified method of slices to determine the minimum factor of safety if seepage is not a factor, and apply the Janbu correction.

21.10. Recalculate the factor of safety in Problem 21.8 using Spencer's method.

**Table 21.3**

Spreadsheet calculation of acting forces in Fig. 21.23

| Slice No. | $H$ (m) | $X$ (m) | $\gamma$ (kN/m³) | $W$ (MN/m) | $\theta$ (deg.) | $S_{ai}$ (MN/m) |
|---|---|---|---|---|---|---|
| 1 | 4.9 | 11 | 17.3 | 0.93 | −40 | −0.60 |
| 2a | 7.3 | 20 | 17.3 | 2.53 | | |
| b | 4.9 | 20 | 19.8 | 1.94 | | |
| Total | | | | 4.47 | −21 | −1.60 |
| 3a | 3.7 | 15 | 17.3 | 0.96 | | |
| b | 15.2 | 15 | 19.8 | 4.51 | | |
| Total | | | | 5.47 | −13 | −1.23 |
| 4 | 24.4 | 15 | 19.8 | 7.25 | −3 | −0.38 |
| 5a | 6.1 | 30 | 18.4 | 3.37 | | |
| b | 26.2 | 30 | 19.8 | 15.56 | | |
| Total | | | | 18.93 | 9 | 2.96 |
| 6a | 3.0 | 30 | 18.4 | 1.66 | | |
| b | 23.8 | 30 | 19.8 | 14.14 | | |
| Total | | | | 15.80 | 28 | 7.41 |
| 7a | 4.3 | 30 | 18.4 | 2.37 | | |
| b | 18.3 | 30 | 19.8 | 10.87 | | |
| Total | | | | 13.24 | 33 | 7.21 |
| 8 surcharge | | | | 0.04 | | |
| a | 8.5 | 30 | 18.4 | 4.69 | | |
| b | 12.2 | 30 | 19.8 | 7.25 | | |
| Totals | | | | 11.95 | 40 | 7.68 |

$$\sum S_a = 21.45$$

**Figure 21.23**

Landslide problem for the method of slices data in Tables 21.3 and 21.4.

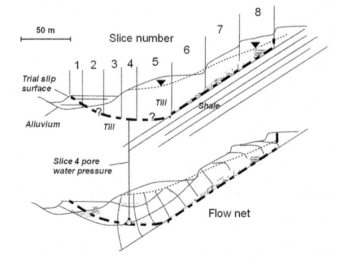

**Table 21.4**

Spreadsheet calculations of resisting forces in Fig. 21.23

| Slice No. | $\Delta L$ (m) | $c$ (kPa) | $c\Delta L$ (MN/m) | $N$ (MN/m) | $u$ (kPa) | $u\Delta L$ (MN/m) | $N'$ (MN/m) | $\phi$ (deg.) | $N' \tan \varphi$ (MN/m) | $S_{ri}$ (MN/m) | $FS_i$ ($S_r/S_a$) |
|---|---|---|---|---|---|---|---|---|---|---|---|
| 1 | 14.4 | 45.5 | 0.655 | 0.712 | 48 | 0.69 | 0.02 | 0 | 0 | 0.655 | −1.09 |
| 2 | 21.4 | 38.3 | 0.819 | 4.17 | 120 | 2.56 | 1.61 | 25 | 0.751 | 1.57 | −0.98 |
| 3 | 15.4 | 38.3 | 0.590 | 5.33 | 185 | 2.85 | 2.48 | 25 | 1.156 | 1.75 | −1.42 |
| 4 | 15.0 | 38.3 | 0.574 | 7.24 | 239 | 3.59 | 3.65 | 25 | 1.702 | 2.28 | −5.99 |
| 5 | 30.4 | 38.3 | 1.164 | 18.70 | 257 | 7.82 | 10.88 | 25 | 5.073 | 6.24 | 2.11 |
| 6 | 34.0 | 38.3 | 1.302 | 13.95 | 233 | 7.93 | 6.02 | 16 | 1.726 | 3.03 | 0.41 |
| 7 | 35.8 | 38.3 | 1.371 | 11.47 | 179 | 6.41 | 5.06 | 16 | 1.451 | 2.82 | 0.39 |
| 8 | 39.2 | 38.3 | 1.501 | 9.15 | 120 | 4.69 | 4.46 | 16 | 1.29 | 2.78 | 0.36 |
| Tot. | 205.6 | | 7.976 | | | | | | | 13.178 | 21.13 |

$$FS = 21.13 \div 21.45 = 0.99$$

*Note*: Negative $FS_i$ for individual slices results when $S_a$ and $S_r$ act in the same direction.

21.11. Repeat the calculations of Problem 21.8 using Bishop's method.

21.12. Set up a spreadsheet, check and if necessary correct the calculations in Tables 21.3 and 21.4 for the slope shown in Fig. 21.23. Now make changes to account for drains that will draw the groundwater table down to a level DD' and determine the effect on factor of safety.

21.13. Drilled lime stabilization of soil in Fig. 21.23 Slice 8 increases cohesion 50% and the friction angle to 25°. Use the spreadsheet to recalculate the factor of safety. (This may be done without a computer by substituting values.)

21.14. Calculate the effect on factor of safety if 25% of the soil in Fig. 21.23 Slice 8 is replaced by Rammed Aggregate Piers having zero cohesion and a friction angle of 50°.

21.15. Set up a table with recommended factors of safety for landslides, roadcuts, road embankments, small earth dams, and large earth dams.

21.16. In Fig. 21.23 what is the most likely explanation for the relatively flat ground surface on Slice 6? (b) How might your explanation be confirmed? (c) Might this subtle observation have any bearing on location of a drain line? Explain.

21.17. Discuss relative merits of circular, log spiral, and cycloid failure surfaces in homogeneous soils.

21.18. Carefully examine soil at and beyond the top of a steep slope for tension cracks. Measure the height and slope angle and prepare a cross-section, and sketch in a possible failure surface.

21.19. The slip surface of an active landslide is planar at the top and thickens to more than a meter downslope. Explain.

21.20. Look to sources to review geological factors that may have influenced the failure at Vaiont Dam.

21.21. Prepare a table with columns for quicksand, dispersive clay, and hydraulic fracturing and entries describing (1) flow rate required, (2) flow direction, (3) factors governing hydraulic pressure, (4) prevention, (5) high, medium, or low potential danger to a dam, and (6) clues that the phenomenon is occurring.

21.22. What should be looked for when inspecting an earth dam? When should inspection be most rigorous, during high or low level of the reservoir? How often should an earth dam be inspected (a) the first 10 years, (b) after 10 years?

## References and Further Reading

Bishop, A. W. (1955). "The Use of the Slip Circle in Stability Analysis of Earth Slopes." *Géotechnique* 5(1), 7–17.

Collin, A. (1846). *Landslides in Clay*. Transl from French by W. R. Schriever. University of Toronto Press, Toronto, 1956.

Creager, W. F., Justin, J. D., and Hinds, J. (1948). *Engineering for Dams*, Vol. III: *Earth, Rock-Fill, Steel and Timber Dams*. John Wiley & Sons, New York.

Duncan, J. M., and Wright, S. G. (2005). *Soil Strength and Slope Stability*. John Wiley & Sons, Hoboken, N.J.

Ellis, H. B. (1973). "Use of Cycloidal Arcs for Estimating Ditch Safety." *ASCE J. Soil Mech. and Foundation Eng.* 99(SM2), 165–180.

Engineering Foundation Conference (1975). *Safety of Small Dams*. ASCE, New York.

Fox, D. E. (1981). "Behavioral Predictions of Compacted Embankments on Rigid Foundations." Unpublished Ph.D. dissertation, Iowa State University, Ames, Iowa.

Frontard, J. (1948). "Sliding Surfaces and Stability Calculations for Soil Masses of Curvilinear Profile." *Proc. 2d Intn. Conf. on Soil Mechanics and Foundation Engineering*, Rotterdam, II,5.

Handy, R. L. (1986). "Borehole Shear Test and Slope Stability." *ASCE In Situ 86*, 161–175.

Handy, R. L. (1995). *The Day the House Fell*. ASCE Press, New York.

Handy, R. L., and Williams, W. W. (1967). "Chemical Stabilization of an Active Landslide." *ASCE Civil Engineering*, 37(8), 62–5.

Huang, Y. H. (1983). *Stability Analysis of Earth Slopes*. Van Nostrand Reinhold, New York.

ICOLD (International Commission on Large Dams) (1995). Ruptures de barrages — Analyse statistique – (Dam Failures — Statistical Analysis). ICOLD Bulletin 99.

Janbu, N. (1973). "Slope Stability Computations." In *Embankment-Dam Engineering: Casagrande Volume*. John Wiley & Sons, New York.

Jumikis, A. R. (1962). *Soil Mechanics*. Van Nostrand, Princeton, N.J.

Koestler, A. (1964). *The Act of Creation*. Macmillan, New York.

Leonards, G. A. (1987). *Dam Failures*. Elsevier, Amsterdam. Reprinted from *Engineering Geology* 26(1–4).

Logani, K. L. (1973). "Dilatancy Model for the Failure of Rocks." Unpublished. Ph.D. dissertation, Iowa State University, Ames, Iowa.

Milligan, V. (1999). "An Historical Perspective of the Development of the Embankment Dam." Canadian Dam Association. Conference, Sudbury, Ontario.

Morgan, A. E. (1971). *dams (sic) and other disasters*. Porter Sargent Publisher, Boston.

Scholtz, C. H. (1967). "Wear and Gouge Formation in Brittle Faulting." *Geology* 15, 493–495.

Seed, H. B. (1968). "Landslides During Earthquakes Due to Soil Liquefaction." Fourth Terzaghi Lecture. *ASCE J.Soil Mech. and Foundation Eng. Div.* 94(SM5), 1053–1122.

Skempton, A. W. (1964). "Long Term Stability of Clay Slopes." *Geotechnique* 14(2), 75–102.

Spencer, E. (1967). "A Method of Analysis of the Stability of Embankments Assuming Parallel Interslice Forces." *Géotechnique* 17(1), 11–26.

Taylor, D. W. (1948). *Fundamentals of Soil Mechanics*. John Wiley & Sons, New York.

Terzaghi, K. (1943). *Theoretical Soil Mechanics*. John Wiley & Sons, New York.

U.S. Department of Interior Bureau of Reclamation (1960). *Small Dams*. U.S. Govt. Printing Office, Washington, D.C.

# 22

# Bearing Capacity of Shallow Foundations

## 22.1 OVERVIEW

### 22.1.1 The Functional Foundation

In civil engineering a foundation links a structure with the material on which it rests. The foundation is designed to transmit the weight of the structure, plus effects of live loads and wind loads, to the underlying material so that it is not stressed beyond its safe bearing capacity. If the supporting material is hard rock, design is simplified, but construction can become more costly because of the requirement for heavy ripping or blasting. Usually, however, the depth to rock is such that the foundation rests on soil.

In this chapter we will deal with shallow foundations, also called spread footings because they spread the load. If the soil is not competent to support the anticipated load, the soil can be reinforced, or the load can be carried deeper to more competent soil or rock by means of piles or piers. The soil or rock support for a foundation is variously referred to as the foundation soil or foundation bed.

### 22.1.2 Site Evaluation

Site evaluation for a small or inexpensive structure may only involve on-site inspection to determine the presence of unstable slopes, landslides, limestone sinks, mine cave-ins, abandoned wells, ravines, trash, pollution, seepage, ponded water, hog confinements, or other more or less obvious deficiencies.

Less obvious but still exceedingly important are clues that can reveal the presence of expansive clays—vertical shrinkage cracks in the soil, heaved and distressed pavements, or cracks in nearby structures. Reference should be made to available documents including soil maps and reports. Hand borings or probings are made as needed for soil identification and evaluation.

Many engineers now are reluctant to make such judgments without detailed on-site borings and tests because of liability issues. All moderately heavy to heavy structures require exploration borings, and many loan agencies require borings and a soil report before approving loans. Soil borings and on-site tests are discussed in Chapter 26.

Regardless of the detail of the investigation, a final step is to inspect open basement and footing excavations for buried topsoil or fill that were not encountered in the borings and which should be removed prior to placing concrete. Any loose soil that may have fallen into open foundation excavations must be scooped out, and the firmness of the bare foundation soil can quickly be verified with a hand probe or penetrometer. Covering up the evidence literally sets the stage for future complications.

### 22.1.3  Building Codes

Many cities and other organizations such as state highway departments have established tables of safe bearing capacities on the basis of experience with the local soils. The Boston Building Code is an example (Table 22.1).

Building codes are a valuable guide, but are oversimplified. For example, they typically specify a value or a narrow range of values for safe bearing capacity for each class of soil without giving any consideration to width, shape, and depth of a foundation, or to the proximity to other structures, and even the relation to a groundwater table. The foundation width is particularly important because wider foundations confine the soil and allow a higher bearing pressure. Building codes therefore may include an option that design can be performed by a registered professional engineer.

## 22.2  FOUNDATION BEARING CAPACITY

### 22.2.1  Two Kinds of Failures

The most ruinous type of foundation failure results when the underlying soil shears and is displaced laterally, allowing the foundation to sink into the ground. This is a classic *bearing capacity failure*. Fortunately, it is rare. The concept is relatively straightforward, downward and sideways. The supporting ability of a soil depends in part on the depth of the foundation; the deeper the foundation, the greater the ability. The overloaded foundation is like a boat, sinking deeper until it finds its stable depth. The distinction from a boat is because soils are much denser than water and possess shear strength. Also eccentric loading and variations in shear strength inevitably cause the structure to tip one way or the other. We can call this a Type A failure. Because the soil is shearing, it becomes weaker through remolding.

**Table 22.1**

Allowable bearing pressures of foundation materials (Sec. 725, Boston Building Code, 1970)

| Class of material | Allowable bearing pressure, tons/ft$^2$ (kPa) |
|---|---|
| 1. Massive igneous rocks and conglomerate, all in sound condition (sound condition allows minor cracks) | 100 (9600) |
| 2. Slate in sound condition (minor cracks allowed) | 50 (4800) |
| 3. Shale in sound condition (minor cracks allowed) | 10 (960)[a] |
| 4. Residual deposits of shattered or broken bedrock of any kind except shale | 10 (960) |
| 5. Glacial till | 10 (960) |
| 6. Gravel, well-graded sand and gravel | 5 (480) |
| 7. Coarse sand | 3 (290) |
| 8. Medium sand | 2 (190) |
| 9. Fine sand | 1 to 2 (95 to 190)[b] |
| 10. Hard clay | 5 (480) |
| 11. Medium clay | 2 (190)[c] |
| 12. Soft clay | 1 (95)[c] |
| 13. Inorganic silt, shattered shale, or any natural deposit of unusual character not provided for herein | [b] |
| 14. Compacted granular fill | 2 to 5 (190 to 480)[b] |
| 15. Preloaded materials | [b] |

Note: [a]The allowable bearing pressures given in this table are intended to prevent bearing capacities from exceeding the soil shear strength. However, these allowable bearing pressures do not ensure that the settlements will be within the tolerable limits for a given structure.
[b]Value to be fixed by the building official.
[c]Alternatively, the allowable bearing pressure shall be computed from the unconfined compressive strength of undisturbed samples, and shall be taken as 1.5 times that strength for round and square footings, and 1.25 times that strength for footings with length-width ratios of greater than four (4); for intermediate ratios interpolation may be used.

The Type B failure also involves sinking but it is slow and not as catastrophic. This is *excessive and/or differential settlement*, which are far more common. In a Type B failure the soil consolidates and becomes stronger instead of shearing and becoming weaker.

An example of Type B is the famous Leaning Tower, which started to lean soon after construction began, so the builders compensated for the lean by using thicker courses of stone on the low side. The correction temporarily relieved the off-center loading, but meanwhile soil underneath the low side had consolidated more than under the high side and was slightly stiffer, so tilting pursued in another direction. The stonemasons played into the game like a juggler balancing a broomstick, so the Tower zig-zagged upward. When it was completed there could be no more corrections, so tilting continued and then slowly speeded up as the load went more off-center. Had the soil become weaker through shearing instead of stronger through consolidation, the Tower would have fallen over.

## 22.2.2 Factors Influencing Type A Bearing Capacity

Bearing capacity failures are relatively rare except when an earthquake temporarily liquefies a sandy or silty foundation soil. During a 1964 earthquake in Niigata, Japan, a cluster of multistory apartment buildings sank and tipped to such a high angle that the occupants reached safety by crawling out of their windows and walking down the outside walls.

Another contributor to low shearing strength is excess pore water pressure caused by rapid loading of a saturated, poorly draining foundation soil. As the load goes on and pore water pressure goes up, frictional strength goes down in accordance with the concept of effective stress. This is unlikely to occur where loading is restricted by a small allowable settlement and therefore is more common where larger settlements can be tolerated, as under road embankments or grain bins. Grain bins commonly are clustered on a single concrete pad, so settlement can be kept on an even keel using the same method as was used for the Tower, distributing more load to the high side.

A famous example of a bearing capacity failure from development of excess pore water pressure is the Tranconia grain elevator in Canada, which tipped about 45°. The story does not end there because after the silos were emptied they were righted by underpinning!

Bearing capacity failures also occur under road, railroad, or dam embankments if the rate of loading exceeds the capacity of the soil to drain away excess pore water pressure. In one case a heavily loaded train making a slow first run over a new railroad track slowed down and laid over on its side as the result of a bearing capacity failure.

Excess pore water pressures are predictable from laboratory consolidation and triaxial tests, and are readily measured in the field with wellpoints, more formally referred to as piezometers. The first stage of failure may be creep, but unlike a landslide where stress remains relatively constant, as a building or other structure tips, soil stresses increase under the low side, so a failure can occur with little warning.

## 22.2.3 Analytical Approach to Bearing Capacity

For mathematical convenience bearing capacity is considered a two-dimensional problem with a unit dimension normal to the page. The analyses therefore directly apply to strip footings, and adjustments are required for them to be applied to square, rectangular, or round footings.

The general procedure is to define an *ultimate bearing capacity*, which is the anticipated failure load, and a *safe bearing capacity*, which is the ultimate capacity divided by a factor of safety. The usual factor of safety based on load is 3 to 5, but as will be shown, the corresponding factor of safety in relation to the soil strength is considerably lower.

The *net bearing capacity* is the bearing pressure minus that of soil that has been removed to make the foundation excavation. The net bearing capacity equals the increase in pressure on the soil at the foundation depth. Bearing capacity is determined at the base of a foundation and therefore includes the weight of the foundation element itself. However, because the unit weight of concrete is only moderately higher than that of the soil removed to make room for the concrete, the small amount of increase in load usually is ignored.

Many bearing capacity theories have been proposed, and reflect the backgrounds and biases of the theorists.

### 22.2.4 Rankine's Analysis

Rankine viewed bearing capacity in terms of active and passive pressures on two sides of a retaining wall. He then left out the wall (Fig. 22.1). The foundation load induces active pressure on one side of the imaginary wall, which in turn induces passive pressure on the other. The active wedge moves downward and sideways, and the passive wedge moves upward and sideways, which is consistent with observations that the ground surface bulges up adjacent to a bearing capacity failure. The development is as follows, and it does not matter whether, as shown in the figure, the failure is one-sided or is symmetrical.

Passive side: $\sigma_h = \gamma D K_p$

Active side: $q_o = \sigma_h K_p$

Substituting for $\sigma_h$,

$$q_o = \gamma D K_p^2$$

where $\sigma_h$ is the horizontal stress, $\gamma$ is the unit weight of the surcharge soil outside of the foundation, $D$ is the depth of the surcharge, $q_o$ is the bearing capacity pressure at failure, and $K_p$ is the passive coefficient. Substituting the value

**Figure 22.1**

Rankine's bearing capacity theory uses a simplified linear failure geometry and ignores vertical shearing resistance along vertical surfaces.

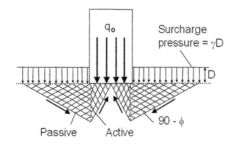

for $K_p$ gives

$$q_o = \gamma D K_p^2 \tag{22.1}$$

$$q_o = \gamma D \left[\frac{1 + \sin\phi}{1 + \sin\phi}\right]^2 \tag{22.1a}$$

This relatively simple equation shows the influence of foundation depth, which is represented by $D$, but does not include soil cohesion. However, if $D = 0$ and there is no cohesion, as in a dry sand, a bearing pressure will sink until $D$ is large enough for the load to be supported.

### Example 22.1
A 100 lb (0.445 kN) person wearing high heels is attempting to walk on a dry sand beach, so you try to explain the situation. $\phi = 25°$ and $\gamma = 90$ lb/ft$^3$ (14.1 kN/m$^3$). The area of each heel is 0.25 in.$^2$ (160 mm$^2$). How far must a heel sink into the sand to reach equilibrium?

*Answer:* For this friction angle $K_p^2 = 6.1$. The bearing pressure is $q_o = 100$ lb $\div$ 0.25 in.$^2 =$ 400 lb/in.$^2$, or 0.445 kN $\div$ 160 mm$^2 = 0.00278$ kN/mm$^2$. Then

$$400 \ \frac{\cancel{lb}}{in.^2} = 90 \ \frac{\cancel{lb}}{\cancel{ft^3}} D \times 6.1 \times \frac{1 \ \cancel{ft^3}}{1728 \ in.^3}$$

$D = 1260$ in. $= 105$ ft, or

$$0.00278 \ \frac{\cancel{kN}}{mm.^2} = 14.1 \ \frac{\cancel{kN}}{\cancel{m^3}} D \times 6.1 \times \frac{1 \ \cancel{m^3}}{(1000 \ mm)^3}$$

$D = 32$ m. And that is why one should not attempt to wear high heels at the beach.

## 22.2.5 Critique of Rankine's Theory

Rankine assumed straight-line boundaries for the shearing soil, which is on the unsafe side if a curved shear surface is more critical. The model also suggests vertical shearing stress along the principal plane separating active and passive wedges, and by definition there can be no shearing stress on a principle plane. A wedge of soil is pushed down but a larger wedge is pushed up, which does not balance out. Rankine's analysis is not reliable for design, but did succeed in pointing out zones of active and passive pressures that became a guide for subsequent analyses.

## 22.2.6 Bell's Equation

Bell added a cohesion term to Rankine's analysis, it can be visualized as two companion triaxial (actually plane strain) test specimens, one oriented vertically and the other laid on its side, as shown in Fig. 22.2. The horizontally oriented specimen on the right has a confining stress $\sigma_3$ equal to the surcharge pressure $\gamma D$,

**Figure 22.2**

Bell's model is similar to Rankine's but includes cohesion. While Bell's equation is useful as a quick check, the angular slip surfaces may give results that are on the unsafe side.

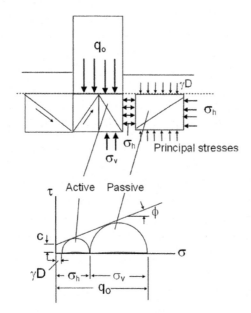

where $\gamma$ is the soil unit weight and $D$ is the depth of the foundation. At failure this value equals $\sigma_1$ for the left-hand specimen. Mohr circles representing this concept are shown at the bottom of the figure. Based on this relationship, in 1915 Bell presented the following equation:

$$q_o = \gamma D \tan^2(45 + \phi/2) + 2c \tan(45 + \phi/2)[\tan^2(45 + \phi/2) + 1] \tag{22.2}$$

where $c$ is the soil cohesion.

A test for any equation is to consider end conditions. If $\phi = 0$, Bell's equation becomes

$$q_o = \gamma D + 4c \tag{22.2a}$$

This relationship is shown by Mohr circles in Fig. 22.3 and also is known as the Casagrande-Fadum illustration. If in addition $D = 0$,

$$q_o = 4c \tag{22.2b}$$

These relationships apply to saturated clay soils with rapid loading so that there is no time for drainage, for example under a moving wheel load.

If the unconfined compressive strength is regarded as an undrained test, as shown by the left-hand Mohr circle in Fig. 22.3, $q = 2c$, and

$$q_o = 2q_u \tag{22.2c}$$

where $q_u$ is the unconfined compressive strength. This equation indicates that if the unconfined compressive strength is used as an estimate for bearing capacity, as sometimes is the case for small jobs, the factor of safety is 2.

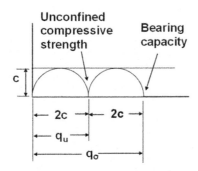

**Figure 22.3**

An adaptation of Bell's model with $\phi = 0$ equates bearing capacity $q_o$ to two times the unconfined compressive strength $q_u$. A surcharge shifts the Mohr circles to the right and increases $q_o$ by the amount of the surcharge.

Bell's model is subject to the same objections as Rankine's with regard to the linear slip surface, and ignores vertical shear and the unit weight of soil below the foundation level. Results can provide a simple check on other more sophisticated analyses.

## 22.2.7 Critical Edge Pressure and Progressive Failure

Theoretically, under the edge of a foundation as the shearing area decreases to zero, vertical shearing stress becomes infinite—except of course the soil shears first. O. K. Fröhlich, a professor of civil engineering at the Technical College in Vienna and a colleague of Terzaghi, applied elastic theory to map zones of shear failure in soil under the edges of a foundation. This development defines a plastic zone that expands as the foundation pressure increases, as shown in Fig. 22.4. As the plastic zones continue to expand under increasing pressure they coalesce in the middle, in effect describing a progressive failure.

If the friction angle is zero, Fröhlich's solution becomes

$$q_o = 3.14c \tag{22.3}$$

where $c$ is the soil cohesion.

## 22.2.8 Positive Gains from Perimeter Shear

Housel (1956), at the University of Michigan, performed plate bearing tests with different diameters of plates, and discovered that the smaller the plate, the higher the bearing pressure at failure. This conflicts with the more widely accepted Terzaghi theory that is discussed below, which indicates the opposite. Housel suggested that there is an additional resistance to loading as a result of shearing resistance at the edges. However, plate load test areas are small compared with foundation areas, and therefore magnify the perimeter effect. Housel's

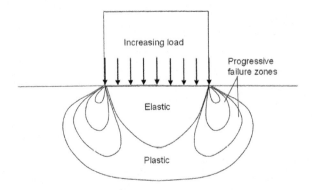

**Figure 22.4**

As shearing stress is infinite at the edges of a foundation, Fröhlich applied elastic theory and theorized progressive development of a plastic zone as a load approaches that will cause a bearing capacity failure. This sketch is for a cohesive soil. (Modified after Jumikis, 1969).

observation nevertheless provides some useful insight into interpretation of plate load test results that without correction will be on the unsafe side for a foundation bearing capacity.

Housel proposed an empirical relationship based on plate bearing test results:

$$W = Pm + An \tag{22.4}$$

where $W$ is the total foundation load, $P$ is the perimeter, $A$ is the bearing area, and $m$ and $n$ are constants. Dividing by $A$ gives

$$q_0 = \frac{P}{A}m + n \tag{22.5}$$

where $q$ is the bearing capacity. For a square or circular bearing area $P/A = 4/B$ where $B$ is the plate width. Then

$$q_0 = \frac{4m}{B} + n = \text{perimeter effect} + \text{area pressure} \tag{22.6}$$

In order to separately evaluate perimeter and area effects, tests are required using two or more different plate sizes. If a bearing capacity is obtained by means other than a plate load test, no correction is made because the influence of perimeter shear diminishes with increasing size of a bearing area. Perimeter shear was observed in model experiments by Vesic (1975), and became the basis for a bearing capacity theory based on vertical punching proposed by Janbu.

### 22.2.9 Loss of Strength from Perimeter Shear

Terzaghi also recognized that perimeter shear must occur, and became concerned about the reduction in strength that occurs durng shearing, particularly in a sensitive soil. The amount of strength lost is difficult to predict as it varies depending on the soil sensitivity and extent of shearing, and Terzaghi suggested an arbitrary one-third reduction in $c$ and in $\tan \phi$ for foundations on sensitive clay. He also applied the same reduction to loose sands because perimeter shear can lead to excessive settlement.

## 22.2.10  When a Plate Bearing Test Should Not Be Relied upon for Design

A plate bearing test essentially is a model footing, but the model is distorted because the soil properties are not appropriately scaled. (This is the basis for conducting bearing tests in a centrifuge.) There are two reasons why plate tests can be unreliable for design: (1) a small plate emphasizes the role of perimeter shear, discussed above, but more importantly, (2) the depth of influence is limited by the small diameter of the plate. The latter effect is shown in Fig. 22.5.

A typical example where a plate load test would give misleading results derives from the tendency for a clay soil to dry out and gain apparent cohesive strength at the ground surface. This objection can be overcome by testing in a pit, but costs go up, and if the test depth is deeper than the footing depth the influence of surcharge pressure becomes exaggerated.

The same objection to the plate load test applies to settlement predictions, as the depth of the pressure bulb is substantially reduced by the small plate width

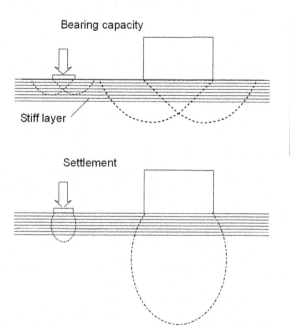

Bearing capacity

Stiff layer

Settlement

**Figure 22.5**

A plate load test does not accurately model a much larger foundation, particularly if there are changes in the soil with depth.

compared with that of a foundation. The most common use of plate bearing tests is to measure the modulus of pavement elements.

## 22.3 TERZAGHI BEARING CAPACITY THEORY

### 22.3.1 Prandtl-Terzaghi Theory

Again demonstrating a benefit from cross-fertilization of ideas, Terzaghi's bearing capacity theory derives from a mathematical analysis for penetration of metal by a punch. Prandtl was a professor at the University of Göttingen in Germany who is most famous for his contributions to flow patterns in aerodynamics. He adapted a similar flow pattern to plastic deformation by defining a curved transition zone to link Rankine's active and passive zones, and noted that it is possible to do so without corrupting principal stresses with shear stresses if the shape is that of a logarithmic spiral. The contours depend on the friction angle (Fig. 22.6(a)): the higher the friction angle, the wider the spiral and the higher the bearing capacity.

Terzaghi modified Prandtl's theory by including a surcharge load (Fig. 22.6(b)). The Prandtl geometry assumes a smooth base for the punch so that the bottom is a principle plane, and Teraaghi hypothesized that because soil tends to expand laterally in accordance with Poisson's ratio, confining friction should be mobilized along the base. This in turn should change the shape of the shear curves, as shown in Fig. 22.6(b). By acting to reduce lateral expansion, the influence of base roughness should tend to increase the bearing capacity even though the shear areas are larger.

The mathematical solution is complex, and Terzaghi took a critical step to make the equations usable by engineers. The simplified equation defines "bearing capacity factors" that depend on the angle of internal friction and are compiled into tables. Terzaghi (1943) emphasized that the solution is not exact, and that it requires a generous factor of safety. He also suggested the arbitrary one-third reduction in bearing capacity for sensitive soils.

Terzaghi's theory has been modified by later workers but still is the most commonly used method for predicting shallow foundation bearing capacity. The equation is as follows:

$$q_0 = \frac{\gamma B}{2} N_\gamma + c N_c + \gamma D N_q \tag{22.7}$$

($q_0$ = width factor + cohesion factor + surcharge or depth factor)

where $q_0$ is the bearing capacity in force per unit area, $\gamma$ is the soil unit weight, $B$ is the foundation width, $c$ is soil cohesion, $D$ is depth of the foundation, and the

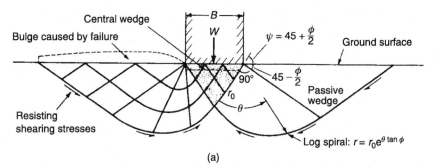

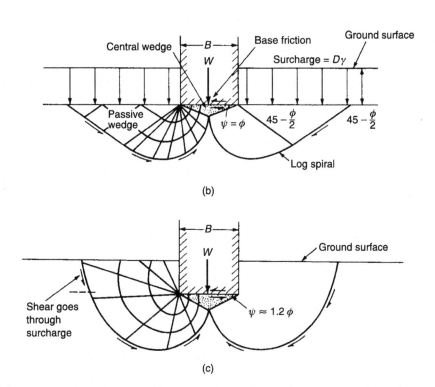

**Figure 22.6**

Bearing capacity failure geometries: (a) Prandtl-Terzaghi smooth base; (b) Terzaghi rough base; (c) Meyerhoff.

$N$ values are bearing capacity factors. An important feature of the equation is that it shows respective contributions from foundation width, cohesion, and surcharge, as indicated in the line under the equation.

Terzaghi bearing capacity factors are given in Table 22.2 and depend only on the soil friction angle and whether the base is smooth or rough. The equations for $N_c$ and $N_q$ are analytical and exact and may be programmed, but the equations for $N_\gamma$ were determined by curve fitting and comparisons to values determined by testing. While commercial programs are available for calculating bearing capacity,

**Table 22.2**

Bearing capacity factors for use in eq. (22.7)

| $\phi$ | Terzaghi rough base | | | Smooth base | | |
|---|---|---|---|---|---|---|
| | $N_\gamma^b$ | $N_c^c$ | $N_q^d$ | $N_\gamma^e$ | $N_c^c$ | $N_q^f$ |
| 0 | 0.0 | 5.7 | 1.0 | 0.0 | 5.1 | 1.0 |
| 5 | 0.5 | 7.3 | 1.6 | 0.5 | 6.5 | 1.6 |
| 10 | 1.2 | 9.6 | 2.7 | 1.2 | 8.3 | 2.5 |
| 15 | 2.5 | 13 | 4.4 | 2.6 | 11 | 3.9 |
| 20 | 5.0 | 18 | 7.4 | 5.4 | 15 | 6.4 |
| 25 | 9.7 | 25 | 13 | 11 | 21 | 11 |
| 30 | 20 | 37 | 22 | 22 | 30 | 18 |
| 35 | 42 | 58 | 41 | 48 | 46 | 33 |
| 40 | 100 | 96 | 81 | 110 | 75 | 64 |
| 45 | 300 | 170 | 170 | 270 | 130 | 130 |
| Local shear: | | | | | | |
| 0 | 0.0 | 5.7 | 1.0 | 0.0 | 5.1 | 1.0 |
| 5 | 0.3 | 6.7 | 1.4 | 0.3 | 6.0 | 1.3 |
| 10 | 0.7 | 8.0 | 1.9 | 0.7 | 7.0 | 1.8 |
| 15 | 1.2 | 9.6 | 2.7 | 1.2 | 8.3 | 2.5 |
| 20 | 2.0 | 12 | 3.8 | 2.1 | 10 | 3.4 |
| 25 | 3.2 | 14 | 5.3 | 3.4 | 12 | 4.6 |
| 30 | 5.0 | 18 | 7.4 | 5.4 | 15 | 6.4 |

Note: $^a q_0 = \dfrac{\gamma B}{2} N_\gamma + c N_c + D N_q$.

$^b$After Bowles (1968). Approximate equation fit by the authors: $N_\gamma = 1.1\,(N_q - 1)$ $\tan 1.3\phi$.

$^c N_c = (N_q - 1)\cot\phi$ (exactly).

$^d N_q = \dfrac{e^{(1.5\pi - \phi)\tan\phi}}{2\cos^2(45 + \phi/2)}$ (exactly).

$^e$Equation fit by Vesic (1975): $N_\gamma = 2(N_q + 1)\tan\phi$.

$^f N_q = e^{\pi\tan\phi}\tan^2(45 + \phi/2)$ (exactly).

it is a simple matter to make a spreadsheet based on these values or their approximate formulas.

Table 22.2 also gives Terzaghi bearing capacity factors that are arbitrarily reduced for local shear, to be used for sensitive clays.

**Example 22.2**

A clay soil has $\phi = 10°$, $c = 19.2$ kPa (400 lb/ft$^2$), and $\gamma = 15.7$ kN/m$^3$ (100 lb/ft$^3$). (a) Assume a rough footing surface and evaluate the relative influences of footing width and depth on supporting capacity of a footing 1.22 m (4 ft) wide and 1.52 m (5 ft) deep. (b) What would be the effect of removing the surcharge on one side of the footing? What would be the influence from using a smooth footing?

*Answer:* (a) From Table 22.2 for a rough footing $N_\gamma = 1.2$, $N_c = 9.6$, and $N_q = 2.7$. Therefore

$$q_o = \frac{\gamma B}{2} N_\gamma + cN_c + \gamma DN_q \tag{22.7}$$

$$q_o = \frac{1}{2}(15.7\,\text{kN/m}^3)(1.22\,\text{m}) \times 1.2 + (19.2\,\text{kN/m}^2) \times 9.6 + (15.7\,\text{kN/m}^3)(1.52) \times 2.7$$
$$= 11.5 + 184 + 64.4 = 260\,\text{kN/m}^2 \text{ or}$$
$$q_u = \frac{1}{2}(100\,\text{lb/ft}^3)(4\,\text{ft}) \times 1.2 + (400\,\text{lb/ft}^2) \times 9.6 + (100\,\text{lb/ft}^3)(5) \times 2.7$$
$$= 240 + 3840 + 1350 = \underline{5430\,\text{lb/ft}^2}$$

It will be noted that these are foundation pressures per unit length of the foundation, and must be multiplied by the width to obtain a supporting capacity in lb/ft or kN/m. The relative proportions of the bearing pressure attributed to foundation width, soil cohesion, and foundation depth are 4.4%, 71%, and 25%, respectively.

(b) Removing the surcharge simply means dropping the third term in the equation, which will reduce the ultimate bearing capacity by 25%.

(c) Substituting bearing capacity factors for a smooth footing gives

$$q_o = \frac{1}{2}(100)(4) \times 1.2 + 400 \times 8.3 + (100)(5) \times 2.5 = \underline{4810\,\text{lb/ft}^2}(230\,\text{kPa})$$

a reduction of 12%.

## 22.3.2 Pore Water Pressure

Calculations normally are on an effective stress basis—that is, the friction angle and cohesion used in the formulas are adjusted to compensate for any development of excess pore water pressure. This is done on the Mohr-Coulomb shear diagram.

### Example 22.3
The soil in the above example is slightly overconsolidated. Assume that prior to shearing $K_o = 0.4$ and the Bishop pore pressure parameter is 0.5 (section 18.9.8). Recalculate the bearing capacity.

*Answer:* The simplified Bishop pore pressure equation states

$$\Delta u = A(\Delta\sigma_1 - \Delta\sigma_3) + \Delta\sigma_3 \tag{18.21}$$

Making appropriate substitutions,

$$\Delta u = (0.5)(\Delta\sigma_1 - 0.4\Delta\sigma_1) + 0.4\Delta\sigma_1 = 0.7\Delta\sigma_1$$

That is, unless loading is very slow or drainage is enhanced, 70% of the bearing pressure is predicted to go to pore water pressure, which should substantially reduce the bearing capacity. Terzaghi's effective stress equation states

$$\tau_f' = c' + (\sigma_n - u)\tan\phi'$$
$$= c' + 0.3(\sigma_n)\tan\phi' \tag{18.7}$$

It can be argued that the cohesion term should be relatively unaffected, in which case the friction angle used in the bearing capacity equation should be reduced by an amount that gives the reduced tangent, or

$$\phi = \tan^{-1}(0.3 \tan 10°) = 3°$$

This value is used to determine new bearing capacity factors, which will be quite low. Drainage can be aided if the foundation is placed on a layer of sand that is allowed to freely drain.

### 23.3.3 Allowable Bearing Capacity

Equation (22.7) gives a bearing capacity at failure, and must be divided by a factor of safety to obtain an allowable bearing capacity that is designated by $q_a$. The factor based on load-carrying capacity normally is 3 to 5. This may appear to be extremely conservative but, as will be shown, it is less conservative than it looks.

### 22.3.4 Meyerhof's Modification

The Prandtl-Terzaghi geometry assumes that the ground surface adjacent to a loaded foundation is a principal plane so that the passive wedge rises at an angle is $45 - \phi/2$ with horizontal. Observations of actual failures in the field in model tests indicate that the surface is not flat but curves upward as shown in Fig. 22.6(c). Meyerhof (1951), at the Technical University of Nova Scotia, recalculated the bearing capacity factors assuming that the log spiral continues to the ground surface, and also varied the basal triangle to obtain the most critical shear angle. The base angle thus determined is about $1.2\phi$ compared with $\phi$ for Terzaghi rough base. While there is only a modest effect on bearing capacity factors, Meyerhoff's analysis probably depicts a more realistic model for what actually happens.

### 22.3.5 Eccentric Loading

Although foundations ideally should be symmetrically loaded, there are situations where eccentric loading is unavoidable, such as for the foundation for a retaining wall. Asymmetrical loading also can occur where a foundation is next to a property line, or is a combined footing supporting two loads such as a column load and a wall load.

With reference to Fig. 22.7, eccentricity $e$ is defined as the distance from the centroid of a bearing area to the central axis of load. As can be seen in the figure, when that distance exceeds $B/6$, where $B$ is the footing width, the contact area is reduced, and if $e = 0.5B$, the load is centered over the edge of the footing so that the contact area in effect becomes a line. Using that as a guide, Hansen suggested that the footing width be reduced as follows for use in bearing capacity calculations:

$$B' = B - 2e \tag{22.8}$$

### Example 22.4

An MSE wall is 0.6 times as thick as it is high, and the friction angle of soil behind the wall is 25°. Assume that the unit weight of the soil in the wall and behind the wall are the same. What is the eccentricity of the foundation load and the effective width?

*Answer:* From the friction angle $K_a = 0.4$. With reference to Fig. 22.7, summing moments around $O$ gives

$$0.6\gamma H^2 \ e = (0.4\gamma H^2/2)(H/3)$$
$$e = 0.11H, \text{ where } B = 0.6H \text{ and } e = 0.185B$$
$$B' = B - 2(0.185B) = 0.44B$$

This will reduce the width factor in the bearing capacity equation.

### Example 22.5

The foundation soil in the above example is silt with $\phi = 25°$ and when saturated, $c = 0$. $\gamma = 15.7 \, \text{kN/m}^3$ ($100 \, \text{lb/ft}^3$). The height of the wall is 20 ft (6.1 m). What is the factor of safety for foundation bearing capacity?

*Answer:* The average bearing pressure is $q = 6.1(15.7) = 95.8 \, \text{kN/m}^2$ ($2000 \, \text{lb/ft}^2$).

$$q_o = \frac{\gamma B}{2} N_\gamma + cN_c + \gamma DN_q$$
$$= \frac{1}{2}(15.7)(0.6)(6.1) \times 9.7 + 0 + 0$$
$$= 279 \, \text{kN/m}^2$$

The factor of safety without considering the eccentric loading therefore is $FS = 279/95.8 = 2.9$, which is marginally acceptable. However, if the eccentric loading is taken into consideration, $B' = 0.44B$ and $q_o = 0.44(279) = 123 \, \text{kN/m}^2$. This reduces the factor of safety to $123/95.8 = 1.3$, which is not acceptable and could invite a progressive failure.

## 22.3.6  Footing on a Slope

It is easy to assume—incorrectly—that a bearing capacity calculated according to the above considerations also applies to footings located on a slope. However, as

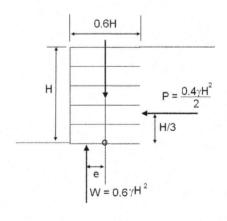

**Figure 22.7**

Example of eccentric foundation loading by an MSE wall.

can be seen in Fig. 22.8, an important effect of a slope is to take away surcharge on the downhill side. There also is a reduced length of the slip surface, which affects the cohesion term. Modified bearing capacity factors were calculated by Bowles (1988) assuming that the slope surface is a principal plane. As a guide, the surcharge term can be omitted and a more generous factor of safety adopted. Another alternative is to analyze the system as a slope stability problem using a method of slices, discussed in the following section.

### 22.3.7  Progressive Failure and Upper- and Lower-Bound Solutions

All of the above solutions approximate "slip-line solutions" based on limit equilibrium, or the maximum load that can be sustained prior to failure. All of the stated geometries are approximations because they do not include the influence of self-weight of the soil on normal stress vectors and therefore the shape of the failure surface. This can be approached by the method of Sokolovski, which shows distorted stress fields. The lowest bearing capacity that can be obtained by adjusting the slip lines is an "upper-bound solution." This means that a pressure in excess of that amount is likely to cause failure.

These solutions also assume simultaneous development of maximum shearing resistance all along the slip surfaces, but failure probably is progressive as shown in Fig. 22.4. This problem is mitigated by applying a generous factor of safety that hopefully will limit shearing to a very small portion of the potential slip surface.

Another approach is to consider energy relationships such that a system *must* fail. This is called a "lower-bound solution." The energy generated equals the weight of a structure times the amount of settlement, and the energy used up depends on stress-stain relationships in the soils. The ultimate achievement is if

**Figure 22.8**

(a) Too high an eccentricity *e* can initiate liftoff and reduce the bearing area. (b) Footing on a slope has a reduced surcharge contribution to bearing capacity. Its effect on slope stability can be analyzed by a method of slices.

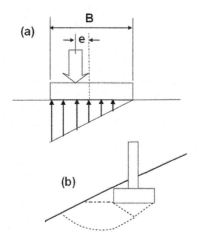

the upper- and lower-bound solutions agree with one another. As a practical matter, inconsistencies and anisotropy of the soil, particularly sedimentary layering, can overshadow the effects of these adjustments.

## 22.3.8 Rectangular, Square, and Circular Footings

Column loads normally are supported on square footings, but both analytical and numerical (finite element) solutions struggle with three dimensions. Qualitatively an equidimensional footing will increase the role of cohesion because the shear surface is increased, and reduce the influence from friction angle for the same reason, by diluting the normal stress on the shear surface. This is shown in Fig. 22.9.

The two-dimensional Terzaghi equation has been extended to three dimensions empirically on the basis of load tests. As anticipated, $N_c$ is increased and $N_\gamma$ is decreased. Tests by De Beer in Belgium indicate that the surcharge factor also is increased, apparently through better lateral confinement. In the case of small footings the Prandtl smooth-base geometry appears most appropriate. Bearing capacity factors suggested by Vesic (1975) are presented in Table 22.3.

Factors for rectangular foundations can be obtained by interpolation of the base width-to-length ratio, $B/L$: for infinitely long footings, $B/L = 0$, with factors given in Table 22.2, and for circular or square footings, $B/L = 1$ (Table 22.3).

**Example 22.6**
Estimate bearing capacity factors for a footing that is twice as long as it is wide, with $\phi = 20°$.

*Answer:* From Tables 22.2 and 22.3 for a smooth base,

|              | B/L | $N_\gamma$ | $N_c$ | $N_q$ |
|--------------|-----|-----------|-------|-------|
| Long         | 0   | 5.4       | 15    | 6.4   |
| Square       | 1   | 3.2       | 21    | 8.7   |
| Interpolation| 0.5 | 4.3       | 18    | 7.6   |

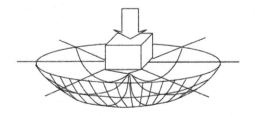

**Figure 22.9**
Square or circular foundations create a larger shear area that exaggerates the influence of cohesion and dilutes the effect of friction angle. These effects are compensated by adjusting the bearing capacity factors.

| $\phi$ | $N_{\gamma_s}^b$ | $N_{c_s}^c$ | $N_{q_s}^d$ |
|---|---|---|---|
| 0 | 0.00 | 6.2 | 1.0 |
| 5 | 0.27 | 8.0 | 1.7 |
| 10 | 0.73 | 11 | 2.9 |
| 15 | 1.6 | 15 | 5.0 |
| 20 | 3.2 | 21 | 8.7 |
| 25 | 6.5 | 31 | 16 |
| 30 | 13 | 49 | 29 |
| 35 | 29 | 79 | 57 |
| 40 | 66 | 140 | 120 |
| 45 | 160 | 270 | 270 |

**Table 22.3**

Bearing capacity factors for circular or square smooth-base foundations, modified after Vesic, (1975)

Note: $^a$ $q_0 = \dfrac{B}{2}N_\gamma + cN_c + \gamma D N_q$.

$^b$ $N_{\gamma_s} = 0.6N_\gamma$.

$^c$ $N_{c_s} = N_c \left(1 + \dfrac{N_q}{N_c}\right)$.

$^d$ $N_{q_s} = N_q(1 + \tan\phi)$.

## 22.4 ANALYSIS USING A METHOD OF SLICES

### 22.4.1 Getting the Big Picture

It is not generally appreciated that bearing capacity and slope stability are two statements of the same problem—bearing capacity equals slope stability with a surcharge and no slope. The same analytical methods therefore can be used for bearing capacity as for slope stability. Influences of soil layering or building on a slope are easily included in the analysis. There also is a justification for the log spiral because self-weight of the soil plays a subordinate role to foundation pressure. The Terzaghi slip geometry therefore can be selected for analysis.

**Example 22.7**

In Example 22.2, (a) what is the design load with a design factor of safety of 3? (b) Determine the factor of safety based on soil shear strength using the simple method of slices.

*Answer:* (a) $q_a = 5430 \div 3 = 1810\,\text{lb/ft}^2$ or
$= 260 \div 3 = 87\,\text{kpa}$

(b) The diagram is shown in Fig. 22.10. Results for individual slices with $q_a$ as above are as follows. Note that the acting forces are negative for slices 4 and 5 because of the

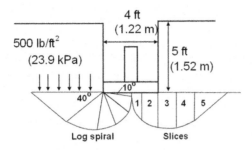

**Figure 22.10**
(Left) Bearing capacity by the Terzaghi method for a rough base, Example 22.2; (right) the same problem solved by the ordinary method of slices, Example 22.7.

reverse slope of the slip surface.

| Slice No. | 1 | 2 | 3 | 4 | 5 | Total |
|---|---|---|---|---|---|---|
| $S_a$, lb | 2910 | 864 | 73 | −377 | −913 | 2557 |
| $S_r$, lb | 1191 | 863 | 769 | 745 | 1456 | 5024 |
| FS | 0.41 | 1.0 | 10.5 | − | − | 2.0 |

## 22.4.2 Will the Real Factor of Safety Please Stand Up?

In the above example there is a substantial disagreement in regard to the factors of safety: 3.0 from a bearing capacity analysis and 2.0 from the ordinary method of slices, which, as previously discussed, tends to overestimate the factor of safety. Who is right? This is not just an academic argument as there are substantial consequences. A resolution is needed.

Both answers are right, because it depends on whether one is up looking down, or down looking up, which is always a source of disagreement. Both calculations assume the same Terzaghi rough base geometry, so how can both be correct? We will pause a moment for quiet contemplation and prayer if it helps.

The explanation is in the definition of a factor of safety. In the bearing capacity equations it represents an *allowable foundation load*, but the foundation load is only part of the acting forces that include the weight of the soil. A slope stability factor of safety is based on the total weight in relation to the shearing strength of the soil. This in effect adds a constant amount to both the numerator and denominator, as the bearing capacity factor of safety is $3 \div 1 = 3$ and the shear strength factor is $(3 + 1) \div (1 + 1) = 2$. Bearing capacity factors of safety are easy to calculate and impress clients; engineers are more concerned with a possibility of failure.

### 22.4.3 Converting Factors of Safety

A bearing capacity factor of safety can be converted to an equivalent factor of safety against shear failure by substituting reduced values for cohesion, $c$, and for tan $\phi$ into the bearing capacity equation and calculating the corresponding $q_o$. The correct answer is obtained by trial and error, dividing $c$ and tan $\phi$ by a trial shear strength factor of safety, determining the allowable, and comparing it with that obtained from the bearing capacity equation. The trial value is correct when the allowables are the same.

**Example 22.8**
In Example 22.7 the load factor of safety is 3 and $q_a = 1810$ lb/ft$^2$ (87 kPa). What is the corresponding shear strength factor of safety?

*Answer:* An answer of 2.0 already has been obtained by the method of slices. This may be tested by dividing $c$ and tan $\phi$ by 2, which gives allowable strength parameters of $\phi_a = 5.04°$ and $c_a = 200$ lb/ft$^2$. Substituting appropriate values in eq. (22.5) gives

$$q_a = 0.5(100)(4) \times 0.5 + 200 \times 7.3(100)(5) \times 1.6$$
$$= 200 + 1460 + 80 = 1740 \text{ lb/ft}^2 \text{ (83 kPa)}$$

This is fairly close to the value obtained with a load factor of safety of 3. Let us therefore try 1.9, noting that the major influence will be in the cohesion term:

$$q_a = 0.5(100)(4) \times 0.5 + 210 \times 7.3 + (100)(5) \times 1.6$$
$$= 200 + 1530 + 80 = 1810 \text{ lb/ft}^2 \text{ (83 kPa)}$$

which agrees with the allowable foundation pressure from the bearing capacity analysis and indicates an equivalency of the two factors of safety, respectively 3 and 1.9. The shear strength factor of safety of 1.9 therefore is the equivalent of a load factor of safety of 3.

### 22.4.4 Importance of Surcharge

The depth of a foundation directly affects the amount and weight of the surcharge, whose function is to keep the soil that is underneath a foundation from shearing and squeezing out. Excavation that inadvertently removes surcharge can have a dramatic and devastating effect, Fig. 22.11.

## 22.5 BEARING CAPACITIES FOR SPECIAL CASES

### 22.5.1 Saturated, Undrained Clay

A special case of bearing capacity that applies only to rapid loading is a "$\phi = 0$" analysis. Because $N_\gamma = 0$ and $N_q = 1$ for both the rough and smooth base geometries, eq. (22.7) reduces to

Rough base: $q = 5.7c + \gamma D$            (22.9a)

Smooth base: $1 = 5.14c + \gamma D$                                       (22.9b)

If $\phi = 0$, the unconfined compressive strength is $q_u = 2c$. Substituting this relationship for $c$ gives

Rough base: $q = 2.85q_u + \gamma D$                              (22.10a)

Smooth base: $q = 2.57q_u + \gamma D$                             (22.10b)

A common procedure for small structures on saturated clay is to assume that the allowable net bearing capacity equals the unconfined compressive strength, which according to the above relationships automatically includes a strength-based net factor of safety between about 2.6 and 2.8. Because the friction angle is zero there is no width factor.

For equidimensional footings on saturated clays with zero drainage, $N_c$ from Table 22.3 is 6.2, or

Equidimensional smooth base: $q_a = 3.1q_u + \gamma D$                    (22.10c)

**Figure 22.11**

Cracks around the window openings testify to settlement of the front wall of this old building when surcharge was removed next to the foundation. The building was temporarily restrained from collapse by cantilevered floor joists, plus some sand hastily piled back against the foundation.

Use of $q_u$ as an allowable net bearing capacity (exclusive of surcharge) for a square or circular footing therefore gives an inferred factor of safety of 3.1. Bearing capacity factors may be interpolated for rectangular footings.

The developed factor of safety will be significantly higher than indicated by eq. (22.10) if partial drainage allows internal friction to become mobilized.

### 22.5.2 Bearing Capacity of Cohesionless Sand

Without the cohesion term, eq. (22.7) reduces to

$$q = \gamma\left[\frac{B}{2}N_\gamma + DN_q\right] \tag{22.11}$$

Substituting the appropriate $N$ values for $\phi = 25°$ gives

$$q = \gamma(3.25B + 16D)$$

For a pneumatic tire $q$ equals the inflation pressure.

### 22.5.3 Other Bearing Capacity Theories

The failure geometries assumed in an assortment of bearing capacity theories are shown in Fig. 22.12, with solutions for a friction angle of zero and no surcharge. The one-sided circular failure suggested by Krey was modified by Wilson, who solved for the most critical location of the circle, and results are close to those obtained with two-sided geometrics of the Prandtl-Terzaghi model. Perhaps most striking is that despite the wide variations in assumed failure geometry, with the exception of the Rankine geometry the results all fall in a range that is less than 20 percent.

### 22.5.4 One-Sided and Two-Sided Models

The Prandtl-Terzaghi-Meyerhoff models are symmetrical, but a foundation loaded to failure inevitably tips one way or the other. This would appear to argue for support from only one side. However, an analogy may be drawn to a board supported by ropes at the ends and loaded in the middle: Both ropes support the load until one breaks.

## 22.6 THE TYPE B PROBLEM, SETTLEMENT

### 22.6.1 Overview

Foundation designs must be checked for bearing capacity as outlined above, but in most cases settlement controls. This will depend on the structure: for example, settlement is not critical for a light pole so long as it does not tilt, whereas settlement

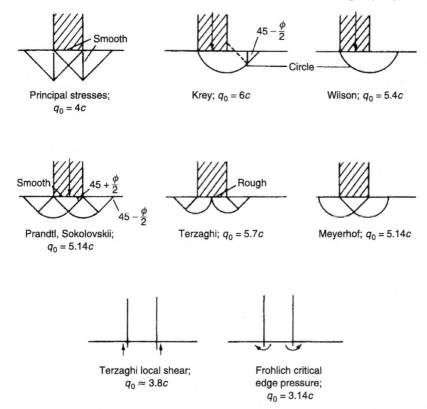

**Figure 22.12**

Assumed failure geometries and solutions from various bearing capacity theories with $\phi=0$ and $D=0$.

should be minimized for a floor supporting sensitive equipment. Differential settlement, where part of a foundation system settles more than another part, can rack or destroy a structure, and is evidenced by uneven floors, cracking and faulting of floors and walls, and doors and windows that stick or drag. Geotechnical engineers frequently are asked to do the forensics on differential settlement problems. The maximum differential settlement occurs when one part of a foundation settles and another does not. Suggestions for allowable maximum and differential settlements for various structures are indicated in Table 16.3.

The larger the area covered by a structure, the more likely it will be to encounter variations in the soils. One way to minimize differential settlement is to tie all foundations together as a continuous mat (or raft) foundation. Such a foundation must be sufficiently reinforced to bridge over soft spots. Another advantage of a mat foundation is that a bearing capacity failure becomes unlikely because of the large width factor. However, a mat generally is considerably more expensive than separated column and/or wall foundations.

Building codes mentioned earlier in this chapter emphasize bearing capacity over settlement.

### 22.6.2  Structures on Clay

Settlement of structures on clay normally is predicted from laboratory consolidation tests by methods discussed in Chapters 16 and 17. Particularly important is the preconsolidation pressure, as foundation pressures that are less than that value will be limited to elastic settlement. The preconsolidation pressure should not be exceeded for foundations on quick clays.

A relatively new procedure involves manipulating the preconsolidation pressure by increasing the lateral stress, which is discussed in Chapter 24.

If foundation pressure exceeds the preconsolidation pressure, a rough estimate of settlement can be made on the basis of an empirical relationship between the compression index and the liquid limit (eq. (16.9a)). A recent trend has been toward the use of in-situ tests to predict settlement, which is discussed in Chapter 26.

The influence of overlapping stress bulbs from adjacent footings should not be overlooked in predicting settlement. This can become critical if a heavy building is founded on shallow footings next to an existing building on shallow footings, and is a reason to select a deep or intermediate foundation for the newer building.

### 22.6.3  Structures on Sand

Sand is very difficult to sample without causing disturbance, so if consolidation tests are performed they use reconstituted samples. This difficulty has led to an increasing use of in-situ tests to predict settlement.

Earlier predictions of settlement on sand were based on Standard Penetration Test (SPT) blow counts ($N$) and width of the footing, and did not consider the distribution of compressive strain with depth. (It will be recalled that clay is divided into arbitrary layers, and the amount of consolidation is predicted for each layer based on an elastic distribution of vertical stress.) The SPT and other in-situ tests will be discussed in relation to prediction of settlement. Details of the test are presented in Chapter 26.

The preferred procedure is that suggested by Schmertmann and discussed in Chapter 16, Section 16.10.

## 22.7  PROPORTIONING FOOTINGS FOR EQUAL SETTLEMENT

### 22.7.1  Overview

A goal is to minimize differential settlement, but column loads usually differ in a single structure, and bearing pressures and the geometry are not the same for column footings and wall footings. Differential settlement is almost assured when part of a structure is supported on shallow foundations and part is on piles. Differential settlement also is unavoidable when parts of a structure on shallow foundations are built at different times, because of the nonlinear time factor controlling the rate of settlement.

A common guide is to limit maximum settlement to one inch (25 mm), which also limits differential settlement to one inch.

## 22.7.2 Proportioning Footing Areas

Differential settlement can be reduced if different foundation bearing areas are proportioned to give approximately the same amount of settlement. The apportionment is based on the dead load plus a percentage of the live and wind loads, depending on whether the live load will be in place over substantial periods of time. For simplicity, settlement often is assumed to be proportional to pressure regardless of the size of the footing—in other words, double the weight means doubling the area of the footing. This is an obvious oversimplification because it omits any depth effect, but normally is adequate so long as loads are not highly variable. If adjacent loads do vary by a large amount, consideration should be given to using a combined footing, discussed in the following paragraphs.

Settlement predictions also assume that footings are free-standing and are not affected by rigidity of the superstructure. This is the *design settlement*, which may or may not be attained by the structure, depending on its flexibility. The designer ordinarily assumes that the full predicted design settlements will occur, then examines the structure to determine if the resulting stresses and differential movements can be tolerated.

## 22.7.3 Settlement of Combined Footings and Rafts

Footings that support more than one column load or column plus wall loads are referred to as *combined footings*. If possible the combined load must act through the centroid of the bearing area, or the footing will tilt. Two unequal column loads may be supported by a footing that is trapezoidal in plan. Combined footings are useful next to property lines so that a footing area extends between adjacent columns instead of extending out symmetrically around each one.

The extreme case of a combined footing is a raft or mat foundation that extends under an entire structure and supports all of the columns and walls. Mats are used on soft soils because bearing pressures are greatly reduced compared to those of individual footings.

Rafts are sometimes incorrectly referred to as "floating foundations," implying that they derive their support from buoyancy, in which case settlement would be zero. A true floating foundation requires deep excavation, so the weight of the structure equals the weight of the soil removed. Partially floating foundations have been used in very soft, compressible soils, as in Mexico City.

Most rafts are reinforced concrete mats with considerable structural rigidity that limits differential settlement to about one-half of the design settlement, so the total allowable settlement often is larger, about 50 mm (2 in.). The raft must have sufficient reinforcement to bridge softer areas. The pressure bulb from a raft extends much deeper because of the larger width, and this also has an averaging effect.

### Example 22.9
Two columns with centerlines 3 ft apart are to be supported by a single footing. One column carries 28 tons and the other one-half of that amount. The allowable bearing capacity is 2 ton/ft². Design a trapezoidal footing.

*Answer:* The first step is to determine the base area: $A = (28 + 14) \div 2 = 21$ ft². This is the area of a square 4.6 ft on a side, which may be used as a guide for selecting an arbitrary length dimension $h$ shown at the right in Fig. 22.13.

Next, the location of the centroid of the load is obtained by summing moments around a reference point exterior to the columns. If the point is selected 1 ft from the heavier column, $m(28 + 14) = 28(1) + 14(4)$, and $m = 2.0$ ft, as shown in the left-hand sketch.

The area of a trapezoid is (Tuma, 1970):

$$A = h(a + b) = 21 \text{ ft}^2$$

Let a trial $h = 6$ ft; then $21 = 6(a + b)$ and $(a + b) = 3.5$ ft. For a rectangular footing $a = b = 1.75$ ft. This configuration is shown in the left half of the right-hand figure.

For a nonrectangular footing, $(a + b) = 3.5$, so a reduction in one means an increase in the other. The centroid of a trapezoid is located a distance $f$ from the longer parallel edge,

### Figure 22.13

Example 22.9: design of a trapezoidal footing.

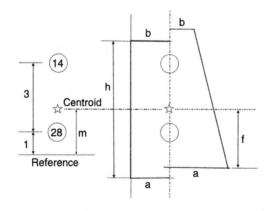

which is defined by

$$f = \frac{h}{3}\frac{a+2b}{a+b} = 2(3.5+b)/3.5$$

A symmetrical footing will extend the same distance from both columns, which is $(6-3)/2 = 1.5$ ft. Adding the distance from one column to the centroid gives $f = 2.5$ ft. Then $b = 0.875$ ft and $a = 2.625$ ft. This configuration is shown at the right in the figure.

### 22.7.4 Minimizing Settlement with Intermediate and Deep Foundations

Deep foundations, piles and piers, minimize settlement by transferring loads downward to stiffer soil layers, as discussed in the next chapter. The supporting capacity of a deep foundation is predicted, and then normally is checked with full-scale load tests of sample production piles or piers, which gives a direct measure of the amount of settlement that can be expected to take place under the load of the structure.

Another approach is to modify the soil to reduce its modulus. This may be done by surcharging to increase the preconsolidation pressure so that it exceeds the anticipated foundation loads, as discussed elsewhere in this book. For granular soils that are remote from other structures, "deep dynamic compaction," or repeatedly dropping a heavy weight, can achieve the same goal. A built-up foundation soil can incorporate lateral restraint by including horizontal tension members, as is done for MSE walls. Another approach that is gaining in popularity is to laterally compact the foundation soil with rammed-in aggregate piers, which in this book is considered an intermediate foundation and is discussed in Chapter 24.

## 22.8 SUMMARY

Shallow foundations are designed to meet two criteria: bearing capacity and settlement. Foundations on or near slopes impose surcharge loads that must not cause a landslide. Another aspect that is discussed in a later chapter is safety from earthquakes. It is axiomatic that the geotechnical investigation that is not made prior to construction often will be conducted later.

## Problems

22.1. Define bearing capacity of a foundation soil. Define allowable bearing value.

22.2. Explain the difference between gross and net bearing capacity.

22.3. What two conditions must be met when determining the bearing capacity of a shallow foundation? Are there other requirements that are discussed in other chapters?

22.4. What is meant by the term "differential settlement"?

22.5. A square footing is to be founded on a stratum of medium clay. What is the allowable bearing capacity according to the Boston Building Code?

22.6. The unconfined compression strength of the clay in the previous problem is $1.5 \, \text{tons/ft}^2$ (144 kPa). Based on this information alone, can you give an approximate bearing capacity? Is this in relation to shear failure, settlement, or both?

22.7. The unconfined compressive strength of a sand is zero. Based on this information alone, can you estimate the bearing capacity?

22.8. A dry dune sand has a friction angle of $25°$ and a unit weight of $95 \, \text{lb/ft}^3$. Use the Rankine formula to estimate the depth of a footprint exerting $200 \, \text{lb/ft}^2$ of pressure.

22.9. The answer in the previous problem is subjected to an experimental proof. Do you expect settlement to be more, less, or exactly as predicted? Explain.

22.10. Explain in simple terms why it is not advisable to wear cowboy boots or high heels when playing beach volleyball.

22.11. Explain two versions of factor of safety with regard to bearing capacity. Which is higher? Which is more scientific?

22.12. Recalculate the bearing capacity in Example 22.1 if the friction angle is $20°$, and compare the allocation to width, cohesion, and surcharge. Give $q_a$ based on a load factor of safety of 3.

22.13. A concrete wall exerting $10,000 \, \text{lb/ft}$ is to be founded at a depth of 4 ft in soil having a cohesion of $400 \, \text{lb/ft}^2$, a friction angle of $16°$, and a unit weight of $100 \, \text{lb/ft}^3$. What width footing will be required for a load factor of safety of 3? Round upward to the nearest foot and calculate the reduction in factor of safety during high water.

22.14. The wall in the previous example is subjected to a horizontal force of 2000 lb at 40% of the wall height. Calculate eccentricity and adjust the footing width as necessary.

22.15. Determine an allowable bearing capacity for a column footing 4 ft square on soil having a unit weight of $100 \, \text{lb/ft}^3$, cohesion of $800 \, \text{lb/ft}^2$, and friction angle of $27°$. The column is interior in a warehouse building so there is no significant surcharge. Why should a bond breaker be incorporated between the outside of the footing and a concrete floor?

22.16. What is the maximum permissible differential settlement for a bell tower 90 ft high with a foundation width of 22 ft? What would be the resulting tilt angle?

22.17. The maximum settlement of a 3 ft × 3 ft footing on sand is not to exceed 1 in. The average STP blow count is 16 blows per foot. Disregard any influence from surcharge and determine the maximum load on this footing.

22.18. The footing of Problem 22.17 is founded 3 ft deep in soil having a unit weight of $110\,lb/ft^3$. What is the maximum load?

22.19. The friction angle in Problem 22.18 is 25°. Determine the factor of safety against a bearing capacity failure and adjust the footing size if necessary for FS = 3.

22.20. When should Schmertmann's $C_2$ correction be used?

22.21. Redesign the trapezoidal footing in Example 22.9 with the larger column load increased 50%.

22.22. The footing of Example 22.9 is on sand having an average STP $N = 24$ blows per foot. Estimate settlement at each end of the footing.

22.23. Calculate the minimum cohesion for a clay soil to support a D8 crawler tractor weighing 227 kN (51,000 lb) with a track width of 560 mm (22 in.) and a total contact area of 3.5 m² (5445 in.²). Assume a soil unit weight of $15.7\,kN/m^3$ ($100\,lb/ft^3$), an undrained friction angle of zero, and zero depth of sinking.

22.24. Construct a computer spreadsheet for bearing capacity identifying contributions from width, cohesion, and surcharge terms.

## References and Further Reading

Bowles, J. E. (1988). *Foundation Analysis and Design*. McGraw-Hill, New York.

Housel, W. R. (1956). "A Generalized Theory of Soil Resistance." *ASTM Special Technical Publication* 26, 13–29.

Holtz, R. D. (1991). "Stress Distribution and Settlement of Shallow Foundations." In H. Y. Fang, ed., *Foundation Engineering Handbook*, 2nd ed., Van Nostrand Reinhold, New York, pp. 166–222.

Jumikis, A. R. (1969). *Theoretical Soil Mechanics*. Van Nostrand Reinhold, New York.

Lambe, T. W., and Whitman, R. V. (1969). *Soil Mechanics*. John Wiley & Sons, New York.

Meyerhof, G. G. (1951). "The Ultimate Bearing Capacity of Foundations." *Géotechnique* 5, 301–332.

Peck, R. B., Hanson, W. B., and Thornburn, T. H. (1974). *Foundation Engineering*, 2nd ed. Wiley, New York.

Robertson, P. K., Campanella, R. G., and Wightman, A. (1983). "SPT-CPT Correlations." *ASCE J. Geotech. Eng.* 109(11), 1449–1459.

Schmertmann, J. H. (1970). "Static Cone to Compute Static Settlement over Sand." *ASCE J. Geotech. Eng.* 96(SM3), 1011–1043.

Schmertmann, J. H., Hartman, J. P., and Brown, P. R. (1978). "Improved Strain Influence Factor Diagrams." *ASCE J. Geotech. Eng. Div.* 104(GT8), 1131–1135.

Terzaghi, K. (1943). *Theoretical Soil Mechanics*. John Wiley & Sons, New York.

Tuma, J. T. (1970). *Engineering Mathematic Handbook*. McGraw-Hill, New York.

Vesic, A. C. (1975) "Bearing Capacity of Shallow Foundations." In H. F. Winterkorn and H. Y. Fang, eds., *Foundation Engineering Handbook, Van Nostrand Reinhold, New York*, Ch. 3.

Wu, T. H. (1976). *Soil Mechanics*. Allyn & Bacon, Boston.

# 23 Deep Foundations

## 23.1 OVERVIEW

### 23.1.1 Pile-Up

Putting structures on posts is not a new idea, but dates back at least 6000 years, to a time when the European climate was warming after centuries of cold, drought, and misery, and lake levels were rising. That was the time of "Ötzi," the Ice Man of the Italian Alps, whose discovery set off a crime investigation that must be the ultimate in cold case files.

Prehistoric builders burned the ends of tree trunks to sharpen them, and probably wielded rocks to drive the posts into soft ground. Today's piles are driven with pile drivers that range from drop hammers to steam or diesel operation.

The most common use of piles is to keep structures from sinking into the ground, but piles also prevent bridge supports from being undercut by scour during periods of rapid current and high water. The famous London Bridge—the one that young children are taught is falling down, thereby fostering an unwarranted disenchantment with civil engineering—was on wooden piles, and it did not fall down; it was pushed down by river ice. It then was repaired and lasted for another 550 years. Somebody should write a song about that.

### 23.1.2 Deep Foundations

Piles are structural columns that extend down into soil. They are either *end-bearing* if they extend all of the way to rock or hard soil, or they are *friction piles* if they are mainly supported by friction along the sides, although friction piles also usually develop some end support. Timber piles are driven top down to take advantage of their taper for increasing friction, and a taper often is incorporated into concrete piles. *Compaction piles* are driven into loose sand to densify it and increase its bearing capacity.

Modern piles are wood, steel, concrete, or composite if they are composed of more than one material. An example of a composite pile is when a wood pile section is used under a groundwater table where it is preserved by reducing conditions, and is connected to a concrete section that extends through aerobic surroundings above the water table.

Whereas piles are driven or can be jacked into the ground, *piers* are large-diameter supports that are placed in pre-bored holes. Caissons are large tubes used for construction below water, as in rivers, and may be pressurized to keep water out. "Caisson foundation" sometimes is applied to bored piers, and piers also are sometimes called shafts or piles, leading to some deep confusion.

Some types of piles and piers are shown in Fig. 23.1.

### 23.1.3 Intermediate Foundations

A growing class of foundations can be referred to as *intermediate foundations*. These are similar to deep foundations because vertical members transfer stress down into soil, but unlike conventional deep foundations they are not rigid columns of wood, concrete, or steel, but are composed of compacted coarse aggregate or stabilized soil. Intermediate foundations include stone columns, Rammed Aggregate Piers, and columns of soil stabilized in situ, and are discussed in the next chapter.

## 23.2 PILE FOUNDATIONS

### 23.2.1 End-Bearing Piles

Settlement of a structure supported on end-bearing piles is limited to compression of hard materials under the ends plus elastic compression of the pile itself.

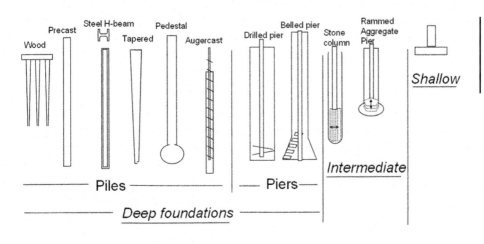

**Figure 23.1**

Deep, intermediate, and shallow foundations.

Although end-bearing piles are long, slender columns, they are not free columns and derive considerable lateral support against buckling from passive resistance of the soil. Failure by buckling is virtually unknown unless the piles extend for long distances through air or water. A different kind of bending of steel pile can occur during driving, if the pile hits a glancing blow on a large boulder deep within the soil. There have been rare occurrences where a driven steel pile has done a 180° bend and emerged at the ground surface.

### 23.2.2  Friction Piles

By transferring a foundation load downward into soil, friction piles also can cause a substantial reduction in settlement, even in a normally consolidated soil, because of the increase in soil density and modulus with depth. In this connection it is helpful to review the *e*-log *p* curve in Fig. 16.9, where because of the logarithmic scale a given increase in pressure results in smaller changes in void ratio with increasing depth.

### 23.2.3  Horizontal Deflections and Batter

Piles may be driven vertically or they may be driven at an angle to better resist horizontal components of a foundation load. Batter piles are sometimes referred to as spur piles. Batter is expressed as a fraction in which the horizontal leg of the slope triangle is the numerator and the vertical leg is the denominator. Thus, a batter of 1:3 means a slope of 1 horizontal to 3 vertical.

Vertical piles also may be designed to resist horizontal loading, particularly when horizontal loading is not constant, as in wharves.

### 23.2.4  Positive and Negative Skin Friction

A pile or pier that moves downward relative to the surrounding soil will derive support from side friction. However, there also can be situations where end bearing results in the soil moving downward relative to the pile. In that case frictional forces are directed downward and add to the weight that must be carried by the pile (Fig. 23.2(b)). Side friction that acts downward is called *negative skin friction*.

Negative skin friction is much more common than might be anticipated because it can be mobilized by even a relatively small lowering of the ground-water table reducing buoyant support for the soil. Friction can be reduced by using smooth piling or coating the sides with a lubricant such as soft asphalt. A failure to recognize a potential for negative skin friction can lead to future difficulties.

Although positive skin friction aids in support of a pile or pier foundation, it does not peak out simultaneously with end support, particularly for long piles, because of compression of the piles. That is, as a pile or pier is loaded, support initially

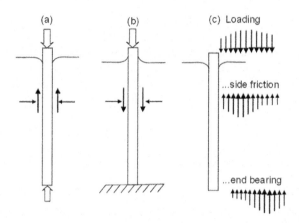

## Figure 23.2

Combined friction and end bearing: (a) ideal development, (b) negative skin friction adds soil weight to the pile; (c) during loading, side friction develops first and decreases, mobilizing end bearing.

derives from skin friction along the upper part, and then as additional load is applied and the pile compresses, more friction is mobilized until finally the load reaches the end of the pile or pier and starts building up end support. This is illustrated in Fig. 23.2(c). It also means that with a factor of safety of 2 little or no load may reach the lower end of the pile or pier.

## 23.2.5 Kinds of Rigid Piles and Piers

The foundation elements shown in Fig. 23.1, and other types that are not shown, have certain advantages and disadvantages, depending on the kind of soil, anticipated load, whether the pile is to be end-bearing or frictional, pile availability, time, and cost.

### Wood Piles

Timber piles were first, and still are commonly used depending on cost and availability. Plain or untreated wood is highly susceptible to decay in aerobic wet conditions, and most wood piles now are chemically treated. Older buildings founded on untreated wood piles may suffer damage because of lowering of the groundwater table. The length and bearing capacity of wood piles also is limited, depending on the kind and size of the trees. In the U.S., southern pine is often used and is limited to a length of about 80 ft (25 m).

Marine borers, both crustaceans and mollusks, are difficult to separate from their appetites, and even wood piles whose surface has been impregnated with creosote may last only 15 to 50 years, depending on climatic factors. The useful life can be extended by periodic maintenance that involves caulking or coating damaged areas.

### Steel H-Beams

Steel H-beam sections are especially useful to penetrate through loose or weathered rock layers to obtain end bearing in solid rock. To minimize end damage and bending they may be fitted with hardened steel shoes that are welded on at the job site. Because of their small cross-sectional area, H-piles can be driven into sand and gravel where it would be difficult to drive displacement piles such as timber or concrete.

Steel H-beams also are used for trestle structures in which the piles serve both as bearing piles and as braced columns.

Side friction on steel H-piles derives about one-half from soil-to-steel friction on the outer surface of the flanges, and the other half from soil friction as soil between the flanges moves with the pile. A driving shoe that is larger than the pile oversizes the hole and reduces side friction, so H-piles intended for side friction should not have such a shoe. However, the lower friction and side area are advantageous for reducing negative skin friction.

When steel piles are exposed to the air or to alternate wetting and drying above a groundwater table they are subject to corrosion, the same as any steel structure under similar conditions. H-beam piles above ground can be protected from corrosion by painting, or they may be encased in concrete for several feet above and below the ground line.

## 23.2.6  Classes of Concrete Piles

Because concrete is readily available, transportation costs often are lower compared with other piles. Concrete piles are in two general classes, cast-in-place and precast. Precast concrete piles are driven, whereas cast-in-place may either use a driven steel shell that is filled with concrete or may involve boring a hole in which to cast the pile. A wide variety of independently developed cast-in-place construction methods is used, leading to some confusion in nomenclature.

Generally, drilled-and-cast piles are now referred to as "drilled shafts" or "drilled piers," but they also are called "drilled caissons," "bored piles," or simply "caissons." These are discussed in the next section.

## 23.2.7  Driven Precast Piles

Precast concrete piles are cast horizontally, and therefore are made square or octagonal instead of circular. They can be tapered uniformly or tapered in segments from tip to butt, but longer lengths normally are tapered only for about 4 or 5 ft (1.2–1.5 m) from the lower end for easier handling. Piles are reinforced with longitudinal bars and transverse steel that may be in the form of separate loops or continuous spirals. The transverse steel is spaced closer together for 3 or 4 ft (1–1.2 m) at each end of the pile because driving stresses are highest near the ends.

Because handling and driving stresses are severe, steel reinforcement often is either pre-tensioned or post-tensioned. Sufficient flexural strength is obtained with only about one-fourth to one-third the usual level of prestressing for more orthodox structural members. Prestressing also can help to prevent piles from breaking in tension during driving, as compression waves going down the pile pass and add to wave displacements echoing up from the bottom.

Whether prestressed or not, precast piles must be carefully handled and lifted by slings under the one-third points to reduce bending stresses to a minimum. Uniformly tapered precast piles are limited in length to about 40 ft (12 m) because of the small cross-sectional area of the lower end. Parallel-sided piles with tapered ends can be over 100 ft (30 m) long. Octagonal precast piles may be up to 36 in. (0.9 m) in diameter, in which case a hole may be cast along the center axis to reduce their weight.

In order to reduce transportation costs, precast piles are usually manufactured in a temporary casting yard near the site where they are to be used. Steel driving shoes may be cast into the concrete to resist hard driving conditions.

## 23.2.8  Driven Shells

Driven steel shells may be tapered or parallel-sided shells or steel pipes. After driving they are filled with concrete.

As an illustration of the procedures that are involved, one type of tapered driven-shell pile is made as follows:

1. A thin corrugated steel shell is closed at the bottom with a steel boot.
2. A steel mandrel or core is placed inside the shell.
3. The mandrel and shell are driven into the ground.
4. When a predetermined driving resistance has been reached, driving stops and the mandrel is withdrawn.
5. The open lined hole is inspected for damage.
6. The shell is filled with concrete.

The driving mandrel protects the shell during driving. This type of shell can be tapered uniformly from the bottom to top, or they may be step-tapered with different-sized shells attached end to end. An advantage of the step-taper is that the driving mandrel distributes driving energy along the length of the shell instead of concentrating it at the bottom.

Another type of driven shell has a scalloped cross-section formed by a series of vertical flutings running the full length of the shell. This shell is driven without a mandrel and is uniformly tapered and closed at the bottom by a steel boot.

The thinnest shells that can be used for this type of pile are about $\frac{1}{8}$ in. (3 mm) thick, but thicknesses of 1.5 to 2 times this value are more common, depending on the length and diameter of the pile.

A parallel-sided, dropped-in shell pile that is designed solely for tip resistance is constructed by driving a heavy steel pipe, usually 16 in. (400 mm) in diameter and ½ in. (12.7 mm) thick, with the aid of a steel mandrel, then pulling out the mandrel and inserting a thin metal shell such as a corrugated-metal culvert pipe, filling it with concrete, and withdrawing the drive pipe.

Another alternative is to drive a steel pipe pile with either a closed end or an open end. The closed-end pipe is filled with concrete. Sometimes a pipe is driven all in one piece, or it may be made of several pieces of pipe either welded together or fitted together with special internal sleeves. The open-end steel pipe is used to obtain better penetration through weathered rock to find hard rock, and then either flushed out with water and air or with a miniature orange-peel dredging bucket. The open pipe then is given a few extra blows, pumped out, and filled with concrete. By sealing into the rock, a pile installed in this manner in effect is a small-diameter caisson and has a high bearing capacity.

### 23.2.9  Pedestal Piles

A modification that produces a bulb of concrete at the bottom of a pile is called the Franki method. The procedure is as follows:

1.  A heavy steel pipe containing a mandrel or core is driven into the ground.
2.  The mandrel is removed and a small charge of concrete is poured into the pipe.
3.  The mandrel is reintroduced and lowered to the top of the concrete.
4.  The pipe is pulled upward 2 or 3 ft ($\frac{2}{3}$ to 1 m).
5.  The mandrel is again driven with the pile-driving hammer, which forces concrete out of the end of the pipe to form a bulb-shaped mass.
6.  The mandrel is removed, and the pipe is filled with concrete up to the top.
7.  The mandrel is reintroduced to hold the concrete down while the steel drive pipe is pulled out of the ground.

### 23.2.10  Composite Piles

Composite piles of timber and concrete combine the low cost of wood piles with the durability of concrete piles. The wood section is driven until its upper end is about at the ground surface, then topped off with a concrete pile that is driven until the wood pile is below the groundwater level. An important element is the joint between the wood and concrete sections so that there is a good bearing surface and the joint has strength to resist tension and bending.

### 23.2.11 Minipiles

Minipiles or micropiles are any small-diameter piles that are installed in borings, jacked in, driven, or vibrated in. They are useful where there is little headroom for installation, for example inside existing buildings. Hollow pipes are used for higher loads and the ends pressure grouted. A simple adaptation used for light underpinning is reinforcing steel bars that are jack-hammered into the soil near a foundation, then bent back underneath a footing and encased in concrete.

## 23.3 PILE DRIVING

### 23.3.1 Types of Hammers

Early pile-driving hammers were simply weights that were repeatedly lifted along guides and dropped on top of a pile. In later innovations the hammers were lifted by steam, which gave the option of either single operation with steam only doing the lifting, or double operation with steam also powering the down cycle. Steam hammers are capable of applying from 50 to 110 blows/min.

Diesel hammers (Fig. 23.3) have a self-contained power source. The ram acts as a vertical piston that when dropped compresses an air-fuel mixture inside a cylinder, causing it to explode and propel the ram back up again. Burned gases exhaust

**Figure 23.3**

Diesel hammer puffing smoke as it drives a steel pipe pile.

through a port, additional fuel is injected into the cylinder, and the cycle repeats automatically until the fuel is shut off. The ram also impacts an anvil that drives the pile. Pile resistance is needed to keep a diesel hammer actuated, so if a pile breaks or goes through very soft soil the hammer may stop.

### 23.3.2 Protecting the Pile Head

Driving a wood pile tends to crush the wood fibers at the top of the pile, which is called "brooming." Wood pile also may split vertically. Damaged portions of the piles are cut off, so piles should have sufficient extra length to allow for the cutoff. A heavy steel ring may be placed over the head of a wood pile to reduce the amount of brooming and splitting.

The head of a precast concrete pile is protected by a metal drive cap or helmet. As shown in Fig. 23.4, a cushion that usually consists of blocks or layers of wood is used to reduce the impact of the hammer. The amount of energy lost becomes apparent when the wood block smolders or catches on fire and must be periodically replaced.

### 23.3.3 Driving Formulas

The efficiency of pile driving in terms of energy per unit of driving depends on the resistance of the soil. Because the load-bearing capacity of the pile also depends on the resistance of the soil, one may expect that there will be a relationship between pile-bearing capacity and driving energy. Unfortunately, energy loss factors are substantial and highly variable, which reduces the reliability of such projections. As a result many empirical pile-driving formulas have been suggested to try and improve the relationships.

**Figure 23.4**

(a) Driver setup showing ram, wood cushion, helmet and pile.
(b) Single-acting steam,
(c) double-acting,
(d) diesel, and
(e) vibratory hammers. (After Vesic, 1977.)

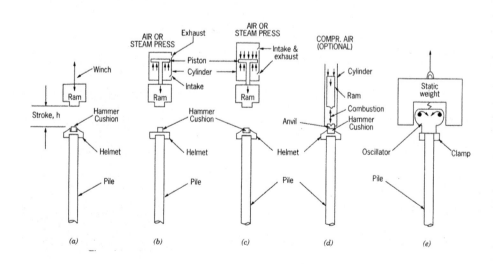

The most common driving formula also is one of the oldest, the *Engineering News* formula that was suggested in 1888 by A. M. Wellington, the editor of that magazine before it became *Engineering News Record*. The formula is based on the principle of conservation of energy—energy in equals energy out—but it also includes a generous and arbitrary energy loss factor. The formula is as follows:

$$Q_p = \frac{1}{6} \frac{WH}{S} \tag{23.1}$$

where $Q_p$ = allowable bearing capacity (tons);

$W$ = hammer weight (tons);

$H$ = height of fall (ft);

$S$ = penetration with one hammer blow (ft, inches in final formulas);

$WH$ = pile-driving energy per blow (ft-tons);

$1/6$ = energy loss factor.

An additional adjustment is made to account for hammer friction losses, which are assumed to be constant, and therefore can be represented by an arbitrary value added to the denominator. For convenience, distance $S$ is converted to inches while the height of fall remains in feet. With these adjustments,

$$Q_p = \frac{2WH}{S+1} \quad \text{for drop hammers} \tag{23.2}$$

$$Q_p = \frac{2WH}{S+0.1} \quad \text{for single-acting steam hammers} \tag{23.3}$$

where $S$ is the settlement per blow in inches. SI equivalents are

$$Q_p = \frac{815 W_{SI} H_{SI}}{S_{SI} + 25} \quad \text{for drop hammers} \tag{23.2a}$$

$$Q_p = \frac{815 W_{SI} H_{SI}}{S_{SI} + 2.5} \quad \text{for single-acting steam hammers} \tag{23.3a}$$

where $W_{SI}$ = hammer weight (kN);

$H_{SI}$ = height of fall (m);

$S_{SI}$ = penetration with one hammer blow (mm).

For double-acting hammers where power is added during the hammer stroke, energy from the manufacturers' specifications is substituted for $WH$. The minimum hammer blow for piles that will carry 25 tons (220 kN) is 15,000 ft-lb (20 kilojoules or kilonewton-meters).

**Figure 23.5**

*Engineering News* and Michigan formulas applied to results of pile load tests from several sources (Spangler and Mumma, 1958).

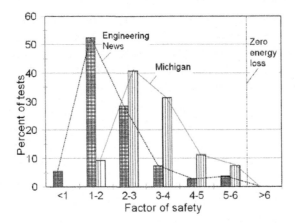

The *Engineering News* formula was developed when practically all piles were of wood and were driven by drop hammers that now are considered light in weight. The formula ignores pile length, weight, cross-section, taper, material, and soil response, which can change with time. A graph comparing the formula with over 100 load tests is shown in Fig. 23.5, where it will be seen that the energy loss factor of 1/6 is valid to the extent that none of the piles had over 6 times the predicted capacity. The test piles included a wide variety of types, lengths, sizes, soil conditions, and geographic locations. Included are 69 load tests by the Michigan State Highway Commission at three sites, one having a hard cohesive soil, another a soft cohesive soil, and the third a deep granular deposit with interbedded organic materials. Pile types included H-section, pipe, flute-tapered monotube, and step-tapered shell.

Despite obvious spread of the data in Fig. 23.5, according to these data the probability of having a factor of safety less than 1.0 with the *Engineering News* formula is less than 1 in 20, and will be further diminished for piles acting in a group. The average factor of safety is about 2. Since the energy loss factor nominally is 6, on the average about two-thirds of the driving energy is lost. However, the wide range in factors of safety leaves room for improvement.

**Example 23.1**

The final penetration resistance for a pile driven with a 30,000 ft-lb (42 kJ or 42 kN-m) drop hammer is 0.6 in./blow (15 mm/blow). Predict the bearing capacity according to the *Engineering News* formula.

*Answer:*

$$Q_p = \frac{2WH}{S+1} = \frac{2 \times 15 \text{ ft-tons}}{0.6 + 1 \text{ (in.)}} = 19 \text{ tons, or}$$

$$= \frac{815 \times 42}{15 + 25} = 860 \text{ kN}$$

Note that there actually is only one significant figure.

## 23.3.4 Michigan Formula

Many modifications have been made to the *Engineering News* formula. A relatively simple change by Hiley was adapted by the Michigan State Highway Commission(1965) to account for the mass of the pile in terms of its weight, $W_p$, and a "coefficient of restitution," $e$, that depends on the pile composition and the cushion.

The coefficient or restitution is defined as the square root of the ratio of energy out to energy in for the driving cushion, and therefore is dimensionless. The value of $e$ was established empirically and varies from 0.25 for timber pile and for a concrete or steel pile with a soft wood cushion, to about 0.55 for a steel pile with no cushion. A value of 0.5 is most commonly used to represent a common practice of driving steel or concrete piles with an oak hardwood cushion. A special Micarta cushion has $e = 0.8$.

The Michigan formula gives a correction factor that is applied to the *Engineering News* formula, and therefore still incorporates the driving energy loss factor of 6. The correction factor is

$$E_m = 1.25 \frac{W_r + e^2 W_p}{W_r + W_p} \tag{23.4}$$

where $E_m$ = a multiplier for the result from the EN formula and is dimensionless;

$W_r$ = weight of the hammer ram;

$W_p$ = weight of the pile;

$e$ = coefficient of restitution.

Results from the Michigan pile study in Fig. 23.5 show considerable improvement over the EN formula by removing evaluations having a low factor of safety, so this modification is recommended.

### Example 23.2

The pile of the previous example is a steel pipe pile 40 ft (12 m) long, 12 in. (0.30 m) in diameter, weighing 17.86 lb/ft (260 N/m). It is driven by a 10,000 lb (44.5 kN) single-acting steam-driven ram operating on an oak cushion with $e = 0.5$. Calculate $E_m$.

*Answer:* The weight of the pile is $W_p = 40$ ft $\times$ 17.86 lb/ft $= 714$ lb, or 12 m $\times$ 260 N/m $= 310$ N.

$$E_m = 1.25[(10{,}000 + (0.5)^2 \times 714]/(10{,}000 + 714) = 1.2, \text{ or}$$
$$= 1.25[(44{,}500 + (0.5)^2 \times 310]/(44{,}500 + 310) = 1.2$$

## 23.3.5 The Wave Equation

The effects of pile mass and a coefficient of restitution that were added in the Michigan formula are more elegantly treated in the wave equation developed in 1960 by E. A. L. Smith of the Raymond Pile Co.

The wave equation treats the pile as a rod that is set into vibration by being hit on the end with a hammer. The behavior can be demonstrated by suspending a pile horizontally on cables, when instrumentation shows that a compression wave travels down the pile from one end to the other; then the far end snaps back and generates another compression wave that goes in the opposite direction. Tension may develop instantaneously as two compression waves traveling in opposite directions pass one another. The method received a considerable boost when some long concrete bridge piles broke in load tests, and when pulled were found to have broken in tension—an obvious conundrum because a hammer can push but not pull.

Smith's model is shown in Fig. 23.6. A pile is divided into an arbitrary series of segments each 8 to 10 ft (2.5 to 3 m) long and having a mass $W$. Soil resistance to driving is represented by individual $R$ values along the sides or at the tip. Each segment is connected to adjoining segments by an elastic connector represented by a spring, $K$.

A ratio of tip to side resistance is assumed, and data for hammer energy and elastic constants and mass of the pile are entered. Equations relate compression and force in each connector spring to displacement, velocity, and accelerating force of each segment. The equations are solved at discrete millisecond time intervals starting when the hammer hits. This generates a record of forces and displacements in time throughout the length of the pile, as shown in Fig. 23.7. Time zero is when the ram hits the pile, and intercepts along the $x$ time-zero axis show the compression wave moving down the pile to the tip section $D12$. After 15 milliseconds the pile tip has moved downward 0.2 in., whereas the top of the pile has moved down 0.86 in. and is bouncing back upward. There is a temporary disconnect between the pile cap $D_2$ and the upper end of the pile $D_3$ at about 12.5 ms.

The pile-bearing capacity is obtained by summing all of the $R$'s that represent both side friction and point bearing. The analysis is repeated with different assumed $R$ values to generate a graph relating pile capacity to blows per foot (0.3 m), as shown in Fig. 23.8.

The wave equation addresses driving variables but it still does not take into account time-related changes in the soil such as drainage of excess pore water pressure and thixotropic hardening. These effects become obvious when driving is stopped for a few hours and the pile has developed "set" or "freeze." Re-driving a test pile after a day or two then gives a new penetration resistance that is divided

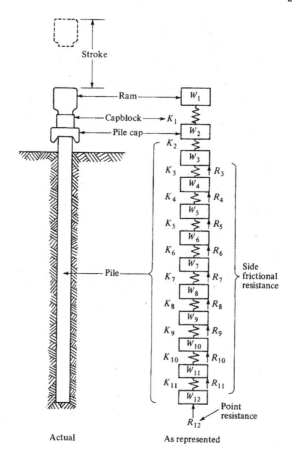

**Figure 23.6**

Smith's wave equation model for driven piles.

Actual                    As represented

by the earlier resistance to give a "setup factor," $B/A$ in Fig. 23.8. The setup factor varies from 1 for sands to 2 or more for clays. The setup factor reduces the number of blows per foot required for a pile to reach a desired bearing capacity.

Because of the many equations and data entries required, the wave equation is most conveniently solved with the aid of a computer, and programs are available.

## 23.4 DRILLED PIERS AND AUGERCAST PILES

### 23.4.1 Installing Drilled Piers

Borings for drilled piers are made by turning a relatively short auger section until the auger is filled, quickly raised it out of the boring, and rapidly rotating it to spin off the soil. This procedure is repeated until the boring reaches the desired depth. Soil is pulled away from the top of the boring with a shovel to prevent it from falling back in. The system requires a drill that can extend to the maximum

**Figure 23.7**

Example of wave equation analysis. (From Smith, 1960, with permission of the American Society of Civil Engineers.)

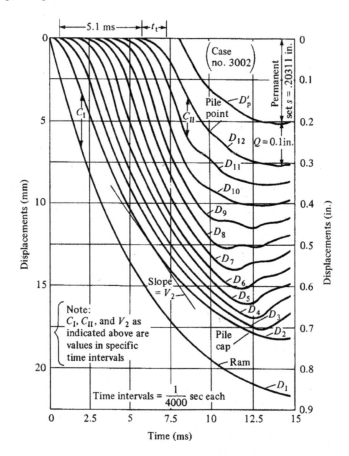

### 23.4.2 Belled Piers

depth of the pier with a square stem that slides up and down through a power sleeve called a "kelly." Holes may be bored dry, or to prevent caving they can be kept full of mud. In clay soils this often is made on-site by mixing soil and water. In this case a casing can be set to restrain the soil while the mud is being removed by bailing. (This probably is how the name became "caisson.") Another procedure is to leave the mud in the hole, lower the steel reinforcing cage, and pour concrete through a canvas chute or tremie that extends to the bottom of the hole. The fluid concrete because of its higher density then replaces the mud from the bottom up.

### 23.4.2 Belled Piers

A special type of drilled pier that can be constructed only in soil that can stand without caving is called a "belled" or under-reamed pier that increases the end-bearing area. The procedure is to bore a hole to a predetermined depth, then insert a reamer with expandable blades (Fig. 23.9). The blades open out to an angle of about 30° to 45°. The hole then is cleaned out and the bottom inspected for loose soil, after which the boring is filled with concrete. Manual inspection and cleanout

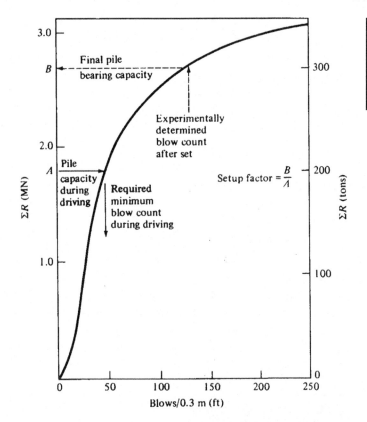

**Figure 23.8**
Evaluating and applying a setup factor, *B/A*, to a wave equation analysis.

**Figure 23.9**
Expandable cutter used for belling out the bottom of drilled piers to increase the end-bearing area. (Photo courtesy of Dr. Lyman C. Reese.)

of the bottoms of belled piers is a colorful occupation that requires a positive outlook, and the popularity of this method is fading.

### 23.4.3 Augercast Piles

Augercast piles were invented in the 1940s by the Intrusion-PrePakt Co. They are installed with a continuous hollow-stem helical auger that equals or exceeds the length of the pile (Fig. 23.10). The auger is bored to the desired depth and then the soil-filled auger is slowly raised while cement mortar is pumped down to fill the space left by the auger. The auger therefore confines the mortar so that it builds up pressure against the soil.

Augercast piles usually are smaller in diameter than drilled piers and more expensive than driven piles, but are adapted for caving soils with difficult groundwater conditions. Augercast piles do not require casing, and installation does not generate the noise and tremors of pile driving. The pumping pressure is monitored to ensure positive pressure as the auger is withdrawn, leaving a continuous column of grout. A cardboard or plastic cylinder is added at the top and filled with grout to the desired elevation, and reinforcing steel can then be lowered into the fluid grout.

The grout used in augercast piles consists of Portland cement, sand, fly ash, and water. Fly ash is a byproduct from burning powdered coal in electric power plants; it is a pozzolan, meaning that it reacts with lime liberated by the hydrating cement to create additional cementitious compounds. The fine spherical particles

**Figure 23.10**

Continuous flight auger and guide for making augercast piles. After boring to full depth, hoses carry liquid grout to the top of the hollow-stemmed auger for pumping to the bottom of the hole as the auger is lifted and confines the grout. The horizontal bar near the top is a torque arm.

of fly ash reduce friction during pumping. Fluidizing agents and an expanding agent such as aluminum powder, which reacts with alkali to give off hydrogen bubbles, also may be added to help aid pumping and maintain positive pressure through expansion until the grout sets.

A recent modification of the augercast method that is intended to eliminate spoil from the boring incorporates a reverse-pitch section of auger, so instead of soil being taken out of the boring it is pushed together and forced to move laterally. This increases lateral stress, but is difficult to accomplish in stiff soils because of the power requirements.

## 23.5   SOIL MECHANICS OF PILES AND PIERS

### 23.5.1   Overview: Side Friction and End Bearing

The total supporting capacity $Q$ of a pile or pier is the sum of side friction and end bearing, and can be expressed by

$$Q = Q_s + Q_b \qquad (23.5)$$

where $Q_s$ derives from side friction and $Q_b$ is from end bearing at the base.

Side friction depends in part on normal stress exerted by soil on a pile or pier, which for the most part has seldom been measured except indirectly through pull tests. The coefficient of friction can be compromised by several factors including remolding.

Prediction of an end bearing of deep foundations from soil mechanics considerations requires a model for bearing capacity failure. Three models are shown in Fig. 23.11. As in the case of shallow foundations, Terzaghi's model assumes that soil above the bottom of the foundation contributes only a surcharge pressure, whereas Meyerhof incorporates shearing resistance of the soil above that base elevation. While this appears more reasonable for deep foundations, the development of shearing resistance may be progressive, so yielding does not occur simultaneously along the entire slip surface, particularly since shearing stresses decrease rapidly with radial distance in a three-dimensional arrangement. Even though the model appears to be unrealistic, the Terzaghi factors corrected for a round base are very close to those determined by Vesic (1963) from model tests.

### 23.5.2   Soil Factors

Factors that have been identified as influencing both driving resistance and pile capacity are as follows:

- *Remolding* obviously must occur in soil adjacent to driven piles and occurs to a lesser degree in the perimeter of borings for drilled piers and augercast pile.

**Figure 23.11**

Models for side
friction and end
bearing of piles
and piers. Other
things being
equal, side
support increases
as the length and
the diameter, and
end bearing as
the square of the
diameter.

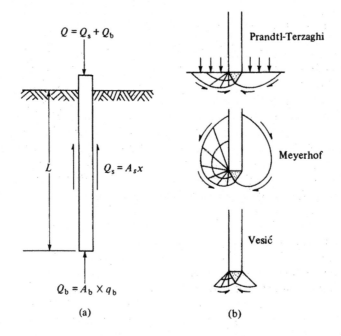

$$Q = Q_s + Q_b$$

$$Q_s = A_s x$$

$$Q_b = A_b \times q_b$$

(a)

Prandtl-Terzaghi

Meyerhof

Vesić

(b)

Side friction has been shown to be increased if boring surfaces are rough or grooved.

- *Temporary high pore water pressures* have been measured in soil adjacent to driven piles, and excess pore pressure also are likely to occur during ramming of the bottom bulb of pedestal piles.

- *Temporary liquefaction* was discovered in soil near Rammed Aggregate Piers, described in the next chapter, and probably occurs near the bulb of pedestal piles. Liquefaction also may occur in weak, saturated soil near tapered piles and would aid driving, but those occurrences have not been verified.

- *Radial tension cracks* have been conjectured to occur near driven piles to explain rapid dissipation of pore water pressure, and have been confirmed in the elastic zone near Rammed Aggregate Piers. Radial cracking may be an important phenomenon for facilitating drainage of excess pore pressure near displacement piles and piers.

- *Lateral stress* directly influences side friction and to some extent end bearing of deep foundations, but for the most part has not been measured except indirectly from pullout resistance. Displacement from pile driving should increase lateral stress, and lateral stress temporarily is decreased by borings and reinstated by fluid pressure of concrete. The contact stress then can change as load is applied. Lateral stress is the least known of the variables affecting pile capacity.

The following discussions follow recommendations of Reese and O'Neil (1989).

### 23.5.3 Side Friction in Clay

Lateral pressure from the fluid pressure from the concrete should enable a calculation of side friction. However, Reese et al. (1976) found that water from the concrete migrates into adjacent clay soil, which should contribute to excess pore pressure during load testing and weaken the soil. A representative series of tests is shown in Fig. 23.12, where side friction does increase with depth down to about 10 to 12 feet (3 to 4 m) and peaks out at about 75 percent of the shearing strength of the unaltered soil. At depths less than about 5 ft (1.5 m), low side friction is attributed to desiccation shrinking of clay away from the pier. Side friction can be seen to decrease during the final load increment, indicating remolding.

Another complicating factor in Fig. 23.12 is a nearly complete loss of side friction near the bottom of the pier. This can be expected if, as the base of the pier settles, it carries a surrounding bulb of soil along with it.

Based on these and similar results, Reese and O'Neil recommend calculating side friction of a pier in clay using an "alpha factor," which represents a portion of the undrained shear strength of the soil acting along a shaft length that excludes the upper 5 ft (1.5 m) and the lower 5 ft (1.5 m) plus the length of a bell. The $\alpha$ factor experimentally was determined to have an average value of 0.55. Hence

$$q_s = 0.55c \quad \text{and} \tag{23.6}$$

$$Q_s = q_s L_s \pi D \tag{23.7}$$

where $q_s$ is the developed cohesion, $L_s$ is the shaft length *after deduction for end effects*, and $D$ is the shaft diameter.

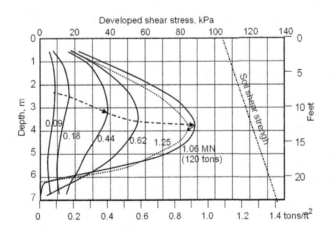

**Figure 23.12**

Development of skin friction during loading of a drilled test pier in expansive Houston clay. (After Reese et al., 1976, with permission of the American Society of Civil Engineers.)

### Example 23.3

Calculate the side friction for a 2.5 ft (0.76 m) diameter drilled pier 15 ft (4.6 m) long penetrating clay with $c = 5$ ton/ft$^2$ (480 kPa). There is no bell.

*Answer:*

$$q_s = 0.55 \times 5 \text{ ton/ft}^2 = 2.75 \text{ ton/ft}^2$$

$$Q_s = 2.75 \text{ ton/ft}^2 \times [(15 - 5 - 5)\pi \times 2.5] \text{ ft}^2 = 108 \text{ tons, or}$$

$$q_s = 5.5 \times 480 \text{ kPa} = 2640 \text{ kPa or } 2640 \text{ kN/m}^2$$

$$Qs = 2640 \text{ kN/m}^2 \times [(4.6 - 1.5 - 1.5)\pi \times 0.76] \text{ m}^2 = 10.1 \text{ MN}$$

## 23.5.4  End Bearing in Clay

The Terzaghi bearing capacity equation for shallow foundations is

$$q_b = \frac{\gamma B}{2} N_\gamma + cN_c + \gamma D N_q \tag{22.7}$$

where $q_b$ is the end-bearing capacity in force per unit area, $\gamma$ is the soil unit weight, $B$ is the foundation width, $c$ is soil cohesion, $D$ is depth of the foundation, and the $N$ values are bearing capacity factors.

If $\phi$ is assumed to be zero in clay to represent undrained conditions, from Table 22.3 the three respective $N$ bearing capacity factors for a round base are 0, 6.2, and 1.0. The first (width) term in the equation therefore is eliminated, the second (cohesion) term is retained, and the third (depth) term equals the amount subtracted to obtain a net bearing capacity, based on the approximate equivalence of the soil and concrete unit weights. The net bearing capacity at the ground surface then becomes

$$q_{bnet} = 6.2c$$

However, back-calculations from load tests indicate a closer approximation if $N_c = 9$, which corresponds to a partially drained friction angle of about 6°. Then

$$q_{bnet} = 9c \tag{23.8}$$

where $q_{bnet}$ is the net end-bearing pressure and $c$ is the average soil cohesion for a depth of two diameters below the base. Multiplying by the base area gives

$$Q_{bnet} = \pi(B/2)^2 c$$
$$Q_{bnet} = 7cB^2 \tag{23.9}$$

where $B$ is the base diameter. In normally consolidated soils, $c'$ increases with depth so there is an increase in end bearing even though $c$ in the equation represents a partially drained shear strength.

### Example 23.4

Calculate allowable total net bearing capacity of the 2.5 ft (0.75 m) diameter, 15 ft (4.6 m) long pier of the previous example on the same overconsolidated clay.

*Answer:*

$Q_{bnet} = (7)(5\,tons/ft^2) \times [\pi(1.25)^2]\,ft^2 = 17\,tons$. Adding side resistance gives

$Q_{net} = 108 + 7 = 115\,tons$. With a factor of safety of 2,

$Q_a = 52\,tons$, of which most is side friction, or

$Q_{bnet} = (7)(480\,kPa) \times [\pi(0.38)^2]\,m^2 = 0.49\,MN$

$Q_{net} = 10.1 + 0.5 = 10.6\,MN;\ Q_a = 5.3\,MN.$

### 23.5.5 Depth Factor for End Bearing in Clay

Reese and O'Neil (1989) suggest a modification to eq. (23.8) to incorporate a moderate depth influence as follows:

$$q_{bnet} = 6[1 + 0.2(L/B)]c \le 40\,tons/ft^2\ (3.8\,MPa) \tag{23.10}$$

where $L$ is the pier length and $B$ is the base diameter, both having the same units.

### Example 23.5
Recalculate the previous example using eq. (23.10).

*Answer:*

$L/B = 15/2.5 = 6.0(4.6/0.75 = 6)$

$q_{bnet} = 6[1 + 0.2(6)]\,5\,ton/ft^2 = 66\,tons/ft^2$; use $40\,tons/ft^2$

$Q_{bnet} = 40\pi(1.25\,ft)^2 = 196\,tons$, compared with $170\,tons$
    from the previous example.

$Q_{net} = 196 + 67 = 263\,tons$. With a factor of safety of 2,

$Q_a = 130\,tons$ instead of $118\,tons$, or

$q_{bnet} = 6[1 + 0.2(6)]\,480\,kPa = 6.3\,MPa$; use $3.8\,MPa.$

### 23.5.6 Settlement on Clay

The larger the base of a pier the deeper the pressure bulb, and the larger the amount of settlement. Reese and O'Neil therefore suggest an empirical bearing pressure reduction factor for piers up to 11 ft (3.5 m) in diameter:

$$F_r = \frac{2.5}{\psi_1 B_b + \psi_2} \le 1 \tag{23.11}$$

where $B_b$ is the base width in inches. The two $\psi$ factors are

$$\psi_1 = \{0.0071 + 0.0021(L/B_b)\} \le 0.015 \tag{23.12}$$

$$\psi_2 = 1.125\sqrt{c}\ \text{and}\ 0.5 \le \psi_2 \le 0.15 \tag{23.13}$$

where $c$ is the undrained cohesive shear strength in kips/ft$^2$. As $\psi_2$ is not dimensionless, SI units should be converted before making this correction.

### Example 23.6

Determine the settlement correction factor to bearing capacity in the preceding example.

*Answer:*

$\psi_1 = \{0.0071 + 0.0021(15/2.5)\} = 0.020$; use 0.015

$\psi_2 = 1.125\sqrt{10}$ kips/ft$^2$ = 3.56; use 0.15

$F_r = 2.5/[0.015(2.5) + 0.15] = 13.3 > 1.0$; use 1.0, no reduction

## 23.5.7   Side Friction in Sand

Side friction of a drilled pier in sand is calculated in a similar manner but uses an empirical beta ($\beta$) factor that incorporates influences of both friction angle and lateral stress. The relationship is

$$q_{si} = \beta\sigma'_{zi} \qquad (23.14)$$

where $q_{si}$ is side friction per unit area of a section of pier passing through a soil layer and $\sigma_{zi'}$ is the vertical effective stress in soil at the middle of the layer. Beta is evaluated from

$$\beta = 1.5 - 0.35\sqrt{z_i}, \quad \text{and } 0.25 \leq \beta \leq 1.20 \qquad (23.15)$$

where the depth zi is in feet, or

$$\beta = 1.5 - 0.63\sqrt{z_i}, \quad \text{and } 0.25 \leq \beta \leq 1.20 \qquad (23.15a)$$

where $z_i$ is in meters. Because, according to eq. (23.15), $\beta$ is not linear with depth, a sand is divided into sublayers for calculations. There is no decrease in side friction at the top or bottom of a pier, and no belling allowed in sand.

### Example 23.7

Calculate the maximum side friction on a 20 ft (6.1 m) long, 2.5 ft (0.76 m) diameter straight pier in sand having a unit weight of 120 lb/ft$^3$ (18.9 kN/m$^3$). The groundwater table is at 10 ft (3.05 m) depth.

*Answer:* Arbitrarily divide the sand into two layers, $z_i = 5$ ft and 15 ft (1.5 and 4.6 m). Respective $\beta$ values are 0.72 and 0.25. Respective $q_{si}$ values are

Upper $q_{si} = 0.72(5\,\text{ft})(120\,\text{lb/ft}^3) = 432\,\text{lb/ft}^2$, or

$\qquad = 0.72(1.5\,\text{m})(18.9\,\text{kN/m}^3) = 20.4\,\text{kPa}$

Lower $q_{si} = 0.25[(10)(120) + 5(120 - 62.4)] = 329\,\text{lb/ft}^2\,(15.8\,\text{kPa})$

$\qquad Q_s = \pi(2.5)(10)(432) + \pi(2.5)(10)(329) = 33,900 + 25,800\,\text{lb} = 30\,\text{tons}$

$\qquad = \pi(0.76)(3.05)(20.4) + \pi(0.76)(3.05)(15.8) = 149 + 115\,\text{kPa} = 264\,\text{kPa}$

## 23.5.8 End Bearing in Sand

If cohesion is zero, eq. (22.7) for bearing capacity becomes

$$q_b = \gamma' \left[ \frac{B}{2} N_\gamma + L N_q \right] \tag{23.16}$$

where $\gamma'$ is the effective soil unit weight, $B$ is the pile diameter, and $L$ the length of the pile.

The bearing capacity factors depend on the friction angle of the sand, which because of the difficulty of sampling for laboratory testing generally is evaluated from in-situ tests. The most satisfactory option would be direct measurement, but most common are penetration tests that are routinely conducted during site evaluations.

Standard Penetration Test (SPT) results are in blows per foot of penetration of the special sampler and are designated by $N$. As discussed in Chapter 26, several correction factors have been proposed depending on the testing depth and on the driving energy. The following formula was developed using *uncorrected* $N$ values:

$$q_0 = 0.6N \leq 45 \, \text{tons/ft}^2, \text{ or}$$
$$q_0 = 0.06 \, N \leq 4.3 \, \text{MN/m}^2 \tag{23.17}$$

where $N$ is the average standard penetration resistance of soil in blows per ft (0.3 m) to a distance $2B$ below the bottom of the pier.

### Example 23.8
Evaluate end-bearing and allowable bearing capacity for the 2.5 ft (0.76 m) diameter drilled pier in the previous example if $N_{av} = 26$ blows/ft at the bearing depth.

*Answer:*

$q_0 = 0.6(26) = 15.6 \, \text{tons/ft}^2$

$Q_0 = \pi(1.25)^2(15.6) = 77 \, \text{tons}$. Adding side friction,

$Q = 30 + 77 = 107 \, \text{tons}$. Dividing by a factor of safety of 2 gives

$Q_a = 53 \, \text{tons}$, or

$q_0 = 0.06(26) = 1.6 \, \text{MN/m}^2$

$Q_0 = \pi(0.38)^2(1.6) = 730 \, \text{kN}$

$Q = 264 + 730 = 994 \, \text{MN}$

$Q_a = 500 \, \text{MN}$

---

*Question:* With the allowable bearing pressure, what percentages of the side and base resistance theoretically will be mobilized?

*Answer:* 100% and about 30%. However, side friction may decrease when the pier is loaded, which will increase the base percentage.

---

### 23.5.9  Soil Mechanics Approach to Side Friction for Sand

Many engineers prefer to compare different methods to obtain answers, which is useful to check for major errors. Also, any empirical approach may not be valid in the universe that exists outside of the test conditions. A logical next step is to develop a rational method that applies to all conditions, but this approach is possible only if relevant soil information is available. A rational approach also has an advantage of pointing out variables that otherwise may remain buried in the empirical correlations.

Critical parameters for side friction are the normal effective stress and the developed friction angle. The normal stress following a concrete pour equals that imposed by the fluid concrete, but it sometimes is assumed that in time soil pressure will be relieved to its initial $K_0$ condition, which is difficult to evaluate. Pier loading can induce an increase in lateral stress because of rotation of the principal stress directions, and dilation or compression of the soil. Finally, there is the developed friction angle between the soil and the pier that because of the movement involved will represent a residual strength without a dilatant component. A typical assumption is that the angle of side friction is defined by $\tan \delta = 0.8 \tan \phi$.

**Example 23.9**

Evaluate side friction using a soil mechanics approach for the pier in the preceding example, the groundwater table being at a depth of 10 ft (3.05 m).

*Answer:* According to a criterion presented later in this book, an uncorrected $N = 26$ corresponds to a friction angle in sand of about 35°. An estimate of the developed side friction is $\tan \delta = 0.8 \tan \phi$, or $\delta = 29°$.

*Option 1:* For a maximum value use $K_0$ for fluid concrete of 1.0 and a unit weight of 150 lb/ft$^3$. The average normal stresses at mid-depths in the layers above and below the groundwater table and corresponding side friction are:

Above gwt: $\sigma_h = 1.0(5)(150) = 750 \text{lb/ft}^2$; $S_s = 750\pi(2.5)(10)\tan 29° = 32{,}600 \text{ lb}$
Below gwt: $\sigma_h = 1.0[10(150) + 5(150 - 62.4)]$

$$= 1940 \text{ lb/ft}^2; \ S_s = 1940\pi(2.5)(10)\tan 29°$$
$$= 84{,}400 \text{ lb}$$

Total $= 58$ tons

*Option 2:* For a minimum value use the soil normally consolidated $K_o = 1 - \sin \phi$ and the unit weight of the soil.

Above gwt: $\sigma_h = (1 - \sin 35°)(5)(120) = 256 \text{ lb/ft}^2$; $S_s = 256\pi(2.5)(10)\tan 29°$
$$= 11{,}100 \text{ lb}$$

Below gwt: $\sigma_h = (1 - \sin 35°)[(10)(120) + 5(120 - 62.4)] = 635 \, \text{lb/ft}^2$;

$$S_s = 635\pi(2.5)(10) \tan 29° = 27,600 \, \text{lb}$$

Total = 20 tons

The two options therefore bracket the answer obtained by the empirical method. The same procedures can be followed using SI units. Obviously for a soil mechanics approach to be valid the friction angle and lateral stress must be accurately evaluated.

## 23.5.10 Settlement of a Pile or Pier in Sand

In sand, particularly in loose sand, settlement may be a more important criterion than a bearing capacity failure. However, the prediction of settlement is complicated because a substantial part of the load is carried by skin friction, so after application of a factor of safety the bottom of a pile or pier may receive very little stress. Settlement therefore is predicted on the basis of stress reaching the bottom of the pier, and therefore is subject to errors in determination of side friction.

Settlement in sand is limited by a correction factor applied to the load-bearing capacity of large pier bases in sand:

$$F_r = \frac{50}{B_b} \qquad \text{when } B_b > 50 \text{ inches, or} \tag{23.18}$$

$$F_r = \frac{1.25}{B_b} \qquad \text{when } B_b > 1.25 \text{ m} \tag{23.18a}$$

## 23.6 LOAD TESTS

### 23.6.1 Compression Load Test

Load tests are more easily arranged for piles and piers than for shallow foundations because no massive dead load is required. Instead, a test is conducted by jacking against a horizontal beam with ends attached to uplift piles or piers, as shown in Fig. 23.13. As the anchor piles or piers are in tension they must be reinforced, and the reinforcing steel is extended upward where it is bolted or welded to the horizontal reaction beam.

For piers having a large base-bearing capacity four anchors may be required to generate a sufficient side friction, in which case the reaction beams are assembled as a letter H. Loads ordinarily are increased incrementally until they reach two times the design capacity, to confirm a factor of safety of 2.

As test loads are applied, settlement is measured with dial gauges or transducers on opposite sides of the pile or pier so that averages may be used to compensate for bending. The gauges are supported on small independently supported beams

**Figure 23.13**

Load test of a
drilled pier using
hydraulic jacks
and two
independently
supported dial
gauges.
Sunshade is to
prevent thermal
warping from
affecting the
readings.

that are transverse to the pile row to avoid influences from ground heave. The
arrangement is shaded to reduce error from thermal expansion and warping of
gauge supports.

Test loads may be applied continuously or in cycles. In the continuous method,
which is most common, a test load is applied in increments until two times the
design load is reached or else a plunging failure occurs. Each load increment is
held constant for 2 hours while settlement is measured after various time intervals
(ASTM Designation D-143). The final load is left on for 48 hours. Unloading also
is in steps, and a final reading is made at least 12 hours after all load has been
removed. The load-settlement curve thus obtained is known as a total-settlement
curve or gross-settlement curve.

In the cyclic method, a load increment is applied, and the settlement is measured.
The load is released and the pile is permitted to rebound. Repeating this process
for several increments will permit data to be obtained for plotting a net or plastic-
settlement curve. The distinction between the two types of tests can be seen by
comparing the initial load curve starting at zero in Fig. 23.14 with the reloading
curve.

### 23.6.2 Defining a Failure Load

Load deflection curves are curvilinear, and a sharp break in a load-settlement
curve is called a "plunge," which normally indicates end-bearing capacity failure.
This occurs after full mobilization and partial reduction of side friction. Plunge
occurs near point $F$ in Fig. 23.14.

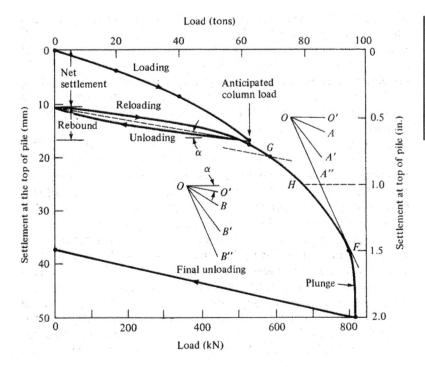

**Figure 23.14**

Pile load test and failure criteria based on settlement, net settlement, and settlement per unit of load.

The maximum test load seldom is sufficient to generate plunge, so many different criteria have been proposed to define acceptability without plunging failure. No single criterion will be found that can apply to all load tests, and a common procedure is to define failure on the basis of several criteria and select the most relevant for a particular situation. For example, if settlement cannot exceed a certain value, that settlement value becomes the criterion for acceptability.

Some common criteria are as follows. In this connection "gross settlement" equals total settlement, and "net settlement" is the gross settlement corrected for rebound during unloading.

- As indicated above, gross settlement may be limited to a specific value, for example 0.5 or 1.0 in. (12 to 25 mm). This is the simplest criterion because it is read directly from the load-settlement curve, as point $H$ in Fig. 23.14, and is based on the amount of settlement that can be tolerated by a structure instead of depending on shifting pile-soil interactions.

- Gross settlement can be defined as a function of rate, for example 0.01, 0.02, or 0.05 in./ton (0.03, 0.06, or 0.15 mm/kN). These criteria are shown by the slopes of lines $OA$, $OA'$, and $OA''$ in Fig. 23.14. A failure load is found by shifting the selected line until it is tangent with the load-settlement curve. A factor of safety of 2 then is applied. The steepest line, $OA''$, probably defines a realistic failure load for this pile. A limitation of this criterion is that the slope may not be attained in the load test because of limits imposed by the jacks and anchors.

- The net settlement rate may be limited to an arbitrary value as in paragraph 1, and a factor of safety of 2 applied. This in effect rotates the allowable slope base line by an angle $\alpha$ in Fig. 23.14 so that it is parallel to the rebound curve, giving slopes $OB$, $OB'$, and $OB''$ in the figure. The line then is shifted until it is tangent to the load curve as in the preceding criterion. Net settlement eliminates most of the elastic compression that depends on pile length and material, but it should be recognized that some compression remains after full removal of load because of residual side forces that act upward on the lower part of the pile and downward on the upper part as the pile rebounds elastically.

- Net settlement may be limited to a specific value, for example 0.25 or 0.5 in. (6 or 12 mm), in which case the failure load is obtained by parallel shifting of the rebound curve so that it intersects the desired settlement value, and reading where this line intersects the load curve, for example $G$ in Fig. 23.14. A factor of safety of 2 usually is applied.

Other criteria such as defining a "break in the curve," or using a point defined by tangents drawn on either side of a break, are more arbitrary and depend on the scale of the plot and the maximum applied load. Another criterion is to define an allowable settlement for a given load increment, in which case that load increment must be defined.

The safe load for a pile or pier is one in which settlement is not excessive for the intended use. A factor of safety is necessary to cover variable soil and pile conditions and to a lesser extent unanticipated loading conditions. Since most pile-soil systems gain strength with time, the eventual mean factor of safety ordinarily will increase over that defined by the load test.

### 23.6.3  Uplift Test

Piles or piers that are intended to resist uplift also can be tested with an arrangement similar to that in Fig. 23.13 except that the center pile is attached to the cross-beam that is jacked against the end piles. A pullout test of course mobilizes only side friction with no end bearing.

### 23.6.4  Two-Way Osterberg Test

A recent innovation by Dr. Jorg Osterberg, an emeritus professor at Northwestern University, pushes up on a pile or pier while pushing down on soil at its base, by use of an expanding hydraulic load cell, called an Ostberg cell or "O-cell." A schematic diagram is shown in Fig. 23.15. Two telltales or vertical rods are attached to opposite sides of the cell and extended to the ground surface

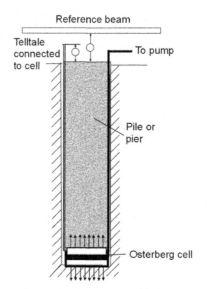

Reference beam

Telltale connected to cell

To pump

Pile or pier

Osterberg cell

**Figure 23.15**
Schematic of the Osterberg cell for pushing upward on a pile or pier.

to provide a base for strain measurements. The O-cell is particularly useful for large-capacity piers that would require a massive reaction beam and jacking capability. A special version of the device can be attached to the ends of driven piles. However, the O-cell is not recovered after a test, which adds to the cost of the test.

As can be seen from the data in Fig. 23.16, the maximum load depends on which yields first, side friction or end bearing. In most cases side friction is the first to peak out, and end bearing is estimated by extrapolation. The data in Fig. 23.16 indicate a failure load of about 7400 plus about 8000 tons (65 plus 55 MN). This is one of the largest capacity piers ever tested.

With the O-cell side friction and bottom bearing are developed simultaneously, whereas when a pile or pier is put into service with top-down loading, side friction initially will carry all of the load prior to activation of bottom bearing. This introduces a depth variable in the test, as side friction generally is higher at depth because of higher normal stresses. If and when side friction is fully mobilized in the test, indicated by side shear movement equaling and exceeding bottom movement, it should make no difference whether shearing is acting upward or downward. The predicted settlement then may equal bottom movement plus elastic compression of the pile or pier, taking into account the cumulative nature of side shear with depth.

## 23.6.5 Electronic Analyzers

An electronic "pile analyzer" consists of transducers and accelerometers attached to the top of a pile in order to monitor the force transmitted to the pile and its

**Figure 23.16**

O-cell data for a
9 ft diameter pier
50 ft deep at
Apalachicola
River, Florida.
(after Osterberg,
2004).

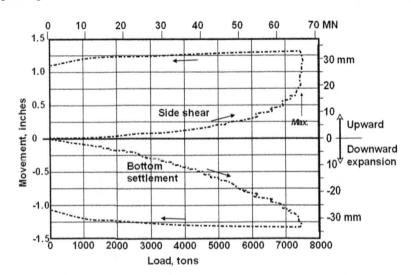

velocity during driving. This bypasses the effect of an important uncontrolled variable, $e$, of the wooden cap blocks. The procedure is more direct than a wave equation analysis but still incorporates an arbitrary soil "damping constant." Testing is repeated after a waiting time to evaluate a setup factor.

### Flaws and Integrity Testing

Flaws in concrete piles or piers can remain undetected until one breaks during a load test. This is indicated by a sharp plunge, and is symptomatic of improperly cured concrete, cracks from handling, tension cracks from driving, or soil falling or pushing into a boring to displace fluid grout or concrete. Failure during a load test is painful for everybody involved, particularly if other foundation elements already are in the ground.

Failure of a single high-capacity pier ordinarily cannot be tolerated, so if a test fails, one alternative is to diamond-bit core for the full length of every pier to ensure its integrity. A far simpler procedure that is less definitive involves attaching a transducer to the top of the foundation element and hitting with a hand-held hammer. A plot of pulse versus time will reveal flaws that are sufficiently large that they will create multiple echoes in addition to an echo emanating from the bottom of a pile or pier. This test is most useful for pile lengths that are less than 30 diameters.

The high cost of pulling and replacing weak or broken piles or piers can result in project abandonment, or in shifting to a slightly different footprint that will allow installation of all new foundation elements. Far better is to get things right the first time.

## 23.7 END BEARING ON ROCK

### 23.7.1 Knowing the Geology

It often is assumed that the safest foundation is on rock, and end bearing can be based on the unconfined compressive strength. However, the rock may be a mix of weak and strong layers of shale, limestone, and sandstone, or it may be deeply weathered. As the rock surface cannot be inspected except in borings, these are only some of many what-ifs. Rock avoids some soil problems but substitutes others, most having to do with competence of the rock.

### 23.7.2 Depth to Sound Rock

A first requirement is to measure the depth to rock by drilling or seismic exploration, in order to determine if end bearing on rock is economically feasible. As most major cities and nearly all bridges are founded on or adjacent to river floodplains, the most common soil cover for rock encountered in engineering is alluvium. The depth to rock obviously depends on thickness of the alluvium, which can be considerable, caused by the $400 + $ ft $(120 + $ m$)$ rise in sea level at the end of the Pleistocene, which caused all graded river valleys to fill with sand and gravel.

Even after solid rock is encountered in exploration borings, they may be continued for a set distance such as 10 ft (3 m) into the rock to ensure that it is not an isolated mass that slid down some primordial slope and happened to be encountered by the test boring. There still remains a possibility for vertical discontinuities such as clay-filled faults or fractures that most likely will be missed by exploration drilling. For example, if vertical discontinuities constitute a 5 percent area coverage, the probability of encounter by a single boring is only 1 in 20, or by 10 borings still only 1 in 2. Such discontinuities often are discovered later in foundation excavations or pier borings, where they may require remedial "dental treatment" or grouting.

### 23.7.3 Gradational Contact

Rock weathers along fractures and bedding planes that allowed infiltration of surface water at the time when the rock was exposed at the ground surface. A common result is a gradual transition with depth from residual soil at the buried former ground surface to soil containing rock fragments to solid rock. Driven piles with hardened steel points are often used to penetrate through the weathered zone into solid rock. For bored piers the boring machine must have sufficient power to penetrate through the weathered zone and create a "rock socket."

### Side Friction

Side friction generally is not included in design of end bearing on rock, but is a consideration for development of negative skin friction.

### Rock Competence

The competence of rock commonly is estimated from the percentage of competent core recovered during exploration drilling. This is the "rock quality designation" or RQD. The modulus of a rock with weathered seams is mainly a function of the thickness and spacing of the seams, which contain material that may not be normally consolidated because of support from arching action. The modulus therefore decreases rapidly and somewhat erratically depending on the RQD, a 30 percent decrease in RQD causing about an 80 percent decrease in modulus. Clues can be obtained from cross-hole seismic testing, and a modulus can be more accurately evaluated with rock pressuremeter tests (Failmezger et al. (2005)).

Experience in local areas often leads to bearing capacity criteria for particular rocks and may become the basis for local building code requirements. Obviously, caution must be used in applying such criteria to other areas and in particular to other rock types, as even a rock name such as "shale" or "limestone" can cover a wide range of strength properties.

One approach to design is to consider only the material in weathered seams as contributing to settlement, and test that material in a consolidation test. Results then are reduced by a percentage that represents the amount of dilution by rock fragments. For example, if the rock is 90 percent sound rock fragments, test data on soil separating the fragments can apply to only 10 percent of the weathered rock mass and therefore can be divided by 10. This method is only approximate and is sensitive to the bearing load because of the influence of arching action. The answer for design usually is in load tests of completed piles or piers, and the final answer is in the performance.

## 23.8   PILE GROUPS

### 23.8.1   Pile Interaction

Piles frequently are used in groups in order to obtain a required bearing capacity. The pile group usually is connected at the top with a concrete pile cap so that loading is simultaneous and stress fields around and under the piles overlap. For example, if two adjacent friction piles are loaded simultaneously, soil between the piles will be subjected to a downward force on both sides and will tend to compress more than if it were adjacent to a single pile.

The bearing capacity of a pile group therefore includes a factor for efficiency:

$$Q_g = EnQ \qquad (23.20)$$

where $Q_g$ = bearing capacity of the pile group;

$E$ = efficiency;
$n$ = number of piles in the group;
$Q$ = bearing capacity of individual pile.

Simultaneous downloading of a group of piles is difficult because of the large forces involved, but loading can be accomplished with Osterberg cells. Another approach is to use scale models, but in order to meet physical requirements as dimensions are scaled down, unit weights should be scaled up. This can be accomplished in a centrifuge.

## 23.8.2  The Feld Rule

A simple procedure to account for pile interactions was introduced by Jacob Feld, and involves reducing the bearing capacity of each pile by one-sixteenth for each adjacent pile, whether on a diagonal or a straight row. The rule derives from the assumption that frictional resistance is compromised along arc distance $a$ in Fig. 23.17. The minimum ratio of spacing to diameter, $S/D$, is approximately 2.5, in which case

$$\tan(0.5\alpha) = (0.5S)/L$$

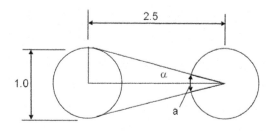

**Figure 23.17**

Derivation and example of the Feld rule.

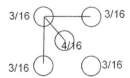

Substituting $S = 2.5D$,

$$\tan(\alpha/2) = (0.5D)/2.5D = 0.20$$
$$a = 22.6°$$
$$22.6/360 = 0.063 = 1/16$$

**Example 23.10**

Calculate efficiency of a 9-pile group in Fig. 23.17 using the Feld rule.

*Answer:* There are 5 piles, and summing the reductions gives 16/16. This is the equivalent of $5 - 1 = 4$ stand-alone piles, for an efficiency of 80%.

*Question:* Would eliminating the center pile increase the efficiency? Would it reduce the load-carrying capacity?

Model tests summarized by Poulos and Davis (1980) indicate that, for a spacing of $2.5D$ in various size groups the efficiency varied from 0.7 to 0.9, averaging about 0.8. With larger spacings the efficiency is slightly higher. Pile group reduction factors are appropriate for friction piles in clay but not for end-bearing or for driven piles in sand, where the efficiency is 1.0 and may exceed 1.0 because of densification of the sand.

## 23.8.3 Bearing Capacity of a Pile Group

Terzaghi and Peck (1967) introduced a concept that a large pile group should be tested for bearing capacity failure as a group that in effect is acting as a single foundation (Fig. 23.18). This is unlikely to occur in soils having internal friction because the depth factor $N_q$ in the bearing capacity equation will be quite large. The evaluation therefore is most appropriate for poorly drained clay, as in eq. (23.8). Applying this equation to a pile group and adding perimeter shear resistance gives

$$Q_g = c(9BA) + c(2BL + 2AL)$$

**Figure 23.18**

Bearing capacity of a pile group.

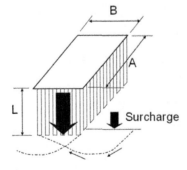

$$Q_g = c\{9BA + 2L(B + A)\} \tag{23.21}$$

where $Q_g$ is the net bearing capacity of the group, $c$ is the undrained soil cohesion, $B$ and $A$ are the plan dimensions of the group, and $L$ is the pile length. Every large pile group should be checked for this failure mode, particularly in soft soil with short pile lengths.

### Example 23.11

Find the maximum allowable net bearing capacity of a pile group with $L = 20$ ft (6.1 m), $D = 1$ ft (0.3 m), $A = 30$ ft (9.1 m), and $B = 20$ ft (6.1 m), and compare with the bearing capacity using the Feld rule. The soil is clay, $c = 500$ lb/ft$^2$ (23.9 kPa), and $\gamma_{sub} = 60$ lb/ft$^3$ (9.4 kN/m$^3$).

*Answer:* From eq. (23.21), for the pile group,

$$Q_g = 500 \text{ lb/ft}^2 [9(30)(20) + 2(20)(30 + 20)] \text{ ft}^2 = 500[5400 + 2000]$$
$$= \underline{1850 \text{ tons}}, \text{ or} = 23.9 \text{ kPa} [9(9.1)(6.1) + 2(6.1)(9.1 + 6.1)] \text{ m}^2 = 16.4 \text{ MN}$$

Single pile:

$$Q_s = cL\pi(B/2)^2 = 500(20)(3.14)(0.5)^2 = 7850 \text{ lb}$$
$$Q_{bnet} = 7cB^2 = 7(500)(1)^2 = 3500 \text{ lb}$$
$$Q = 7850 + 3500 = 11{,}350 \text{ lb} = 5.68 \text{ tons} (50.5 \text{ kN})$$

Feld rule: Assume a spacing of $2.5D$, or 2.5 ft center to center. The group will be $(30/2.5 - 1) = 11$ piles in the $A$ direction and $(20/2.5 - 1) = 7$ piles in the $B$ direction, or 77 piles total. This will include:

4 corner piles each having an efficiency of 13/16, giving $4 \times 13/16 = 3.25$ pile equivalencies.

$2(7-2) + 2(11-2) = 45$ edge piles, each with 9/16, which gives 25.3 equivalencies.

$77 - 45 - 4 = 29$ center piles, each with 8/16, which gives 14.5 equivalencies for a total of $3.25 + 25.3 + 14.5 = 43$, or an efficiency of $43/77 = 56\%$.

$Q = 0.56(77)(45) = \underline{1940 \text{ tons}}$ (17.3 MN) or approximately the same as the group bearing capacity.

## 23.8.4   What Is an Appropriate Factor of Safety?

It will be recalled from the previous chapter that the factor of safety that is based on a foundation load is not the same as the factor of safety against shear failure in the soil, which is smaller. Global factors of safety for pile, pier, and pile group bearing capacities are a mix, as they include both a load criterion for punching shear and a shear strength criterion for side friction. For example, at the right in Fig. 23.16 when side friction is fully mobilized and its factor of safety is 1.0, end-bearing is not fully mobilized and its factor of safety is greater than 1.0.

Because of reliance on load tests, a factor of safety of 2 for load is generally considered acceptable for deep foundations even though in terms of shear failure it is lower than that.

## 23.9 SETTLEMENT OF PILE GROUPS

### 23.9.1 Overview

The larger the base area, the deeper the pressure bulb and the more soil will be involved in consolidation settlement. A procedure suggested by Terzaghi and Peck (1967) is widely used and is somewhat similar to that used for bearing capacity of a pile group but with an important difference: the center of action is assumed to be at two-thirds of the depth of the pile (Fig. 23.19). In other words, for settlement calculations it is as if the pile group is replaced by a shallow foundation having the same size and extending to a depth equal to two-thirds of the lengths of the piles.

Fellenius (1991) suggested a rationale for this procedure, that settlement of soil adjacent to the pile creates negative skin friction on the upper part of the pile. As this is opposed by positive skin friction on the lower part, there exists a neutral plane where there is no shearing stress and no shearing movement between the soil and the pile. The load from the pile group therefore can be conceptually replaced with a foundation at the elevation of the neutral plane. For friction piles the neutral plane is at about the one-third point, and for end-bearing piles it is at the bottom of the piles. The position of the neutral plane dictates the level of the equivalent foundation.

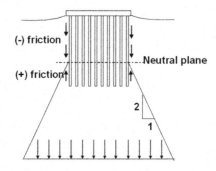

**Figure 23.19**

Terzaghi-Peck equivalent foundation method for predicting settlement of a pile group. Negative skin friction can be caused by surface loading of the soil and/or lowering of the groundwater table.

**Example 23.12**

A 1000 ton (8.9 MN) surface load is supported by a 24 ft pile group 20 × 40 ft (6.1 × 12.2 m), on soil having a cohesion of 400 lb/ft² (19.2 kPa). Estimate the vertical stress at a depth of 8 ft below the bottom of the piles.

*Answer:* A depth of 8 ft below the pile tips is $8 + (24/3) = 16$ ft (4.9 m) below the neutral plane. With a 2:1 slope this extends the base dimension $16/2 = 8$ ft in four directions, giving an expanded area of $(20 + 16)(40 + 16) = 2016$ ft² (57.1 m²). The pressure at this level is $1000/2016 = 0.5$ ton/ft² $= 1000$ lb/ft² (48 kPa).

## 23.10 SPECIAL TOPICS

### 23.10.1 Soils that Can Lift and Pull Piles

Deep foundations often are used to isolate a structure from the effects of clay expansion, where a structure founded on shallow foundations would be racked and destroyed. During the expansion cycle of such a clay, the normal stress against the sides of a pile can be multiplied by as much as 6 or 8 times in the active zone of the soil, which can extend from 7 to 10 ft (2 to 3 m) below the ground surface. This adds friction to cohesion so expansion also can lift the pile itself. In fact, in areas of severe expansive clay problems bell-bottomed piers may be used, not for their larger bearing area but to prevent uplift.

Side shear can be reduced by using protective sheaths that can slip on the pile, or coating the pile with soft mastic layers. Foundation elements of course must be reinforced to withstand tensile forces.

Where structures are supported on piles, floors that normally would be in contact with the soil also must be supported or they can be lifted by clay expansion. This is called a "structural floor," and is supported on beams connected to the pile. It also is critical that a sufficient clear space be left under the floor to allow for clay expansion.

A similar problem can occur as a result of freezing of soil to the surface of a pile or pier, followed by frost heave of the soil. This mainly occurs in frost-susceptible silty soils with access to water. In permafrost areas, foundations can be extended into the permafrost where they may be presumed to remain permanently frozen, subject to climatic influences.

### 23.10.2 Lateral Loading

Lateral loads commonly are imparted to pile foundations from the following causes:

- Wind.
- Earthquakes.

- Bridge pier supports, from current, floating debris, and ice.
- Wharves, piers, and offshore structures, from wind, waves, ice, moored ships, and ships coming to berth. Unbraced piles intended to absorb ship impacts are called fender piles.
- Railway bridges, from train sway, traction, braking, and centrifugal force on curves.
- Machinery vibrations, especially large, horizontal compressors.

Lateral forces can be provided for in design by combining vertical and nonvertical or "battered" piles in the same pile group, the latter to act as braces. In some applications, such as for bridges, lateral forces can be expected to come from one direction, whereas in other situations, such as an earthquake, they may come from any direction. Design with batter piles is simply a matter of drawing force diagrams with forces parallel to the pile directions, or of calculating the horizontal component of batter pile resistance and comparing to anticipated lateral loads. If appropriately pinned and reinforced, batter pile may act in tension as well as in compression.

### Example 23.13

A bridge pier will exert a vertical force of 440 tons (3.9 MN). Select an array of 50 ton (443 kN) piles with 70% efficiency due to group action, to support this load plus a maximum anticipated lateral load of 44 tons (390 kN) caused by current, ice, and debris.

*Answer:* Each pile capacity is $0.7 \times 50 = 35$ tons (310 kN). The horizontal capacity of this pile driven at 20° batter is $35 \sin 20° = 12$ tons (106 kN), To meet the required 44 tons, four piles in the group must be battered at that angle. The next step is to determine the vertical capacity of the four battered piles, which is $4 \times 35 \cos 20° = 131$ tons (1165 kN). The required number of vertical pile therefore is $(440 - 121)/35 = 8.8$, so the pile group will have 9 vertical piles plus 4 battered at 20°.

## 23.10.3 Pile Design for Lateral Loads

Another option is to design vertical piles to resist lateral forces, the two criteria being deflection and maximum resistance to a lateral load. The first criterion, the amount of allowable deflection, can be approached by assuming ideal elastic behavior, but a more common procedure is to use the same modulus as used in pavement design, called the modulus of subgrade reaction. Test data show this modulus to increase with depth, so the interaction between soil resistance and bending moment is defined by a "$p$-$y$ curve," where $p$ indicates pressure and $y$ indicates depth. Determination of a $p$-$y$ curve is beyond the scope of this book and usually is accomplished with a computer program. For descriptions of analytic methods see Prakash and Sharma (1990).

## 23.10.4 Wind, Signs, and Light Poles

Shallow foundations are not practical for resisting wind forces on elevated signs because of the large pad area and weight required to develop a large resisting moment. One solution is to simply insert the poles deep enough into the soil so they will not tip and tear out a chunk of soil, like a spade when the handle is pushed down. In this case the amount of deflection is less important than the lateral bearing capacity. A simple procedure was devised by Rutledge (1947) to determine a minimum embedment depth. (The authors are indebted to Dr. John Schmertmann for pointing out this method.)

Rutledge based his procedure on lateral pull-tests using a 1.5-inch diameter steel auger, and significantly, signposts that have endured a hurricane generally are bent instead of being ripped out of the ground. The assumed stress distribution is shown at the right in Fig. 23.20, which also includes Rutledge's nomograph that can be used for a solution. A factor of safety is incorporated into the soil strength data, and a deeper embedment is required in sands because of the dependence of frictional strength on overburden pressure. It is recommended that SI units be converted to ft-lb-sec for use of this method.

A diagram of wind pressure versus velocity is shown in Fig. 23.21. Arrows indicate pressures from a 100 mph (160 km/hr) wind.

**Example 23.14**
Determine the embedment depth in soil having an unconfined compressive strength of 2000 lb/ft$^2$ for a 20 ft long, 1.5 ft diameter pole supporting a sign having an area of 20 ft$^2$, to resist a wind of 100 mph.

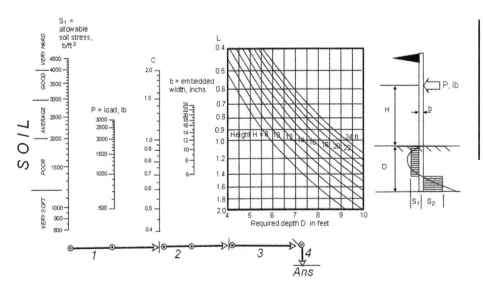

**Figure 23.20**

Nomograph for calculating required embedment depth for pole structures. (Redrawn from Rutledge, 1947.)

**Figure 23.21**

Wind pressure as
a function of
velocity. Arrows
show pressures
from 199 mph
(320 km/hr) wind.
(Drawn from data
of Farr, 1980.)

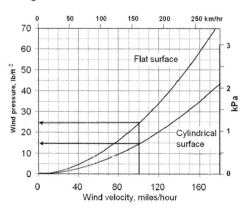

*Answer:* With a factor of safety of 2 the allowable horizontal stress is $1000\,\mathrm{lb/ft^2}$, which is entered on the left scale. The wind force is $1.5 \times 20 \times 16 + 20 \times 25 = 980\,\mathrm{lb}$, which is entered on the $P$ scale. A line through these two points reads 0.97 on the $C$ scale. Then draw a line through 0.97 and 18 in. on the $b$ scale and read 0.68 on the $L$ scale. A horizontal line from 0.68 intersects the $H = 10\,\mathrm{ft}$ mid-height to give $5\,\mathrm{ft}$ embedment.

### 23.10.5 Increase in Pile Friction from Lateral Load

Lateral loading can substantially increase the bearing capacity of piles by increasing normal stress and therefore the side friction as the pile tends to rotate about a point near the center (John Schmertmann, personal communication). This has been confirmed by pullout tests and may be an important factor that can aid design to withstand hurricane-force winds, but does not appear to have been incorporated into design. The concept is relatively straightforward: a lateral force at the top of a pile will develop an opposing couple from a triangular distribution of pressure that is a maximum at the ground surface and at the bottom of the pile. These forces multiplied by a coefficient of friction equals the increase in side resistance. This will be accompanied by decreases in normal stress and friction on the opposite sides of the pile, but the increase should be substantially larger than the decrease. A trial calculation indicates as much as a 100 percent increase in pullout capacity, depending on the soil, which may have saved more than a few structures from tipping over in a strong wind.

### Problems

23.1. What are friction piles? End-bearing piles? Compaction piles? Describe the soil requirements for each. What distinguishes between a pile, a pier, and a caisson?

23.2. What are the principal materials in deep foundation piles?

23.3. What are batter or spur piles and why are they used?

23.4. A bridge pier in a navigable stream is supported on bearing piles. What functions do the piles perform in addition to increasing the load-carrying capacity of the pier?

23.5. What precautions can be taken to increase the useful life of wood pile?

23.6. What is a bored-and-belled pier and how is it installed?

23.7. Cite some advantages of steel H-beam piles for end bearing. Can steel H-beams be used as friction pile? Explain.

23.8. Join appropriate pairs:

| | |
|---|---|
| A. Franki | a. negative skin friction |
| B. end-bearing | b. pier |
| C. bored and cast-in-place | c. taper |
| D. driven pile | d. pedestal |
| E. steel shell | e. mandrel |

23.9. A pile is driven by a drop hammer weighing 17.8 kM (4000 lb). The height of fall is 3 m (10 ft) and the average penetration of the last few blows is 1.27 mm (0.05 in.). What is the bearing capacity of the piles according to the *Engineering News* formula?

23.10. The pile of Problem 23.15 is concrete, 254 × 254 mm (10 × 10 in.) × 9.1 m (30 ft) long. The driving cushion is oak. What is the bearing capacity by the Michigan formula?

23.11. A pile is driven by a single-acting steam hammer which weighs 8.9 kM (2000 lb). The height of fall is 1.2 m (4 ft) and the average penetration under the last few blows is 8.4 mm (0.33 in). What is the bearing capacity of the pile by the *Engineering News* formula?

23.12. Explain how driving forces can induce tension in a pile.

23.13. A second test pile is driven under the same conditions as in Fig. 23.8 until the driving resistance is 50 blows/0.3 m (ft). Two weeks later the pile is re-driven with the same hammer, and 16 blows were found to be necessary to advance the pile 25 m (1 in.). What is the setup factor? How is this used in design?

23.14. As a professional engineer you review a foundation design with piers appropriately belled out to rest on a top of a layer of stiff clay, and above the clay is a saturated sand. Is this design acceptable? Why (not)? What is meant by "buildable?"

23.15. An 0.9 m (3 ft) diameter straight bored pier is founded at a depth of 10.4 m (34 ft) in clay which has an average unconfined compressive strength of 47.9 kPa (1000 lb/ft$^2$). Calculate end bearing, side friction, and the allowable bearing capacity with a factor of safety of 2.0.

23.16. Check the previous answer for settlement.

23.17. The pier of Problem 23.15 was installed and tested, and settled excessively at 180% of the design load. Revise the design with a longer pier that will provide a factor of safety of 2.2 and explain why you require 2.2 instead of the target 2.0.

23.18. Design a pier that will carry 20 tons in end bearing on sand with an average blow count of 12 blow/ft and check for settlement. The factor of safety will be 2.0.

23.19. Make a preliminary estimate of side friction and end bearing of a 15 ft, 1 ft diameter pile driven in soil with an unconfined compressive strength of $800 \, \text{lb/ft}^2$. What assumptions are made in your analysis?

23.20. A 10 ft, 0.5 ft diameter pipe pile is driven in sand having an internal friction angle of $30°$. Calculate side friction assuming that $K = 1.0$, $\gamma = 120 \, \text{lb/ft}^3$, and contact friction is 80% of the internal friction.

23.21. A group of 24 piles 0.3 m (1 ft) in diameter is driven in four rows of six piles each. They are spaced 1.1 m (3.5 ft) center to center. If each individual pile has a bearing capacity of 214 kN (24 tons), determine the bearing capacity of the group and indicate the criterion.

23.22. If the piles in Problem 23.21 are driven 6.1 m (20 ft) into cohesive soil having an unconfined compression strength of 19.2 kPa (400 lb/in.$^2$), check to determine stability of the whole group.

23.23. What area should be used to determine the average vertical stress in soil at the bottom of the piles in Problems 23.21 and 23.22?

23.24. Calculate vertical soil stress imposed by the pile group in Problems 23.21 and 23.22 at a depth 10 ft below the bottom. How would this value be used to predict settlement?

23.25. Sketch the arrangement for a pile load test, indicating jacks, reinforced anchor piles, and dial gauges. What loading sequence should be used for a pile with a design load of 120 tons?

23.26. What are the failure loads and working loads of the pile in Fig. 23.14 according to the 29 mm/N (0.01 in./ton) gross settlement and net settlement criteria? Are these reasonable criteria for this test?

23.27. Deep foundation elements are to be installed in a soil with the groundwater table at a depth of 8 ft. Soil samples show mottled gray and brown colors to a depth of 18 ft. What water table depth should be used in design? How will this affect a design?

23.28. Give reasons why the natural groundwater table generally is lowered in cities.

23.29. Calculate the maximum tensile and net uplift forces on a 0.3 m (12 in.) diameter end-bearing pile extending through 6.1 m (20 ft) of expansive clay, one-third of which is below the permanent water table. Assume $c' = 14 \, \text{kPa} \, (2 \, \text{lb/in.}^2)$.

23.30. Design a vertical and batter 6-pile system to sustain bidirectional lateral loads equal to 30% of the vertical load, all piles being equal in length.

23.31. Telephone poles averaging 20 inches in diameter with an above-ground height of 30 ft long are designed for a wind speed of 120 mph. Ignore wind pressure on the cross-bars and lines and determine the embedment depth in soil having an unconfined compressive strength of $1600\,lb/ft^2$.

# REFERENCES

Failmezger, R. R. A., Zdinak, A. L., Darden, J. N., and Fahs, R. (2005). "Use of the rock pressuremeter for deep foundation design." *Pressio 2005/ISP5 International Symposium* 1:469–478. Presses de l'ENPC/LCPC, Paris.

Farr, H. H. (1980). *Transmission Line Design Manual.* U.S. Dept. of Interior, Denver. U.S. Govt. Printing Office.

Fellenius, B. H. (1990). "Pile Foundations." In H.-Y. Fang, ed., *Foundation Engineering Handbook*, 2nd ed., pp. 511–536.

Fox, L. (1948). "Computations of Traffic Stresses in a Simple Road Structure." *Proc. 2d Int. Conf. on Soil Mechanics and Foundation Engineering* 2, 236–246.

Hirsch, T. J., Carr, L., and Lowery, L. L. Jr. (1976). *Pile Driving Analysis—Wave Equation Users' Manual.* USDOT Rept. No. FHWA-1P-76-13, National Tech. Inf. Service, Springfield, Va.

Kulhawy, F. (1991). "Drilled Shaft Foundations." In H.-Y. Fang, ed., *Foundation Engineering Handbook,* 2nd ed., pp. 537–552.

Michigan State Highway Commission (1965). "A Performance Investigation of Pile Driving Hammers and Piles." Research Project 61 F-60, Lansing, Michigan.

Osterberg, Jorg (1999). "The Osterberg Load Test for Bored and Driven Piles—the First Ten Years." *Proc. 7th Int. Conf. on Deep Foundations.* The Deep Foundation Institute, Vienna, Austria.

Poulos, H. G., and Davis, E. H. (1980). *Pile Foundation Analysis and Design.* John Wiley & Sons, New York.

Prakash, S., and Sharma, H. D. (1990). *Pile Foundations in Engineering Practice.* John Wiley & Sons, New York.

Rabe, W. H. (1946). "Dynamic Pile-Bearing Formula with Static Supplement." *Engineering News Record (*Dec. 26).

Reese, L. C., and O'Neil, M. W. (1989). "New Design Method for Drilled Shafts from Common Rock and Soil Tests." In F. H. Kulhawy, ed., *Foundation Engineering: Current Principles and Practices,* 2, 1026–1039. ASCE, New York.

Reese, L. C., Touma, F. T., and O'Neil, M. W. (1976). "Behavior of Drilled Piers under Axial Loading." *ASCE J. Geotech. Eng. Div.* 10(GT5), 493–510.

Rutledge, P. C. (1947). "Chart for Embedment of Posts with Overturning Loads." *Outdoor Advertising Assoc. Amer.* Quoted by Donald Pattterson in *Civil Engineering* 527, 69, July 1958.

Smith, E. A. L. (1960) "Pile-Driving Analysis by the Wave Equation." *ASCE Soil Mech. and Foundation Eng. Div.* 86(SM4), 35–61.

Spangler, M. G., and Ill. Mumma, F. (1958). "Pile Test Loads Compared with Bearing Capacity Calculated by Formulas." *Proc. Highway Res. Bd.* 37.

Terzaghi, K., and Peck, R. B. (1967). *Soil Mechanics in Engineering Practice,* 2nd ed. John Wiley & sons, New York.

Vesic, A. B. (1963). "Bearing Capacity of Deep Foundations in Sand." *Highway Research Record* 39, 112–153.

# 24

# Intermediate Foundations and Ground Improvement

## 24.1 OVERVIEW

### 24.1.1 Definitions

"Intermediate foundations" are similar in purpose to deep foundations, but are sufficiently different that in this book they are treated as a separate category and described in a separate chapter. Intermediate foundations generally are intermediate in cost, depth, and bearing capacity between deep and shallow foundations. The target is soils that are too weak to develop adequate support from shallow foundations, but not so weak as to require deep foundations. Like rigid piles and piers, intermediate foundations can markedly reduce settlement, but the mechanisms by which this is accomplished can be quite different,

"Ground improvement" has become somewhat of a catch-all term that presumably might include any process that improves the ground, from drainage to soil nailing. In this textbook ground improvement includes processes such as grouting, where a foreign material is introduced into soil without depending on discrete structural elements. "Soil reinforcement" strengthens soil by introducing structural elements.

Intermediate foundations will be discussed first.

### 24.1.2 Contrasts

The strength of the chain depends on the weakest link, and in the case of a foundation system the weakest link is the soil. It therefore is permissible to dilute the strength of stronger components without adversely affecting overall performance. Stone columns and Rammed Aggregate Piers substitute compacted coarse aggregate for concrete, and the aggregate columns still are stronger than the soil. Lateral confining pressure increases stability but the aggregate columns are

AGGREGATE COLUMNS    STABILIZED SOIL    GROUT COLUMNS

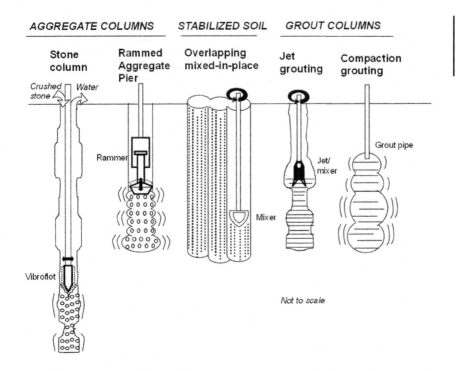

**Figure 24.1**
Some kinds of
intermediate
foundations.

not rigidly cemented, and allow adjustments to occur during loading. The method of installation can create much higher lateral stresses than is possible with hydrostatic pressure from fluid concrete, which changes the responsiveness of the system.

An example of a difference in behavior is when upper parts of uncemented aggregate columns compress and bulge outward under load, thereby increasing side friction and causing stress to be more quickly transferred into the surrounding soil. As soil around and between the foundation elements is compressed laterally its strength also is increased, analogous to the behavior of soil in a triaxial test.

Other kinds of intermediate foundations incorporate cementitious materials so the end product can behave more like a pile, except that the methods of installation tend to create irregular surfaces that increase side friction. Schematic sketches of some intermediate foundations are shown in Fig. 24.1.

## 24.2 STONE COLUMNS

### 24.2.1 It Started with Vibroflotation

A method for penetrating and strengthening soil called *Vibroflotation* was developed in Germany in the 1930s for deep compaction of sand. A long, horizontally vibrating probe is lowered into a sandy soil. Penetration can be aided with water jets that liquefy and wash loose soil up to the ground surface. In the original process, sand was used to fill the void left as the probe was withdrawn, so the process in essence was one of ground improvement. However, when crushed stone replaced sand as a backfill, the result is structural columns of compacted coarse aggregate. Open-graded aggregate is used to facilitate drainage.

Stone columns are best adapted for moderately weak soils, but they should not be so weak that they do not hold an open hole. The vibroflot is lowered into the soil and either makes its own hole or water is flushed out the end to help advance the hole.

After the full depth is reached, crushed stone can either be dumped in at the ground surface or introduced at the bottom through a feed tube. The final column diameter is 2.5 to 4.5 ft (0.75 to 1.4 m), depending in part on the strength of the soil. Typical replacement precentages are 2 to 3.5 percent, and columns as deep as 100 ft (30 m) have been installed.

### 24.2.2  Applications of Stone Columns

Stone columns are best adapted for stabilizing large areas, for example under embankments or mat foundations. They also are used for support of individual columns and walls, but are limited to about 20 to 50 tons (180 to 450 kN) per column (Bachus and Barksdale, 1989).

The compacted stone has a moderately high shear strength with a friction angle that can be expected to be around 36°. In typical installations the probe diameter is 12 to 18 in. and the spacing is 6 to 9 ft on a triangular pattern. An important design parameter is the area replacement ratio, $R_a$, which is defined as the ratio of the cross-sectional area of the stone columns, $A_c$, to the total area, $A$. The relation between diameter, spacing, and area coverage is shown in Fig. 24.2, from which

$$R_a = \frac{A_c}{A} = \frac{\pi}{2\sqrt{3}}\left(\frac{d}{s}\right)^2 \tag{24.1}$$

**Figure 24.2**

Area ratio is determined from a "unit cell" that contains the complete repeating pattern.

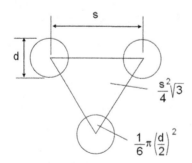

$$\frac{s^2}{4}\sqrt{3}$$

$$\frac{1}{6}\pi\left(\frac{d}{2}\right)^2$$

where $d$ and $s$ are as shown in Fig. 24.2. It will be noted that $d$ represents an expanded diameter after the aggregate is vibrated into place.

### 24.2.3 Shear Strength

Bearing capacity and slope stability depend on shearing resistance of the soil-stone column composite, which usually is calculated on an area replacement basis. As the column area increases the soil area decreases, so the net effect on shearing strength is

$$\Delta S = R_c(\tan \phi_{sc} - \tan \phi) \tag{24.2}$$

This ignores stress concentration, which puts more weight on the columns and can substantially increase their horizontal shearing resistance, as discussed in the following paragraphs.

### Example 24.1

The shear strength in the slip zone of an active landslide is determined to be $\phi = 12°$ and $c = 0$. Without considering stress concentration, what will be the contribution of stone columns to a factor of safety of a landslide stabilized with stone columns having $\phi_c = 38°$, an average expanded diameter $d = 24$ in. (0.6 m), and a spacing $s = 10$ ft (3 m)?

*Answer:*

$$R_a = \frac{\pi}{2\sqrt{3}}\left(\frac{2}{10}\right)^2 = 0.036, \quad \text{or} \quad \frac{\pi}{2\sqrt{3}}\left(\frac{0.6}{3}\right)^2 = 0.036$$

$$\Delta S = 0.036\,(\tan 38° - \tan 12°) = 0.0205$$

The factor of safety in an active landslide is 1.0, so

$$FS = (1 + 0.0205)/1 = 1.02$$

This ignores stress concentration in the columns.

### 24.2.4 Stress Concentration

Stress is concentrated in the stone columns as a result of downdrag by the surrounding soil as it compresses under its own weight or its weight in combination with a surface load. This is expected because of the higher modulus of the stone column compared with that of the soil. As vertical stress is increased in the columns, it is compensated by a decrease in vertical stress in the soil, but as the soil has a lower friction angle the net result is a substantial gain in shear strength. The soil weight will receive more support close to the columns, but the average should remain the same.

The stress concentration factor $n$ may be defined as the ratio of vertical stress in the stone column to that in the soil, and is reported by Barksdale and Bachus (1983) to be in the neighborhood of 3.5 to 5. Studies reported by Kirsch and

Sondermann (2003) indicate values between 2 and 3, and they describe field measurements in which $n$ was 2.6 and 2.8.

With stress concentration, the column weight per total unit area is multiplied by $n$. This is compensated by a reduction in net weight of the soil. As the column normal force is increased by $n$ and its shearing resistance is increased by $n \tan \phi_{sc}$, the soil normal force is decreased by $n$ and its shearing resistance is decreased by $\tan \phi$ for the soil. The net effect is

$$\Delta S = nR_c \left( \tan \phi_{sc} - \tan \phi \right) \tag{24.3}$$

Equation (24.2) may be overly conservative because it does not include benefits from compaction of the soil.

**Example 24.2**
Correct the factor of safety in the above example for a stress concentration ratio $n = 2.5$ without and with a reduced contribution from the soil.

*Answer:* Without a reduction for the decrease in soil area,

$$\Delta S = 2.5(0.036)(\tan 38°) = 0.090(0.78) = 0.70 \text{ and FS} = 1.07$$

With a soil reduction,

$$\Delta S = 0.090(\tan 38° - \tan 12°) = 0.05 \text{ and FS} = 1.05$$

## 24.2.5 Foundation Bearing Capacity

A simple calculation of bearing capacity of a stone column-supported foundation on clay may be made by assuming that the controlling factor is passive yielding of the clay that will allow the column to shear. Bearing capacity then is expressed in terms of a modified cohesion factor $N_c$ in the bearing capacity equation:

$$q_0 = cN_c'' \tag{24.4}$$

where $c$ is cohesion of the matrix soil and $N_c''$ is the cohesion factor modified for the presence of stone columns. Limited experimental evidence indicates that $N_c''$ is between 18 and 25, which according to data in Table 22.2 corresponds to an effective combined friction angle for the soil and columns of about 28° to 31°. Part of the apparent better effectiveness of stone columns for foundation reinforcement compared with landslide stabilization probably relates to higher stresses and better lateral confinement induced under a foundation. For undrained clay soil $c$ is one-half of the unconfined compressive strength.

**Example 24.3**
Estimate the influence of stone columns on the net bearing capacity of a square foundation on clay having $\phi = 0$.

*Answer:* From Table 22.3, without stone columns $N_c'' = 6.2$. With stone columns and $N_c'' = 18$ to 25, bearing capacity is estimated to be increased about 3 to 4 times.

## 24.2.6 Bearing Capacity and Settlement from Soil Mechanics

A more rigorous solution for bearing capacity will use all terms in the bearing capacity equation, with the bearing capacity factors defined by composite friction angles that are adjusted for stress concentration and area replacement. The same approach can be used to predict settlement. For long columns an expression by Barksdale and Bachus (1983) for estimating maximum settlement with stone columns is as follows:

$$I_c = 1 + (n+1)/R_c \tag{24.5}$$

where $I_c$ is an improvement factor, or ratio of settlement without the columns to settlement with the columns. A graph of this equation is shown in Fig. 24.3. The higher the stress concentration factor, the less stiff the soil and the greater the need for improvement.

### Example 24.4

A mat foundation is predicted to settle 1.5 inches. Estimate the $R_c$ stone column replacement ratio to reduce settlement to 1.0 inch.

*Answer:* Let $n = 3$. The required improvement factor is $I_c = 1.5/1 = 1.5$. If the anticipated stress concentration factor $n = 3$, from Fig. 24.3, $1/R_c = 3.7$ and $R_c = 0.27$.

The stress concentration factor is a consequence of negative skin friction, and therefore may be limited by the shearing resistance of the soil. The factor therefore may decrease and become less significant below a depth that depends on the soil shearing strength and the developed lateral stress.

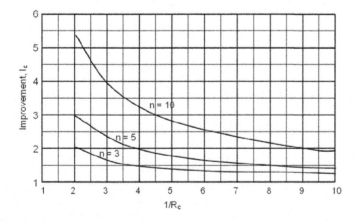

### Figure 24.3

Settlement factor in relation to replacement ratio $R_c$ and stress concentration factor $n$ (modified from Barksdale and Bachus, 1983).

### 24.2.7 Summary

Stone columns are used to improve soil bearing capacity and to stabilize slopes, where their effectiveness depends on an area replacement ratio and on a stress concentration factor. Soil settlement around the columns can increase vertical stress in the columns by about a factor of 2.5, which increases the shearing resistance of the columns but decreases that of the soil.

In order for the soil to be suitable it must be weak enough to allow efficient penetration by the vibrator, and strong enough to resist caving during installation. Stone columns are best adapted to soils having a shearing strength of 200 to 1000 lb/ft$^2$ (10 to 50 kPa) (Bachus and Barksdale, 1989). In stiff soils columns can be installed in pre-drilled holes or by driving an open pipe and filling and compacting stone with a drop hammer as the pipe is withdrawn (Broms, 1991a), but these procedures increase installation costs.

## 24.3 MIXED-IN-PLACE PIERS

### 24.3.1 Concept and Use

Soil-cement was developed in the 1930s as a base material for secondary roads, and is particularly useful in areas where aggregate is scarce. Soil-cement is a mixture of Portland cement, soil, and water that is compacted like soil instead of being poured like concrete. The same concept was later extended to roller-compacted concrete in order to decrease the water/cement ratio to a minimum that is required to hydrate the cement.

In the 1950s a technique was developed by the Intrusion Prepakt Co. in the U.S. to make a continuous wall by mixing Portland cement with soil in overlapping boreholes. The method was not widely used until it was independently developed in Asia and in Scandinavia in the 1970s, where it is used to stabilize soft bay muds in coastal areas and to construct diaphragm walls to prevent seepage. The same mixing equipment is used to introduce unslaked lime (quicklime) or fly ash into soil, and can be employed to stabilize or confine hazardous wastes and sludges.

### 24.3.2 Construction Methods

Modern Deep Soil Mixing (DSM) is accomplished with a special cutter or a hollow-stemmed auger. As the cutter is slowly advanced into the soil, cement slurry is injected under pressure through holes near the tip. Additional mixing paddles may be attached along the shaft of an auger. Soil and slurry are mixed as the cutter advances and more slurry is added as the mixer is removed. In very wet soils dry cement or quicklime can be injected using compressed air. A modification used in Japan has up to eight simultaneous and overlapping mixing units to increase productivity and help ensure continuity of a wall.

### 24.3.3  Mix Design

Because the main variable is the soil, it is critical that trial mixes be prepared, cured, and tested. For example, lime reacts only with certain soil clay minerals (smectite) and is nonreactive with sand. Bay muds and deltaic deposits even in the tropics often contain smectites that have been carried in from remote areas. Sulphate soils can react with lime to create a highly expansive mineral called ettringite, which at least in theory can decrease the integrity of the column. Cement can react poorly with organic soils that contain humic acids. Fly ash is the relatively inexpensive byproduct from burning powdered coal, and Type C (for calcium) ash contains appreciable amounts of lime both as the oxide and as cementing compounds. Type C ash is produced from coal that contains limestone. If Type F (for iron) low-calcium fly ash is used, it must be mixed with lime to obtain a reaction.

For soil-cement columns the amount of cement required depends on the soil, and usually is in a range of 5 to 15 percent by weight of the soil. After setting, soil-cement columns have a compressive strength of about 100 to 300 lb/in.$^2$ (700 to 2100 kPa) in silts and clays, and two or three times that in sands. Lime introduced into a clay soil flocculates the clay and substantially increases the soil permeability, whereas hydrating cement tends to fill pores and decrease it. Lime columns therefore can function as drains, while cement columns are more satisfactory for diaphragm walls. Both additives increase the soil compressive strength, but the pozzolanic reaction between clay and lime requires a longer curing time, so soil-lime columns are more susceptible to creep and may develop internal fractures that decrease the compressive strength.

### 24.3.4  Use for Foundation Support

Design methods for lime or cement columns are similar to those used for drilled shafts and for pile groups (Broms, 1991b), so long as the stabilized soil column has sufficient strength to support the foundation load.

## 24.4  RAMMED AGGREGATE PIERS™ (RAP)

### 24.4.1  Why They Are Different

Rammed piers are similar to stone columns in being composed of crushed stone, but differ because instead of being horizontally vibrated into place, the stone is densely compacted by vertical ramming in 1-foot (0.3 m) layers. A particularly important feature of the installation is the beveled shape of the rammer, which drives the stone sideways and builds up lateral pressure in the soil (Fig. 24.4). The rammer is driven by a high-energy hydraulic hammer so there is uniform compaction and lateral stress along the length of the pier.

Unlike rigid piles and piers, stone columns and Rammed Aggregate Piers (RAPs) are compressible. RAPs are load tested like piles, and pulling resistance can be measured by pulling upward on a steel plate installed at the bottom and connected to steel rods that extend upward along the sides to the ground surface. Telltales installed in RAPs indicate that a plate load applied at the top is quickly transferred to soil around the upper parts of the piers (Fig. 24.5). This is in sharp contrast to the stress distribution along rigid piers such as shown in Fig. 23.12.

Under wider loaded areas such as embankments, pier-soil interaction can carry stresses deeper.

## 24.4.2 New Insights

Because lateral stress is uniform in the RAP system, it is well suited for research investigations of lateral in-situ stress. The $K_o$ Stepped Blade, which is discussed in Chapter 26, has been extensively employed in RAP research and gives new insights that also should be applicable to other foundation systems. Soil behaviors that will be discussed include temporary liquefaction, radial compaction, and radial tension cracks.

**Figure 24.4**

Rammed Aggregate Pier$^{TM}$ (or Geopier® soil reinforcement) are constructed by ramming layers of graded coarse aggregate into an open boring. The downhole position of the rammer helps to contain the impact noise. The bevelled face of the rammer pushes stone outward to create a high lateral stress in the matrix soil.

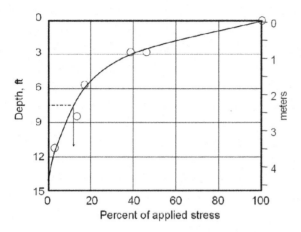

**Figure 24.5**

Unlike rigid piles and piers, the upper part of Rammed Aggregate Piers and stone columns compresses and quickly transfers load to the soil. Arrows show transfer of 90 percent of the load to the upper half of the pier. For comparison see Fig. 23.12. Data are from Lawton and Merry (2000).

## 24.4.3  Radial Influences from Ramming

Approximately one-half of the contact area of the rammer used for compacting RAPs is beveled at 45° in order to direct a substantial part of the ramming energy laterally as well as downward. The rammer is attached to a hydraulically operated rammer (Fig. 24.4) that is a large version of a jack hammer except that it is operated with hydraulic oil instead of compressed air.

A hole is bored to the full depth of the pier, the most common diameter being 2.5 ft (0.75 m). The hole preferably is self-supporting, but it also can be held open with a steel casing that is withdrawn as pier layers are compacted from the bottom up. The first layer of aggregate then is dumped in and rammed into place. The first layer is open-graded, and is given extra compaction energy to create a bulb that is a base for compaction of the next layer. The open grading does not affect infiltration by soil because only a small fraction of the foundation pressure reaches the lower end of the pier.

Subsequent layers are composed of graded stone, and each layer is apportioned to be approximately 1 ft (0.3 m) thick after ramming. The amount of compaction energy is adjusted to obtain a uniform product. A schematic diagram of finished piers is shown in Fig. 24.6.

## 24.4.4  Temporary Liquefaction During Ramming

In Fig. 24.7 radial stresses measured with the $K_o$ Stepped Blade are plotted versus radial distance from the pier center, and indicate that the same stress exists

**Figure 24.6**

Lateral stress
from pier ramming
creates an elastic
upper zone in the
soil. At the right is
a detail for an
uplift pier. The
high skin friction
allows them to be
used for
tie-downs in pier
load tests.

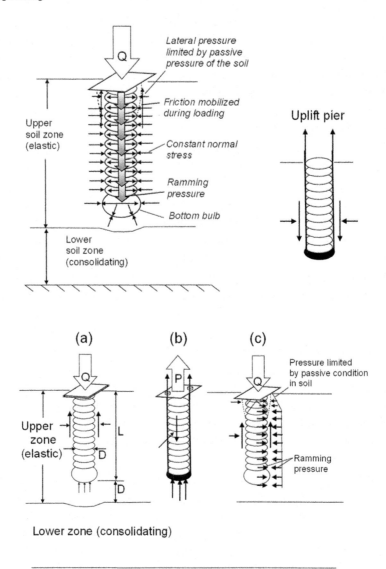

at B, approximately 2 m (6 ft) from the center of the pier, as at A along the
pier surface. This behavior is consistent in soft soil that is below a groundwater
table, and is possible only if the soil behaved as a liquid during ramming. There
are important implications, shown in Fig. 24.8: as liquefied soil increases stress in
all directions, it also must act vertically, prestressing the soil to the lateral
ramming pressure that in this case is 2500 lb/ft$^2$ (120 kPa). Vertical prestressing
can be effective only if the soil drains, which is discussed in the following
paragraphs.

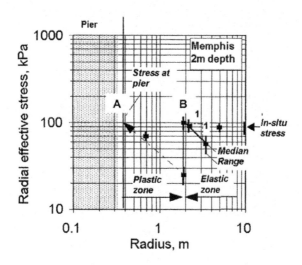

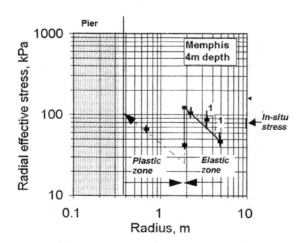

**Figure 24.7**
Transfer of radial stress from A to B in soil near Rammed Aggregate Piers indicates temporary liquefaction to a distance of about 6 ft (2 m) from the center of the pier. (From Handy and White, 2006, with permission of the American Society of Civil Engineers.)

Liquefaction can occur only if the soil is saturated. A second requirement is that the soil structure must break down as a result of lateral pressure from ramming, in which case the governing factor appears to be the soil unconfined compressive strength, which must not exceed about 2500 lb/ft² (120 kPa) for liquefaction to occur, although this will vary depending on the impact forces developed by the rammer.

### 24.4.5 Radial Tension Cracks

Outside of the remolded zone there is a consistent 1:1 downward slope of the stress line as shown in Fig. 24.7. This is not in accordance with elastic theory unless there is tangential stress relief from radial tension cracking. Radial cracking previously has been hypothesized to explain a rapid decline in pore water pressure

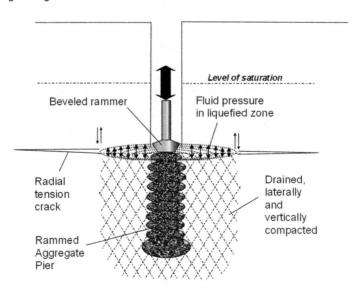

**Figure 24.8**

Stress measurements indicate that temporary liquefaction can vertically prestress soil to approximately 2500 lb/ft² (120 kPa) and inject it into peripheral tension cracks. The prestressed zone is approximately 3.5 m (12 ft) in diameter. (Courtesy of the Geopier Foundation Co., Inc.)

measured near driven piles. In addition to allowing rapid drainage, injection of liquefied soil into the cracks can prop them open and prevent stress relief from rebound as the liquefied soil drains. The stress measurements indicate that radial tension cracks extend about 3 m (9 ft) beyond the diameter of the remolded zone to act as a drainage gallery.

Radial cracking is prevented if the induced radial stress, which is of the order of 2500 lb/ft² (120 kPa), is lower than the existing lateral in-situ stress.

**Example 24.5**

Calculate the depth below which radial cracking will not occur in a normally consolidated soil having a friction angle of 25° and a unit weight of 124 lb/ft³ (19.5 kN/m³), (a) above the groundwater table, (b) below the groundwater table.

*Answer:* Assume $K_0 = 1 - \sin 25° = 0.58$.

(a) $2500 \, \text{lb/ft}^2 = 0.58(124 \, \text{lb/ft}^3) \, z$; and $z = 35 \, \text{ft}$, or $120 \, \text{kN/m}^3 = 0.58(19.5 \, \text{kN/m}^2) \, z$; and $z = 10.6 \, \text{m}$.

(b) $2500 = 0.58(62) \, z$, and $z = 70 \, \text{ft}$ (21 m).

This probably puts a depth limit on radial drainage, which is a reason for these to be classified as intermediate foundations.

## 24.4.6 Recovery

Liquefied soil obviously has zero shear strength, but as it drains and is simultaneously stressed by ramming from above it can consolidate and gain strength. The slope of the stress line in the plastic zones in Fig. 24.7 can be analyzed by the theory of plasticity, which indicates that the soil friction angle after ramming is about 38°. This was confirmed by borehole shear tests at a different site (Fig. 24.9). Depending on the friction angle prior to ramming, there can be a 40 to 60 percent increase in tan $\phi$, which can be beneficial for slope stabilization.

A schematic representation of soil conditions around RAPs with different soil conditions is shown in Fig. 24.10. Radial tension cracks have been observed at the ground surface in hard, unsaturated soils.

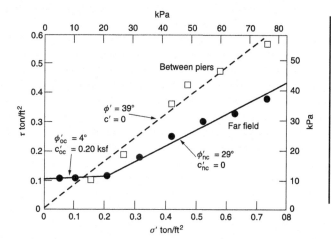

**Figure 24.9**

Shear tests performed in situ before and after RAP installation show effects of radial compaction. (Courtesy of Geopier Foundation Co. Northwest, Portland, Oreg.)

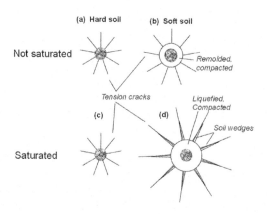

**Figure 24.10**

Different soil responses near Rammed Aggregate Piers: (a), (c) tension cracks if tangential tensile stress >$K_0$ in-situ stress; (b) radial compaction if radial ramming pressure <soil $q_u$; (d) temporary liquefaction of saturated soil with drainage and wedging into open cracks.

### 24.4.7 Liquefaction and Earthquakes

As discussed in Chapter 27, a major cause of damages to structures in earthquakes is from temporary liquefaction of saturated sandy soils. Induced liquefaction such as by pier ramming or other artificially induced ground vibrations, which in the case of Rammed Aggregate Piers is followed by forced drainage, therefore can protect against earthquake liquefaction. This has been confirmed with case histories.

## 24.5 SETTLEMENT OF FOUNDATIONS SUPPORTED BY RAMMED AGGREGATE PIERS

### 24.5.1 Boosting the Preconsolidation Pressure

Structures supported on Rammed Aggregate Piers show surprisingly little settlement, and measurements of settlement indicate that the soil penetrated by the piers is not consolidating but is behaving elastically. This can be explained in part by the prestressing from liquefaction illustrated in Fig. 24.8, but also appears to occur in soil above a groundwater table.

One hypothesis that may help to explain this is illustrated by Mohr circles in Fig. 24.11. The circle at the left represents a normally consolidated soil with a lateral stress at $A$ and a vertical stress at $C$. The circle diameter is such that the top is on the $K_0$ line.

Now ramming increases the lateral stress to $B$, and before the soil can consolidate, the vertical stress must increase to $D$, so $D$ is a new quasi-preconsolidation pressure. It can be seen that the influence of an increase in lateral stress is magnified, as line $AB$ is shorter than $CD$.

A physical model for this hypothesis is shown at the bottom of Fig. 24.11. At the left the particles are on the verge of slipping if there is any increase in the vertical load, but at the right, if a lateral stress is applied, the vertical load must be larger in order to initiate slipping. Note that no slipping need actually occur for friction to change at the grain contacts. This concept is supported by recent laboratory tests (Pham, 2005).

Another indication of elastic response is from uplift load tests that are conducted using a bottom plate and vertical steel rods as shown at the right in Fig. 24.6. When the uplift load is removed there is essentially a 100 percent strain recovery.

The depiction in Fig. 24.11 does not include the intermediate principal stress, but stress measurements in the field show that lateral stress is more or less homogeneous near a pier group. As a limiting case the friction arrows can be reversed by a lateral stress, and then again must reverse to allow consolidation.

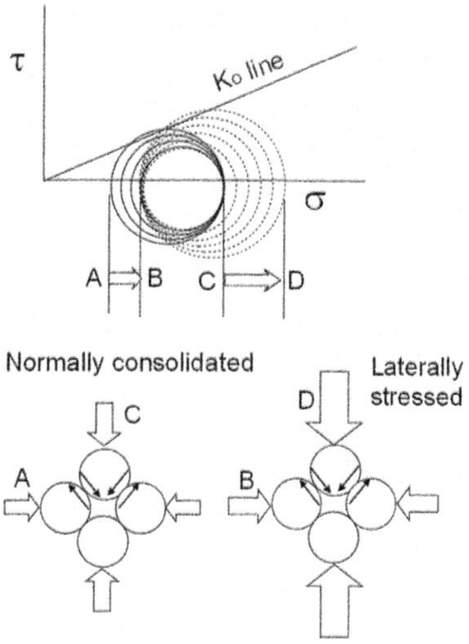

Figure 24.11

(Top) Increasing lateral stress from A to B by ramming means that vertical stress must increase from C to D to initiate consolidation. The reason is a reversal of the friction arrows (below). The net result is an increase in the preconsolidation pressure.

If the lateral stress is taken to the passive limit and the vertical stress is subsequently taken to the limit defined by $K_0$, it can be shown that the maximum multiplier for vertical stress is (Handy, 2001):

$$M = \frac{1 + \sin\phi}{(1 - \sin\phi)^2} \qquad (24.6)$$

Some solutions for this equation are as follows:

| $\phi$: | 0 | 10 | 20 | 25 | 30 |
|---|---|---|---|---|---|
| $M$: | 1 | 1.7 | 3.1 | 4.3 | 6.0 |

The factor also is limited by lateral stress already existing in the soil, which generally increases with depth. For example, if the existing lateral stress equals or exceeds that induced by pier ramming there can be no stress-induced change to elastic behavior. Some depths at which this occurs are calculated in Example 24.5. Lateral stress also can be high in an expansive clay, but then the problem usually is not settlement under load, but settlement as a result of desiccation.

## 24.5.2 Elastic vs. Consolidation Settlement

Elastic compressibility is much lower than that which occurs during consolidation as soil grains actually slip over one another. This is illustrated by the graph in Fig. 16.14, where the curve is assumed to be flat at pressures lower than the preconsolidation pressure.

In practice, pier load tests are used to confirm a pier design and to predict settlement from compression of the elastic zone. As in the case of stone columns, it is convenient to define an area ratio and a modulus ratio (Fox and Cowell, 1998). However, in the case of RAPs the modulus is measured from a load test. In a load test where the influence depth is not fixed by a specimen length, a modulus is determined from the slope of the deflection vs. pressure diagram. Whereas the modulus of elasticity, $E$, is the slope of a stress-strain curve and has units of stress, the pier modulus, $k$, has units of stress per unit of compression. A common designation is in *pounds per square inch per inch*, which is the equivalent of pounds per cubic inch, or the equivalent in $kN/m^3$. It should be emphasized that these are *not* unit weights, but indicate compressibility under a surface load. An analogous *modulus of subgrade reaction* is used in pavement design, but in that case the test involves loading on a laterally continuous layer instead of being concentrated on a pier diameter.

Let $R_a$ = area ratio; ratio of the pier area to the gross footing contact area (dimensionless);

$R_k$ = modulus ratio; ratio of the pier modulus to that of the matrix soil from load tests (dimensionless).

The following equation can be derived from a consideration of these ratios:

$$q_{RAP} = \frac{qR_k}{R_a R_k - R_a + 1} \qquad (24.7)$$

where $q_{RAP}$ is the pier bearing stress at the top of the pier element and $q$ is the average footing bearing pressure on a total area basis. Settlement from compression of the upper zone is

$$S_1 = \frac{q_{RAP}}{k_{RAP}} \qquad (24.8)$$

where $k_{RAP}$ is the pier subgrade modulus measured from a modulus load test.

The pier modulus test using a circular plate exerts a concentrated bearing pressure that probably is an extreme condition compared with that which will exist in service, because of partial support of a foundation by the surrounding soil. The measured $k$ modulus therefore should be on the conservative side. This has been confirmed from performance evaluations.

### Example 24.6

An 80 ton column load is to be supported on footing 6 ft × 6 ft (1.8 × 1.8 m) square footing. $R_a$ is 0.33, $R_k = 20$, and $k_{RAP} = 260 \, lb/in.^3$ (0.271 N/cm$^3$). Predict settlement from compression of the upper layer.

*Answer:* The bearing pressure is $q = 160,000/6 \times 6 = 4440 \, lb/ft^2$ (412 m$^2$). From eq. (24.7),

$$q_{RAP} = \frac{(4440)(20)}{(0.33)(20 - 0.33 + 1)} = 12,215 \, lb/ft^2 \ (585 \, kN/m^2)$$

From eq. (24.8),

$$S_1 = \frac{12,215\,\text{lb/ft}^2 \div (12\,\text{in./ft})^2}{260\,\text{lb/in.}^2} = 0.33\,\text{in. (8.3 mm)}$$

## 24.5.3 Stress Transfer to the Lower Layer

Compaction of soil below the bottom of a pier extends the upper layer to an approximate depth equal to the pier length plus one pier diameter. Telltale plates typically are installed at the bottom of test piers to ensure that all or nearly all of the load is transferred to soil in the upper, elastic zone. This is important to eliminate a "mattress effect" of uneven stress distribution to the lower layer. As in the design of flexible pavements, a stiff elastic upper layer acts to spread a surface load out on a supporting weaker layer.

An example of the influence of an elastic zone on stress distribution is shown in Fig. 24.12. The Boussinesq distribution for a uniform modulus is shown on the left, and a corresponding two-layer distribution is shown on the right, where the modulus ratio $E_1/E_2 = 10$ and Poisson's ratio is reduced 50 percent in the upper layer. The maximum stress transmitted to the top of the lower layer is reduced about 50 percent.

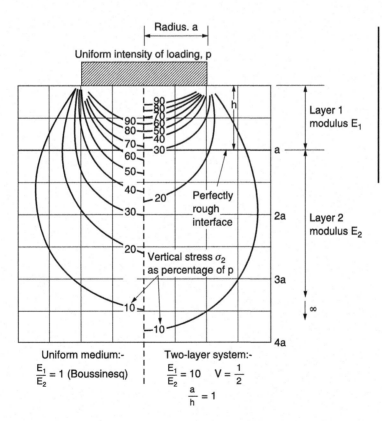

**Figure 24.12**

Influence from creating an increased elastic modulus $E_1$ in an upper layer. (After L. Fox, reproduced in Poulos and Davis, 1974.)

Several options may be considered for transfer of stress to the lower layer (Fox and Cowell, 1998):

(a) A Westergaard distribution, on the basis that a high lateral stress and compaction in the upper layer may reduce Poisson's ratio so that lateral bulging is not a factor. The Westergaard distribution is shown in Fig. 15.13.

(b) A Boussinesq stress distribution with results multiplied by 0.80.

(c) A 1.67 vertical:1 horizontal rule similar to the 2:1 rule used for shallow foundations.

(d) Schmertmann's layer strain method, which is most appropriate for sand.

In any of these methods, stress is assumed to be distributed from the surface pad. The center of stress transfer actually is deeper, but not so deep as with rigid piles because of compressibility of the upper part of the pier.

Compression of the lower layer is analyzed using Schmertmann's layer method or by conventional consolidation theory and tests.

### Example 24.7

Estimate the percent of vertical stress transmitted to the lower layer with a Rammed Aggregate Pier supporting a 1 m (3.3 ft) square pad at the ground surface. The pier is 8 ft (2.44 m) long and 2.5 ft (0.76 m) in diameter.

*Answer:* The depth to the lower layer is the pier length plus one diameter, or 10.5 ft (3.2 m). The projected square at that depth is $3.3 + 2 \times (10.5/1.67) = 16.7$ ft, or $1 + 2 \times (3.2/1.67) = 4.83$ m. The projected area is $(4.83)^2 = 23.4$ m$^2$ so the reduction is $1:23.4 = 4\%$, or $16.7^2 = 279$ ft$^2$ for a stress reduction of $(3.3)^2/279 = 4\%$.

In addition to spreading the load, by carrying it deeper the soil normally is stiffer, having already been consolidated under the overburden pressure.

## 24.5.4 Compression of the Lower Layer

As the lower layer is not modified, its compression is analyzed using conventional consolidation theory, and the amount of settlement is added to that realized from compression of the upper layer. As in the case of friction piles or piers, transferring stress deeper into a soil deposit can reduce consolidation settlement by an order of magnitude. A simple way to predict this effect is to analyze settlement of a shallow foundation by dividing the soil into layers with one layer break being at the boundary between the upper and lower layers.

### Example 24.8

A water tower in Example 16.4 was predicted to settle 3.6 in. (90 mm). How much would the settlement be reduced if the tower were supported by RAPs so that the layer boundary is at the top of the C layer in Fig. 16.11?

*Answer:* The length of 2.5 ft (0.76 m) diameter piers will be $(20 - 2.5) = 17.5$ ft (5.33 m). Try 4 piers 3 ft (0.91 m) center to center and a pad diameter of 12 ft (3.66 m). The weight of the tank full is 22,540 lb (100 kN).

*Upper layer:* The area ratio is $R_a = 4\pi (1.25)^2/\pi (6)^2 = 0.174$. The average footing pressure is $q = 22,540/\pi (6)^2 = 200$ lb/ft$^2$ (9.58 kPa). With a modulus ratio $R_k = 20$ and $k_{RAP} = 200$ lb/in.$^3$ (43.9 MN/m$^3$), from eq. (24.7) the pressure on the pier is

$$q_{RAP} = \frac{qR_k}{R_aR_k - R_a + 1} = \frac{200 \text{ lb/ft}^2(20)}{(0.174)(20) - 0.174 + 1} = 929 \text{ lb/ft}^2 = 6.45 \text{ lb/in.}^2 \text{ (44.5 kPa)}$$

From eq. (24.28),

$$S_1 = \frac{q_{RAP}}{k_{RAP}} = \frac{6.45 \text{ lb/in.}^2}{200 \text{ lb/in.}^3} = 0.03 \text{ inch, which is negligible.}$$

Consolidation of the A and B layers totaling 2.6 in. therefore is substituted by elastic compression, which is negligible, reducing the total settlement to 1 in. (25.4 mm).

## 24.5.5 Uplift Piers

Rammed Aggregate Piers are readily converted to resist uplift by attaching steel rods to a plate that is placed on top of the first lift. The rods run up along opposite sides of the boring and are attached to a surface element to resist uplift from wind or other lateral loads such as from earthquakes, and to serve as anchor piers for load tests. Design involves equating side friction to uplift force, assuming that cohesion is zero:

$$P = \pi DL\sigma_h \tan \phi \tag{24.9}$$

where $P$ is the uplift force, $D$ is the nominal pier diameter, $L$ is its length, $\sigma_h$ is the radial stress at the interface, and $\phi$ is the friction angle of the soil. The increase in pier diameter from ramming usually is not included as it is on the safe side.

### Example 24.9

Calculate the maximum uplift resistance of a 30 in. (0.76 m) diameter RAP 10 ft (3 m) long in soil having a friction angle after compaction of 35° and a uniform lateral ramming pressure of 2500 lb/ft$^2$ (120 kPa).

*Answer:* $P = \pi(2.5 \text{ ft})(10 \text{ ft})(2500 \text{ lb/ft}^2) \tan 35° = 137,000 \text{ lb} = 69$ tons (610 kN). A factor of safety of 2.5 to 3 normally is applied. The 100% strain recovery commonly realized with uplift piers when the pulling force is released is fairly unique in the annals of measured soil responses.

## 24.5.6 Shear Failure

A bearing capacity failure is unlikely in an RAP designed to control settlement, and RAPs have been successfully used in peat. The modulus in peat has been measured in a range from 75 to 125 lb/in.$^3$, compared with 125–300 for clays and

165–360 for sands. Lateral containment by peat appears to be related to fibers that act as a binding. Because side friction is essentially zero, RAPs in peat are strictly end-bearing, and transfer all load to a lower, more substantial layer.

**Example 24.10**

Evaluate the failure load on a 30 in. (0.76 m) diameter Rammed Aggregate Pier with $\phi = 50°$ and a lateral pressure of 2500 lb/ft². Allow for an arbitrary 20% relaxation of lateral pressure after ramming.

*Answer:* The principal stress ratio is (eq. (19.2))

$$\frac{\sigma_1}{\sigma_3} = \frac{1 + \sin\phi}{1 - \sin\phi} = 7.5$$

With $\sigma_3 = 0.8(2500 \text{ lb/ft}^2)$, the pier bearing capacity is $7.5 \times 0.8 \times 2500 = 15,000 \text{ lb/ft}^2$. Without any enlargement of the pier diameter from ramming, the failure load in this case would be $Q = 15,000\pi(1.25)^2] = 74,000 \text{ lb} = 37 \text{ tons}$ (330 kN). With a factor of safety of 3, $Q_a = 12 \text{ tons}$ (110 kN), which would be transferred in full to the underlying layer. This would have to be confirmed with a load test.

# 24.6  PRESSURE GROUTING

## 24.6.1  Permeation Grouting

Classic pressure grouting consists of injecting a setting fluid into cracks and voids in rock or soil. The intent may be to strengthen the soil or to reduce or prevent seepage, for example in rocks under dams and in the abutments.

The most common permeation grouts are thin mortars consisting of Portland cement, fly ash, and water, with or without the addition of various other chemicals. Particulate grouts are limited to relatively open void spaces such as in fractured rock, gravel, or coarse sand. The reason is that they are not ideal Newtonian fluids, but have a yield stress that must be overcome in order to initiate flow. The distribution of shearing velocities of Newtonian and non-Newtonian fluids is shown schematically in Fig. 24.13, which indicates that grout particles need not exceed the pore diameter for plugging to occur.

In 1925 H. Joosten, a Dutch engineer, developed a two-solution chemical that can penetrate finer-grained soils. In the Joosten process a solution of sodium silicate ("water glass") is injected in one boring and one of calcium chloride is injected into the next. Where the two solutions meet and intermingle in the soil they immediately react to form a calcium silicate gel that plugs the voids. The process was used for many years but now has been largely replaced by other "one-shot" processes that go under various trade names. Some use sodium silicate plus a setting agent. Acrylamide grouts ("AM-9") showed great promise but are highly toxic in high concentrations and have been replaced by nontoxic acrylate gels. Finer soils can be penetrated but costs are higher than for sodium silicate

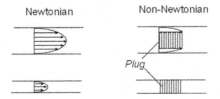

**Figure 24.13**

Shearing stress is zero in the middle of a flowing grout mass and a maximum at the boundaries, so if there is a yield stress a particulate grout will plug voids that are larger than the grout particle diameter.

grouts, which in turn cost many times more than cement grouts, which are the least expensive.

The groutability of soils is related in a general way to their hydraulic conductivity. Soils having $k \geq 10^{-1}$ cm/sec can be grouted with a particulate grout, whereas if $k = 10^{-1}$ to $10^{-3}$, a soil is groutable with commonly used chemical grouts. Soils that cannot be permeated by grout still may be suitable for compaction grouting, discussed below.

An approximate guide to the volume of grout was suggested by Baker (1982), based on the volume and porosity of the treatment zone, minus an amount for incomplete void filling, and plus an amount for grout loss to surrounding zones:

$$\frac{V_g}{V} = nF(1 + L) \tag{24.10}$$

where $V_g/V$ is the volume of grout per unit volume of soil grouted, $n$ is porosity, $F$ is a void filling factor, and $L$ is a loss factor. For example, if the porosity of a loose sand is 40 percent, the void filling is 75 percent, and the grout loss is 25 percent, the ratio of grout volume to treated volume is

$$\frac{V_g}{V} = 0.4(0.75)(1 + 0.25) = 0.4$$

If the sand is very loose it may be displaced instead of permeated. This is referred to as *compaction grouting*, which is discussed later in this section.

## 24.6.2  Pattern Grouting

Grouting normally starts at the bottom of a boring, then after grout has been injected, the grout pipe is pulled up a preset distance and the operation repeated. The grout penetration approximates a sphere, and ideally the spheres will overlap. Where grouting is used to seal off water flow a common technique is to inject a line of borings, then inject in new holes halfway between the borings, called secondary grouting, and then if necessary repeat the process, called tertiary grouting. The final spacing may be only a meter, depending on the desired result.

Grouting across a large area to strengthen a foundation soil normally is done from the outside in, to reduce the loss of grout into surrounding areas. Grouting is first performed around the perimeter of a project area to create an underground containment, and then grouting can proceed across the entire area. If that is not done, far more grout can be lost than will perform a meaningful purpose.

Another indication of grout loss is if the pump pressure suddenly decreases and the volume pumped increases. This can occur if the pressure is sufficient to fracture the soil or lift and penetrate horizontally. Grouting contractors often are paid by the volume of grout pumped, and refer to this as "take." What the engineer views as excessive they like to think of as normal.

It always is important to remember that grout under pressure will seek out the least resistant path.

### 24.6.3  Measuring Success

Careful records are kept of every injection to determine the volume and rate of grout pumped and the pumping pressure. Borings or in-situ tests such as the cone penetration test, or boring and pressure testing with water, can be conducted to determine the extent to which the soil has been penetrated and/or solidified, or rendered impermeable. A follow-up is recommended because in some instances no grout will be found because it was washed away by flowing groundwater or simply sank into voids deep in a gravel stratum or karst limestone. One of the most difficult grouting projects is in karst; another is in random uncompacted fill that often will occupy a buried stream valley where dumping occurred. Much grout can be lost under these circumstances. In the case of rapid groundwater flow, such as underneath a dam, fibrous materials or asphalt may be pumped to try and stem the flow. The other option of course is to draw down the reservoir to decrease the pressure head on the water.

As grout contractors often are paid on the basis of volume pumped, a sharp increase in "take" during pumping can be viewed as a positive sign by the contractor and a negative sign by the engineer. A sudden increase in "take" often means that the grout pressure is sufficient to lift, fracture, and invade soil strata, so there is no telling where or how far it will go. In some instances grout has been found in the house next door or across the street, or has emerged at the ground surface on the other side of a hill. Grouting is as much an art as a science, and requires experienced and knowledgeable operators.

### 24.6.4  Jet Grouting

Jet grouting is a relatively new process that creates mixed-in-place columns of stabilized soil. A cement slurry is injected immediately above an auger tip using fluid pressures of the order of 5000 lb/in.$^2$ (3500 kPa), which destroy the soil structure and aid mixing. Because some soil layers will be more resistant to

---

**Case History**

A house on expansive clay was grouted in an attempt to relieve structural distress because the doors would not close. A surveyor's level was set up to monitor the amount of uplift of sagging areas. Pumping began but there was no measurable movement, so pumping was stopped and an inspection was made for grout leaks.

*Investigation.* The bathroom doors could not be opened as the rooms were full of concrete. Fortunately nobody was in there at the time.

*Lesson learned.* Grout under pressure will find a path of least resistance, including a leaky or broken sewer pipe.

---

erosion than others, the result is a somewhat irregular column of hardened grout, as illustrated in Fig. 24.1. The column typically is about 15 to 45 in. in diameter, depending not only on erodibility but also on nozzle diameter and time spent at each grouting elevation. Air and water also may be injected above the grout nozzle to aid in breaking down the soil structure.

Whereas most grouting methods are restricted to particular kinds of soil, jet grouting can be used with soils from soft clay to gravel so long as the structure can be disrupted. In permeable soils bentonite may be added to reduce penetration by the grout.

On-site trials of jet grouting are recommended to help determine a suitable grout pressure, nozzle size, rotation rate, and rate of withdrawal. A set grout column can be partially excavated to determine its size and compressive strength so that adjustments may be made as needed. These investigations normally are the responsibility of the contractor.

Jet grouting is convenient for underpinning, as the grout column allows the introduction of reinforcing steel that can be bent back underneath an existing foundation. Jet grouting also is used to create diaphragm walls, but there will be "windows" where grout columns do not meet.

## 24.6.5 Compaction Grouting and Mudjacking

Whereas permeation grouting pushes grout into soil pores and jet grouting mixes grout with soil, compaction grouting *displaces* soil.

Compaction grouting is an outgrowth of "mudjacking," which has been used in the U.S. since the 1940s to jack up sagging pavements (Fig. 24.14(a)). A mudjacked pavement is readily detected from a pattern of 2–3 in. holes that are filled with concrete. As only low pressures are required, mudjacking can be performed with a hand pump, but usually powered pumps are used.

**Figure 24.14**

(a) Mudjacking to raise sagging pavements,
(b) quicklime columns to tame expansive clay.

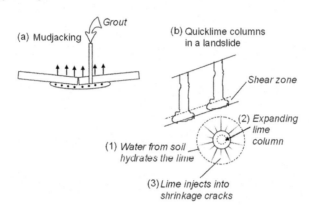

(a) Mudjacking

Grout

(b) Quicklime columns in a landslide

Shear zone

(2) Expanding lime column

(1) Water from soil hydrates the lime

(3) Lime injects into shrinkage cracks

Much higher pressures and deeper probing are required for compaction grouting. Compaction grout is viscous so it does not invade soil pores, but compacts soil laterally. As the grout pipe is incrementally raised and pumped it leaves a continuous grout column, as shown in Fig. 24.1.

Low-density soils are most readily compacted by compaction grouting, and should be relatively free-draining. Compaction grouting can prevent liquefaction of sand or collapse of loess, and is used to strengthen soil to bridge over tunnels and sink-holes (Schmertmann and Henry, 1992). Compaction grouting also can be used in a manner similar to mudjacking to reinforce soils under sagging tanks or foundations.

A compaction grout composition can be silty sand with or without Portland cement and fly ash, depending on the strength required. It should have a low slump, about 2 in. (50 mm) maximum. If the grout pressure is too high it can cause hydraulic fracturing, or the development of vertical radial tension cracks similar to those developed during ramming of RAPs. Cracks then can bleed away grout so that the procedure is less effective.

A special case of grouting involves filling abandoned underground coal mines in an attempt to control fires and acid mine water from oxidation of pyrite (iron sulfide) to form sulfuric acid. Large volumes of grout are required, the bulk of which is fly ash that in a sense is going home, as it is the power plant ash from burning pulverized coal.

### 24.6.6    Quicklime Columns (Drilled Lime)

A chemical soil treatment involving the use of quicklime was introduced in the 1960s to stabilize landslides involving soils containing active clay minerals (smectite). Later the same process was adapted to reduce clay expansion under and around foundations.

Instead of applying a fluid pressure that would further destabilize a landslide, borings are made filled with quicklime (CaO) that immediately reacts with water from the soil, temporarily drying it and stopping the landslide. The main and long-lasting benefit is not the drying, but the reaction between hydrated lime $(Ca(OH)_2)$ and expansive clay minerals. As the quicklime hydrates it approximately doubles in volume, so the expanding column of lime creates and injects into radial tension cracks, greatly aiding distribution into the soil. The target amount of lime is 1 to 3 percent by weight of the soil, depending on the clay content (Handy and Williams, 1967).

The function of the lime is to increase the soil plastic limit so that it is higher than the natural moisture content, changing the soil from plastic to solid even though the moisture content may remain the same. The liquid limit is less affected, so a soil that is close to the liquid limit is not suitable for treatment by this method, nor are soils in which the clay mineral is nonexpansive. A simple test is to mix some wet soil from the landslide shear zone with a few percent hydrated lime in a sealed plastic bag, knead the mixture by hand, and determine if after a few minutes the soil becomes crumbly. A second bag can contain untreated soil as a reference. More exact is to measure the plastic limit with and without lime and compare to the natural moisture content. Excess lime reacts with the clay pozzolanically to increase both the soil cohesion and internal friction. Drains also can be installed in soil that prior to stabilization would be impossible to trench.

---

### Case History

A shallow landslide was bearing against the back wall of a house, breaking in the wall and pushing the house off its foundation. After contacting his attorney, the owner of the adjacent property denied access to a drilling machine.

*Resolution.* Lime was introduced using hand-held concrete vibrators to make a hole. The house was saved.

*Lesson.* An attorney's first obligation is to his client.

---

The drilled quicklime method is one of the most economical ways to stop a shallow landslide by treating the active, upper part of the slide. Figure 21.1 shows a landslide prior to stabilization by this method.

## Problems

24.1. Differentiate between shallow, deep, and intermediate foundations.

24.2. What is meant by "ground improvement?"

24.3. Connect all appropriate matches:

| Clay | Stone column |
| Silt | Mixed-in-place pier |
| Sand | Jet grouting |

| Gravel | Permeation grouting |
| All soft soils | Compaction grouting |
| Pavement | Rammed Aggregate Piers |
| Fractured rock | Mudjacking |

24.4. What evidence supports liquefaction of soil close to Rammed Aggregate Piers? What conditions are required for liquefaction to occur?

24.5. As part of a research project both RAPs and stone columns are used to support different areas of embankment on soft clay soil. Which do you expect will settle the least? Why?

24.6. What conditions are required for stress concentration in a stone column and what is the significance for horizontal shearing?

24.7. An active landslide involving soil having a friction angle of 12° and cohesion from remolding of zero is to be treated with stone columns having a friction angle of 35°. With a stress concentration factor of 3, what area replacement ratio will give a factor of safety of 1.05?

24.8. What intermediate treatments are most satisfactory for creating an underground cutoff wall to prevent seepage through horizontal sand seams in clay? Evaluate stone columns, compaction grouting, mixed-in-place piers, jet grouting, and permeation grouting.

24.9. Recalculate Example 24.4 for a depth of 15 ft and a lateral ramming stress of $2200 \, \text{lb/ft}^2$.

24.10. On a Mohr diagram show how preconsolidation pressure influences cohesion.

24.11. A landslide slip zone has a friction angle of 12°. One-half of the landslide area will be reinforced with stone columns having a friction angle of 44°. The area ratio in the reinforced area is 0.3. Predict the factor of safety (a) with no stress concentration, (b) with a stress concentration factor of 2. Note that concentrating vertical stress on the columns decreases vertical stress on the matrix soil.

24.12. What is the thickness of the upper layer in Example 24.7?

24.13. In Example 24.7 the column load is increased to 100 tons. Calculate compression of the upper layer. How does the increase in load affect the thickness of the upper layer? Explain.

24.14. In Example 24.7 calculate the added stress at a depth of 15 ft. How is this affected by the length of the pier? Explain.

24.15. An RAP will be subjected to a load test using two anchor piers that have the same dimensions as the test pier. Do you expect that the anchor will be adequate?

24.16. A proposal is made to mudjack and raise a sidewalk next to a basement wall. Do you anticipate any problems with the wall? What should be monitored during jacking? What can reduce the likelihood of problems?

24.17. A landslide involves slip along a bedding plane in illitic shale. Is drilled lime applicable? Why (not)?

## References and Further Reading

Bachus, R. C., and Barksdale, R. D. (1989). "Design Methodology for Foundations on Stone columns." In F. Kulhawy, ed., *Foundatrion Engineering: Current Principles and Practices*, pp. 244–257. ASCE, New York.

Baker, W. H. (1982). "Planning and Performing Structural Chemical Grouting." In *Grouting in Geotechnical Engineering*, pp. 515–539. ASCE, New York.

Barksdale, R. D., and Bachus, R. C. (1983). *Design and Construction of Stone Columns*, Vol. 1. FHWA/RD-83/026. (Available in PDF format on the FHWA website.)

Broms, B. B. (1991a). "Deep Compaction of Granular Soils." In H.-Y. Fang, ed., *Foundation Engineering Handbook*, 2nd ed., Van Nostrand Reinhold, New York, pp. 814–832.

Broms, B. B. (1991b). "Stabilization of Soil with Lime Colunns." In H.-Y. Fang, ed., *Foundation Engineering Handbook*, 2nd ed., Van Nostrand Reinhold, New York, pp. 833–855.

Fox, L. (1948). "Computations of Traffic Stresses in a Simple Road Structure." *Proc. 2d Int. Conf. on Soil Mechanics and Foundation Engineering* 2, 236–256.

Fox, N. S., and Cowell, M. J. (1998). *Geopier$^{TM}$ Foundation and Reinforcement Manual*. Geopier Foundation Company, Blacksburg, Va.

Graf, E. D. (1992). "Compaction Grout, 1992." *Grouting Soil Improvement and Geosynthetics*, ASCE Spec. Publ. 30, 275–287.

Handy, R. L. (2001). "Does Lateral Stress Really Affect Settlement?" *ASCE J. Geotech. and Geoenviron. Eng.* 127(7), 623–626.

Handy, R. L., and White, D. J. (2006). "Stress Zones near Rammed Aggregate Piers: I. Plastic and Liquefied Behavior; II. Radial Cracking and Wedging." *ASCE J. Geotech. and Geonenviron Eng.* 132(1), 54–71.

Handy, R. L., and Williams, W. W. (1967). "Chemical stabilization of an active landslide." *ASCE Civil Engineering* 37(8), 62–5.

Karol, R. H. (1990). *Chemical Grouting*, 2nd ed. Marcel Dekker, New York.

Kirsch, F., and Sondermann, W. (2003). "Field measurements and numerical analysis of the stress distribution below stone column supported embankments and their stability." *Intern. Workshop on Geotechnics of Soft Soils.* VGA Publishing House GmbH, Essen, Germany.

Lawton, E. C., and Merry, S. M. (2000). "Performance of Geopier® Supported Foundations during Simulated Seismic Tests on Northboutnd Interstate 15 Bridge over South Temple, Salt Lake City Utah." Rept. No. UUCVEEN 00–03, University of Utah.

Pham, Ha (2005). "Support mechanisms of rammed aggrgate piers." Unpublished Ph.D. thesis, Iowa State University Library, Ames, Iowa.

Poulos, H. G., and Davis, E. H. (1974). *Elastic Solutions for Soil and Rock Mechanics*. John Wiley & Sons, New York.

Schmertmann, J. H., and Henry, J. F. (1992). "A Design Theory for Compaction Grouting." *Grouting Soil Improvement and Geosynthetics*, ASCE Spec. Publ. 30, 215–228.

Westergaard, H. M. (1939). "A Problem of Elasticity Suggested by a Problem of Soil Mechanics: Soft Material Reinforced by Numerous Strong Horizontal Sheets." In *Sixtieth Anniversery Volume of S. Timoshenko*. Macmillan, New York.

# 25

## Underground Conduits

## 25.1 INTRODUCTION

### 25.1.1 Uses

Underground conduits—or buried pipes—have served humanity since ancient times. Underground channels and pipes were used in the Middle East to convey scarce rainwater for storage in cisterns. Around 1720 B.C.E. the Minoan palace at Knossos was rebuilt after an earthquake and had an elaborate system of drains, conduits, and pipes, but things went wrong and it didn't last. Romans made extensive use of conduits to feed their baths and to connect to their famous aqueducts, but things went wrong and that didn't last. It was not until centuries later that city life was made safer and more pleasant by getting the sewage out of the streets. Civil engineers' dedication to improving sanitation has saved more human lives than has any other profession. We hope it lasts.

Although water and sewage systems are far from perfect, and there still are occasional warnings not to drink the water. It was extensive failures of field drain tile that led to the first definitive research on soil pressures on buried pipe. The "Experiment Station" approach was initiated, which involved testing and determining quantitative relationships that have broad and useful applications. The Marston-Spangler theory of pressure on buried conduit resulted from research starting in 1906 and extending into the 1940s, and is the basis for most current buried pipe design.

Another wide application for underground conduits began in the 1920s when it was discovered that automobiles are far less intelligent than horses for getting around obstacles, and a campaign was initiated to "get out of the mud." Roads were graveled or paved, and wooden bridges over creeks gradually were replaced with concrete culverts covered with soil. As embankments went higher in order to reduce hillside grades, increased soil pressures on culverts led to a modified theory. "Projecting conduits" were so named because they project upward into an embankment instead of being buried in a ditch.

The introduction of deformable metal culverts required another modification of Marston's theory as the culverts tend to bulge under a soil load, emphasizing the need for a lateral restraining pressure. The general theory that covers all kinds of conduits now is known as the Marston-Spangler theory of soil loads on buried conduit.

### 25.1.2 Marston's Classification of Underground Conduits

A *ditch conduit* is defined as one that is installed in a relatively narrow ditch and covered with earth backfill. Examples are sewers, drains, and water and gas mains.

A *positive projecting conduit* is installed in shallow bedding with its top projecting above the surface of the natural ground and then is covered with an embankment. Railway and highway culverts generally are installed in this manner.

In addition to the two major classes of conduit there are two subclasses:

A *negative projecting conduit* is designed to reduce soil pressure on pipe under an embankment, by placing it in a shallow ditch before it is covered with the embankment. The top of the pipe must be lower than the surrounding natural ground surface in order that the soil column over the conduit will settle and become partially supported by friction. This method of construction is most effective for minimizing the load if soil backfill on top of the conduit is loose and uncompacted.

The *imperfect ditch* or *induced trench conduit* uses a concept that is similar to that of a negative projecting conduit, only in this case trenches are cut into the embankment over the conduit and backfilled with compressible material. This is effective for reducing the soil load on a pipe, but cannot be used in embankments that serve as water barriers because the loosely placed backfill will allow channeling of seepage water through the embankment. The essential elements of these four main classes of conduits are shown in Fig. 25.1.

## 25.2 SOIL LOADS ON DITCH CONDUITS

### 25.2.1 The Most Favorable Geometry

As trench backfill settles downward its weight becomes partially carried by friction on the walls of the trench. The action is similar to arching action in soil behind a fascia-type retaining wall, where frictional support comes from both the inner surface of the wall and the exposed surface of a rock outcrop. As in the case

**Figure 25.1**

Marston's classification of underground conduit: (a) ditch, (b) positive projecting, (c) negative projecting, (d) imperfect ditch.

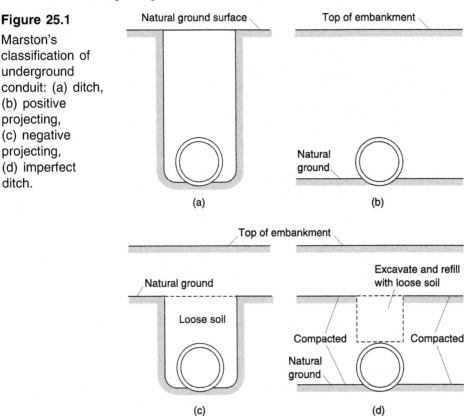

Natural ground surface

Top of embankment

Natural ground

(a)

(b)

Top of embankment

Natural ground

Loose soil

(c)

Excavate and refill with loose soil

Compacted

Compacted

Natural ground

(d)

of retaining walls, cohesion between the backfill material and sides of the ditch is assumed to be negligible because the soil has been disturbed and cohesive bonds broken. This assumption is justified on the basis that considerable time must elapse for cohesion to redevelop, and a solution with zero cohesion gives the maximum soil load on the conduit immediately after backfilling.

## 25.2.2 Analysis of Load on a Ditch Conduit

Frictional support for soil in a trench is the product of lateral pressure from the soil and a coefficient of friction between soil and the trench walls. The following notation is employed in the mathematical derivation of a formula for loads on ditch conduits:

$W_c$ = load on conduit, in N/m or lb/ft;

$\gamma$ = total unit weight (including water) of the backfill material, in $N/m^3$ or $lb/ft^3$;

$V$ = average vertical stress on any horizontal plane in backfill, in N/m or lb/linear ft;

$B_c$ = horizontal diameter or width of the conduit, in m or ft;

$B_d$ = horizontal width of the ditch at the top of the conduit, in m or ft;

$H$ = height of backfill above the top of conduit, in m or ft;

$h$ = distance from the ground surface down to any horizontal plane in backfill, in m or ft;

$C_d$ = a load coefficient for ditch conduits;

$\mu'$ = coefficient of friction between the fill material and sides of the ditch on an effective stress basis;

$K$ = ratio of lateral pressure of soil at the sides of the ditch to average vertical pressure;

$e$ = base of natural logarithms.

In the initial development of Marston's theory, $K$ was assumed to equal the Rankine coefficient of active earth pressure, $K_a$, a view that has persisted in the literature for many years. It qualifies as an oxymoron because a principal stress by definition cannot induce friction, and if it does induce friction it cannot be a principal stress (Handy 2004). As discussed in the preceding chapter, rotation of the principal stress directions increases $K$ and will have the effect of increasing supporting friction, so the actual value is approximated by $K_0$. Frictional support theoretically should relate to $K_0 \tan \delta$, where $\delta$ is the coefficient of sliding friction. The preferred designation is $K\mu$, where $K$ is the lateral stress ratio and $\mu$ is the coefficient of soil-wall sliding friction.

A section of a ditch and ditch conduit 1 unit in length is shown in Fig. 25.2. A thin horizontal element of fill material of thickness $dh$ is located at a depth $h$ below the ground surface. The forces acting on this element at equilibrium are: $V$ on the top of the element; $V + dV$ on the bottom of the element; the weight of the element, which equals $\gamma B_c\, dh$; and friction on the sides of the element. The latter equals two times the lateral pressure times the thickness of the element times the coefficient of sliding friction, or $2K(V/B_d)\mu\, dh$.

Equating the upward and downward vertical forces on the element gives

$$V + \gamma B_c\, dh = V + dV + 2K(V/B_d)\mu\, dh \tag{25.1}$$

This is a linear differential equation, the solution for which is

$$V = \gamma B_c^2 \frac{1 - e^{-2K\mu(h/B_d)}}{2K\mu} \tag{25.2}$$

If $h$ is set equal to the height of soil over the top of the conduit, $H$, then $V$ is the total load on a horizontal plane across the top of the conduit. The portion carried by the conduit depends on its rigidity compared with that of the fill material

**Figure 25.2**

Forces involved in the development of Marston's equation.

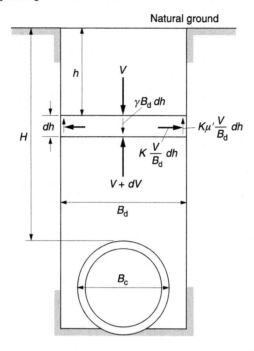

between the sides of the conduit and sides of the ditch. In the case of a rigid pipe, the side fills typically are compressible so the pipe will carry practically all of the load. On the other hand, if the pipe is flexible and soil is thoroughly tamped along the sides, the stiffness of the side fills may approach that of the conduit, so the load is spread over the entire width of the ditch.

### 25.2.3 Maximum Loads on Ditch Conduits

If we assume relatively compressible side fills alongside the pipe, the load on the conduit can be described by

$$W_c = C_d \gamma B_d^2 \tag{25.3}$$

where

$$C_d = \frac{1 - e^{-2K\mu(H/B_d)}}{2K\mu} \tag{25.4}$$

Evaluation of $W_c$ is made easy by use of the graph in Fig. 25.3. To use this diagram, calculate $H/B_d$ and read a corresponding value of $C_d$ for the appropriate fill material. Comparisons between actual weighed loads on rigid pipes in ditches and calculated loads in two of Marston's early experiments are shown in Fig. 25.4.

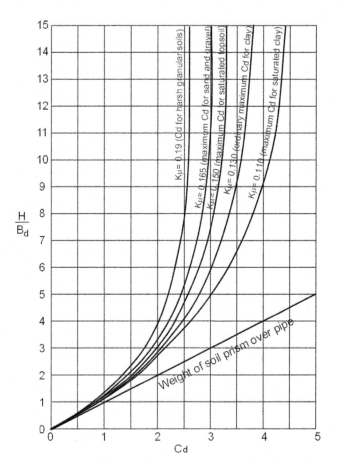

**Figure 25.3**
Graph for estimating $C_d$ for use in eq. (25.3).

## 25.2.4 Comparison with Measurements

Marston's arrangement for measuring soil loads on pipes in ditches was as simple as it is remarkable, having been devised a century ago. A section of pipe was supported inside an open trench by a beam running through the pipe, and the beam was suspended from a lever, one end of which rested on a platform scale. Soil was shoveled in on top of the pipe and confined at the ends with boards that moved with the pipe so that there was no end friction, and the system was calibrated with known loads. Some results from experiments in two different widths of ditches are shown in Fig. 25.4.

**Example 25.1**
A rigid sewer pipe with an outside diameter of 610 mm (24 in.) is to be laid in a ditch 1.1 m (3.5 ft) wide at the top of the pipe, and is to be covered with 8.5 m (28 ft) of clayey soil backfill that weighs $18.9\,\text{kN/m}^3$ ($120\,\text{lb/ft}^3$). Determine the load on the sewer.

## Figure 25.4

Some results from Marston and Anderson's (1913) experiments. The most important factor determining load on pipe in a ditch is not the pipe diameter but the ditch width, which is squared in the formula.

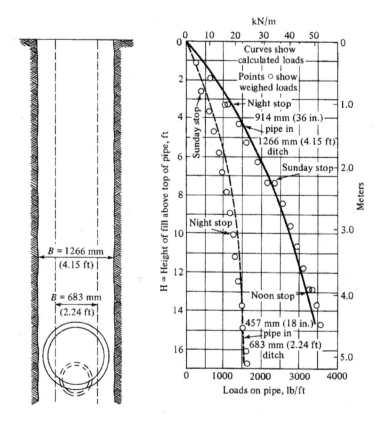

*Answer:*

$H = 8.53\,\text{m}$ (28 ft)  $\quad \gamma = 18.9\,\text{kN/m}^3$ (120 lb/ft$^3$)

$B_d = 1.1\,\text{m}$ (3.5 ft)  $\quad B_c = 1.14\,\text{m}$ (12.25 ft)

$H/B_c = 8.0$

From $H/B_c$ and the curve in Fig. 25.3 marked "ordinary maximum for clay," $C_d$ is 3.3. Substituting in eq. (25.3) gives

$$W_c = C_d\gamma B_d^2 = 3.3 \times 120 \times (3.5)^2 = 4900\,\text{lb/ft of pipe}$$

or

$$W_c = 3.3 \times 18.9 \times (1.1)^2 = 75\,\text{kN/m}$$

It will be seen that the pipe diameter does not enter into this determination.

### 25.2.5  Flexible Conduits in Ditches

If soil alongside a flexible pipe is thoroughly tamped it may have essentially the same degree of stiffness as the pipe itself. In this case the load on flexible pipe becomes

$$W_c = C_d \gamma B_d B_c \qquad (25.5)$$

Since $B_c < B_d$, a consequence of the pipe deflection is to divert part of the load to the soil and therefore decrease the load on the pipe. It must be emphasized that for eq. (25.5) to be applicable, side fills must be compacted sufficiently to have the same resistance to vertical deformation as the pipe itself. This equation should not be used simply because the pipe is a flexible type.

It is probable that the load on a flexible pipe lies somewhere between that indicated by eqs. (25.3) and (25.5), depending on the relative rigidity of the pipe and the side fill columns of soil. This was confirmed by laboratory tests by a private research agency, which indicated that the load imposed on a 250 mm (10 in.) diameter flexible sewer pipe under a dropped-in sand backfill with no compaction of the side fills was about 70 to 90 percent of the backfill load on a similarly installed rigid pipe.

## 25.2.6  Design to Prevent Cracking

The theoretical loads from eqs. (25.3) and (25.5) are working values that should be used in the design of sewers, drains, and other ditch conduits to prevent cracking of the pipe. Obviously, cracking cannot be tolerated for pressurized water or gas pipes. Other cracked pipes may continue to function without collapsing, but they will leak. Leaky sewer pipes pollute groundwater, and during periods when the groundwater table is high, flow may be into instead of out of the pipe, adding to the burden placed on sewage treatment facilities. Wetting from a leaky pipe can weaken soil along the sides of a pipe so more weight is transferred to the pipe, eventually causing it to fail. Small sewer pipes such as those that serve residential housing can readily become plugged with soil at a break, causing a sewer backup.

Pressure imposed by a trench backfill may increase as the soil settles in as a result of creep and changes in the soil moisture content, and in some cases there is as much as a 20 to 25 percent increase in total load in a matter of years. It is for this reason that sewers and other conduits that are observed to be structurally sound immediately after construction sometimes are found to be cracked months or years later.

## 25.2.7  Ditches with Sloping Sides

Worker safety is of paramount importance because of ever-present dangers of trench cave-ins. Many trenches therefore are now cut with sloping sides if the ditches are not too deep and space is available. Experimental evidence indicates that in this case $B_d$ should be the width of the ditch at or slightly below the *top* of the conduit, not at the bottom. This is shown in Fig. 25.5. In order to reduce the effective width, the conduit can be laid in a relatively narrow subditch at the

## Figure 25.5

Wide trenches reduce the danger from cave-ins but can greatly increase the soil load on pipe. (a) shows the correct width measurement, (b) (c) are options to reduce the effective trench width, and (d) shows an unacceptable geometry used in the case history in Example 25.2.

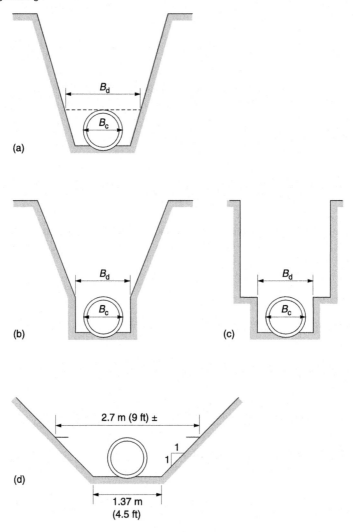

bottom of the wider ditch, which will cause a considerable reduction in the load that must be carried by the conduit.

### Example 25.2

*Case history.* A 24 in. (610 mm) sewer line was laid in a ditch, as shown at the bottom of Fig. 25.5. The specifications called for a ditch width $B_d = 4$ ft (1.22 m), which the contractor assumed was the width at the bottom of the excavation. The trench was cut with 1:1 side slopes and the actual width at the top of the pipe was $B_d = 8$ ft (2.4 m). The trench was 10 ft (3.0 m) deep. The pipe cracked and failed. Calculate the increase in load on the pipe attributable to this construction error.

*Answer:* As designed, $H = 10 - 2 = 8$ ft (2.4 m), and $H/B_d = 8/4 = 2.0$ (or 2.4/1.2 = 2). From Fig. 25.3, $C_d = 1.5$. As built, $H/B_d = 8/8 = 1.0$ and $C_d = 0.8$. Although $C_d$ is reduced, this is more than offset by the increase in load as the square of the trench width (eq. (25.3)).

Substituting values in that equation indicates that the load on the pipe was increased by a factor of $0.8(8)^2\gamma/1.5(4)^2\gamma = 2.1$, which is much higher than the normal factor of safety, and the pipe failed. An additional check is shown in Example 25.7.

## 25.2.8 Worker Safety in Trenches

Even partial burial of a person as a result of a trench cave-in can be fatal because the impact from caving soil expels air from the lungs, and in order to inhale, the victim must overcome passive resistance from soil that is composed of clumps and has a very high friction angle. Oxygen will be needed unless the person can be quickly freed. Prevention of blood circulation in the legs can lead to future medical problems. Rescuers who enter the ditch run the risk of being caught in another cave-in, compounding the tragedy. The only safe procedure is prevention, and regulations now require that workers in trenches with vertical walls be protected by shoring or by a steel trench box that is pulled along as pipe sections are installed.

### Example 25.3

Is it a fair statement to say that the passive pressure at a depth of 1 ft (0.3 m) on one's torso is like having an elephant sit on your chest? Assume the soil $\phi = 45°$ and $\gamma = 100\,\text{lb/ft}^3$ $(15.7\,\text{kN/m}^3)$.

*Answer:* $K_p = 5.8$ so the passive force will be $P = Kp\gamma\ H^2 = 5.8(100)(1)^2 = 580\,\text{lb}$, or $(5.8)(15.7)(0.3) = 27\,\text{kPa}$, so at a time like this nobody is going have time to worry about an elephant.

## 25.3 PIPE JACKING AND TUNNELING

### 25.3.1 Definition and Uses

Pipes frequently must go under existing roads, railroads, canals, etc., where open trenching would interrupt service and is not a viable option. Another procedure is to excavate at ends of the crossing, then use a horizontal auger to bore a hole underneath the structure. As the hole is drilled, jacks advance sections of a casing pipe through the hole. Finally an inner "carrier pipe" is installed inside the casing to facilitate replacement without re-boring. Large borings may be hand-excavated by workers inside the casing pipe as it is jacked forward. Once started, the jacking operation must be continuous to prevent "freezing" of the pipe so that it cannot be moved.

### 25.3.2 Soil Loads on Jacked Pipe

As the boring for a jacked pipe is slightly oversize, soil over the pipe tends to settle and become partly supported by friction. The geometry therefore is similar to that of pipe in a ditch, but soil over a jacked pipe or tunnel is not disturbed and therefore retains its cohesion. There also is a higher lateral in-situ stress than in the case of soil backfill in a trench, but that is difficult to evaluate. The Marston formula modified for cohesion acting on both ends of the arching element is as follows:

$$W_t = C_d \gamma B_c (\gamma B_c - 2c_t) \geq 0 \tag{25.6}$$

where $W_t$ is the load per unit length of pipe in a tunnel, $C_d$ is from Fig. 25.3, $B_c$ is the width of the conduit, and $c_t$ is the adjusted soil cohesion. Because cohesion can be affected by moisture content, shrinkage cracking, and creep, a recommendation has been made that $c_t$ should be one-third of the measured cohesion, $c$.

**Example 25.4**

A 30 in. diameter pipe is to be jacked at a depth of 15 ft in clay soil with $\phi = 11°$, $c = 600 \, \text{lb/ft}^2$, and $\gamma = 135 \, \text{lb/ft}^3$. Estimate load on the pipe.

*Answer:* $H/B = 15/2.5 = 6$, so from Fig. 25.3, $C_d = 3.3$ for saturated clay. Then $W_t = 3.3(2.5)$ $(135 \times 2.5 - 2 \times 200) \geq 0 = 0$; there should be essentially no soil load on the pipe.

### 25.3.3   New Austrian Tunneling Method

The NATM was developed for tunneling through rock, but uses principles from soil mechanics. Traditionally it was assumed that for reasons of safety a tunnel lining should be installed immediately as a tunnel is being advanced, thereby substituting lining support for high stresses existing in situ in deeply buried rock. The NATM allows a relaxation of these stresses before the lining is installed. Stress measurements in the lining indicate a substantial reduction in the amount of support required, and therefore in cost. Relaxation of stress occurs by elastic rebound and mobilization of supporting friction along fractures and bedding planes. The method involves careful monitoring of the rock expansion in order to select a proper time for installation of the liner. The liner plates then are assembled inside the tunnel and grouted to fill the space between the rock and the plates.

## 25.4   POSITIVE PROJECTING CONDUIT

### 25.4.1   Of Conduits and Culverts

Positive projecting conduits commonly are called "culverts," and are installed underneath soil embankments. Culverts have a wide variety of shapes including circular, rectangular, elliptical, etc., and are composed of concrete, corrugated metal, or plastic. If the culvert is more rigid that the soil, friction on the soil column over the culvert acts downward and adds to pressure, instead of acting upward and subtracting from pressure. Rigid culverts therefore must be made strong enough to resist this additional force. Another option is to use a flexible metal culvert that can yield as the soil load is applied, in which case the soil pressure may be approximately the same as the weight of the column of overburden. This is illustrated in Fig. 25.6, where measurements also confirm the long-lasting nature of soil frictional forces as they influence loads on pipe.

## 25.4.2 Geometry of Positive Projection

The geometric relationships of the soil prism over a projecting conduit are shown in Fig. 25.7, where it will be seen that the friction arrows point downward. However, tests show that friction may not be mobilized through the entire weight thickness of the overlying soil. An adjustment therefore is made based on a hypothesized "plane of equal settlement" existing in the soil at some level $H_e$ above the top of the conduit, and frictional forces are assumed to be mobilized only through this height. $H_e$ then was back-calculated from the measured loads. This concept was later confirmed from settlement measurements at different levels over a conduit.

## 25.4.3 Settlement Ratio

The level of the plane of equal settlement also is affected by settlement of the conduit itself. This led to definition of a *critical plane,* which defines settlement of soil adjacent to the conduit and initially at the same level as the top of the conduit, and a *settlement ratio,* a complex factor that with reference to Fig. 25.7 is defined as:

$$r_{sd} = \frac{(s_m + s_g) - (s_f + d_c)}{s_m} \tag{25.7}$$

where $r_{sd}$ = settlement ratio;

$s_m$ = compression strain of the side columns of soil of height $PB_c$;

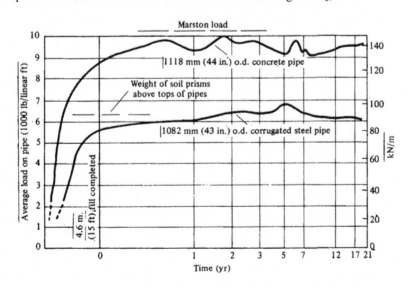

**Figure 25.6**

Loads measured over two decades on rigid and flexible pipes under an embankment (Spangler, 1973.)

**Figure 25.7**

Geometry of the positive projection condition.

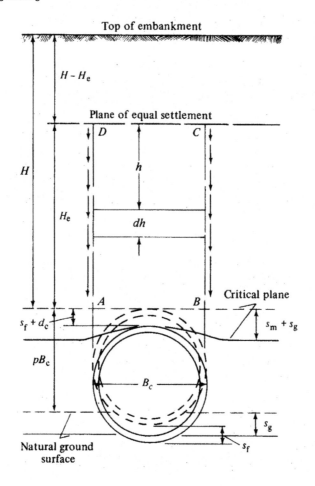

$s_g$ = settlement of the natural ground surface adjacent to the conduit;

$s_f$ = settlement of the conduit into its foundation;

$d_c$ = shortening of the vertical height of the conduit.

The settlement ratio defines the projection condition; if the ratio is positive, as shown in Fig. 25.7, projection is positive. However, the settlement ratio also can be negative if the critical plane settles less than the top of the conduit, in which case the load is less than the weight of the soil column, similar to a ditch condition. If the settlement ratio is zero, the load on the pipe exactly equals the weight of the soil column.

Although allowing a pipe underneath an embankment to settle can reduce the settlement ratio and load on the pipe, it introduces another problem shown in Fig. 25.8(a): the pipe settles most where the embankment pressure is highest, near

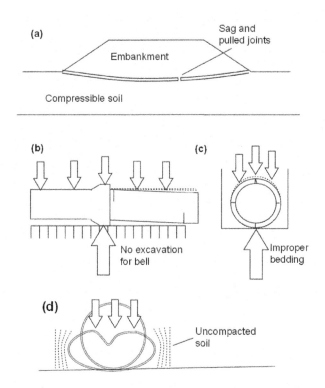

**Figure 25.8**

Common causes of distress in buried conduits: (a) sag under an embankment; (b) no excavation for pipe bells; (c) load concentration at the bottom due to improper bedding; (d) flexible pipe with insufficient side pressure from soil.

the middle. The resulting sag not only holds water, it stretches the pipe and can pull joints apart. Where this is anticipated, conduit can be constructed with an initial camber and with joints that can move with compression.

Another approach is to allow the pipe to flex and deform sufficiently so that $r_{sd}$ becomes negative. There is a limit to this approach because excess vertical compression of the pipe diameter can allow the top to cave in (Fig. 25.8(d)). This problem will be investigated later in this chapter.

## 25.4.4 Projection Ratio and Plane of Equal Settlement

The height of the plane of equal settlement, $H_e$, was shown analytically to depend on the product of the settlement ratio and a *projection ratio*, which is defined as

$$p = \frac{H}{B_c} \tag{25.8}$$

where $H$ is the vertical distance from the ground surface to the top of the pipe and $B_c$ is the pipe diameter. Analytical solutions for $H_e$ are shown by intersections of

**Figure 25.9**

Load coefficients
for positive
projecting
conduits. The
theoretical height
of the plane of
equal settlement,
$H_e$, is determined
from the
intersections of
the ray lines with
the outer curved
lines.

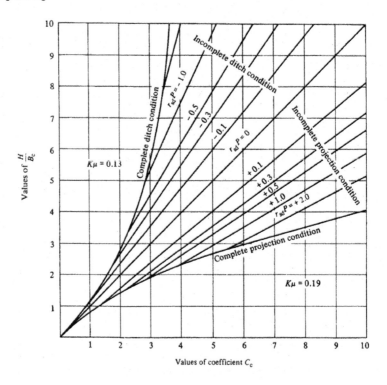

the linear arrays of lines in Fig. 25.9 with the outer curved lines. The existence of a plane of equal settlement has been confirmed experimentally.

**Example 25.5**

The height of the plane of equal settlement for the rigid conduit in Fig. 25.6 with $H = 15$ ft and $B_c = 3.67$ ft was measured to be approximately 8 ft. How does this compare with the theoretical height?

*Answer:* $H/B_c = 4.1$. In Fig. 25.9 the intersection of a line drawn from $H/B_c = 4.1$ on the y-axis and the line labeled "Complete projection condition" indicates a theoretical $H_e = 10$ ft.

It will be seen in Fig. 25.9 that the line for "Complete projection condition" is straight for higher values of $H/B_c$, so higher values are easily obtained by extrapolation.

### 25.4.5 Loads on Positive Projecting Conduit

The vertical load on a positive projecting conduit can be expressed by a formula that is similar to that for the ditch condition:

$$W_e = C_c \gamma B_c^2 \tag{25.9}$$

where

$$C_d = \frac{e^{\pm 2K\mu(H/B_c)} - 1}{\pm 2K\mu} \tag{25.10}$$

The plus signs are used for the complete projection conditions in which $H_e$ extends to the ground surface, and the minus signs are used for a complete ditch condition under an embankment, in which $H_e$ also extends to the ground surface.

In most cases $H_e$ does not extend all of the way to the ground surface, which would create either a bump or a sag in the roadway. This is called an incomplete ditch or incomplete projection condition. Equation (25.10) then is modified:

$$C_d = \frac{e^{\pm 2K\mu(H/B_c)} - 1}{\pm 2K\mu} + \frac{H - H_e}{B_c} e^{\pm 2K\mu(H/B_c)} \tag{25.11}$$

The plus signs are used for the incomplete projection condition, and the minus signs for the incomplete ditch condition.

In eqs. (25.9) to (25.11),

$W_c$ = load on conduit, in lb/linear ft (newtons per meter);

$\gamma$ = unit weight of embankment soil, in $lb/ft^3$ ($kN/m^3$);

$B_c$ = outside width of conduit, in ft (m);

$H$ = height of fill above *the top of* the conduit, in ft (m);

$H_e$ = height of the plane of equal settlement above the top of the conduit, in ft (m);

$K$ = lateral stress ratio;

$\mu$ = tan $\phi$ where $\phi$ is the angle of internal friction of the embankment material;

$e$ = base of natural logarithms

Solutions for these equations are shown in Fig. 25.9. It will be noted that $C_c$ depends both on the ratio of the height of fill to the width of the conduit, $H/B_c$, and on the product of the settlement ratio and the projection ratio, or $r_{sd}p$.

Results also are influenced by the coefficient of internal friction, but as the influence is relatively minor in soils that are acceptable for embankment construction, it is not considered necessary to differentiate between various soils as in the case of ditch conduits. Therefore, in constructing Fig. 25.9 it was assumed that $K\mu = 0.19$ for projection conditions, in which shearing forces are directed downward, and 0.13 for a ditch condition under an embankment, in which the shearing forces are directed upward. This diagram gives reasonable accuracy based on the assumptions used in the analysis.

| **Table 25.1** | Conditions | Settlement ratio |
|---|---|---|
| Design values of settlement ratio | Rigid culvert on foundation of rock or unyielding soil | +1.0 |
| | Rigid culvert on foundation of medium to stiff soil | +0.5 to +0.8 |
| | Rigid culvert on foundation of material that yields with respect to adjacent natural ground | 0 to +0.5 |
| | Flexible culvert with poorly compacted side fills | −0.4 to 0 |
| | Flexible culvert with well-compacted side fills | −0.2 to +0.2 |

### 25.4.6 Nature of the Settlement Ratio

Although the settlement ratio, $r_{sd}$, is a rational quantity, it is difficult to predict. It has been determined from measurements of actual culverts under embankments, which results in the guidelines given in Table 25.1.

#### Example 25.6

Calculate the load on rigid and flexible pipes in Fig. 25.6 founded on stiff ordinary soil. The total unit weight of soil in the embankment is approximately $120\,\text{lb/ft}^3$.

*Answer:* (a) *Rigid pipe*: From Table 25.1, $r_{sd} = 0.8$ for rigid pipe founded on relatively stiff soil. With $H/Bc = 4.1$, from Fig. 25.9 with $r_{sd} = 0.8$, $C_c$ is approximately 6.6. Then

$$W_e = C_c \gamma B_c^2 = 6.6(120)(3.67)^2 = 10{,}700\,\text{lb/ft}$$

In Fig. 25.6 the maximum measured load was $10{,}000\,\text{lb/ft}$.

(b) *Flexible pipe*: $B_c$ for the flexible pipe is $3.58\,\text{ft}$, so $H/B_c = 4.2$. From Table 25.1, $r_{sd} = -0.2$ to $+0.2$, the average being 0. In that case the load on the pipe equals the weight of the prism of soil over the pipe, and $C_c = 4.2$. Hence

$$W_e = C_c \gamma B_c^2 = 4.2(120)(3.58)^2 = 6500\,\text{lb/ft}$$

In Fig. 25.6 the maximum measured load was $6900\,\text{lb/ft}$.

### 25.4.7 Wide Ditches vs. Positive Projection

The increasing use of wide ditches for safety reasons indicates a need to determine at what point a ditch conduit becomes a projecting conduit. Studies by Schlick (1932) indicate that the ditch-conduit formula can be used for all ditch widths below that which gives a load equal to the load indicated for a positive projecting conduit. The width at which this occurs is called the *transition width*. Solutions therefore can be made for both conditions and the lower value used, or the transition width can be obtained from Fig. 25.10 to indicate which formula should be used. If $B_d/B_c$ is less than that indicated in the diagram, the formula for a ditch conduit is used, or if greater, the formula for a projecting conduit is used. In the absence of specific information, $r_{sd} = 0.5$ is suggested as a working value for rigid conduit.

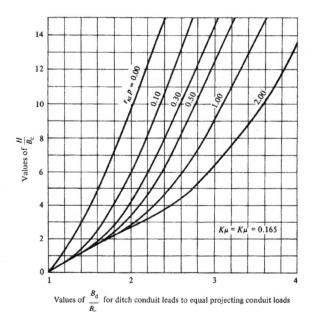

**Figure 25.10**

Transition widths from wide trench to positive projecting conduit installations.

## Example 25.7.

Does the conduit in Fig. 25.5(d), Example 25.2, qualify as a ditch conduit or a positive projecting conduit? If projecting, what is the increase in load on the conduit?

*Answer:* $H/B_c = 8/2 = 4.0$. With $r_{sd}$ estimated at 0.5, from Fig. 25.9 the transition width is $B_d B_c = 2.1$. The actual value is $B_d B_c = 8/2 = 4$, so the installation is positive projecting. From Fig. 25.10. $C_d = 6$, and the increase in load on the pipe is $6(8)^2\gamma/1.5(4)^2\gamma = 16$ times, which is many times higher than if the conduit had remained in a trench condition.

# 25.5 MARSTON'S IMPERFECT DITCH

## 25.5.1 Reducing Soil Loads with the "Imperfect Ditch"

The "imperfect ditch" or "induced trench" method of construction was suggested by Marston in the 1920s to reduce loads on conduit, and is particularly useful where fill is to be placed over an existing buried conduit. The action is similar to the negative projecting conduit except that the ditch is over instead of under the conduit.

After a conduit has been installed in a positive projecting condition and fill layers are extended over the top, a trench is cut into the fill directly over the conduit and backfilled with compressible material such as straw, leaves, or expanded Styrofoam. Then the fill is extended to the desired depth. If the prism of soil over the conduit settles more than the soil alongside, the prism is partially supported by friction.

### 25.5.2  Design Loads with the Imperfect Ditch

The equation for loads using either the negative projecting conduit or imperfect ditch method is as follows, where $C_n$ values are shown in Fig. 25.11 for a situation where the thickness of the compressible layer equals the width of the conduit:

$$W_n = C_n \gamma B_c^2 \tag{25.12}$$

The desired thickness of the compressible layer also depends on its compressibility, but this is not critical because relatively little movement is necessary to develop supporting friction.

### Example 25.8

A circular reinforced concrete pipe is to be installed using the imperfect ditch method of construction. The outside diameter of the pipe is 1830 mm (6.0 ft) and when installed in a Class C bedding it will project 1.28 m (4.2 ft) above the adjacent subgrade. The imperfect ditch will be excavated to a depth of 1.83 m (6.0 ft) and refilled with very loose soil. Assume a settlement ratio of −0.3. The height of fill will be 15.25 m (50 ft) and the soil weighs 18.9 kN/m³ (120 lb/ft³). Assume the soil $\phi = 30°$. Calculate the reduction in load on the pipe.

**Figure 25.11**

Theoretical values of $C_n$ for the imperfect ditch condition with a compressible fill thickness equal to $B_c$. The line for $r_{sd} = -0.5$ is suggested for general use.

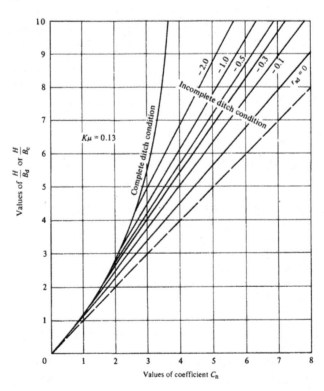

*Answer:*

| | |
|---|---|
| $H = 15.25\,\text{m}$ (50 ft) | $B_c = 1.83\,\text{m}$ (6.0 ft) |
| $r_{sd} = -0.3$ | $H/B_c = 8.33$ |
| $\gamma = 18.9\,\text{kN/m}^3$ (120 lb/ft³) | $C_n = 5.2$ (Fig. 25.11) |

$$W_c = C_n \gamma B_c^2 = 5.2 \times 18.9 \times 1.83^2 = 329\,\text{kN/m, or}$$

$$5.2 \times 120 \times 6^2 = 22{,}500\,\text{lb/ft with the imperfect ditch.}$$

Without the ditch method, assume a settlement ratio of $+0.5$ and a projection ratio of 0.8: $r_{sd}p = +0.4$. With reference to Fig. 25.9, $C_c =$ is off the chart and is estimated to be 12. The reduction in load therefore is $(12 - 5.2)/12$ or about 60%.

## 25.5.3 Adding to the Height of Fill over an Existing Buried Conduit

An important application of the imperfect ditch method is where additional fill is to be placed over an existing conduit, for example when the height of an embankment is to be increased as shown in Fig. 25.12. In this case the engineer is faced with a choice of either excavating and replacing the conduit, reinforcing it with a liner, or employing the imperfect ditch method. The older embankment may respond elastically, so the ditch acts similarly to a foundation except that, instead of adding to the surcharge pressure, it subtracts from it. The imperfect ditch procedure has been successfully used to solve some difficult problems. The complexity of the relationships has led investigators to apply finite element modeling to the imperfect ditch problem (Kim and Yoo, 2005).

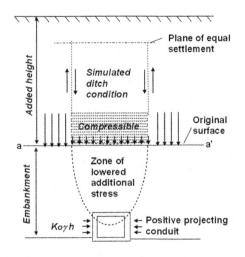

**Figure 25.12**

Application of the imperfect ditch method for increasing the fill height over an existing conduit.

## 25.6  SURFACE LOADS

### 25.6.1  Static Point Loads on Soil

Buried conduits often are subjected to loads at the ground surface from highway, railway, or other traffic. What minimum thickness of soil cover may be required to protect a pipe?

Experimental evidence favors use of the Boussinesq solution to determine the influence of point or limited area loads on buried conduit, and the presence of a rigid pipe in a mass of soil does not significantly increase the contact pressure.

The influence of a point surface load on a buried horizontal area is the reverse of that discussed in relation to settlement of foundations and uses the same elastic solutions. Vertical stress is highest when a load is centered over a culvert and then decreases, whereas horizontal stress increases and reaches a peak value as the load moves a short distance away.

### Example 25.9

An 18,000 lb (80 kN) truck wheel load passes over a 2 ft (0.61 m) diameter culvert that has a soil cover 3 ft (0.9 m) deep. (a) What is the maximum vertical stress transferred to the culvert? (b) What is the horizontal stress at mid-height on the culvert when the wheel load is directly over an edge? (c) When it is 1 ft (0.3 m) from the edge?

*Answer:* (a) According to the Boussinesq equation the vertical stress is

$$\sigma_z = \frac{3P}{2\pi} \frac{z^3}{R^5} \tag{15.1}$$

where $R = \sqrt{(x^2 + z^2)}$. Directly under the load $z = R$. Hence

$$\sigma_z = \frac{3P}{2\pi} \frac{1}{z^2} = \frac{3 \times 18{,}000 \text{ lb}}{2\pi(3)^2 \text{ ft}^2} = 950 \text{ lb/ft}^2, \text{ or}$$

$$\sigma_z = \frac{3 \times 80 \text{ kN}}{2\pi(0.9)^2 \text{ m}^2} = 47 \text{ kPa}$$

Note that because of the inverse relation to $z^2$, it is a simple matter to calculate pressures for other depths of burial. In this example, if the soil cover layer were only half as thick, the pressure would be 4 times as high, which is an effective argument for incorporating a sufficient cover. If there were no cover, the pressure would equal the truck tire pressure.

(b) The Boussinesq relationship for horizontal stress with Poisson's ratio equal to 0.5 is

$$\sigma_x = \frac{3P}{2\pi} \frac{x^2 z}{R^5}$$

Directly over an edge $x=0$, so the lateral stress $\sigma_x=0$. When the load is 1 ft (0.3 m) from the edge, $R = \sqrt{[(1)^2+(4)^2]} = 4.1$ ft (1.26 m). Then

$$\sigma_x = \frac{3 \times 18,000}{2\pi} \frac{(1)^2(4)}{(4.1)^5} = 30 \text{ lb/ft}^2, \text{ or}$$

$$\sigma_x = \frac{3 \times 80}{2\pi} \frac{(0.3)^2(1.2)}{(1.26)^5} = 1.3 \text{ kPa}$$

## 25.6.2  Impact Loads

Although a rigid pavement spreads loads, this influence can be nullified by impact loads, which depend on many factors including vehicle weight and speed, pavement roughness, and nature of the soil response which, as discussed in the next chapter, can include resonance and damping. A maximum amplification factor appears to be about 2, but that will decrease with increasing depth to a conduit. Factors such as this that are not readily quantifiable are taken into account by increasing the design factor of safety.

# 25.7  SUPPORTING STRENGTH OF CONDUIT

## 25.7.1  Strength Measurement

Underground conduits are constructed in a wide variety of shapes and of many different structural materials. Monolithic reinforced-concrete structures such as arch and box culverts may be satisfactorily designed by procedures used for analysis of rigid-frame structures, using a load that is uniformly distributed over the width of the conduit. Such structures also may be subjected to lateral pressure similar to that on retaining walls. A culvert at the toe of a landslide can be subjected to very high passive pressures that can be calculated using Rankine theory.

The supporting strength of rigid circular or elliptical pipes is routinely evaluated by compressing a section of pipe in a laboratory compression testing machine, as shown in Fig. 25.13. The "three-edge bearing test" has been standardized and is the simplest and easiest to perform, although some engineers prefer the sand-bearing test because of the wider distribution of both the applied load and the reaction base. Nevertheless, for comparison purposes the strength of pipe often is expressed in terms of its equivalent three-edge bearing strength.

Based on numerous tests of rigid pipe, the conversion factors to three-edge bearing strength are shown in Fig. 25.13. For example, to convert results from a sand-bearing test to a three-edged bearing test, divide the measured strength by 1.5. These are "load factors," designated by $L_f$. It will be noted that these are empirically derived and carry no more that two significant figures.

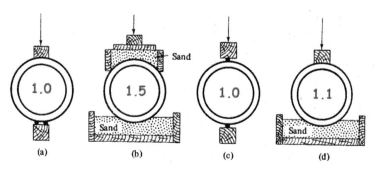

**Figure 25.13**

(a) Three-edged bearing test; (b) sand bearing; (c) two-edged bearing; (d) Minnesota bearing. Numbers indicate load factors for each type of test.

### 25.7.2 Bedding

The different load factors listed above illustrate the significance of bedding, which functions to spread the load over larger sectors of the pipe. Bedding is one of the most important factors governing pipe strength in field installations, as a lateral distribution of the bottom reaction decreases the bending moment in the pipe walls. The principle involved is essentially the same as that of a simple beam supported at two ends: if a concentrated load, $P$, is applied at the center of the span, the maximum bending moment is at the load point and is equal to $0.25PL$, where $L$ is the length of the beam. On the other hand, if the same load is uniformly distributed over the span length, the maximum bending moment is reduced one-half, to $0.125PL$.

In the field a distribution of the bottom contact stress is achieved by shaping a bedding material such as a layer of sand by use of a template cut to fit the contour of the pipe. Another practice, which is more labor-intensive and therefore less effective, is to tamp soil into the spaces under the haunches of a pipe.

Backfilling practices also affect lateral pressures on the sides of the pipe. Lateral pressures create bending moments in the opposite direction from those produced by the vertical load and reaction, and therefore decrease the bending stress in the pipe wall and increase its supporting strength. Maximum bending moments are shown in Fig. 25.14, where in (a) a circular pipe is placed on a flat surface without any attempt to shape the bedding to fit the contour of the pipe, and without any applied lateral pressure. This load condition causes the bottom reaction to be highly concentrated and gives a maximum bending moment of $0.294rW_c$, where $r$ is the radius of the pipe and $W_c$ is the vertical load. This is called *Class D, impermissible bedding*.

In Fig. 25.14(b) the bedding is preshaped to fit the pipe over a width of 90°, which distributes the reaction approximately uniformly over a width equal to

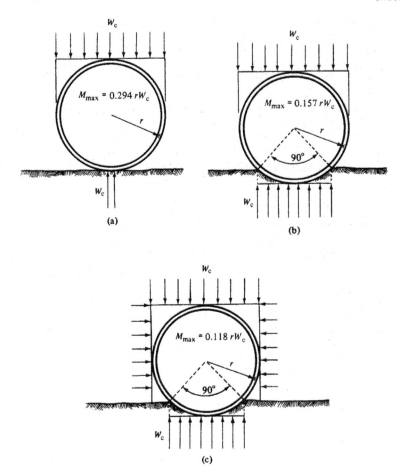

**Figure 25.14**
Importance of
bedding on pipe
strength. $M_{max}$
values are the
calculated
maximum bending
moments in the
walls of the pipes.

70 percent of the pipe diameter. The maximum bending moment is reduced
to $0.157rW_c$, so pipe (b) theoretically can support 1.87 times the load on
pipe (a) without increasing the wall stress. This improvement is taken into
account by introduction of the load factor, which is further discussed in the next
section.

Fig. 25.14(c) shows the effect of lateral pressure on stress on bending
moment in the pipe wall. If it is assumed that the pipe bedding is the same
as in (b) and that lateral pressure is one-third of the vertical pressure of the
overlying soil, the bending moment is further reduced to $0118rW_c$, and the
supporting strength is increased to 2.49 times that of pipe (a) and 1.33 times
that of pipe (b).

These illustrations indicate the very great influence of bedding and backfilling
practices on the structural performance of a pipe in the ground. The quality of

these procedures and the care taken in the installation of a pipe can mean the difference between success and structural failure of a buried conduit.

The supporting strength of this type of structure therefore depends on three factors: (1) the inherent strength of the pipe; (2) the quality of the bedding as it affects the distribution of the bottom reaction; and (3) the magnitude and distribution of lateral pressures that may act on the sides of the pipe.

### 25.7.3  Classification of Pipe Bedding

Several classes of pipe bedding have been defined to represent construction practices used in the field. Both the distribution of the bottom pressure and the influence of lateral pressure are taken into account in the bedding class. Bedding classes and corresponding load factors are shown in Fig. 25.15; the better the bedding, the higher the load factor and the larger the load that can be sustained by the pipe. It will be noted that because of the empirical nature of the load factors, answers will carry no more than two significant figures.

**Impermissible, or Class D Bedding: Load Factor = 1.1**
This bedding class describes a situation where no effort has been made to shape the foundation to fit the lower part of the conduit and no attempt has

**Figure 25.15**

Classes of pipe bedding. Italicized numbers are bedding load factors for ditch conduits.

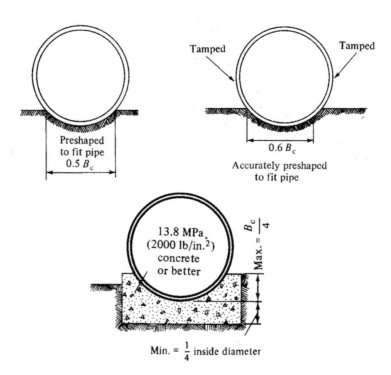

been made to fill the spaces under and around the conduit. Nevertheless there will be some side pressure from backfill, and the load factor determined experimentally for this bedding is 1.1, which is multiplied times the strength from a three-edged bearing test to obtain an estimated strength at failure in the field.

### Ordinary or Class C Bedding: Load Factor = 1.5

This class is used to indicate "ordinary care" in preshaping the earth foundation to fit the lower part of the conduit exterior for a width of at least 50 percent of the conduit breadth, and in which the remainder of the conduit is surrounded to a height of at least 0.15 m (0.5 ft) above its top by granular materials that are shovel-placed and shovel-tamped to completely fill all spaces under and adjacent to the conduit. This type of installation must be under the general direction of a competent engineer.

### First-Class or Class B Bedding: Load Factor = 1.9

In this class a conduit is carefully laid on fine granular materials that have been carefully preshaped by means of a template to fit the lower part of the conduit exterior for a width of at least 60 percent of the conduit breadth, and the remainder of the conduit is entirely surrounded to a height of at least 0.3 m (1.0 ft) above its top by granular materials that must fill all spaces under and adjacent to the conduit, and are thoroughly tamped on each side and under the conduit in layers not exceeding 0.15 m (0.5 ft) in thickness. Fill at the sides of positive projecting conduits is tamped to a distance at least equal to the conduit width. All work must be done under the direction of a competent engineer, who is represented by an inspector who is constantly present during the operation.

### Concrete-Cradle (Class A) Bedding: Load Factor 2.8 to 3.4

This bedding involves settting the lower part of a conduit in plain or reinforced concrete of suitable thickness and extending upward on each side of the conduit for a distance that is *not greater than* one-fourth of the outside diameter of the conduit. Pipe laid in concrete bedding usually shows initial cracking at the top.

In all cases where pipe sections are joined by means of a widened flange or bell at one end, the bell *must* be accommodated by excavation to prevent loading the bell. A failure to excavate for the bell easily can reduce strength by a factor of 2. The strength can be estimated by assuming the pipe acts as a simple beam supported at the ends, but with proper installation procedures it should not be necessary to make this estimation—except in preparation for trial.

## 25.7.4 Increased Load Factors for Positive Projecting Conduits

Increased efficiency of compaction of soil alongside positive projecting conduits results in higher load factors compared to ditch conduits. A formula that takes this into account is

$$L_f = \frac{A}{N - xq} \tag{25.13}$$

where $A$ = shape parameter, 1.43 for circular conduit, or 1.34 for a horizontal elliptical shape;

$N$ = bedding factor (Table 25.2);

$x$ = side area function, which depends on $p$, the projection ratio for circular pipe, or $m$, the projection ratio from the vertical diameter of a horizontal elliptical pipe;

$$q = \frac{pK}{C_c}\left(\frac{H}{B_c} + \frac{p}{2}\right) \tag{25.14}$$

where $p$ = projection ratio (proportion of the pipe diameter projecting above the bearing surface);

$K$ = lateral stress ratio;

$C_c$ = load coefficient;

$H$ = height of fill above the conduit;

$B_c$ = horizontal outside dimension of conduit.

### 25.7.5  Factor of Safety

Design that is based on a failure load must incorporate a factor of safety. Rigid underground conduits may be selected according to the following criterion:

$$S_{eb} = \frac{W_c \times FS}{L_f} \tag{25.15}$$

**Table 25.2**

Values of $N$ and $x$ for use in eq. (25.13)

| $p$ or $m$ | $x$ | Bedding B, C, D | $N$ | Bedding A | |
|---|---|---|---|---|---|
| | | | | $x$ | $N$ |
| Circular | | | | | |
| 0.3 | 0.217 | D | 1.310 | 0.743 | — |
| 0.5 | 0.423 | C | 0.840 | 0.856 | — |
| 0.7 | 0.594 | B | 0.707 | 0.811 | — |
| 0.9 | 0.655 | | | 0.678 | — |
| 1.0 | 0.638 | | | 0.638 | 0.505 |
| Horizontal elliptical: | | | | | |
| 0.3 | 0.146 | C | 0.763 | | |
| 0.5 | 0.268 | B | 0.630 | | |
| 0.7 | 0.369 | | | | |
| 0.9 | 0.421 | | | | |

where $W_c$ = design load;

$S_{eb}$ = three-edge bearing strength;

FS = factor of safety;

$L_f$ = load factor.

Factors of safety vary widely in practice. For unreinforced concrete pipe, suggested values based on minimum test strength of a representative number of test specimens are from 1.3 to 1.5. For reinforced concrete pipe, the suggested factor of safety is 1.2. As such pipes do not fail by collapsing, the failure load is defined as the load causing one or more 0.25 mm (0.010 in.) wide cracks measured at close intervals throughout a pipe section length of 0.3 m (1 ft) or more. This is usually expressed as a "D-load," which is the test strength per linear meter (foot) of pipe divided by the *internal* diameter in meters (feet).

The 0.25 mm (0.010 in.) crack width can be measured with a mechanic's "feeler gauge" and is not to be regarded as defining a failure situation in the field. All reinforced concrete structures can be expected to crack as the tensile stress is transferred to the reinforcement, because the modulus of elasticity of steel is much greater than that of concrete, and the steel will stretch more than the concrete at working stresses. Unless cracks open up a sufficient amount to permit or promote oxidation and corrosion of the steel, reinforced concrete pipes will continue to perform their load-carrying function indefinitely, even though fine cracks have developed.

## Example 25.10

Determine the required three-edge bearing strength for the pipe in Example 25.8 with ordinary Class C bedding and a factor of safety of 1.0 based upon the 0.010 in. (0.025 mm) crack strength. $W_c$ = 329 kPa (22,500 lb/ft$^2$). The internal diameter $D$ = 1.52 m (5 ft).

*Answer:* Expressed in terms of D-load, $W_c$ = 329/1.52 = 216$D$ kPa, or 22,500/5 = 4500$D$ lb/ft.

$\phi = 30°$; $K = 0.33$      $A = 1.43$

$p = 4.2/6.0 = 0.7$      $N = 0.840$ (Table 25.2)

$X = 0.594$ (Table 25.2)

$$q = (0.7 \times 0.33)/5.2(8.33 + 0.7/2) = 0.385 \qquad \text{(eq. (25.14))}$$

$$L_f = 1.43/[0.840 - (0.594 \times 0.386)] = 2.34$$

$$\text{(eq. (25.13))}$$

Required three-edge strength = $(3.29 \times 1)/2.34 = 140$ kN/m,

or $(22,500 \times 1)/2.34 = 9600$ lb/ft

D-load strength = $140/15.25 = 9.18D$ kN/m$^2$, or $9600/5 = 1920D$ lb/ft$^2$

ASTM Specification C76 Class IV pipe: $95.8D$ kN/m$^2$ or $2000D$ lb/ft$^2$ at 0.25 mm (0.01 in.) crack criterion.

Factor of safety with Class IV pipe: $(95.8D \times 2.34)/216D = 1.04$, or $(2000D \times 2.34)/4500D = 1.04$, which is marginal.

### 25.7.6 Conduit Repairs

Buried conduits are checked for signs of distress visually, in the case of a large conduit, or by pulling through a special TV camera. If cracks or other signs of deterioration are encountered, the pipe can be "slip-lined" with smaller pipe or by unrolling a plastic sleeve. The space between the pipe and the lining then is pressure grouted, usually with a cement or fly ash-based grout, and the pipe is returned to service. Although the internal diameter is decreased, the smoothness normally is increased so the transmission capability is not greatly affected.

## 25.8 ADDITIONAL REQUIREMENTS

### 25.8.1 Need for Continuous Support

A criterion that is easily overlooked is the critical need for continuous support along the length of a pipe. For example, if a pipe crosses from resistant to soft soil, a transition is needed to prevent bending and shearing stresses at the boundary. The transition can be as simple as a supporting beam crossing from one soil to the other. Most violations of this principle are from ignorance and can be regarded as construction errors. It therefore is instructive to review some case histories.

### 25.8.2 Case History: Pipe Crossing a Trench Backfill

A basic premise of Marston's theory is that soil backfill in a trench tends to settle and develop side friction, so crossing through a backfill with another pipe is to invite localized bearing pressures. In one instance a perimeter drain had been installed around a large one-story building, and some months later a gas pipe was installed at a higher level so that the pipe crossed through the backfilled trench. This situation *demanded* that the gas pipe be supported where it crossed the trench. The situation was aggravated because the earlier installation had not yet settled to reach equilibrium. The unsupported pipe ruptured, so gas went through the perimeter drain and into the building. The explosion and fire killed over 40 people.

### 25.8.3 Case History: Class D Bedding and Frozen Backfill Soil

The importance of bedding and tamped side-fill soil has been emphasized. In this case the weather was freezing, and the trench bottom apparently was not shaped to receive a gas pipe. Large clods of frozen backfill soil were dumped into the trench and created pressure points on top of the pipe, which broke. Gas passed through the loose trench backfill that led directly into the building. The house exploded and collapsed, killing the sole occupant.

## 25.8.4  Case History: Water Hammer at an Elbow

Most people are familiar with "water hammer," when a pipe receives a severe knock if the flow of water suddenly is shut off. The cause is the momentum in the mass of flowing water and Newton's Law, $f=ma$. Sudden changes in water velocity also can occur in water mains, so pipe bends are required to have vertical bearing blocks to resist hammering. In this case the pipe was laid in saturated loose sand, hammering densified the sand, and the pipe settled, went out of alignment, and was in danger of rupturing.

## 25.8.5  Case History: Improper Bend

A plastic gas pipe was installed with a relatively sharp bend that was not within manufacturer's guidelines. Residents of a large apartment complex smelled gas, called the gas company, and evacuated of their own volition. A gas company repairmen was sent to check for an inside leak, and did not shut off gas to the building. The leak was outside the building and came in through the trench, and several hours later the building burned to the ground.

## 25.8.6  Case Histories: Landslide and Settlement Problems

Any movement of soil relative to an adjacent structure will induce shearing stresses and bending moments in utility pipes. Most gas companies are aware of the dangers and routinely install flexible connecting pipes if there is any danger or evidence of movement. Water and sewer pipes are less readily accessible, and an early indication of trouble can be a sewer backup. If a water main is broken it can contribute to a landslide

These examples should serve to demonstrate the importance of proper installation and maintenance of underground conduit. Unfortunately, explosions and leaks still continue to make headlines.

## 25.8.7  Case History: Why Not Drink the Water

Water and sewer lines often run parallel to one another, so there is a danger of cross-contamination if the pipes should leak. Contamination of the water supply is prevented if there is a continuous positive internal pressure that exceeds the groundwater pressure throughout the system, but this becomes less likely in older systems, particularly during times of peak usage.

## 25.9  INTERNAL PRESSURE

### 25.9.1  Overview

Many buried conduits are subjected to both external loads from soil and to internal fluid pressure that stresses the pipe in circumferential tension. These applications include lines carrying water, natural gas, and petroleum products. Internal pressure increases stress in the pipe wall, so a conduit will not carry as much external load, and conversely a pipe subjected to external load will not withstand so much internal pressure.

Two general classes of pipe material will be considered: cast-iron pipe, which is relatively brittle, and steel, ductile iron, or plastic pipe, which is more yielding. Cast-iron pipe was used extensively in municipal water distribution systems, but in recent years has largely been superseded by ductile or nodular iron.

### 25.9.2 Pressurized Steel Pipe

Pipelines of these kinds are usually laid in trenches and then backfilled before they are pressurized. The pipe deflects an amount $\Delta x$ under the influence of the earth load and assumes an elliptical shape with the major axis horizontal and the minor axis vertical, in effect increasing the upper and lower boundary areas. When internal pressure is introduced into the pipe, pressure on the enlarged area will tend to push the pipe back to a circular shape. Bending stresses in the pipe wall corresponding to this equilibrium deflection are assumed to be algebraically additive.

On the basis of the foregoing hypothesis, the maximum combined stress in a pipe due to earth load and internal pressure may be expressed by the formula

$$S = S_1 + S_2$$
$$S = \frac{P(D - 2t)}{2t} + 0.117 \times \frac{C_d \gamma B_d^2 Etr}{Et^3 + 2.592 pr^3} \tag{25.16}$$

where $S$ = maximum combined stress, in $\text{lb/in.}^2$ (MPa);

$S_1$ = hoop stress due to internal pressure, in $\text{lb/in.}^2$ (MPa);

$S_2$ = bending stress due to fill load, in $\text{lb/in.}^2$ (MPa);

$C_d$ = load coefficient for fill load on ditch conduits;

$\gamma$ = unit weight of soil backfill, in $\text{lb/ft}^3$ ($\text{kN/m}^3$);

$B_d$ = width of ditch at level of top of pipe, in ft (m);

$E$ = modulus of elasticity of pipe material, in $\text{lb/in.}^2$ (MPa);

$D$ = outside diameter of pipe, in inches (mm);

$t$ = thickness of pipe wall, in inches (mm);

$r$ = mean radius of the pipe = $(D - t)/2$, in inches (mm);

$p$ = internal pressure, in $\text{lb/in.}^2$ (MPa).

If the pipeline is subjected to vertical loads from surface traffic, the stress from this source should be added to that indicated by eq. (25.16).

A typical diagram showing maximum and minimum stresses in the wall of a 914 mm by 9.5 mm (36 in. by $\frac{3}{8}$ in.) steel pipeline, subjected to earth load and to

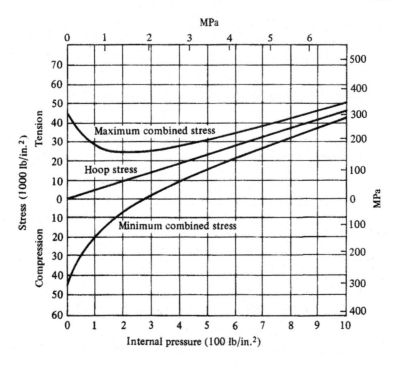

**Figure 25.16**
Example of wall
stresses in a
$36 \times \frac{3}{8}$ in.
$(914 \times 9.5\,\text{mm})$
pipe calculated
from a soil load
plus variable
internal pressure.

various internal pressures, is shown in Fig. 25.16. It will be noted that the combined tensile stress decreases to a minimum value as internal pressure is increased, and then increases with further increase in pressure.

### 25.9.3  Design of Pressure Pipe

For manufacture, design, and installation of pressure pipes the reader is referred to standards developed by the American Water Works Association (AWWA). Kinds of pipe include prestressed and pretensioned reinforced concrete, polyvinyl chloride (PVC), polyethylene (PE), and polybutylene (PB).

## 25.10  FLEXIBLE CONDUIT

### 25.10.1  Overview

All underground conduits derive their ability to support the soil above them from two sources, the inherent strength of the pipe and the lateral pressure from soil at the sides of the pipe. Rigid pipes derive most of their supporting strength from their inherent strength. In flexible pipes, such as corrugated-metal culverts and thin-walled steel or plastic conduits, the situation is reversed: the pipe itself has relatively little inherent strength, and a large part of its ability to support vertical load is derived from side pressures induced as the sides move outward against the soil. Soil at the sides of the pipe therefore is an integral part of the structure. The analysis therefore becomes more complex, particularly as soil is more variable and less predictable than the various manufactured pipe products.

### 25.10.2 Failure Modes of Flexible Conduit

Flexible conduits often fail by vertical deflection instead of rupture of the pipe walls. The outward movement of the sides of the pipe is resisted by increasing passive pressure of the soil. As an embankment is built higher and vertical pressure increases, the top of the pipe may become approximately flat and eventually may develop a reverse curvature so that the top is concave upward. As soon as that occurs, the sides of the pipe are pulled inward so that side support is lost, and the top of the pipe caves in. This mode of failure is shown at the bottom of Fig. 25.8. If the sides are restrained and loads are sufficient, the pipe may fail by ring compression at the 3 and 9 o'clock positions. External fluid pressure also can contribute to this type of failure.

### 25.10.3 Development of the Iowa Formula

The deflection mode of failure of flexible pipe was analyzed by Spangler, who developed an equation that generally is known as the "Iowa formula." The analysis was based on measurements of soil pressure using "friction ribbons," stainless steel tapes that were installed lengthwise in contact with soil around the perimeter of a pipe. After a soil load was in place, the tapes were pulled and pulling forces converted to soil pressures by the use of calibrations. This deceptively simple system has an important advantage over load cells—pressures are averaged over the entire length of a pipe section.

The following conclusions were based on the results of full-scale experiments:

1. The vertical load may be determined by Marston's theory of loads on conduits and is distributed approximately uniformly over the breadth of the pipe.
2. The vertical reaction at the base equals the vertical load and is distributed approximately uniformly over the width of bedding of the pipe.
3. The horizontal pressure on each side of the pipe is distributed parabolically over the middle $100°$ of the pipe as the pipe deflects and sides move outward.
4. The maximum side pressure occurs at the ends of the horizontal diameter of the pipe, and equals the modulus of the fill material multiplied by one-half the horizontal deflection of the pipe.

This pattern of loading is shown in Fig. 25.17. The formula for deflection as modified by Watkins is as follows:

$$\Delta x = D_L \frac{K W_c r^3}{EI + 0.061 E' r^3} \tag{25.17}$$

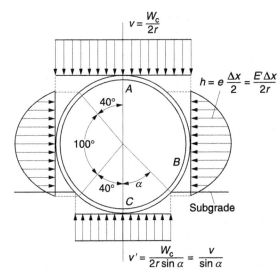

**Figure 25.17**
Distribution of soil pressure determined from full-scale experiments on flexible pipe (Spangler, 1941.)

where $\Delta x$ = horizontal deflection of the pipe in mm (in.), which may be considered the same as the vertical deflection;

$D_L$ = deflection lag factor;

$K$ = a bedding constant that depends on the angle $\alpha$ in Fig. 25.17;

$W_c$ = vertical load per unit length of the pipe in lb/in. (N/m);

$r$ = mean radius of the pipe in inches (mm);

$E$ = modulus of elasticity of the pipe material in lb/in.$^2$ (kPa);

$I$ = moment of inertia of cross-sectional area of the pipe wall in lb/in.$^2$/in. (kPa/mm);

$e$ = modulus of the enveloping soil in lb/in.$^2$/in. (kPa/mm);

$E'$ = $er$ = modulus of soil reaction in lb/in.$^2$ (kPa).

Values of the bedding constant $K$ for various values of the bedding angle are shown in Table 25.3. The bedding angle $\alpha$ is defined as one-half the angle subtended by the arc of the pipe ring that is in contact with the pipe bedding, and over which the bottom reaction is distributed, as shown in Fig. 25.17.

## 25.10.4 Deflection Lag Factor

Measurements of field installations indicate additional yielding with time, particularly with fine-grained soils. The estimate of deflection therefore should be increased by an empirical "deflection lag factor." This factor cannot be less than 1.0 and has been observed to range upward toward a value of 2.0. A well-graded dense soil will permit very little, if any, residual deflection, and the lag

| **Table 25.3** | Bedding angle, $\alpha$ (deg) | Bedding constant, $K$ |
|---|---|---|
| Values of bedding constant | 0 | 0.110 |
| | 15 | 0.108 |
| | 22.5 | 0.105 |
| | 30 | 0.102 |
| | 45 | 0.096 |
| | 60 | 0.090 |
| | 90 | 0.083 |

factor can safely be ignored, but loosely placed soil may allow a relatively large deflection lag. A deflection lag factor of 1.5 is recommended for general design purposes, and for best results the backfill soil should be compacted for a width of one or two pipe diameters on each side of the pipe.

Although eq. (25.17) was developed for pipes made of elastic material such as steel, research at Utah State University indicates that the equation also can apply to plastic pipes. However, the modulus for plastic is many times lower than that of steel, wall thicknesses typically are larger, and a larger deflection can be tolerated before the top of the pipe dips into a reverse curve.

### 25.10.5  Approximate Method for Predicting Deflection

Difficulties in evaluating the constants, in particular $E'$, in eq. (25.17), led to the use of a simplified method for predicting deflection by Watkins at Utah State University and Howard at the U.S. Bureau of Reclamation. This analysis assumes that the resistance of the pipe to compression is the same as that of the soil on both sides, as shown in Fig. 25.18. The approximate method is supported by full-scale load tests.

Application of the simplified method is simple. Consolidation tests are performed on soil that is compacted according to the specification density and moisture content, and the decrease in thickness $T$ is determined from the amount of compression under the anticipated embankment pressure $\sigma_v$.

It also is important to apply a deflection lag factor to the results. It might be expected that the lag factor would be larger for plastic than for steel pipe, but tests indicate that lag is governed more by the soil than by the composition of the pipe.

The pipe strength also must be evaluated, either analytically or empirically from tests in a test box. An argument favoring this approach is that, unlike the situation with rigid pipe, the governing criterion for flexible pipe is not the actual soil load on the pipe, but the amount of deflection.

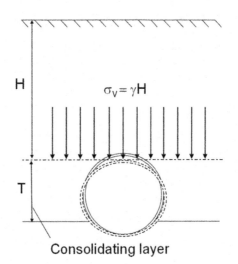

**Figure 25.18**
Approximate
method for
predicting
deflection of
flexible conduit.

$$\sigma_v = \gamma H$$

H

T

Consolidating layer

## 25.10.6    Permissible Deflections

Experimental studies have shown that flexible steel pipe may develop a reverse curvature in the top section when the deflection is about 20 percent of the nominal pipe diameter. A maximum allowable deflection of 5 percent therefore is recommended for design purposes, giving a deflection factor of safety of 4. Deflection measurements also are used to evaluate the condition of flexible steel culvert pipes in the field, and can be repeated over time to determine if the deflection rate is constant, increasing, or decreasing. If it is decreasing, FORE, discussed earlier in this book, may be used to predict a final amount.

A somewhat larger deformation is required to develop the reverse curvature in the top section of plastic pipe, about 30 percent, so the design maximum may be specified as 7.5 percent.

## 25.10.7    Finite Element Analysis

The complexity of flexible pipe response to loading and difficulties in assessing parameters for use in the Iowa formula have led to the use of finite element analysis (FEA). This usually is based on a hyperbolic stress-strain relationship for soil proposed by Duncan and Chang (1970), based on analyses of triaxial test data by Kondner. Iteration is required as the modulus changes both in response to compression under vertical stress and confinement under a changing lateral stress. An advantage of FEA is that it can incorporate elastic properties of the conduit and of the reinforcement. However, the assumptions made in the analysis still suggest that results should be correlated with results from physical tests. For more details of the method as it applies to underground conduit, see Moser (1990).

### 25.10.8 Strength Criteria for Flexible Pipe

While deflection amount is the major criterion for design of flexible pipe, the pipe also must meet strength requirements. These functions are from engineering mechanics and are as follows:

**Wall Crushing**

Pipe walls at the 3 and 9 o'clock positions act like short columns that must support the weight of the overlying soil. Compressive stress at these positions in the pipe is

$$S_c = \frac{W_c B_c}{2A} \tag{25.18}$$

where $W_c$ is the load per unit length of pipe, $B_c$ is the pipe diameter, and $A$ is the cross-sectional area of a unit length of wall.

**Bending Stress**

Bending stress is in addition to compressive stress, but is relatively small in thin-walled pipe:

$$S_b = \frac{Mt}{2I} \tag{25.19}$$

where $M$ is the bending moment, $t$ the wall thickness, and $I$ the moment of inertial of the cross-sectional area.

**Buckling**

Caving-in of the top of a flexible pipe is a form of buckling, but buckling also can occur anywhere in the perimeter as a result of high external pressure, for example in a deeply buried pipe subjected to hydrostatic pressure from groundwater. The external pressure for buckling is

$$P_{cr} = \frac{3EI}{R^3} \tag{25.20}$$

where $P_{cr}$ is the critical external pressure or the pressure differential if the pipe holds a vacuum, $E$ is the elastic modulus of the pipe, $I$ is the moment of intertia of a unit length of the pipe wall, and $R$ is the nominal radius of the pipe.

## Problems

25.1. Sketch cross-sections showing a ditch conduit, a positive projecting conduit, and an imperfect ditch conduit, labeling the plane of equal settlement if it is appropriate and showing the friction arrows according to Marston's theory

25.2. Compute the load on a sewer pipe having an outside diameter of 533 mm (21 in.) installed in a ditch with a width of 1.14 mm (3 ft 9 in.) at the elevation of the top of the pipe. The height of backfill over the top of the pipe is 5.5 m (18 ft). Assume the soil is a clayey material having a coefficient of friction of 0.2 and unit weight of 18.9 kN/m³ (20 lb/ft³).

25.3. Suppose the pipe in Problem 25.2 is installed in a ditch that is 1450 mm (4 ft 9 in.) wide. What will be the percentage increase in load?

25.4. A 1524 × 1524 mm (5 × 5 ft) reinforced concrete box culvert with 254 mm (10 in.) sidewalls is installed with its top 914 mm (3 ft) above the adjacent natural ground level. What is the projection ratio for this installation?

25.5. A circular pipe culvert is installed as a positive projecting conduit. The following settlement measurements were made after the embankment was constructed. Calculate the settlement ratio for this installation.

| Element | Settlement |
|---|---|
| Critical plane | 165 mm (0.54 ft) |
| Top of pipe | 55 mm (0.18 ft) |
| Natural ground at sides of pipe | 27 mm (0.09 ft) |

25.6. An elliptical concrete pipe has outside dimensions of 2135 mm (7.0 ft) on the major axis and 1250 mm (4.1 ft) on the minor axis. When installed in a culvert with the major axis horizontal, the top of the pipe is 915 mm (3 ft) above the adjacent ground surface. What is the projection ratio?

25.7. A 2134 mm (84 in.) diameter flexible pipe is installed with its top 1525 mm (5 ft) above the adjacent ground level. It is covered with an embankment extending 5.5 m (18 ft) above its top. The soil weighs 18.5 kN/m³ (120 lb/ft³). Assume a settlement ratio of −0.2. Estimate the load on the pipe.

25.8. A reinforced concrete pipe culvert having an outside diameter of 1778 mm (5.83 ft) is installed in such a manner that it projects 1403 mm (4.6 ft) above the ground level. The foundation soil is described as very unyielding in character. If the culvert is covered with a fill 7.5 m (24 ft) high that has a unit weight of 18.5 kN/m³ (120 lb/ft³), what is the load on the structure?

25.9. Estimate the height of the plane of equal settlement in Problem 25.8.

25.10. A rigid pipe with an outside diameter of 1778 mm (5.83 ft) is installed as a positive projecting conduit. Soil is thoroughly compacted to a distance of 3.66 m (12 ft) on each side and to an elevation 2.1 m (7 ft) above the top of the pipe. A trench 1778 mm (5.83 ft) wide is excavated directly over the pipe to a depth of 1.8 m (6 ft) and refilled with loose compressible soil. Then an embankment is constructed to a height of 17 m (56 ft) above the top of the pipe. Assume the unit weight of the soil is 19.6 kN/m³ (125 lb/ft³) and the settlement ratio is −0.35. Estimate the load on the pipe and the height of the plane of equal settlement.

25.11. A 457 mm (18 in.) o.d. pipe is installed in a ditch 1524 mm (60 in.) wide at the top of the pipe. Assume $r_{sd}p = 0.3$ and the height of fill is to be 6.1 m (20 ft). Should the load on this pipe be calculated by the ditch conduit theory or the projecting conduit theory?

25.12. A concrete box culvert 1220 mm (4 ft) wide is covered with 610 mm (2 ft) of fill that weighs $19.6 \, kN/m^3$ ($125 \, lb/ft^3$). A truck wheel weighing 44.5 kN (10,000 lb) stops directly over the center of the culvert. Determine the total dead and live load on a 0.91 m (3 ft) length of the culvert centered directly under the wheel.

25.13. Five specimens of concrete sewer pipe are tested in three-edge bearing. The minimum test strength is found to be 26.6 kN/m (1820 lb/ft). Applying a factor of safety of 1.4, what will be the design load for these pipes when installed in a ditch with ordinary (Class C) bedding? When installed with first-class (Class B) bedding?

25.14. What will be the required minimum three-edge bearing strength of the culvert pipe in Problem 25.8 if it is installed with first-class (Class B) bedding? Assume the friction angle of the fill material is 30° and use a factor of safety of 1.0 based on the 0.25 mm (0.01 in.) crack strength.

25.15. A 457 mm (18 in.) nonreinforced concrete pipe is to be installed under 3.7 m (12 ft) of cover in a ditch that is 914 mm (36 in.) wide at the elevation of the top of the pipe. Assume a clay soil weighing $18.9 \, kN/m^3$ ($120 \, lb/ft^3$), Class C bedding, and a factor of safety of 1.5. Determine the required minimum three-edge bearing strength of the pipe.

25.16. A 1524 mm (60 in.) reinforced concrete culvert pipe having 152 mm (6 in.) sidewalls is to be installed as a positive projecting conduit in a Class C bedding and covered with an embankment 7.32 m (24 ft) high. Assume the projection ratio is 0.5, the settlement ratio is +0.8, and the unit weight of the soil is $18.9 \, kN/m^3$ ($120 \, lb/ft^3$). Also assume the lateral earth pressure ratio $K = 0.33$, and that the lateral earth pressure is effective over the full projection of the pipe. Determine the required minimum three-edge bearing strength for a crack no wider than 0.25 mm (0.01 in.) with a factor of safety of 1.0.

25.17. The same size pipe as in Problem 25.16 is to be installed as an imperfect ditch conduit on a Class C bedding under 15.25 m (50 ft) of fill. Assume $p' = 1$, $r_{sd} = 0.4$, $\gamma = 18.9 \, kN/m^3$ ($120 \, lb/ft^3$), $K = 0.25$. Determine the required minimum 0.25 mm (0.001 in.) crack three-edge bearing strength of the pipe. The factor of safety is 1.0.

25.18. A 1524 mm (60 in.) 12-gauge corrugated steel pipe is to be installed as a projecting conduit in 60° bedding and covered with an embankment 9.15 m (30 ft) high. Assume the projection ratio is 0.7, the settlement ratio is zero, the unit weight of soil is $18.9 \, kN/m^3$ ($120 \, lb/ft^3$). (a) Calculate the appropriate consolidation pressure to predict deflection of this pipe. (b) Assume that the soil is compressed 3% at this pressure. What is the predicted long-time deflection of the pipe, assuming the deflection lag factor of 1.25. Is this design adequate?

25.19. A 762 mm (30 in.) o.d. by 9.5 mm ($\frac{3}{8}$ in.) steel line pipe that will carry natural gas at a maximum operating pressure of 4140 kPa ($600 \, lb/in.^2$) is to be installed in a 1.07 m (3.5 ft) wide trench under 1.83 m (6 ft) of cover.

Assuming the soil weighs $18.9 \, \text{kN/m}^3$ ($120 \, \text{lb/ft}^3$) and $E = 207 \, \text{GN/m}^2$ ($30 \times 10^6 \, \text{lb/in.}^2$), locate and determine the maximum tensile stress in the pipe wall.

## References and Further Reading

Duncan, J. M., and Chang, C. Y. (1970). "Non-linear Analysis of Stress and Strain in Soils." *ASCE J. Soil Mech. and Foundation Eng. Div.* 96(5), 1629–1653.

Handy, R. L. (2004). "Anatomy of an Error." *ASCE J. Geotech. and Geoenviron. Eng.* 130(7), 768–771.

Howard, A. (1996). *Pipeline Installation*. Relativity Publ., Lakewood, Colo.

Kim, K., and Yoo, C. H. (2005). "Design Loading on Deeply Buried Box Culverts." *ASCE J. Geotech. and Geoenviron. Eng.* 131(1), 20–27.

Larson, N. G. (1962). "A Practical Method for Constructing Rigid Conduits under High Fills." *Proc. Highway Res. Bd.* 41, 273.

Marston, A. (1930). "The Theory of External Loads on Closed Conduits in the Light of the Latest Experiments." Bull. 96, Iowa Engineering Experiment Station, Ames, Iowa.

Marston, A., and Anderson, A. O. (1913). "The Theory of Loads on Pipes in Ditches and Tests of Cement and Clay Drain Tile and Sewer Pipe." Bull. 31, Iowa Engineering Experiment Station, Ames, Iowa.

Moser, A. P. (1990). *Buried Pipe Design*. McGraw-Hill, New York.

Schlick, W. J. (1932). "Loads on Pipe in Wide Ditches." Bull. 108, Iowa Engineering Experiment Station, Ames, Iowa.

Schlick, W. J. (1952). "Loads on Negative Projecting Conduits." *Proc. Highway Res. Bd.* 31, 308.

Spangler, M. G. (1941). "The Structural Design of Flexible Pipe Culverts." Bull. 153, Iowa Engineering Experiment Station, Ames, Iowa.

Spangler, M. G. (1950a). "A Theory of Loads on Negative Projecting Conduits." *Proc. Highway Res. Bd.* 29, 153.

Spangler, M. G. (1950b). "Field Measurements of the Settlement Ratios of Various Highway Culverts." Bull. 170, Iowa Engineering Experiment Station, Ames, Iowa.

Spangler, M. G. (1958). "A Practical Application of the Imperfect Ditch Method of Construction." *Proc. Highway Res. Bd.* 37, 271.

Spangler, M. G. (1973). "Long Time Measurement of Loads on Three Pipe Culverts." *Highway Res. Rec.* 443.

Watkins, R. K. (1975). "Buried Structures." In H. F. Winterkorn and H.-Y. Fang, ed., *Foundation Engineering Handbook*, pp. 649–672. Van Nostrand Reinhold, New York.

Watkins, R. K., and Moser, A. P. (1971). "Response of Corrugated Steel Pipe to External Loads." *Highway Res. Rec.* 373, 86. Also discussion by M. G. Spangler.

# 26 In Situ Soil and Rock Tests

## 26.1 OVERVIEW

### 26.1.1 Probing for Answers

The first attempts at in-situ testing of soils probably involved jabbing the ground with a pointed stick, and may have been the origin of the phrase, to "get the point." The stick eventually became standardized, modified, and improved, evolving into a cone-tipped rod that became the "cone penetration test" or CPT. This was further improved by adding a separate sleeve behind the tip to measure side friction. More recent embellishments include measuring pore water pressure and using wireless data transmission. It is difficult to comprehend the scientific advances that are possible for a pointed stick. The multichanneled attack had developed extensive empirical correlations that relate cone data to soil type and bearing capacity.

As the cone test was being developed for relatively soft soils, particularly in Holland, similar procedures were being developed in the U.S. for testing harder soils. A sampler was devised that is driven into the soil instead of being pushed in. The sampler and driving effort were standardized to become the Standard Penetration Test or SPT, which still is the most widely used soil test in the U.S. An advantage over the cone is that it measures penetration resistance and simultaneously extracts a soil sample for identification and classification, although the sample is too disturbed for meaningful strength tests. A disadvantage is that the sampler must be removed from the boring after each test, a boring instrument inserted, and the boring advanced by drilling to the next test depth. A test is conducted by counting the number of hammer blows required to drive the sampler prescribed distances.

Because both CPT and SPT test results are influenced by combined influences from shearing, remolding, and compression of soil, other tests have been devised to try and isolate and measure these separate parameters. For example, modulus tests might be directly used to predict settlement, and shear tests to predict bearing capacity and slope stability.

Another type of in-situ testing is nondestructive and substitutes electrical, acoustic, or microwave (radar) energy for pushing force or driving energy. Many of these tests were developed in the petroleum exploration industry for subsurface mapping, and more recently adapted to measure properties such as a dynamic modulus. They are discussed in Chapter 27.

The reliability of in-situ tests is evaluated by comparisons with more traditional test methods, and by comparisons with field performance.

## 26.1.2 Exploration Drilling

At one time water wells were drilled with "cable tool rigs" that alternately lift and drop a heavy drill stem with a percussion cutter attached. Drill cuttings were intermittently bailed from the hole. Cable tool rigs are slow and simple, but can drill through both soil and rock. They still are used for drilling shallow wells, particularly in less developed areas.

More efficient drilling methods are required for drilling deep wells including oil wells. Rotary drilling employs a continuous flow of water or a thin mud that is circulated down the drill stem and carries cuttings back up between the stem and the perimeter of the hole. There the fluid is directed to a "mud pit" where cuttings settle out and the mud is siphoned off and pumped back down the hole.

The main component of drilling mud besides water is bentonite, or relatively pure smectite, because the mix is thixotropic. Then if there is an interruption in drilling, the mud holds cuttings in suspension instead of letting them settle to the bottom of the boring and lock up the bit.

Rock drilling usually is performed with a Hughes "tricone bit," which has three rollers with teeth that impact the rock. In soft rocks and soils a "fishtail" or similar hardened, steel bit is adequate to advance the boring. Rock cores are obtained with a hollow diamond-laced coring bit.

The first step in exploration drilling is to plan and stake out boring locations and do an elevation survey that will enable cross-sections to be drawn. Every geotechnical report will have a disclaimer pointing out that identifications of soils between borings are based on interpolation and can be expected to vary from the interpretations—a matter of considerable importance where variations are expected, for example in karst areas.

### 26.1.3 Undisturbed Sampling

Soil samples are obtained with the Standard Penetration Test by pounding a thick-walled sampler into the ground, and can hardly be considered "undisturbed." An optional sampling method uses a thin-walled steel tube called a Shelby tube, which is pushed into soil at the bottom of a boring. This procedure allows laboratory tests to be performed on the soil samples, including unconfined compression, triaxial, and consolidation tests. Special "piston samplers" draw the soil into the tube with partial vacuum, and "sleeve samplers" have been developed that encase the soil in a Nylon tube to reduce side friction.

A measure of sample disturbance is the "area ratio," defined by Hvorslev (1949) to represent the ratio of soil displaced to that which is sampled. The area ratio is the cross-sectional area of the sample divided by the doughnut-shaped area of the sampler. The larger the area ratio, the larger the degree of sample disturbance:

$$\text{Area ratio} = \frac{(\text{o.d.})^2 - (\text{i.d.})^2}{(\text{i.d.})^2} \tag{26.1}$$

### Example 26.1
Calculate area ratios for an STP sampler, and for a nominal 3 in. (75 mm) diameter 16 Gauge Shelby tube having a wall thickness of 0.065 in. (1.65 mm).

*Answer:*

$$\text{STP: Area ratio} = (2^2 - 1.5^2)/(1.5)^2 = 0.78$$

$$\text{Shelby tube:} \quad (3^2 - 2.87^2)/2.87^2 = 0.093$$

Each of the two sampling operations, SPT and Shelby tube, has advantages that depend on the kind of soil, SPT being most useful for evaluating cohesionless soils, and Shelby tube and laboratory testing most useful for cohesive soils. Because of the unknown character of soil ahead of a boring, the two procedures often are performed alternately, typically at 5 ft (1.5 m) depth intervals. A hole is bored to the first testing depth, either STP or Shelby tube sampling performed, the hole is drilled out to the next test depth, the alternate method used, and so on through the depth of the boring. Boring depths are selected on the basis of the application—deep enough to allow a realistic analysis for consolidation settlement, foundation bearing capacity, and slope stability, deeper than the longest anticipated piles or piers, or a sufficient distance into rock to ensure that the rock is in place and not part of an old rockfall.

Shelby tube samples of sand are compressed during sampling and are not suitable for laboratory tests, which has led to a continuing dependence on in-situ testing. Even the relatively tranquil Shelby tube sampling of clay soils does not generate a truly undisturbed sample because in-situ soil stress is not preserved. The confining stresses existing in the field can be reinstated in a laboratory triaxial test in the "Lambe stress path method," but only if the lateral stress that exists in the field is known.

STP and Shelby tube sampling can be conducted in a boring made with or without drilling mud. Another option to hold a boring open while avoiding the extra setup and mud disposal is to use a "hollow-stemmed auger." The auger has a central plug that remains in place during drilling; then at the desired sampling or testing depth the plug is removed and the sampler introduced down the center of the auger.

## 26.1.4 Determining the Character of Soil Between Sampling Depths

Sampling for laboratory tests or SPTs normally includes about 30 percent of a soil column and leaves the rest for interpolation. It therefore is of critical importance that the driller carefully examine the remaining 70 percent. An advantage of some in-situ tests, in particular cone tests, is that they can be continuous with depth.

Most soil drilling is performed with a helical screw-type "flight auger." Drillers whose main objective is to "make hole" (and money) will prefer to rotate the auger rapidly so that pieces of soil are cut loose at the tip and pushed along up the auger to emerge at the ground surface. In this case the only way to identify the depth from which the soil was removed is to stop advancing the auger and wait for cuttings from the bottom to come up.

Much more satisfactory for geotechnical exploration drilling is to rotate the bit slowly while advancing the auger so that it literally screws itself into the ground like a wood screw. The auger then is pulled vertically to shear off the soil and hold it in place between flights of the auger. The soil then is exposed by cutting away the surface with a knife, examined, and boundaries determined and depths measured on the auger. This information is critical for a geotechnical evaluation. For example, a settlement analysis requires knowing the thickness of the consolidating clay layers and whether they are sandwiched between sand layers. When using this procedure the driller must gauge each auger advance so that it does not exceed the capability of the drilling machine to pull it out.

## 26.1.5 Groundwater Table

An important step in exploration drilling is measuring the depth to the groundwater table. This conveniently is done with a line that completes an electric circuit on contact with free water, and ideally should be done after completion of drilling and 24 hours later. This cannot be done in mudded holes, which is another reason to prefer auger-type drilling where possible.

## 26.2    STANDARD PENETRATION TEST (SPT)

### 26.2.1    Redefining the SPT

The Standard Penetration Test (ASTM Designation D-1586) uses a standardized sampler shown in Fig. 26.1, and a driving effort that is standardized using a 140 lb (63.5 kg) weight dropping 30 in. (0.76 m) for each blow. The number of blows is counted and recorded for each of three 6 in. (152 mm) increments of advance. The first count usually is lower than the last two, which are added to give $N$ blow counts per foot (per 0.3 m).

A "doughnut hammer" is a hollow cylindrical weight that fits over the drill rod and hits on an expanded section of the rod. The hammer is lifted by looping a rope two turns around a smooth pulley called a "cathead" that is continually turning. When the end of the rope is pulled it grabs onto the cathead so that friction raises the hammer. Then at the moment when the required height is reached, the rope is forcibly released and the hammer drops. This constitutes one blow. The drill stem is marked with chalk at 6 in. (152 mm) intervals, so the number of blows can be determined for each interval. A typical rate is 40 to 60 blows per minute.

The "safety hammer" fits over the exposed end of a drill stem (Fig. 26.2) and is the current standard. Measurements indicate that whereas with a rope release the doughnut hammer has an efficiency of about 45 percent, the safety hammer has an efficiency of about 60 percent. Results obtained with a doughnut hammer are corrected to $N_{60}$ on the basis of the ratio of efficiencies.

### Example 26.2

The blow count for 3 driving invervals with a doughnut hammer are 10, 14, and 16. What is $N_{60}$?

*Answer:* Adding the last two blow counts gives $N_{45} = 30$ blows per ft (0.3 m). The energy correction gives $(45/60) \times 30 = 22$ blows/ft (0.3 m).

**Figure 26.1**

Standard Penetration Test sampler or "split spoon" is driven vertically by repeatedly dropping a large weight.

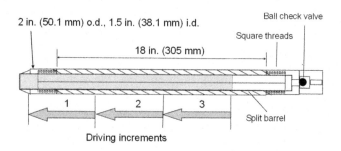

2 in. (50.1 mm) o.d., 1.5 in. (38.1 mm) i.d.

Ball check valve

Square threads

18 in. (305 mm)

1    2    3

Split barrel

Driving increments

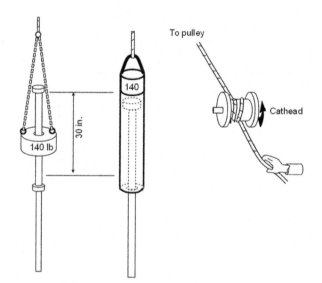

**Figure 26.2**

Hammers used to drive a Standard Penetration Test sampler. (Left) doughnut hammer, (middle) safety hammer, (right) rotating cathead and rope for lifting and manual release. Modern "trip hammers" have an automatic release mechanism.

The rope-and-cathead lifting procedure requires operator skill and introduces variables that can be on the unsafe side. For example, if the weight is prematurely released or the rope drags on the cathead, the count is increased and the soil is reported to be stronger than it actually is. Calibrations are used to measure hammer efficiency. These variables are eliminated by use of a "trip hammer" that is raised and automatically released when gripping levers ride up on a trip cone. The efficiency of trip hammers is considerably higher, and data are provided by the manufacturer. Detailed investigations also will include a correction for increasing inertia of the drill rod as the test proceeds deeper.

## 26.2.2 Overburden Pressure

A second correction that is applied to SPT data involves overburden pressure. The sensitivity of blow count to overburden pressure was first reported by Gibbs and Holtz at the U.S. Bureau of Reclamation. Soil confined in a box was loaded through a steel plate that supported various surcharge loads, and SPTs were performed through holes in the plate. While the concept was valid, the test arrangement may have included boundary conditions that exaggerated the development of horizontal pressure on the sampler and gave correction factors that are now believed to be unconservative, and factors suggested by Liao and Whitman (1986) are based on studies by several investigators and are shown in Fig. 26.3. The recommended correction factors may be calculated from the following empirical formula:

$$C_N = \sqrt{(1/\sigma'_V)} \qquad\qquad (26.2)$$

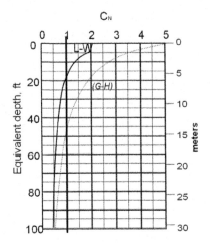

**Figure 26.3**

Liao-Whitman (L-W) overburden correction to SPT blow count data. The earlier Gibbs-Holtz correction is shown for comparison and should not be used. The equivalent depth is the sum of the depth above the groundwater table plus one-half of the depth below the groundwater table, based on $\gamma = 125 \, lb/ft^3$ (19.6 kN/m³). For a different soil unit weight, multiply depth by the appropriate ratio.

where $C_N$ is the correction factor and $\sigma'_v$ is the effective overburden stress in tons/ft². Alternatively,

$$C_N = \sqrt{(1/100\sigma'_v)} \qquad (26.2a)$$

where $\sigma'_v$ is in kPa.

In Fig. 26.3 the equivalent depth is based on a soil unit weight of $125 \, lb/ft^3$ (19.6 kN/m³). The effective thickness is defined as thickness above a groundwater table plus one-half of the thickness below a groundwater table. The Gibbs-Holtz (G-H) corrections are shown for comparison.

**Example 26.3**

The $N$ data in Example 26.2 were obtained at a depth of 30 ft (9.1 m) and a groundwater table exists at a depth of 8 ft (2.4 m). What is the blow count corrected for overburden pressure?

*Answer:* The hammer correction remains valid and gave $N = 22$. The effective depth is $8 + (30 - 8)/2 = 19$ ft [or $(2.4 + (9.1 - 2.4)/2 = 5.8$ m] From Fig. 26.3, $C_N = 1$ and the blow count remains at 22.

### 26.2.3 What Does the SPT Measure?

It sometimes is assumed that the resistance to driving the SPT sampler is mainly from end resistance, but the incremental increase in resistance for successive sections suggests that friction both on the inside and on the outside of the sampler is at least equally important. Studies by Schmertmann (1979) indicate that friction contributes from 45 to 85 percent of the resistance, the percentage being highest in cohesive soils.

Other investigators have found that the $N$ blow count relates more closely to side friction measured with the cone than to cone end resistance. The SPT mainly measures soil-to-steel friction and therefore is affected by lateral confining stress and surface roughness of the sampler. The measured "friction" is higher for cohesive than for granular soils, indicating a considerable influence from adhesion of clay to the sampler. Side friction can be evaluated by subtracting the second from the third increment, or a more precise measurement can be obtained by simply twisting the sampler with a torque wrench (Kelley and Lutenegger, 2004).

## 26.2.4 Relationships to Bearing Capacity and Settlement of Shallow Foundations on Sand

Generally the higher the penetration resistance of sand, the higher the friction angle, since part of the penetration resistance of sand comes from end bearing that in turn depends on the relative density, and hence the angle of internal friction. Excess pore water pressure is assumed to be negligible in sand and coarser-grained soils. An approximation to the friction angle can be obtained from Fig. 26.4, but it should be emphasized that these values derive from empirical correlations and are only approximate. Because a

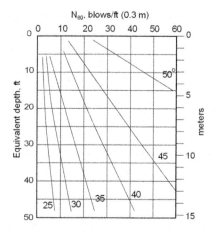

## Figure 26.4

Approximate friction angles of sand from SPT data, adapted from data of de Mello (1971). $N$ values should be corrected to $N_{60}$ but should *not* be corrected for overburden pressure. The equivalent depth incorporates a groundwater correction as in Fig. 26.3.

depth factor is included, $N$ values should *not* be corrected for overburden pressure.

The friction angle and unit weight may be used to calculate the bearing capacity and settlement of shallow foundations on sand according to the principles discussed in Chapter 23.

### Example 26.4

An SPT performed at 10 ft (3 m) depth in sand above the groundwater table gives $N=3$, 3, 4. Estimate the bearing capacity of a column footing $2 \times 2$ ft (0.6 m) at a depth of 1 ft (0.3 m).

*Answer:* $N=3+4=7$. From Fig. 26.6, $\phi=35°$. From Fig. 26.4 for a square footing the bearing capacity factors $N_\gamma$ and $N_q$ are 29 and 57, respectively. Then from eq. (22.7),

$$q_o = (\gamma B/2)N_\gamma + \gamma D N_q = (125 \, \text{lb/ft}^3 \times 2 \, \text{ft}/2)(29)$$
$$+ \, 125(1)(57) = 3625 + 7125 = 10{,}750 \, \text{lb/ft}^2 (515 \, \text{MPa})$$

With a factor of safety of 3, $q_a = 3500 \, \text{lb/ft}^2$ (170 MPa). Multiplying by the base area gives

$$Q_a = 4 \, \text{ft}^2 \times 3500 \, \text{lb/ft}^2 = 14{,}000 \, \text{lb} = 7 \, \text{tons} \, (62 \, \text{kN})$$

This is the bearing capacity against shear failure and does not ensure that there will not be excessive settlement.

Settlement of foundations on sand is discussed in section 16.10.3. Schmertmann's method is shown in Fig. 16.17. SPT blow counts for sand may be converted to approximate cone equivalents with Meyerhoff's (1956) approximation, $q_c = 4N$, where $q_c$ is in tons/ft². As both cone and SPT values are affected by overburden pressures, this relationship should apply to both corrected and uncorrected $N$ values.

### Example 26.5

Predict settlement of the footing in the preceding example.

*Answer:* In Fig. 16.17 for a square base, settlement is

$$S = 0.25 B p / q_c$$

where $B$ is the base width, $p$ is the foundation pressure, and $q_c$ is the cone bearing value.

Let

$$q_c = 4N = 4(7) = 28 \, \text{tons/ft}^2 (250 \, \text{kN})$$
$$S = 0.25(2 \, \text{ft})(7 \, \text{ton/ft}^2)/28 \, \text{ton/ft}^2 = 0.125 \, \text{ft} = 1.5 \, \text{in.} \, (38 \, \text{mm})$$

The allowable bearing pressure therefore must be reduced if settlement is to be less than 1 in. (25 mm).

## 26.2.5 Deep Foundations on Sand

Meyerhof (1956, 1976) showed empirical relationships between SPT blow counts and pile end-bearing capacity and side friction. Representative data are as shown in Fig. 26.5. It is not mentioned if the $N$ values were corrected, so it must be assumed that they were not. Depth correction factors in Fig. 26.3 vary at most by a factor of 2 and probably average close to 1 for most tests.

Most of the data in Fig. 26.5 cluster around the line $q_p = 4N$, where $q_p$ is the point bearing pressure in tons/ft$^2$ (100 kPa). This relationship is exactly the same as that which relates cone bearing pressure $q_c$ to $N$ value, and is consistent with the concept that the cone is a model pile.

From these and similar observations Meyerhof suggests the following formulas for end-bearing and for side friction on piles driven in sand:

$$q_p = 0.4N\frac{D}{B} \leq 4N \tag{26.3}$$

$$f_s = N_{av}/50 < 1\text{ton/ft}^2(100\text{kPa}) \tag{26.4}$$

where $q_p$ and $f_s$ are end-bearing and side friction stresses in tons/ft$^2$ (100's of kPa), $D$ is depth to the bottom and $B$ is the bottom width, and $N_{av}$ is the average blow count for the length of the pile. An H-pile is assumed to have areas between the flanges filled with soil. Taper can increase side friction by as much as 1.5.

### Example 26.6

SPT data for a sand give an average blow count of 18 blows/ft (0.3 m). Estimate bering capacity for a 20 ft (6.7 m) pile 1 ft (0.3 m) in diameter.

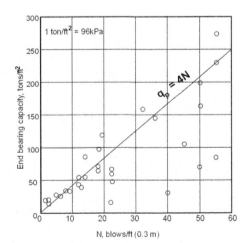

**Figure 26.5**

Data for end-bearing capacity of piles versus SPT blow count, summarized by Meyerhoff (1976) from many sources.

*Answer:* End bearing cannot exceed $4N = 4(18) = 72$ tons/ft2 (7 MPa)

$q_p = 0.4(18)(20\,\text{ft})/(1\,\text{ft}) = 144\,\text{tons/ft}^2(14\,\text{MPa})$ so

$q_p = 72\,\text{tons/ft}^2(7\,\text{MPa})$ end bearing

Side friction: $fs = 18/50 = 0.36\,\text{tons/ft}^2 < 1\,\text{ton/ft}^2$.

Multiplying by the resective areas,

$Q = 72\,\text{ton/ft}^2 \times (\pi)(0.5\,\text{ft})^2 + 0.36\,\text{ton/ft}^2 \times (\pi)(1\,\text{ft})(20\,\text{ft}) = 56.5 + 7.2$
$= 63\,\text{tons}\,(567\,\text{kN})$

With a factor of safety of 2,

$Q_a = 31\,\text{tons}\,(280\,\text{kN})$

### 26.2.6  SPT in Clay

Dynamic penetration resistance in saturated clay varies depending on sensitivity of the clay and excess pore water pressures developed during driving, so SPT data are considered even less definitive than for sand. The data therefore may correlate to driving resistance of piles but not the final bearing capacity, which depends on the setup factor.

For a preliminary estimate of end bearing in clay, Meyerhoff suggests eq. (26.3) with a maximum value of $3N$ instead of $4N$, and for side friction eq. (26.4).

## 26.3  CONE PENETRATION TESTS

### 26.3.1  Push vs. Drive-Cone Tests

Cone penetration tests measure the resistance of soil to penetration by a cone-tipped rod. An advantage of cone tests over SPTs is that cone tests do not require a breakdown and reassembly of the drill string after every test. Another advantage is that instead of being driven with a hammer, the cone is pushed, which better simulates a field loading condition. However, cone tests can be difficult to perform from the back of a truck-mounted drill without the use of screw-in ground anchors to hold the truck down. A drive cone can be driven with an SPT hammer, but the test is not standardized and should not be used without extensive correlation data. A hand-operated drive cone is useful for inspecting footing excavations to ensure that they are in firm soil.

### 26.3.2  Static Cone Tests

Cone penetration tests (ASTM Designation D-3441) are most readily performed with a hydraulic pushing apparatus located near the center of gravity of a truck so that the reaction base is the entire weight of the truck. The system also can be enclosed, which is amenable to the use of electronic and computer facilities.

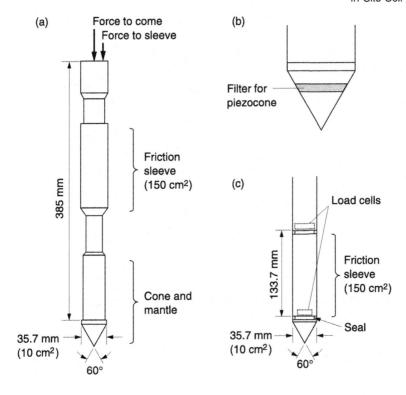

**Figure 26.6**
Schematic representations of cone apparatus: (a) mechanical cone, (b) detail of piezocone, (c) electrical cone. Tip and friction sleeve dimensions are standardized.

Cones have been standardized to have a 60° apical angle, which simulates the cone of soil with a friction angle of 30° that is pushed ahead of a flat foundation during a bearing capacity failure. The diameter is standardized at 35.7 mm (1.4 in.) in order that the cone base area is 1000 mm² (1.54 in.²).

Static cone penetrometers have an independently supported "friction sleeve" that measures side friction on the sleeve. The standard sleeve area is 15,000 mm² (23.25 in.²). As the sleeve is the same diameter as the cone, this corresponds to a sleeve length of 133.7 mm (5.26 in.). Diagrams of mechanical and electronic cones are shown in Fig. 26.6.

The standard rate of advance is 20 mm/s, and measurements are made by mechanical, hydraulic, or electrical means. Electrical readouts allow continuous testing whereby the cone and sleeve are pushed simultaneously, whereas mechanical units require alternate pushing of the cone and of the sleeve. Cone penetration, sleeve friction, and friction ratios can be plotted versus depth as a test proceeds (Fig. 26.7).

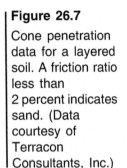

**Figure 26.7**

Cone penetration data for a layered soil. A friction ratio less than 2 percent indicates sand. (Data courtesy of Terracon Consultants, Inc.)

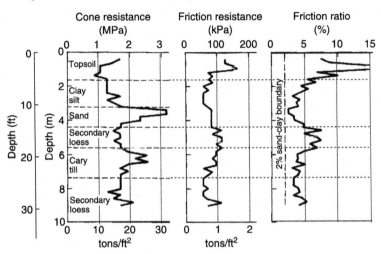

### 26.3.3 Influence of Depth and Lateral Stress

A gradual increase in cone resistance with depth is apparent in the left-hand graph of Fig. 26.7, similar to that which occurs with the Standard Penetration Test. Robertson and Campanella (1983) employed a depth function in relating friction angle of sand to cone bearing value (Fig. 26.8). Note that this correlation applies only to sand.

Combined influences of depth and lateral stress were examined theoretically by Durgunoglu and Mitchell (1975) (Fig. 26.9). In most instances $K$ is between 0.5 and 1.0, but it can be many times higher in overconsolidated soils and much lower in underconsolidated, collapsible soils. It can be seen that a high $K$ should increase the cone bearing value and a low $K$ should decrease it.

### 26.3.4 Relation of Soil Type to Cone Data

The friction ratio is the ratio of side friction to point stress on a unit area basis. Because clay increases side frictional resistance, cone data are an indicator of soil type. For sands the friction ratio is less than 2 percent, whereas for clays it is from 2 to 10 percent. Intermediate soil types are not as well differentiated and there is a considerable overlapping of soil types, as shown in Fig. 26.10. The logarithmic scale for cone bearing suppresses but does not preclude the role of a depth function. The overlapping of the various fields is indicative of the accuracy of the determinations. Computer programs used to interpret cone data can be misleading by specifically identifying soil types, even though the correlations are not that specific, and the soil identifications from cone data should be indicated as "suggested based on average soil responses."

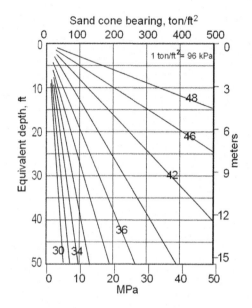

**Figure 26.8**
Friction angle of sand related to cone bearing value. The equivalent depth is depth above the groundwater table plus one-half of the depth below the groundwater table. (From data of Robertson and Campanella, 1983.)

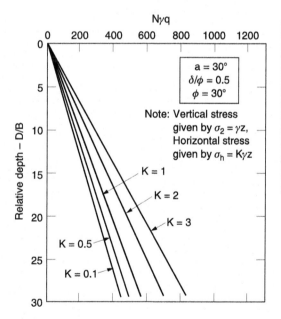

**Figure 26.9**
Relation of cone bearing value to depth and to lateral stress ratio, *K*. (From Durgunoglu and Mitchell, 1975, reprinted with permission of the American Society of Civil Engineers.)

**Example 26.7**
In Fig. 26.7, a sand layer identified from samples has a cone bearing from 1.5 to 3 MPa and a friction ratio of 2.5 to 5. How would this soil be identified from cone data?

*Answer:* Silty sand or sandy silt.

**Figure 26.10**

Soil identifications from cone data. (Modified from Robertson and Campanella, 1983.)

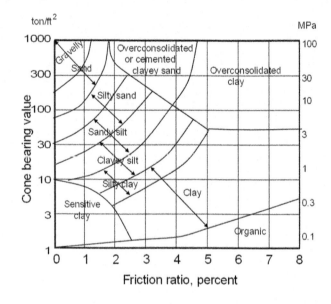

### 26.3.5 Thin Layers

It sometimes is assumed that because of the short length of the cone tip, 30.9 mm (1.2 in.), the cone test will detect thin layers of dissimilar soils. However, the soil zone influencing the cone bearing value extends over several tip diameters, and layers thinner than about 100 mm (4 in.) may remain undetected (Meigh, 1987), which is unfortunate because thin layers of sand can have a huge influence on the rate of primary consolidation settlement. Such layers can be reliably detected only by continuous sampling.

The friction cone measures an average response over the length of the sleeve, which is 133 mm (5.2 in.). It also should be noted that the center of the friction sleeve is about 100 mm (4 in.) above the cone tip with the electric cone, or 245 mm (9 in.) with the mechanical cone, so measurements are not simultaneous for a particular depth.

Stick-slip occasionally may produce a harmonic response or "chatter" in the results. This tendency will increase as tests go deeper and the rod acts more like a spring. Audible stick-slip is sometimes observed in the Borehole Shear test.

### 26.3.6 Other Uses of Cone Data

Cone data are used for predicting settlement of structures on sand in accordance with procedures outlined in section 16.10.3, and to gain an approximation for the friction angle of sand as indicated in Fig. 26.8, with a possible modification for the influence of lateral in-situ stress (Fig. 26.9).

Cone tests are especially useful for quality control of compacted fill, being complementary to near-surface density tests made during construction. An illustration of how cone data were used to diagnose excessive settlement is shown in Fig. 26.11.

As a rough guide, to prevent excessive settlement the allowable load on clay is sometimes taken as 1/10 of the cone bearing capacity, and on sand 1/30 of the cone value (Broms and Flodin, 1988).

The cone can be regarded as a model pile and used to predict pile bearing capacity, particularly that part of the bearing capacity that is attributable to skin friction. However, accuracy is influenced by scaling effects and inconsistent pore water pressures, and is of the order of ±50 percent (Schmertmann, 1975).

The cone resistance in saturated clays is a measure of the undrained or partially drained shear strength, being influenced by the rate of advance and soil sensitivity. By assuming an undrained condition a value for cohesion may be obtained for use in the bearing capacity equation:

$$c_u = \frac{q_c - \sigma_v}{N_K} \tag{26.5}$$

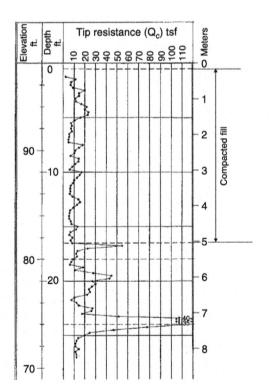

**Figure 26.11**
Cone bearing values showing layering in a 13 ft (4 m) section of compacted fill. The regular variation is described by Allender as the "Oreo cookie effect," but some of these cookies had too much filling. Under the fill is alluvium and glacial till. (Courtesy of Allender and Butzke Engineers.)

where $c_u$ is the undrained cohesion, $q_c$ the cone bearing value, $\sigma_v$ the *total* overburden pressure, and $N_K$ is a cone factor that varies from about 10 to 20, averaging about 15. $N_K$ often is evaluated from local correlations with vane shear measurements of cohesion (discussed below).

An important use of cone data is to determine the potential of sand for liquefaction during an earthquake, which is discussed in the next chapter.

### 26.3.7  The Piezocone

Sensitivity of the cone test to layering is improved by monitoring pore water pressures, which are higher in poorly draining soils. The pore water pressure generated by probing with the cone depends on the position of the filter, being highest when it is near the middle as shown in Fig. 26.6(b). The cone also can be left at one depth to obtain an equilibrium pore water pressure, but this information is more conveniently obtained with piezometers.

Pore pressure affects both the cone bearing and friction values, and bearing values may be corrected to be on an effective stress basis (Robertson et al., 1986). Empirical soil classification schemes also have been devised that incorporate pore pressure data.

Use of the piezocone and correction for excess pore pressure also improves the relationship between pile skin friction and cone data (Takesue et al., 1998). However, this use requires careful de-airing of the system, and the porous stone is susceptible to plugging.

## 26.4   VANE SHEAR

### 26.4.1   A Direct Approach

An alternative to a heavy reliance on empirical correlations is development of other methods that more directly target specific soil properties. For example, to determine the shear strength of soil, one can measure it directly instead of relating it to driving resistance or friction against a steel surface.

The vane shear test was developed in Sweden in the 1940s and uses a vertical rod with flat plates attached radially in the shape of an X (Fig. 26.12). The instrument is pushed into the ground or into the bottom of a boring and twisted, and the torque that is required to cause shearing around the cylindrical surface defined by the blades is measured. The soil shear strength is calculated from the torque measurement and the size and shape of blades of the vane.

A unique feature of the vane shear test is that after a peak strength value has been attained, rotation can be continued to obtain a measure of the residual shear

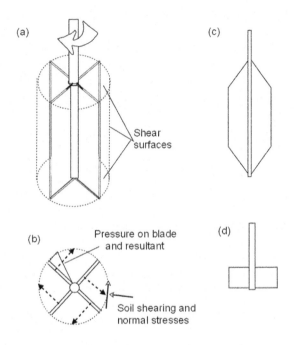

**Figure 26.12**

Vane shear: (a) and (c) standard blade shapes; (b) force being directed to the perimeter and tending to mobilize soil internal friction; (d) an experimental shape used to emphasize shearing on horizontal surfaces.

strength and therefore of the sensitivity. In the laboratory this is possible only in a direct shear test by a time-consuming cyclical reversal of the shear box.

## 26.4.2 Dimensions

The standardized height-to-diameter ratio is 2.0 (ASTM Designation D-2573). Because, as shown in Fig. 26.12(a), part of the shearing surface is vertical and part is horizontal, ends may be cut at $45°$ to prevent shearing along horizontal bedding planes (Fig. 26.12(c)). Other specialized shapes such as Fig. 26.12(d) have been used to vary the ratio of vertical and horizontal shearing resistance.

---

*Question:* Disregarding the shaft dimensions, what is the ratio of vertical to horizontal shear areas for a standard rectangular vane?

*Answer:* With diameter $D$ and length $2D$, the ratio is $(\pi D \times 2D)/2\pi(D/2)^2 = 4$.

---

## 26.4.3 What Does Vane Shear Measure?

It often is assumed that the vane test measures only soil cohesion, but as shown in Fig. 26.12(b), rotation causes a triangular distribution of pressure on a blade such that the resultant force pushes outward on the soil and will mobilize friction if there is any drainage. The vane is most suitable for testing slow-draining soft to medium clay such as occurs in deltas or offshore. The relevant pore water pressures are at the ends of the vanes where they are difficult to measure. By assuming full drainage,

Farrant in Australia showed that one should be able to evaluate the friction angle by varying the number of blades, but the method was found to be imprecise.

Comparisons with field situations indicate two additional discrepancies that have contributed to a declining use of vane shear (Schmertmann, 1975). First, failure is progressive, starting at the leading edge of each vane and then extending until failure involves the entire circumference. Therefore by the time shearing is complete, part of the soil has lost strength due to remolding, with the amount of strength lost depending in part on compressibility of the soil.

Second, the measured shearing strength is sensitive to the rate of application of torque. This can be attributed to several factors: development of negative pore pressure from dilatancy, development of positive pore pressure from compression, and viscous behavior during measurement of a residual strength. Some of these conditions give answers that are on the unsafe side.

The formula for reducing torque to shear strength for a vane of rectangular section and a height/diameter ratio of 2 is

$$\tau = \frac{3T}{28\pi r^3} \tag{26.6}$$

where $\tau$ is the shearing resistance in lb/in.$^2$ (N/m$^2$ or Pascals), $T$ is the torque in inch-pounds (newton-meters), and $r$ is the vane radius in inches (meters). The equation is dimensionally homogeneous.

## 26.5  BOREHOLE SHEAR TEST

### 26.5.1  Controlling Normal Stress

The Borehole Shear Test was devised in the early 1960s to perform a direct shear test on the sides of a borehole. The hole is 3 in. (75 mm) in diameter, preferably made by pushing a Shelby tube, but it also can be hand-augered.

As shown in Fig. 26.13, two opposing, sharply grooved plates of the shear head are pushed into soil along the sides of a borehole with a controlled constant pressure, and after a time for consolidation the device is pulled upward and the maximum pulling force measured. The normal and shearing stresses are calculated from the respective forces and the plate areas:

$$\sigma_n = \frac{N}{A} \tag{26.7}$$

$$\tau_f = \frac{F}{2A} \tag{26.8}$$

where $\sigma_n$ is the normal stress, $N$ the normal force on each shear plate, $A$ is the area of one plate, $\tau_f$ is the shearing stress at failure, and $F$ is the pulling force acting on two shear plates.

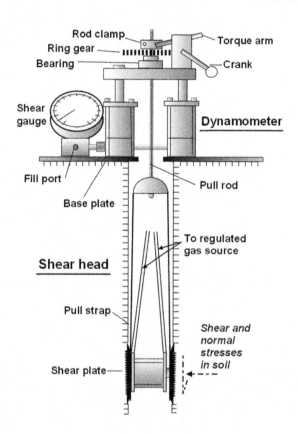

**Figure 26.13**

Schematic diagram of the borehole shear tester (BST). The shear head is expanded with a constant normal stress, time is allowed for consolidation, and the shear head is pulled upward. (Courtesy of Handy Geotechnical Instruments, Inc.)

## 26.5.2  The Question of Pore Water Pressure

The circumferential grooves and thin shear zone contribute to rapid dissipation of excess pore water pressure, as shown in Fig. 26.14. The standard test procedure allows 15 minutes consolidation time for the first data point and 5 minutes for each successive data point; however, additional drainage occurs during application of the shearing stress. Pore pressures in Fig. 26.14 were measured above the midpoint of the shear plate where they were found to be highest. Because drainage is unimpeded at the edges of the shear plates, pore pressure should decline to near zero, so the average should be about one-third to one-half of the values in the figure.

Pore pressures again increase as shearing stress is applied due to the increase in major principal stress, so shearing rate is slow, 0.002 in. (0.05 mm)/sec, and is further slowed down manually as the shearing stress approaches a maximum value. Pore water pressures can be monitored during the test, or their effect can be more conveniently determined by alternating shorter and longer consolidation times to see if they influence the shear envelope (J. Schmertmann, personal communication).

**Figure 26.14**

Maximum pore water pressures during consolidation phases of a Borehole Shear test in marine clay (from data of Lutenegger and Tierney, 1986).

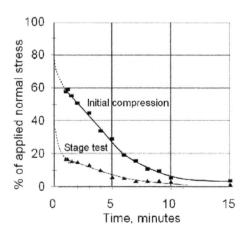

### 26.5.3   Borehole Shear Tests in Unsaturated Soils

Pore water pressures in unsaturated soils are negative and referred to as matric suction. This usually is interpreted as moving the origin of a Mohr-Coulomb failure envelop to the left, increasing cohesion and leaving the friction angle unchanged (Fredlund and Rahardjo, 1993). However, Miller et al. (1998) show that with a decreasing moisture content and increased suction prior to testing, cohesion decreases and the friction angle significantly increases. Tests were conducted in a 300 mm (12 in.) diameter rigid test chamber that may accentuate changes in matric suction, which becomes more negative with increasing strain. The phenomenon has not been observed in field tests, and in Miller's tests the uplift capacity of model drilled shafts was closely predicted from the Borehole Shear test results.

Borehole shear tests preferably are conducted in holes left by a 75 mm (3 in.) diameter thin-walled (Shelby) sampling tube. In caving soils sampling and testing are conducted through a hollow-stemmed auger, Fig. 26.15.

### 26.5.4   Serendipity Will Out

Tests revealed an important feature of the Borehole Shear test, that it does not require repositioning of the shear head between test data points. The reasons appear to be the radial geometry and concentration of normal stress close to the shear plates, coupled with rapid drainage of excess pore water pressure. As a result the consolidated-drained shear strength after shearing and remolding normally exceeds that of the surrounding undisturbed soil, so the shear surface moves outward with each successive cycle of shearing. If this does not occur, as in a quick clay, very low normal stress increments are used and drainage time is increased. A complete Mohr-Coulomb shear envelope with five or more data points normally is obtained in 30 minutes to an hour.

This test procedure is called a "stage test," and results in high degree of reproducibility and linearity because essentially the same depth of soil is repeatedly tested.

A representative data plot is shown in Fig. 26.16. The bilinear failure envelope is similar to those obtained with a laboratory direct shear test (Fig. 18.8), and it will be seen that the normally consolidated leg of the shear envelope extrapolates

**Figure 26.15**
Borehole Shear tester shear head being lowered into a hollow-stemmed auger for testing at the bottom of a boring. (Photo courtesy of R. Failmezger, In Situ Testing, L.C., and Dr. David White, Iowa State University.)

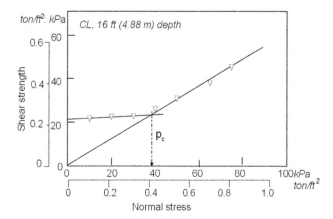

**Figure 26.16**
Borehole shear test data. This test was continued beyond the preconsolidation pressure to establish a bilinear failure envelope. (Courtesy of Dr. Evert Lawton, University of Utah.)

**Figure 26.17**

Rock Borehole
Shear test
comparative data.
RBST strengths
are slightly lower
because of
chipping of the
rock by the shear
plate teeth. The
test can be
performed in rock
that is too
fractured to core.
(Data courtesy of
the Oyo
Corporation,
Japan.)

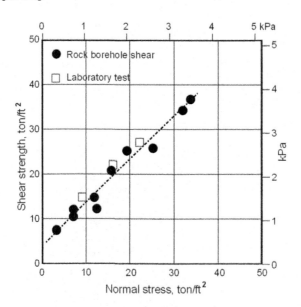

through the origin. Because of its speed and resolution compared with routine triaxial testing, the test sometimes is adapted for laboratory use.

### 25.5.5  Rock Borehole Shear Test

In the 1970s the principle of the Borehole Shear test was adapted for testing sedimentary rocks because of the difficulty of obtaining intact cores of fractured rocks for laboratory testing. This device uses smaller shear plates in order to administer normal pressures up to 12,000 lb/in.$^2$ (80 MPa). This is a single-point test as the shear plates engage and break off a chip of rock, after which the shear head is removed from the boring, rotated 45°, and the test repeated at a different normal stress. As in the case of the BST, suppression of the sampling variable results in highly linear failure envelopes that are closely comparable to results from laboratory tests, as shown in Fig. 26.17.

The test went full circle when the rock tester shear plate was adapted to the soil tester for testing intermediate rocks such as shale or soft limestone, where it often can be used as a stage test instead of a single-point test (Handy et al., 1976).

## 26.6  PRESSUREMETERS

### 26.6.1  Applying Only a Normal Stress

A pressuremeter consists of an expanding rubber bladder that fits inside a borehole. The bladder is expanded with water, and the expansion pressure and

**Figure 26.18**
Advances in pressure-expansion devices, or "pressuremeters."

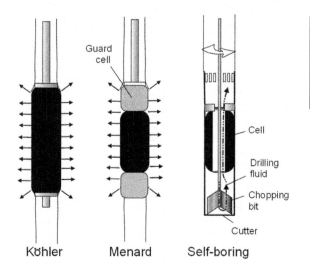

Köhler  Menard  Self-boring

corresponding increases in volume are measured. The pressure at failure is the "limit pressure."

A predecessor to the pressuremeter was invented in the 1930s by Köhler, and is shown in Fig. 26.18. However, inflation was with gas pressure, and results were difficult to interpret because the soil compresses vertically as well as laterally. In the 1950s, Loius Menard in France cleverly remedied the latter situation by adding two "guard cells" that are inflated at the same time, thereby giving essentially a horizontal plane-strain expansion of the center cell. The Menard pressuremeter and its derivatives are in common use today.

An important modification developed simultaneously in England and in France is a "self-boring" feature that is designed to slide the instrument into a boring without allowing relaxation of the soil pressure. Data obtained with self-boring pressuremeters give additional insight into the expansion mechanics of the Menard pressuremeter.

## 26.6.2 Theory

Lamé's theory for expansion of a hole in an elastic medium indicates that radial compressive stress should decrease with the square of radial distance. Tangential stress is tensile, and increases (algebraically) with radial distance, shown at the left in Fig. 26.19.

In an ideal plastic material, when the radial pressure reaches the proportional limit, the material in theory will continue to expand so long as the pressure is maintained. This is the basis for the concept of a pressuremeter limit pressure. Soils, however, have internal friction that increases as a result of compression,

**Figure 26.19**

Ideal elastic zone stresses are shown on the left, and changes that occur when a limit pressure is exceeded are shown on the right. Tension cracks reduce tangential stress to zero. The rate of change in radial stress in the plastic zone depends on the angle of internal friction.

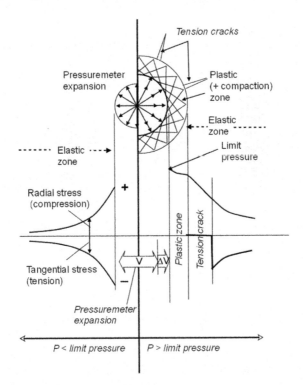

so the limit pressure is arbitrarily defined as the pressure at which the hole volume is doubled.

The relation of a plastic zone where the limit pressure has been exceeded and a surrounding elastic zone is shown at the right in Fig. 26.19. Radial tension cracking can occur in the elastic zone if tangential tension exceeds the lateral in-situ soil stress. The existence of these stress zones was confirmed by lateral stress measurements near Rammed Aggregate Piers shown in Fig. 24.7.

### 26.6.3 Instrumentation

In the Menard pressuremeter, gas pressure is applied to water in a standpipe and the water level monitored to indicate changes in the expansion cell volume. Simultaneously a second standpipe shows inflation of the two guard cells at close to and slightly below the main cell pressures.

In self-boring pressuremeters instead of the volume being monitored, the diameter is measured with internal flat springs that carry strain gauges and bear against external walls of the cell. Another alternative is load cells attached to the exterior of the device and bearing directly against the soil. Self-boring pressuremeters normally do not have guard cells, as there should be no initial gap between the device and walls of the boring.

A "push-in pressuremeter" is similar to the self-boring devices except that the hole is cut with a Shelby tube-like cutter and soil remains in the center core of the device until a test is completed and the pressuremeter is removed from the boring. Another type of pressuremeter is an expansion cell that is arranged to follow a cone to supplement the cone data with lateral expansion data, but in this case soil disturbance becomes a major factor.

An external shield of overlapping metal strips is commonly used to protect the exposed pressuremeter membrane from abrasion or puncturing.

## 26.6.4  Performing a Test

All pressuremeters must be calibrated by expanding inside a steel tube having known expansion characteristics. This step is essential to account for expansion of the hoses, standpipes, etc., under pressure.

A hole is bored and the Menard device lowered to the desired test depth, or the self-boring or push-in devices are taken to the desired depth. Pressure is raised in increments with each pressure held constant for 1 minute. The volume change is measured and pressure-volume expansion curves developed as shown in Fig. 26.20.

The initial upward curve obtained with the Menard device is sensitive to hole disturbance and does not occur with the self-boring pressuremeter. This part of the curve is attributed to repair of disturbed soil with the perimeter of the boring. French geotechnical engineers who make extensive use of pressuremeter tests recommend testing in hand-augered borings made with drilling mud that maintains a positive pressure on the sides of the hole. The use of a circulating mud

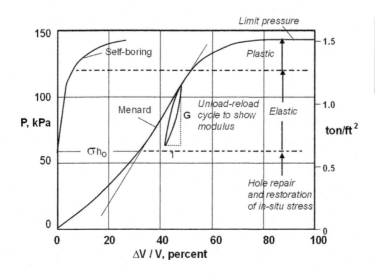

**Figure 26.20**

Self-boring and Menard pressuremeter curves.

means that the auger is not withdrawn to empty it, and depths of 60 to 90 ft (20 to 30 m) are possible.

### 26.6.5  In-Situ Stress

The starting point of the linear pressure-volume relationship may reasonably be assumed to represent the pre-existing lateral stress, and as can be seen in Fig. 26.20, the starting point from a self-boring device generally is more definitive than the reading with the Menard device because of hole disturbances.

### 26.6.6  Modulus

The linear portions of pressure-volume curves generally are well defined, particularly for a reloading cycle, so it is reasonable that they should be useful for predicting settlement of foundations. Difficulties are that the drainage state is not known, compression is horizontal instead of vertical, and is radial instead of uniaxial.

According to elastic theory,

$$G = V\frac{\Delta P}{\Delta V} \tag{26.9}$$

where $G$ is the shear modulus, $V$ the volume of the expanding hole, and $\Delta P/\Delta V$ is the slope of the pressure-volume curve after the calibration corrections are applied. The volume used in eq. (26.9) is the average of $V_0$ and $V_1$ that bracket the linear response range. With the Menard device the modulus determination may be made from an unload-reload cycle prior to reaching a limit pressure.

The shear modulus is converted to Young's modulus $E$ by

$$E = 2(1 + v)G \tag{26.10a}$$

where $v$ is Poisson's ratio. Menard recommended assuming a standard value of $v = 0.33$, which gives

$$E_M = 2.66G \tag{26.10b}$$

where $E_M$ is the "Menard modulus." However, the modulus differs from that obtained with a self-boring device, and is horizontal instead of vertical. Menard developed a complex, semiempirical equation to predict settlement from this modulus while taking into account footing dimensions and rheological and shape factors.

### 26.6.7  Limit Pressure

The limit pressure is a measure of soil strength, but does not differentiate between internal friction and cohesion. By making certain assumptions the limit pressure can be used to estimate cohesion by assuming that the friction angle is zero, which represents an undrained shear strength. Alternatively, a limit pressure can be used to estimate friction angle by assuming that cohesion is zero, as for a sand. However, the limit pressure is influenced by simultaneous radial compression and tangential tension, which is appropriate for soil around driven piles but not under shallow foundations. Stress may or may not be partly relieved by radial tension cracks, which are difficult to detect.

Menard related bearing capacity to limit pressure through an empirical bearing capacity factor $k$ based on grain size and depth of embedment. Other $k$ factors are defined for pile foundations. Comparisons between predicted and measured bearing capacities are shown by Briaud (1986) to be variable, with a factor of safety of 3 required to be on the safe side. Data reduction programs are provided by the manufacturers.

### 26.6.8  Role of Empiricism

Most in-situ test methods rely for their interpretation on empirical relation-ships developed from data. One danger is that there is a world of conditions that are outside of those used to establish the relationships and that remain untested. On the other hand, design that is based on experience should not be discounted if there is a long and continuous record of success in a particular area.

## 26.7  THE DILATOMETER

### 26.7.1  Overview

The soil Dilatometer, shown in Fig. 26.21, was developed in the late 1970s by Dr. Silvano Marchetti at the University of Turin in Italy. The device is spade-shaped, 14 mm (0.55 in.) thick, and on one face has a 60 mm (2.4 in.) diaphragm that is expanded by gas pressure for a distance of 1 mm (0.039 in.). The Dilatometer is pushed in to the next test depth, pressure readings are obtained with the diaphragm relaxed and expanded, and the instrument pushed to the next test depth where the operation is repeated.

The diaphragm is expanded with gas pressure, and the closed and open positions of the diaphragm are indicated by an electrical contact and buzzer. Because the diaphragm has an innate stiffness, reference pressure readings are made after the diaphragm is "exercised" to obtain a uniform response. Subsequent readings are corrected by reference to these values.

The device is pushed downward for a distance of 100–200 mm (4–8 in.) and a reading taken, which is the "A" reading. The diaphragm then is expanded 1 mm, and the expansion pressure is the "B" reading. After correction the two readings are referred to as $p_0$ and $p_1$. A "C" reading may be taken when the diaphragm is closed as an indication of static pore water pressure.

**Figure 26.21**

The Marchetti soil Dilatometer. Two pressure readings are obtained, one with the diaphragm flat and one with it expanded 1 mm into the soil.

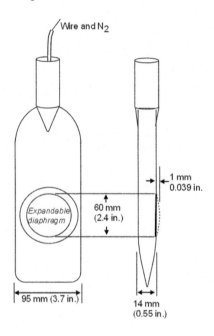

## 26.7.2 Evaluating $K_0$

Marchetti defines two dimensionless parameters plus one that has the dimensions of a modulus. The first parameter is in the nature of a lateral stress ratio:

$$K_D = \frac{p_0 - u_0}{\sigma_v'} \tag{26.11}$$

where $K_D$ is the Dilatometer modulus, $u_0$ is the static pore water pressure that is calculated from the depth below the groundwater table, and $\sigma_v'$ is the vertical effective stress that is calculated from the test depth and groundwater level. If the $p_0$ reading were an undistorted lateral in-situ stress, $K_D$ would equal $K_0$. However, insertion of the device increases the lateral stress, so a correction is necessary. On the basis of correlations in clays for a range of $K_0 = 0.4$ to 2.8, Marchetti suggested the following empirical relationship:

$$K_0 = (K_D/1.5)^{0.47} - 0.6 \tag{26.12}$$

Because the response to lateral displacement of different soils is not uniform, this expression has been modified by different investigators as shown in Fig. 26.22, and according to the various correlations the ratio of $K_0$ to $K_D$ value varies from about 1/20 to 3. A correlation for sands by Baldi et al. (1986) gave a standard error of estimate of 0.119. This means that if $K_0 = 0.5$, about two-thirds of the test values will be in a range 0.38 to 0.62. Dilatometer $K_0$ values appear reasonable as they are based on averaged soil responses, and they are useful if this level of reliability for $K_0$ can be tolerated. As discussed in the next section, the

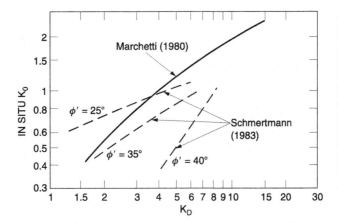

**Figure 26.22**

Some empirical relationships between $K_o$ and the Dilatometer $K_D$ (From Baldi et al., 1986, reprinted with permission of the American Society of Civil Engineers.)

Dilatometer thickness/width ratio is such that $p_0$ probably represents a limit pressure similar to that which is obtained with the pressuremeter.

### 26.7.3    Material Index

A second dimensionless parameter defined by Marchetti is the "material index," which depends on the change in pressure from expansion of the diaphragm in relation to pressure on the expanded diaphragm. The index is described as a ratio of soil stiffness to soil strength:

$$I_D = \frac{\Delta p}{p_1 - u} \qquad (26.13)$$

Low $I_D$ values (0–0.5) indicate clayey soils, medium values (0.5–1.5) silty soils, and high values (> 1.5) sandy soils. The material index also is used to help define other soil parameters including overconsolidation ratio, which also depends on $K_D$, and in turn is used to define preconsolidation pressure. A value for cohesion is obtained from $K_D$ if the friction angle is zero, or of internal friction from $I_D$ if cohesion is zero. The various measures are readily calculated and printed out, but their number should not defy the principle of "degrees of freedom," that the number of answers cannot exceed the number of input parameters. In this case there are two measurements, $p_0$, $p_1$, plus two other fixed variables, overburden pressure and pore pressure, plus optional additional measurements such as penetration resistance and the "C" pressure. If the number of answers exceeds the number of variables the answers cannot be independent but must be interrelated.

### 26.7.4    Predicting Settlement

The most successful use of the Dilatometer appears to be to predict settlement. The Dilatometer modulus, $E_D$, is based only on the increase in pressure needed to expand the diaphragm:

$$E_D = 38.2\Delta p \qquad (26.14)$$

where the constant was derived on the basis of elastic theory for stress near a circular hole (Marchetti, 1980). A vertical compression modulus $M$ is defined from elastic theory as

$$M = E/(1 - \mu^2) \tag{26.15}$$

where $E$ is the elastic modulus and $\mu$ is Poisson's ratio. As the limits on $\mu$ for an elastic material are 0 and 0.5, the corresponding limits on $M/E$ are 1.0 to 1.33. However, when the analogous equation is written using the Dilatometer modulus and evaluated experimentally, the range is much larger (Fig. 26.23). Let

$$M = R_M E_D \tag{26.16}$$

where $R_M$ was found to vary with $K_D$, which appears to be influenced by the limit pressure, and to a much lesser degree on the material index. Except for very soft soils the Dilatometer modulus underestimates the elastic modulus, perhaps related to disturbance from insertion of the instrument. Additional corrections that further increase $M$ with depth for normally consolidated soils are suggested by Schmertmann (1986). Settlement is

$$\Delta H = \frac{\Delta \sigma_v H}{M} \tag{26.17}$$

where $\Delta H$ is the change in thickness of a layer having a thickness $H$ and vertical compression modulus $M$, as a result of a change in vertical stress $\Delta \sigma_v$.

In 15 comparisons with measured settlements, the discrepancy from Dilatometer predictions varied from $-29$ percent to $+67$ percent, with an average over-prediction of about 10 percent (Schmertmann, 1986), lending credibility to the empirical approach. The soils in that investigation varied from a highly overconsolidated clay (where the overprediction was 67 percent) to silty clay, silt, silty sand, sand, and peat. The last two showed the largest variability. (The prediction for a quick clay is not included in this summary as settlement was overpredicted because of remolding.)

**Figure 26.23**

$R_M$ values to convert Dilatometer modulus $E_D$ to a compression modulus $M$. (Plotted from equations presented in *Dilatometer Manual* by Marchetti and Crapps.)

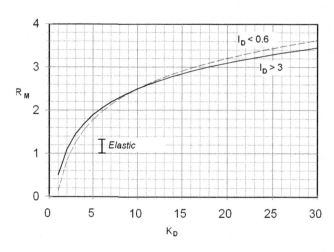

## Example 26.8

A layer of silty clay 2.5 m (8.2 ft) thick will sustain an added load of 2 bars ($2 \, \text{ton/ft}^2$, 191 kPa). Adjusted A and B readings give $p_0 = 1.01$ bar and $p_1 = 1.43$ bar on a total stress basis. The average pore water pressure is 0.14 bar, and effective overburden pressure at the middle of the layer is 0.46 bar. Predict compression of this layer.

*Answer:* On an effective stress basis, $p_0' = 0.87$ bar and $p_1' = 1.29$ bar. $\Delta p = 0.42$ bar.

$$K_D = p_0'/\sigma_v' = 0.87/0.46 = 1.90.$$
$$I_D = \Delta p/p_0' = 0.42/0.87 = 0.48$$
$$E_D = 34.7\Delta p = 34.7(0.42) = 14.6 \, \text{bar}$$

From eq. 26.27, $R_M = 0.9$ so $M = 0.9(14.6) = 13.1$ bar.

Then $\Delta H = \Delta \sigma_v \, H/M = 2$ bar $(2.5 \, \text{m})/13.1$ bar $= 0.4 \, \text{m}$ (1.2 ft).

# 26.8 LATERAL STRESS WITH THE $K_0$ STEPPED BLADE

## 26.8.1 The Challenge

Lateral in-situ stress is one of the most difficult soil properties to measure, while being one of the most important in engineering, not only as the basis for retaining wall design, but also for foundation bearing capacity and skin friction on piles. The most accurate means for measuring lateral stress in situ in soils probably is with a self-boring pressuremeter, but the test is expensive and time-consuming and the instrument is easily clogged or damaged.

Another approach is to push in a device such as a cone or Dilatometer that displaces soil and increases lateral stress according to shear or compression behavior, and attempt an empirical correction. A modification is to push in a thin load cell such as a "Gloetzl" cell and leave it in place until the soil pressure equilibrates, but there is no assurance that the final measured pressure represents that which existed prior to insertion of the instrument.

A load cell also can be buried in soil or attached to a solid surface to measure changes in pressure in response to some type of loading. This approach has been widely used in retaining wall and underground conduit research, but caution still must be used in interpreting the data because of a difference in modulus between the load cell and the soil, such that pressures may tend to concentrate on the load cell.

A fourth approach is to pump in a fluid and cause hydraulic fracturing, indicated by a sudden reduction in pressure, but in an overconsolidated soil the fractures may occur horizontally so that the pressure is measured vertically.

In the 1970s the U.S. Federal Highway Administration sponsored research to develop a better method for measuring lateral in-situ stress in soils.

The "$K_0$ Stepped Blade" introduces different levels of disturbance and extrapolates measurements to obtain a hypothetical pressure on a zero-thickness blade. A similar approach is used in chemical analysis and X-ray diffraction—to add a measured amount of a chemical or mineral and make two measurements.

## 26.8.2 Operation of the Blade

As shown in Fig. 26.24, the Stepped Blade has four steps incrementally increasing in thickness from 3 to 7.5 mm. Each step carries a pneumatic pressure cell that is flush with the surface and has a cover membrane that is in contact with the soil. The cell has two chambers that are inflated with slightly different gas pressures so that when the soil pressure is reached, the membrane is lifted and allows a cross-leak that is indicated by a differential pressure gauge, at which time the gas pressure is read. As the gas pressure equals the soil pressure, no additional calibration is needed. The cell then is immediately vented to prevent bulging of the membrane.

A test is conducted by boring to the test depth, then removing the drilling instrument. The first, thinnest step of the blade is pushed into the bottom of the boring and the pressure is read. Then the second step is pushed in, and the

**Figure 26.24**

Lateral pressures are measured sequentially *at the same depths* with the $K_0$ Stepped Blade and extrapolated to obtain a hypothetical pressure on a zero-thickness blade. (Courtesy of Handy Geotechnical Instruments, Inc.)

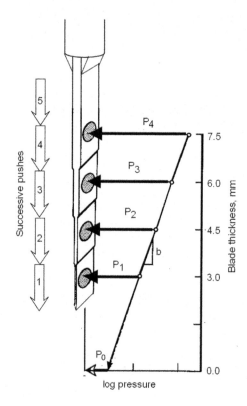

pressure is read on both the second and first steps, in that order. Then the third step is pushed in and the pressure read on cells 3, 2, and 1, and so on, until all steps have been pushed and read. An additional blank section of the blade is pushed to obtain an additional set of readings. The result is 4 pressure readings at the shallowest depth, 4 at the next, 3 at the next, 2 at the next, and 1 at the deepest depth, giving 14 pressure readings in all. This operation normally requires about 15 minutes. The instrument then is removed and the hole advanced to the next test depth, where the operation is repeated.

The device measures total stress, so it is important that the level of the groundwater table be determined in order to reduce the data to effective stresses.

## 26.8.3 Interpretation

The logarithm of pressure was found to relate to blade thickness, which is analogous to a consolidation $e$-log $P$ plot. This implies that for a test to be valid the soil must be consolidating. The sequence of reading is intended to allow approximately the same drainage time for each step. If consolidation is prevented by poor drainage, all steps should measure a limit pressure.

## 26.8.4 Results

Some representative test results are shown in Fig. 26.25. An important step in interpretation is to discard data that do not show an increase in pressure with increasing step thickness. The two most common causes are (1) in stiff soils the first step often creates an elastic response because the soil structure is not broken down, which gives a high reading that is a measure of in-situ stress plus a

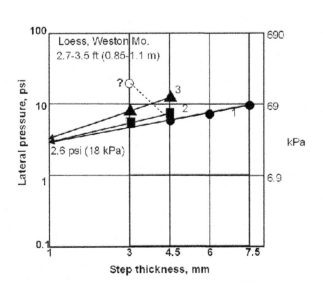

**Figure 26.25**

$K_0$ Stepped Blade test data at three subdepths showing extrapolation to obtain lateral in-situ stress. The question mark indicates elastic response prior to the soil structure breaking down.

modulus; (2) in soft soils the thicker steps can cause a lateral bearing capacity failure that is analogous to the limit pressure in the pressuremeter test. Data that do not meet the requirement that pressures must increase with increasing thickness therefore are omitted from the interpretation.

In Fig. 26.25 one measurement is omitted because of an elastic response and several are omitted because the pressure reached a limiting pressure and stayed constant or decreased. The change in slope of the data relates to drainage times, and excess pore pressures also appear to extrapolate out and give the same zero-thickness extrapolation for total lateral stress.

### 26.8.5 Accuracy of the Stepped Blade Test

A comparison of Stepped Blade test data with lateral stresses measured with a self-boring pressuremeter in a saturated, highly plastic expansive clay is shown in Fig. 26.26. In this series 44 Stepped Blade determinations were made that allowed calculation of a statistical confidence interval, discussed later in this chapter. Another series of tests conducted in a passively displaced silt indicated a test

**Figure 26.26**

Blade averages (open circles) and 90 percent confidence limits from 44 tests bracket six determinations with a self-boring pressuremeter. Pressuremeter data are courtesy of Dr. Michael O'Neil, University of Houston. (From Handy et al., 1982.)

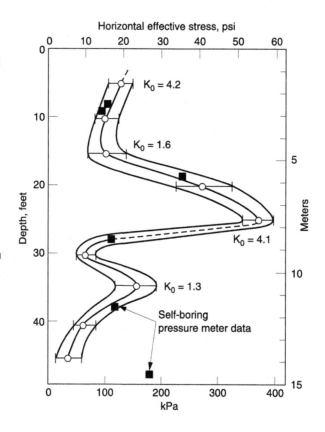

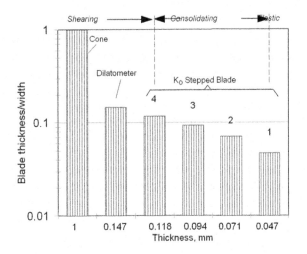

**Figure 26.27**
Thickness/width ratios of various push-in soil measurement devices, and (top) soil responses identified from Stepped Blade tests.

precision of ±5 percent, with pressures that were consistent with Rankine theory (Handy and White, 2006).

### 26.8.6 Relation to Thickness/Width Ratio

Relationships between various push-in soil-testing instruments are shown in Fig. 26.27, with soil responses identified from Stepped Blade tests. The thickness/width ratio of a Gloetzl cell varies from about 0.05 to 0.025, so insertion normally will cause either an elastic or consolidating response. An elastic response will not result in a time-related relief of pressure, and a consolidating response eventually should reach an equilibrium pressure after completion of primary and secondary consolidation.

## 26.9 STATISTICAL TREATMENT OF DATA

### 26.9.1 Overview

Statistics can be applied to any data, but statistical methods are particularly appropriate for soils because of their large variability, and for in-situ tests because of the abundance of data that can be generated in a relatively short time. These two characteristics are complementary, because a large variability requires many evaluations to establish a reliable average that may be used for design. For example, if the strength of a soil deposit varies between 20 and 40, it is very unlikely that a single measurement will give an answer that is representative of the true average. If two tests are performed and averaged, the estimate will be improved, and if 10 tests are performed and averaged, the estimate will be improved even further. A statistical analysis provides a means for establishing reliability of an average (or mean), based on variability and the number of evaluations.

The engineer often must make a choice between performing a few highly sophisticated tests or performing more tests that are less accurate and less costly. Which approach is best, or can they be combined, a survey test to establish an average and more detailed tests on a sample that is representative of the average?

Variability is expressed by a "standard deviation," which is determined by summing departures of data from an average. However, as the average will vary depending on the number of tests, the standard deviation is only estimated and depends on the quality of the average. The standard deviation, which once required hours to determine with a hand-crank calculator, now is determined almost instantly with a computer or a small electronic calculator that has statistics functions.

The standard deviation calculated from test results actually represents the sum of two standard deviations, one representing the soil variability and the other a test variability. This relationship is as follows:

$$s^2 = s_s^2 + s_t^2 \qquad (26.18)$$

where $s$ is the standard deviation that is measured, $s_s$ is that of the soil, and $s_t$ that of the test.

This equation is easily visualized because it describes the hypotenuse of a right triangle (Fig. 26.28). Soil variability is represented by the long leg of the triangle, test variability by the shorter leg. The objective of testing should be to minimize the length of the hypotenuse, but it can be seen that if the soil is highly variable, little can be gained by using highly refined test procedures that have a low variability. This may help explain the continuing popularity of rapid test procedures, such as SPT, CPT, and unconfined compressive strength, even though they only indirectly address basic soil parameters. However, these procedures can be valuable to screen and select representative samples that can be subjected to more detailed tests, thus giving the best of both worlds.

The relationship in Fig. 26.28 also indicates that a way to reduce variability is to target the soil variability by subdividing the soil on the basis of distinguishing

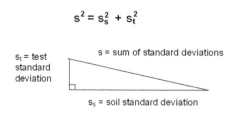

$$s^2 = s_s^2 + s_t^2$$

$s_t$ = test standard deviation

$s$ = sum of standard deviations

$s_s$ = soil standard deviation

**Figure 26.28**

Statistical variability $s$ is the square root of the sum of squares of test variability $s_t$ and soil variability $s_s$. For most efficient evaluations $s$ should be minimized, so the test should fit the soil variability.

characteristics. For example, if sand layers are interbedded with clay layers, it would be foolish to determine critical engineering properties based on a single composite sample, as there should be at least two samples.

## 26.9.2 Measuring Variability

The simplest statistic is the average or arithmetic mean, which is readily calculated and easily understood. However, the mean of 20, 25, 30 is the same as the mean of 5, 25, and 45, so an expression also is needed to describe variability. The standard deviation includes the number of samples or determinations:

$$s = \sqrt{\frac{\sum (X - \bar{X})^2}{n}} \tag{26.19}$$

where $s$ is the estimate for the standard deviation, $X$ represents each individual value, $\bar{X}$ is the mean of the $X$ values, and $n$ is the number of $X$ values. (An alternative procedure is to divide by $n-1$ instead of $n$ in recognition of the degrees of freedom.) The estimate of $s$ becomes more reliable with an increasing $X$ or number of measurements, but there should be no systematic change in $s$ related to the number of measurements. (In statistical literature $s$ usually is represented by $\sigma$, but the Roman version is used here to avoid confusion with stress. Also, $s$ can signify an *estimate* of the standard deviation.)

One use of the standard deviation is to calculate how much of a population probably exceeds certain limits. This usually is done by assuming an ideal normal or Gaussian distribution of data, which is illustrated in Fig. 26.29. As indicated in the figure, 67 percent of the area under the curve, which represents 68 percent of the population, is bounded by $\bar{X} \pm s$.

### Example 26.9
The following Standard Penetration Test blow counts were obtained in 10 tests:

$N = 4, 12, 18, 9, 16, 6, 13, 10, 8, 14$

Calculate the standard deviation (without using a program) and, assuming that this is an accurate representation of the true population spread, estimate the $N$ value exceeded by 95% of the deposit. (Use 90% area boundaries since these will define 5% "tails" at both ends of the distribution.)

*Answer:* The sum is 110 so the mean is $\bar{X} = 110/10 = 11.0$.

Values of $(X - \bar{X})$ are as follows: $-7$, $1$, $7$, $-2$, $5$, $-5$, $2$, $-1$, $-3$, $3$. Therefore $(X - \bar{X})^2 = 49 + 1 + 49 + 4 + 25 + 25 + 4 + 1 + 9 + 9 = 176$. Then

$$s = \sqrt{\frac{176}{10}} = 4.2$$

Therefore $\bar{X} = 11.0 \pm 4.2$ where the $\pm$ entry signifies one standard deviation. If these are accurate estimates, from Fig. 26.29, 90% of the values will be between $11.0 + 1.645(4.2)$,

**Figure 26.29**

A statistical normal distribution relating the number of tests (vertical axis) to deviations from a mean value (horizontal axis). The *s* values in this case are standard deviations (*s* in the text).

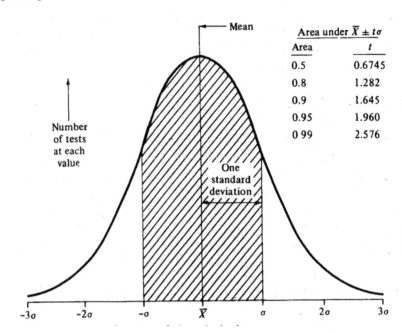

| Area under $\bar{X} \pm t\sigma$ | |
| Area | t |
| --- | --- |
| 0.5 | 0.6745 |
| 0.8 | 1.282 |
| 0.9 | 1.645 |
| 0.95 | 1.960 |
| 0 99 | 2.576 |

or 4.1 to 17.9. This means that if the distribution is symmetrical, 5% will be less than 4.1 and 95% will exceed 4.1.

### 26.9.3 Certainty of an Average

Because engineering decisions often are based on average properties instead of the spread, a more useful application for standard deviation is to establish the reliability of an average. This procedure establishes *confidence limits on the mean* and is a function of *s* and *n*. This range refers only to the mean, and not to the population or "spread." The larger the sample, the closer the confidence limits. This is analyzed according to a "*t* distribution" (Table 26.1).

**Example 26.10**

Calculate the 95% confidence limits on the mean in the previous example.

*Answer:* The 95% confidence limits are $X \pm 0.75(s) = 11.0 \pm 0.75(4.2) = 11.0 \pm 3.2$, or between 7.8 and 14.2.

If the average value of 11.0 is used for design without a factor of safety, the likelihood that the true average is lower than that value is 50 percent, so if the design procedures are accurate the failure rate would be 50 percent, which is unacceptable. If the lower value of 7.8 is used for design, the failure rate is reduced to 2.5 percent, which represents one "tail" of the distribution

| No. of observations | 90% confidence | 95% confidence | Table 26.1 |
|---|---|---|---|
| 4 | 1.36 | 1.84 | Confidence limits on a mean value, $\bar{X} \pm a \times s$ |
| 5 | 1.06 | 1.39 | |
| 6 | 0.90 | 1.15 | |
| 7 | 0.79 | 1-00 | |
| 8 | 0.72 | 0.89 | |
| 9 | 0.66 | 0.82 | |
| 10 | 0.61 | 0.75 | |
| 12 | 0.54 | 0.66 | |
| 14 | 0.49 | 0.60 | |
| 16 | 0.45 | 0.55 | |
| 18 | 0.44 | 0.51 | |
| 20 | 0.40 | 0.48 | |
| 25 | 0.35 | 0.42 | |
| >25 | $1.645 \div \sqrt{(n-3)}$ | $1.96 \div \sqrt{(n-3)}$ | |

Adapted from ASTM (1951).

and normally would be unacceptable even though the nominal "factor of safety" is $11 \div 7.8 = 1.4$.

The statistical "$t$-test" therefore can allow one to evaluate the reliability of a factor of safety. The lower the standard deviation represented in Fig. 26.28, the better the reliability.

## 26.9.4 How Many of Anything Is Enough?

In the above example if the number of tests had been 20 instead of 10 and had generated a comparable set of numbers, the 95 percent confidence limits on the mean would have been $11.0 \pm 0.48(4.2) = 11.0 \pm 2.0$ instead of $\pm$ 3.2, thereby reducing the range by more than one-third. The larger the number of tests, the greater the certainty of the mean value. However, *the tests must be real and not fictitious, which would be a fraudulent misrepresentation of data.*

A guide to the number of tests is presented in ASTM Designation E-122:

$$n = (3s/E)^2 \tag{26.20}$$

where $n$ is the number of tests, $s$ is the standard deviation, and $E$ is the acceptable error. In theory this still leaves a 3 in 1000 chance of being out of bounds based on a normal distribution. However, in engineering the normal distribution has its wings clipped because it is impossible to have negative values for blow counts, cone resistance, factor of safety, etc., so 3 in 1000 usually can be regarded as virtual certainty.

### Example 26.11

How many tests will be required for an allowable error of 20% of the mean of the preceding example? (This allowable error would be covered by increasing the factor of safety by 20%.)

*Answer:* The estimate of $E = 0.20(11.0) = 2.2$. Then $n = (3 \times 4.2/2.2)^2 = 33$ tests.

As this calculation is based on preliminary evaluations of $s$ and $\bar{X}$ from 10 tests, those estimates could be repeated after 20 tests to determine if there are any changes.

## 26.9.5 Two Populations

The most effective way to reduce data variability is to avoid the fruit salad of mixed data, such as not discriminating between sand and clay layers. In some cases differences may not be obvious until the data are plotted as a *bar graph*, which is readily accomplished with a computer spreadsheet.

A bar graph shows the number of data points falling within prescribed data intervals, and in theory will approach a statistical normal distribution if there is sufficient data. Figure 26.30 shows a bar graph for 123 standard penetration tests conducted in glacial till at a hospital site. To plot such a graph an arbitrary interval is assigned on the horizontal axis, and each test value is added to the bar having the corresponding interval. In Fig. 26.30 three tests have $N$ values from 10 to 14, seven from 15 to 19, and so forth.

**Figure 26.30**

Standard Penetration Test blow counts in a glacial deposit consisting of brown (oxidized) till over gray (unoxidized) till. (The heights of the bars only reflect the number of tests and not the strength.)

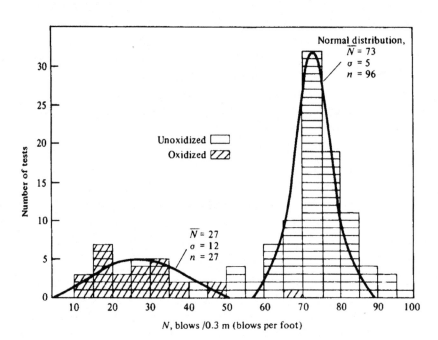

Even though the soil is glacial till, the distribution in this case is obviously *bimodal* because it has two modes, or population peaks. The modes correlate with the field description, one soil being gray and unoxidized and the other brown and oxidized. At one time the depth of oxidation was attributed to weathering time, but this graph indicates a more fundamental reason, that the gray till is harder and more dense so it restricted entry of oxygen-charged water. The gray till probably is "subglacial," meaning that it was compressed under the weight of the glacial ice, whereas the brown till probably was deposited during the final retreat of the ice front. This information obviously can be helpful for interpreting data from other similar sites.

The two populations therefore are treated separately to calculate means and standard deviations, which show that not only is the brown till layer much weaker than the underlying gray till ($N = 27$ compared to 73), it also is more variable ($s = 12$ compared to 5). This indicates a greater likelihood that the brown variation may contain weaker zones that were not detected in the testing. The building therefore was founded on deep foundations that extend into the unoxidized gray till.

## 26.9.6 Other Statistical Measures

### t-Tests
A t-test is used as illustrated above to define confidence limits on the mean, and also may be used to compare means of two samples to determine the likelihood that they represent two distinct populations.

### Variance
Variance is the square of the standard deviation. An "analysis of variance" (ANOVA) is used to compare means of three or more samples.

### Standard Error
The standard error sometimes is included as a $\pm$ notation after the mean to relate the reliability of the mean to the number of observations. Standard error is smaller than the standard deviation, and is defined as

$$s_e = s \left[ 1 \pm \frac{1}{\sqrt{(2n)}} \right] \qquad (26.21)$$

### Coefficient of Variation
How can one compare apples with oranges? Statistics at least allow a comparison of the *variability* of two populations. This is useful to compare variability of different test methods. The coefficient of variation is

$$C_v = \frac{s}{\bar{X}} \qquad (26.22)$$

where s is the standard deviation and $\bar{X}$ is the mean. The coefficient of variation often is expressed as a percent. In order to determine the coefficient of variation for a particular test method, the test should be repeated many times in a uniform soil.

### Example 26.12
Calculate the coefficient of variation in the above example.

*Answer:* $C_v = 4.2/11.0 = 0.38$ or 38%. This includes contributions from both soil and test variability.

### Linear Regression and Correlation
Linear regression, or fitting a straight line to $X$ and $Y$ data, is frequently used in engineering and can readily be performed with a computer or programmable calculator. In applying this method it is *critical* to assign $X$ values to an *independent variable that has little or no error*, because is it assumed that all random error is in the $Y$ variable. If that is not possible, the regression can be repeated, first assuming that the $X$ variables are independent, and then that the $Y$ variables are independent, discussed below.

Linear regression investigates the data and fits a linear relationship that minimizes the sums of squares of deviations in the $Y$ direction. The result has two coefficients, $a$ and $b$, for a linear equation having the form

$$Y = aX + b \tag{26.23}$$

Most computer or calculator programs also will give a value for $R^2$, which is the *correlation coefficient*. If all data are on line, $R = 1$ or $-1$, and there is no deviation of data from the line. A negative value only means that $X$ increases as $Y$ decreases. Squaring $R$ takes away the sign. $R^2$ also is affected by the number of data points, $n$, which therefore should be stated.

### No Independent Variable
In engineering both $X$ and $Y$ data may have random variability, in which case a simple regression of one on the other is not correct and will give the wrong $a$ and $b$ coefficients. If both random errors are unavoidable and approximately equal, two regressions, $Y$ on $X$ and $X$ on $Y$, will give two lines that intersect at the mean values of both variables, and an average can be sketched in. Both relationships will have the same $R^2$ and the lines will bracket the correct relationship. If $R^2 = 0$, the two regression lines will be at right angles and there is not a relationship.

There are many other statistical methods that can be useful in engineering, for example to determine confidence limits for a linear regression, and engineers are encouraged to take a course in statistics, if only to gain a better appreciation of the inevitable variability of data and the presence of experimental error.

## Example 26.13

The following data were obtained from calibration of a test instrument:

| Dynamometer: | 0 | 3.4 | 6.5 | 9.5 | 13 | 16.4 | 19.5 | 22.6 | 25.6 | 29 |
| Gauge pressure: | 0 | 20 | 40 | 60 | 80 | 100 | 120 | 140 | 160 | 180 |

Determine the regression coefficients and correlation coefficient with the dynamometer as the independent variable.

*Answer:* $a = -0.5230$, $b = 6.222$, $R^2 = 0.9998$, $n = 10$.

## Beta Distribution

The Gaussian distribution reaches into prohibited territory, for example by indicating a positive probability for a negative factor of safety, which is physically impossible. This has led to increasing use of a *beta distribution*, which is more versatile because end points can be defined. Usually a beta distribution is selected that is closest to a normal distribution but is subject to an experimental validation. A beta distribution has two shape functions, $\alpha$ and $\beta$, that are equal if the distribution is symmetrical. Some examples of beta distributions are shown in Fig. 26.31. R. Failmezger (personal communication) applied a beta distribution to show a linear relationship between standard deviation and a required factor of safety.

## Normalizing Data

In statistics to "normalize" literally means to make data more normal, that is, to more closely fit a normal distribution such as shown in Figs. 26.29 and 26.30. Plotting particle sizes to a logarithmic instead of a linear scale is an example of this kind of normalization, and is justified because sedimentation rate depends on the square of the particle diameter and not on the diameter, and most engineering soils are sediments. Any use of a logarithmic scale to obtain a linear relationship, as in interpretation of consolidation tests or Stepped Blade data, is in fact

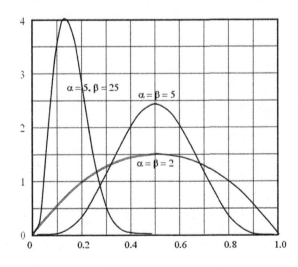

**Figure 26.31**

Examples of statistical beta distributions. Advantages of beta are that it is adjustable and can avoid any implication of negative values.

normalizing the data. This definition can be broadened to include any rearrangement of data that will either yield or take advantage of a linear relationship.

Another kind of transform is if data variability increases with size—the larger the size, the larger the variability. The coefficient of variation is an example of this kind of normalization, and allows comparisons between entirely different objects or occurrences.

In computer science, data are normalized to fit within a prescribed range that is easier to compute. "Floating point" arithmetic uses numbers expressed with an exponential notation, such as $12 \times 10^{-5}$ instead of 0.00012.

Researchers sometimes normalize data by expressing it in dimensionless terms. For example, pressure can be expressed relative to atmospheric pressure and expressed in "bars." An advantage is that this eliminates the system of units, as a dimensionless term is the same regardless whether units are in SI, English, or some other system. The Reynolds number is a dimensionless term that empirically has been shown to separate laminar from turbulent flow of fluids. Dimensionless terms are the "pi terms" of dimensional analysis.

Sometimes a transform to a dimensionless number can introduce unintended bias, which defeats the purpose of normalization. Two examples that are common in geotechnical engineering are the overconsolidation ratio (OCR) and coefficient of lateral stress ($K_0$). Both of these ratios involve vertical stress that is zero at the ground surface. It therefore is impossible to calculate an OCR or $K_0$ value for soil at the ground surface because it means dividing by zero. It also is not appropriate to describe a soil as having a particular value or even an average value of OCR or $K_0$, any more than it is to describe a soil as having a particular overburden pressure. The popularity of these terms may relate to their use for laboratory triaxial tests where the reference vertical stress is known or controlled.

## 26.9.7  Rejection of Data

It is not permissible to omit data simply because points do not fit a relationship. *Bad data can be worse than no data, so care is required in making measurements.* Data points that are far off the mark are called *outliers*. An outlier should be rechecked to make certain there has been no mistake in calculating or entering the numbers. Nevertheless outliers are unavoidable, as when a penetration test hits a rock or a void.

One guideline is the probability of occurrence of an outlier, for example that the probability should not be less than $1/2n$. For example, if the number of observations is 10, an outlier point may be rejected and labeled with a question mark if its probability of occurrence is less than 1/20 or 0.05. This and other probabilities are presented in ASTM Designation E-178.

A simple calculation is required to test an outlier; its departure from the estimated mean value is divided by the standard deviation:

$$D = \frac{|\bar{X} - X|}{s} \qquad (26.24)$$

The criteria for rejection are shown in Fig. 26.32.

Considerable caution should be used in omitting outliers, particularly if they are on the unsafe side, because they could be real. Preferably the test should be repeated. A low CPT or SPT value could flag a void or a soft zone.

### Example 26.14

The following SPT blow counts are obtained from tests in a uniform clay. Estimate the mean and standard deviation, examine for outliers, and if necessary recalculate the mean.

$N = 18, 12, 46, 13, 16$

*Answer:* The mean and standard deviation are $21.0 \pm 14.2$, $n = 5$. The questionable data entry is 46, which is on the high side and will raise the mean value. From eq. (26.24),

$D = |21.0 - 46| \div 14.42 = 1.73$

From Fig. 26.32, with $n = 5$ there is less than a 5% probability that this is from the same population as the rest of the values. It therefore is reasonable that this data point should be rejected. The recalculated mean and standard deviation are $15 \pm 2.8$ instead of $21.0 \pm 14.2$, giving a lower and therefore a safer and more reliable estimate that can be used for design.

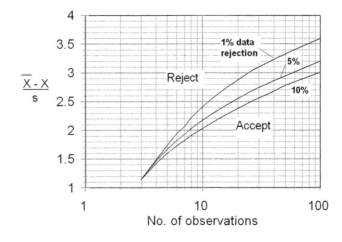

**Figure 26.32**

Criteria for rejection of data (adapted from ASTM Designation E-178).

**Figure 26.33**

The borehole shear test is one of the few in-situ tests that can be performed without a drilling machine, and is the only field test that measures soil cohesion and friction angle on an effective stress basis. The shear head is being expanded into the soil with air pressure and will be pulled upward to cause shearing.

## 26.10 SUMMARY

### 26.10.1 Tests with an Attitude

The pioneers of geotechnical (then called soil) engineering had specific objectives in mind when they devised tests to measure soil properties such as particle sizes, plasticity, compressibility, and friction angle and cohesion (Fig. 26.33). This is in contrast to the early developments in in-situ testing, where primary goals were speed and ease of use with a drilling machine. Tests such as the standard penetration test are effective as survey tests, but do not discriminate between compressibility and shear strength. Important improvements include development of cones that are pushed instead of driven, and separate determinations of side friction and end bearing, Pore water pressures can be measured as an aid to soil identification since there is no sample. Nevertheless the ambiguity of empirical correlations has opened journal pages to hundreds of papers that delve into the details. A unique and positive aspect of the standard penetration test is that it provides a sample.

### 26.10.2 The Next Stage

The next generation of in-situ tests was more targeted and includes vane shear, the pressuremeter and Dilatometer. However, these tests still measure combined factors. Users also should be cautioned that if many different parameters can appear on the printed page, they cannot be independent if their number exceeds the number of measurements. Thus a soil may be evaluated for friction or for cohesion but not for both, which leads to overdesign. This is permissible for small structures but can become very expensive for large ones, so a trend continues to develop in-situ tests that are directed towards specific objectives such as internal friction and cohesion and lateral in-situ stress.

## 26.10.3 A Useful Combination

Soils are so variable that many tests are required for statistical reliability. The speed and simplicity of tests such as the cone and SPT therefore are appropriate for initial survey purposes, and can be followed with more definitive tests that focus on particular soil properties. For both safety and economy this can be a winning combination.

## Problems

26.1. Prepare a table showing advantages and disadvantages of the cone penetration test compared with the Standard Penetration Test. Include level of skill of the technicians.

26.2. How does geotechnical exploration drilling differ from drilling for water? For oil?

26.3. Assume that a soil loses one-half of its strength by remolding. A standard cone push rod is 36 mm (1.4 in.) o.d. and 16 mm (0.63 in.) i.d. Calculate the amount of extension of 10 m (30.5 ft) of rod as a soil goes from a peak strength of 2 MPa (20 tons/ft$^2$) to its residual strength. The modulus of elasticity of steel is $200(10)^6$ kPa ($30 \times 10^6$ lb/in.$^2$). Explain the significance.

26.4. What is the vane shear strength of sand? Explain.

26.5. As a geotechnical engineer you are asked to observe exploration drilling and testing at a job site where they are performing a Standard Penetration Test. What do you look for in terms of the test, samples, and characterization of soil between the test depths?

26.6. The following SPT data were obtained at a depth of 5 ft in soil above the groundwater table with a doughnut hammer: 15, 23, 26. What is $N$ in blows per foot (0.3 m) What is $N_{60}$ with a depth correction?

26.7. The soil in the preceding problem is a sand. Which value of $N$ is used to estimate the friction angle? What is the estimated friction angle?

26.8. Use the answer from the preceding problem to estimate the bearing capacity of the footing in Example 26.4.

26.9. Predict settlement for the footing in Problem 26.8.

26.10. From averaged cone data in Figs. 26.7 and 26.10, estimate the grain size of the Cary till and of the sand. Which identifications are more reliable?

26.11. The apical angle of the cone test simulates what soil friction angle?

26.12. Estimate total settlement from compression of the weaker layers of fill in Fig. 26.11 under a broad load of $1\,ton/ft^2$, assuming that the layers behave as a sand.

26.13. The stiff layer immediately underneath the fill in Fig. 26.11 is a clay paleosol. Estimate its cohesion and bearing capacity. How does $N_T$ compare with the bearing capacity $N_c$ for a circular foundation? How do you explain the differences, if any?

26.14. The maximum torque measured with a standard rectangular vane is 35 ft-lb (47.5 N-m). What is the shear strength?

26.15. The test in Fig. 26.16 was conducted at a depth of 16 ft (4.88 m). The soil unit weight is $105\,lb/ft^3$ ($16.5\,kg/m^3$). The groundwater table is at 4 ft (1.2 m) depth. What is the OCR?

26.16. A Borehole Shear test develops the following data in unsaturated loess:

| Normal stress, lb/in.$^2$ (kPa): | 5.5 (38.5) | 12 (84) | 17.7 (124) | 24 (168) |
|---|---|---|---|---|
| Shear failure  "       " | 4 (28) | 4.6 (32) | 11.6 (81) | 14.7 (1.3) |

(a) Plot a Mohr-Coulomb graph and failure envelope, and explain if any data are omitted. Estimate cohesion and angle of internal friction. (b) Perform a linear regression for linear portions of the graph and determine the correlation coefficient, cohesion, and friction angle.

26.17. The following Stepped Blade measurements were obtained at two subdepths:

| Step thickness (mm): | 3 | 4.5 | 6 | 7.5 |
|---|---|---|---|---|
| Pressure lb/in.$^2$ (kPa): | 32 (224) | 24 (168) | 26.5 (186) | 29 (200) |
| | 45 (315) | 29 (186) | 34 (238) | 30 (210) |

(a) Plot the data with pressure on a log scale (or plot log pressure), examine the data for acceptability, and extrapolate to obtain lateral in-situ stresses. (b) Perform linear regressions on the thickness-log pressure data with thickness as the independent variable, calculate the correlation coefficient, lateral in-situ stress, and slope of the lines.

26.18. $K_0$ Stepped Blade measurements in a basal stratum of loess give $K_0 = 1.0$. Explain. What are the engineering implications for the overlying loess soil?

26.19. Give two explanations why lateral stress measurements in a residual soil consistently give $K_0 > 1$. Will $K_0$ be constant with depth? Explain. (*Hint*: What is $K_0$ at the ground surface?)

26.20. In Example 26.8 a clay layer 1 m (3.3 ft) thick gives a Dilatometer A reading $= 1.14 + 0.22$ bar and $B = 2.39 - 0.43$ bar. The pore water pressure is 0.475 bar, and the effective overburden pressure is 0.70 bar. Calculate settlement from compression of this layer.

26.21. In Fig. 26.30 there are several data points that occur in the wrong population. Perform a statistical test to determine if they should be omitted.

26.22. In the glacial till in Fig. 26.30, what laboratory test might give an estimate of the thickness of the glacier? What in-situ test?

26.23. In Example 26.9 the standard deviation is 4.2. (a) What is the coefficient of variation? (b) Estimate the limits on the mean that will encompass 50% of the data (known as the "probable error"). (c) What is the standard error?

26.24. In Example 26.9 predict how many tests will be required for an acceptable error of $\pm 10\%$.

26.25. In Fig. 26.30, which curve most nearly simulates a symmetrical normal distribution? How could one confirm that this is the best curve?

26.26. SPTs are performed in a glacial till that contains gravel particles. (a) Calculate the mean and standard deviation for $N = 17$, 16, 25, 20, 18, 37, 45, 26, 18, 22 blows/ft (0.3 m). (b) Plot a bar graph of the data. Would you be justified in deleting any value(s)? If so, suggest a possible explanation and recalculate part (a). (c) The tests start at a depth of 5 ft (1.5 m) and are performed at 5 ft (1.5 m) depth intervals. Apply a depth correction as necessary. (d) Calculate 95% confidence intervals on the mean.

26.27. What tests are most appropriate to determine soil variability across and under a site? What tests are most appropriate to establish design parameters such as compressibility and effective stress internal friction and cohesion?

## References and Further Reading

Almar, S., Baguelin, F., Canépa, Y., and Frank, R. (1998). "New Design Rules for the Bearing Capacity of Shallow Foundations Based on Menard Pressuremeter tests." In *Geotechnical Site Characterization* 2, 727–733. Balkema, Rotterdam.

American Society for Testing and Materials (Annual Book of Standards). "Standard Practice for Dealing with Outlying Observations," Designation E-178.

American Society for Testing and Materials (1951). *ASTM Manual for Quality Control of Materials*, Spec. Tech. Publ. 15-C. ASTM, Philadelphia.

Baldi, G., Bellotti, G., Ghionna, V., Jamiolkowski, M., Marchetti, S., and Pasqualini, E. (1986). "Flat Dilatometer Tests in Calibration Chamber." *Use of In Situ Tests in Geotechnical Engineering*, ASCE Geotech. Spec. Pub. No. 6, 431–446. ASCE New York.

Briaud, J. L. (1986). "Pressuremeter and Foundation Design." *Use of In Situ Tests in Geotechnical Engineering*. 74–115. ASCE, New York.

Broms, B., and Flodin, N. (1988). "History of Penetration Testing." In *Penetration Testing 1988*. Balkema, Rotterdam.

Brouwer, J. J. M. (2002). "Guide to Cone Penetration Testing," http://www.conepenetration.com.

de Mello, V. (1971). "The Standard Penetration Test—a State-of-the-Art Report." *Proc. 4th Int. Conf. on Soil Mechanics and Foundation Engineering* 1, 1–86.

Deacon, J. (2006). "The Really Easy Statistics Site" http://helios.bto.ed.ac.uk/bto/statistics/tress3.html

Durgunoglu, H. T., and Mitchell, J. K. (1975). "Static Penetration Resistance of Soils." *In Situ Measurement of Soil Properties* I, 151-188. ASCE, New York.

Failmezger, R. A., and Anderson, J B., ed. (2006). *Flat Dilatometer Testing.* Proc. of the Second Intern. Conf. on the Flat Dilatometer. In-Situ Soil Testing, L. C., Lancaster, Va.

Fredlund, D. G., and Rahardjo, H. (1993). *Soil Mechanics for Unsaturated Soils.* John Wiley & Sons, New York.

Gibbs, H. J., and Holtz, W. G. (1957) "Research on Determining the Density of Sands by Spoon Penetration Testing." *Proc. 4th Int. Conf. on Soil Mechanics and Foundation Engineering* **1**, 35.

Handy, R. L., Pitt, J. M., Engle, L. E., and Klockow, D. E. (1976). "Rock Borehole Shear Test." *Proc. 17th U.S. Symp. on Rock Mechanics* 486, 1–11.

Handy, R. L., Remmes, B., Moldt, S., Lutenegger, A. J., and Trott, G. (1982). "In Situ stress determination by the Iowa Stepped Blade." *ASCE J. Geotech. Eng. Div.* 108(11), 1405–1422.

Handy, R. L., and White, D. J. (2006). "Stress Zones Near Rammed Aggregate Piers: I. Plastic and Liquefied Behavior; II. Radial Cracking and Wedging." *ASCE J. Geotech. and Geonenviron Eng.* 132(1), 54–71.

Hvorslev, M. J. (1949). *Subsurface Exploration and Sampling of Soils for Civil Engineering Purposes.* Waterways Experiment Station, U.S. Army Corps of Engineers, Vicksburg, Miss.

Kelley, A. M., and Lutenegger, A. J. (2004). "Unit Skin Friction from the Standard Penetration Test Supplemented with the Measurement of Torque." *ASCE J. Geotech. and Geoenviron. Eng.* 130(5), 540–3.

Liao, S., and Whitman, R. V. (1986). "Overburden correction factors for SPT in sand." *ASCE J. Geotech. Eng.* 112(3), 373–7.

Lutenegger, A. J., and Tierney, K. F. (1986). "Pore Pressure Effects in Borehole Shear Testing." *Use of In Situ Tests in Geotechnical Engineering*, ASCE Geotech. Spec. Pub. No. 6, 752–764. ASCE, New York.

Marchetti, S. (1980). "In situ by flat Dilatometer"*J. Geotech. Eng. Div. ASCE.* 106(3), 299–321.

Meigh, A. C. (1987). *Cone Penetration Testing.* Butterworths, London.

Meyerhoff, G. G. (1956). "Penetration Tests and Bearing Capacity of Cohesionless Soils." *ASCE J. Soil Mechanics and Foundation Eng. Div.* 182(SM1), 1–19.

Meyerhoff, G. G. (1976). "Bearing Capacity and Settlement of Pile Foundations." *ASCE J. Geotech. Eng. Div.* 102(GT3), 195–228.

Miller, G. A., Azad, S., and Hassell, C. E. (1998). "Iowa Borehole Shear Testing in Unsaturated soil." In *Geotechnical Site Characterization* 2, 1321–1326. Balkema, Rotterdam.

Robertson, P. K., and Campanella, R. G. (1983). "Interpretation of Cone Penetration Tests." *Canadian Geotech. J.* 20, 718–745.

Robertson, P. K., Campanella, R. G., Gillespie, D., and Greig, J. (1986). "Use of Piezometer Cone Data." *Use of In Situ Tests in Geotechnical Engineering.* ASCE Geotech. Spec. Pub. No. 6, 1263–1280. ASCE, New York.

Schmertmann, J. (1975). "Measurement of in Situ Shear Strength." *In Situ Measurement of Soil Properties* II, 57–138. ASCE, New York.

Schmertmann, J. (1979). "Statics of SPT." *ASCE J. Geotech. Eng. Div.* 105(5), 655–670.

Schmertmann, J. H. (1986). "Dilatometer to compute foundation settlement." *Use of In Situ Tests in Geotechnical Engineering*, 303–321. ASCE, New York.

Takesue, K., Sasao, H., and Matsumoto, T. (1998). "Correlation Between Ultimate Pile Skin Friction and CPT Data." *Geotechnical Site Characterization* 2, 1177–1182. Balkema, Rotterdam.

# 27

# Introduction to Soil Dynamics

## 27.1 DYNAMIC LOADING CONDITIONS

### 27.1.1 Overview

Dynamic conditions relevant to geotechnical engineering fall into several categories: (1) earthquakes, (2) machinery vibrations, and (3) other human-made disturbances such as blasting, pile driving, soil compaction, rail or truck traffic, water hammer in pipes, etc. In addition, (4) dynamic instrumentation such as a portable seismograph or ground-penetrating radar are used as nondestructive investigative tools.

### 27.1.2 Earthquakes

Earthquakes are recurring and inevitable in many areas. The shaking induces strong, reciprocating horizontal and vertical forces due to inertia of soil and rock masses and supported structures. Inertial forces act through the center of gravity of a structure, causing tilting of tall structures if foundations should fail, as shown in Fig. 27.1. The soil itself also is subjected to back-and-forth shearing stresses that can contribute to landslides and collapse of retaining walls. Earthquake-induced ground vibrations also can cause liquefaction of susceptible soils, which can allow a sudden and dramatic sinking of structures into the ground. The extent of damages to structures therefore depends on the response of the soil or rock supporting the structures in addition to the nature of the structures and strength of the earthquakes. Vibrations can reinforce in some areas and partially annul in others, so destruction becomes selective. Earthquake problems therefore tend to be soil problems, related at least in part to the composition and density of the soil or rock.

**Figure 27.1**

Chimney inertia, and let's hope that nobody decides to use the broom. (Photo by James L. Ruhle, Orange, Cal.)

### 27.1.3 Machinery Foundations

Vibrations from rotating or reciprocating machines, while not usually destructive, can be troublesome and a source of annoyance. There also can be dangers from metal fatigue and ruptured water or gas service lines. The problems are most severe when a machine creates a resonant response in the foundation and supporting soil, sometimes ringing church bells a kilometer (half a mile) away. Managing machine vibrations often is a matter of managing the soil response to decrease or dampen resonance.

### 27.1.4 Blasting and Pile Driving

Vibrations from blasting, and razing of the old structures and pile driving for new ones, also are an annoyance and are potentially damaging to nearby structures. Vibrations seek a soil natural frequency that may or may not resonate and reinforce in a nearby structure. Geotechnical and structural engineers may be called upon to investigate, photograph, and make an inventory of damages existing prior to blasting or pile driving in order to determine if additional damage has occurred, and accelerometers can be installed to monitor the intensity of vibrations. Much more difficult is to try and assign damages afterwards, which becomes a forensic investigation based on evidence such as whether horizontal mortar joints may have puffed dust.

### 27.1.5 Useful Vibrations

Ground vibrations are intentionally induced in dynamic soil compaction, in deep dynamic compaction by repeatedly dropping huge weights, and in the manufacture of Rammed Aggregate Piers. Another category of vibrations is those that are introduced for nondestructive testing such as by the use of portable engineering seismographs and ground-penetrating radar. Some fundamental relationships and uses are discussed in this chapter.

## 27.2 EARTHQUAKES

### 27.2.1 Plate Tectonics

One of the broader implications of continental drift, or more formally plate tectonics, is stick-slip at plate margins, indicated by dots in Fig. 2.3. As plates tend to move past one another, shearing stress increases along an array of intersecting faults near plate margins until stress locally exceeds the holding power of friction, and one or more of the faults slip. It is the sudden elastic rebound of rock on each side of the fault that creates an energy burst that emanates outward as waves that are an earthquake. After the first slip occurs there usually are a series of "aftershocks" that are less intense but still can be destructive, and are evidence for a stick-slip nature of dislocations along the fault.

About 90 percent of earthquakes are associated with plate margins, but the other 10 percent are not, and occur in plate interiors. Fault-line earthquakes in central plate areas tend to be less frequent but stronger than those at plate margins. Focal points, or hypocenters, of severe earthquakes tend to be relatively shallow, at a depth less than 32 km (20 miles). Earthquakes also can be associated with volcanism. Earthquakes near or under the sea can create large swells that as they approach land compress and intensify into enormously destructive waves called tsunamis. The most important defense against a tsunami is to go to the high ground if there is any, which also requires an effective system that warns people to get out of the way before the wave gets there.

### 27.2.2 Pluses and Minuses of Pore Water Pressure

The role of pore pressure in generating shallow earthquakes was discovered in the 1950s when water was injected into an active fault zone in Colorado and resulted in a series of small earthquakes. When the injections stopped, the earthquakes stopped. Some consideration has been given to injection as a possible stress-relief mechanism that might substitute many small earthquakes for the occasional large one, but there are some major liability issues, partly because stress relief in one place will lead to a stress intensification in others.

Pore water pressures also can decrease naturally as shearing stress in a fault zone causes the fractured rock mass to dilate or increase in volume. The low void ratio of the system intensifies the reaction because only a slight increase in volume can create a substantial decline in pore pressure that draws down the groundwater level in nearby water wells. A temporary increase in effective stress helps to postpone movement, but that only stores more energy for release in an earthquake.

Dilatancy also is detected from a reduction in electrical resistivity of the fault zone due to the enlargement of rock fissures. Another clue is an emission of the colorless, odorless but radioactive gas radon into deep wells. Radon is continuously generated by radioactive decay of certain elements that occur in igneous rocks, and the release is caused by improved flow and interchange of groundwater in the dilated zone.

Considerable attention is given to clues to an impending earthquake, called precursors, but predictions still remain only approximate. In keeping with the stored energy model, in general the longer the time interval between the precursor and the quake, the more intense the earthquake, as shown in Fig. 27.2, but all that means is that the longer one waits, the worse it may get.

### 27.2.3  Whose Fault?

Geologists classify faults according to the angle of dip and the relative movement. An upward or downward movement of one side relative to the other defines a "normal" or "reverse" fault depending on the dip angle. A normal fault pulls a rock layer apart and a reverse fault compresses it. Normal and reverse faults create fault scarps, so recent activity can be estimated from the degree of weathering and erosion, dissection by streams, and waterfalls.

Many faults that cause earthquakes are "strike-slip" faults, meaning that one side moves horizontally relative to the other, tearing the ground and creating offsets in streams, roads, and fences. The grinding action of faults is such that the thickness of the fault zone relates to the amount of movement (Scholte, 1987). The average thickness is about 1 percent of the slip distance, but thickness also depends on the rock hardness and on normal stress on the fault surface.

**Figure 27.2**

Average time delays from precursor to earthquake: the longer the wait, the wilder the ride. (Plotted from data of Scholz et al., 1973.)

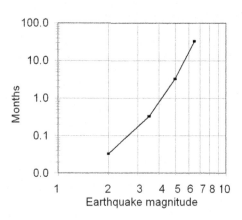

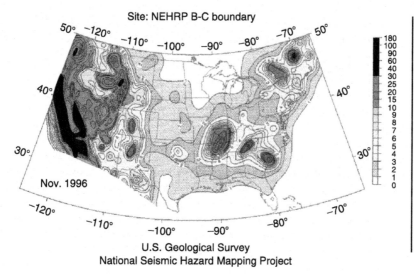

Site: NEHRP B-C boundary

U.S. Geological Survey
National Seismic Hazard Mapping Project

**Figure 27.3**

Peak accelerations with a 50 percent probability of being exceeded in 50 years, expressed as a percent of gravitational acceleration. Dark shaded areas are 0.1 to 0.3*g* and subject to liquefaction black areas 0.3 to 1*g*.

The thickness therefore generally increases with depth. As previously noted, a similar thickening has been qualitatively observed in slip zones under landslides.

The pulverized rocks in fault zones are easily eroded and therefore affect stream drainage patterns. A sharp zig-zag offset in a drainage pattern is a clue that it is crossed by an active fault. A similar offset of roads and fences is an even more dramatic clue to recent activity. A geological assessment of fault activity is an important consideration for determining the suitability and safety of an area for a proposed structure, particularly for dams, although putting a building across an active fault generally is not a good idea.

A concept of the amount of energy that can be released when a fault slips is indicated by localized rock melting, which can either act as a lubricant or as a viscous brake (Toro et al., 2006).

## 27.2.4  Intensities and Magnitudes

The Modified Mercalli scale is used as a semiquantitative descriptor of earthquake intensities and is based on observed damages:

XII. Damage total. Rolling ground, impossible for a person to stand.

XI. Few masonry structures remain standing. Bridges destroyed, railroad rails bent.

X. Most masonry and frame structures destroyed. Ground cracked, rails bent, landslides.

IX. Damage considerable; structures out of plumb or shifted off foundations.

VIII. Damage slight to great, depending on design of structure. Chimneys topple, auto drivers disturbed. Sand boils and changes in well flow.

VII. Damage negligible to considerable, depending on design of structure. Some chimneys broken.

VI. Damage slight: some falling plaster, damaged chimneys.

V or less. No damage; felt in varying degrees.

A more scientific measure of earthquake energy is the *magnitude*, which is indicated by amplitude of swings of the pointer on a seismograph. The scale was devised by C. F. Richter and utilizes a horizontal pendulum with a magnification of 2800. The magnitude $M = \log A$, where $A$ is the amplitude in micrometers, empirically corrected to a standard distance of 100 km (62 miles) from the epicenter. The logarithmic scale means that each increment of magnitude is by a factor of 10.

The strongest recorded earthquake in the U.S., magnitude 9.2, occurred on March 28, 1964, in Prince William Sound, Alaska. The quake caused many landslides and major structural damage, and vertical displacements up to 15 m (50 ft). The largest earthquake in the last century was 9.5 in Chile in 1960. The San Francisco earthquake of 1906 had an estimated magnitude of 7.8. An earthquake of magnitude 7.3 occurred in Charleston, S. C., in 1886.

The strongest earthquakes in historical times in the contiguous U.S. were in the New Madrid area in southeast Missouri and northeast Arkansas, December 1811 and January 1812. Four earthquakes all had magnitudes between 7.0 and 8.0, and created disturbances of intensity "V" over much of the eastern half of the U.S. The New Madrid quake was predicted by native American Indian legends that undoubtedly reflect earlier occurrences. Radioactive dating of buried geological features such as sand boils has been used to establish a recurrence interval, which is a matter of a few hundred years but is highly irregular. The repeat could have devastating consequences in cities like Memphis and St. Louis.

The 2004 Indian Ocean earthquake, which created a devastating tsunami, had a magnitude of 9.15, a near record. The northern Pakistan earthquake of October 2005, which left tens of thousands dead and millions homeless, registered 7.6, much of the difficulty being related to the steep rock slopes and mountainous terrain, and the use of unreinforced masonry for houses. Wood structures generally are lighter, more flexible, and less susceptible to collapse, but wood suitable for structures can be a scarce commodity.

In what part of the U.S. are earthquakes most frequent? Of those of magnitude 7.0 or larger in the last 130 years, 87 were in Alaska—many in the Aleutian Islands. Sixteen were in California, and 14 in other states. None were in Iowa.

## 27.2.5  Acceleration

Ground motions during earthquakes are periodic but highly irregular due to subsurface wave refractions and reflections, annulments, and reinforcements. The vibrations have both horizontal and vertical components, but generally it is the horizontal components that are most damaging, because of induced horizontal shearing forces. The magnitude of these forces can be estimated if the acceleration is known, from Newton's Second Law: $f = ma$. A peak acceleration of $0.5g$ therefore will induce a force that is equal to one-half of the weight of an object and may act in any direction. Ground accelerations having a 50 percent probability of occurring in the next 50 years in the U.S. are indicated in Fig. 27.3.

## 27.2.6  Amplification

Ground surface accelerations depend not only on the magnitude of the earthquake, but also on the nature of the soils. Soft clay in particular tends to amplify horizontal vibrations, since the soil acts both as a mass and a spring (Fig. 27.4). Structures built on thick clay layers therefore may sustain considerable damage while those nearby founded on rock are not seriously affected. The problem can be further magnified in sedimentary basins that tend to focus earthquake waves on narrow strips, or can trap waves so they move back and forth as surface waves, like water sloshing in a bucket.

At certain frequencies successive shakes are additive as the system is at resonance, but amplitudes still are in part controlled by viscous damping. Whether or not amplification occurs depends on how closely the natural vibrational frequency of the system is a multiple of externally applied vibration frequency. As illustrated in

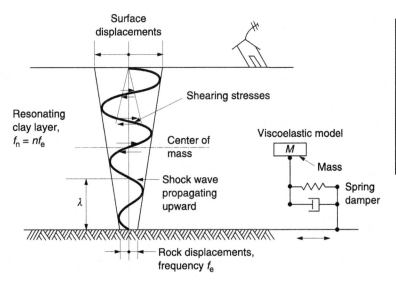

**Figure 27.4**

Soft clay layers can amplify earthquake vibrations. An undulating rock surface also can focus vibrations, like a parabolic mirror.

the next section dealing with machine vibrations, viscous damping reduces surface displacements, particularly if frequencies do not match.

The natural frequency of vibrations of a clay deposit is related to the velocity of upward propagating shear waves in relation to the layer thickness. The velocity of shear waves is

$$V_s = \sqrt{\frac{G}{\rho}} \tag{27.1}$$

where $G$ is the shear modulus and $\rho$ is the density of soil. For shear waves to reinforce one another, the layer thickness must equal an odd number of wavelengths. The wavelength is

$$\lambda = V_s/f_n \tag{27.2}$$

where $\lambda$ is the wavelength and $f_n$ is the natural frequency. Then

$$D = O_{dd}\lambda = b(V_s/f_n)$$

where $D$ is the layer thickness and $O_{dd}$ is an odd number or its reciprocal, and

$$f_n = \frac{O_{dd}}{D}\sqrt{\frac{G}{\rho}} \tag{27.3}$$

Therefore the thicker the soil layer $D$ and the lower its shear modulus $G$, the lower the natural frequency of the soil. The trouble begins when the natural frequency of the soil is close to that of a structure resting on the soil, which is a major consideration for construction in earthquake-prone areas. As a result tall, flexible buildings may survive because their natural frequency is low, where an adjacent shorter structure may be shaken into a pile of rubble.

### 27.2.7 Damping

Viscous damping can occur in thick clay layers and is beneficial because it decreases accelerations at the ground surface. Measurements made during the 1957 San Francisco earthquake ($M = 5.3$) indicated that the frequency of peak accelerations in underlying rock of 3.3 Hz (cps) was reduced to 0.4 Hz at the surface of a clay layer 500 m (1600 ft) thick. Horizontal ground accelerations varied from 0.12$g$ on rock to 0.05$g$ on clay. For a detailed discussion of damping see Newmark and Rosenblueth (1971).

### 27.2.8 Acceleration and Stress

Measurements during the 1940 El Centro earthquake in California ($M = 7.1$) indicated a maximum acceleration of 0.32$g$, which would create a body force that is almost one-third the pull of gravity. As will be discussed later, any ground accelerations exceeding about 0.10$g$ can cause liquefaction of susceptible soil.

For comparison, the Pasadena, California, earthquake of 1952, which had a comparable magnitude ($M = 7.7$), gave a much lower maximum ground acceleration, $0.06g$. The July 28, 1957, Mexico City earthquake on soft clay gave a maximum acceleration of only 0.05 to $0.10g$, but nevertheless caused collapse of several multistory buildings. Maximum accelerations of 0.05, 0.10, or $0.15g$ have been assumed in various building codes, depending on earthquake severity experienced in an area. For retaining walls, values of 0.1 to $0.3g$ are used because of lower damping capability. A simple procedure for design is to add a horizontal body force that equals the weight of a mass times the anticipated maximum acceleration.

Horizontally oriented shearing stresses are the most damaging to structures, and can exceed the horizontal shearing strength of the soil. This situation is aggravated if the soil is sensitive, losing strength on shearing. The New Madrid earthquake caused extensive landslides in banks along the Mississippi River and in the river itself, as riverboat pilots later discovered that many islands were simply gone, either from sinking or from having slipped off into adjacent channels. Reelfoot Lake in Tennessee was formed at the same time by faulting and sinking, and the Mississippi River itself is reported to have temporarily flowed backward in some areas.

Slopes are designed to withstand earthquake forces by adding a horizontal component to each slice in a method of slices. However, the first priority is to evaluate the potential of a soil for liquefaction—and to do something about it

## 27.2.9  Liquefaction

The most devastating action of an earthquake occurs when earthquake-induced horizontal shearing stresses exceed the strength of a loose, saturated sand so that it collapses. The sudden loss of soil structure allows the sand particles to temporarily become suspended in water, a condition that is the equivalent of quicksand, but the quick condition normally is limited to a deeper zone within the soil instead of occurring at the ground surface. As sand particles settle, displaced pore water moves upward through cracks and channels, lifting and carrying sand to the surface where the liquid mass erupts as sand boils that resemble miniature volcanoes.

The 1964 earthquake at Niigata, Japan ($M = 7.3$), developed maximum accelerations of about $0.12g$ in rock and $0.16g$ in soil, and caused extensive liquefaction, ground subsidence, sand boils, springs, tilting of bridge piers, and a dramatic tipping of several multistory apartment buildings. The buildings remained intact but were so severely tilted that occupants evacuated by climbing out of windows and walking down the outside walls.

Earthquakes in hilly loess areas of northern China have triggered severe, extensive, and tragic landslides as the relatively dry loess moved en masse for several kilometers before coming to an abrupt stop (Close and McCormack, 1922).

As the soil was not saturated, A. Casagrande suggested that pore air pressure contributed to the liquefaction. Another possibility is that flow down a valley trapped air like a hovercraft, a process that has been inferred from movement of prehistoric landslides over long distances on slopes of only a few degrees (Shreve, 1968).

### 27.2.10  Conditions for Liquefaction

Extensive research by H. B. Seed and his co-workers at the University of California, Berkeley, laid some important ground rules for liquefaction. Two requirements are that the soil must be saturated and it must decrease in volume when shaken or sheared.

Because horizontal shearing stresses derive from inertia of the overlying soil mass, shearing stress increases linearly with depth, but shearing strength also increases with depth. The result is a zone of liquefaction, illustrated in Fig. 27.5.

The minimum acceleration required for liquefaction is about $0.10g$, with an earthquake magnitude of 5 or above (Day, 2002). The dark shaded areas in Fig. 27.3 indicate a 50 percent probability of an earthquake with $a > 0.10g$ in 50 years.

**Figure 27.5**

Much devastation from earthquakes can be attributed to horizontal shearing stresses causing liquefaction of loose, saturated sandy soils.

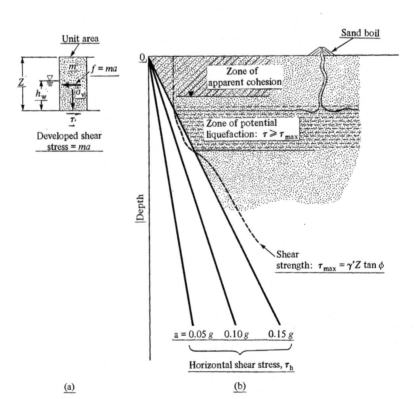

## 27.2.11 Predicting Liquefaction

In-situ tests, in particular the Standard Penetration Test and cone penetration test, are commonly used to evaluate a soil potential for liquefaction. A guide suggested by Seed et al. (1985) is as follows, using blow counts corrected to standard $N_{60}$ energy and for overburden pressure:

| SPT blow count, $N_{60}$ | Potential damage |
| --- | --- |
| 0–20 | High |
| 20–30 | Intermediate |
| >30 | No significant damage |

A closer approximation can be obtained by calculating a *cyclical stress ratio (CSR)*, also called a *seismic shear stress ratio*.

Consider a rigid soil prism (Fig. 27.6) with a mass $m$ exposed to a horizontal acceleration $a$. The horizontal shearing stress over a unit area is

$$\tau_h = f = ma$$

For a prism height $h$ the mass $m$ is

$$m = \frac{\gamma h}{g}$$

where $\gamma$ is the unit weight of the soil and $g$ is the acceleration of gravity. Substituting the vertical stress for $\gamma h$ gives

$$\tau_h = \sigma_v \frac{a}{g} \tag{27.4}$$

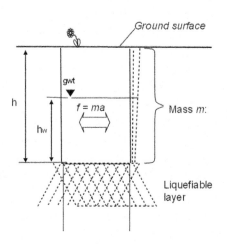

**Figure 27.6**
Analysis of horizontal stress acting at the base of a soil element.

Dividing both sides by the effective vertical stress at depth $h$ gives

$$\frac{\tau_h}{\sigma_v'} = \frac{\sigma_v}{\sigma_v'} \frac{a}{g} \tag{27.5}$$

The maximum shear stress is multiplied by 0.65 to give an approximate average shear stress or cyclical stress ratio (Seed et al., 1985):

$$CSR = 0.65 \frac{\sigma_v}{\sigma_v'} \frac{a}{g} \tag{27.6}$$

This is compared with a *cyclical resistance ratio (CRR)*, which is the shearing resistance on a horizontal plane, $\sigma_v' \tan \phi_{rep}'$:

$$CRR = \tan \phi_{rep}' = \frac{\tau_f}{\sigma_v'} \tag{27.7}$$

Studies by Seed et al. (1985) indicate that this value is approximately $N_{60}/100$ (corrected for depth) with an $M = 7.5$ earthquake (Fig. 27.7). A factor of safety may be defined as CRR/CSR. The presence of up to 35 percent fines (silt and clay) can substantially reduce the liquefaction potential for a particular value of $N_{60}$, with no liquefaction occurring when $N_{60} > 20$ instead of, as shown in Fig. 27.7, $N_{60} > 30$. Engineers working in earthquake areas should refer to the original articles or a reference such as Day (2002).

**Example 27.1**
Estimate the liquefaction potential for a zone in sand with $N_{60} = 25$, $a/g = 0.30$. Assume that the groundwater table is at the ground surface so that $\sigma_v/\sigma_v' = 2$.

*Answer:* CSR $= 0.65(2)(0.30) = 0.39$. CRR corresponding to $N_{60} = 28$ is $0.30 < 0.39$; unsafe.

## 27.2.12 Adjustment for Magnitude

A correction factor for earthquake magnitude by Seed et al. (1985) is shown in Fig. 27.8, to reduce the liquefaction potential for lower magnitude earthquakes.

**Example 27.2**
Recalculate the preceding example for an $M = 5$ earthquake.

*Answer:* The correction factor is 0.64, so CSR $= 0.64(0.39) = 0.25$. FS $= $ RR/SR $= 0.30/0.25 = 1.2$, which is marginal.

## 27.2.13 Use of the Cone Penetration Test

Because of the longer experience record with SPT than cone penetration tests, cone data often are converted to equivalent SPT values for liquefaction analysis. It also can be argued that the dynamic SPT may be more suitable than a quasi-static

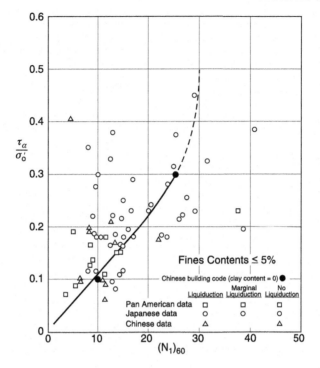

**Figure 27.7**

Liquefaction case histories for clean sand in relation to developed seismic stress ratio (CSR) and $N_{60}$ corrected SPT blow counts. (From Seed et al., 1985, reprinted with permission of the American Society of Civil Engineers.)

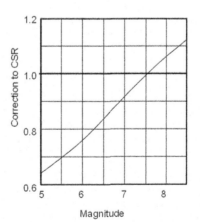

**Figure 27.8**

Correction to CSR for earthquake magnitude. (From data of Seed et al., 1985.)

cone test. A relationship developed from case histories by Stark and Olson (1995) is shown in Fig. 27.9 with cone data corrected for overburden pressure. As in the case of the SPT, the presence of fines reduces the liquefaction potential. These authors also document cases of liquefaction in gravel. Figure 27.8 can be used to correct for magnitude.

**Figure 27.9**

Cone penetration test liquefaction criteria for clean sand. (Modified from Stark and Olson, 1995, reprinted with permission from the American Society of Civil Engineers.)

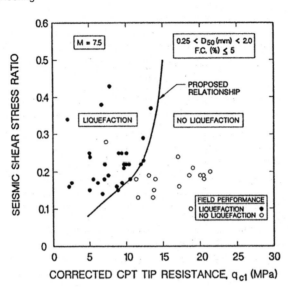

## 27.2.14  Geology, Liquefaction, and Lateral Spread

After liquefaction already has occurred and the soil grains settle into a more dense packing, the friction angle is increased and the soil becomes more resistant to future episodes of shaking. The most likely candidates for liquefaction therefore are recently deposited shallow sedimentary soils that have not yet sustained an equivalent episode of severe shaking. The phenomenon therefore is most common in geologically young deposits such as deltaic or alluvial sands, or in loose sandy fill soil that has become saturated.

A particularly damaging aspect of liquefaction is called "lateral spread," as the area involved expands toward an open face such as a river bank or down a gentle slope. In such cases the slip surface is flat, and the soil involvement far exceeds the amount that would slide under the influence of gravity alone.

Lateral spread played a prominent role in the 1906 San Francisco earthquake, toppling many buildings. The 1964 Turnagain Heights landslide in Anchorage, Alaska, involved a horizontally sliding soil mass about 18 m (60 ft) thick by 400 m ($\frac{1}{4}$ mile) wide. Since the shearing stresses work alternately in opposite directions, such slide areas often exhibit crevasses and faulting that break the ground into a loosely assembled mass of tilted blocks. Empirical equations fit to data in the U.S. and in Japan indicate that the amount of displacement depends on earthquake magnitude, distance from the seismic source, thickness, fines content, and particle size of the liquefiable layer (Bartlett and Youd, 1995).

## 27.2.15 Preventing Liquefaction

The simplest methods for reducing the likelihood of liquefaction are to permanently lower the groundwater table, or else induce liquefaction and allow the ground to settle prior to construction. Liquefaction can be induced with deep dynamic compaction, which involves repeatedly dropping a heavy weight as shown in Fig. 27.10. In more restricted areas Rammed Aggregate Piers, driven piles, or vibrating probes (Vibroflot process) are successfully used. Grouting, especially compaction grouting, may be appropriate to avoid damaging nearby existing structures. Surcharge loading over a broad area can increase lateral stress in the soil so that it resists liquefaction, but narrow loads such as building foundations are less effective because stress alternately is increased and decreased by rocking action.

### Example 27.3

A sand layer at a depth of 7 m (23 ft) has $\gamma = 117.3 \, kN/m^3$ ($110 \, lb/ft^3$) and $\varphi = 25°$. Evaluate the effect on liquefaction potential from lowering the groundwater table from the ground surface to a depth of 3 m (10 ft). The average soil unit weight above the water table is $15 \, kN/m^3$ ($95.5 \, lb/ft^3$).

*Answer:* Lowering the groundwater table should reduce the cyclical stress ratio (eq. (27.6)):

$$SR = 0.65 \frac{\sigma_v}{\sigma'_v} \frac{a}{g}$$

**Figure 27.10**

Deep dynamic compaction of a gravelly sand foundation soil prior to reconstruction of Jackson Lake Dam, Wyoming. After 8 blows from a 100 ton (900 kN) weight dropping over 105 ft (32 m), the gravelly sand was densified to a depth of approximately 45 ft (15 m) underneath the crater.

With the groundwater table at the ground surface, $\sigma_v = 23(110) = 2530$ lb/ft$^3$ and $\sigma_v = 23(110 - 62.4) = 1095$ lb/ft$^3$, giving a ratio $\sigma_v/\sigma'_v = 2.31$. After lowering of the groundwater table the values are respectively 2380 and 1756 lb/ft$^3$, giving a ratio 1.36, so the critical $a/g$ is reduced by a factor of $1.36/2.31 = 1.7$. Drainage therefore can be an effective way to prevent liquefaction.

## 27.2.16  Design of Slopes to Resist Earthquakes

Even without liquefaction many slopes subjected to earthquake stresses may fail due to the inertia-induced horizontal forces. The simplest method for including such forces is to assume a maximum acceleration relative to gravity, $a/g$, which is particularly well suited to the method of slices because $a/g$ is simply multiplied times the weight of the slice to give each horizontal force. Thus, an earthquake acceleration of $0.10g$ will result in a horizontal inertial force equal to 0.1 times the weight of the soil in each slice. An analysis that includes moment equilibrium is complicated by the fact that seismic forces act through the centroid of each slice, but many slope stability computer programs incorporate this capability.

**Example 27.4**
A design acceleration of $0.10g$ is assumed for a slice weighing 890 kN (100 tons). Find the horizontal force and resolve into components acting parallel and normal to the slip plane, which is inclined at an angle of $10°$.

*Answer:* The horizontal body force is $0.1 \times 890 = 89$ kN (10 tons). The component parallel to the slip surface is $89 \cos 10° = 87.6$ kN (9.8 tons), to be added to the downslope acting force. The normal component is $89 \sin 10° = 15.5$ kN (1.7 tons), which should be subtracted from the normal force.

The above procedure is not adequate for large earth structures such as large earth dams that respond elastically to earthquake vibrations, causing complex annulments and reinforcements at different levels. The response of such structures can be modeled by finite element analysis, in which the embankment is represented by interconnected linear elements whose response to forces simulates the response of soils.

## 27.2.17  Retaining Walls

As for slope stability analyses, the simplest method for including earthquake forces in retaining wall design is to add horizontal force components acting through the centroid of the active soil wedge and the mass of the wall itself. This ignores possible resonant effects and reduced damping because of rigidity of the wall. Also, if granular backfill remains in the active state, cyclical tilting of the wall will be cumulative. Assumed accelerations for retaining wall design therefore are higher than for other structures.

## 27.2.18 Foundations and Bridge Piers

The two most important seismic-induced forces in structures are rocking and horizontal shear. The tendency for sliding of a foundation is easily resisted by base friction and passive restraint from soil on the sides of foundations and pile. On the other hand, rocking can significantly increase maximum bearing pressures, particularly under narrow structures such as water towers, storage tanks, and bridge piers. These forces can be calculated based on the mass and height of the center of mass of the structure and acceleration of the earthquake. The anticipated increases in bearing stress are compensated by increasing the factor of safety against a static bearing capacity failure. A common specification for ordinary structures is to increase the bearing capacity requirement by 30 percent (Day, 2002).

Rocking forces can be particularly devastating for bridge piers, where the analogy to dominoes standing on edge cannot be easily dismissed. Bridges and overpasses are frequent casualties of earthquakes if tilting is sufficient to remove support from the spans, and consequences can be dire. Buildings can be of even more concern because so many lives can be endangered.

Failure of buildings often starts at the ground floor because of inertia of the overlying mass, and because ground floors may have larger windows and door openings that decrease lateral stability. In recent years a considerable effort has been directed toward retrofitting reinforcement for columns and walls supporting buildings and bridges. Sometimes a building will remain intact but rock sufficiently to jam into adjacent buildings. Quite often the relatively rigid floors in one building are not even with floors on the adjacent building, leading to battering of the walls.

Structural design involves cross-bracing against lateral stress and avoiding resonance. As previously mentioned, the latter depends on stiffness and amplification by the soil. Natural resonant frequencies vary from about 3 Hz for rock to 0.25 Hz for soft soil. The natural frequency of multistory buildings in hertz is approximately $10/N_s$, where $N_s$ is the number of stories (Seed et al., 1991), so a three-story building may be resonant on rock whereas a 40-story building would be resonant on soft clay. As previously pointed out, a soft clay layer can greatly lower the frequency at the ground surface.

Various isolation procedures that have been used and perfected with machines also have been applied to entire buildings. These methods are discussed in the next section.

## 27.3 INDUCED VIBRATIONS

### 27.3.1 Overview

Machine vibrations that resonate with the foundation system natural frequency can cause problems ranging from minor annoyance to damage to structures or to the machine itself. The purpose of this review is not to provide a basis for design, but to present important principles that may prove useful for remedial purposes.

Vibrating machine systems consist of the machine and its foundation acting as a mass, resting on soil that acts as a spring. Since most machines involve rotation of an eccentric mass, or rotation-coupled linear movement of a mass such as a piston, the movement is sinusoidal and can be described by the following equations:

$$\text{Displacement:} \quad z = A_m \sin \omega t \tag{27.8}$$

where $A_m$ is the amplitude, or maximum displacement in one direction, $\omega$ is the rotational frequency in radians per second, and $t$ is time in seconds. Differentiating this equation with respect to time gives

$$\text{Velocity:} \quad dz/dt = \omega A_m \cos \omega t \tag{27.9}$$

$$\text{Acceleration:} \quad d^2z/dt^2 = -\omega^2 A_m \sin \omega t \tag{27.10}$$

The relationship between angular frequency $\omega$ and cyclical frequency $f$ is

$$f = \frac{1}{2\pi} \omega \tag{27.11}$$

where $f$ is in hertz, or cycles per second. From Hooke's Law the static displacement from a weight on a spring is

$$z_s = \frac{W}{K} \tag{27.12}$$

where $z_s$ is deflection of the spring, $W$ is the weight, and $k$ is the spring constant in units of force per unit length. If the weight is bouncing, at any instant it is opposed by an acceleration force consistent with Newton's Second Law. Solution of the resulting differential equation gives a natural vibration frequency $f_n$:

$$f_n = \frac{1}{2\pi} \sqrt{\frac{k}{m}} = \frac{1}{2\pi} \sqrt{\frac{kg}{W}} \tag{27.13}$$

where $m$ is the bouncing mass and $g$ is the acceleration of gravity. Thus the larger the mass or the softer the spring, the lower the natural frequency. Substituting eq. (27.12) in eq. (27.13) gives

$$f_n = \frac{1}{2} \sqrt{\frac{g}{z_s}} \tag{27.14}$$

where $g = 32.2 \, \text{ft/s}^2 \, (9.81 \, \text{m/s}^2)$.

## Example 27.5

Grandma's antique vertical-plunger Maytag sits on a springy porch floor. The washer full weighs 80 lb (356 N) and depresses the floor 0.02 in. (0.5 mm). Estimate the natural frequency of the system.

*Answer:*

$$f_n = \frac{1}{2}\sqrt{\frac{32.2\,\text{ft/s}^2}{(0.02/12)\,\text{ft}}} = 70\,\text{Hz, or}$$

$$f_n = \frac{1}{2}\sqrt{\frac{9.81\,\text{m/s}^2}{(0.5/1000)\,\text{m}}} = 70\,\text{Hz}$$

## 27.3.2 Resonance

The development of eq. (27.14) for a mass freely bouncing on a spring assumes no actuating force other than an initial displacement. Such vibrations are called natural or free vibrations, and occur as a result of impact, earthquake tremors, blasting, etc. On the other hand if the actuating force is repeated at the natural frequency of the system, amplitudes add like kids jumping on a trampoline.

Most machine-induced vibrations are repetitive, and if the machine frequency matches the natural frequency of the system, the amplitudes theoretically would become infinite except for energy losses from damping. If the machine frequency and natural frequency are not precisely equal, the reinforcement is reduced to give a maximum steady amplitude that is defined by

$$A_m = \frac{P}{m(f_n^2 - f^2)} \tag{27.15}$$

where $A_m$ = the amplitude measured from a steady-state position (one-half of the maximum vibration movement);

$P$ = the actuating force;

$m$ = the vibrating mass;

$f_n$ = natural frequency of the system;

$f$ = machine frequency.

The frequency of the forced vibrations is that of the machine. The *dynamic magnification factor* is defined as

$$M_g = \frac{1}{[1 - (f/f_n)^2]} \tag{27.16}$$

Some solutions for this equation are as follows

| $f/f_n$: | 0 | 0.5 | 0.75 | 0.095 | 1.0 | 1.05 | 1.414 | 2 |
|---|---|---|---|---|---|---|---|---|
| $M_g$: | 1 | 1.33 | 2.29 | 10.26 | $\infty$ | −9.76 | −1.0 | −0.33 |

As the two frequencies close, the magnification factor rapidly increases, then reverses as the minus sign signifies a 180° phase shift. A graph of eq. (27.16) that ignores the phase shift is shown by the upper curves in Fig. 27.11.

A simple demonstration of magnification factor is a ball suspended from a paddle on a rubber band. When the up-and-down motion is synchronized and resonates, amplitude of bouncing increases, but when the paddle is totally out of synch, bouncing becomes sharply reduced, particularly, as shown in Fig. 27.11, if $f$ is faster than $f_n$.

## 27.3.3 Damping

Damping, shown by $D$ in Fig. 27.11, reduces vibrational amplitude. *Attenuation damping* converts energy to heat by friction and/or viscosity. The automobile shock absorber is an attenuation damper and absorbs "shock" by forcing a fluid

**Figure 27.11**

Amplification as an activating frequency approaches the natural frequency of a system, and influence from damping.

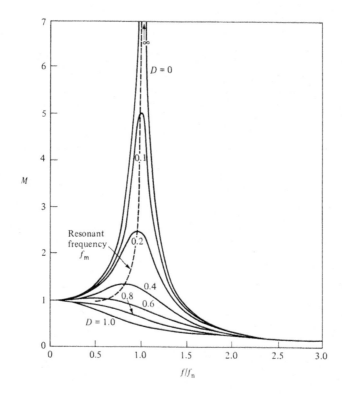

through an orifice. As shown in Fig. 27.11, increased damping also reduces the natural frequency, resulting in a smoother ride on a bouncy pavement.

Another mechanism of damping is *geometrical damping*, which results from a geometrical spreading of wavefronts as they propagate outward from the source. Geometrical damping therefore is governed by the type of oscillations and characteristics of the system.

By assuming a damping force as proportional to momentary displacement velocity $dz/dt$, one can develop an expression for magnification factor analogous to eq. (27.16):

$$M = \frac{1}{\sqrt{\{1 - (f/f_n)^2\}^2 + \{2D(f/f_n)\}^2}} \qquad (27.17)$$

where $D$ is the *damping ratio*. When $f = f_n$ eq. (27.17) becomes

$$M = \frac{1}{2D} \qquad (27.18)$$

The resonant frequency or frequency that gives the maximum $M$ also changes, defined by the formula

$$f = f_n\sqrt{1 - 2D^2} \qquad (27.19)$$

Higher values of $D$ greatly reduce vibrational amplitudes, and when $D = 2$ there is no resonant frequency and $M < 1$.

## 27.3.4 Soil as a Spring

Foundation soils behave elastically under repeated loading, and may be envisioned as a bed of individual coil springs. The larger the loaded area, the more springs that are covered, and the stiffer the system. Damping also occurs and depends on the foundation area, and the vibrating mass includes an undefined amount of soil, which complicates the problem. Damping can be evaluated experimentally from the shape of the frequency response curve (Richart et al., 1975). Geometrical damping also relates to the type of vibrational waves and directions of the vibrations.

The relationship of spring constant to foundation area is taken into account by defining $k'$ for soils such that

$$k = k'A \qquad (27.20)$$

where $k$ is the equivalent of a spring constant, $A$ is the foundation contact area, and $k'$ is the repeated-load version of the modulus of subgrade reaction used in pavement design.

This aspect of soil elastic behavior is important since it means that $k'$, and therefore the natural frequency $f_n$, can be changed by changing the dimensions of the foundation. $k'$ is evaluated from reloading phases of plate-load tests. The plate must be loaded, unloaded, and reloaded in order to eliminate consolidation effects, and the load range used should be representative of actual loading conditions.

### Example 27.6

A plate-load test gave the following data upon loading, unloading, and reloading. Calculate $k'$.

| Test mode | Load, kPa (lb/in.) | Settlement, mm (in.) |
|---|---|---|
|  | 0 | 0 |
| Loading | 69 (10) | 15.2 (0.598) |
| Unloading | 0 | 10.0 (0.393) |
| Reloading | 69 (10) | 15.4 (0.606) |
| Unloading | 0 | 10.4 (0.409) |

*Answer:* From the reloading cycle, $15.4 - 10.0 = 5.4$ mm (loading), and $15.4 - 10.4 = 5.00$ mm (rebound). The difference represents hysteresis, or energy loss due to friction in the systems. Since friction acts one way upon loading and the opposite upon unloading, one justifiably may use the average deflection, 5.2 mm (0.204 in.). The subgrade reaction modulus with repeated loading is $k' = (10\,\text{lb/in.}^2)/0.204\,\text{in.} = 49\,\text{lb/in.}^3$ (13.3 MN/m³).

Some estimates of $k'$ are given in Table 27.1. Plate-load tests give only $k$ in the vertical direction; for horizontal movements Barkan (1962) recommended that $k'$ be reduced one-half, or for rocking action $k'$ should be doubled.

### 27.3.5   Natural Frequency of Soil-Foundation Systems

Equation (27.14) for natural frequency was modified by Tschebetarioff to include the soil modulus, $k'$, and a vibrating soil mass, $m_s$, which moves with the machine and its foundation:

$$fn = \frac{1}{2\pi}\sqrt{\frac{k'A}{m + m_s}} = \frac{1}{2\pi}\sqrt{\frac{k'Ag}{W + W_s}} \qquad (27.21)$$

where $f_n$ = natural frequency of the soil-foundation system, in hertz (or cycles per second);

$k'$ = dynamic soil modulus, in newtons per cubic meter (lb/ft³);

$A$ = foundation base area, in square meters (ft²);

| Soil | Vertical $k'$, MN/m³ (ton/ft³) | Table 27.1 |
|---|---|---|
| | | Representative values for $k$, adapted from Barkan (1962) |
| Weak clays, silty clays, clayey and silty sands | <30 (<95) | |
| Medium-strength clays and sandy-silty clays close to plastic limit, sands | 30–50 (95–155) | |
| Strong clays and sandy-silty clays, gravel, gravelly sand, loess (well-drained) | 50–100 (155–310) | |
| Rock | >100 (>310) | |

$m$, $m_s$ = masses of the machine-foundation system and of vibrating soil, respectively, in kilograms (slugs);

$W$, $W_s$ = weights of the machine-foundation system and of vibrating soil, respectively, in newtons (lb);

$g$ = acceleration of gravity = 9.81 m/s² (32.2 ft/s²).

Dimensional homogeneity is required, so if $m$ is in kilograms and $A$ is square meters, $k'$ must be in newtons per cubic meter, since 1 N = 1 kg × m/s². That is, 1 N force will give 1 kg an acceleration of 1 m/s². For engineering applications weight terms are used.

The assumption of a soil mass $m_s$ (or soil weight $W_s$) vibrating in phase with a foundation is a gross simplification since the outer boundary of vibrating soil cannot be discrete. Thus an accurate prediction of $f_n$ from eq. (27.16) is not possible without an evaluation of the soil mass from theory of elasticity. Procedures are available in Richart et al. (1975). For preliminary estimations, the vibrating soil mass or weight usually is about 10 to 30 percent of that of the machine and its foundation, the higher value for foundations on pile. Note that this equation only predicts $f_n$, not amplitude, which requires a more rigorous analysis.

### Example 27.7
A 26.7 kN (6000 lb) vertical compressor-foundation system is operated at 40 Hz. The soil foundation is a medium stiff clay, and the foundation bearing area is 2.25 m² (8 ft²). Predict the natural frequency, $f_n$, and magnification factor, $M$, (a) without and (b) with a damping factor $D = 0.4$.

*Answer:* (a) Assume $W_s = 0.2W$ and $k'_z = 5$ MN/m³. From eq. 27.21,

$$f_n = \frac{1}{2\pi}\sqrt{\frac{50(10)^6(2.25)(9.81)}{26.7(10)^3(1.2)}} = 30\,\text{Hz}$$

From eq. 27.17,

$$M_g = \frac{1}{\sqrt{\{1-(40/30)^2\}^2+0}} = 1.3$$

(b) With damping,

$$M_g = \frac{1}{\sqrt{\{1 - (40/30)^2\}^2 + \{(2)(0.4)(40/30)\}^2}} = 0.70$$

From eq. 27.19, the natural frequency with damping is predicted to be

$$f = 30\sqrt{1 - 2(0.4)^2} = 25\,\text{Hz}$$

## 27.3.6 Wave Transmission, Damping, and Kinds of Waves

Vibrations in an infinite elastic medium may be classified as P-waves (primary or compression waves) and S-waves (secondary or shear waves), depending on relative motions of individual particles. Near the surface a third type of wave occurs, the R-wave or Rayleigh wave, where an interaction between compression-expansion and shear produces an elliptical motion analogous to water waves, but with the long axis of the ellipse vertical instead of being inclined forward by wind drag.

P-waves are the first to arrive on seismograph records and present horizontal oscillations; these are followed shortly by S-waves. These carry most of the earthquake energy and supply a vertical component that, while significant, is usually ignored in building codes because of the preponderance of damage from horizontal forces. Nevertheless vertical accelerations have been recorded that momentarily exceed $1g$. R-waves, as well as more complicated types such as Love waves, arrive later, during early adolescence. The first-arrival P-waves are used in seismic refraction surveying, discussed in the next section.

Vertically vibrating machinery generates waves that propagate radially downward and outward as a hemispherical wavefront. Because of the rapidly expanding area of the wavefront, its energy flux is diluted as the square of the distance outward. This dilution is the cause of geometrical damping, and results in high values of $D$, of the order of 0.3 to 0.6, for vertically vibrating machinery. For horizontal vibrations $D$ is somewhat lower, 0.2 to 0.4.

The Rayleigh wave carries most of the energy from machine vibrators, and being confined to a surface layer, propagates as a cylindrical wavefront with its axis vertical, diminishing as the square root of the radial distance. R-waves therefore propagate farther with less geometrical damping than P- and S-waves. This correlates with a relatively low $D$ measured for rocking motions, of the order of 0.01 to 0.1. Because of the low damping coefficient, foundations should be designed to minimize rocking action. This can be done by increasing the foundation mass, and making foundation and its contact area long in the direction of potential rocking.

### 27.3.7  Combined Motions

Unless the reciprocating part of the machine mass happens to be aligned with the center of gravity, most machines will develop two separate motions, usually vertical plus rocking, or horizontal plus rocking. Such a system has two degrees of freedom, or two ways to move, namely translation and rotation. As many as six degrees of freedom are possible, along and about the three coordinate axes.

Two degrees of freedom give two natural frequencies because of interaction or coupling. These will be larger and smaller than frequencies for the individual motions considered separately.

Horizontal vibrations may be approximately analyzed by the methods outlined above but with a reduced $k'$ (Table 27.1). Rocking and torsional vibrations involve moment of inertia calculations and a modification of the procedure.

### 27.3.8  Reducing Machine Vibrations

Geotechnical engineers often are called on for advice on how to reduce troublesome vibrations from machinery, trains, trucks, etc. In the case of machines, the exciting frequency changes during startup and may or may not be precisely controlled. Two machines operating independently will oscillate into and out of phase, periodically doubling and partially annulling the vibrations.

In general, with eq. (27.21) and Fig. 27.11 as guides, one should strive to increase damping, for example by increasing the foundation embedment. Machines often are supported on rubber bearing pads, rubber having a high mechanical hysteresis and damping effect.

Another approach is to lower the natural frequency, with eq. (27.21) as a guide. For example, in the case of moderate to high-frequency machines, one might attempt to reduce $f_n$ by increasing mass and reducing the foundation bearing area.

Damping will occur if a machine can be supported on springs or yokes that have a different vibrational frequency from that of the machine.

*Dynamic damping* is possible in a machine that operates at a constant frequency, by attaching an accessory mass by a spring such that it has the same natural frequency as the machine. Thus as the machine moves one way the damper moves the other and automatically gets out of phase and acts to annul the vibrations. However, a low frequency requires a large mass and a strong but soft spring. Careful synchronizing is required or the system will pulse and amplify instead of reducing vibrations.

In the case of heavy, slower reciprocating machinery, a higher $f_n$ can be sought by using lightweight reinforced mat foundations with a large contact area, or by increasing $k'$ by grouting or by underpinning with piles.

Isolation trenches can be used to interrupt horizontal P-waves, but a trench wall becomes a secondary source for return vibrations that can interact and develop a periodicity. Experiments have shown that such trenches must extend at least $0.3\lambda$ below the foundation level, where $\lambda$ is the wavelength. The wavelength may be calculated from $\lambda = V/f$, where $V$ is the wave velocity that can be measured with an engineering seismograph.

## Example 27.8

Troublesome vibrations occur during startup of a large horizontal natural-gas compressor with foundations on clay. Self-igniting diesel operation prevents phasing, so when two compressors are in phase the vibration amplitude is approximately doubled. The machine operates at approximately 18 Hz but vibrations are worse during startup. Evaluate the following options, noting whether they are likely to increase or decrease the natural frequency. They are listed in approximate order of increasing cost:

(a) Isolate the foundations from floors and walls.
(b) Increase the foundation mass.
(c) Grout to harden the soil underneath the foundation.
(d) Enlarge the foundation contact area.
(e) Add buttress walls and lateral anchorages.
(f) Raise the level of the soil around the foundation.
(g) Underpin the foundation with pile.
(h) Install a row of sheet pile outside the foundation area.
(i) Isolate with a peripheral trench.
(j) Other methods.

*Answer:* An important clue is that vibrations are worst during startup, which means that the natural frequency must be less than the operating frequency. The treatment therefore should attempt to increase damping and/or increase $f_n$ and move to the right in Fig. 27.11. Let us examine the alternatives:

(a) Detach any rigid contacts such as with floors. This should tend to decrease both $k'$ and $W_s$. As these two effects work in opposite directions, feasibility depends on which influence dominates. A trial might be recommended with vibrations monitored with accelerometers.
(b) Increasing the foundation mass without increasing the area will decrease $f_n$, probably not a good idea.
(c) Grouting is not feasible in clay.
(d) (d)–(i). It is suggested that the student examine the other possibilities for their feasibility: what is each likely to do? There also are practical matters that may prevent use of a particular solution; for example, large compressors servicing a natural-gas pipeline cannot all be shut down at the same time.

# 27.4   PILE DRIVING AND BLASTING

## 27.4.1   Sorry, but Sometimes a Little Noise Is Not Avoidable

Pile driving and blasting create ground vibrations that can be minimized but not altogether eliminated. A pre-blast survey normally is made with owners' permissions and documented with descriptions, dates, and photographs showing all prior cracking and settlement of structures closer than 150 to 300 m (500 to 1000 ft) from the blast, farther if the blast will be large. Vibrations from pile driving can become unexpectedly ampified if the pile tip encounters a hard layer such as a buried pavement.

Rayleigh waves from a blast are attenuated as the square root of the distance from the blast, so doubling the distance reduces these vibrations only by a factor $\sqrt{2} = 0.7$. Damage from blasting vibrations has been shown to relate to maximum vibrational velocity, no damages occurring when $V < 50$ mm/s (2 in./s). Only minor cracking occurs when $V$ is twice this value, but even the lower velocity will be considered by nearby residents to be "very unpleasant," occasionally bordering on unprintable.

Solving eq. (27.11) for $\omega$ and substituting in eq. (27.19) for velocity gives

$$V_m = 2\pi f_n A_m \tag{27.22}$$

where $V_m$ is the maximum velocity and $A_m$ is the maximum amplitude. Thus from a velocity standpoint, a high frequency can be as damaging as a large amplitude.

Accelerations also should be limited to less than $0.1g$, and from eq. (27.11) the maximum acceleration is

$$a_m = 4\pi^2 f_n{}^2 A_m \tag{27.23}$$

further emphasizing the importance of frequency. Since frequencies are not too predictable, they are measured by trial. The limits suggested above are indicated in Fig. 27.12.

### Example 27.9

One of your clients is being sued for damages to a 140-year-old building that was not considered worthy of a pre-blast survey (which is even more important for old buildings). The clients' independent consultant recorded a maximum acceleration of $0.08g$ near the building and an amplitude of 0.1 mm. Can you help your client?

*Answer:* Always remember there are two ways that can be equally helpful to a client: "In my opinion the evidence supports (does not support) your side of the argument." In this case the amplitude is troublesome and the acceleration is barely within permissible limits, and the building is over a century old.

**Figure 27.12**

Suggested
permissible ranges
of velocities and
accelerations from
blasting or pile
driving. (Modified
from Richart et al.,
1970.)

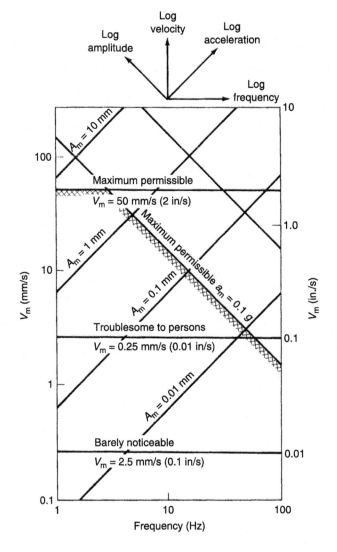

Blasting vibrations are controlled by limiting the amount of explosives
set off at any one time. This is accomplished by detonating a series of
smaller explosives at preset time intervals called "delays," with each delay
being long enough for preceding vibrations to damp out. It is important
to prevent reinforcement from successive delayed changes, which can be
expected with random shot timing. Heavy, overcast weather tends to confine
and intensify the noise. Many times permanent seismograph records will
be made of ground and building vibrations at the time of a blast, as
evidence relevant to allegations of damage. This also is recommended close
to pile driving.

## 27.5  VIBRATORY COMPACTORS

Vibrating rollers are used for rapid compaction of embankment or fill soils in layers up to 600 mm (24 in.) thick, depending on the soil and on the compactor. Vibrating compactors are relatively lightweight because of the ramming effect from the vibration. For example, if the vibrator momentarily is lifted so that zero weight is applied to the ground, the weight will be approximately doubled on the down cycle and will be further increased if the compactor leaves the ground.

Vibration is most effective for densifying granular soils, but heavy vibratory compactors also have been found to perform efficiently with cohesive soils. Small plate vibrators are adapted for work in small areas and may be hand-operated.

Vertical vibration is achieved with two counter-rotating weights as shown in Fig. 27.13, so that horizontal forces oppose and cancel out. In plate vibrators the two counter-rotating weights are slightly out of phase, so an oscillating horizontal component is developed that carries the machine forward.

A number of studies have shown an advantage in operating vibratory compactors at or near a resonant frequency for the system, particularly for granular soils. This is reasonable since effectiveness depends in part on how well successive vibrations reinforce one another and propagate into the soil. Equation (27.11) can be rewritten for frequency of a soil-vibrator system:

$$f = \sqrt{\frac{c}{W}} \tag{27.24}$$

where c is a constant and $W$ is the weight of the vibrator. Weight therefore can be added to a machine to decrease its resonant frequency, or the frequency itself can be changed to match that of the soil. If adjustments can be made it will increase the efficiency of compaction. However, the soil also is changing, increasing $k'$ as it compacts, so tuning should be to the end product.

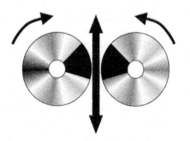

**Figure 27.13**
Counter-rotating weights to create vertical motion.

As previously indicated, if the maximum acceleration exceeds that of gravity, the machine will hop off of the ground. This appears to be a minimum requirement for effective densification, and acceleration commonly exceeds 1.5 to 2 for commercial vibrating compactors. Tamping action is especially effective for compacting clay, where impact force may be more important than resonance for smashing the clay structure. Equation (27.10) describes the maximum acceleration from a sinusoidal vibration, and emphasizes the importance of frequency.

A current area of research is to determine if a measured response to vibratory compaction can be used to determine if a soil has reached a specified density.

Air-driven rammers are used to compact soil in close quarters. These can be machine- or hand-operated, and do not cause harmonic motion. Instead the force delivered depends on the deceleration after impact, $f = ma$, which depends on hardness of the soil. As the soil compacts, the deceleration rate and force gradually increase.

Dynamic compaction is not a substitute for moisture content control, and the same dangers exist with regard to overcompaction on the wet side of the optimum moisture content or creating a collapsible soil on the dry side, as discussed in detail in Chapter 13.

## 27.6  SEISMIC SURVEYS

### 27.6.1  Types of Surveys

Seismic surveying has been used in the petroleum production industry for many decades, and a similar electronic instrumentation has been adapted for geotechnical engineering uses. Portable hand-operated seismographs create P-waves and measure arrival times to rapidly determine and map depths to different strata such as bedrock. More recently, downhole and cross-hole methods have been devised to measure S-wave velocities.

### 27.6.2  Refraction Seismographs

The classic engineering seismograph consists of (1) a sound source, typically a hammer hitting a steel plate or ball placed on the ground, and carrying an impulse trigger that starts a timer; (2) one or more detectors called geophones; and (3) a timer accurate to $10^{-3}$ seconds or better. For deep surveys a heavy thumping machine or small explosive charge can be used.

A survey is conducted by either laying out a line of geophones or using a single geophone and moving it along a line to different locations and repeating the source impulse for each new location. P-wave first arrival times are measured at

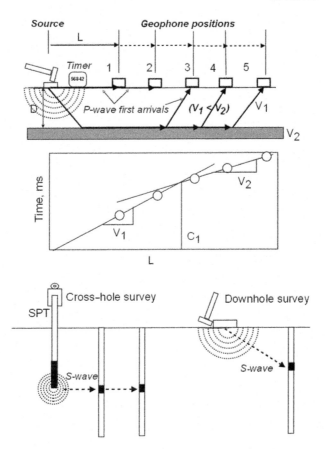

**Figure 27.14**

(Top) Geophysical P-wave refraction survey and (below) S-wave survey methods.

each location. As surveys often are made near construction sites or around traffic, it is important that the seismograph have a capability to add pulses from repeated hammer hits in order to eliminate the influence of random noise. Early seismographs were set to stop recording and give a reading upon first arrival; later models often show a complete trace, which is useful to distinguish signal from noise.

Pulse arrival times are plotted versus distance as shown in Fig. 27.15. The reciprocal of the slope of the plot is velocity in meters (or feet) per second. The line may not go through the origin because of a starting time error that should remain constant and thus will not affect the slope.

As shown in Fig. 27.14, as the survey distance is increased, the first arrival will divert through a buried higher-velocity layer. That is, instead of following a direct route from source to geophone, a faster path will be to detour down and across through the deeper layer. The graph therefore breaks at a distance $C_1$ from the

**Figure 27.15**

Refraction
seismogram
showing two layer
thicknesses
(Example 27.10).

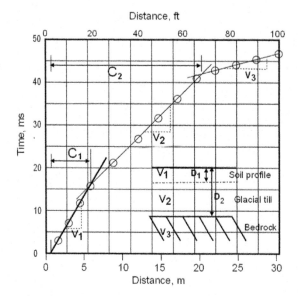

sound source. This is converted to an equivalent depth $D_1$ through the following formula:

$$D_1 = \frac{C_1}{2} \sqrt{\frac{V_2 - V_1}{V_2 + V_1}}$$

(27.25)

where $V_1$ and $V_2$ are velocities determined for the upper and lower layer, respectively.

If more than one layer is involved, its depth $D_2$ is approximated by

$$D_2 = \frac{C_2}{2} \sqrt{\frac{V_3 - V_2}{V_3 + V_2}} + 0.85 D_1$$

(27.26)

**Example 27.10**

Determine velocities and thicknesses of the layers in Fig. 27.15.

*Answer:* Measured from the initial time delay the line intersections are at $C_1 = 18$ ft (5.5 m) and $C_2 = 71$ ft (21.6 m). Velocities are $V_1 = 1200$ ft/s (366 m/s), $V_2 = 2000$ ft/s (610 m/s), and $V_3 = 2900$ ft/s (884 m/s).

$$D_2 = \frac{18}{2} \sqrt{\frac{2900 - 2000}{2900 + 2000}} + 0.85(4.5) = 7.7 \text{ ft, or}$$

$$D_1 = \frac{5.5}{2} \sqrt{\frac{610 - 366}{610 + 366}} = 1.4 \text{ m}$$

$$D_2 = \frac{5.5}{2} \sqrt{\frac{884 - 610}{884 + 610}} + 0.85(1.4) = 2.4 \text{ m}$$

## 27.6.3  Precautions

The velocity of sound in porous soils such as loess is approximately the same as in air, in which case the geophone must be covered to prevent triggering by an air wave.

If the refracting stratum is inclined with respect to the ground surface, the velocities will be correct but the $C$ intersections will differ depending on the direction of the traverse. Where a dipping stratum is suspected, the survey is conducted in both directions and the determinations are averaged. As the traverse is continued the amount of dip can be ascertained.

Erroneous results will be obtained if seismic refraction follows a buried pipe or detours laterally through a nearby building foundation, street, or other structure. While seismic refraction is useful for a rapid area survey, determinations always should be spot-checked by drilling.

If a low-velocity layer is underneath a higher-velocity layer it will not be detected. A seismic refraction survey cannot be conducted through frozen ground.

A refraction survey will not detect small underground openings such as caverns because the sound waves go around.

## 27.6.4  Seismic Velocity and Rippability

According to eq. (27.1) the velocity of a shear wave is

$$V_s = \frac{G}{\rho}$$

where $G$ is the shear modulus and $\rho$ is density. According to theory of elasticity the shear modulus is related to the compression modulus as follows:

$$G = \frac{E}{2(1+v)} \qquad (27.27)$$

where $v$ is Poisson's ratio. A seismograph measures velocity of the compression wave. From theory of elasticity the compression wave velocity is directly proportional to the shear wave velocity according to

$$V_p = V_s \sqrt{\frac{2(1-v)}{1-2v}} \qquad (27.28)$$

(It will be seen that this relationship breaks down if $v = 0.5$.) By combining equations it will be seen that

$$V_p = k\frac{E}{\rho} \qquad (27.29)$$

**Table 27.2**

Representative seismic P-wave velocities. Velocities lower than those in parentheses are tippable with a large tractor

| Material | ft/s | | m/s | |
|---|---|---|---|---|
| Topsoil | 800–1800 | (3000) | 240–550 | (910) |
| B–horizon clay | 1100–6000 | (6100) | 240–1800 | (1900) |
| Loessial silt | 1100–1400 | – | 340–430 | – |
| Glacial till | 1500–8000 | (6000) | 460–2400 | (1800) |
| Sand or gravel below water table | 1500–4000 | – | 460–1200 | – |
| Caliche | 4000–7000 | (6300) | 1200–2100 | (1900) |
| Clay, soft shale | 3000–7000 | (6100) | 910–2100 | (1900) |
| Hard shale | 6000–10,000 | (7500) | 1830–3040 | (2300) |
| Sandstone | 5000–10,000 | (7400) | 1520–3040 | (2300) |
| Limestone, weathered | 4000–8000 | (7600) | 1200–2400 | (2300) |
| Limestone | 8000–18,000 | (7600) | 2400–5500 | (2300) |
| Basalt | 8000–13,000 | (7500) | 2400–3960 | (2300) |
| Granite | 10,000–20,000 | (6800) | 3050–6100 | (2100) |
| Frozen soil | 4000–7000 | Var. | 1200–2100 | Var. |
| Ice | 10,000–12,000 | | 3050–3700 | |
| Liquid water | 4800 | | 1460 | |
| Air | 1130 | | 344 | |

where $E$ is the elastic modulus, $\rho$ is density, and $k$ is a constant that depends on Poisson's ratio with the proviso that it cannot equal 0.5—that is, that the soil cannot be incompressible. Since density and modulus tend to increase together, their effects might tend to cancel out, but generally the modulus prevails, and seismic velocity is more an indicator of modulus than of density.

Tests by the Caterpillar Tractor Company (1994) indicate that seismic velocity can be an indicator of rippability if soils are properly identified. Any soils and rocks with seismic velocities lower than about 6000 ft/s (1800 m/s) can be ripped with a large (D9N) crawler tractor. A compact glacial till having a velocity above 7000 ft/s (2100 m/s) probably will require blasting, but layered sedimentary rocks probably can be ripped. Representative P-wave velocities and approximate relationships to rippability are shown in Table 27.2.

### Example 27.11

A small contractor is experiencing excavation costs that have increased by a factor of 10 over his bid because of a need for blasting. The rock is a mix of shale and soft sandstone having an average velocity of 4500 ft/s (1400 m/s). How do you advise?

*Answer:* Get a bigger tractor.

## 27.6.5 Reflection Seismographs

In seismic reflection profiling, sound waves simply travel down to a discontinuity and then bounce back to the ground surface. Reflection profiling has a long history in petroleum exploration, and with increased speed of timers has been adapted for shallow exploration and engineering uses.

Typically a linear array of 24 geophones is used with the source in the middle. For deeper penetration the source may be a shotgun blast or rifle shot into the ground. As shown in Fig. 27.16, travel paths are slightly longer to outer geophones, resulting in hyperbolic reflections in unprocessed data (Fig. 27.17). Direct sound paths such as through the air can be discerned from a linear relationship to distance, and in the case of air waves can be confirmed from the travel velocity. A portable computer is used to record the data, which then are processed to remove distortion and other irrelevant information. Many echoes are recorded, so different discontinuities can be detected, and a major advantage is that unlike refraction surveys it is not necessary for velocities to increase with depth. A reflection seismograph also can be used for refraction surveys.

## 27.6.6 Sonar

Water depths can be measured with relatively simple "fish finder" recorders that cost a few hundred dollars. These belong in a class called "pingers" because they use piezoelectric transducers that emit a high-frequency pulse; after a time the echo then re-excites the transducer, causing an electrical pulse. Because the velocity of sound in water is known, the time between sending and arrival is easily translated into an equivalent depth. Depth surveys may be conducted from boats, or from the upper surface of frozen lakes.

Higher-energy acoustical "boomers" or electrical "sparkers" are used to obtain deeper penetration of bottom sediments. Transducers offer a wide range of

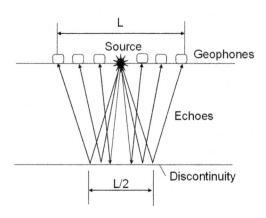

**Figure 27.16**

Principle of the reflection seismograph.

**Figure 27.17**

Unprocessed field seismogram showing curved response caused by different echo distances shown in Fig. 27.16. Linear responses are from direct transmission through air or soil and are ignored. (After Steeples and Miller, 1990.)

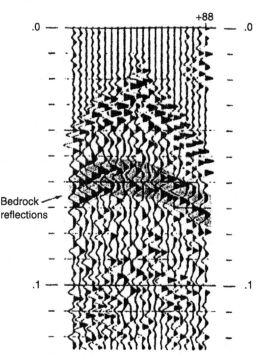

operating frequencies; the higher the frequency the shorter the wavelength and the higher the resolution, but the lower the energy and the lower the penetration. For engineering purposes the frequency is around 1 kHz. As the velocity of sound in loose sediments is very close to that in water, approximate depths may be scaled directly on the profile. These instruments are useful to supplement borings for bridge piers, etc., and to quickly measure scour depths where a rapid current prevents accurate soundings.

### 27.6.7  Ground-Penetrating Radar

Radio frequency (RF) microwaves may be used for shallow-depth continuous profiling on land or water. A radar "bowtie" antenna emits very high frequency (VHF) waves in about a 90° cone, and therefore must be close to the ground to reduce losses. Upon entering the soil and successively denser soil layers, the outer waves entering at an angle are refracted into a narrower cone, which aids penetration. Charts showing reflections appear much like sonar or sound profiling, but the propagation velocity and therefore the depth scale are variable and highly dependent on the moisture content and dielectric constant as conducting elements (molecules, clay particles, etc.) oscillate polarity in the applied a.c. electrical field. Water has a high dielectric constant that increases the microwave velocity and decreases the travel time, compressing the recorder depth scale. However, a high dielectric constant also increases attenuation, or energy loss, particularly at high frequencies.

The maximum penetration depth for ground-penetrating radar normally is in the range 3 to 30 ft (1 to 10 m). The microwave exposure to personnel is minimal, being less than that of a cell phone. Ground-penetrating radar is used to locate objects such as buried pipes, drums of hazardous waste, concrete rebars, defective railroad ties, sinkholes, and unmarked graves.

## Problems

27.1. Name three harmful effects of soil vibrations. Name three beneficial ones.

27.2. Explain how an earthquake precursor works.

27.3. (a) Estimate the intensity of the earthquake in Fig. 27.1. (b) Draw a free-body diagram and estimate the compressive force on the prop. Can the chimney be righted and underpinned?

27.4. Why are so many earthquakes concentrated in Alaska? How may this relate to the direction of plate movement?

27.5. Why is a "window" where there are no earthquakes along an active fault considered a danger sign?

27.6. What is meant by earthquake intensity? Earthquake magnitude?

27.7. A 10 m (33 ft) thick drained fill is planned on a soft, saturated deltaic sand with $\gamma=14$ kN/m$^3$ (89 lb/ft$^3$) and $\phi=23°$. If the unit weight of the fill is 18 kN/m$^3$ (115 lb/ft$^3$), predict the effect on earthquake sensitivity at a depth of 10 m (33 ft) in the sand.

27.8. Calculate static, dynamic, and total soil force on a unit length of 12 m (39.4 ft) high retaining wall, $a=0.2g$, $\phi=30°$, and $\gamma=16$ kN/m$^3$ (102 lb/ft$^3$). Ignore wall friction.

27.9. A water tower with the center of mass 18.9 m (62 ft) above the base is designed to withstand a horizontal acceleration of 0.10$g$. Support is from a symmetrical nine-pile (3 × 3) group with piles 1 m (0.3 ft) center-to-center. Calculate the increase in load to each pile. *Answer.* 31% increase to outer pile.

27.10. A machine weighing 400 N (90 lb) becomes resonant at 100 Hz. If there is no damping, what are the static displacement and the actuating force?

27.11. What maximum displacement may be expected from a vibrator operating at 60 Hz if the natural frequency is 10 Hz and the actuating force is 5 N (1.12 lb)?

27.12. The base of a machine-foundation system weighing 4 kN (900 lb) measures 2 × 4 m (6.6 × 13.1 ft). The machine will generate horizontal oscillations and will rest on a dense sand. Predict $f_n$.

27.13. Explain the relation between vibration direction, wave type, and damping coefficient.

27.14. What may be done to increase damping of a vertically vibrating machine and foundation? What can be done to lower the natural frequency?

27.15. A machine gives two resonant frequencies, one for vertical vibrations and one for rocking. Explain. Which frequency is most likely the higher?

27.16. In Problem 27.15 the observed frequencies are 9 and 22 Hz. The machine is to be operated at 15 Hz. Estimate damping for each vibration, and find the magnification ratio.

27.17. How should the foundation for the machine of Problem 27.15 be laid out to minimize rocking vibrations? Why might emphasis be placed on rocking?

27.18. Sketch an arrangement for sheet piling to reduce vibrations from a horizontal compressor.

27.19. A blast gives a ground response of 40 Hz and an acceleration of $0.01g$. Estimate the maximum particle velocity and amplitude. Is this vibration level noticeable, troublesome, or destructive?

27.20. The static load from a vibratory compactor on the vibrating roller is 8.9 kN (1 ton). What is a desirable frequency and acceleration?

## References and Further Reading

Barkan, D. D. (1962). *Dynamics of Bases and Foundations*. Trans. from Russian by L. Drashevska. McGraw-Hill, New York.

Bartlett, S. F., and Youd, T. L. (1995). "Empirical Prediction of Liquefaction-Induced Lateral Spread." *ASCE J. Geotech. Eng.* 121(4), 316–329.

*Caterpillar Performance Handbook*. (1994). Caterpillar, Inc., Peoria, Ill.

Close, U., and McCormack, E. (1922). "Where the Mountains Walked." *National Geographic Magazine* XLI(5), 445–464.

Coffman, J. L., and von Hake, C. A., ed. (1973). *Earthquake History of the United States*. U.S. Dept, of Commerce Publ.41–1. U.S. Gov. Printing Office, Washington, D.C.

Day, R. W. (2002). *Geotechnical Earthquake Engineering Handbook*. McGraw-Hill, New York.

Newmark, N.M., and Rosenblueth, E.(1971). *Fundamentals of Earthquake Engineering*. Prentice-Hall Inc., Englewood Cliffs, N. J.

Richart, F. E., Jr. (1975). "Foundation Vibrations." In H. Winterkorn and H. Y. Fang, ed., *Foundation Engineering, Handbook* 1st ed., pp. 673–699. Van Nostrand Reinhold, New York.

Scholz, C.H., Sykes, L.R., and Aggarwal, Y.P. (1973). "Earthquake Prediction: A Physical Basis." *Science* 181(4102), 803–810.

Seed, H. B., and Idriss, I. M. (1971). "Simplified Procedure for Evaluating Soil Liquefication Potential." *ASCE J. Soil Mech. and Foundation Eng.* 97(SM9), 1249–1273.

Scholte, C. H. (1987). "Wear and Gouge Formation in Brittle Faulting." *Geology* 15, 493–495.

Seed, H. B., Tokimatsu, K., Harder, L. F., and Chung, R. M. (1985). "Influence of SPT Procedures in Soil Liquefaction Resistance Evaluations." *ASCE J. Geotech. Eng.* 111(12), 1425–1445.

Seed, H. B., Chaney, R. C., and Pamukcu, S. (1991). "Earthquake Effects on Soil-Foundation Systems." In H. Y. Fang, ed., *Foundation Engineering Handbook*, 2nd ed., pp. 594–672. Van Nostrand Reinhold, New York.

Shreve, R. L. (1968). "The Blackhawk Landslide." *Geol. Soc. Amer. Special Paper 108.*

Stark, T. D., and Olson, S. M. (1995). "Liquefaction Resistance Using CPT and Field Case Histories." *ASCE J. Geotech. Eng.* 121(12), 856–869.

Steeples, D. W., and Miller, R. D. (1990). "Seismic Reflection Methods Applied to Engineering, Environmental, and Ground-water Problems." *Volumes on Geotechnical and Environmental Geophysics* 1, 130.

Toro, G. D., Hirose, T., Nielson, S., Pennacchione, G., and Schimamoto, T. (2006). "Natural and Experimental Evidence Of Melt Lubrication of Faults During Earthquakes". *Science* 311, 647–649.

Winterkorn, H. R., and Fang, H. Y., eds. (1975). *Foundation Engineering Handbook.* Van Nostrand Reinhold, New York.

# 28
# Geotechnical Investigation and Report

## 28.1 OVERVIEW

### 28.1.1 Why Write when Straight Talk Is Easier?

Some students know deep down in their soul that courses in writing are a waste of time, and that there are better ways to waste time. However, writing is communication, and an engineer who cannot communicate effectively is going to have a difficult time of it. Effective writing requires effective thinking. That is why writing is called a discipline.

As they advance in their profession, engineers and particularly geotechnical engineers often discover that they spend more time using a computer as a word processor than as an instrument for calculations, as calculations now are performed almost instantly while writing takes time. Writing time can be reduced by using stock paragraphs to describe such things as the general geology or standardized test procedures, but the heart of a report will depend on the specifics.

### 28.1.2 Writing and Engineering

Many engineers write as they analyze a problem in order to have a "paper trail" as well as to have a head start on a final report. A first draft may not be much more than a sentence outline ready for additions and revisions. The final draft may be the product of many revisions. Even after it is completed, a report should be thoroughly checked, preferably by an associate, to minimize the possibility for misinterpretations, misstatements, or errors.

The tangled relationship between writing and thinking often can allow an engineer to arrive at new ideas and concepts during the writing process, which is one reason why advanced degrees, even in engineering, often require a written thesis or dissertation. Prospective employers will want to know if an advanced degree is with or without a thesis or other written report, because that can indicate if a student has successfully followed a rational thought process and procedures from beginning to end. A thesis usually must be defended in an oral examination. This is a test of verbal skills that also will be needed in the future.

Writing is invention, and invention can be fun. Geotechnical engineers who are not born writers soon become writers, as witness the copious geotechnical engineering literature that includes an ever-broadening range from case histories to speculations. Every geotechnical problem involves discovery and therefore is a new adventure, as every architectural design can be a new adventure in art, and every structural design a new adventure in applied mechanics and materials.

## 28.1.3   Writing and Geotechnical Engineering

Geotechnical engineers not only deal with the world's most abundant construction materials, but the materials also are the most variable and changeable in response to treatment and to the environment. Soils are not selected by number from a mill catalog, nor are they mixed according to some standard recipe like concrete. Soils are whatever is there and is available. Geotechnical engineers start each project by identifying and characterizing their materials, and then consider and evaluate the options for design and construction. It is no accident that students and engineers become intrigued with this kind of problem-solving, partly because it takes them out of the office and into the field. Architects and structural engineers mainly communicate through graphical means; geotechnical engineers do so with written reports.

## 28.1.4   The Importance of Style

Style in writing has nothing to do with a fashion statement, and involves more than simply selecting words and phrases that can captivate their readers. Style in writing also involves a more fundamental relationship, the order in which information is presented.

People intuitively write like they think, in a narrative 1-2-3, style, describing things in the order in which they occurred or were investigated or perceived. While this approach is logical, it can create an immediate disconnect with readers who, as indicated in Fig. 28.1, want answers up front. Their reaction when they read a report is, "Give me some answers and spare me the details." Nevertheless, details are essential in order to defend the conclusions. Geotechnical engineers' recommendations cost money, and sooner or later when money talks people do listen.

Engineering and scientific writing traditionally follows a narrative style, starting with (1) goals or requirements of the investigation, (2) a review of literature and other available information, (3) details of the procedures used, (4) results of the

**Figure 28.1**

(a) Engineers think and write in narrative style, but (b) clients want answers now.

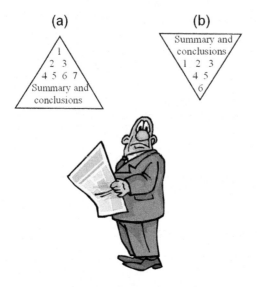

investigation, (5) analysis of the results, and (6) recommendations and conclusions. While all of this information may seem fascinating to the person doing the writing, it may begin to look like "boilerplate" to the person doing the reading.

A simple and effective compromise is to include an abstract or "executive summary" at the beginning of a report, where it can spur some interest and quickly convey the most relevant information. Some engineers prefer not to include details of bearing capacity as incentive to read the rest of the report, or if they are included there may be a qualifier such as: "... can vary depending on the width and depth of the foundation element; details are in the body of the report." The body of the report then can meet requirements of objectivity and thoroughness.

In contrast with the narrative style, a newspaper story starts with a headline that can bash the reader into submission, followed by a quick summary of what happened in the first paragraph. The rest of the story fills in the details in descending order, in recognition that the nether end of an article may be lopped off to make room for the lottery results.

## 28.2 PROPOSAL: WHERE WRITING COUNTS

### 28.2.1 Overview

Except in isolated circumstances where there is mutual trust and understanding, the days are gone when a geotechnical investigation could initiate with a letter or phone call. The proposal serves two functions, to outline the problem as it was presented to the geotechnical engineer and on which the recommendation will be based, and to allow the client to compare costs, timing, and other issues including qualifications and method of approach to a problem. Unfortunately in a competitive market some clients will be swayed by low cost figures, but the professional engineer is obligated to perform an adequate investigation, and may become liable if the investigation is not in keeping with current standards of the profession. The client not only is buying a service, he or she is depending on the experience, expertise, and competence of the engineer to do a proper job.

A written proposal includes a brief job description and location, structural loads and other requirements provided by the structural engineer and/or architect, and scope of the work. The geotechnical proposal may include a tentative boring plan, testing program, and cost estimate, and should be carefully written because if accepted it will be part of a legally binding contract. Any unusual conditions that may be anticipated based on the engineer's experience should be noted, not as a "scare tactic," but as a gentle reminder of the importance of having a competent geotechnical investigation.

## 28.2.2  Site Plan and Boring Locations

The scope and intent of a project form the basis for the investigation: is it for a building, dam, or parking ramp, and how many stories? Included should be anticipated final grades and floor elevations, and a preliminary estimate of dead and live loads prepared by a structural engineer. A topographic site survey normally is conducted independently and the information is provided to the geotechnical engineer.

A considerable part of the cost of a site investigation is for borings, and geotechnical consulting firms may calculate a total cost simply on the basis of the boring number and depths. A boring plan therefore is required that will include locations, depths, and tests that are anticipated to be performed in the field and in the laboratory.

## 28.2.3  Boring Depths

Anticipated boring depths will depend on the site geology, loads, and footprint of the proposed structure, and relationships to topography and other structures. On large projects at least one boring may be deep enough to encounter hard material or bedrock. Other boring depths must be sufficient to extend below the bottom of a bearing capacity failure, which generally is about 1.5 times the width of the bearing area.

A second consideration in addition to bearing capacity is settlement. Because of the decrease in added stress with depth below a foundation, one guideline that is based on elastic theory is to extend borings to a depth such that the added

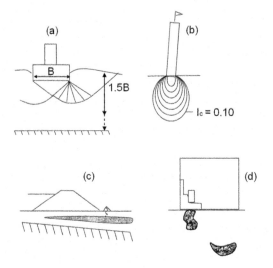

**Figure 28.2**

Some criteria for selecting boring depths and spacings: (a) bearing capacity can depend on the soil, footing width, or depth to a firm layer; (b) settlement depends on pressure, soil type, and the pressure bulb; (c) leakage can readily find paths under an earth dam; (d) sinkholes, debris-filled ravines, and collapsed mine workings are difficult to locate and pose dangers.

pressure is less than 10 percent of the bearing pressure. Some factors affecting appropriate boring depths are illustrated in Fig. 28.2; the broader the loaded area, the deeper should be the borings.

### 28.2.4 Boring Spacing

Borings normally are located at all exterior corners of buildings to ensure that the same soils exist across a building site, and they may be spaced at no more than 50–100 ft (15–30 m) intervals on a grid pattern. A center boring often is extended to greater depth to ensure that there are no radical changes that could adversely affect a construction project. Not only are soils identified; equally important may be the location of the groundwater table.

The boring pattern should be adjusted to reflect geological features observed on the ground or from airphotos. For example, borings on a floodplain should be arranged to show locations of point bars, natural levees, clay plugs, and terraces, as otherwise important features can be missed by an arbitrary grid pattern.

In-situ tests often are employed, and applications of geophysical methods should not be overlooked.

---

### Case History

One corner of a new school building located on a floodplain settled several inches. Borings revealed that the corner was located on top of a clay-filled oxbow, whereas the remainder of the building was on point-bar sand. The original site investigation had reported a surficial layer of clay, but borings did not go deep enough to detect the oxbow clay.

*Lesson learned*: Do not ignore the geology.

---

## 28.2.5 Contract and Performance

After a contract is negotiated the investigation should be performed in a timely manner, weather permitting. If unusual or adverse site conditions are encountered, that information should immediately be relayed to the client, the structural engineer, and/or the architect, as it could mean a change of plans or of location. The drilling and other field investigations are conducted, samples are taken to the laboratory for identification and analysis, and a report is prepared.

## 28.2.6 Organization of the Report

There are no set rules for organization, and most firms will have developed their own style and preferences. A logical order is as follows:

### Title Page
This should include the title and date and identify the firm conducting the investigation. Also included are the name and title of the person preparing the report, and the signature and seal of the engineer who will bear the ultimate responsibility for the report.

### Executive Summary
An executive summary may include (1) general project description and location, (2) a few words to explain the geology and primary geotechnical concerns, which can include a description of special problems such as expansive clay or random fill, and (3) general recommendations that often will include alternative approaches that can be considered in relation to cost and feasibility.

### Body of the Report

- Location and general description of the project. This should include structural loads and floor levels that form a basis for the report.
- Site geology and soils. Even though this sometimes is referred to as "boilerplate" because it can be standardized from files, it is an essential part that shows that the engineer knows the territory. It may be prepared with the

assistance of a geologist, and should refer to relevant publications such as state geological reports and soil surveys.

- A map showing boring locations.

- Boring logs that include soil identifications and groundwater table, and usually will include Standard Penetration Test and unconfined compressive strength data, moisture contents, liquid and plastic limit data, geological identification, soil color and gradation, and engineering classification.

- Cross-sections showing boring logs at their proper elevations, geological identifications, groundwater table. Strata are joined by straight lines that often are dashed to convey that they are interpolations.

- Additional test data and graphs, plotted versus depth where appropriate. The data will include results from both in-situ field tests and laboratory tests.

- Special considerations: a discussion of groundwater, quick conditions, slopes and landslides, expansive clay, collapsible soils, liquefaction potential, sinkholes, mines, hazardous wastes, and other potential problems.

- Summary.

- Conclusions and recommendations.

- An indication of availability for questions, discussions, and additional supplementary investigations.

- *Disclaimer:* A legal clause indicating that the interpretations and conclusions in the report are based on the available data and boring information, and cannot guarantee conditions that may exist between the borings. Usually prepared by an attorney.

- *Every report must be checked!* A final step before a report is submitted is to have it carefully checked word for word by a senior engineer. Clarity and accuracy are requirements, and correct spelling is nonnegotiable. Many offices require that two engineers sign every report.

- *Invoice for services.* This often is mailed separately out of consideration for the client's sensibilities.

- *Dinner and movie.* Optional.

### 28.2.7  ASFE

Liability claims against geotechnical engineers reached an all-time high in the 1960s, to the extent that insurance companies refused to insure. ASFE was founded in 1969 as the American Association of Foundation Engineers to attempt to remedy this situation through a combination of education of both clients and members, and by initiating peer reviews of activities and reports. (The formal name now is ASFE.) A limited liability was written into contracts, and the Association wrote the insurance. The result was a phenomenal success, and the concept of a limited liability has served as a model for many other design and environmental professionals. ASFE membership includes having methods and materials reviewed and critiqued by experienced peers, and ASFE publications are an invaluable guide for the professional geotechnical engineer. ASFE "case histories" prepared by members are useful and instructive for students.

## 28.3 THE GEOTECHNICAL ENGINEER AS AN EXPERT WITNESS[*]

### 28.3.1 Overview

Geotechnical engineers may be asked to testify as expert witnesses in lawsuits, an activity that normally is voluntary and for which the engineer is duly compensated. This is in contrast to a subpoena, which is served by an officer of the court and *requires* that a witness will appear and testify at a specified time and date. On the other hand, an invitation from a valued client also can be persuasive, so long as the client understands that facts will not be distorted or concealed in order to favor one side or the other in a case. Expert witnesses soon come to appreciate the ability of many attorneys to grasp even relatively complex technical details, and probe into their possible significance.

### 28.3.2 Steps Leading up to the Courtroom

1. Upon being approached by a client or the client's attorney, the engineer can ask sufficient questions concerning the case to determine if it is within his or her area of expertise, and if the claims may have merit. The expert then can indicate a willingness to serve without prejudice, and emphasize that he or she will call it as they see it. Most clients and attorneys will respect this position because they want the facts of the matter. Full and honest disclosures result in most cases being settled out of court.

2. The engineer will supply the attorneys with a vita showing education and experience. Many federal courts require that the expert supply opposing counsel with a list of all cases, including attorneys' names, in which the witness has given testimony in the last five years. There sometimes is a rush to a particular expert to prevent his or her being retained by the other side.

3. At the request of the client's counsel and prior to giving testimony, the expert will review all available information including reports and legal depositions of others, and if requested prepare a written report analyzing and explaining the information. The attorney may ask that anything in writing carry a written notation, "Privileged and Confidential Attorney-Client Work Product," because preliminary studies can lead to new areas of investigation, and as additional evidence is compiled opinions may change. As many of the same documents will be used by both sides there can be a huge duplication of materials, thanks to the copy machine.

4. Discussions with the client or client's counsel may lead to additional testing to try and help resolve the conflict. The client should be aware that test results can work either way.

---

[*]These are observations and not legal opinions.

5. The witness will discuss his or her findings with the attorney and client. A written report will include various documents, drawings, calculation sheets, maps, etc., used in rendering an opinion. When these are finalized they will become formal "exhibits" and given exhibit numbers. The report itself also will become a court document.

6. In contrast to popularized versions of courtroom trials, attempts will be made to ferret out all relevant information for review by both sides prior to trial so that there will be no surprises. The most common method for accomplishing this is to require depositions under oath before attorneys representing all sides of a conflict. The deposition is requested and paid for by the opposing side(s). Opinions expressed in the deposition should be documented and based on the evidence and not on opinions or preferences of the client.

7. If matters are not settled, testify at trial.

### 28.3.3   Mediation and Arbitration

If an agreement is not reached, in order to save costs of a trial both sides may agree to hire an independent mediator or team of mediators, which will include both engineering and legal representatives, to try and make recommendations that are equally dissatisfying to both sides. The recommendations may or may not be accepted.

A more stringent approach is called "binding arbitration." In this case a single independent arbitrator or a team of three or more, always an odd number to prevent a tie, will reach a decision that is binding. Arbitrators are nominated by each side, and at least one will be an independent investigator who will analyze the various arguments and may carry most of the weight of the decisions. The arbitration procedure may require a formal record and testimony under oath, or it may simply involve a review of the evidence and exhibits by the expert or panel of experts.

### 28.3.4   Testifying under Oath

Of the various steps, the legal deposition can be the most challenging because attorneys will be probing for soft spots and inconsistencies. Prior to the deposition, the client's attorney will discuss the deposition protocol and give hints as to what the witness may expect. He will offer suggestions on how to give a good deposition, such as waiting until an entire question has been asked and not simply shaking one's head to answer a question. The discussion will focus on the issues, and the well-informed attorney will know the answers to questions before they are asked.

A deposition is a formal question and answer procedure that begins with swearing-in of the witness, who will be asked to state his or her name and affiliation. The opposing attorney(s) then ask questions of the expert, which is the "direct examination." Attorneys on both sides will alternate questioning to seek additional clarifications.

An official record of every word spoken is typed in shorthand except for moments that are requested by an attorney to be "off the record." Answers should be carefully considered based on the evidence, and never hurried. An answer can carry the qualifier, "It is my opinion that...."

At trial the order of questioning is reversed, with the attorney for the client doing the direct examination and the opposing attorney the cross-examination.

In many ways a deposition and trial have elements of a game, so the witness should be relaxed and remember it is only a game. Each attorney is ethically bound to represent the best interests of his or her client, so while the evidence that is presented must be truthful, counsel may prefer that not all of the evidence be presented. In theory, with equally competent representations, truth will out and the right side will win.

### 28.3.5 Witness Demeanor

Expert witnesses have some special privileges, the most important being that they are not be required to give a yes or no answer but are free to explain the basis for their opinions. *The expert witness should be a teacher, not an advocate or a combatant.* Graphics often are used and are prepared ahead of time. During trial the judge can ask for clarifications.

Some trial attorneys resort to simple tricks, like asking the same question over and over again to test if the witness is consistent with the answers, or may be overly consistent by giving the same answer word for word, implying that it has been coached and rehearsed. A report may be read sentence by sentence and the witness asked if he agrees with each statement, even though it may be the omissions that are relevant. The expert inevitably will be asked if he or she is being paid for his or her testimony, which is intended to demonstrate if the witness is a "hired gun." The appropriate answer is that one is being compensated for one's time, the same as the attorneys.

Evasive answers, bluffing, name-dropping, inconsistencies, and technical jargon destroy witness credibility. Lying under oath is perjury, can impeach a witness, and carries stiff penalties. Opinions should be carefully phrased; for example, it would be unwise to slam water witching if the township's leading water-witcher may be on the jury and it has nothing to do with the case. (More accurate would be to say that one is not aware of any scientific validation for water witching so its reliability is open to question.) During trial the witness will be asked to explain any apparent deviations from his or her earlier reports and depositions.

Attorney smirking or ridicule can be expected, but only indicate that the opposing attorney is not comfortable with the way things are going.

The expert often is asked if in his or her opinion an investigation or report meets professional standards that were in place at the time that the report was written, which in many jurisdictions can form a basis for a decision.

### 28.3.6 Attitudes

At trial the expert who really is expert will speak slowly and clearly, use simple terms that are easily understood by lay persons, and explain technical terms as necessary. In an ideal world the experts would not take sides and would come to the same conclusions based on the same evidence, but this seldom happens. However, the disagreements generally are over interpretations and relevance of particular information or data. In rare cases experts on both sides will be asked to work together and issue a joint opinion as a basis for settlement.

During court recesses, witnesses are not permitted to talk to members of the jury, or there may be a mistrial.

---

### *Case History*

A large 50 ft (15 m) deep sewage lagoon was being tested by filling with water when a breach occurred that cut down through the embankment and drained the entire lagoon. Had the test not been conducted, the consequences would have been devastating.

*Investigation*: Three deficiencies were discovered: (1) The clay lining had been cut through to bury inlet pipes, and the trench had not been properly backfilled. This is where the breach occurred. (2) Even though the lagoon was only partly filled, the clay lining had penetrated into the gravelly soil that supported the lagoon, such that leakage was immanent. The lining was only one-third as thick as originally recommended by the geotechnical engineer, who then relaxed the requirement because of a shortage of clay. (3) Unexcavated random fill was discovered at one corner of the lagoon, but had yet not been reached by the level of filling.

*Resolution*: A report was issued prorating responsibility to the design firm, the geotechnical firm, and the contactor on the basis of estimated probability and likely consequences. The firm responsible for item (1) did not accept the recommendation and the case went to binding arbitration. The arbitration upheld the experts' recommendations. Repair was with a plastic liner.

*Lesson learned*: Many cases involve multiple contributing factors, and all should be examined instead of focusing on one and ignoring the rest.

---

## Case History

A landslide threatened a house uphill from the landslide scarp, and the homeowner sued the county claiming that road maintenance operations had plugged a tile drain and caused the soil to become saturated.

*Investigation*: The alleged tile drain was not actually observed; its location was determined by water witching.

*Resolution*: The case was thrown out of court.

*Lesson learned*: It is not advisable to rely on questionable data.

---

## Problems

28.1. Identify the writing style of this chapter and compare with that of a newspaper article.

28.2. What is meant by "executive summary?"

28.3. Develop a boring plan for a $50 \times 80$ ft ($15 \times 24$ m) warehouse on an alluvial floodplain.

28.4. Develop a boring plan for a 15-story brick building on a thick alluvial clay.

28.5. Develop a boring plan for a gas station on peat.

28.6. Develop a boring plan for a 100 ft (30 m) square concrete pad to support a cluster of four grain elevators on sandy alluvium.

28.7. As a class exercise examine a building on campus that is showing some distress, sketch and describe the crack patterns, indicate possible causes, prepare a written report, and invite the campus architect for an oral presentation.

28.8. Examine some additions to older buildings on campus and determine if there has been differential settlement. If so, explain the cause and write a brief report.

28.9. Have your professor contact a local geotechnical firm and ask for boring logs and data from a completed project. The class then will write a report and invite engineers from the firm to hear an oral presentation in class and ask questions. (Many firms welcome this opportunity.)

28.10. Answer the call from a homeowner in distress, and as a class project examine the nature and cause of the distress and write a report.[*]

---

[*]Consulting geotechnical firms often are reluctant to investigate homeowner complaints because the financial compensation is not in keeping with liability issues, making this a fertile area for a class investigation—so long as it is understood that it is a student training exercise and carries no guarantees or liabilities, and that the most that the class can expect is a grade and a couple of pizzas.

28.11. Examine and hand-probe a small landslide and write a brief report indicating probable causes and suggesting remedies.

28.12. Examine a tilted retaining wall and write a brief report indicating probable causes and suggesting remedies.

28.13. Examine an MSE wall for bulges or inclinations.

28.14. Walk and hand-probe some steep terrain and sketch the extent of old landslides on a map or airphoto.

28.15. Develop a boring plan for a 40 ft (12 m) high earth dam including abutments.

28.16. Recommend a boring and testing program if the earth dam in the preceding problem is in an area of limestone.

28.17. Trek through an old cemetery and determine if tombstone weathering and tilting may relate to composition and time of interment. Write a happy report.

28.18. What special privilege is accorded an expert witness, aside from going to the bathroom as required?

28.19. How should one answer, "Are you being *paid* for your testimony?"

## Further Reading

ASFE Website. Various publications including case histories.

Hunt, R. E. (1984). *Geotechnical Engineering Investigation Manual.* McGraw-Hill, New York.

Hvorslev, M. J. (1949). *Subsurface Exploration and Sampling of Soils for Civil Engineering Purposes.* Reprinted by Engineering Foundation, United Engineering Center, New York.

# Appendix
# Elastic Solutions for Horizontal Pressures on Retaining Walls Resulting from Surface Loads

*Note*: These solutions assume elastic behavior of the soil and are for rigid, unyielding walls that mirror and double lateral stresses compared with those in a semi-infinite half-space.

## 1. Point Load

$$\sigma_h = \frac{3P}{\pi} \frac{x^2 z}{R^5} \tag{A.1}$$

where $P$ is the point load and $R = \sqrt{x^2 + y^2 + z^2}$. The $x$, $y$, $z$, and $R$ dimensions are indicated in Fig. 19.30.

## 2. Long Line Load Parallel to Wall (Fig. A.1)

$$\sigma_h = \frac{4p}{\pi} \frac{x^2 z}{R^4} \tag{A.2}$$

where $p$ is the load per unit length and $R = \sqrt{x^2 + z^2}$.

## 3. Long Line Load Perpendicular to Wall (Fig. A.2)

$$\sigma_h = \frac{p}{\pi} \frac{z}{R^2} \tag{A.3}$$

where $p$ is the load per unit length and $R = \sqrt{y^2 + z^2}$.

## 4. Long Area Load Parallel to Wall (Fig. A.3)

$$\sigma_h = \frac{2q}{\pi} \left[ \tan^{-1}\left\{\frac{x}{z}\right\} - \frac{yz}{(x^2 + z^2)} \right]_{x_1}^{x_2} \tag{A.4}$$

where $q$ is the area contact pressure and the angle is expressed in radians.

**Figure A.1**
Line load parallel
to wall.

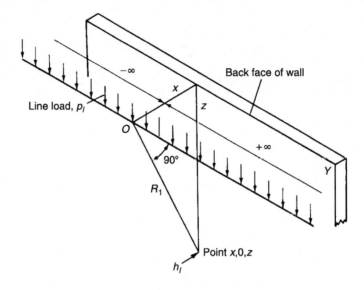

**Figure A.2**
Line load
perpendicular to
wall.

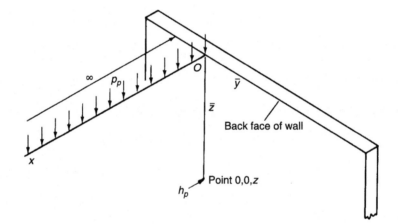

## 5. Any Area Load (Fig. A.4)

$$h_a = p_a \int_{x_1}^{x_2} \int_0^{y_1} \frac{x^2 z}{R^5} \, dy \, dx$$

or

$$h_a = \frac{p_a}{\pi} \left[ \tan^{-1}\left(\frac{x_2 y_1}{z R_{x_2}}\right) - \frac{x_2 y_1 z}{(x_2^2 + z^2) R_{x_2}} - \tan^{-1}\left(\frac{x_1 y_1}{z R_{x_1}}\right) - \frac{x_1 y_1 z}{(x_1^2 + z^2) R_{x_1}} \right] \tag{A.5}$$

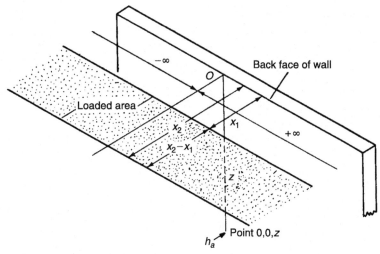

where

$h_a$ = horizontal unit pressure on a retaining wall at any depth $z$ opposite *one end*
   of a uniformly distributed area load on the backfill surface

$p_a$ = surcharge load per unit area

$x_1$ = distance from wall to near side of load

$x_2$ = distance from wall to far side of load

$x_2 - x_1$ = width of load in direction normal to wall

$y_1$ = length of loaded area

$$R_{x_1} = \sqrt{(x_1^2 + y_1^2 + z^2)}$$

$$R_{x_2} = \sqrt{(x_2^2 + y_1^2 + z^2)}$$

Since maximum pressures occur at points opposite the center of an area load, the area can be divided into two parts each having a length $y_1/2$ and the results from one of the areas doubled.

**Figure A.4**

Area load (a) to a point even with the end of the loaded area; (b) to a point even with any location in the loaded area; (c) to a point even with a location outside of the loaded area.

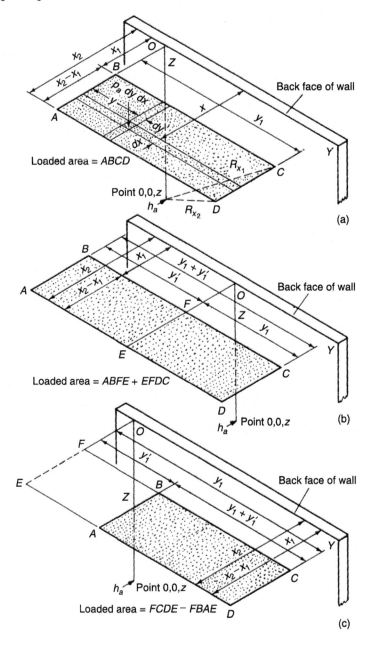

# Index

CPSIA information can be obtained
at www.ICGtesting.com
Printed in the USA
BVOW11*1940101217
502139BV00004B/14/P